HARCOURT BRACE JOVANOVICH COLLEGE [illegible] SERIES

PHYSICAL CHEMISTRY

J. Edmund White

Department of Chemistry
Southern Illinois University at Edwardsville
Edwardsville, Illinois

Books for Professionals
Harcourt Brace Jovanovich, Publishers
San Diego   New York   London

Permissions
Harcourt Brace Jovanovich, Publishers
Orlando, Florida 32887

Printed in the United States of America

Library of Congress Cataloging-in-Publication Data

White, J. Edmund.
Physical chemistry.

(Books for professionals) (Harcourt Brace Jovanovich college outline series)
Includes index.
1. Chemistry, Physical and theoretical. I. Title. II. Series. III. Series: Harcourt Brace Jovanovich college outline series.
QD453.2.W49 1987 541.3 86-31979
ISBN 0-15-601657-5

First edition

A B C D E

PREFACE

Do not read this Outline—use it. You can't learn chemistry of *any* variety simply by reading it: You have to *do* it, in the laboratory and on paper. Solving specific, practical problems is the best way to master—and to demonstrate your mastery of—the theories, laws, and definitions that make up physical chemistry. Outside the laboratory, you need three tools to do physical chemistry: a pencil, paper, and a calculator. Add a fourth tool, this Outline, and you're all set.

This HBJ College Outline has been designed as a tool to help you sharpen your problem-solving skills in physical chemistry, skills that are vital to success in a "p-chem" course. Although some of your examinations may contain a few questions involving explanations, derivations, or diagrams, most of the questions you'll be asked in your course will be problems to be solved. Your knowledge of concepts, terms, and laws will be measured by your success in applying them to the solution of problems.

Each chapter of this Outline covers a major topic, whose fundamental principles are broken down into outline form for easy reference. The text is heavily interspersed with worked-out examples, so you can see immediately how each new idea is applied in solving a problem. Each chapter also contains a Summary and a 'flowchart' of the major equations which, taken together, will enable you to review the primary principles of the chapter and the connections between the equations expressing those principles.

Most important, this Outline gives you plenty of problems for practice. Work the Solved Problems before you read the answers; then check yourself against the step-by-step solutions provided. Show your mastery of the material in each chapter by doing the Supplementary Exercises. (In the Supplementary Exercises, you're given answers only—the details of the solution are up to you.) Finally, test yourself on the topics covered in the Outline by taking the two Mid-Semester Exams and the two Final Exams. The solutions to all exam questions are given in detail, so you can use them to diagnose your own strengths and weaknesses.

Remember that this is an *outline*, to be used as a complement to your textbook or for review. This Outline *does not* contain every topic found in a typical physical chemistry textbook, but it does contain the fundamental topics common to all textbooks. Be sure you understand these fundamentals before you try to learn the more advanced aspects of the topics that are included in your course. To relate this Outline to your textbook, use the Textbook Correlation Table, which correlates chapters of the Outline with pages in several of the leading textbooks of physical chemistry.

The choice of topics to include—and the emphasis to give each—was based partly on a report of a committee of the Division of Chemical Education of the American Chemical Society (ACS) and partly on analysis of examinations shared with me by seventeen other professors. Statistical mechanics, for example, was treated lightly in these materials, so Chapter 18 is quite brief. I have omitted or merely introduced some advanced topics that have "trickled down" into the beginning physical chemistry textbooks, but I have included some topics that seem to have "trickled out." I believe that the latter could be helpful to many students whose difficulties with a p-chem course or textbook may be due to lack of familiarity with basic concepts.

You may find differences between the symbols used in this Outline and those used in your textbook. Probably no two books are exactly alike in this regard; each author must make his/her own choices for symbols—and for units as well. The SI system of units, which I have used extensively, now bears official approval, but other systems still exist. So I have mixed in other units that have been used regularly in the past because they still turn up in older tables, books, and (especially) journals. At first, you may be confused by the variety of symbols and units; but, as you work with these symbols and units, converting them from one to another, they will become familiar, and you will be prepared for almost any situation.

I wish to acknowledge support from Southern Illinois University at Edwardsville: I wrote most of the first draft while I was on sabbatical leave, the Office of Research and Projects provided some help on typing, and the Department of Chemistry provided some supplies, copying, and assigned time. Colleagues Thomas Bouman and David Rands helped on some of the physical chemistry, and Antony Wilbraham was supportive and helpful from his experience as a successful textbook author. Special thanks are due to Dr. Gordon R. Franke of Northeast Missouri State University, who read every word of the manuscript, caught errors, and offered many valuable suggestions which greatly improved both text and problems. My HBJ editors, Charles Arthur and Philip Unitt, also made useful suggestions and caught errors. Any remaining errors are, of course, the responsibility of the author. Finally, I thank my wife, Betty Evans White, for her encouragement to undertake the project, for her constant support, and for her patient acceptance of various deprivations during the long process of completing it.

Southern Illinois University at Edwardsville
Edwardsville, Illinois

J. EDMUND WHITE

CONTENTS

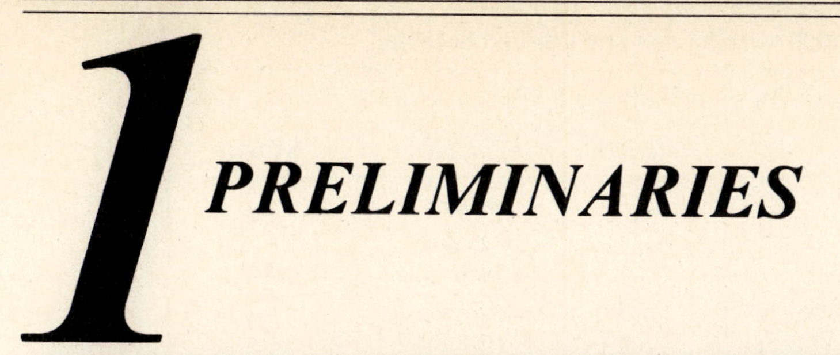

1 PRELIMINARIES

THIS CHAPTER IS ABOUT

- ☑ **Dimensions, Units, and SI**
- ☑ **Constants and Conversion Factors**
- ☑ **Essentials of Logarithms**
- ☑ **Essentials of Calculus**

1-1. Dimensions, Units, and SI

A. Dimension and unit

The terms *dimension* and *unit* have different meanings and can't be used interchangeably. A **dimension** is a concept or abstraction that expresses a type of measurement, e.g., length or time. A **unit** is a specific amount of a dimension that serves as a reference for actual measurement, e.g., meters or seconds. **Dimensional analysis** is a mathematical approach for analyzing the dimensions involved in a problem and its answer without regard to numerical values or units.

EXAMPLE 1-1: For a problem that requires division of an area by a volume, show that the answer will contain the dimension of length (actually, the reciprocal).

Solution: Since area and volume both contain the dimension of length (one squared and the other cubed), we write

$$\frac{(\text{Length})^2}{(\text{Length})^3} = \frac{l^2}{l^3} = l^{-1}$$

Note that we did not use any units.

EXAMPLE 1-2: List some commonly used units of the dimension *length.*

Solution: Inch, foot, mile, furlong, rod, meter, and kilometer.

B. Physical quantities

A number by itself has no physical meaning; it is dimensionless. If a number is followed by a unit, the combination is a measurement of a **physical quantity**.

EXAMPLE 1-3: The number 6.022×10^{23} by itself has no meaning, although you will recognize it as the number of objects represented by Avogadro's number, N_A. When a unit is added, it becomes Avogadro's constant, $L_A = 6.022 \times 10^{23}$ [objects] mol^{-1}. This constant contains the dimension *amount of substance.*

C. International system of units

By international agreement, seven dimensionally independent physical quantities have been selected as the foundation of a system of units, known as SI from the French *Le Système*

TABLE 1-1: Defined and Derived SI Units

Base Quantities and Units			
Physical quantity	Symbol	Unit	Symbol
Length	l	meter	m
Mass	m	kilogram	kg
Amount of substance	n	mole	mol
Thermodynamic temperature	T	kelvin	K
Time	t	second	s
Electric currrent	I	ampere	A
Luminous intensity	I_v	candela	cd

Quantities with Derived Units			
Physical quantity	Unit	Symbol	Definition in base units
Force	newton	N	kg m s^{-2}
Pressure	pascal	Pa	$\text{kg m}^{-1}\text{ s}^{-2}$
Energy	joule	J	$\text{kg m}^{2}\text{ s}^{-2}$

International d'Unités. A **base unit** has been designated for each quantity. These seven quantities are listed in Table 1-1 along with their base units and symbols.

note: The **amount of substance** is a new concept of physical quantity, and the meaning of its symbol conflicts with prior usage. The symbol is n, and it requires *both a number and a unit*. The unit is the **mole**. Thus it is not correct to say "n moles"; rather, "$n = y$ moles," where y is the number of moles.

EXAMPLE 1-4: Consider an analogy between length and amount of substance. Length is designated by l, and let's say that we have $l = 5$ m. The l is not just the number of meters, "5," or the unit, "meters"; the l is the whole expression, "5 meters." Thus for an amount of substance, n, we might have $n = 5$ mol. Both the number and the unit are required to describe n completely.

EXAMPLE 1-5: If we have 64 grams of sulfur, we could say that we have 2 moles of sulfur, but how would we write it?

Solution: We should not write "$n = 2$." The correct use of n is "$n = 2$ mol," a number and a unit. Since the substance is known, "2 mol S" is even better.

D. Numerical values of physical quantities

We obtain the numerical value of a quantity by dividing the quantity by the unit.

EXAMPLE 1-6: For 2 moles of sulfur, we would write

$$\text{Number of moles} = \frac{n}{\text{mol}} = \frac{2\text{ mol S}}{\text{mol S}} = 2$$

We emphasize the distinction between a quantity and its numerical value because it is important in certain situations.

EXAMPLE 1-7: A table of thermodynamic data has a column headed $\Delta H_v/\text{kJ mol}^{-1}$. **(a)** Why is this designation used? **(b)** How should we read the table?

Solution:

(a) The table contains a list of numbers, not physical quantities. Let's use a physical quantity of

$\Delta H_v = 41$ kJ mol^{-1}. Its numerical value, as illustrated in Example 1-6, is

$$\text{Numerical value} = \frac{\Delta H_v}{\text{kJ mol}^{-1}} = \frac{41 \text{ kJ mol}^{-1}}{\text{kJ mol}^{-1}} = 41 \qquad \textbf{(1-1)}$$

The column heading tells us that the physical quantity was divided by its units and the numerical value was entered in the table.

(b) To read the table (i.e., to find a ΔH_v), we simply find the number for a particular compound and write the units after it: 41 kJ mol^{-1}. More precisely, we are applying Eq. (1-1):

$$41 = \frac{\Delta H_v}{\text{kJ mol}^{-1}}$$

Cross-multiplying gives $\quad \Delta H_v = 41 \text{ kJ mol}^{-1}$

E. Derived quantities and units

Other physical quantities are derived from the seven SI base quantities. Their units are expressed as combinations of the defined (base) units. Some of the derived units (e.g., the pascal) were named for famous scientists. A few of the derived quantities and units are included in Table 1-1.

EXAMPLE 1-8: The dimensions of velocity are length and time (reciprocal), or, in symbols, $l \times t^{-1}$. The base units of length and time are meters and seconds, respectively. Thus the derived unit for velocity is, in symbols, m s^{-1}.

EXAMPLE 1-9: Newton's first law is Force = Mass × acceleration. **(a)** What are the dimensions of force? **(b)** What are the units of force?

Solution:

(a) The dimensions of acceleration are $l \times t^{-2}$; thus force equals $m \times l \times t^{-2}$.

(b) The base units of m, l, and t are kg, m, and s, respectively, so the unit of force is kg m s^{-2}; this unit is called the *newton*, N. (See Table 1-1.)

F. Auxiliary units

An auxiliary unit is a combination of a prefix and a base unit. Prefixes are expressed in multiples or submultiples of one thousand. You are already familiar with milli- and kilo- for 10^{-3} and 10^{3}, respectively. The newer prefixes such as femto- for 10^{-15} seldom appear in physical chemistry. Strictly speaking, the quantity 10 cm should be written as 10^{-1} m or 10^{2} mm, but the well-established prefixes centi-, deci-, and deca- are still allowed by SI rules.

1-2. Constants and Conversion Factors

A. Precision of constants

The numerical values of fundamental physical constants are usually known to six or seven significant figures. This degree of precision is far greater than the experimental error in most undergraduate experiments and isn't needed for solving most practice problems. We usually round off constants to four significant figures; but, for practice problems, we may further shorten the values to three significant figures. Some examples of frequently used constants are given in Table 1-2.

TABLE 1-2: Some Fundamental Constants

Avogadro constant	L_A	$6.022\,05 \times 10^{23}$ mol^{-1}	Planck constant	h	$6.626\,18 \times 10^{-34}$ J s
Boltzmann constant	k	$1.380\,66 \times 10^{-23}$ J K^{-1}	Electron charge	e	$1.602\,19 \times 10^{-19}$ C
Faraday constant	$\mathcal{F}$	$9.648\,46 \times 10^{4}$ C mol^{-1}	Electron mass	m_e	$9.109\,53 \times 10^{-31}$ kg
Gas constant	R	$8.314\,41$ J K^{-1}mol^{-1}	Speed of light	c	$2.997\,93 \times 10^{8}$ m s^{-1}

B. Conversions

The data for solving a problem may not be compatible with each other or with the available constants, making conversions necessary. In your general chemistry course, you learned to include the units when you set up a calculation and then to make sure that you canceled or multiplied them properly to end up with the correct units for the answer. This is even more important in physical chemistry because the problems are more complex. Verifying units is a good way to check your work.

1. The simplest conversion is the addition of a constant.

EXAMPLE 1-10: Convert 30.00 °C to the Kelvin scale.

Solution: One degree on the Celsius scale is equal to one degree on the Kelvin scale, so we convert from Celsius to Kelvin by adding 273.15, the number of degrees between absolute zero and the freezing point of water. Recalling the point made in Example 1-7 with Eq. (1-1), we should write

$$\frac{T_K}{K} = \frac{T_C}{°C} + 273.15$$

If $T_C = 30.00\ °C$,

$$\frac{T_K}{K} = \frac{30.00\ °C}{°C} + 273.15 = 303.15 \qquad \text{or} \qquad T_K = 303.15\ K$$

---

2. Most conversions require multiplication by a conversion factor.

EXAMPLE 1-11: To convert mm to m, we need the relationship 10^3 mm = 1 m. You could probably work out in your head that 50 mm = 5×10^{-2} m, but how would you write a mathematical equation to get this answer?

Solution:

$$(50\ \text{mm})\left(\frac{1\ \text{m}}{10^3\ \text{mm}}\right) = 50 \times 10^{-3}\ \text{m} = 5.0 \times 10^{-2}\ \text{m}$$

Note in Example 1-11 that (1 m)/(10^3 mm) = 1. *A conversion factor is unity.* Multiplying by 1 does not change a quantity, but it does change the units in which the quantity is measured.

C. Non-SI units

The use of certain well-established units (e.g., liter, atmosphere, calorie, and angstrom) may eventually end. Until then, you will continue to encounter them in tables, books, and articles; you will need to know how to convert them to SI units. Several of these conversion factors are

$$\begin{aligned} 1\ \text{atmosphere} &= 1.013\,25 \times 10^5\ \text{Pa} \\ 1\ \text{liter} &= 10^{-3}\ \text{m}^3 = 1\ \text{dm}^3 \\ 1\ \text{bar} &= 10^5\ \text{Pa} \\ 1\ \text{calorie} &= 4.184\ \text{J} \end{aligned}$$

EXAMPLE 1-12: A new table of thermodynamic properties from the National Bureau of Standards uses 1 bar as the standard-state pressure. (The *bar* is not an SI unit but is a simple multiple of one, the pascal.) Find the value of the pressure–volume gas constant, $R = 0.082\,05$ L atm $K^{-1}mol^{-1}$, for this new pressure unit.

Solution: You know that

$$\begin{aligned} 1\ \text{atm} &= 1.013\,25 \times 10^5\ \text{Pa} \\ 1\ \text{Pa} &= 10^{-5}\ \text{bar} \\ 1\ \text{atm} &= 1.013\,25\ \text{bar} \end{aligned}$$

So $$R = 0.082\,05 \frac{\text{L atm}}{\text{K mol}}\left(\frac{1.013\,25\text{ bar}}{1\text{ atm}}\right) = 0.083\,14 \frac{\text{L bar}}{\text{K mol}}$$

Note that the conversion from atmospheres to bars causes a change of about 1.3% in the numerical value, which is close to experimental error in many cases. For practice problems, we will use atmosphere or bar interchangeably, as if they were identical. For a real situation involving three or four significant figures, you should convert atmospheres to bars.

1-3. Essentials of Logarithms

A. Exponential nature

The **logarithm** of a number to a certain base is the power to which that base must be raised to give the number. If a is the base,

$$u = a^{\log u} = a^v \tag{1-2}$$

where $v = \log_a u$.

The usual bases are 10 and e ($e = 2.718\,28\ldots$). The abbreviation log is used routinely to represent logarithms to the base 10, and ln is used for logarithms to the base e. Thus the symbol $\ln x$ may be read "the *natural* logarithm of x to the base e." The relationship between the two systems of logarithms is

$$\ln u = 2.303 \log u$$

which may be expressed as a conversion factor:

$$\frac{\ln u}{2.303 \log u} = 1$$

The exponential character of logarithms makes them useful for expressing exponential relationships between physical quantities. In general, if the dependence of one quantity on a second quantity is proportional to the value of the second quantity, the dependence is exponential. Specific cases include distribution functions, first-order reactions, and the barometric formula. Treating such cases mathematically usually requires calculus and the expression $du/u = \ln u$.

EXAMPLE 1-13: The dependence of u on v is proportional to u. Express this as an exponential and as a logarithm.

Solution: The dependence can be written as

$$\frac{du}{dv} = ku$$

where k is a proportionality constant. Rearranging gives

$$\frac{du}{u} = k\,dv$$

We now have to use calculus. Letting $u = u_0$ when $v = 0$, we integrate between the limits u and u_0 and v and 0:

$$\int_{u_0}^{u} d\ln u = k\int_0^v dv$$

$$\ln\frac{u}{u_0} = kv$$

$$\frac{u}{u_0} = e^{kv}$$

B. Properties of logarithms

We need to manipulate logarithms in some derivations before doing any calculations. Some commonly encountered relationships (valid for any base) are

$$\log_a a = 1 \quad \textbf{(1-3)}$$

$$\log nm = \log n + \log m \quad \textbf{(1-4)}$$

$$\log \frac{n}{m} = \log n - \log m \quad \textbf{(1-5)}$$

$$\log 1 = 0 \quad \textbf{(1-6)}$$

$$\log \frac{1}{n} = -\log n \quad \textbf{(1-7)}$$

$$\log n^p = p \log n \quad \textbf{(1-8)}$$

C. Logarithms as a computational tool

In the past, physical-chemistry textbooks always included a table of logarithms. In this age of calculators with their log keys, you may never need to use such a table; modern texts do not include them. You should remember that logarithms to the base 10 are useful in multiplying and dividing and in raising to powers or taking roots, especially to large or fractional values.

EXAMPLE 1-14: Raise the number 234.56 to the 1.62 power using logarithms to the base 10.

Solution: First, find the logarithm of the number, using the following three-step procedure:

1. Convert the number to *scientific notation*, i.e., a number with one digit in front of the decimal times 10 to the appropriate power:

$$234.56 = 2.3456 \times 10^2$$

2. Look up the *mantissa* of the logarithm in a table:

$$\text{Mantissa of } 2.3456 = 0.3703$$

3. Insert the *characteristic* of the logarithm, i.e., the power of 10 (2 in this case), in front of the mantissa:

$$\log(2.3456 \times 10^2) = 2.3703$$

Next, use the following steps to raise the number to the power:

1. Multiply the logarithm of the number by the power to get the logarithm of the result; i.e., use Eq. (1-8):

$$\log n^p = p \log n$$

$$\log[(2.3456 \times 10^2)^{1.62}] = 1.62 \log(2.3456 \times 10^2) = 1.62(2.3703) = 3.8399$$

2. Find the antilogarithm of the mantissa by looking up 0.8399 in a table:

$$\text{antilog}(0.8399) = 6.917$$

The characteristic gives the power of 10:

$$\text{antilog}(3.8399) = 6.917 \times 10^3$$

Alternative Solution: Start with the scientific notation form; then use Eq. (1-2):

$$(2.3456 \times 10^2)^{1.62} = (10^{\log 2.3456} \times 10^2)^{1.62} = (10^{0.3703} \times 10^2)^{1.62} = (10^{2.3703})^{1.62}$$

Multiply the exponents:

$$(2.3456 \times 10^2)^{1.62} = 10^{3.8399} = 10^{0.8399} \times 10^3$$

The antilogarithm of 0.8399 is 6.917, so

$$(2.3456 \times 10^2)^{1.62} = 6.917 \times 10^3$$

EXAMPLE 1-15: Find x in $\log(x/4) = 6.5432$.

Solution: Use Eq. (1-5) and write $\log(x/4)$ as

$$\log x - \log 4 = 6.5432$$

$$\log x = 6.5432 + \log 4 = 6.5432 + 0.6021 = 7.1453$$

$$x = \text{antilog}(7.1453) = 1.397 \times 10^7$$

Or, we could let $y = x/4$, take the antilogarithm of 6.5432, and then solve for x:

$$y = \text{antilog}(6.5432) = 3.493 \times 10^6$$
$$x = 4y = 1.397 \times 10^7$$

Be careful if you have a negative logarithm and want to take its antilogarithm. The mantissas in tables are positive; thus a negative logarithm must be changed to a form that has a positive mantissa.

EXAMPLE 1-16: The result of a calculation is $\log x = -2.6789$; find the value of x.

Solution: The mantissa is -0.6789 (which can't be found in a table), so we have to convert it to a positive mantissa. We simply add 1 and subtract 1, which doesn't change the value of the logarithm. We can see how this works if we visualize the logarithm as two separate parts (-2 and -0.6789) and then add 1 to the mantissa and subtract 1 from the characteristic:

$$\begin{array}{rr} -2 & -0.6789 \\ -1 & +1. \\ \hline -3.0 & +0.3211 \end{array}$$

A logarithm with a negative characteristic is often written in the form $\bar{3}.3211$ to show that the characteristic is negative, while the mantissa is positive.

---

note: In the rest of the book, we don't include steps for solving logarithms in the detailed solutions. If you don't already have access to a hand-held scientific calculator, you should buy one and learn how to use it. Be sure you know how to raise to powers, to take roots, and to find logarithms and antilogarithms, especially for negative cases.

1-4. Essentials of Calculus

A. Derivatives

A few of the simpler derivative forms will do for most applications in physical chemistry. Those used most often (where a is a constant and u and v represent functions) are

$$\frac{d}{dx}(x) = 1 \qquad \textbf{(1-9)}$$

$$\frac{d}{dx}(a) = 0 \qquad \textbf{(1-10)}$$

$$\frac{d}{dx}(au) = a\frac{d}{dx}(u) \qquad \textbf{(1-11)}$$

$$\frac{d}{dx}(uv) = u\frac{d}{dx}(v) + v\frac{d}{dx}(u) \qquad \textbf{(1-12)}$$

$$\frac{d}{dx}(u^n) = nu^{n-1}\frac{d}{dx}(u) \qquad \textbf{(1-13)}$$

$$\frac{d}{dx}(\ln u) = \frac{1}{u}\frac{d}{dx}(u) \qquad \textbf{(1-14)}$$

$$\frac{d}{dx}(e^u) = e^u\frac{d}{dx}(u) \qquad \textbf{(1-15)}$$

note: These equations may be written as differentials, i.e., without the denominator: The differential for $\frac{d}{dx}(x) = 1$ is $d(x) = 1$, etc.

EXAMPLE 1-17: Differentiate the expression x^2y with respect to x.

Solution: We use Eq. (1-12) for the product of two terms and Eq. (1-13) for the x^n term. First, substitute into Eq. (1-12):

$$\frac{d}{dx}(x^2y) = x^2\frac{d}{dx}(y) + y\frac{d}{dx}(x^2)$$

Now substitute into Eq. (1-13), to complete the second term:

$$\frac{d}{dx}(x^2y) = x^2\frac{dy}{dx} + y\left[2x\frac{d}{dx}(x)\right] = x^2\frac{dy}{dx} + 2xy$$

B. Characteristics of derivatives

You should know the following properties of derivatives:

1. The reciprocal relationship:

$$\frac{dy}{dx} = \left(\frac{dx}{dy}\right)^{-1} \quad \text{or} \quad 1\bigg/\frac{dx}{dy} \tag{1-16}$$

2. If u is a function of y and y is a function of x, then

$$\frac{d}{dx}(u) = \frac{d}{dy}(u)\frac{d}{dx}(y) \tag{1-17}$$

3. Further differentiation is possible. The second derivative with respect to the same variable is

$$\frac{d}{dx}\left(\frac{dy}{dx}\right) \quad \text{or} \quad \frac{d^2y}{dx^2}$$

The second derivative may be taken with respect to a different variable:

$$\frac{d}{dy}\left(\frac{du}{dx}\right) \quad \text{or} \quad \frac{d^2u}{dy\,dx}$$

C. Integrals

The *indefinite* integral (antiderivative) always includes a constant of integration. The *definite* integral requires limits of integration. The definite integral represents the area bordered by the graph of a function, the horizontal axis, and vertical lines that intersect the graph at two selected endpoints (limits of integration). Some basic forms of the indefinite integral are

$$\int dx = x + C \tag{1-18}$$

$$\int a\,dx = a\int dx = ax + C \tag{1-19}$$

$$\int (u + v)\,dx = \int u\,dx + \int v\,dx + C \tag{1-20}$$

$$\int u^n\,du = \frac{1}{n+1}u^{n+1} + C \qquad (\text{if } n \neq -1) \tag{1-21}$$

$$\int u^{-1}\,du = \int \frac{1}{u}\,du = \int d\ln u = \ln u + C \tag{1-22}$$

$$\int e^u\,du = e^u + C \tag{1-23}$$

EXAMPLE 1-18: Integration is the reverse of differentiation. Demonstrate this by differentiating x^5 and then integrating the result between the endpoints 0 and x.

Solution: Use Eq. (1-13) as a differential:

$$d(u^n) = nu^{n-1}\,du$$

Substitute x^5 for u^n:

$$d(x^5) = 5x^4\,dx$$

To integrate, first use Eq. (1-19) to remove the constant from the integral:

$$\int_0^x 5x^4\,dx = 5\int_0^x x^4\,dx$$

Then use Eq. (1-21):

$$5\int_0^x x^4\,dx = 5(\tfrac{1}{5}x^5)\Big|_0^x = x^5 + 0$$

We can handle expressions that don't fit any of the basic integration forms by using integration by parts, transformations (or substitutions), or partial fractions. If one of these methods is needed, we will describe it.

D. Partial derivatives

In the basic-form derivatives, u is a function only of x, i.e., $u = f(x)$. If u is a function of two variables, x and y, i.e., $u = f(x, y)$, we can use the basic expressions by holding x constant and taking the derivative with respect to y. The result is the **partial derivative** of u with respect to y (x constant) and is written $\left(\frac{\partial u}{\partial y}\right)_x$. The **total differential** of a function is the sum of its partial derivatives, each multiplied by the infinitesimal change dx or dy:

$$du = \left(\frac{\partial u}{\partial x}\right)_y dx + \left(\frac{\partial u}{\partial z}\right)_x dy \tag{1-24}$$

EXAMPLE 1-19: Write the expression for the total differential of $M(x, y, z)$.

Solution: We want an expression in the form of Eq. (1-24), but we must hold two variables constant for each partial derivative:

$$dM = \left(\frac{\partial M}{\partial x}\right)_{y,z} dx + \left(\frac{\partial M}{\partial y}\right)_{x,z} dy + \left(\frac{\partial M}{\partial x}\right)_{x,y} dz$$

Partial derivatives exhibit the three properties of derivatives listed in Section 1-4B. These properties, combined with appropriate mathematical manipulations, enable us to obtain expressions for additional partial derivatives. One approach is illustrated in the following example. We write the total differential for the variable [e.g., u in Eq. (1-24)] which will be the constant subscript in the desired partial derivative. Then we differentiate with respect to the variable (either x or y) which will yield the desired partial derivative. When u is made constant, its derivative equals zero, and any remaining derivatives become partial derivatives with u as the constant subscript. (Later, we'll use such techniques to write expressions for calculating the rate of change of one physical property with respect to another.)

EXAMPLE 1-20: Derive an expression for $\left(\frac{\partial x}{\partial y}\right)_u$ from Eq. (1-24).

Solution: We need to introduce the term $\left(\frac{dx}{dy}\right)$ into Eq. (1-24). We do this by dividing both sides of the equation by dy. Or, in proper mathematical terminology, we differentiate the expression with respect to y:

$$\frac{du}{dy} = \left(\frac{\partial u}{\partial x}\right)_y \frac{dx}{dy} + \left(\frac{\partial u}{\partial y}\right)_x \frac{dy}{dy}$$

Of course, $\frac{dy}{dy} = 1$, and if we hold u constant, $\frac{du}{dy} = 0$. Now, we must write $\frac{dx}{dy}$ as a partial derivative with u held constant:

$$0 = \left(\frac{\partial u}{\partial x}\right)_y \left(\frac{\partial x}{\partial y}\right)_u + \left(\frac{\partial u}{\partial y}\right)_x$$

Solving for the desired quantity, we have

$$\left(\frac{\partial x}{\partial y}\right)_u = -\frac{(\partial u/\partial y)_x}{(\partial u/\partial x)_y} = -\left(\frac{\partial u}{\partial y}\right)_x \left(\frac{\partial x}{\partial u}\right)_y$$

We can obtain an important relationship by rearranging the result of Example 1-20:

EULER'S CHAIN RELATION

$$\left(\frac{\partial x}{\partial y}\right)_u \left(\frac{\partial y}{\partial u}\right)_x \left(\frac{\partial u}{\partial x}\right)_y = -1 \qquad \textbf{(1-25)}$$

Note that each of the three variables appears once in the numerators, once in the denominators, once as constants, and only once in each term.

E. Exact differentials

For certain functions, the change in value between two points is independent of the path. (In the chapter on thermodynamics, we call these *state functions* because they depend only on the state of the system.) The differentials of such functions are said to be **exact** (or **perfect**) **differentials**. A differential is exact if and only if the following conditions are satisfied:

1. The *line integral* is independent of the path: The actual change in value must equal the integral of the differential change.
2. The *cyclic integral* equals zero: If the system returns to the original point, or state, it has completed a *cycle*; i.e., the value of the state function returns to the original value. Thus the net change is zero.
3. There exists a function that can be differentiated to give the differential. (The importance of this will become clear later.)

If any of these conditions is satisfied for a differential, the others must hold and the differential is exact. If any of these conditions does not hold for a differential, then the differential is **inexact**.

F. Euler's reciprocity theorem

To test whether a differential is exact, we use a procedure credited to Euler. If a function depends on more than one variable, e.g., $L = f(x, y)$, the total differential of the function may be expressed as $dL = M\,dx + N\,dy$, where M and N are *partial derivatives* of L, as in Eq. (1-24). The partial derivatives are functions of x and y:

$$M(x, y) = \left(\frac{\partial L}{\partial x}\right)_y \quad \text{and} \quad N(x, y) = \left(\frac{\partial L}{\partial y}\right)_x$$

If dL is an exact differential, we have

EULER'S RECIPROCITY THEOREM

$$\frac{dM}{dy} = \frac{dN}{dx} \qquad \textbf{(1-26)}$$

These quantities are second derivatives. We see this when we substitute for M and N:

$$\left[\frac{\partial}{\partial y}\left(\frac{\partial L}{\partial x}\right)_y\right]_x = \left[\frac{\partial}{\partial x}\left(\frac{\partial L}{\partial y}\right)_x\right]_y$$

or

$$\frac{\partial^2 L}{\partial y\,\partial x} = \frac{\partial^2 L}{\partial x\,\partial y} \qquad \textbf{(1-27)}$$

Note that the order of differentiation does not matter for an exact differential.

EXAMPLE 1-21. If $dL = y\,dx + x\,dy$, is dL an exact differential?

Solution: Applying Eq. (1-26) and letting $M = y$, $N = x$, we have

$$\frac{dM}{dy} = \frac{dy}{dy} = 1 \quad \text{and} \quad \frac{dN}{dx} = \frac{dx}{dx} = 1$$

Being equal, these derivatives satisfy the reciprocity theorem, and so dL is exact.

EXAMPLE 1-22 In Example 1-21, dL is exact. Thus there must be a function that can be differentiated to give y dx + $x\,dy$. What is it?

Solution: The expression given in Example 1-21 is the same as that in Eq. (1-12) for the derivative of a product: $d(uv) = v\,du + u\,dv$. Thus $L = xy$.

EXAMPLE 1-23. If $dR = y^2\,dx + x^2\,dy$, is dR exact?

Solution: Again, applying Eq. (1-26), we have

$$\frac{dM}{dy} = \frac{dy^2}{dy} = 2y \quad \text{and} \quad \frac{dN}{dx} = \frac{dx^2}{dx} = 2x$$

Because $2y \neq 2x$, dR is not exact; we say that the differential is inexact or imperfect.

G. Inexact differentials

Various symbols are used to show that a differential is inexact, including $đ$, $\delta\!\!\!{}^-$, and δ. Some authors simply use d for both exact and inexact differentials. In the chapter on the first law of thermodynamics, we'll see that, in general, the expressions for heat and work are inexact differentials. They will be designated $đq$ and $đw$, respectively.

SUMMARY

1. A dimension is a concept or abstraction that expresses a type of measurement. A unit is a specific amount of a dimension.
2. The International System of Units (SI) is based on seven dimensionally independent physical quantities. A base unit is designated for each of the seven physical quantities. Units for all other physical quantities are derived from the seven base units.
3. Auxiliary units are expressed as multiples of 10^3, although centi-, deci-, and deca- are retained in SI.
4. Conversion factors must be used to change units into an SI-compatible form, although some well-established non-SI units are accepted in physical chemistry.
5. Exponential relationships between variables may be written using logarithms.
6. An exact differential satisfies three significant conditions: The line integral is independent of the path; the cyclic integral equals zero; and a function exists that can be differentiated to give the differential.
7. Euler's reciprocity theorem is the test for whether a differential is exact.

SOLVED PROBLEMS

Dimensions, Units, and SI

PROBLEM 1-1 What are the dimensions of density? Name some units of density.

Solution: Density is the ratio of mass to volume; thus it has dimensions of mass and reciprocal volume: $m \times V^{-1}$. Since volume has dimensions of length cubed, you also could say that density has the dimensions,

$m \times l^{-3}$. Many different units of density can be used because there are many units of mass, volume, and length. Some combinations are g L^{-1}, lb gal^{-1}, g cm^{-3}, or kg m^{-3}.

PROBLEM 1-2 Define the unit *pascal* in SI base units. (See Table 1-1.) Remember that pressure is force per unit area.

Solution: From Table 1-1, you find that the unit of force is the newton; thus

$$\text{Pressure} = \frac{\text{Force}}{\text{Area}} \qquad \text{becomes} \qquad \frac{\text{N}}{\text{m}^2}$$

In SI base units, the newton is kg m s^{-2}; therefore

$$\text{Pressure} = \frac{\text{kg m s}^{-2}}{\text{m}^2} = \text{kg m}^{-1}\,\text{s}^{-2} = \text{pascal}$$

PROBLEM 1-3 The unit *liter* is not an SI unit. How should you describe a volume of one liter in SI units?

Solution: You must use the dimension of length (cubed). Remember that

$$1\ \text{L} = 10^3\ \text{cm}^3$$

Since

$$1\ \text{cm}^3 = (10^{-2}\ \text{m})^3 = 10^{-6}\ \text{m}^3$$

you have

$$1\ \text{L} = 10^{-3}\ \text{m}^3$$

You may also use decimeters:

$$1\ \text{dm} = 10\ \text{cm}$$
$$1\ \text{dm}^3 = 10^3\ \text{cm}^3 = 1\ \text{L}$$

This permits a simple replacement of L with dm^3 in traditional units, e.g., those of the gas constant, R.

Constants and Conversion Factors

PROBLEM 1-4 Convert a density of 1.43 g mL^{-1} to the proper SI units, kg m^{-3}.

Solution: You must find a conversion path through various units. For the numerator, the conversion of grams to kilograms is clear. For the denominator, you must convert volume to length. Recall that 1 mL = 1 cm^3 and that the meter is the unit required.

You now need to use the numerical relationships between the units to complete the conversion:

$$x = 1.43\,\frac{\text{g}}{\text{mL}}\left(\frac{1\ \text{kg}}{10^3\ \text{g}}\right)\left(\frac{1\ \text{mL}}{1\ \text{cm}^3}\right)\left(\frac{100\ \text{cm}}{1\ \text{m}}\right)^3 = 1.43 \times 10^3\ \text{kg m}^{-3}$$

PROBLEM 1-5 Convert the energy gas constant, $R = 8.314$ J $K^{-1}mol^{-1}$, to the pressure–volume (L atm) gas constant.

Solution: You will need to use some numbers from Sections 1-1 and 1-2 and Table 1-1. Using Table 1-1, you can set up a conversion factor from the units:

$$\frac{\text{J}}{\text{Pa}} = \frac{\text{kg m}^2\,\text{s}^{-2}}{\text{kg m}^{-1}\,\text{s}^{-2}} = \text{m}^3 \qquad \text{or} \qquad \frac{\text{J}}{\text{m}^3\,\text{Pa}} = 1$$

From Problem 1-4,

$$1\ \text{L} = 10^{-3}\ \text{m}^3 \qquad \text{or} \qquad \frac{1\ \text{L}}{10^{-3}\ \text{m}^3} = 1$$

From Section 1-1F,

$$1\ \text{atm} = 101\ 325\ \text{Pa} \qquad \text{or} \qquad \frac{1\ \text{atm}}{101\ 325\ \text{Pa}} = 1$$

Now you can multiply R by these conversion factors (inverting when necessary):

$$R = \left(\frac{8.314\ \text{J}}{\text{K mol}}\right)\left(\frac{\text{m}^3\ \text{Pa}}{\text{J}}\right)\left(\frac{\text{L}}{10^{-3}\ \text{m}^3}\right)\left(\frac{\text{atm}}{101\,325\ \text{Pa}}\right)$$

$$= \frac{8.314}{101\,325 \times 10^{-3}}\,\frac{\text{L atm}}{\text{K mol}} = 0.082\,05\ \text{L atm K}^{-1}\text{mol}^{-1}$$

Essentials of Logarithms

PROBLEM 1-6 Use logarithms to the base 10 to find x in $x = \dfrac{46.789(0.000\,345\,67)}{2.4353}$.

Solution: You must apply Eqs. (1-4) and (1-5) and rewrite the numbers in scientific notation:

$$\log x = \log(4.6789 \times 10^1) + \log(3.4567 \times 10^{-4}) - \log(2.4353)$$

Next, find the mantissas from a table (or a calculator); you should have

$$\log x = 1.6701 + \bar{4}.5387 - 0.3865 = \bar{3}.8223$$

Then

$$x = \text{antilog}(\bar{3}.8223) = 6.642 \times 10^{-3}$$

PROBLEM 1-7 Raise 3 to the 3.5 power using logarithms to the base 10.

Solution:

$$x = 3^{3.5}$$
$$\log x = 3.5 \log 3 = 3.5(0.4771) = 1.6699$$
$$x = \text{antilog}(1.6699) = 46.76$$

PROBLEM 1-8 Use a calculator to solve the following equations:

(a) $x = \log 0.074$ **(b)** $y = 4^{15}$ **(c)** $x = e^{4.2}$

Solution: Different makes of calculators may have different key labels and procedures. Be sure you know how to use yours.

(a) Enter 0.074; push [log] key:

$$x = -1.131$$

(b) Enter 4; push [y^x] key; enter 15, push [=] key:

$$y = 1.07 \times 10^9$$

(c) Rearrange the expression to $\ln x = 4.2$. Enter 4.2; push [inv] key; push [ln] key:

$$x = 66.7$$

PROBLEM 1-9 In the chapter on kinetics, you will encounter the equation $kt = \ln\left(\dfrac{a}{a-x}\right)$. Solve for x if $a = 15$ mol, $t = 30$ min, and $k = 0.046\ \text{min}^{-1}$.

Solution: In general, do all the mathematical manipulation you can before substituting numbers into an equation, in order to avoid confusion. In this problem, you must solve for $a - x$ to obtain x. Note that you don't need to write units into the fractional logarithmic term because they cancel: The quantities a and $a - x$ may have any unit of concentration so long as it is the same for both terms.

Substituting y for $a - x$ and applying Eq. (1-5), you get

$$kt = \ln\frac{a}{y} = \ln a - \ln y$$

Rearrange terms:

$$\ln y = \ln a - kt$$

Substitute numbers for a, t, and k:

$$\ln y = \ln 15 - 0.046\ \text{min}^{-1} \times 30\ \text{min}$$

Use your calculator to find ln 15 = 2.708, substitute, and solve for x:

$$\ln y = 2.708 - 1.380 = 1.328$$
$$y = 3.77 = a - x$$
$$x = a - y = 15 - 3.77 = 11.23 \text{ mol}$$

An alternative method emphasizes the exponential character of logarithms. Change the given expression to

$$e^{kt} = \frac{a}{a - x}$$

Rearrange terms:

$$a - x = ae^{-kt}$$
$$a - ae^{-kt} = x$$
$$x = a(1 - e^{-kt}) = (15 \text{ mol})[1 - e^{-(0.046)(30)}] = 15(1 - e^{-1.38})$$

Use your calculator to find the inverse ln of -1.38. Then:

$$x = 15(1 - 0.2516) = 11.23 \text{ mol}$$

Essentials of Calculus

PROBLEM 1-10 Differentiate the expression $y = e^{4-x^2}$.

Solution: The derivative $\frac{dy}{dx} = \frac{d}{dx}(e^{4-x^2})$ is of the form of Eq. (1-15):

$$\frac{d}{dx}(e^u) = e^u \frac{d}{dx}(u) \qquad \textbf{(a)}$$

Thus you need to differentiate the exponent, using Eqs. (1-10) and (1-13):

$$\frac{d}{dx}(4 - x^2) = 0 - 2x$$

Now you apply Eq. (1-12) for u^n to find

$$-\frac{d}{dx}(x^2) = -2x \qquad \textbf{(b)}$$

Substituting eq. (b) into eq. (a), you get

$$\frac{d}{dx}(e^{4-x^2}) = -2xe^{4-x^2}$$

PROBLEM 1-11 Differentiate the expression $y = \frac{1}{x^3} + \ln x^3$.

Solution: This is really two independent problems:

$$\textbf{(a)}\ d\left(\frac{1}{x^3}\right) \qquad \text{and} \qquad \textbf{(b)}\ d(\ln x^3)$$

(a) Using Eq. (1-13):

$$d\left(\frac{1}{x^3}\right) = d(x^{-3}) = -3x^{-4} = -\frac{3}{x^4}$$

(b) Using Eq. (1-14):

$$d(\ln x^3) = \frac{1}{x^3}\,d(x^3) = \frac{1}{x^3}(3x^2) = \frac{3}{x}$$

Combining parts (a) and (b):

$$dy = -\frac{3}{x^4} + \frac{3}{x}$$

PROBLEM 1-12 Integrate the expression $\frac{dy}{dx} = \frac{a}{x} + bx + cx^2$ without limits.

Solution: First you must write the expression in the proper form with dx on the right-hand side of the equation:

$$\int dy = \int \left(\frac{a}{x} + bx + cx^2\right) dx$$

The integral of a sum is the sum of the integrals, and constants may be moved outside the integral sign as in Eq. (1-19):

$$\int dy = a \int x^{-1}\, dx + b \int x\, dx + c \int x^2\, dx$$

For the first term, you use Eq. (1-22) and, for the second and third terms, Eq. (1-21):

$$y = a \ln x + \frac{1}{2} bx^2 + \frac{1}{3} cx^3 + C$$

PROBLEM 1-13 Evaluate the integral $\int x^{-2}\, dx$ between the limits $x_2 = 25$ and $x_1 = 15$.

Solution: You are asked to find the definite integral

$$\int_{15}^{25} x^{-2}\, dx$$

Use Eq. (1-21) to carry out the integration:

$$\int_{15}^{25} x^{-2}\, dx = \frac{1}{-1} x^{-1} \Big|_{15}^{25} = -\frac{1}{x}\Big|_{15}^{25}$$

You subtract the value of the term at the lower limit from the value at the upper limit:

$$\int_{15}^{25} x^{-2}\, dx = -\frac{1}{25} - \left(-\frac{1}{15}\right) = -\frac{1}{25} + \frac{1}{15} = 0.0267$$

PROBLEM 1-14 **(a)** If the variable z is a function of both x and y, i.e., $z = f(x, y)$, write the total differential of z. **(b)** If the relation of z, x, and y is $z = 7\frac{x^2}{y}$, evaluate the partial derivatives of z and write the total differential of z.

Solution:

(a) You need to write an expression like Eq. (1-24):

$$dz = \left(\frac{\partial z}{\partial x}\right)_y dx + \left(\frac{\partial z}{\partial y}\right)_x dy$$

(b) Begin by doing each partial derivative separately:

$$\left(\frac{\partial z}{\partial x}\right)_y = \frac{7}{y}\frac{\partial}{\partial x}(x^2) \qquad \left(\frac{\partial z}{\partial y}\right)_x = 7x^2 \frac{\partial}{\partial y}\left(\frac{1}{y}\right)$$

Use Eq. (1-13):

$$\frac{\partial}{\partial x}(x^2) = 2x\frac{\partial x}{\partial x} = 2x \qquad \frac{\partial}{\partial y}\left(\frac{1}{y}\right) = -y^{-2}\frac{\partial y}{\partial y} = -\frac{1}{y^2}$$

Thus

$$\left(\frac{\partial z}{\partial x}\right)_y = \frac{14x}{y} \qquad \left(\frac{\partial z}{\partial y}\right)_x = -\frac{7x^2}{y^2}$$

and the total differential is

$$dz = \frac{14x}{y} dx - \frac{7x^2}{y^2} dy$$

PROBLEM 1-15 In Example 1-21, we saw that $dL = y\,dx + x\,dy$ is exact. Is this true if the sign of x is negative?

Solution: You must apply Euler's reciprocity theorem, Eq. (1-26), to test whether the differential is exact. For $dR = y\,dx - x\,dy$,

$$M = y \quad \text{and} \quad \frac{dM}{dy} = \frac{dy}{dy} = 1$$

$$N = -x \quad \text{and} \quad \frac{dN}{dx} = \frac{d}{dx}(-x) = -1$$

Because these expressions are not equal, dR is not exact and should be written $đR$.

PROBLEM 1-16 Is $du = \dfrac{dx}{y} - \dfrac{x\,dy}{y^2}$ an exact differential?

Solution: Again you must apply Euler's theorem, Eq. (1-26). Let $M = 1/y$ and $N = -x/y^2$. Take the derivative of M with respect to y:

$$\frac{dM}{dy} = \frac{d}{dy}\left(\frac{1}{y}\right) = -\frac{1}{y^2}$$

Then take the derivative of N with respect to x:

$$\frac{dN}{dx} = \frac{d}{dx}\left(-\frac{x}{y^2}\right) = -\frac{1}{y^2}\frac{dx}{dx} = -\frac{1}{y^2}$$

Since the two are equal, du is exact.

Supplementary Exercises

PROBLEM 1-17 The dimensions of velocity are ________ and ________ (reciprocal). The SI unit for velocity is ________. The SI base units for temperature, mass, and current are ________, ________, and ________, respectively. The SI unit, symbol, and definition (in base units) for pressure are ________, ________, and ________, respectively. There are ________ mm^3 in 1 m^3 and ________ dm^3 in 1 kL. The derivative of x^4 is ________. The method used for testing for an exact differential is ________.

PROBLEM 1-18 Solve the following for x: **(a)** $x = 6.53^{2.5}$; **(b)** $\log x = -5.678$.

PROBLEM 1-19 Differentiate $y = 6x^3 - 4x^2 + 5$ with respect to x.

PROBLEM 1-20 Differentiate $y = x^2e^{-3x}$ with respect to x.

PROBLEM 1-21 Determine the indefinite integral of $6e^{3x}\,dx$.

PROBLEM 1-22 Evaluate $\displaystyle\int_{10}^{20}\left(\frac{1}{x}\right)dx$.

PROBLEM 1-23 If J is a function of K and L, derive an expression for $\left(\dfrac{\partial K}{\partial L}\right)_J$.

PROBLEM 1-24 **(a)** Is dZ an exact differential if $dZ = (y - 3x^2)\,dx + (x + 2y)\,dy$?
(b) What is the value of the partial derivatives obtained from testing dZ?

Answers to Supplementary Exercises

1-17 length; time; m s^{-1}; kelvin, kilogram, ampere; pascal, Pa, kg m^{-1} s^{-2}; 10^9, 10^3; $4x^3$; Euler's reciprocity theorem **1-18** (**a**) 1.09×10^2 (**b**) 2.10×10^{-6} **1-19** $18x^2 - 8x$

1-20 $xe^{-3x}(2 - 3x)$ **1-21** $2e^{3x} + C$ **1-22** 0.693 **1-23** $-\left(\frac{\partial J}{\partial L}\right)_K \left(\frac{\partial K}{\partial J}\right)_L$

1-24 (**a**) yes (**b**) 1

2 GASES

THIS CHAPTER IS ABOUT

- ☑ The Ideal Gas Laws
- ☑ The Gas Constant
- ☑ Other Laws Concerning Ideal Gases
- ☑ Properties of Real Gases
- ☑ Equations of State for Real Gases
- ☑ The Kinetic Molecular Theory of Ideal Gases

2-1. The Ideal Gas Laws

A substance may be in a gaseous, liquid, or solid state. Its state is determined primarily by the pressure and the absolute (Kelvin) temperature. However, we must also specify the volume and amount of substance if we are to describe a gas completely. In order to determine the relationships among the four variables, P, V, T, and n, we must hold two of them constant and measure the other two as they change. Six relationships (some of which bear famous names) are possible; all are special cases of the ideal gas equation.

A. If n and T are constant, $P \propto 1/V$.

This relationship between P and V is known as **Boyle's law**. It may be stated as

BOYLE'S LAW $$P \propto \frac{1}{V} \quad \text{or} \quad P = k_A\left(\frac{1}{V}\right) \quad \text{or} \quad PV = k_A \qquad \textbf{(2-1)}$$

where k_A is a proportionality constant.

We can illustrate this $P-V$ relationship with a hand pump used for filling balls or bicycle tires. If we close the outlet and push down on the handle, the air in the cylinder is compressed into a smaller volume at the higher pressure. When we place the end of the pump hose on the air needle or inner-tube stem, we effectively close the outlet with the ball or tire, forcing the air at the higher pressure to inflate the ball or the tire.

EXAMPLE 2-1: If a certain quantity of gas occupies 5.0 dm^3 at 300 K and 2.0 atm pressure, what will the volume be at 3.0 atm?

Solution: We apply Eq. (2-1) to calculate k_A.

$$k_A = PV = (2.0\ \text{atm})(5.0\ \text{dm}^3) = 10\ \text{atm dm}^3$$

At the new pressure,

$$V_2 = k_1\left(\frac{1}{P_2}\right) = (10\ \text{atm dm}^3)\left(\frac{1}{3.0\ \text{atm}}\right) = 3.3\ \text{dm}^3$$

A shorter solution is possible if we recognize that the PV products for all sets of conditions are equal:

$$P_1V_1 = k_A = P_2V_2 = P_3V_3 = \cdots \qquad \textbf{(2-1a)}$$

Thus

$$P_2 = \frac{P_1 V_1}{V_2} = \frac{(2.0\ \text{atm})(5.0\ \text{dm}^3)}{3.0\ \text{atm}} = 3.3\ \text{dm}^3$$

B. If n and P are constant, $V \propto T$.

This relationship between V and T is known as **Charles' law** (sometimes as **Gay-Lussac's law**). It may be stated as

CHARLES' LAW $\qquad V \propto T \quad \text{or} \quad V = k_B T \quad \text{or} \quad \frac{V}{T} = k_B \qquad$ **(2-2)**

We can demonstrate this V–T relationship if we blow up and tie a balloon at room temperature and then place it in a refrigerator. Its volume will decrease.

EXAMPLE 2-2: If a balloon has a volume of 500 mL at 25 °C and 1 atm pressure, what will be its volume at the boiling point of liquid nitrogen (−196 °C)?

Solution: We use Eq. (2-2) to calculate k_B (remembering to use absolute temperatures).

$$\frac{V}{T} = k_B = \frac{500\ \text{mL}}{(273 + 25)\ \text{K}} = 1.68\ \text{mL K}^{-1}$$

At the new temperature,

$$V_2 = (1.68\ \text{mL K}^{-1})(273 - 196\ \text{K}) = 129\ \text{mL}$$

We could have used

$$\frac{V_1}{T_1} = \frac{V_2}{T_2} \quad \text{or} \quad V_2 = \frac{V_1 T_2}{T_1} \qquad \textbf{(2-2a)}$$

C. If n and V are constant, $P \propto T$.

This relationship between P and T doesn't have a name. It may be stated as

$$P \propto T \quad \text{or} \quad P = k_C T \quad \text{or} \quad \frac{P}{T} = k_C \qquad \textbf{(2-3)}$$

We can demonstrate this P–T relationship by measuring the pressure of our automobile tires both before and after a long drive. The pressure will be higher and the tires will feel hotter to the touch after the drive.

EXAMPLE 2-3: If the tire pressure is 28 lb in^{-2} before a trip, and the air temperature is 24°C, what will be the pressure after the friction of driving has raised the temperature to 32 °C?

Solution: We use Eq. (2-3).

$$k_C = \frac{28\ \text{lb in}^{-2}}{297\ \text{K}} = 0.094\ \text{lb in}^{-2}\text{K}^{-1}$$

At the new temperature,

$$P_2 = k_C T_2 = (0.094)(305) = 29\ \text{lb in}^{-2}$$

D. If P and T are constant, $V \propto n$.

This relationship between V and n is called **Avogadro's law**, although he didn't do the studies to establish it as did Boyle, Charles, and Gay-Lussac for the laws that carry their names. It may be stated as

AVOGADRO'S LAW $V \propto n$ or $V = k_D n$ or $\frac{V}{n} = k_D$ **(2-4)**

We can demonstrate this V–n relationship by blowing air into a flattened plastic bag. The bag will fill out as more air enters.

EXAMPLE 2-4: If a nonelastic balloon has a volume of 700 m^3 when it contains 20.0 mol of helium, what volume will it have at the same temperature and pressure if we inject an additional 5.0 mol of helium into it?

Solution: We use Eq. (2-4).

$$k_D = \frac{V}{n} = \frac{700 \text{ m}^3}{20.0 \text{ mol}} = 35.0 \text{ m}^3 \text{ mol}^{-1}$$

At the new amount of substance,

$$V_2 = k_D n_2 = (35.0 \text{ m}^3 \text{ mol}^{-1})(25.0 \text{ mol}) = 875 \text{ m}^3$$

E. If V and T are constant, $P \propto n$.

This relationship may be stated as

$$P \propto n \quad \text{or} \quad P = k_E n \quad \text{or} \quad \frac{P}{n} = k_E \qquad \textbf{(2-5)}$$

We can illustrate this V–T relationship by a familiar laboratory activity, the use of a cylinder of gas. As gas is removed from a cylinder, the needle of the cylinder's pressure gauge drops.

EXAMPLE 2-5: A steel reaction vessel contains 0.50 mol H_2 and shows a pressure of 9.0 lb in^{-2}. How many moles of H_2 must be added to raise the pressure to 15.0 lb in^{-2}?

Solution: We use Eq. (2-5).

$$k_E = \frac{P}{n} = \frac{9.0}{0.50} = 18 \text{ lb in}^{-2} \text{ mol}^{-1}$$

At the new pressure,

$$n_2 = \frac{P_2}{k_E} = \frac{15.0}{18} = 0.83 \text{ mol}$$

The amount to be added is $n_2 - n_1 = 0.83 - 0.50 = 0.33$ mol.

F. If P and V are constant, $T \propto 1/n$.

This relationship between T and n has no name, and it is unfamiliar and difficult to visualize. It can be stated as

$$T \propto \frac{1}{n} \quad \text{or} \quad T = k_F\left(\frac{1}{n}\right) \quad \text{or} \quad nT = k_F \qquad \textbf{(2-6)}$$

We can illustrate this $T - n$ relationship by imagining a rigid vessel containing a certain quantity of gas. Then we ask ourselves what will happen if we add more gas. Our first impulse probably would be to say that the pressure would rise. Remember, however, that we have imposed a condition of constant pressure. The only way to add more gas to a fixed volume without a change in pressure is to *cool the system*.

EXAMPLE 2-6: A steel flask contains 5.0 mol of a gas at 400 K and a certain pressure. We want to put an additional 3.0 mol of gas into the flask while maintaining the same pressure. What must we do?

Solution: We must cool the flask. We calculate the needed temperature by using Eq. (2-6).

$$k_F = nT = (5.0 \text{ mol})(400 \text{ K}) = 2.0 \times 10^3 \text{ mol K}$$

For the new quantity, 8.0 mol,

$$T_2 = \frac{k_F}{n_2} = \frac{2.0 \times 10^3 \text{ mol K}}{8.0 \text{ mol}} = 250 \text{ K}$$

G. If none of the variables is constant, $PV \propto nT$.

From Eq. (2-1), $PV = k_A$, but from Sections 2-1B and 2-1C, both P and V independently are directly proportional to T. Thus the product PV also must be directly proportional to T.

$$PV \propto T \qquad \text{or} \qquad PV = k_G T \qquad \text{or} \qquad \frac{PV}{T} = k_G \tag{2-7}$$

You have probably used this relationship in the form

$$\frac{P_1 V_1}{T_1} = \frac{P_2 V_2}{T_2} \tag{2-7a}$$

Equations (2-7) and (2-7a) are valid only if n is constant. From Sections 2-1D and 2-1E, we know that P and V are directly proportional to n. From Section 2-1F, T is inversely proportional to n or, conversely, $1/T$ is directly proportional to n. Thus

$$\frac{PV}{T} \propto n \qquad \text{or} \qquad \frac{PV}{T} = k_H n \tag{2-8}$$

If we replace k_H with the usual symbol R and rearrange, we have

IDEAL GAS EQUATION

$$PV = nRT \tag{2-9}$$

We can now see that Eqs. (2-1) to (2-7) are special cases of Eq. (2-9), obtained by holding the desired variables constant. We use Eq. (2-9) in calculations in which the amount of substance is known or is to be determined.

EXAMPLE 2-7: What is the pressure exerted by 4.00 mol of an ideal gas contained in a 15.0-L flask at 250 K?

Solution: We need to rearrange Eq. (2-9), then solve.

$$P = \frac{nRT}{V} = \frac{(4.00 \text{ mol})(0.0821 \text{ L atm K}^{-1}\text{mol}^{-1})(250 \text{ K})}{15.0 \text{ L}} = 5.47 \text{ atm}$$

If the number of moles is constant, Eq. (2-7a) is more convenient for calculations. A practical application of this expression is to change data concerning a gas from experimental conditions to reference-state conditions.

EXAMPLE 2-8: In an experiment, we fill a 1-L flask with a gas at an atmospheric pressure of 751 mm Hg and a room temperature of 26 °C. Usually gas data are referred to STP, i.e., standard temperature and pressure, which are 273 K and 1 atm, respectively. What volume would this gas occupy at STP?

Solution: It can be helpful to list the information given and the unknown:

$$\begin{aligned} P_1 &= 751 \text{ mm Hg} & P_2 &= 760 \text{ mm Hg} \\ V_1 &= 1.00 \text{ L} & V_2 &= \text{Unknown} \\ T_1 &= (26 + 273) \text{ K} & T_2 &= 273 \text{ K} \end{aligned}$$

We rearrange Eq. (2-7a) and solve for V_2.

$$V_2 = \frac{P_1 V_1}{T_1}\left(\frac{T_2}{P_2}\right) = \frac{(751 \text{ mm Hg})(1.00 \text{ L})}{299 \text{ K}}\left(\frac{273 \text{ K}}{760 \text{ mm Hg}}\right) = 0.902 \text{ L}$$

2-2. The Gas Constant

The constant in the ideal gas equation is a universal constant. By rearranging Eq. (2-9), we see that several physical quantities are incorporated in R.

$$R = \frac{PV}{nT} = \frac{PV_m}{T} \qquad \textbf{(2-9a)}$$

where V_m is the volume of one mole. The numerator is an energy term and is expressed in different units for different applications, as shown in Table 2-1.

TABLE 2-1: Values of the Gas Constant *R*

Precise value and units	Value to four significant figures	Areas of application
1.987 19 cal $K^{-1}mol^{-1}$	1.987	Heat and work
0.082 056 8 L atm $K^{-1}mol^{-1}$	8.206×10^{-2}	P in atm; V in L
8.314 41 J $K^{-1}mol^{-1}$	8.314	P in Pa; V in m^3
8.314 41 Pa m^3 $K^{-1}mol^{-1}$	8.314	P in Pa; V in m^3
8314. 41 Pa dm^3 $K^{-1}mol^{-1}$	8.314×10^3	P in Pa; V in dm^3 (or L)
0.083 144 1 L bar $K^{-1}mol^{-1}$	8.314×10^{-2}	P in bar; V in dm^3 (or L)

You should memorize the first three values to four significant figures; you can obtain the second three if you know the relationships between units.

2-3. Other Laws Concerning Ideal Gases

Two important laws usually are not included in the gas laws—Dalton's law and Graham's law.

A. Dalton's law of partial pressures

Dalton's law states that the total pressure of a mixture of gases is the sum of the pressures that each quantity of gas would exert if that quantity occupied the same volume by itself. This law has great practical importance when a gas is collected over water and thus is mixed with water vapor. Mathematically, it may be expressed as

DALTON'S LAW

$$P_{\text{total}} = P_A + P_B + P_C + \cdots \qquad \textbf{(2-10)}$$

EXAMPLE 2-9: We collect some oxygen by bubbling it through water and into a gas buret. When the water levels equalize, the volume is 33.6 mL. The room temperature is 24.6 °C, and the barometric pressure is 752.6 mm Hg. We want to calculate the amount of oxygen present.

Solution: We should recognize that the oxygen will have water vapor in it. In order to calculate the pressure of oxygen alone, we use Eq. (2-10) and rearrange:

$$P_{O_2} = P_{\text{total}} - P_{H_2O}$$

We must look up the vapor pressure of water at 24.6 °C. It is 23.2 mm Hg; so

$$P_{O_2} = 752.6 - 23.2 = 729.4 \text{ mm Hg} = (729.4 \text{ mm Hg})\left(\frac{1 \text{ atm}}{760 \text{ mm Hg}}\right) = 0.960 \text{ atm}$$

Then

$$n = \frac{PV}{RT} = \frac{(0.960 \text{ atm})(33.6 \text{ mL})}{(82.1 \text{ mL atm K}^{-1}\text{mol}^{-1})(297.8 \text{ K})} = 1.32 \times 10^{-3} \text{ mol}$$

B. Graham's law of diffusion or effusion

Diffusion is the movement of one gas through another, and **effusion** is the process whereby gas escapes through tiny holes. Both processes slow down as the weights of molecules increase. **Graham's law** states that the rate of movement of a gas is inversely proportional to the square root of the molar mass M_m. For two gases at the same temperature and pressure,

GRAHAM'S LAW

$$\frac{\text{rate}_A}{\text{rate}_B} = \frac{\sqrt{M_{m,B}}}{\sqrt{M_{m,A}}} = \sqrt{\frac{M_{m,B}}{M_{m,A}}} \qquad \textbf{(2-11)}$$

EXAMPLE 2-10: Flasks of H_2S and NH_3 are opened at the same distance from you. (**a**) Which gas would you smell first? (**b**) How much faster does that gas move than the other?

Solution:

(**a**) The faster moving gas will reach you first; it is the lighter one, NH_3.
(**b**) We rearrange Eq. (2-11) and solve:

$$\text{rate}_{NH_3} = \left(\sqrt{\frac{M_{m,H_2S}}{M_{m,NH_3}}}\right)(\text{rate}_{H_2S})$$

$$\sqrt{\frac{M_{m,H_2S}}{M_{m,NH_3}}} = \sqrt{\frac{34}{17}} = \sqrt{2.0} = 1.4$$

Thus $\text{rate}_{NH_3} = 1.4(\text{rate}_{H_2S})$.

2-4. Properties of Real Gases

A. Critical point

When a liquid is warmed at constant pressure, a point is reached at which the boundary between the liquid and the gas phases disappears. This point is known as the **critical point**, and the substance is said to be in the **critical state**. The temperature at this point, the **critical temperature**, T_c, often is defined as that temperature above which a gas cannot be liquefied by raising the pressure. The volume and pressure corresponding to T_c are the **critical volume**, V_c, and the **critical pressure**, P_c. The latter may be defined as the pressure that causes liquefaction when the temperature of the substance is T_c. The critical constants, T_c, P_c, V_c, for gases vary widely.

EXAMPLE 2-11: Note that the critical constants of water, carbon dioxide, nitrogen, and oxygen are very different, as shown in Table 2-2. Values in different reference books may vary but should be close to these.

TABLE 2-2: Critical Constants for Four Gases

Gas	T_c (°C)	P_c (atm)	V_c ($cm^3\ mol^{-1}$)
H_2O	374	218	59
CO_2	31	73	94
N_2	−147	34	90
O_2	−118	50	78

B. Compressibility factor

The **compressibility factor** Z (also called the *compression factor* in some textbooks) is merely the ratio of PV to nRT for a gas:

COMPRESSIBILITY FACTOR

$$Z = \frac{PV}{nRT} \qquad \textbf{(2-12)}$$

For one mole of an ideal gas, $Z = 1$ at all pressures; for most gases, $Z = 1$ only at low pressures. Thus Z measures deviation from the ideal. Typical experimental data are plotted for

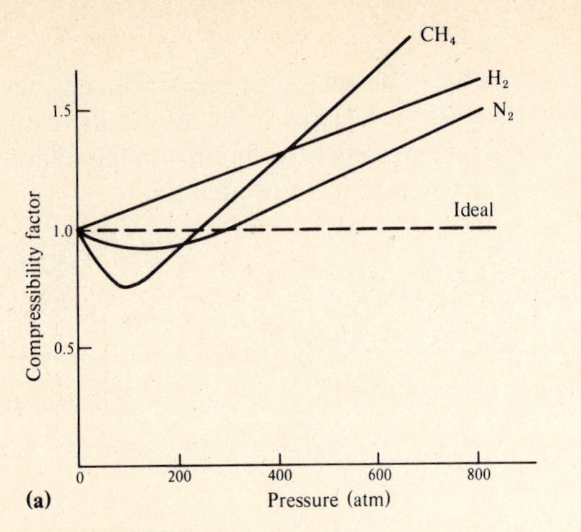

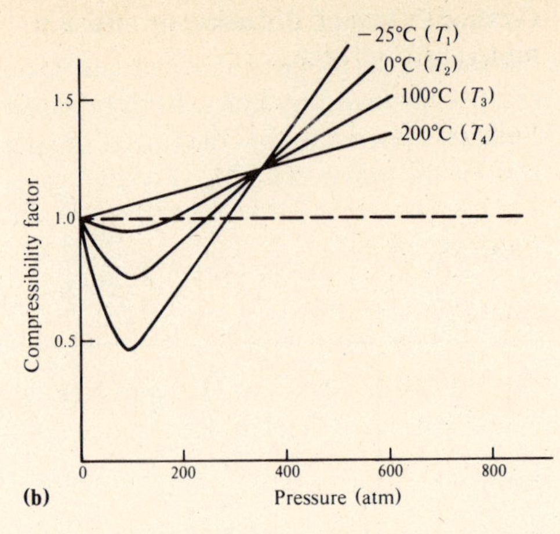

FIGURE 2-1. Compressibility Factors: (**a**) For several gases at 273 K; (**b**) for methane at several temperatures.

several gases at the same temperature in Figure 2-1a and for one gas at several temperatures in Figure 2-1b. Numerically, the value of Z for methane at 0 °C and 150 atm is about 0.76. For light gases, such as hydrogen and helium at temperatures of 0 °C or above, Z is always greater than 1.

EXAMPLE 2-12: We find that 1 mol of a certain gas occupies 0.124 L at 60 °C and 200 atm. Calculate the compressibility factor.

Solution: We substitute in Eq. (2-12) and solve for Z:

$$Z = \frac{PV}{nRT} = \frac{(200 \text{ atm})(0.124 \text{ L})}{(1 \text{ mol})(0.0821 \text{ L atm K}^{-1} \text{mol}^{-1})(333 \text{ K})} = 0.907$$

C. Boyle temperature

The **Boyle temperature,** T_B, is the temperature at which a real gas behaves ideally over a significant range of pressures. In Figure 2-1b, we can see that, between T_3 and T_4, there must be a temperature for which the curve will begin tangent to the $Z = 1$ line, i.e., the initial slope is zero; the curve will remain close to that line for some distance and then rise slowly. Above the Boyle temperature, a gas will deviate positively from ideal behavior.

EXAMPLE 2-13: The Boyle temperatures of hydrogen and oxygen are 117 K and 432 K, respectively. Thus we can predict that at 0 °C hydrogen gas will deviate positively from the ideal at all pressures, whereas oxygen will deviate negatively at first and then positively as the pressure reaches several hundred atmospheres.

2-5. Equations of State for Real Gases

At pressures greater than a few atmospheres, most gases do not behave as predicted by the ideal gas equation, Eq. (2-9). A number of more complex equations of state have been proposed to describe the behavior of real gases. Each has a range of conditions within which it does a better job of predicting behavior than it does in other regions. Each contains parameters with values that were selected to give a good fit to the curve obtained from experimental data.

A. Van der Waals equation

The most useful of the equations of state for real gases is the **van der Waals equation,** in which factors designed to account (1) for attractive forces between molecules and (2) for the volume occupied by the molecules are added to the pressure and volume terms of Eq. (2-9). For n moles, Eq. (2-9) becomes

VAN DER WAALS EQUATION $$\left(P + \frac{an^2}{V^2}\right)(V - nb) = nRT \qquad \textbf{(2-13)}$$

The parameters a and b are known as the van der Waals constants. Values for them can be found in tables.

If we solve Eq. (2-13) for P, we can see that it is essentially the ideal gas equation plus a term in V^2:

$$P = \frac{nRT}{V - nb} - \frac{an^2}{V^2} \qquad \textbf{(2-14)}$$

Note that this equation looks like the beginning of a power series in $1/V$.

EXAMPLE 2-14: Use the van der Waals equation to calculate the pressure exerted by 2 mol CO_2 in a 20.0-L container at 373 K and compare your answer to the pressure calculated for ideal behavior. For CO_2 $a = 3.64 \times 10^5$ Pa L^2 mol^{-2} and $b = 4.27 \times 10^{-2}$ L mol^{-1}.

Solution: We use Eq. (2-14):

$$\begin{aligned} P &= \frac{nRT}{V - nb} - \frac{an^2}{V^2} \\ &= \frac{(2\ \text{mol})(8314\ \text{Pa dm}^3\ \text{K}^{-1}\text{mol}^{-1})(373\ \text{K})}{(20.0\ \text{L}) - (2\ \text{mol})(0.0427\ \text{L mol}^{-1})} - \frac{(3.64 \times 10^5\ \text{Pa L}^2\ \text{mol}^{-2})(2\ \text{mol})^2}{(20.0\ \text{L})^2} \\ &= 3.08 \times 10^5\ \text{Pa} \end{aligned}$$

For ideal behavior

$$P = \frac{nRT}{V} = \frac{(2\ \text{mol})(8314\ \text{Pa dm}^3\ \text{K}^{-1}\text{mol}^{-1})(373\ \text{K})}{20.0\ \text{L}} = 3.10 \times 10^5\ \text{Pa}$$

The van der Waals constants of a gas can be used to approximate the critical constants, or vice versa, from the relationships

$$a = 3P_cV_c^2, \qquad b = \frac{V_c}{3} \qquad \textbf{(2-15)}$$

or

$$V_c = 3b, \qquad P_c = \frac{a}{27b^2}, \qquad T_c = \frac{8a}{27Rb} \qquad \textbf{(2-16)}$$

EXAMPLE 2-15: Calculate the values of a and b for CO_2 from the critical constants in Example 2-11 (Table 2-2) and compare the results to the values given in Example 2-14.

Solution: We use Eqs. (2-15):

$$a = 3P_cV_c^2 = 3(73\ \text{atm})(0.094\ \text{L mol}^{-1})^2\left(\frac{1.01 \times 10^5\ \text{Pa}}{1\ \text{atm}}\right) = 1.95 \times 10^5\ \text{Pa L}^2\ \text{mol}^{-2}$$

$$b = \frac{V_c}{3} = \frac{0.094\ \text{L mol}^{-1}}{3} = 3.1 \times 10^{-2}\ \text{L mol}^{-1}$$

We see that the discrepancy is rather large; in fact, a and b vary somewhat with pressure and temperature. In addition, a molecule of CO_2 is not spherical. Thus it is customary to obtain values of a and b from experimental data, rather than from critical constants.

B. Corresponding states

A useful general equation of state may be obtained by replacing the variables in the van der Waals equation by the ratios of each variable to the corresponding critical constant. The ratios are called **reduced variables** and are defined by

REDUCED VARIABLES
$$P_r = \frac{P}{P_c}, \qquad V_r = \frac{V}{V_c}, \qquad T_r = \frac{T}{T_c} \tag{2-17}$$

EXAMPLE 2-16: Derive a form of the van der Waals equation that contains the reduced variables instead of *P*, *V*, and *T*. We will need the expressions in Eq. (2-17).

Solution: We introduce the reduced variables into Eq. (2-13) by substituting $P = P_cP_r$, etc., from Eq. (2-17). We let $n = 1$ mol, so

$$\left(P_cP_r + \frac{a}{(V_cV_r)^2}\right)(V_cV_r - b) = RT_cT_r$$

Now, substituting from Eq. (2-16), we obtain

$$\left(\frac{a}{27b^2}P_r + \frac{a}{(3b)^2V_r^2}\right)(3bV_r - b) = R\frac{8a}{27Rb}T_r$$

Factoring and canceling the *R*'s, we have

$$\frac{a}{9b^2}\left(\frac{P_r}{3} + \frac{1}{V_r^2}\right)(b)(3V_r - 1) = \frac{8a}{27b}T_r$$

The *a*'s, *b*'s, and the 9 cancel out to yield

$$\left(\frac{P_r}{3} + \frac{1}{V_r^2}\right)(3V_r - 1) = \frac{8T_r}{3}$$

We now divide by $(3V_r - 1)$ and transpose $1/V_r^2$:

$$\frac{P_r}{3} = \frac{8T_r}{3(3V_r - 1)} - \frac{1}{V_r^2}$$

We then multiply by 3:

$$P_r = \frac{8T_r}{3V_r - 1} - \frac{3}{V_r^2} \tag{2-14a}$$

Note that Eq. (2-14a) has the same form as Eq. (2-14) but has no empirical constants. Thus it is independent of the nature of the gas. In other words: *All gases should obey the same equation of state when their states are described by reduced variables.* This is one statement of the **principle of corresponding states**. Another statement is that all gases at the same P_r and T_r have the same V_r. This rule becomes less reliable as the molecules of gases become more complex—perhaps polar or nonspherical—but more reliable at conditions close to the critical point.

EXAMPLE 2-17: Using the data in Example 2-11, calculate and compare the reduced variables for 1 mol CO_2 at 144 atm and 390 K and for 1 mol H_2O at 439 atm and 845 K. Do these sets of conditions seem to be corresponding states?

Solution: Since we are given pressures and temperatures, we can use Eq. (2-17) to calculate the reduced pressures and temperatures.

For CO_2: $\quad T_r = \dfrac{T}{T_c} = \dfrac{390\text{ K}}{304\text{ K}} = 1.28 \quad$ and $\quad P_r = \dfrac{P}{P_c} = \dfrac{144\text{ atm}}{73\text{ atm}} = 1.97$

For H_2O: $\quad T_r = \dfrac{845\text{ K}}{647\text{ K}} = 1.31 \quad$ and $\quad P_r = \dfrac{439\text{ atm}}{218\text{ atm}} = 2.01$

The solution of Eq. (2-14a) for V_r is rather complicated, so let's merely compare T_r's and P_r's. The two T_r's are almost the same, as are the two P_r's. Thus the two V_r's must also be almost identical and satisfy Eq. (2-14a). We can say that the widely disparate conditions for the two gases actually are *corresponding states.*

C. Virial equation

Sometimes called the *Kamerlingh Onnes equation,* the **virial equation of state** may be written as a power series in P. However, the following form is more common.

VIRIAL EQUATION

$$\frac{PV}{nRT} = 1 + \frac{B(T)n}{V} + \frac{C(T)n^2}{V^2} + \frac{D(T)n^4}{V^4} + \cdots \tag{2-18}$$

Note the missing odd powers of V (except the first); experience has shown that this form agrees better with experimental results. The parameters, B, C, D, etc, are called the **virial coefficients** and are functions of temperature.

Solving Eq. (2-18) for P, we obtain

$$P = \frac{nRT}{V} + \frac{RTB(T)n^2}{V^2} + \frac{RTC(T)n^3}{V^3} + \cdots \tag{2-19}$$

We see that the first two terms look very much like Eq. (2-14). Each succeeding term will make a smaller contribution to the total correction. When we compare Eqs. (2-12) and (2-18), we note that the virial equation expresses the deviation of the compressibility factor from its ideal value of 1.

EXAMPLE 2-18: For CO_2 at 373 K, the value of B is $-72.2\ \text{cm}^3\ \text{mol}^{-1}$. Use only the first two terms of Eq. (2-19) and calculate the pressure for the conditions in Example 2-14; compare the two results.

Solution: From Eq. (2-19)

$$\begin{aligned} P &= \frac{nRT}{V} + \frac{RTB(n)^2}{V^2} \\ &= \frac{(2\ \text{mol})(8.314 \times 10^3\ \text{Pa L K}^{-1}\text{mol}^{-1})(373\ \text{K})}{20.0\ \text{L}} \\ &\quad + \frac{(8.314 \times 10^3\ \text{Pa L K}^{-1}\text{mol}^{-1})(373\ \text{K})(-0.0722\ \text{L mol}^{-1})(2\ \text{mol})^2}{(20.0\ \text{L})^2} \\ &= 3.10 \times 10^5\ \text{Pa} - 0.0224 \times 10^5\ \text{Pa} = 3.08 \times 10^5\ \text{Pa} \end{aligned}$$

The two equations of state give identical answers in this situation.

At higher pressures and at temperatures near the critical temperature, the van der Waals equation doesn't give satisfactory results; you should use the virial equation or one of those listed next, in Section 2-5D.

D. Other equations of state

Three more well-known equations of state are listed here without examples. Calculations are similar to those in Sections 2-5A and 2-5C when values of the various constants are available.

1. The **Berthelot equation** is very accurate at pressures of about one atmosphere or lower.

BERTHELOT EQUATION

$$P = \frac{RT}{V_m - b} - \frac{A}{TV_m^2} \tag{2-20}$$

At low pressures, the equation can be expressed in terms of reduced variables.

$$P = \frac{nRT}{V}\left[1 + \left(\frac{9}{128}\right)\left(\frac{P_r}{T_r}\right)\left(1 - \frac{6}{T_r^2}\right)\right] \tag{2-20a}$$

2. The **Dieterici equation** fits the data well near the critical point.

DIETERICI EQUATION $$P = \frac{RT}{V_m - b} e^{-a/RTV_m} \tag{2-21}$$

3. The **Beattie–Bridgeman equation** is accurate over wide ranges of pressure and temperature. Five constants are involved, but this consolidated form shows its similarity to the previous equations.

BEATTIE–BRIDGEMAN EQUATION $$P = \frac{RT}{V_m} + \frac{D}{V_m^2} + \frac{E}{V_m^3} + \frac{F}{V_m^4} \tag{2-22}$$

Here, D, E, and F represent expressions containing the five Beattie–Bridgeman constants, temperature, and R.

To conclude this section, we emphasize that the ideal gas equation of state is a **limiting law**: It fits the data well only at low pressures. Real gases deviate from ideal behavior at higher pressures; a number of modifications to the ideal gas equation have been proposed in attempts to predict more accurately the behavior of nonideal gases.

2-6. The Kinetic Molecular Theory of Ideal Gases

A. Properties of the model

The fundamental postulates of the kinetic molecular theory are as follows:

1. A gas consists of tiny particles called molecules (atoms in a few cases).
2. Molecules are treated as points, having no volume.
3. Molecules are so far apart that any attractive forces may be ignored.
4. Molecules move continuously and randomly.
5. Molecules collide with each other, changing direction and velocity.
6. Collisions with the walls of the container exert a force, which causes pressure.
7. Collisions are elastic (no loss of total kinetic energy).
8. Absolute temperature is a measure of the average kinetic energy of the molecules, and the two are directly proportional.

B. Derivation of gas laws

The usual derivation starts with one molecule in a container and develops the following expressions and conclusions.

1. Single molecule

For a single molecule, the change, Δ, of momentum upon collision with the wall is

$$\frac{\Delta\text{Momentum}}{\text{Collision}} = 2mu_x \tag{2-23}$$

where m is the molecular mass and u_x is the velocity component along the x-axis. The number of collisions in a unit of time is

$$\frac{\text{Collisions}}{\text{Time}} = \frac{u_x}{2x} \tag{2-24}$$

The force exerted is the product of Eqs. (2-23) and (2-24):

$$\text{Force} = \frac{\Delta\text{Momentum}}{\text{Time}} = \frac{mu_x^2}{x} \tag{2-25}$$

The pressure (force per unit area) exerted in the x-direction by a single molecule is

$$P_x = \frac{mu_x^2}{V} \tag{2-26}$$

2. M molecules

For a collection of M molecules, the u_x^2 term is replaced by its average over all particles, $\overline{u_x^2}$.

In three dimensions,

$$\overline{u^2} = \overline{u_x^2} + \overline{u_y^2} + \overline{u_z^2} \tag{2-27}$$

The three components are equal; therefore $\overline{u_x^2} = \frac{1}{3}\overline{u^2}$. Substituting in Eq. (2-26) and multiplying by M, we get

$$P = \left(\frac{1}{3}\right)\left(\frac{Mm\overline{u^2}}{V}\right) \tag{2-28}$$

or

$$PV = \tfrac{1}{3}Mm\overline{u^2} \tag{2-29}$$

The new velocity term is called the **mean-square velocity**. Its square root, $(\overline{u^2})^{1/2}$, is called the **root-mean-square velocity**.

note: Sometimes confusion exists concerning the terms *speed* and *velocity*. **Velocity** is a vector quantity having magnitude and direction; the magnitude alone is the **speed**. Speed is a scalar quantity, defined as the positive square root of u^2. For our applications in this chapter, the direction of motion is not important, so it is proper to use speed, mean-square speed, and root-mean-square speed. Some authors use velocity; some alternate the terms. Various symbols are used to represent these quantities, including c, u, and v. We'll use u for both speed and velocity.

EXAMPLE 2-19: Suppose that we have 50 molecules: 15 have $u = 2.0 \times 10^2$ m s^{-1}, 30 have $u = 4.0 \times 10^2$ m s^{-1}, and 5 have $u = 8.0 \times 10^2$ m s^{-1}. Calculate the root-mean-square speed.

Solution: For M molecules, we may replace Eq. (2-27) with

$$\overline{u^2} = \frac{\sum_{i=1}^{M} u_i^2}{M} = \frac{n_1u^2 + n_2u_2^2 + n_3u_3^2}{n_1 + n_2 + n_3} \tag{2-30}$$

Sometimes, such an expression is called a **weighted average**.

We can conveniently set up the calculations in columns:

	n	u	u^2	nu^2
	15	2.0×10^2	4.0×10^4	60×10^4
	30	4.0×10^2	16.0×10^4	480×10^4
	5	8.0×10^2	64.0×10^4	320×10^4
Total	50			860×10^4

So $$\overline{u^2} = \frac{860 \times 10^4 \text{ molecules m}^2\text{ s}^{-2}}{50 \text{ molecules}} = 17.2 \times 10^4 \text{ m}^2\text{ s}^{-2}$$

and $$(\overline{u^2})^{1/2} = 4.15 \times 10^2 \text{ m s}^{-1}$$

---

3. Boyle's law

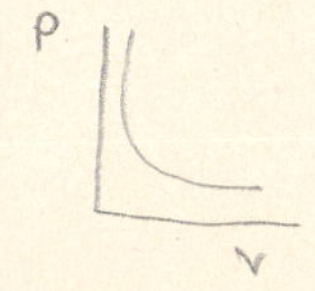

Boyle's law is obtained by replacing M in Eq. (2-29) with nL_A (i.e., Molecules = Moles × Avogadro's constant),

$$PV = \tfrac{1}{3}nL_Am\overline{u^2} \tag{2-31}$$

to obtain Boyle's law:

$$PV = k \qquad \textbf{[Eq. (2-1)]}$$

4. Absolute temperature and average kinetic energy

One postulate of the kinetic molecular theory is that the absolute temperature of a gas is directly proportional to the average kinetic energy of the molecules, $\overline{\text{KE}}$, which equals the product of the number of molecules and the average kinetic energy of a single molecule, $\overline{\text{ke}}$. For one mole,

$$\overline{\text{KE}} = L_A \overline{\text{ke}} = L_A(\tfrac{1}{2}m\overline{u^2}) \tag{2-32}$$

Also

$$\overline{\text{KE}} = CT \tag{2-33}$$

where C is the proportionality constant. Combining and rearranging these equations, we have

$$2CT = L_A m\overline{u^2} \tag{2-34}$$

Substituting into Eq. (2-31), we get

$$PV = \tfrac{2}{3}CnT \tag{2-35}$$

5. **Charles' law**

   At constant pressure, Eq. (2-35) becomes Charles' law:

$$\frac{V}{T} = k \qquad \text{[Eq. (2-2)]}$$

6. **Kinetic energy and temperature**

   Substituting Eq. (2-33) into Eq. (2-35) leads to a specific relationship between kinetic energy and temperature:

$$PV = \tfrac{2}{3}n\text{KE} \tag{2-36}$$

   If we let the constant C be $\tfrac{3}{2}R$, we see that Eq. (2-36) is the ideal gas equation. Equation (2-33) becomes

$$\text{KE} = \tfrac{3}{2}RT \tag{2-37}$$

   and Eq. (2-36) becomes the ideal gas equation:

$$PV = \tfrac{2}{3}n(\tfrac{3}{2}RT) = nRT \qquad \text{[Eq. (2-9)]}$$

   Obtaining the empirical expression by deduction from the theory adds validity to the assumptions of the theory.

   Dividing both sides of Eq. (2-37) by L_A gives

$$\overline{\text{ke}} = \tfrac{3}{2}kT \tag{2-38}$$

   where $k = R/L_A$ ($= 1.3807 \times 10^{-23}$ J K^{-1}), the **Boltzmann constant,** or the gas constant per molecule. This relationship will turn up again when we deal with the internal energy of molecules, where it may be called the **translational energy** or **kinetic energy of translation,** E_k.

7. **Temperature and mean-square speed**

   If we combine Eq. (2-32) with Eq. (2-37), we get a relationship between temperature and mean-square speed:

$$3RT = L_A m\overline{u^2}$$

   or

$$\overline{u^2} = \frac{3RT}{L_A m} \tag{2-39}$$

EXAMPLE 2-20: Assume that the gas in Example 2-19 is nitrogen and calculate its temperature.

Solution: We rearrange Eq. (2-39) to

$$T = \frac{L_A m\overline{u^2}}{3R} \tag{2-40}$$

The product, $L_A m$, is the weight of 1 mol: 28.0 g or 0.0280 kg; thus

$$T = \frac{(0.0280 \text{ kg mol}^{-1})(17.2 \times 10^4 \text{ m}^2 \text{ s}^{-2})}{3(8.314 \text{ J K}^{-1} \text{mol}^{-1})}\left(\frac{1 \text{ J}}{1 \text{ kg m}^2 \text{ s}^{-2}}\right) = 193 \text{ K}$$

C. Collision frequency and mean free path

We assume that two molecules have collided if they merely touch. When two molecules just touch, the distance between their centers is $r_A + r_B$ or, for identical molecules, $2r(= d)$. A single moving molecule (assuming that all other molecules are identical and stationary) is viewed as moving along the axis of a cylinder of radius d, colliding with any molecule that overlaps the cross-sectional area (πd^2) of the cylinder. The quantity πd^2 is called the **collision cross-section,** σ, of the moving molecule. The number of molecules hit per second by the moving one is $\bar{u}\sigma M/V$. (The average speed, $\bar{u}$, will be described in Section 2-6D.) When all molecules are moving, we must use the average relative speed of the two molecules that collide, or $\sqrt{2}\bar{u}$. Thus the **collision frequency** of one molecule, i.e., the collisions per second per unit volume, is given by

$$Z_A = \frac{\sqrt{2}\bar{u}\sigma M}{V} \tag{2-41}$$

For M molecules, there will be $(M/V)(\bar{u}\sigma M/V)$ collisions, but all are counted twice—once for each of the two colliding molecules. Now the total collisions per second per unit volume is given by

$$Z_{AA} = \left(\frac{1}{2}\right)\left(\frac{\sqrt{2}\bar{u}\sigma M^2}{V^2}\right) = \frac{\bar{u}\sigma M^2}{\sqrt{2}V^2} \tag{2-42}$$

If molecules of two sizes, A and B, are present, the average relative speed is given by the geometric average $(\bar{u}_A^2 + \bar{u}_B^2)^{1/2}$, and the internuclear distance is the sum of the two radii, $d_{AB} = \frac{1}{2}d_A + \frac{1}{2}d_B$. The equation for this case is

$$Z_{AB} = \left(\frac{1}{2}\right)(\bar{u}_A^2 + \bar{u}_B^2)^{1/2}\pi d_{AB}^2\left(\frac{M_A M_B}{V^2}\right) \tag{2-43}$$

EXAMPLE 2-21: The effective atomic radius of argon is found to be 1.9 angstrom units (Å). Calculate the number of collisions per second per liter if 1.0 mol of argon is confined in a 5.0-L vessel. Assume that $\bar{u} = 400 \text{ m s}^{-1}$.

Solution: First, let's get the radius into proper units:

$$1 \text{ Å} = 10^{-8} \text{ cm} = 10^{-10} \text{ m}$$

Thus $r = 1.9 \times 10^{-10}$ m, and $d = 3.8 \times 10^{-10}$ m. For 1 mol, $M = L_A = 6.02 \times 10^{23}$. Next, we substitute into Eq. (2-42):

$$Z_{AA} = \frac{(400 \text{ m s}^{-1})(\pi)(3.8 \times 10^{-10} \text{ m})^2(6.02 \times 10^{23})^2}{\sqrt{2}(5.0 \text{ L})^2} = 1.9 \times 10^{30} \text{ m}^3 \text{ s}^{-1} \text{ L}^{-2}$$

We want collisions per liter, so we must remove m^3.

$$1 \text{ L} = 1000 \text{ cm}^3 = 1000(10^{-6} \text{ m}^3/\text{cm}^3) = 10^{-3} \text{ m}^3$$

or

$$1 \text{ m}^3 = 10^3 \text{ L}$$

Then

$$Z_{AA} = 1.9 \times 10^{33} \text{ s}^{-1} \text{ L}^{-1}$$

The average distance a molecule travels between collisions is called the **mean free path** (mfp). If we know how far the molecule travels and how many times it collides in one second, we simply divide to determine the distance per collision:

MEAN FREE PATH
$$\text{mfp} = \frac{\text{Velocity}}{\text{Frequency}} = \frac{\text{m s}^{-1}}{\text{Collisions s}^{-1}} = \frac{\text{m}}{\text{Collision}} \tag{2-44}$$

Using Eq. (2-41) for the collisions of a single molecule, we get

$$\text{mfp} = \frac{\bar{u}}{\sqrt{2}\bar{u}\sigma M/V} = \frac{V}{\sqrt{2}\sigma M} \tag{2-45}$$

EXAMPLE 2-22: To get an idea of the magnitude of mean free paths, calculate the mfp for the argon in Example 2-21.

Solution: We substitute πd^2 for σ in Eq. (2-45):

$$\text{mfp} = \frac{5.0 \times 10^{-3}\,\text{m}^3}{\sqrt{2}(\pi)(3.8 \times 10^{-10}\,\text{m})^2(6.02 \times 10^{23})} = 1.3 \times 10^{-8}\,\text{m}$$

D. Velocity distributions

The actual kinetic energy of a molecule (and hence its velocity) changes continually, owing to collisions with other molecules and the walls of the container. For a huge number of molecules at constant temperature, the number n_i having a particular energy ε_i statistically will remain constant and is given by

$$n_i = Ae^{-\varepsilon_i/kT} \tag{2-46}$$

where k is the Boltzmann constant. (See Eq. 2-38.) We will encounter other equations of this form, which relate the probability of an event to the temperature and to an energy term.

Maxwell showed that, if the number of molecules having speeds between u and $u + du$ is represented by dn_i, the fraction du_i divided by the total number of molecules is equal to an expression similar to Eq. (2-46).

MAXWELL DISTRIBUTION LAW
$$\frac{dn_i}{n} = 4\pi\left(\frac{m}{2\pi kT}\right)^{3/2} e^{-mu^2/2kT}u^2\,du \tag{2-47}$$

A similar expression for the distribution of molecular energies is obtained if we let the kinetic energy of one molecule ($\frac{1}{2}mu^2$) equal ε and replace u^2 and du in Eq. (2.47):

MAXWELL–BOLTZMANN DISTRIBUTION LAW
$$\frac{dn_i}{n} = \frac{2\pi}{(\pi kT)^{3/2}} e^{-\varepsilon/kT}\varepsilon^{1/2}\,d\varepsilon \tag{2-48}$$

Plots of these equations have similar shapes and probably are already familiar to you from your general chemistry course.

EXAMPLE 2-23: For practice with the units and math, evaluate the expression $\frac{1}{n}\frac{dn_i}{du}$, using Eq. (2-47) for nitrogen at 298 K where the speed is 600 m s^{-1}.

Solution: First, we divide Eq. (2-47) by du:

$$\frac{1}{n}\frac{dn_i}{du} = 4\pi\left(\frac{m}{2\pi kT}\right)^{3/2} e^{-mu^2/2kT}\,u^2$$

We calculate the main terms separately and then combine them.

$$-\frac{mu^2}{2kT} = -\frac{(M_m)u^2}{2RT} = -\frac{(28.0\,\text{g mol}^{-1})(600\,\text{m s}^{-1})^2}{2(8.314\,\text{J K}^{-1}\text{mol}^{-1})(298\,\text{K})} = -2.03 \times 10^3$$

One joule equals 1 kg m^2 s^{-2}, so we'll need the molecular weight in kg. (Remember that kg is the SI base unit.) If we divide by 1000 g kg^{-1} the units cancel and the exponent becomes -2.03, a pure number, as it should be. Then $e^{-2.03} = 0.131$.

$$\frac{m}{2\pi kT} = \frac{(0.0280 \text{ kg mol}^{-1})/(6.022 \times 10^{23} \text{ molecules mol}^{-1})}{2\pi(1.3806 \times 10^{-23} \text{ J K}^{-1} \text{ molecule}^{-1})(298 \text{ K})}$$
$$= 1.80 \times 10^{-6} \text{ kg J}^{-1} = 1.80 \times 10^{-6} \text{ m}^{-2}\text{ s}^2$$

$$\left(\frac{m}{2\pi kT}\right)^{3/2} = (1.80 \times 10^{-6} \text{ m}^{-2}\text{ s}^2)^{3/2} = 2.41 \times 10^{-9} \text{ m}^{-3}\text{ s}^3$$

Substituting these values in the original expression, we obtain

$$\frac{1}{n}\frac{dn_i}{du} = 4(3.1416)(2.413 \times 10^{-9} \text{ m}^{-3}\text{ s}^3)(0.1308)(600 \text{ m s}^{-1})^2 = 1.43 \times 10^{-3} \text{ m}^{-1}\text{ s}$$

note: This is a good place to consider the meaning of the term we have evaluated. The $\frac{dn_i}{n}$ term represents the fraction of molecules having speeds between u and $u + du$. When we divide by du and let $u = 600$, we have the fraction of molecules in a small range of about 600 m s^{-1}.

A typical plot of Eq. (2-47) is shown in Figure 2-2, with $\frac{1}{n}\frac{dn_i}{du}$ as the y axis. The height and position of the maximum value of $\frac{1}{n}\frac{dn_i}{du}$ will vary with temperature and with the gas (more precisely, with the mass of the molecule). The arrows show the relative magnitudes and the relationship to the curve of three ways of expressing an *average*, or representative, speed of a gas in three dimensions.

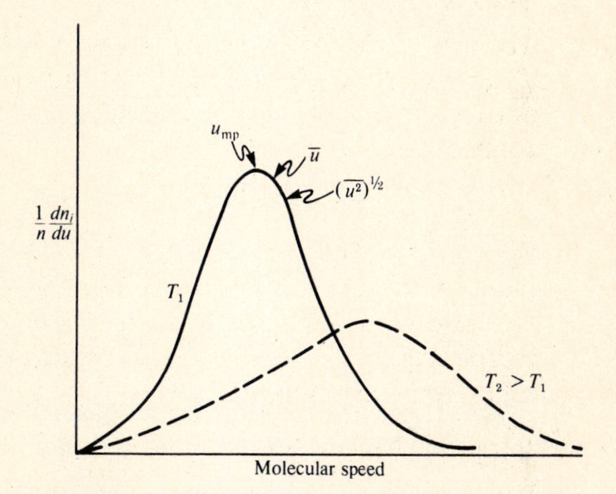

FIGURE 2-2. Qualitative Representation of Distribution of Molecular Velocities.

1. The *root-mean-square speed* was defined previously (Eq. 2-29). From Eq. (2-39), written for molecular mass,

ROOT-MEAN-SQUARE SPEED $$(\overline{u^2})^{1/2} = \left(\frac{3kT}{m}\right)^{1/2} \tag{2-49}$$

2. The *average speed* is the usual average: the sum of all speeds divided by the total number. It turns out to be related to the rms speed:

AVERAGE SPEED $$\bar{u} = \left(\frac{8kT}{\pi m}\right)^{1/2} = 0.921\left(\frac{3kT}{m}\right)^{1/2} = 0.921(u^2)^{1/2} \tag{2-50}$$

3. The *most probable speed* is the speed that occurs more often than any other and thus is the maximum point of the plot. It is calculated from

MOST PROBABLE SPEED $$U_{mp} = \left(\frac{2kT}{m}\right)^{1/2} \tag{2-51}$$

We could express these three equations in terms of moles by using R instead of k and molecular weight instead of the mass of one molecule.

EXAMPLE 2-24: Calculate the three representative speeds—(**a**) rms, (**b**) average, (**c**) most probable—for SO_2 gas at 25 °C and (**d**) determine the ratios of the largest two to the smallest. (To avoid the 10^{23} terms, rewrite the equations for moles.)

Solution:

(a) For rms speed, Eq. (2-49) becomes

$$(\overline{u^2})^{1/2} = \left(\frac{3RT}{M_m}\right)^{1/2} \qquad \textbf{(2-49a)}$$

$$\overline{u^2} = \left(\frac{3(8.314 \text{ J K}^{-1}\text{mol}^{-1})(298 \text{ K})}{0.0640 \text{ kg mol}^{-1}}\right)\left(\frac{1 \text{ kg m}^2 s^{-2}}{1 \text{ J}}\right) = 11.6 \times 10^4 \text{ m}^2 s^{-2}$$

$$(\overline{u^2})^{1/2} = 3.41 \times 10^2 \text{ m s}^{-1}$$

(b) For average speed, Eq. (2-50) becomes

$$\bar{u} = \left(\frac{8RT}{\pi(M_m)}\right)^{1/2} \qquad \textbf{(2-50a)}$$

$$= \left(\frac{8(8.314 \text{ kg m}^2 \text{ s}^{-2} \text{ K}^{-1}\text{mol}^{-1})(298 \text{ K})}{\pi(0.0640 \text{ kg mol}^{-1})}\right)^{1/2} = 3.14 \times 10^2 \text{ m s}^{-1}$$

(c) For most probable speed, Eq. (2-51) becomes

$$U_{\text{mp}} = \left(\frac{2RT}{M_m}\right)^{1/2} \qquad \textbf{(2-51a)}$$

$$= \left(\frac{2(8.314 \text{ kg m}^2 \text{ s}^{-2} \text{ K}^{-1}\text{mol}^{-1})(298 \text{ K})}{0.0640 \text{ kg mol}^{-1}}\right)^{1/2} = 2.78 \times 10^2 \text{ m s}^{-1}$$

(d) Now the ratios are

$$\frac{\text{rms}}{\text{mp}} = \frac{341}{278} = 1.22$$

$$\frac{\text{Average}}{\text{mp}} = \frac{314}{278} = 1.13$$

These ratios are the same for all gases, and are often written as proportions (see Problem 2-34). To four significant figures, they are

$$u_{\text{mp}}\text{: } \bar{u}\text{:}(\overline{u^2})^{1/2} = 1\text{:}1.128\text{:}1.225$$

SUMMARY

Table S-2 (at the end of the chapter) summarizes the major equations of this chapter and shows connections between them. You should determine the conditions that are necessary for each equation to be valid or the conditions that are imposed to make the connections. Other important things you should know are listed below.

1. The proportionalities of volume, pressure, temperature, and amount of substance are as follows:

Held constant	Proportionality	Held constant	Proportionality
n, T	$P \propto 1/V$ (Boyle's law)	P, T	$V \propto n$ (Avogadro's law)
n, P	$V \propto T$ (Charles' law)	V, T	$P \propto n$
n, V	$P \propto T$	P, V	$T \propto 1/n$

2. The four variable factors are related by the ideal gas equation of state: $PV = nRT$.
3. The commonly used values of the gas constant are 1.987 cal $\text{K}^{-1}\text{mol}^{-1}$, 0.082 06 L atm $\text{K}^{-1}\text{mol}^{-1}$, and 8.314 J (or Pa m^3) $\text{K}^{-1}\text{mol}^{-1}$.
4. Dalton's law of partial pressures may be expressed as $P_{\text{total}} = P_A + P_B + \cdots$.
5. Graham's law of diffusion may be stated as follows: The rate of diffusion is inversely proportional to the square root of the molecular weight of the gas.

6. Reduced variables are the actual pressure, temperature, and volume of a gas divided by the corresponding critical constants.
7. When two gases have the same values for the reduced variables, the gases are in corresponding states.
8. The compressibility factor of a gas provides a measure of ideality: $Z = PV/nRT$.
9. At the Boyle temperature, a gas behaves ideally over a range of pressures.
10. The constants in the van der Waals equation of state account for attractive forces between molecules and for the volume occupied by the molecules.
11. The virial, Berthelot, and Beattie–Bridgeman equations of state utilize power series; the Dieterici equation uses an exponential function.
12. A gas is assumed to consist of tiny particles relatively far apart, moving at a high speed.
13. Absolute temperature is a measure of the average kinetic energy of gas molecules, which is determined by the mean-square velocity.
14. The gas laws can be derived from the kinetic molecular theory:
    - Boyle's law becomes $PV = \frac{1}{3}nL_A m\overline{u^2}$.
    - The ideal gas law becomes $PV = \frac{2}{3}CnT$.
15. For an ideal gas, $\overline{\mathrm{KE}} = E_k = \frac{3}{2}RT$; for one molecule, $\overline{\mathrm{ke}} = \frac{3}{2}kT$.
16. Collision frequencies can be calculated
    - as collisions per second per unit volume per molecule: $Z_A = \sqrt{2}\bar{u}\sigma M/V$;
    - for M identical molecules: $Z_{AA} = \dfrac{\bar{u}\sigma M^2}{\sqrt{2}V^2}$; and
    - for molecules A and B: $Z_{AB} = \frac{1}{2}(\bar{u}^2 + \bar{u}_B^2)^{1/2}\pi d_{AB}^2 \dfrac{M_A M_B}{V^2}$.
17. The mean free path of a particle is the average distance traveled between collisions: $\mathrm{mfp} = V/\sqrt{2}\sigma M$
18. Three ways of expressing an average velocity of a group of particles are the root-mean-square velocity, the average velocity, and the most probable velocity. Their relative magnitudes are 1.225:1.128:1.00, respectively.

SOLVED PROBLEMS

The Ideal Gas Laws

PROBLEM 2-1 If the pressure on 15 L of an ideal gas is tripled, at constant temperature, what is the new volume?

Solution: This is the type of elementary problem that you can do in your head. You know that volume is inversely proportional to pressure [Boyle's law, Eq. (2-1)]. If the pressure increases by a factor of 3, the volume must decrease by a factor of 3, so $V = 15/3 = 5$ L.

[Note that, since the actual pressures are not given, you effectively assume that $P_1 = 1$ and $P_2 = 3$. You could have used $P_1 = 2.5$ and $P_2 = 7.5$, or any other set of pressures in the ratio of 3:1.]

PROBLEM 2-2 A certain quantity of a gas is confined in an 8.0-dm³ cylinder under a pressure of 4.0×10^5 Pa. What volume would it occupy at standard pressure?

Solution: This problem is mathematically simple, and your inclination probably is to start multiplying and dividing. Let's go more slowly, however. Only pressures and volumes are involved, which should make you think of Boyle's Law and Eq. (2-1a). Also, you must recall the value of the standard pressure (1 atm, or 101 325 Pa). Then,

$$V_2 = \frac{P_1 V_1}{P_2} = \frac{(4.0 \times 10^5 \text{ Pa})(8.0 \text{ dm}^3)}{1.0 \times 10^5 \text{ Pa}} = 32 \text{ dm}^3$$

Actually, you first should ask yourself whether Boyle's law is applicable. The problem doesn't state that the temperature is constant or that the gas is ideal. On the other hand, from the information given you could not solve the problem unless these two conditions hold. So you should *assume* that they hold and make the calculation. Speaking precisely, you should say that the final volume will be 32 dm^3 *if* the gas has behaved ideally and *if* its temperature is the same as the initial temperature. This qualification would not be needed if the problem were stated more completely, as in Problem 2-3.

PROBLEM 2-3 A certain quantity of an ideal gas occupies a volume of 6.0 mL at a pressure of 700 mm Hg. If it expands isothermally to a pressure of 300 mm Hg, what is the volume?

Solution: This time you are told that the gas is ideal and the temperature is constant, so it is clear that Boyle's law applies. Could you use the ideal gas equation to calculate the answer? You can't, for two reasons: You are not given a value for the temperature or the amount of substance. From Eq. (2-1a),

$$V_2 = \frac{P_1 V_1}{P_2} = \frac{(700 \text{ mm Hg})(6.0 \text{ mL})}{300 \text{ mm Hg}} = 14 \text{ mL}$$

PROBLEM 2-4 You collect some gas over mercury and find that it occupies 36.0 mL at a barometric pressure of 752.2 mm Hg and a temperature of 24.3 °C. What volume will the gas occupy at STP?

Solution: Your first thought probably was "STP means 1 atm and 0 °C." Your second should be that this is a typical gas law problem. Then you should remember that you can set up the expression for calculating the answer in several different ways. The desired answer, of course, is the new volume. Again you must assume ideal behavior to get anywhere.

Method A: Use the combined equation (2-7a).
Rearranging, you have

$$V_2 = \frac{P_1 V_1}{T_1}\left(\frac{T_2}{P_2}\right) = \frac{752.2(36.0)}{297.5}\left(\frac{273.2}{760}\right) = 32.7 \text{ mL}$$

Method B: Use conversion factors.
This approach requires that you decide how each change (in P and T) affects the original volume, if each change occurs independently. Then you write a fraction that gives the proper effect mathematically. Symbolically,

$$V_2 = V_1(\text{Pressure fraction})(\text{Temperature fraction})$$

In this problem, $P_1 = 752.2$ and $P_2 = 760.0$. The increase in pressure will cause a decrease in volume [Eq. (2-1a)]; thus the pressure fraction must be less than 1, or 752.2/760.0. Since the temperature decreases (24.3 °C to 0 °C), the volume must decrease [Eq. (2-2)]; thus the temperature fraction must be less than 1, or 273.2/297.5. Now,

$$V_2 = (36.0 \text{ mL})\left(\frac{752.2}{760.0}\right)\left(\frac{273.2}{297.5}\right) = 32.7 \text{ mL}$$

Method C: Use the ideal gas equation (2-9).
This method is cumbersome for this problem, because you must calculate n first. Rearranging Eq. (2-9), you obtain

$$n = \frac{P_1 V_1}{RT_1} = \frac{752.2(0.0360)}{0.0821(297.5)} = 1.11 \text{ mol}$$

Rearranging again, you get

$$V_2 = \frac{nRT_2}{P_2} = \frac{1.11(0.0821)(273.2)}{760} = 0.0328 \text{ L}$$

PROBLEM 2-5 What pressure would be exerted by 50 g of hydrogen gas confined in a 10-L cylinder at 25 °C?

Solution: When the amount of gas is given, you should think of the ideal gas equation but should check to see whether the information is needed. There may be enough information so that the weight can be ignored and the answer obtained by one of the methods used in Problem 2-4. This is not possible in this problem, so you must assume ideal behavior and use Eq. (2-9):

$$n = (50\ \mathrm{g}\ H_2)\left(\frac{1\ \mathrm{mol}}{2.0\ \mathrm{g}}\right) = 25\ \mathrm{mol}$$

Rearranging Eq. (2-9) yields

$$P = \frac{nRT}{V} = \frac{25(0.0821)(298)}{10} = 61\ \mathrm{atm}$$

PROBLEM 2-6 How many molecules of an ideal gas are present if 15 dm^3 of the gas at 300 °C exert a pressure of 5.5×10^4 Pa?

Solution: When faced with a question about molecules, you should think of Avogadro's number. Usually, you must find moles and then convert to molecules. In this problem, you can calculate the number of moles using the ideal gas equation (2-9). Rearrange the equation first:

$$n = \frac{PV}{RT} = \frac{(5.5 \times 10^4\ \mathrm{Pa})(15\ \mathrm{dm}^3)}{(8.314 \times 10^3\ \mathrm{Pa\,dm^3\,K^{-1}mol^{-1}})(573\ \mathrm{K})} = 0.17\ \mathrm{mol}$$

Thus

$$\text{No. of molecules} = (0.17\ \mathrm{mol})(6.02 \times 10^{23}\ \text{molecules mol}^{-1}) = 1.0 \times 10^{23}$$

PROBLEM 2-7 What is the density of CO_2 (in g L^{-1}) at 100 °C and 0.50 atm, if the gas behaves ideally?

Solution: You might think that not enough data are given. However, density is a ratio, and you can rearrange the ideal gas equation to give the ratio of moles to volume; then moles can be changed to grams. Rearranging Eq. (2-9), you have

$$\frac{n}{V} = \frac{P}{RT} = \frac{0.50}{0.082(373)}\ \mathrm{mol\,L^{-1}}$$

Then multiply by the molecular weight (44 g mol^{-1}) to obtain

$$d = \frac{0.50(44)}{0.082(373)} = 0.72\ \mathrm{g\,L^{-1}}$$

PROBLEM 2-8 Estimate the molar weight of a liquid if 510 mL of its vapor, measured at 752 mm Hg and 98.5 °C, weighs 2.45 g when it is condensed at room temperature.

Solution: Because the problem asks for an estimated molar weight, you may rearrange and use the ideal gas equation (2-9) to obtain the number of moles: $n = PV/RT$. Put P in atm and T in K, so $P = 752/760$ atm and $T = 273.2 + 98.5 = 371.7$ K. Then

$$n = \frac{752(0.510)}{760(0.0821)(371.7)} = 0.0165\ \mathrm{mol}$$

The weight of 0.0165 mol is given as 2.45 g. Thus

$$M_m = \frac{2.45\ \mathrm{g}}{0.0165\ \mathrm{mol}} = 148\ \mathrm{g\,mol^{-1}}$$

Other Laws Concerning Ideal Gases

PROBLEM 2-9 If you mix 6.0 g of SO_2 and 6.0 g of CO_2 in a 2.0-L flask at 20 °C, what will be the pressure in the flask? Assume that the gases behave ideally.

Solution: With a mixture of gases, you need to use Dalton's law of partial pressures, Eq. (2-10), which means that you need to know the partial pressure of each gas. Each gas acts as if it occupies the entire volume by itself, and each has a different n. The variables T and V are constant; thus

$$\frac{RT}{V} = \frac{0.0821(293)}{2.0} = 12\ \mathrm{atm\,mol^{-1}} \quad \text{and} \quad P = n(RT/V)$$

So
$$P_{SO_2} = \frac{6.0}{64}(12) = 1.1$$
$$P_{CO_2} = \frac{6.0}{44}(12.0) = \underline{1.6}$$
$$\text{Total pressure} = 2.7 \text{ atm}$$

PROBLEM 2-10 Hydrogen gas can be produced by treating a metal with an acid and collected by displacement of water. If the reading on the gas buret is 26.7 mL at a room temperature of 23.4 °C and a barometric pressure of 750.8 mm Hg, how many grams of hydrogen can be collected? Assume ideal behavior.

Solution: You should think immediately of Dalton's law of partial pressures. The gas buret will contain the desired gas plus water vapor, and the pressure exerted by the mixture will be the sum of the pressures of each gas. Solve Eq. (2-10) for the pressure of hydrogen.

$$P_{H_2} = P_{total} - P_{H_2O}$$

The vapor pressure of water at 23.4 °C is 21.6 mm Hg, so

$$P_{H_2} = 750.8 - 21.6 = 729.2 \text{ mm Hg}$$

In order to calculate grams, you will first need to obtain the number of moles.

$$n = \frac{PV}{RT} = \frac{(729.2/760)(0.0267)}{0.0821(296.6)} = 1.05 \times 10^{-3} \text{ mol}$$

When you multiply moles times molecular mass, you get

$$\text{mass}_{H_2} = 1.05 \times 10^{-3}(2.02) = 2.12 \times 10^{-3} \text{ g}$$

PROBLEM 2-11 Which gas will diffuse more rapidly, CO_2 or NO_2?

Solution: This is another problem that you can do in your head. Remember that the heavier gas moves more slowly. Since the mass of an NO_2 molecule is greater than that of a CO_2 molecule, the CO_2 must diffuse more rapidly.

PROBLEM 2-12 If SO_2 diffuses through a certain tube in 12.2 min and an unknown gas diffuses in 6.3 min, what is the molar mass of the unknown gas?

Solution: Diffusion always suggests Graham's law. In this case, time of diffusion is given rather than its rate. Time is inversely proportional to rate and thus directly proportional to the square root of molecular mass. Equation (2-11) becomes

$$\frac{t_2}{t_1} = \sqrt{\frac{M_{m,2}}{M_{m,1}}}$$

Squaring and substituting, you have

$$\left(\frac{t_{unk}}{t_{SO_2}}\right)^2 = \frac{M_{m,unk}}{M_{m,SO_2}}$$

Rearranging gives

$$M_{m,unk} = M_{m,SO_2}\left(\frac{t_{unk}}{t_{SO_2}}\right)^2 = 64\left(\frac{6.3}{12.2}\right)^2 = 17$$

PROBLEM 2-13 The density of nitrogen at 0 °C is 1.251 g L^{-1}. Calculate the density of an unknown gas if it diffuses 0.84 times as fast as nitrogen under the same conditions.

Solution: Again, Graham's law is needed. You should recognize that density is directly proportional to molar weight. Thus Eq. (2-11) becomes

$$\sqrt{\frac{d_2}{d_1}} = \frac{r_1}{r_2} \quad \text{or} \quad \frac{d_{unk}}{d_{N_2}} = \left(\frac{r_{N_2}}{r_{unk}}\right)^2 \quad \text{or} \quad d_{unk} = d_{N_2}\left(\frac{r_{N_2}}{r_{unk}}\right)^2$$

You are given that $r_{unk} = 0.84 r_{N_2}$, so

$$d_{unk} = 1.251\left(\frac{r_{N_2}}{0.84 r_{N_2}}\right)^2 = 1.8 \text{ g L}^{-1}$$

Properties of Real Gases

PROBLEM 2-14 At a pressure of 400 atm and 0 °C, the density of H_2 is found to be 24.4 g L^{-1}. Calculate the compressibility factor. Does the answer agree with the value for these conditions in Figure 2-1a?

Solution: The compressibility factor, Eq. (2-12), contains V and n. From the density given, you can calculate the volume of any amount of substance. For 1 mol:

$$V = \frac{\text{Mass}}{\text{Density}} = \frac{2.02 \text{ g mol}^{-1}}{24.4 \text{ g L}^{-1}} = 0.0828 \text{ L mol}^{-1}$$

Now, substitute in Eq. (2-12):

$$Z = \frac{PV}{nRT} = \frac{(400 \text{ atm})(0.0828 \text{ L mol}^{-1})}{(0.0821 \text{ L atm K}^{-1}\text{mol}^{-1})(273 \text{ K})} = 1.48$$

It is difficult to read Figure 2-1a to two decimal places, but you probably would say that $Z = 1.45$ at 400 atm.

Equations of State for Real Gases

PROBLEM 2-15 The van der Waals constants for water are $a = 5.46$ atm dm^6 mol^{-2} and $b = 3.05 \times 10^{-2}$ dm^3 mol^{-1}. Calculate the values of the critical constants and compare them to the values $V_c = 59$ cm^3 mol^{-1}, $P_c = 218$ atm, and $T_c = 374$ °C given in Table 2-2. Explain the discrepancies.

Solution: Using the expressions in Eqs. (2-16), you obtain

$$V_c = 3b = 3(3.05 \times 10^{-2} \text{ dm}^3 \text{ mol}^{-1})\left(\frac{1000 \text{ cm}^3}{1 \text{ dm}^3}\right) = 91.5 \text{ cm}^3 \text{ mol}^{-1}$$

$$P_c = \frac{a}{27b^2} = \frac{5.46 \text{ atm dm}^6 \text{ mol}^{-2}}{27(3.05 \times 10^{-2} \text{ dm}^3 \text{ mol}^{-1})^2} = 217 \text{ atm}$$

$$T_c = \frac{8a}{27Rb} = \frac{8(5.46 \text{ atm dm}^6 \text{ mol}^{-2})}{27(0.0821 \text{ atm dm}^3 \text{ K}^{-1}\text{mol}^{-1})(3.05 \times 10^{-2} \text{ dm}^3 \text{ mol}^{-1})} = 646 \text{ K} = 373 \text{ °C}$$

The values of P_c and T_c agree well with Table 2-2, but that of V_c doesn't. A similar discrepancy was found in Example 2-15. It seems that $3 \times b$ ALWAYS gives too large a value for V_c. This may be because b values are given for 25 °C, but b changes with temperature, or it may be caused by the invalidity of the van der Waals equation near the critical point.

PROBLEM 2-16 **(a)** Calculate the pressure exerted by 1.00 mol CO_2 in a 15.00-dm^3 vessel at 20 °C, using the van der Waals equation. **(b)** Is the pressure exerted by 4.00 mol CO_2 under the same conditions four times as much?

Solution: You must find the values of the van der Waals constants a and b in your textbook or some other source. For CO_2, they are $a = 3.592$ atm dm^6 mol^{-2} and $b = 4.267 \times 10^{-2}$ dm^3 mol^{-1}. Using Eq. (2-14), the rearranged form of the van der Waals equation, you get

$$P = \frac{nRT}{V - nb} - \frac{an^2}{V^2}$$

(a) For 1 mol:

$$P_1 = \frac{1(0.0821)(293)}{15.00 - 1(0.0427)} - \frac{3.592(1)^2}{(15.00)^2} = 1.61 - 0.0160 = 1.59 \text{ atm}$$

(b) For 4 mol:

$$P_2 = \frac{4(0.0821)(293)}{15.00 - 4(0.0427)} - \frac{3.592(4)^2}{(15.00)^2} = 6.49 - 0.26 = 6.23 \text{ atm}$$

Comparing: $\frac{P_2}{P_1} = \frac{6.23}{1.59} = 3.92$ Nearly four times as much.

PROBLEM 2-17 Repeat Problem 2-16(b), assuming that the gas behaves ideally, and compare the results.

Solution:

$$P_2 = \frac{nRT}{V} = \frac{4(0.0821)(293)}{15.00} = 6.41 \text{ atm}$$

The van der Waals result was 6.23 atm; thus the deviation from the ideal is small at pressures up to 6 atm: only 0.18 atm, or 2.8%.

PROBLEM 2-18 Derive an expression for the compressibility factor at the critical point for a van der Waals gas.

Solution: The compressibility factor is given by Eq. (2-12), and the expressions for the critical constants are listed as Eqs. (2-16). You should write the expression for Z in terms of critical constants and substitute. Take V_c to mean the molar volume, V/n.

$$Z = \frac{P_cV_c}{RT_c} = \frac{1}{RT_c}P_cV_c = \frac{27Rb}{R(8a)}\left(\frac{a}{27b^2}\right)(3b) = \frac{3}{8}$$

PROBLEM 2-19 A useful expression for the compressibility factor can be obtained from the van der Waals equation under the condition of low pressure. Make the assumption in this case that the volume in the pressure correction is the ideal volume and show that $Z = 1 + \frac{b}{RT}P - \frac{a}{(RT)^2}P$. Also assume that the product ab is negligibly small and that V is the molar volume, so that $n = 1$.

Solution: First, you must multiply the two terms on the left-hand side of the van der Waals equation. Let $n = 1$, so that Eq. (2-13) becomes $\left(P + \frac{a}{V^2}\right)(V - b) = RT$, then multiply.

$$PV + \frac{a}{V} - bP - \frac{ab}{V^2} = RT$$

Neglect the ab term and rearrange.

$$PV = RT - \frac{a}{V} + bP$$

You want Z, which is $\frac{PV}{RT}$, so you must divide through by RT.

$$\frac{PV}{RT} = Z = 1 - \frac{a}{RTV} + \frac{bP}{RT}$$

Now use the approximation that, at low pressure, the V in the correction term is the ideal volume; substitute RT/P for V.

$$Z = 1 - \frac{a}{RT}\left(\frac{P}{RT}\right) + \frac{bP}{RT} = 1 + \frac{b}{RT}P - \frac{a}{(RT)^2}P \qquad \textbf{(2-52)}$$

PROBLEM 2-20 The van der Waals constants for hydrogen are $a = 0.244\ \text{L}^2\ \text{atm mol}^{-2}$ and $b = 0.0266\ \text{L mol}^{-1}$. Use Eq. (2-52) to calculate Z for hydrogen at 400 atm and 0 °C, and compare your answer to the value from Problem 2-14.

Solution: This is a straightforward substitution exercise.

$$Z = 1 + \frac{0.0266(400)}{0.0821(273)} - \frac{0.244(400)}{[0.0821(273)]^2} = 1 + 0.475 - 0.195 = 1.28$$

This Z-value is much lower than the 1.48 of Problem 2-14 and Figure 2-1a. The difference is probably the result of the approximations made in deriving the expression for Z.

PROBLEM 2-21 Calculate the pressure exerted by 12.0 g water vapor confined in a 100-cm^3 steel vessel at 200 °C. **(a)** Assume ideal behavior. **(b)** Use the reduced-variable form of the van der Waals equation.

Solution:

(a) You merely need to substitute in the ideal gas equation.

$$P = \frac{nRT}{V} = \frac{(12.0/18.0)(0.0821)(473)}{0.100} = 259 \text{ atm}$$

(b) The equation specified is Eq. (2-14a).

$$P_r = \frac{8T_r}{3V_r - 1} - \frac{3}{V_r^2}$$

Calculate the reduced variables from Eqs. (2-17), using critical constants from Table 2-2. For water, $T_c = 374$ °C, $V_c = 59$ cm^3 mol^{-1}, and $P_c = 218$ atm. Thus

$$T_r = \frac{T}{T_c} = \frac{473}{647} = 0.731$$

$$V_c = \frac{V}{V_c} = \frac{(100 \text{ cm}^3)(0.667 \text{ mol})}{59 \text{ cm}^3 \text{ mol}^{-1}} = 2.5$$

$$P_r = \frac{8(0.731)}{3(2.5) - 1} - \frac{3}{(2.5)^2} = 0.90 - 0.48 = 0.42$$

And

$$P = P_r P_c = 0.42(218) = 92 \text{ atm}$$

PROBLEM 2-22 Use Eq. (2-52) to find a general expression for the Boyle temperature.

Solution: Remember that the Boyle temperature is that temperature for which the plot of Z versus P is horizontal, i.e., tangent to the $Z = 1$ line. (See Figure 2-1b.) Or, stated another way, the slope is zero, i.e., at the Boyle temperature, $\frac{dZ}{dP} = 0$, or better yet, $\left(\frac{\partial Z}{\partial P}\right)_{T_B} = 0$. Factoring the P in Eq. (2-52) gives

$$Z = 1 + \left(\frac{b}{RT} - \frac{a}{(RT)^2}\right)P$$

Differentiating yields

$$\left(\frac{\partial Z}{\partial P}\right)_T = 0 + \frac{b}{RT} - \frac{a}{(RT)^2}$$

At the Boyle temperature, T_B, the partial derivative equals 0, so

$$\frac{b}{RT_B} - \frac{a}{(RT_B)^2} = 0$$

Now multiply by RT_B and transpose the negative term:

$$b = \frac{a}{RT_B}$$

Thus

$$T_B = \frac{a}{Rb} \quad \textbf{(2-53)}$$

PROBLEM 2-23 Calculate the Boyle temperature for CH_4, using Eq. (2-53), and compare your answer to Figure 2-1b. The van der Waals constants are $a = 2.253$ L^2 atm mol^{-2} and $b =$ 0.0428 L mol^{-1}.

Solution: Again, you merely have to substitute numbers in an equation.

$$T_B = \frac{a}{Rb} = \frac{2.253 \text{ L}^2 \text{ atm mol}^{-2}}{(0.0821 \text{ L atm K}^{-1}\text{mol}^{-1})(0.0428 \text{ L mol}^{-1})} = 641 \text{ K}$$

According to Figure 2-1b, T_B should fall between 373 K and 573 K. The difference emphasizes that you must be skeptical of values calculated from expressions that are based on assumptions and approximations.

The Kinetic Molecular Theory of Ideal Gases

PROBLEM 2-24 Calculate the rms speed of helium atoms at 1 atm and 273 K, assuming ideal behavior.

Solution: You can solve this problem by using either Eq. (2-29) or Eq. (2-39); do it both ways for the experience with units.

(a) To use Eq. (2-29) you need the volume, which you can take as 22.4 L because of the assumption of ideal behavior and the temperature of 0 °C. At another temperature, you would have to calculate the volume or make a substitution that would lead to Eq. (2-39). Now, solve for $\overline{u^2}$ and substitute:

$$\overline{u^2} = \frac{3PV}{Mm} = \frac{3(1\text{ atm})(22.4\text{ L mol}^{-1})}{4.00\text{ g mol}^{-1}} = 16.8\text{ L atm g}^{-1}$$

Note that Mm (in the denominator) is the mass of all molecules present, or the weight of 1 mol in this case. Velocity should be stated in m s^{-1}, which you can obtain by converting the pressure to newtons:

$$1\text{ atm} = 101\,325\text{ N m}^{-2} = (1.01 \times 10^5\text{ kg m s}^{-2})(\text{m}^{-2})$$

Also you need to change liters to cubic meters (1 L = 10^{-3} m^3) and grams to kilograms. Thus

$$\overline{u^2} = (16.8\text{ L atm g}^{-1})\left(\frac{1.01 \times 10^5\text{ kg m}^{-1}\text{s}^{-2}}{1\text{ atm}}\right)\left(\frac{10^{-3}\text{ m}^3}{1\text{ L}}\right)\left(\frac{10^3\text{ g}}{1\text{ kg}}\right) = 1.697 \times 10^6\text{ m}^2\text{s}^{-2}$$

$$(\overline{u^2})^{1/2} = 1.30 \times 10^3\text{ m s}^{-1}$$

(b) Use Eq. (2-39):

$$\overline{u^2} = \frac{3RT}{L_A m}$$

Recall that L_A was introduced for M in deriving this expression, so $L_A m = 4.00 \times 10^{-3}$ kg mol^{-1}, as in **(a)**. You must use a value for R that leads to units of speed. Of the usual values, the joule is the easiest to manipulate.

$$1\text{ J} = 1\text{ Pa m}^3 = 1\text{ N m}^{-2}(\text{m}^3) = 1\text{ kg m s}^{-2}(\text{m}) = 1\text{ kg m}^2\text{s}^{-2}$$

$$8.314\text{ J} = 8.314\text{ kg m}^2\text{s}^{-2}$$

$$\overline{u^2} = \frac{3(8.314\text{ kg m}^2\text{s}^{-2}\text{ K}^{-1}\text{mol}^{-1})(273\text{ K})}{4.00 \times 10^{-3}\text{ kg mol}^{-1}} = 1.702 \times 10^6\text{ m}^2\text{s}^{-2}$$

$$(\overline{u^2})^{1/2} = 1.30 \times 10^3\text{ m s}^{-1}$$

PROBLEM 2-25 Calculate the average kinetic energy of 1 mol He at 0 °C. Use any results from Problem 2-24 that you need.

Solution: The reference to Problem 2-24 suggests that you use Eq. (2-32) for 1 mol:

$$\overline{\text{KE}} = L_A(\tfrac{1}{2}m\overline{u^2}) = \tfrac{1}{2}L_A m\overline{u^2}$$

For $\overline{u^2}$, use an approximate average of the two values in Problem 2-24: 1.70×10^6 m^{2}s^{-2}. Again, $L_A m =$ 4.00×10^{-3} kg. Now, substitute and solve.

$$\overline{\text{KE}} = \tfrac{1}{2}(4 \times 10^{-3}\text{ kg})(1.70 \times 10^6\text{ m}^2\text{s}^{-2}) = 3400\text{ kg m}^2\text{s}^{-2} = 3400\text{ J} = 3.40\text{ kJ}$$

A more direct approach is to use Eq. (2-37):

$$\overline{\text{KE}}\text{ (or }E_k) = \tfrac{3}{2}RT$$

At $T = 273$ K,

$$E_k = 1.5(8.314\text{ J K}^{-1})(273\text{ K}) = 3.40\text{ kJ}$$

PROBLEM 2-26 How would the answer to Problem 2-25 differ if argon were used instead of helium? Explain your answer.

Solution: The answer would be exactly the same. For heavier atoms or molecules, the term, $L_A m$, would be larger, but this term appears in the calculation of $\overline{u^2}$ (in Problem 2-24) and would cause the value of $\overline{u^2}$ to be correspondingly smaller.

Thus $\overline{\text{KE}}$ of 1 mol of any ideal gas at 0 °C and 1 atm equals 3.40 kJ.

PROBLEM 2-27 Derive an expression that states the mean-square speed of an ideal gas as a function of the density of the gas.

Solution: Density equals mass/volume, so you need an expression that includes these units. When you rearrange Eq. (2-29), you get

$$\overline{u^2} = \frac{3PV}{Mm}$$

The denominator is the mass of gas, so density $= \rho = (Mm)/V$. Thus

$$\overline{u^2} = \frac{3P}{\rho} \tag{2-29a}$$

PROBLEM 2-28 Calculate the rms speed of nitrogen at STP, using Eq. (2-29a).

Solution: You need the density and should recall that the density of a gas can always be calculated from the molar weight and the molar volume. At STP,

$$\rho_{N_2} = \frac{28.0 \times 10^{-3}\ \text{kg mol}^{-1}}{22.4 \times 10^{-3}\ \text{m}^3\ \text{mol}^{-1}} = 1.25\ \text{kg m}^{-3}$$

$$1\ \text{atm} = 1.01 \times 10^5\ \text{Pa} = 1.01 \times 10^5\ \text{kg m}^{-1}\ \text{s}^{-2}$$

Thus

$$\overline{u^2} = \frac{3P}{\rho} = \frac{3(1.01 \times 10^5\ \text{kg m}^{-1}\text{s}^{-2})}{1.25\ \text{kg m}^{-3}} = 2.42 \times 10^5\ \text{m}^2\,\text{s}^{-2}$$

and

$$(\overline{u^2})^{1/2} = 4.92 \times 10^2\ \text{m s}^{-1}$$

PROBLEM 2-29 A mixture of 0.50 mol of neon and 0.25 mol of krypton is contained in a 20-L flask. The effective atomic radii are 1.6×10^{-10} m for Ne and 2.0×10^{-10} m for Kr; the average velocities are 535 m s^{-1} and 263 m s^{-1}, respectively. Calculate the number of collisions per second per cubic meter.

Solution: For the case of two sizes of atoms, you must use Eq. (2-43):

$$Z_{AB} = \left(\frac{1}{2}\right)(\bar{u}_A^2 + \bar{u}_B^2)^{1/2}\pi d_{AB}^2\left(\frac{M_A M_B}{V^2}\right)$$

For the velocity term

$$(\bar{u}_A^2 + \bar{u}_B^2)^{1/2} = (535^2 + 263^2)^{1/2} = 596\ \text{m s}^{-1}$$

For the collision cross-section term

$$d_{AB}^2 = (\tfrac{1}{2}d_A + \tfrac{1}{2}d_B)^2 = (r_A + r_B)^2 = (1.6 \times 10^{-10} + 2.0 \times 10^{-10})^2 = 1.3 \times 10^{-19}\ \text{m}^2$$

For the number of particles term

$$\frac{M_A M_B}{V} = \frac{0.50(6.02 \times 10^{23})(0.25)(6.02 \times 10^{23})}{(20 \times 10^{-3}\ \text{m}^3)^2} = 1.13 \times 10^{50}\ \text{m}^{-6}$$

Substitute these terms in Eq. (2-43) and solve.

$$Z_{AB} = \tfrac{1}{2}(596\ \text{m s}^{-1})(3.142)(1.30 \times 10^{-19}\ \text{m}^2)(1.1 \times 10^{50}\ \text{m}^{-6}) = 1.3 \times 10^{34}\ \text{s}^{-1}\,\text{m}^{-3}$$

PROBLEM 2-30 Use the information from Problem 2-29 to calculate the mean free path of neon if the same amount of that gas occupied the flask by itself.

Solution: The expression for the mean free path is Eq. (2-45):

$$\text{mfp} = \frac{V}{\sqrt{2}\sigma M} = \frac{V}{\sqrt{2}\pi d^2 M}$$

Substituting, you get

$$\text{mfp} = \frac{20 \times 10^{-3}\ \text{m}^3}{\sqrt{2}(3.142)(3.2 \times 10^{-10}\ \text{m})^2(0.50)(6.02 \times 10^{23})} = 1.5 \times 10^{-7}\ \text{m}$$

PROBLEM 2-31 What would be the mean free path of the neon in Problem 2-30 if the pressure were increased by 10 times? First derive an expression for the mfp of an ideal gas that includes the pressure.

Solution: Start with Eq. (2-45) and replace V with nRT/P and M with nL_A to get

$$\text{mfp} = \frac{nRT}{\sqrt{2}\sigma MP} = \frac{RT}{\sqrt{2}\sigma L_A P} = \frac{kT}{\sqrt{2}\sigma P}$$

where k is the Boltzmann constant, $1.381 \times 10^{-23}\ \text{J K}^{-1}$.

For this problem, the significant point is that mfp is proportional to $1/P$. Thus, if $P_2 = 10\,P_1$, mfp decreases by 1/10, or

$$\text{mfp} = 1.5 \times 10^{-8}\ \text{m}$$

PROBLEM 2-32 If neon is contained in a 1-cm cube, how low would the pressure have to be for the mean free path to be about the same as the length of one side of the container? (See Problems 2-29 through 2-31 and assume 0 °C.)

Solution: This involves a rearrangement of Eq. (2-45a) and a conversion of units:

$$P = \frac{kT}{\sqrt{2}\,\sigma(\text{mfp})} = \left(\frac{(1.381 \times 10^{-23}\ \text{J K}^{-1})(273\ \text{K})}{\sqrt{2}(3.142)(3.20 \times 10^{-10}\ \text{m})^2(1 \times 10^{-2}\ \text{m})}\right)\left(\frac{1\ \text{Pa m}^3}{1\ \text{J}}\right) = 0.829\ \text{Pa}$$

Convert to atmospheres:

$$P = (8.29 \times 10^{-1}\ \text{Pa})\left(\frac{1\ \text{atm}}{1.0133 \times 10^5\ \text{Pa}}\right) = 8.18 \times 10^{-6}\ \text{atm}$$

PROBLEM 2-33 Show that the ratios of the three representative speeds can be obtained from the equations for each and thus are the same for all gases.

Solution: You must write ratios with Eqs. (2-49) and (2-50) as the numerator and with Eq. (2-51) as the denominator.

(a) For rms to mp,

$$\frac{(\overline{u^2})^{1/2}}{u_{\text{mp}}} = \frac{\left(\dfrac{3kT}{m}\right)^{1/2}}{\left(\dfrac{2kT}{m}\right)^{1/2}} = \left(\frac{3}{2}\right)^{1/2} = 1.225$$

(b) For average to mp,

$$\frac{\bar{u}}{u_{\text{mp}}} = \frac{\left(\dfrac{8kT}{\pi m}\right)^{1/2}}{\left(\dfrac{2kT}{m}\right)^{1/2}} = \left(\frac{8}{2\pi}\right)^{1/2} = 1.128$$

PROBLEM 2-34 Obtain an appropriate expression for and then calculate the collision frequency of one molecule of oxygen when $\frac{1}{2}$ mol of oxygen is confined in a 5.00-L container at 298 K. Take the effective diameter of an oxygen molecule to be 3.57×10^{-10} m.

Solution: The average velocity isn't given, so you can't use Eq. (2-41) directly. However, you can substitute the value for $\bar{u}$ from Eq. (2-50).

$$Z_A = \sqrt{2}\left(\frac{8kT}{\pi m}\right)^{1/2}\sigma\frac{M}{V} = 4\left(\frac{kT}{\pi m}\right)^{1/2}\sigma\frac{M}{V}$$

First, find the value of m:

$$m = \frac{0.0320\ \text{kg}}{6.02 \times 10^{23}} = 5.32 \times 10^{-26}\ \text{kg}$$

Then

$$\frac{kT}{\pi m} = \frac{(1.381 \times 10^{-23}\ \mathrm{kg\,m^2\,s^{-2}\,K^{-1}})(298\ \mathrm{K})}{3.142(5.32 \times 10^{-26}\ \mathrm{kg})} = 2.462 \times 10^4\ \mathrm{m^2\,s^{-2}}$$

$$\left(\frac{kT}{\pi m}\right)^{1/2} = (2.462 \times 10^4)^{1/2} = 1.57 \times 10^2\ \mathrm{m\,s^{-1}}$$

And

$$Z_A = 4(1.57 \times 10^2\ \mathrm{m\,s^{-1}})(3.142)(3.57 \times 10^{-10}\ \mathrm{m})^2\left(\frac{0.5(6.02 \times 10^{23})}{5.00 \times 10^{-3}\ \mathrm{m^3}}\right) = 1.51 \times 10^{10}\ \mathrm{s^{-1}}$$

Supplementary Exercises

PROBLEM 2-35 If the pressure on 15 L of an ideal gas is tripled, the new volume is ____________. For a gas collected over water, ____________ probably will be needed in calculations. CO_2 will diffuse ____________ rapidly than SO_2. The van der Waals b constant attempts to correct for ____________. The virial equation of state is a ____________ series. The expression for the average molar kinetic energy of an ideal gas is ____________. The collision frequency ____________ if the volume is increased. If the radius of atom A is twice the radius of atom B, the mean free path of atom A is ____________ than that of atom B. The average speed of a molecule is ____________ than its rms speed.

PROBLEM 2-36 What is the temperature of 7.0 g N_2 if it occupies 4.0 L under a pressure of 1.5 atm? (Assume ideal behavior.)

PROBLEM 2-37 We have 5 mol of an ideal gas in a rigid vessel under a pressure of 3.0 atm and a temperature of 30 °C. If we remove 1 mol of gas, what must we do to keep the pressure at 3.0 atm?

PROBLEM 2-38 If a 2-L flask filled with N_2 at 400 K is found to have a pressure of 1.5 atm, how many grams of N_2 are in the flask?

PROBLEM 2-39 What is the true molecular formula of ironIII chloride in the vapor state if 20.0 g of the compound at 600 K in a 2.0-L flask exerts a pressure of 1.53 atm?

PROBLEM 2-40 Flask A has a volume of 3.0 L and contains CO_2 at a pressure of 4.0 atm. Flask B contains N_2 at a pressure of 8.0 atm. The two are connected, and, after mixing of the gases, the mole fraction of N_2 was found to be 0.30. Calculate the volume of Flask B.

PROBLEM 2-41 Which would effuse more rapidly through a given small opening, methane or ethane? For your choice, calculate how much faster.

PROBLEM 2-42 If it takes 5.0 min for 1 mol of neon to effuse through a certain pinhole, how long will it take for 1 mol of argon to effuse through the same hole?

PROBLEM 2-43 What pressure and temperature for N_2 would be the *corresponding state* of N_2 to water vapor at 100 atm and 400 K?

PROBLEM 2-44 The compressibility factor of methane is 0.88 at 60×10^5 Pa and 273 K. What is the molar volume under these conditions?

PROBLEM 2-45 The van der Waals constants for ammonia are $a = 4.17$ L^2 atm mol^{-2} and $b = 3.70 \times 10^{-2}$ L mol^{-1}. Calculate the temperature of 4.0 mol NH_3 confined in a 2.0-L vessel under a pressure of 50.0 atm.

PROBLEM 2-46 Calculate the van der Waals constants for oxygen using the critical constants in Table 2-2.

PROBLEM 2-47 When the pressure is in atm and V is in liters, the virial coefficients at 500 °C for H_2 are $B = 1.80 \times 10^{-2}$ and $C = 1.00 \times 10^{-6}$. (**a**) Calculate the pressure exerted by 2.0 mol H_2 in a 5.0-L container at 500 °C. (**b**) Compare the pressure calculated to that obtained from the ideal gas equation.

PROBLEM 2-48 The density of HBr gas was measured to be 2.4220 g L^{-1} at 0 °C and 0.666 67 atm. Using five significant figures, calculate the molecular weight of HBr using the Berthelot equation. The critical constants are $T_c = 90$ °C and $P_c = 84.44$ atm.

PROBLEM 2-49 Calculate the root-mean-square speed of HBr for the conditions in Example 2-23. (See Problem 2-27.)

PROBLEM 2-50 For CO_2, the effective diameter of a molecule is 4.1×10^{-10} m, and its average velocity at 298 K is 380 m s^{-1}. If a 4-L flask contains 2.0 mol CO_2, calculate (**a**) the collision frequency and (**b**) the mean free path.

PROBLEM 2-51 Calculate the average kinetic energy of the CO_2 in Problem 2-50.

PROBLEM 2-52 For methane at 300 K, calculate (**a**) the root-mean-square speed, (**b**) the average speed, (**c**) the ratio of (a) to (b), and (**d**) the ratio you would expect. (See Example 2-24.)

Answers to Supplementary Exercises

2-35 5 L; Dalton's law of partial pressures; more; volume of molecules; power; KE $= \frac{3}{2}RT$; decreases; smaller; smaller **2-36** 2.9×10^2 K **2-37** Increase temperature to 106 °C **2-38** 2.6 g **2-39** Fe_2Cl_6 **2-40** 0.64 L **2-41** Methane; 1.37 times as fast **2-42** 7.0 min **2-43** 15.6 atm and 78.0 K **2-44** 0.33 L **2-45** 376 K **2-46** $a = 0.913$ L^2 atm mol^{-2}; $b = 2.6 \times 10^{-2}$ L mol^{-2} **2-47** (**a**) 25.56 atm (**b**) 25.38 atm **2-48** 80.851 g mol^{-1} **2-49** 2.9×10^2 m s^{-1} **2-50** (**a**) 1.29×10^{37} s^{-1} m^{-3} (**b**) 4.45×10^{-9} m **2-51** 3.7 kJ **2-52** (**a**) 6.8×10^2 m s^{-1} (**b**) 6.3×10^2 m s^{-1} (**c**) 1.086; (**d**) 1.086

TABLE S-2: Summary of Important Equations

For ideal gases		For real gases	
$PV = nRT$	(2-9; Ideal gas equation)	$Z = \dfrac{PV}{nRT}$	(2-12; Compressibility factor)
→ $PV = k_A$	(2-1; Boyle's law)	$\left(P + \dfrac{an^2}{V^2}\right)(V - nb) = nRT$	(2-13; van der Waals equation)
→ $\dfrac{PV}{T} = k_G$	(2-7)		
→ $R = \dfrac{PV_m}{T}$	(2-9a; Gas constant)	→ $P = \dfrac{nRT}{V - nb} - \dfrac{an^2}{V^2}$	(2-14)
$P_{\text{total}} = P_A + P_B + P_C + \cdots$	(2-10; Dalton's law)	$V_c = 3b \quad P_c = \dfrac{a}{27b^2} \quad T_c = \dfrac{8a}{27Rb}$	(2-16)
		$P_r = \dfrac{P}{P_c} \quad V_r = \dfrac{V}{V_c} \quad T_r = \dfrac{T}{T_c}$	(2-17; Reduced variables)
$\dfrac{\text{rate}_A}{\text{rate}_B} = \sqrt{\dfrac{M_{m,B}}{M_{m,A}}}$	(2-11; Graham's law)	→ $P_r = \dfrac{8T_r}{3V_r - 1} - \dfrac{3}{V_r^2}$	(2-14a)
→ $\dfrac{r_A}{r_B} = \sqrt{\dfrac{d_B}{d_A}}$	(see Problem 2-13)	$\dfrac{PV}{nRT} = 1 + \dfrac{B(T)n}{V} + \dfrac{C(T)n^2}{V^2} + \dfrac{D(T)n^4}{V^4} + \cdots$	(2-18; Virial equation)

Kinetic Theory			
$\overline{u^2} = \overline{u_x^2} + \overline{u_y^2} + \overline{u_z^2}$	(2-27; single molecule)	$\text{mfp} = \dfrac{V}{\sqrt{2}\sigma M}$	(2-45; mean free path)
$\overline{u^2} = \dfrac{\sum_{i=1}^{M} u_i^2}{M}$	(2-30; *M* molecules)	$n_i = Ae^{-\varepsilon_i/kT}$	(2-46; energy distribution)
$PV = \frac{1}{3}Mm\overline{u^2}$	(2-29)	→ $\dfrac{dn_i}{n} = 4\pi\left(\dfrac{m}{2\pi kT}\right)^{3/2} e^{-mu^2/2kT}u^2\,du$	(2-47; Maxwell distribution)
$\overline{\text{KE}} = CT$	(2-33)		
$PV = \frac{2}{3}CnT = \frac{2}{3}n\overline{\text{KE}}$	(2-35; 2-36)	→ $\dfrac{dn_i}{n} = \dfrac{2\pi}{(\pi kT)^{3/2}} e^{-\varepsilon/kT}\varepsilon^{1/2}\,d\varepsilon$	(2-48; Maxwell–Boltzmann)
→ $\overline{\text{KE}} = \frac{3}{2}RT$	(2-37)	$(\overline{u^2})^{1/2} = \left(\dfrac{3kT}{m}\right)^{1/2}$	(2-49; root-mean-square speed)
$Z_{AA} = \dfrac{\bar{u}\sigma M^2}{\sqrt{2}\,V^2}$	(2-42; Collision frequency, single species)	$\bar{u} = \left(\dfrac{8kT}{\pi m}\right)^{1/2}$	(2-50; average speed)
$Z_{AB} = \left(\dfrac{1}{2}\right)(\bar{u}_A^2 + \bar{u}_B^2)^{1/2}\pi d_{AB}^2\left(\dfrac{M_A M_B}{V^2}\right)$	(2-43; Collision frequency, two species)	$u_{\text{mp}} = \left(\dfrac{2kT}{m}\right)^{1/2}$	(2-51; most probable speed)

3 THE FIRST LAW OF THERMODYNAMICS

THIS CHAPTER IS ABOUT

- ☑ **Temperature and the Zeroth Law of Thermodynamics**
- ☑ **More Fundamental Concepts**
- ☑ **Heat, Work, and Internal Energy**
- ☑ **State Functions and Exact Differentials**
- ☑ **Enthalpy**
- ☑ **Ideal Gases**
- ☑ **Real Gases**

Thermodynamics deals with the relationships between thermal energy and other forms of energy and is based on three fundamental principles: the first, second, and third laws of thermodynamics. All are true laws because they were derived empirically, that is, by countless observations and experiments over many years.

There are two approaches to the study of thermodynamics:

- *Classical thermodynamics*, which doesn't require any theory of matter. It consists of defined terms and mathematical relationships derived from experimental data; and
- *Statistical thermodynamics*, which provides an explanation of the three laws based on the theories of molecular structure and motion. Statistical methods are used to describe molecular behavior because of the large number of objects involved.

The two approaches to thermodynamics may be described, respectively, as the macroscopic and the submicroscopic views of matter, i.e., the properties of bulk quantities of matter, on the one hand, and the properties of individual molecules, on the other. In this and the next three chapters, we avoid mentioning atoms and molecules as much as possible. You should be aware, however, that some authors prefer to start with the statistical approach and that others treat the two approaches together.

3-1. Temperature and the Zeroth Law of Thermodynamics

Definitions of familiar concepts, such as temperature, often seem awkward or unnecessarily complex. The precision of science, though, requires unambiguous, noncircular definitions of concepts.

A. Temperature

We choose **temperature** as the most fundamental concept of thermodynamics. Subjectively, we say that, if object A feels hotter than object B, then object A has a higher temperature than object B. Thus temperature is a measure of how hot or cold an object is. An object feels hot because it transmits heat to us. From this viewpoint, temperature may be defined as a measure of the tendency of an object to lose heat. This definition isn't entirely satisfactory because we haven't defined heat.

B. Zeroth Law

The **zeroth law of thermodynamics** is a necessary principle that seems self-evident, and some authors do not even mention it. Nevertheless, it provides a fundamental definition of temperature and underlies the use of thermometers, as well as much of thermodynamics. One version of this law states that, if object C and object D are both in thermal equilibrium with a third object E, then C and D are in thermal equilibrium with each other.

This implies that these three objects share a property, which, by being the same for all three objects, shows that thermal equilibrium exists. We call this property **temperature**. Now we have a definition that depends only on the observation that no change of state occurs when two objects are in close contact thermally. When we have a means of measuring temperature, we can say that, if two objects are in direct contact thermally and have the same temperature, they are in thermal equilibrium.

C. Thermometers

Any system or device that changes in a regular, observable manner when its temperature changes may serve as a **thermometer**, i.e., an instrument for measuring temperature. Examples of physical properties that may be used as a basis for thermometers are the height of a liquid in a tube (alcohol or mercury), the volume of a gas (argon), and the electrical resistance of a wire (platinum). Each thermometer must be *calibrated* (determination of a base point and a graduated scale) by comparing it to systems in reproducible states, such as water at its triple point or silver at its melting point. Under prescribed conditions, the systems are at equilibrium at these points and are assigned a numeric value for their temperatures. With proper graduations, a thermometer then may be used to measure the temperature of any other system.

3-2. More Fundamental Concepts

A. Definitions

A **system** is some particular portion of the universe in which we are interested; it usually is some form of matter that is undergoing a change. The **surroundings** are the rest of the universe. The interface between a system and its surroundings is often called a **boundary**.

If matter can pass between a system and its surroundings across their boundary, the system is *open*; if not, the system is *closed*. If a closed system is not in thermal contact with its surroundings, it is *isolated*. In other words, if a system is isolated, neither matter nor energy can be exchanged with its surroundings. For a system that is not isolated, an exchange of energy usually is in the form of heat or work.

If two systems or objects are at different temperatures and are placed in thermal contact, one temperature will fall and the other will rise. We say that **heat** flows from the system having the higher temperature to that having the lower temperature. **Work** is defined as the mechanical transfer of energy. If a weight is raised or lowered by some mechanical process (e.g., with a rope and pulley), work is being done.

Energy is defined as the capacity of a system to do work or to transfer heat, and we often say that a system contains energy. In thermodynamics, this property of a system is called the **internal energy**, E. (Some authors prefer the symbol U.) We don't need to know how this energy is contained or its total amount, because thermodynamics is concerned with changes, not with absolute values.

B. The first law of thermodynamics

If heat is applied to a system and no work is involved, the internal energy of the system will increase. This can be achieved by inserting a hot object into a system or by applying a flame to its container. Similarly, the energy of a system may be increased by doing work on the system in the absence of heat; for example, by turning a paddlewheel, rubbing two objects together, or passing an electric current through a resistor. Joule employed all these techniques to show that a certain amount of work always produces the same amount of heat, or in modern terms, the same amount of energy. This result is known as the **mechanical equivalence of heat**.

If heat and work processes occur simultaneously, the net change in the internal energy of a system is the sum of the energy gained or lost as heat, q, and the energy gained or lost because of the work done on or by the system, w. This relationship provides one statement of the **first law of thermodynamics**:

FIRST LAW
$$\Delta E = q + w \tag{3-1}$$

where the symbol Δ (delta) represents a noninfinitesimal change in a property of a system, that is, the value in the final state minus the value in the initial state, or $\Delta E = E_f - E_i$. Note that

Eq. (3-1) applies to a system and leads to the conclusion that E is constant in an isolated system (since no energy can cross the boundary).

When a system is not isolated and loses energy in the form of heat, its surroundings must gain an equal amount of energy. Similarly, if work is done on a system, something in the surroundings had to do the work and use up some of its own energy. The net gain or loss of energy by a system must equal the net loss or gain by its surroundings:

$$-\Delta E(\text{system}) = +\Delta E(\text{surroundings}) \tag{3-2}$$

or

$$\Delta E(\text{surroundings}) + \Delta E(\text{system}) = 0$$

and

$$\Delta E(\text{universe}) = 0 \tag{3-3}$$

In words, the energy of the universe is constant. This is another expression of the First Law, better known perhaps as the **law of conservation of energy**. Let's look at some other equivalent statements of the first law:

1. Energy cannot be created or destroyed.
2. The total energy of an isolated system is constant.
3. The sum of all of the energy changes for all systems participating in a process must be zero.
4. It is impossible to construct a perpetual motion machine of the first kind (one that would create energy from nothing).
5. If work is produced during a cyclic process, an equivalent quantity of heat must be consumed.

C. Sign convention for heat and work

If no work is involved and heat is added to a system, the internal energy increases, and ΔE has a positive sign. Since $w = 0$, Eq. (3-1) becomes $\Delta E = q$. If ΔE is positive, q must be positive. Conversely, if ΔE is negative, q must be negative. Thus the sign convention is

- If heat is added to the system, q is $+$, and ΔE is $+$.
- If heat is lost by the system, q is $-$, and ΔE is $-$.

If no heat is involved and work is done *on* a system, the internal energy increases, and ΔE is positive. Since $q = 0$, Eq. 3-1 becomes $\Delta E = w$. If ΔE is positive, w must be positive. Conversely, if ΔE is negative, w must be negative. These statements are the basis of the sign convention for work recommended by the International Union of Pure and Applied Chemistry:

- If work is done *on* a system: w is $+$, and ΔE is $+$.
- If work is done *by* a system: w is $-$, and ΔE is $-$.

Unfortunately, a convention for work using the opposite signs was established when thermodynamics was used primarily by engineers concerned with steam engines, and some physical chemistry textbooks still use it. You can spot the difference easily because Eq. (3-1) becomes $\Delta E = q - w$. Check your text to see how the first law is expressed. If it is $\Delta E = q - w$, then you will have to keep in mind, while using this outline, that values of work will have signs opposite from those in your text.

3-3. Heat, Work, and Internal Energy

A. Heat

Heat gained or lost is directly proportional to the mass of an object times the temperature change:

$$q \propto (\text{mass})(t_f - t_i) \qquad \text{or} \qquad q = (\text{sp heat})m\,\Delta t \tag{3-4}$$

where sp heat is the **specific heat** of a substance, that is, the amount of heat required to raise the temperature of 1 gram of substance 1 °C, usually expressed in units of cal $g^{-1}deg^{-1}$ (in older texts) or J $g^{-1}deg^{-1}$ (in newer texts). If Δt is positive, then q is positive, and heat has been gained by the object or system, an **endothermic process**. If Δt is negative, then q is negative, and heat has been lost, an **exothermic process**.

EXAMPLE 3-1: Calculate the amount of heat required to heat 15 g of iron from 20 °C to 60 °C. The specific heat of iron is 0.107 cal $g^{-1}deg^{-1}$.

Solution: Use Eq. (3-4):

$$q = (\text{sp heat})m\,\Delta t = (0.107 \text{ cal g}^{-1}\text{deg}^{-1})(15 \text{ g})[(60 - 20) \text{ deg}] = 64 \text{ cal}$$

When two objects are involved, the range of possible unknown quantities is large. If two objects, A and B, at different initial temperatures are placed in contact and come to thermal equilibrium (same temperature), the heat lost by one must equal the heat gained by the other:

$$-q_A = -q_B \tag{3-5}$$

We use Eq. (3-4) and substitute for q_A and q_B:

$$(\text{sp heat})_A(-m_A)(t_f - t_i)_A = (\text{sp heat})_B m_B(t_f - t_i)_B \tag{3-6}$$

In the resulting equation, there are seven quantities that could be the unknown in a problem.

We can readily extend this concept from two objects to a system and its surroundings. The negative sign is needed because the system (B) gains energy while its surroundings (A) lose energy.

EXAMPLE 3-2: In an experiment, a 50.0-g sample of silver was heated to 98.0 °C and then dropped into 50.0 g of water at 24.0 °C. The final temperature was 27.9 °C. Calculate the specific heat of silver.

Solution: Use Eq. (3-6) and let A = Ag and B = H_2O:

$$(-50.0 \text{ g Ag})(\text{sp heat})(27.9 - 98.0) = (50.0 \text{ g H}_2\text{O})(1 \text{ cal g}^{-1}\text{deg}^{-1})[(27.9 - 24.0) \text{ deg}]$$

$$-50.0(-70.1)(\text{sp heat}) = 50.0(3.9)$$

$$\text{sp heat} = \frac{50.0(3.9)}{50.0(70.1)} = 0.056 \text{ cal g}^{-1}\text{deg}^{-1}$$

B. Mechanical work: constant force

When an object is moved some distance by an applied force, the work is defined as the applied force multiplied by the distance moved. In symbols,

$$w = F\,\Delta l \tag{3-7}$$

More generally,

$$dw = F\,dl \tag{3-8}$$

EXAMPLE 3-3: Calculate the work that must be done to lift a box weighing 20.0 kg to a height of 40.0 m.

Solution: The force required to overcome the pull of gravity is given by the mass times the acceleration due to gravity:

$$F = mg = (20.0 \text{ kg})(9.81 \text{ m s}^{-2}) = 196 \text{ kg m s}^{-2}$$

We integrate Eq. (3-8):

$$w = \int_{l_1}^{l_2} F\,dl \tag{3-9}$$

and, since the force is constant,

$$w = F\int_{l_1}^{l_2} dl = F(l_2 - l_1)$$

$$= (196\ \text{kg m s}^{-2})[(40.0 - 0)\ \text{m}] = 7.84 \times 10^3\ \text{kg m}^2\,\text{s}^{-2} = 7.84 \times 10^4\ \text{J} = 7.84\ \text{kJ}$$

EXAMPLE 3-4: If the box in Example 3-3 is lowered slowly back to its starting position, how much work is done?

Solution: As in Example 3-3, the distance moved is 40.0 m, this time in the opposite direction, so $\Delta l = -40.0$ m, and $w = 196(-40.0) = -7.84$ kJ.

EXAMPLE 3-5: The two steps in Examples 3-3 and 3-4 constitute a **cyclic process**. What is the total work done in this cyclic process?

Solution:

$$w_1 + w_2 = 7.8 - 7.8 = 0$$

The box is back in its original position or "state."

C. Mechanical work: variable force

Sometimes the force is not constant and cannot be removed from the integration.

EXAMPLE 3-6: In the stretching of a spring the force changes with the amount of displacement from the rest position $r = 0$. According to Hooke's law, the restoring force exerted by the spring is negative.

HOOKE'S LAW $$F = -kr \qquad \textbf{(3-10)}$$

Calculate the work required to stretch a spring from its rest length of 5.0 cm to 7.0 cm. The Hooke's law constant of this spring is 80 dyn cm^{-1}.

Solution: The displacement is $l - l_0 = r = 7.0 - 5.0 = 2.0$ cm. We use Eq. (3-9), replacing l with r:

$$w = \int_0^r F\,dr$$

The force *on* the spring has the opposite sign of the force exerted *by* the spring; that is, $F = +kr$. Thus

$$w = \int_0^2 kr\,dr$$

Now we apply Eqs. (1-19) and (1-21) but in the form of definite integrals:

$$w = k\left(\frac{r^2}{2}\right)\Bigg|_0^2 = (80\ \text{dyn cm}^{-1})\left(\frac{1}{2}\right)(2^2 - 0)\ \text{cm}^2 = 160\ \text{dyn cm} = 160\ \text{erg}$$

D. Expansion against constant pressure

The expansion of a gas can be envisioned as the mechanical process of pushing back a piston. If we assume that the piston is weightless and frictionless (a necessary, often unstated condition), the force opposing the expansion of the gas is the force of the pressure that holds the piston in place.

Pressure is force per unit area ($P = F/A$), e.g., lb/in^2 or N/m^2. Thus the force is the pressure times the area of the piston: $F = PA$. This pressure, which opposes the expansion, is called the *external pressure*, whereas the pressure exerted by the gas is called the *internal pressure*. In symbols,

$$P_{\text{op}} = P_{\text{ext}} \qquad \text{and} \qquad P_{\text{gas}} = P_{\text{int}}$$

If $P_{int} > P_{ext}$, the gas should expand, pushing the piston through some distance, dl. The work done once again is given by $\int F\,dl$. Here $F = -P_{ext}A$, where the negative sign shows that work is done *by* the gas, not *on* the gas.

$$w = -\int P_{ext}A\,dl = -\int P_{ext}\,dV \tag{3-11}$$

EXAMPLE 3-7: Calculate the work done by a gas in a cylinder fitted with a piston when the gas expands from a volume of 4.0 L to a volume of 7.0 L under an external pressure of 4.0 atm at constant temperature.

Solution: Although the condition is not stated explicitly, the wording of the problem implies that the external pressure is constant. Thus

$$P_{op} = P_{ext} = \text{Constant} = 4.0 \text{ atm}$$

Substituting in Eq. (3-11), we get

$$w = -\int_{V_1}^{V_2} P_{op}\,dV = -P_{ext}\int_{V_1}^{V_2} dV = -P(V_2 - V_1) \tag{3-12}$$

$$= -(4.0 \text{ atm})(7.0 \text{ L} - 4.0 \text{ L}) = -12.0 \text{ L atm}$$

This example, involving three dimensions, is analogous to Example 3-3, in one dimension, where the integral for w became simply $F\,dl$; here it becomes $P\,\Delta V$, because P_{op} is constant.

E. Reversible expansion or compression

Now consider a case analogous to Example 3-6, in which the force was not constant. In this case, P_{op} varies with volume, and the integration to obtain w requires a mathematical expression for the volume dependence of P_{op}. To obtain the relationship, we may use an idealized procedure known as a **reversible process**. In such a process, the essential requirement is that the external pressure must differ only infinitesimally from the internal pressure. Thus they can be set equal for calculation purposes, or

$$P_{op} = P_{ext} = P_{int} = P_{gas} \qquad \text{(for reversible case only)}$$

If the pressure–volume dependence of the gas can be expressed by a mathematical expression, the amount of work can be calculated. Such expressions can be obtained from equations of state, such as the ideal gas equation and the van der Waals equation.

EXAMPLE 3-8: Calculate the work done if 2.00 mol of an ideal gas, initially occupying a volume of 30.0 L at 280 K, is allowed to expand isothermally (constant temperature) and reversibly until the volume is 50.0 L.

Solution: Note that pressure is not mentioned. The initial pressure could be calculated, of course, but the value is not needed.

Since the gas is ideal and the process is reversible,

$$P_{gas} = \frac{nRT}{V} = P_{ext} = P_{op}$$

We substitute in Eq. (3-11):

$$w = -\int_{V_1}^{V_2} P_{op}\,dV = -\int_{V_1}^{V_2} P_{gas}\,dV = -\int_{V_1}^{V_2}\left(\frac{nRT}{V}\right)dV$$

Then

$$w = -nRT\int_{V_1}^{V_2}\frac{dV}{V} = -nRT\ln\frac{V_2}{V_1} \tag{3-13}$$

$$= (-2.00 \text{ mol})(8.314 \text{ J K}^{-1}\text{mol}^{-1})(280 \text{ K})\left(\ln\frac{50.0}{30.0}\right) = -2.38 \text{ kJ}$$

F. Constant-pressure work and reversible work

In Example 3-8, we can change "reversibly" to "at constant pressure," but in order to compare the results, we must be certain that the constant pressure is no greater than the final pressure calculated from the volume data in the example, namely,

$$P_f = \frac{nRT}{V_f} = \frac{(2.00 \text{ mol})(0.0821 \text{ L atm K}^{-1}\text{mol})(280 \text{ K})}{50 \text{ L}} = 0.920 \text{ atm}$$

EXAMPLE 3-9: Calculate the work done in a nonreversible expansion for the ideal gas in Example 3-8 under a constant pressure of 0.920 atm. Compare the result to that of Example 3-8.

Solution:

$$w = -\int P_{\text{op}}\, dV = -\int P_{\text{ext}}\, dV$$

$$= -0.920 \text{ atm} \int_{30}^{50} dV = -0.920(50 - 30) \text{ L atm}$$

$$= -18.4 \text{ L atm} = (-18.4 \text{ L atm})(101.3 \text{ J L}^{-1}\text{atm}^{-1}) = -1.86 \text{ kJ}$$

More work was obtained from the reversible process. In fact, 2.38 kJ is the maximum work possible for this change under these conditions, i.e., expansion from 30 L to 50 L at 280 K.

Let's consider a variation on the constant-pressure problem in which the initial and final pressures are given.

EXAMPLE 3-10: If 2.00 mol of an ideal gas, initially at 1.53 atm and 280 K, is allowed to expand isothermally under a constant external pressure of 0.920 atm until the final internal pressure is 0.920 atm, what is the work done?

Solution: Again $w = -\int P_{\text{op}}\, dV = -P_2(V_2 - V_1)$, but we have to calculate the volumes or substitute an expression for them. The latter is easier:

$$w = -P_2\left(\frac{nRT}{P_2} - \frac{nRT}{P_1}\right)$$

$$w = -nRT\left(1 - \frac{P_2}{P_1}\right) \qquad \textbf{(3-14)}$$

$$= -(2.00 \text{ mol})(8.314 \text{ J K}^{-1}\text{mol}^{-1})(280 \text{ K})\left(1 - \frac{0.920}{1.53}\right) = -1.86 \text{ kJ}$$

G. Changes in internal energy

Placing restrictions on processes that a system may undergo creates special cases that simplify the calculation of ΔE. To obtain mathematical expressions for these special cases, we may use Eq. (3-1) directly ($\Delta E = q + w$) or, with the substitution of Eq. (3-11), as

$$\Delta E = q - \int P_{\text{ext}}\, dV \qquad \textbf{(3-15)}$$

1. Constant volume

At constant volume, $dV = 0$ and no work is done. Thus $\Delta E = q$. For example, chemical reactions may be carried out in stainless-steel pressurized vessels. Since the walls cannot move, the volume is fixed, and any heat produced must increase the internal energy by raising the temperature.

2. **Zero pressure**

   At zero pressure, $P_{ext} = 0$ and no work is done. Again, $\Delta E = q$. If a gas is allowed to expand into an evacuated chamber, there is no opposing pressure between the gas and its surroundings. After a portion of the gas has been transferred, it may exert an opposing pressure against the gas still being transferred, but such work is all *within* the system, so it is not of concern. Such a process is called a **free expansion**. Applying Eq. (3-11), we obtain $w = -\int P_{op}\, dV = 0$.

3. **Constant pressure**

   At constant pressure, work is done. We use Eq. (3-15) and get $\Delta E = q - P\,\Delta V$.

EXAMPLE 3-11: A gas is contained in a cylinder with a piston, which is held in place by an opposing pressure of 10 atm and is in thermal equilibrium with the surroundings. If the piston is pulled out at such a speed that the opposing pressure is always 5 atm, what happens?

Solution: The gas will expand against the constant pressure of 5 atm. The temperature of the gas will drop, and heat from the surroundings will flow into the cylinder. If the amount of heat can be measured and the initial and final volumes are known, ΔE can be calculated.

---

4. **Variable pressure**

   If the pressure isn't constant, work will be done, and $\Delta E = q - \int P_{op}\, dV$. To calculate the work, we need either an expression relating pressure to volume for substitution in the integral or enough data to plot P against V and integrate graphically.

EXAMPLE 3-12: If the gas involved is ideal and the process is reversible, obtain an expression for w.

Solution: We can use the ideal gas equation of state, substitute in Eq. (3-11), and solve for P. If $P = nRT/V$, then

$$w = -\int_{V_1}^{V_2} \frac{nRT}{V}\, dV = -nRT\left(\ln \frac{V_2}{V_1}\right) \qquad \text{[Eq. (3-13)]}$$

---

5. **Adiabatic conditions**

   If $q = 0$, that is, the system is insulated so that no heat can be gained or lost (**adiabatic conditions**), then $\Delta E = w$.

6. **Cyclic process**

   In a cyclic process, the system is returned to the state from which it started. Think about the energy changes when a system changes from state A to state B and then back to state A. For A to B, $\Delta E_1 = E_B - E_A$. For B to A, $\Delta E_2 = E_A - E_B$. The total $\Delta E = \Delta E_1 + \Delta E_2 = (E_B - E_A) + (E_A - E_B) = 0$.

   Any property of a system that undergoes no net change in a cyclic process is called a **state function**, since its value is a function only of the state of the system. Some authors call such functions *thermodynamic properties*.

3-4. State Functions and Exact Differentials

A. Cyclic integration

The cyclic integral of the differential of a state function equals zero. This is another way of expressing the definition of a state function. In symbols, $\oint dE = 0$, where the circle on the integration symbol means that the integration is carried out around a closed path; i.e., the final point coincides with the initial point.

B. Line integration

The line integral of the differential of a state function is independent of the path. If the limits of integration are two points that do not coincide, the value of the integral $\int dE$ is constant. The shape of the curve between the points does not matter. Using finite changes, the ΔE for a change from state A to state B is the same regardless of the nature of the change.

EXAMPLE 3-13: Assume that the temperature of a given system is 30 °C in state A and 150 °C in state B. Describe two different paths for getting from state A to state B. What can be said about the ΔE for each path?

Solution: One possible path between the states is to heat the system to 500 °C, cool it to 100 °C, heat it to 300 °C, and cool it to 150 °C. The line integral, $\int dE$ equals $E(150°) - E(30°) = \Delta E$. Another path is to cool the system to 10 °C and then heat it to 150 °C. Again, $\Delta E = E(150°) - E(30°)$. The value of $\int dE$ is the same for both paths—and for any other possible path.

C. State-function differential

The differential of a state function is an exact differential. Either of the two preceding conditions is sufficient to establish the exactness of a differential. Thus we know that state functions have the three properties of exact differentials listed in Section 1-4E.

The converse of the third property is a useful way of identifying a new state function. If a differential is shown to be exact, a state function must exist and can be differentiated to give that differential.

A state function may also be defined as a function that yields a certain exact differential.

EXAMPLE 3-14: Show that a function corresponding to our definition of internal energy must exist.

Solution: In Examples 3-8 and 3-9, we found that the work done was different for two paths between the same initial and final states. In other words, $\int dw$ depends on the path, violating one of the requirements for an exact differential. Thus we know that no function can be differentiated to give dw and that $\oint dw \neq 0$. The same conclusions can be drawn about the heat. Both dq and dw are inexact. We show inexactness by the symbol $đ$ and can now write

$$\oint đw \neq 0 \quad \text{and} \quad \oint đq \neq 0$$

In a cyclic process, $\Delta E = 0$. Then, by the first law, if work is done by a system, an equivalent amount of heat must be added to the system: $-w = q$. For infinitesimal changes,

$$\oint đw = -\oint đq \quad \text{or} \quad \oint đq + \oint đw = 0 \quad \text{or} \quad \oint (đq + đw) = 0 \tag{3-16}$$

Since the cyclic integral of the quantity in the parentheses equals zero, this quantity must be an exact differential. We can write $\oint dL = 0$, where $dL = đq + đw$. The function L must be a state function. If we give it the name *internal energy* and the symbol E, we have

$$dE = đq + đw \tag{3-1a}$$

This equation is another statement of the first law of thermodynamics.

This approach is satisfying, because there is no need to invent arbitrarily a property such as internal energy. We see that there must be such a property because of our experience with heat and work quantities in cyclic processes.

D. Heat and work differentials

Under special conditions, the heat and work differentials may become exact. We've seen that no work is done if a process is carried out at constant volume; i.e., $dE = đq_V$. Since the heat differential is identical to an exact differential, it must be exact and we could change the symbol to dq. In other words, in a constant-volume process, the heat involved becomes a state function. We may also write $\Delta E = q_V$.

3-5. Enthalpy

Most chemical reactions in the laboratory take place under constant pressure and with a volume that isn't constant. Reaction vessels are usually open to the atmosphere, and thus any exchanges of energy as heat or work occur under pressure exerted by the atmosphere. To achieve constant volume would require a sealed container.

Because the exchange of heat under constant pressure is so common, we need to ask: Is q_P, like q_V, a state function? In answering this question, we'll use the properties of exact differentials.

Let's start by rearranging our new differential expression for the first law, Eq. (3-1a), to read $đq = đE - đw$. Imposing the condition that only pressure–volume work is done, rewriting Eq. (3-12) as $đw = -P\,dV$ (since P is constant), and substituting, we have

$$đq = dE + P\,dV \qquad \textbf{(3-17)}$$

We need to substitute for $P\,dV$. Instead of trying to derive a term, let's take the simpler, if less satisfying, approach of working backwards from the result we want, which is $d(PV)$. Recall the standard expression for the differential of a product, Eq. (1-12). Applying it to the product PV, we get

$$d(PV) = V\,dP + P\,dV$$

Rearranging

$$P\,dV = d(PV) - V\,dP$$

and substituting in Eq. (3-17) yields

$$đq = dE + d(PV) - V\,dP$$

If the pressure is constant, $dP = 0$, so the third term equals 0, and

$$đq_P = dE + d(PV) = d(E + PV) \qquad \textbf{(3-18)}$$

Since E, P, and V are state functions, the quantity $E + PV$ must also be a state function. This function has its own name and symbol: **enthalpy** H, where $H = E + PV$. Now,

$$đq_P = dH = dq_P = d(E + PV) = dE + d(PV) \qquad \textbf{(3-19)}$$

The heat differential is identical to the exact differential, dH, so it must also be exact. In a constant-pressure process, the heat involved becomes a state function, and we may write

$$\Delta H = q_P \qquad (P\text{–}V \text{ work only}) \qquad \textbf{(3-20)}$$

Other approaches to developing the concept of enthalpy can be found in various textbooks. Some authors define it directly as $H = E + PV$. Because it is impossible to measure absolute values of H or E, it is useful to define properties as changes, which can be measured.

EXAMPLE 3-15: Calculate the enthalpy change when 1 mol Fe is heated from 100 °C to 150 °C at atmospheric pressure. (The specific heat of Fe is 0.107 cal $g^{-1}deg^{-1}$.)

Solution: This should remind you of specific-heat problems. Note that P is constant and only P–V work is involved. Combining Eqs. (3-4) and (3-20) gives

$$\Delta H = q_P = (\text{sp heat})m\,\Delta t$$

Thus

$$\Delta H = (55.8 \text{ g})(0.107 \text{ cal g}^{-1}\text{deg}^{-1})(50 \text{ deg}) = 3.0 \times 10^2 \text{ cal}$$

A. Relationships between *E* and *H*

We start with the definition of H in absolute terms, $H = E + PV$, and consider the general case in which neither V nor P is constant. For the final state,

$$H(\text{final}) = E(\text{final}) + PV(\text{final})$$

For the initial state,

$$H(\text{initial}) = E(\text{initial}) + PV(\text{initial})$$

1. For a change between the two states,

$$\Delta H = -H(\text{final}) - H(\text{initial}) = \Delta E + \Delta(PV) \tag{3-21}$$

2. Imposing the condition of ideal gas behavior, we can substitute nRT for PV, or

$$\Delta H = \Delta E + \Delta(nRT) = \Delta E + R\,\Delta(nT) \tag{3-22}$$

3. For constant n,

$$\Delta H = \Delta E + nR\,\Delta T \tag{3-23}$$

4. For constant T,

$$\Delta H = \Delta E + RT\Delta n \tag{3-24}$$

Equation (3-24) may be applied to a chemical reaction, but remember that ideal behavior was assumed and that Δn refers to amounts of gases only:

$$\Delta n = \Sigma n_{(\text{gas, products})} - \Sigma n_{(\text{gas, reactants})} \tag{3-25}$$

EXAMPLE 3-16: Calculate the difference between ΔE and ΔH at 1000 °C for the reaction

$$2NO(g) + O_2(g) \longrightarrow 2NO_2(g)$$

Solution: We assume ideal behavior and rearrange Eq. (3-24):

$$\Delta H - \Delta E = RT\,\Delta n$$

From Eq. (3-25),

$$\Delta n = 2\text{ mol} - 3\text{ mol} = -1\text{ mol}$$

Now

$$\Delta H - \Delta E = (8.314\text{ J K}^{-1}\text{mol}^{-1})(1273\text{ K})(-1\text{ mol}) = -1.058 \times 10^4\text{ J}$$

B. Heat capacity

We can rewrite Eq. (3-4) as

$$q = C\,\Delta T \tag{3-26}$$

where $C = m \times$ (sp heat) and is called the **heat capacity**. The heat capacity of a substance varies with temperature, although for temperature ranges up to about 100 °C, it often is assumed to be constant. It also varies with the other conditions under which the process is carried out, and we must specify constant volume, C_V, or constant pressure, C_P.

These symbols represent the total heat capacity of a system or of a quantity of a substance and are extensive quantities. We represent *molar heat capacity* by $C_{V,m}$ or $C_{P,m}$. This is an intensive property, having the units J (or cal) $K^{-1}mol^{-1}$.

1. Constant volume

If we impose the condition of constant volume and make various substitutions, we obtain from Eq. (3-26) several useful ways of looking at C:

$$C_V = \frac{q_V}{\Delta T} = \frac{dq_V}{dT} = \frac{\Delta E}{\Delta T} = \frac{dE}{dT} = \left(\frac{\partial E}{\partial T}\right)_V \tag{3-27}$$

Perhaps the most useful of these (rewritten) is

$$dE = C_V \, dT \tag{3-28}$$

EXAMPLE 3-17: Assuming that $C_{V,m}$ is constant and equal to 24.8 J K^{-1}mol^{-1}, calculate ΔE of 2.0 mol NH_3 heated from 300 K to 370 K at constant volume.

Solution: In order to obtain an expression for ΔE, we must integrate Eq. (3-28):

$$\int_{E_1}^{E_2} dE = \int_{T_1}^{T_2} C_V \, dT = C_V \int_{T_1}^{T_2} dT$$

$$E_2 - E_1 = nC_{V,m}(T_2 - T_1)$$

$$\Delta E = (2.0 \text{ mol})(25 \text{ J K}^{-1}\text{mol}^{-1})(370 \text{ K} - 300 \text{ K}) = 3.5 \times 10^3 \text{ J}$$

---

2. Constant pressure

If we impose the condition of constant pressure and make various substitutions, we obtain from Eq. (3-26) another set of relationships:

$$C_P = \frac{q_P}{\Delta T} = \frac{dq_P}{dT} = \frac{\Delta H}{\Delta T} = \frac{dH}{dT} = \left(\frac{\partial H}{\partial T}\right)_P \tag{3-29}$$

Perhaps the most useful of these (rewritten) is

$$dH = C_P \, dT \tag{3-30}$$

EXAMPLE 3-18: If a system containing 1.00 mol CO_2 gas ($C_{P,m}$ = 9.80 cal K^{-1}mol^{-1}) at 400 K is heated by 500 cal at constant pressure, what is the final temperature?

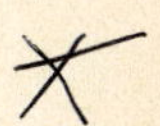

Solution: Remember that $q_P = \Delta H$ and that we can get an equation for ΔH by integrating Eq. (3-30).

$$\int_{H_1}^{H_2} dH = \int_{T_1}^{T_2} C_P \, dT = C_P \int_{T_1}^{T_2} dT$$

$$H_2 - H_1 = C_P(T_2 - T_1)$$

$$T_2 - T_1 = \frac{\Delta H}{nC_{P,m}}$$

$$T_2 = \frac{500 \text{ cal}}{(1.00 \text{ mol})(9.80 \text{ cal K}^{-1}\text{mol}^{-1})} + 400 \text{ K} = (51.0 + 400) \text{ K} = 451 \text{ K}$$

---

3. Relationship between C_P and C_V

Using Eq. (3-1a) as a statement of the first law and substituting for dE the expression for the total differential [see Eq. (1-24)], we have

$$\text{đ}q + \text{đ}w = dE = \left(\frac{\partial E}{\partial T}\right)_V dT + \left(\frac{\partial E}{\partial V}\right)_T dV \tag{3-31}$$

The first partial derivative equals C_V in Eq. (3-27). We make that substitution and also substitute $-P\,dV$ for $\text{đ}w$ to get

$$\text{đ}q - P\,dV = C_V\,dT + \left(\frac{\partial E}{\partial V}\right)_T dV$$

Imposing the condition of constant pressure, we obtain $\text{đ}q = dq_P = C_P\,dT$ and

$$C_P\,dT - C_V\,dT = \left(\frac{\partial E}{\partial V}\right)_T dV + P\,dV = \left[\left(\frac{\partial E}{\partial V}\right)_T + P\right] dV$$

Dividing by dT gives

$$C_P - C_V = \left[\left(\frac{\partial E}{\partial V}\right)_T + P\right]\frac{dV}{dT} = \left[\left(\frac{\partial E}{\partial V}\right)_T + P\right]\left(\frac{\partial V}{\partial T}\right)_P \tag{3-32}$$

We write the last term as a partial derivative in order to state explicitly that the pressure is constant.

The value of C_P will be greater than the value of C_V. You could have predicted this by remembering that the work of expansion must be done in a constant-pressure process. This requires energy; thus the total energy needed to raise the temperature one degree is greater than that needed to achieve the same temperature change in a constant-volume process.

From Eq. (3-32), you can see that the first partial derivative, $\left(\frac{\partial E}{\partial V}\right)_T$, must represent a pressure since it is to be added to P. It is also related to forces between molecules and usually is called the **internal pressure**.

EXAMPLE 3-19: For an ideal gas, the internal pressure equals zero. Evaluate $C_P - C_V$ for an ideal gas.

Solution: Since the internal pressure is zero, Eq. (3-32) becomes

$$C_P - C_V = P\left(\frac{\partial V}{\partial T}\right)_P$$

For an ideal gas, $V = nRT/P$, and

$$\left(\frac{\partial V}{\partial T}\right)_P = \frac{nR}{P}$$

Then

$$C_P - C_V = nP\left(\frac{R}{P}\right) = nR \tag{3-33}$$

Alternatively,

$$C_{P,m} - C_{V,m} = R \tag{3-33a}$$

EXAMPLE 3-20: Assuming ideal behavior, calculate **(a)** $C_{P,m}$ of NH_3 (see Example 3-17), and **(b)** $C_{V,m}$ of CO_2 (see Example 3-18).

Solution: In both cases, we must rearrange Eq. (3-33a).

(a) $C_{P,m} = C_{V,m} + R$
$= (24.8 + 8.314)\ \text{J K}^{-1}\text{mol}^{-1}$
$= 33.1\ \text{J K}^{-1}\text{mol}^{-1}$

(b) $C_{V,m} = C_{P,m} - R$
$= (9.80 - 1.987)\ \text{cal K}^{-1}\text{mol}^{-1}$
$= 7.81\ \text{cal K}^{-1}\text{mol}^{-1}$

3-6. Ideal Gases

Three conditions must be met for ideality:

1. The gas obeys the ideal gas equation of state:

$$PV = nRT \tag{3-34}$$

2. The internal pressure of the gas equals zero:

$$\left(\frac{\partial E}{\partial V}\right)_T = \left(\frac{\partial E}{\partial P}\right)_T = 0 \tag{3-35}$$

In terms of enthalpy, this condition becomes

$$\left(\frac{\partial H}{\partial V}\right)_T = \left(\frac{\partial H}{\partial P}\right)_T = 0 \tag{3-36}$$

3. The heat capacities of the gas are constant:

$$C_V = \left(\frac{\partial E}{\partial T}\right)_V = \text{Constant} \tag{3-37}$$

$$C_P = \left(\frac{\partial H}{\partial T}\right)_P = \text{Constant} \tag{3-38}$$

A. Isothermal expansion

For either reversible or irreversible conditions, the isothermal internal energy change of an ideal gas is zero, as defined by Eq. (3-35). Then $q + w = 0$, and $q = -w$. If work is done by an ideal gas at constant temperature, an equivalent amount of heat must be supplied. Recall Eq. (3-12):

$$w = -\int P_{\text{op}}\, dV$$

Thus

$$q = +\int P_{\text{op}}\, dV \tag{3-39}$$

In Examples 3-7 and 3-8, we have already considered the work done in the irreversible and the reversible cases. We can repeat the expressions obtained there and assign the proper sign convention to the work *done* to obtain q for both cases.

1. Reversible case

Using Eq. (3-13), we get

$$q = -w = nRT \ln \frac{V_2}{V_1} = nRT \ln \frac{P_1}{P_2} \tag{3-40}$$

2. Irreversible case

Using Eqs. (3-12) and (3-14), we get

$$q = -w = P_2(V_2 - V_1) \tag{3-41}$$

$$q = nRT\left(1 - \frac{P_2}{P_1}\right) \tag{3-42}$$

EXAMPLE 3-21: Compare the quantities of heat required to carry out the isothermal expansion of 1.00 mol of an ideal gas, initially occupying 28.0 L at a pressure of 0.880 atm and a temperature of 300 K, to a volume of 40.0 L (**a**) reversibly, and (**b**) irreversibly against a pressure equal to the final pressure of the gas.

Solution: For both cases, the conditions of ideal behavior and constant temperature require that $\Delta E = 0$. Thus $q = -w$ and $\Delta H = 0$. For both cases, Eq. (3-39) applies:

$$q = -w = \int P_{\text{op}}\, dV$$

(**a**) For the reversible case, P_{op} changes continuously and is given by nRT/V. Thus

$$q = \int_{28.0}^{40.0} nRT \frac{dV}{V} = nRT \ln \frac{40.0}{28.0}$$

$$= (1.00 \text{ mol})(8.314 \text{ J K}^{-1}\text{mol})(300 \text{ K}) \ln \frac{40.0}{28.0} = 8.90 \times 10^2 \text{ J}$$

Note that we applied Eq. (3-40) but derived it when needed, rather than substituting into it blindly.

(**b**) For the irreversible case, P_{op} is constant and is the final pressure, which we can calculate from either (1) Boyle's law or (2) the ideal gas equation.

(1) $P_{op} = P_2 = \dfrac{V_1 P_1}{V_2}$

$$q = P_{op} \int dV = P_{op}(V_2 - V_1)$$

$$= \frac{(28.0 \text{ L})(0.880 \text{ atm})}{40.0 \text{ L}}(40.0 - 28.0) \text{ L} = 7.39 \text{ L atm} = 7.49 \times 10^2 \text{ J}$$

(2) $P_{op} = P_2 = \dfrac{nRT}{V_2}$

$$q = P_{op}(V_2 - V_1) = \frac{nRT}{V_2}(V_2 - V_1) = nRT\left(1 - \frac{V_1}{V_2}\right) \qquad \textbf{(3-43)}$$

Equation (3-43), a variation of Eq. (3-42), could be derived by substituting Boyle's law in Eq. (3-42). We use Eq. (3-42) now because we already know the volumes and will not have to calculate P_2. As in (a), we derived the needed equation, rather than substituting numbers in a memorized equation. Then

$$q = (1.00 \text{ mol})(8.314 \text{ J K}^{-1}\text{mol}^{-1})(300 \text{ K})\left(1 - \frac{28.0}{40.0}\right) = 7.48 \times 10^2 \text{ J}$$

Note that the reversible process requires the most heat. This agrees with the fact that a reversible process does the maximum amount of work.

B. Adiabatic expansion

Since $q = 0$ by definition, $\Delta E = w$ [Eq. (3-1)]. This expresses the fact that, if heat cannot enter or leave, the energy needed to do any work can come only from the internal energy of the system. Thus the temperature must drop.

Integrating Eq. (3-28) to get $\Delta E = \int C_V\, dT$ and using Eq. (3-11), $w = -\int P_{op}\, dV$, we obtain three useful expressions:

$$\Delta E = -\int P_{op}\, dV \qquad \textbf{(3-44)}$$

$$w = \int C_V\, dT \qquad \textbf{(3-45)}$$

$$\int C_V\, dT = -\int P_{op}\, dV \qquad \textbf{(3-46)}$$

1. Reversible case

Since $P_{op} = P_{int} = P_{gas}$, we can substitute for P in Eq. (3-46) its value from the ideal gas equation and rearrange to obtain

$$\int C_V \frac{dT}{T} = -\int nR\frac{dV}{V} \qquad \textbf{(3-47)}$$

Since C_V is independent of temperature [Eq. (3-37)], integrating gives

$$C_V \ln\frac{T_2}{T_1} = -nR \ln\frac{V_2}{V_1} = nC_{V,m} \ln\frac{T_2}{T_1} \qquad \textbf{(3-48)}$$

This relationship may be rearranged into several useful forms:

$$\left(\frac{T_2}{T_1}\right)^{C_{V,m}} = \left(\frac{V_2}{V_1}\right)^{-R} = \left(\frac{V_1}{V_2}\right)^{R} \qquad \textbf{(3-49)}$$

$$\frac{T_2}{T_1} = \left(\frac{V_1}{V_2}\right)^{R/C_{V,m}} \qquad \textbf{(3-50)}$$

or $$TV^{R/C_{V,m}} = \text{Constant} \tag{3-51}$$

Sometimes the exponent is expressed in terms of γ (gamma), defined as the ratio $C_{P,m}/C_V$:

$$\frac{R}{C_{V,m}} = \frac{C_{P,m} - C_{V,m}}{C_{V,m}} = \frac{C_{P,m}}{C_{V,m}} - \frac{C_{V,m}}{C_{V,m}} = \gamma - 1$$

Thus

$$TV^{\gamma-1} = \text{Constant} \tag{3-52}$$

EXAMPLE 3-22: Calculate the final temperature if 1.00 mol of an ideal gas, initially occupying 10.0 L at 300 K, expands adiabatically and reversibly to 20.0 L. ($C_{V,m} = \frac{5}{2}R$)

Solution: We rearrange Eq. (3-50) to get T_2:

$$T_2 = T_1\left(\frac{V_1}{V_2}\right)^{R/C_{V,m}}$$

The exponent is $$\frac{R}{C_{V,m}} = \frac{R}{(5/2)R} = \frac{2}{5} = 0.40$$

Then $$T_2 = (300\text{ K})\left(\frac{10.0\text{ L}}{20.0\text{ L}}\right)^{0.40} = 227\text{ K}$$

An expression involving P and V is useful because the temperatures are not needed. Substituting $T = PV/R$ in Eq. (3-51), we get

$$\frac{PV}{R}V^{\gamma-1} = \text{Constant} \quad \text{or} \quad PV^{(1+\gamma-1)} = R\,(\text{Constant})$$

So $$PV^{\gamma} = \text{Constant} \tag{3-53}$$

Note the similarity between Eq. (3-53) and Boyle's law (for the isothermal case). Equation (3-53) is one expression of the reversible adiabatic equation of state for an ideal gas. From it, you should be able to derive other forms, e.g., Eq. (3-52).

EXAMPLE 3-23: Start with Eq. (3-53) and derive an expression similar to Eq. (3-52), which contains P instead of V.

Solution: In order to remove V and insert P in Eq. (3-52), we substitute $V = RT/P$, since the ideal-gas restriction already applies, and get

$$P\left(\frac{RT}{P}\right)^{\gamma} = \text{Constant} \quad \text{or} \quad P\frac{T^{\gamma}}{P^{\gamma}} = \frac{\text{Constant}}{R^{\gamma}} \quad \text{or} \quad P^{(1-\gamma)}T^{\gamma} = \frac{\text{Constant}}{R^{\gamma}}$$

Then we take the γth root:

$$TP^{(1-\gamma)/\gamma} = \text{Constant} \tag{3-54}$$

EXAMPLE 3-24: If an ideal gas having $C_{P,m} = \frac{7}{2}R$ expands from 15 L at a pressure of 5.0 Pa in an adiabatic reversible manner to a volume of 25 L, what is the final pressure?

Solution: In order to use Eq. (3-53), we need to know the value of γ. Recall that $C_{P,m} - C_{V,m} = R$ [Eq. (3-33a)], or $C_{V,m} = C_{P,m} - R$. Thus

$$C_{V,m} = \frac{7}{2}R - R = \frac{5}{2}R \quad \text{and} \quad \frac{C_{P,m}}{C_{V,m}} = \frac{7}{5} = 1.4$$

Rearranging Eq. (3-53) and substituting:

$$\frac{P_2}{P_1} = \left(\frac{V_1}{V_2}\right)^{\gamma}$$

$$P_2 = (5\text{ Pa})\left(\frac{15\text{ L}}{25\text{ L}}\right)^{1.4} = 2.4\text{ Pa}$$

The work done by an adiabatic process may be calculated using either Eq. (3-11) or Eq. (3-45), whichever is appropriate.

EXAMPLE 3-25: Refer to Example 3-22 and calculate the work done.

Solution: From Example 3-22, we obtain $T_1 = 300$ K, $T_2 = 227$ K, $C_{V,m} = \frac{5}{2}R$, and $n = 1.00$ mol. Because we have the initial and final temperatures, we use Eq. (3-45). Since $C_{V,m}$ is constant, Eq. (3-45) becomes

$$w = nC_{V,m}\,\Delta T$$

Substituting:

$$w = \tfrac{5}{2}(8.314)(227 - 300) = -1.52\text{ kJ}$$

---

2. Irreversible case

No equations similar to the adiabatic equations of state for the reversible case can be obtained because there is no expression to substitute for P_{op} in Eq. (3-46). The equation is still valid, but we can use it only for the constant-pressure case. If the adiabatic expansion of an ideal gas is against a constant opposing pressure, $P_{op} = P_2$, Eq. (3-46) becomes

$$C_V \int_{T_1}^{T_2} dT = -P_2 \int_{V_1}^{V_2} dV$$

and

$$nC_{V,m}(T_2 - T_1) = -P_2(V_2 - V_1) \tag{3-55}$$

The volumes can be replaced by nRT/P. Canceling the n's, we get

$$C_{V,m}(T_2 - T_1) = -R\left(T_2 - \frac{T_1P_2}{P_1}\right) \tag{3-56}$$

EXAMPLE 3-26: Using Eq. (3-56), calculate the final temperature if 1.0 mol of an ideal gas ($C_{V,m} = 5.0$ cal K^{-1}mol^{-1}), initially at 5.0 atm and 400 K, is allowed to expand adiabatically against a constant pressure of 2.0 atm until the gas pressure is 2.0 atm.

Solution: First, we rearrange the equation to get T_2:

$$T_2 - T_1 = -\frac{R}{C_{V,m}}\left(T_2 - \frac{T_1P_2}{P_1}\right)$$

$$T_2 + \frac{R}{C_{V,m}}T_2 = T_1 + \frac{R}{C_{V,m}}\left(\frac{T_1P_2}{P_1}\right)$$

$$T_2 = \frac{1}{1 + R/C_{V,m}}\left[T_1 + \frac{R}{C_{V,m}}\left(\frac{T_1P_2}{P_1}\right)\right]$$

Then substitute:

$$T_2 = \frac{1}{1 + (2.0/5.0)}\left[400 + \frac{2.0}{5.0}\left(\frac{400(2.0)}{5.0}\right)\right] = 331\text{ K}$$

3-7. Real Gases

A. Heat capacity as a function of temperature

Because the heat capacity of a nonideal gas varies with temperature, its value must be determined experimentally. Empirical equations are used to consolidate the data for a wide range of temperatures. Two forms of these equations are

$$C_{P,m} = a + bT + \frac{c'}{T^2} \quad (3\text{-}57)$$

$$C_{P,m} = a + bT + cT^2 + dT^3 \quad (3\text{-}58)$$

Most textbooks contain a table of the coefficients a, b, and c for Eq. (3-58), omitting the T^3 term. A short version of such tables, which also illustrates the two common energy units, is given in Table 3-1. When C_P is dependent on temperature, the calculation becomes complicated but not difficult.

TABLE 3-1: Heat Capacities of Some Gases
Applicable Range: 300–1500 K at 1 atm

Substance	a	$b \times 10^3$	$c \times 10^7$
		cal $K^{-1}mol^{-1}$	
$Cl_2(g)$	7.576	2.424	−9.650
$SO_3(g)$	6.077	23.537	−96.87
		J $K^{-1}mol^{-1}$	
$H_2(g)$	29.07	−0.836	20.12
$HCl(g)$	28.17	1.81	15.47

EXAMPLE 3-27: Calculate the heat required to raise the temperature of 1 mol $H_2(g)$ from 400 K to 1000 K at constant pressure.

Solution: From the various possibilities in Eq. (3-29), we choose the expression $dq_P = C_P\,dT$, substitute $C_P = nC_{P,m}$, and integrate:

$$q_P = n\int_{T_1}^{T_2} C_{P,m}\,dT$$

Then we replace $C_{P,m}$ by its value from Eq. (3-58), substitute $n = 1$, and separate the integral:

$$q_P = \int_{T_1}^{T_2} a\,dT + \int_{T_1}^{T_2} bT\,dT + \int_{T_1}^{T_2} cT^2\,dT = a(T_2 - T_1) + \frac{b}{2}(T_2^2 - T_1^2) + \frac{c}{3}(T_2^3 - T_1^3)$$

Finally, we substitute the values for H_2 from Table 3-1:

$$q_P = 29.07(1000 - 400) - \left(\frac{0.836 \times 10^{-3}}{2}\right)(1000^2 - 400^2) + \left(\frac{20.12 \times 10^{-7}}{3}\right)(1000^3 - 400^3)$$

$$= 17.7 \text{ kJ}$$

B. Joule–Thomson effect

One condition of ideality [Eq. (3-35)] is that the internal pressure must be zero, or $\left(\frac{\partial E}{\partial V}\right)_T = 0$. Joule was unable to measure a temperature change when a gas expanded, so ΔE appeared to be zero for real gases. Thomson suggested an alternative experimental procedure: Force a gas slowly through a restriction, such as a silk handkerchief or some other porous material, and let it expand against a piston on the other side. When this was done, an energy change was detected. The apparatus was insulated from the surroundings, so the process was adiabatic, and occurred at constant enthalpy.

EXAMPLE 3-28: Show that the Joule–Thomson expansion is isenthalpic. (Consider the work done on each side of the restriction.)

Solution: We assume that 1 mol of gas is moved from the high-pressure (initial) side to the low-pressure (final) side. As given by Eq. (3-12), the work done under constant pressure is $-P\,\Delta V$. Then

$$w_i = -P_i(0 - V_i) = P_iV_i \quad \text{and} \quad w_f = -P_f(V_f - 0) = P_fV_f$$

Since $q = 0$,

$$w_{\text{net}} = \Delta E = w_i + w_f = P_iV_i - P_fV_f = E_f - E_i$$

Now we collect terms with the same subscript:

$$E_f + P_fV_f = E_i + P_iV_i$$

And, since $H = E + PV$,

$$H_f = H_i \quad \text{or} \quad \Delta H = 0$$

Thomson's experimental procedure also revealed a change of temperature when the pressure changed. The ratio of temperature change to pressure change is known as the **Joule–Thomson coefficient**. It is usually written as a partial derivative and designated μ_{JT}:

JOULE–THOMSON COEFFICIENT

$$\mu_{\text{JT}} = \left(\frac{\partial T}{\partial P}\right)_H \qquad \textbf{(3-59)}$$

If μ is positive, then T decreases when P decreases, or the gas cools upon expansion. This relation applies at about room temperature for all gases except hydrogen and helium. A good illustration is provided by a carbon dioxide fire extinguisher. When the gas is released, there is a large drop in pressure with a consequent drop in temperature. The gas actually solidifies, forming a white powder. At high temperatures and pressures, μ becomes negative. Then expansion causes heating.

The combination of temperature and pressure at which $\mu = 0$ (sign changes from $+$ to $-$) is called the **inversion point**. For a given pressure, the temperature at which $\mu = 0$ is called the **inversion temperature**. For nitrogen gas at 300 °C, μ is 0.0048 at 33 atm and -0.0013 at 60 atm. By rough interpolation, we can estimate that the inversion point is about 55 atm at 300 °C. At a pressure of 60 atm, μ is 0.037 at 200 °C and -0.0013 at 300 °C. We can estimate that the inversion temperature at 60 atm is about 295 °C.

EXAMPLE 3-29: Show that $\mu_{\text{JT}}C_P = -\left(\frac{\partial H}{\partial P}\right)_T$.

Solution: Recall that μ_{JT} requires constant H and that C_P can be written as a partial derivative of H. The partial derivatives suggest that we start by writing the total differential of H as a function of P and T:

$$dH = \left(\frac{\partial H}{\partial P}\right)_T dP + \left(\frac{\partial H}{\partial T}\right)_P dT$$

For constant H, $dH = 0$. Then dividing by dP, we have

$$0 = \left(\frac{\partial H}{\partial P}\right)_T + \left(\frac{\partial H}{\partial T}\right)_P \left(\frac{\partial T}{\partial P}\right)_H$$

The last two partial differentials are special terms and have their own symbols:

From Eq. (3-29):

$$\left(\frac{\partial H}{\partial T}\right)_P = C_P$$

From Eq. (3-59):

$$\left(\frac{\partial T}{\partial P}\right)_H = \mu_{\text{JT}}$$

Finally, we transpose and substitute:

$$\mu_{\text{JT}}C_P = -\left(\frac{\partial H}{\partial P}\right)_T \qquad \textbf{(3-60)}$$

In order for a gas to be liquefied by increased pressure, its temperature must be below the critical temperature. For many common gases, the critical temperatures are far below room temperature; e.g., for O_2, -119 °C, and for H_2, -240 °C. A Joule–Thomson expansion can be used to decrease the temperature, if μ is positive. For hydrogen μ is negative at room temperature, and other types of cooling must be used to lower the temperature below the inversion temperature; then expansion would cause further cooling. In the Linde process for liquefying air, an initial compression removes water, and ordinary cooling lowers the temperature. Then successive Joule–Thomson expansions lower the temperature until liquid forms.

SUMMARY

Table S-3A summarizes the major equations of this chapter and shows connections between them. You should determine the conditions that are necessary for each equation to be valid or the conditions that are imposed to make the connections.

Table S-3B summarizes the applications of the first law under various specified conditions. It shows how to calculate the values for the four basic quantities, q, w, ΔE, and ΔH, in the various cases. You should derive these expressions for yourself and become alert to the conditions under which each is valid.

TABLE S-3A: Summary of Important Equations

$\Delta E = q + w$ (3-1; First law)

$dE = đq + đw$ (3-1a)

→ $dE = C_V\, dT$ (3-28)

→ $\Delta E = -\int P_{op}\, dV$ (3-44)

$\Delta E(\text{universe}) = 0$ (3-3)

$q = (\text{sp heat}) m\, \Delta t$ (3-4)

$dw = F\, dl$ (3-8)

→ $w = -\int P_{ext}\, dV$ (3-11)

→ → $w = -P\, dV$ (3-12)

→ → $w = -nRT \ln \dfrac{V_2}{V_1}$ (3-13)

$dq_V = dE$ (3-27)

→ $dE = C_V\, dT$ (3-28)

→ → $w = \int C_V\, dT$ (3-45)

(3-13) and (3-28) → $C_V \ln \dfrac{T_2}{T_1} = -nR \ln \dfrac{V_2}{V_1}$ (3-48)

→ $TV^{R/C_{V,m}} = \text{Constant}$ (3-51)

→ $PV^{\gamma} = \text{Constant}$ (3-53)

→ $TP^{(1-\gamma)/\gamma} = \text{Constant}$ (3-54)

$dq_P = d(E + PV) = dH$ (3-19)

→ $dH = C_P\, dT$ (3-30)

→ $\Delta H = \Delta E + \Delta(PV)$ (3-21)

→ → $\Delta H = \Delta E + nR\, \Delta T$ (3-23)

→ → $\Delta H = \Delta E + RT\, \Delta n$ (3-24)

$C_{P,m} = a + bT + cT^2 + dT^3$ (3-58)

$\mu_{JT} = \left(\dfrac{\partial T}{\partial P}\right)_H$ (3-59; Joule–Thomson coefficient)

→ $\mu_{JT} C_P = -\left(\dfrac{\partial H}{\partial P}\right)_T$ (3-60)

TABLE S-3B: Basic Information for First-Law Calculations

P, V, T, and n relationships	q	w	ΔE	ΔH
ISOTHERMAL				
n and T constant; P or V variable; $PV = K$	$-w$	$-\int P\,dV$	0	0
ADIABATIC				
n constant; P, V, or T variable $PV^{\gamma} = K'$ $TV^{\gamma-1} = K''$ $TP^{(1-\gamma)/\gamma} = K''' = \dfrac{T}{P^{(\gamma-1)/\gamma}}$	0	ΔE or $-\int P\,dV$ or $\int C_V\,dT$	$\int C_V\,dT$	$\int C_P\,dT$
SPECIAL CASES				
n and T constant; P or V variable	$-w$	$-\int P\,dV$	0	0
n and V constant; P or T variable	ΔE	0	$\int C_V\,dT$	$\Delta E + \Delta(PV)$
n and P constant; V or T variable	ΔH	$-P\,\Delta V$ or $-nR\,\Delta T$	$\Delta H - \Delta(PV)$	$\int C_P\,dT$
P and T constant; n or V variable (for phase transitions and chemical reactions)	ΔH	$-P\Delta V$ or $-RT\Delta n$	$\Delta H - \Delta(PV)$ or $\Delta H - RT\Delta n$	$\Delta H_{\text{transition}}$

note: P–V work only. Expressions restricted to ideal gases are shaded.

Other important things you should know are listed below.

1. A thermometer may be based on almost any physical property that changes in a regular and measurable manner when the temperature changes.
2. If two objects are at the same temperature as a third object, they are at the same temperature as each other (zeroth law).
3. In a reversible expansion (**a**) the external pressure changes continuously but is always equal to the pressure of the gas, and (**b**) the amount of work done is the maximum possible.
4. The properties of exact differentials may be used to establish the existence of state functions.
5. An ideal gas must
   (**a**) obey the ideal gas equation: $PV = nRT$
   (**b**) have an internal pressure of zero: $(\partial E/\partial V)_T = 0$
   (**c**) have a constant heat capacity: $C =$ constant
6. The Joule-Thomson coefficient tells us that, for all gases except hydrogen and helium, a drop in pressure at room temperature will cool a gas.

SOLVED PROBLEMS

Heat and Work

PROBLEM 3-1 In an experiment a student found that when a 40.0-g piece of zinc was warmed by 50.0 cal of heat, the temperature rose 13.4 °C. Calculate the experimental value of the specific heat of zinc.

Solution: Note first that the problem deals with a change of temperature and second that only *one* object is involved. Thus you can apply Eq. (3-4):

$$q = (\text{sp heat})m\,\Delta t$$

Rearranging:

$$\text{sp heat} = \frac{q}{m\,\Delta t}$$

Substituting:

$$\text{sp heat} = \frac{50.10\ \text{cal}}{(40.0\ \text{g})(13.4\ \text{deg})} = 0.0935\ \text{cal g}^{-1}\text{deg}^{-1}$$

PROBLEM 3-2 Calculate the amount of heat needed to heat a 20.0-g piece of zinc from 25 °C to 150 °C. Use the value of the specific heat calculated in Problem 3-1.

Solution: Again you can use Eq. (3-4) but without rearranging:

$$q = (\text{sp heat})m\,\Delta t = 0.0935(20.0)(150 - 25) = 234\ \text{cal}$$

PROBLEM 3-3 In an experiment to measure the specific heat of lead, 80.0 g of lead was heated to 98.0 °C and quickly added to 80.0 g of water, which was at 23.0 °C. The temperature of the water rose to 25.2 °C. Calculate the experimental value of the specific heat of lead.

Solution: This problem differs from Problems 3-1 and 3-2 in that two objects are involved, one gaining heat and one losing heat. The final temperatures of the two objects must be the same, namely, 25.2 °C, so the heat lost by the lead must equal the heat gained by the water. Assume that there were no losses of heat to the air or to the container holding the water during transfer of the metal. Giving the negative sign to the heat lost by the lead, you can write the equality

$$-q(\text{Pb}) = q(\text{H}_2\text{O})$$

or

$$-\text{mass(Pb)} \times \text{sp heat(Pb)} \times \Delta t(\text{Pb}) = \text{mass(H}_2\text{O)} \times \text{sp heat(H}_2\text{O)} \times \Delta t(\text{H}_2\text{O})$$

Rearranging and substituting:

$$\text{sp heat(Pb)} = \frac{-(80.0\ \text{g})(1.00\ \text{cal g}^{-1}\text{deg}^{-1})(25.2 - 23.0\ \text{deg})}{(80.0\ \text{g})(25.2 - 98.0\ \text{deg})} = 0.0302\ \text{cal g}^{-1}\text{deg}^{-1}$$

PROBLEM 3-4 Calculate the heat required to raise the temperature of 100.0 g of H_2 from 25.0 to 35.0 °C at constant pressure. The heat capacity of hydrogen can be taken as constant at 29.1 J K^{-1}mol^{-1} over a small range of temperature.

Solution: Here heat capacity is used rather than specific heat, so Eq. (3-26) is applicable. [For constant pressure, you could also use Eq. (3-29).] Thus,

$$q = (\text{sp heat})m\,\Delta t = C_P\,\Delta t = nC_{P,m}\,\Delta t$$

Substituting:

$$q = (100.0\ \text{g})\left(\frac{1\ \text{mol}}{2.016\ \text{g}}\right)(29.1\ \text{J K}^{-1}\text{mol}^{-1})(35.0 - 25.0\ \text{K}) = 1.44 \times 10^4\ \text{J} = 14.4\ \text{kJ}$$

PROBLEM 3-5 You lift a 50.0-kg weight from the ground to above your head. (**a**) If the height is 2.40 m, how much work do you do? (**b**) If you hold the weight in this position for 5 min, how much work do you do? (**c**) If you lower the weight slowly back to the ground, how much work do you do?

Solution:

(**a**) Work equals force times distance moved. Recall that the force of gravity equals mass times the gravitational constant: $F = mg$. Substitute this in Eq. (3-9) to get

$$w = \int_{l_1}^{l_2} mg\,dl = mg\,\Delta l = (50.0\ \text{kg})(9.81\ \text{m sec}^{-2})(2.40\ \text{m}) = 1.18 \times 10^3\ \text{J} = 1.18\ \text{kJ}$$

(**b**) Because the weight doesn't move, you've done no work, even though you may be very tired.

(c) Lowering the weight is the opposite of raising it, so you did the same absolute amount of work as in part (a). Now, however, the sign is negative: $w(c) = -1.18$ kJ. Thus, for the entire process, the net work equals 0.

PROBLEM 3-6 A certain spring has a spring constant of 50 dyn cm^{-1}. Calculate the work required to compress this spring by 6.0 cm.

Solution: Recall that $dw = F\,dl$ [Eq. (3-8)] and that Hooke's Law for springs is $F = -kr$ [Eq. (3-10)], where F is the force exerted by the spring. Since the spring undergoes compression, F(compression) $= +kr$, and the force *on* the spring is $-kr$. Change dl to dr and keep in mind that $l_1 = 0$ and that $l_2 = r$. Thus

$$w = -\int_0^r kr\,dr = -k\frac{r^2}{2}\bigg|_0^6 = -(50\text{ dyn cm}^{-1})\left(\frac{(6.0\text{ cm})^2}{2}\right) = -9.0 \times 10^2\text{ dyn cm} \quad\text{or}\quad -9.0 \times 10^2\text{ erg}$$

PROBLEM 3-7 If 1.30 mol of a gas expands from a volume of 4.0 L to a volume of 6.0 L against a constant pressure of 5.0 atm at a constant temperature of 280 K, what is the amount of work done?

Solution: The critical factor is the constant pressure. This means that Eq. (3-11) becomes

$$w = -P\int_{V_1}^{V_2} dV = -P(V_2 - V_1) \qquad \text{[Eq. (3-12)]}$$

Thus

$$w = -5.0(6.0 - 4.0) = -10\text{ L atm}$$

Note that the number of moles and the temperature given in the statement of the problem are excess information.

PROBLEM 3-8 Given the same starting conditions as in Problem 3-7, calculate the work done by the gas undergoing the same volume change in a reversible process. Assume that the ideal gas equation holds.

Solution: In a reversible process, the pressure is not constant but is a function of volume. To replace P in the integration, you need an expression for its dependence on V; thus it is necessary for the problem to tell you which equation of state to use.

For an ideal gas, $P = nRT/V$, so n and T are needed, although they were not needed in Problem 3-7. Now the substitution in Eq. (3-11) gives

$$w = -\int_{V_1}^{V_2} P_{op}\,dV = -\int_{V_1}^{V_2}\frac{nRT}{V}\,dV = -nRT\int_{V_1}^{V_2} d(\ln V) = -nRT\ln\frac{V_2}{V_1} \qquad \text{[Eq. (3-13)]}$$

When you substitute numerical values,

$$w = -1.30(0.0821)(280)\left(\ln\frac{6.0}{4.0}\right) = -12\text{ L atm}$$

Note that more work is done in the reversible process.

PROBLEM 3-9 Calculate the maximum work that could be obtained if 2.00 mol of an ideal gas, initially at STP, is allowed to expand to 100 L isothermally.

Solution: The maximum work is obtained in a reversible expansion; you should make the calculation for that special case. Eq. (3-13) is appropriate:

$$w = -nRT\ln\frac{V_2}{V_1}$$

Since the molar volume of an ideal gas at STP is 22.4 L, the 2.00 mol occupies 44.8 L at the start. Substituting:

$$w = -2.00(0.0821)(273)\left(\ln\frac{100.0}{44.8}\right) = -36.0\text{ L atm}$$

PROBLEM 3-10 Convert the answer to Problem 3-9 to joules and calories.

Solution: To avoid memorizing conversion factors for these units, you can use the values of R, which you should have memorized. From Section 2-2, the ones you need here are

$$R = 0.082\,06\ \text{L atm K}^{-1}\text{mol}^{-1} = 8.314\ \text{J K}^{-1}\text{mol}^{-1} = 1.987\ \text{cal K}^{-1}\text{mol}^{-1}$$

You can make conversions by setting up ratios of R values, which will make the units come out right. You are given L atm and want joules; thus

$$w = -(36.0\ \text{L atm})\left(\frac{8.314\ \text{J K}^{-1}\text{mol}^{-1}}{0.082\,06\ \text{L atm K}^{-1}\text{mol}^{-1}}\right) = -3.65\ \text{kJ}$$

Similarly, to obtain calories,

$$w = -(36.0\ \text{L atm})\left(\frac{1.987\ \text{cal K}^{-1}\text{mol}^{-1}}{0.082\,06\ \text{L atm K}^{-1}\text{mol}^{-1}}\right) = -872\ \text{cal}$$

PROBLEM 3-11 If 5.00 mol of an ideal gas is heated at constant pressure from 300 K to 500 K, how much work is done by the expansion of the gas?

Solution: When work of expansion is requested, you probably think immediately of $-\int P\,dV$, which in the constant pressure case becomes $-P\,\Delta V$ [Eq. (3-12)]. In this problem, however, the pressure isn't given, so it's not possible to solve the problem as in Example 3-10 or Problem 3-7.

What you can do is replace $P\,\Delta V$ with $nR\,\Delta T$, since n is constant, and obtain

$$w = -nR\,\Delta T = -5.00(8.314)(500 - 300) = -8.31\ \text{kJ}$$

PROBLEM 3-12 A certain gas is found to obey the equation of state $(P + nA)V = nRT$. Derive an expression for the work done in an isothermal reversible expansion of this gas.

Solution: Because the process is reversible, the opposing pressure equals the pressure of the gas. Rearranging the equation of state gives

$$P_{\text{gas}} = \frac{nRT}{V} - nA = P_{\text{op}}$$

Substituting in Eq. (3-11):

$$w = -\int_{V_1}^{V_2}\left(\frac{nRT}{V} - nA\right)dV = -\int_{V_1}^{V_2}\frac{nRT}{V}\,dV + \int_{V_1}^{V_2} nA\,dV$$

$$= -nRT\int_{V_1}^{V_2} d(\ln V) + nA\int_{V_1}^{V_2} dV = -nRT\ln\frac{V_2}{V_1} + nA(V_2 - V_1)$$

Internal Energy and the First Law

PROBLEM 3-13 Calculate the change in internal energy of a quantity of gas that absorbs 1000 cal of heat while expanding from 15 L to 25 L at a constant pressure of 4.0 atm.

Solution: The process is at constant pressure, so you obtain the work from Eq. (3-12). The heat is given; thus the first law, $\Delta E = q + w$, becomes

$$\Delta E = q - P(V_2 - V_1) = (1000\ \text{cal}) - (4\ \text{atm})(25 - 15\ \text{L})\left(\frac{1.987\ \text{cal K}^{-1}\text{mol}^{-1}}{0.082\,06\ \text{L atm K}^{-1}\text{mol}^{-1}}\right) = 31\ \text{cal}$$

PROBLEM 3-14 Calculate w, q, ΔE, and ΔH if 1 mol of an ideal gas, initially at 20 °C and 8.0 atm, undergoes an isothermal expansion against a constant pressure of 2.0 atm until the final pressure is 2.0 atm.

Solution: For an isothermal process and ideal gas, you know that $\Delta E = \Delta H = 0$, from Eqs. (3-35) and (3-36). Since $\Delta E = 0$, you know that $q = -w$, from the first law. You calculate work using Eq. (3-11), which for a constant-pressure process becomes Eq. (3-12), or

$$w = -P_2(V_2 - V_1)$$

and, for an ideal gas,

$$w = -P_2\left(\frac{nRT_2}{P_2} - \frac{nRT_1}{P_1}\right)$$

Since $T_2 = T_1 = T$,

$$w = -nRT\left(1 - \frac{P_2}{P_1}\right) = -1(8.314)(293)\left(1 - \frac{2.0}{8.0}\right) = -1.8 \text{ kJ}$$

$$q = 1.8 \text{ kJ}$$

PROBLEM 3-15 Repeat Problem 3-14 for a reversible isothermal expansion.

Solution: Again, $\Delta E = \Delta H = 0$ and $q = -w$. For the reversible case, Eq. (3-11) becomes Eq. (3-13):

$$w = -nRT \ln \frac{V_2}{V_1} = -nRT \ln \frac{P_1}{P_2} = -1(8.314)(293)\left(\ln \frac{8.0}{2.0}\right) = -3.4 \text{ kJ}$$

$$q = 3.4 \text{ kJ}$$

Enthalpy

PROBLEM 3-16 Calculate the amount of heat required to raise the temperature of 1.00 mol H_2 from 300 K to 1200 K at constant pressure. Don't assume that heat capacity is constant.

Solution: Note that the problem could have asked for the enthalpy change for this process, since $\Delta H = q_P$. From Eq. (3-29), you should choose the expression $dq_P = C_P\,dT$. Since C_P isn't constant, substitute the analytical expression, $C_{P,m} = a + bT + cT^2$ [Eq. (3-58)], and find the values of a, b, and c from Table 3-1.

Integrating:

$$q_P = \int nC_{P,m}\,dT$$

For 1 mol:

$$q_P = \int (a + bT + cT^2)\,dT$$

Separating terms:

$$q_P = \int a\,dT + \int bT\,dT + cT^2\,dT$$

Integrating with limits:

$$q_P = a(T_2 - T_1) + \frac{b}{2}(T_2^2 - T_1^2) + \frac{c}{3}(T_2^3 - T_1^3)$$

Substituting:

$$q_P = 29.07(1200 - 300) + \frac{-0.836 \times 10^{-3}}{2}(1200^2 - 300^2) + \frac{20.12 \times 10^{-7}}{3}(1200^3 - 300^3)$$

$$= 26\,163 - 564 + 1141 = 26\,740 \text{ J} = 26.7 \text{ kJ}$$

Ideal Gases

PROBLEM 3-17 **(a)** Derive the exponent in Eq. (3-54) from $-R/C_{P,m}$. **(b)** Rearrange Eq. (3-54) so that the exponent of P is $\dfrac{\gamma - 1}{\gamma}$.

Solution: Recall that $\gamma = C_{P,m}/C_{V,m}$.

(a) You need to substitute the value of R from Eq. (3-33a):

If

$$R = C_{P,m} - C_{V,m}$$

then

$$-\frac{R}{C_{P,m}} = -\frac{C_{P,m} - C_{V,m}}{C_{P,m}} = -\frac{C_{P,m}}{C_{P,m}} + \frac{C_{V,m}}{C_{P,m}} = -1 + \frac{1}{\gamma} = \frac{-\gamma + 1}{\gamma} = \frac{1 - \gamma}{\gamma}$$

(b) Part (a) provides an expression that is close to the one you want. If you factor a minus sign from the right-hand term, you get

$$\frac{1 - \gamma}{\gamma} = \frac{-\gamma + 1}{\gamma} = -\frac{\gamma - 1}{\gamma}$$

Then the term in Eq. (3-54) becomes

$$TP^{-(\gamma-1)/\gamma} \quad \text{or} \quad \frac{T}{P^{(\gamma-1)/\gamma}}$$

PROBLEM 3-18 For a certain ideal gas, $C_{V,m} = \frac{3}{2}R$. If 1 mol of this gas undergoes an adiabatic reversible expansion from STP to a volume of 40.0 L, what is the final temperature?

Solution: You know the two volumes: $V_1 = 22.4$ L and $V_2 = 40.0$ L. Thus you should use Eq. (3-49), Eq. (3-50), or Eq. (3-51). Equation (3-50) probably is the most convenient.

Rearranging, you obtain

$$T_2 = T_1\left(\frac{V_1}{V_2}\right)^{R/C_{V,m}}$$

$$\frac{R}{C_{V,m}} = \frac{R}{\frac{3}{2}R} = \frac{2}{3} = 0.667$$

Substituting numbers yields

$$T_2 = 273\left(\frac{22.4\text{ L}}{40.0\text{ L}}\right)^{0.667} = 185\text{ K}$$

PROBLEM 3-19 An ideal gas, initially at 400 K and 4.00 atm, is allowed to expand reversibly and adiabatically until the temperature has dropped to 300 K. What is the final pressure? ($C_{V,m} = \frac{5}{2}R$.)

Solution: For pressure and temperature problems involving adiabatic and reversible processes, you start with Eq. (3-54):

$$TP^{(1-\gamma)/\gamma} = \text{Constant}$$

Then

$$T_1P_1^{(1-\gamma)/\gamma} = T_2P_2^{(1-\gamma)/\gamma}$$

or

$$\frac{T_2}{T_1} = \left(\frac{P_1}{P_2}\right)^{(1-\gamma)/\gamma}$$

When evaluating the exponent, you should recall that $\gamma = C_{P,m}/C_{V,m}$. Since $C_{V,m} = \frac{5}{2}R$, $C_{P,m} = \frac{7}{2}R$ [Eq. (3-34)], and

$$\gamma = \frac{\frac{7}{2}R}{\frac{5}{2}R} = \frac{7}{5}$$

Then

$$\frac{1-\gamma}{\gamma} = \frac{1}{\gamma} - 1 = \frac{5}{7} - 1 = -\frac{2}{7}$$

Now

$$\frac{T_2}{T_1} = \left(\frac{P_1}{P_2}\right)^{-2/7} = \left(\frac{P_2}{P_1}\right)^{2/7}$$

or

$$\frac{P_2}{P_1} = \left(\frac{T_2}{T_1}\right)^{7/2}$$

Rearrange and substitute numbers to obtain

$$P_2 = 4.00\left(\frac{300}{400}\right)^{7/2} = 1.46\text{ atm}$$

PROBLEM 3-20 Change the conditions of Problem 3-19 to an irreversible expansion against a constant pressure of 1.46 atm until the pressure is 1.46 atm. What would be the final temperature?

Solution: Under these conditions, the integration of Eq. (3-46) yields Eq. (3-55):

$$nC_{V,m}(T_2 - T_1) = -P_2(V_2 - V_1)$$

When you replace the volumes using the ideal gas equation, you get Eq. (3-56):

$$nC_{V,m}(T_2 - T_1) = -nR\left(T_2 - \frac{T_1P_2}{P_1}\right)$$

Solve for T_2:

$$C_{V,m}T_2 - C_VT_1 = -RT_2 + \frac{RT_1P_2}{P_1}$$

$$C_{V,m}T_2 + RT_2 = C_{V,m}T_1 + \frac{RT_1P_2}{P_1}$$

$$T_2(C_{V,m} + R) = \left(C_{V,m} + \frac{RP_2}{P_1}\right)T_1$$

$$T_2 = \frac{1}{C_{V,m} + R}\left(C_{V,m} + \frac{RP_2}{P_1}\right)T_1$$

Now substitute numbers:

$$C_{V,m} + R = \tfrac{5}{2}R + R = \tfrac{7}{2}R$$

$$T_2 = \frac{2}{7R}\left(\frac{5}{2}R + \frac{RP_2}{P_1}\right)T_1$$

The R's cancel:

$$T_2 = \frac{2}{7}\left(\frac{5}{2} + \frac{1.46}{4.0}\right)(400) = 327 \text{ K}$$

Compare this solution with that for Example 3-26.

PROBLEM 3-21 Show that $w = nR\,\Delta T/(\gamma - 1)$ for the adiabatic reversible expansion of an ideal gas.

Solution: Since P, V, and T aren't constant in an adiabatic reversible change, you can't make a simple substitution in Eq. (3-11) as you did for the isothermal case. However, you can obtain an expression for $P\,dV$ by differentiating Eq. (3-53).

$$PV^\gamma = \text{Constant}$$

Let $P = u$ and $V^\gamma = v$ in $d(uv) = v\,du + u\,dv$:

$$V^\gamma\,dP + P\gamma V^{\gamma-1}\,dV = 0$$

Divide by $V^{\gamma-1}$ and rearrange:

$$V\,dP = -\gamma P\,dV$$

To get an expression for dV, differentiate the ideal gas equation:

$$V\,dP + P\,dV = nR\,dT$$

Substitute for $V\,dP$ from the last expression:

$$-\gamma P\,dV + P\,dV = nR\,dT$$

$$P\,dV(-\gamma + 1) = nR\,dT$$

$$P\,dV = \frac{nR}{1-\gamma}\,dT$$

Substitute for $P\,dV$ in Eq. (3-11):

$$w = -\int_{V_1}^{V_2} P\,dV = -\int_{T_1}^{T_2} \frac{nR}{1-\gamma}\,dT = \frac{nR}{\gamma - 1}\,\Delta T \tag{3-61}$$

PROBLEM 3-22 Continue Problem 3-19 by calculating the work done if the process involved 2.00 mol of gas.

Solution: The initial and final temperatures are given, so you might use Eq. (3-61). You will need γ.

$$\gamma = \frac{C_P}{C_V} = \frac{\frac{7}{2}R}{\frac{5}{2}R} = \frac{7}{5} = 1.4$$

$$w = \frac{2.00(8.314)(300 - 400)}{1.4 - 1} = -4.16 \text{ kJ}$$

The usual approach, and the one you normally should choose, is integration of Eq. (3-45). For n mol,

$$w = n \int_{T_1}^{T_2} C_{V,m}\, dT = nC_{V,m}\, \Delta T$$

Substituting, you get

$$w = 2.00(\tfrac{5}{2})(8.314)(300 - 400) = -4.16 \text{ kJ}$$

Real Gases

PROBLEM 3-23 Calculate the heat required to raise the temperature of 200 g SO_3 from 300 K to 500 K. Assume no dissociation.

Solution: For this range of temperature, the heat capacity can't be taken as constant, so the expression for $C_{P,m}$ as a function of T must be used. Values of the coefficients for SO_3 are in Table 3-1. Simplify the equations by calculating for 1 mol, then convert to the mass given.

Start with the basic equation, $dq_P = C_{P,m}\, dT$, and integrate:

$$q_P = \int_{T_1}^{T_2} C_{P,m}\, dT = \int_{T_1}^{T_2} a\, dT + \int_{T_1}^{T_2} bT\, dT + \int_{T_1}^{T_2} cT^2\, dT$$

$$= a(T_2 - T_1) + \frac{b}{2}(T_2^2 - T_1^2) + \frac{c}{3}(T_2^3 - T_1^3)$$

Then substitute the values from Table 3-1 and for the T's:

$$q_P = 6.077(500 - 300) + \left(\frac{23.537 \times 10^{-3}}{2}\right)(500^2 - 300^2) - \left(\frac{96.87 \times 10^{-7}}{3}\right)(500^3 - 300^3)$$

$$= 2.78 \text{ kcal mol}^{-1}$$

For the mass given, 200 g,

$$q = \left(\frac{200 \text{ g}}{80.0 \text{ g mol}^{-1}}\right)(2.78 \text{ kcal mol}^{-1}) = 6.95 \text{ kcal}$$

PROBLEM 3-24 For CO_2, $\mu_{JT} = 1.290$ K atm^{-1} at 0 °C and 1.0 atm ($C_P = 6.2$ cal K^{-1}mol^{-1}). Assuming that both of these remain constant, calculate the enthalpy change if the pressure on 1 mol CO_2 is increased to 20 atm isothermally.

Solution: The quantities involved are related according to Eq. (3-60). You should rewrite it for large-scale changes, or

$$\Delta H = -\mu_{JT} n C_{P,m}\, \Delta P$$

Substituting, you get

$$\Delta H = -(1.290 \text{ K atm}^{-1})(1 \text{ mol})(6.2 \text{ cal K}^{-1}\text{mol}^{-1})(20 - 1 \text{ atm}) = -1.5 \times 10^2 \text{ cal}$$

Supplementary Exercises

PROBLEM 3-25 For an endothermic process, the sign given to the heat involved is ______________, and the sign of the enthalpy change is ______________. The fundamental principle that serves as the basis for the measurement of temperature is ______________. The SI equivalent of one atmosphere is ______________. For constant volume, adiabatic processes, and constant temperature, ΔE equals ____, ____, and ____, respectively. For an ideal gas, if $C_{P,m} = \frac{5}{2}R$, then $C_{V,m} =$ ____.

PROBLEM 3-26 For the following reactions, predict whether the numeric value of ΔH is more negative than, less negative than, or the same as the value of ΔE, which is negative.

(a) $C_2H_5OH(l) + 3O_2(g) \longrightarrow 2CO_2(g) + 3H_2O(g)$
(b) $CO(g) + H_2O(g) \longrightarrow CO_2(g) + H_2(g)$

PROBLEM 3-27 The Joule–Thomson coefficient of He is -6.2×10^{-3} at 0 °C. If a quantity of He at 0 °C is allowed to expand, will the temperature rise or fall?

PROBLEM 3-28 In an experimental determination of the specific heat of nickel, an 8.78-g piece of the metal was heated in boiling water long enough to be thoroughly heated to the metal's boiling temperature of 99.3 °C. Then it was transferred rapidly to 9.54 g of water at 24.2 °C. The temperature of the water rose to 31.1 °C. Calculate the experimental value of the specific heat of nickel.

PROBLEM 3-29 Calculate the work done in raising a 0.500-kg weight to a height of 4.00 m above the floor.

PROBLEM 3-30 Calculate the work done in compressing a spring from its unstressed (rest) length of 12.0 cm to 9.0 cm, if the spring constant, k, is 60 dyn cm^{-1}.

PROBLEM 3-31 A frictionless piston having an area of 10.0 cm^2 is pushed in 15.0 cm by a pressure of 800 torr. Calculate the work done in L atm, calories, and joules.

PROBLEM 3-32 Calculate the work done if a gas expands from 3.0 L to 8.0 L under a constant pressure of 3.0 atm.

PROBLEM 3-33 Calculate the work done if 2.00 mol of an ideal gas expands reversibly from 2.00 L to 4.00 L at a constant temperature of 350 K.

PROBLEM 3-34 If 0.50 mol of an ideal gas undergoes a reversible compression from 2.0 atm pressure to 6.0 atm at a constant temperature of 300 K, how much work is done?

PROBLEM 3-35 Electrolysis of a solution of $FeCl_3$ causes the reaction

$$2FeCl_3(aq) \longrightarrow 2Fe(s) + 3Cl_2(g)$$

If 1 mol of $FeCl_3$ is consumed, at 1 atm and 300 K, how much work is done on the atmosphere?

PROBLEM 3-36 A certain quantity of an ideal gas is contained in a 2.0-L cylinder with a frictionless piston under a pressure of 10.0 atm and at a temperature of 300 K. Calculate the work done if the gas is allowed to expand isothermally **(a)** under a constant pressure of 2.0 atm until the pressure becomes 2.0 atm, **(b)** into an evacuated 2.0-L container, and **(c)** reversibly until the final pressure becomes 2.0 atm.

PROBLEM 3-37 What are the values of q, ΔE, and ΔH for the process in Problem 3-36(c)?

PROBLEM 3-38 What is the final temperature if 1 mol of an ideal gas, initially at 400 K and 5.0 L, expands adiabatically and reversibly to a volume of 15 L? $C_{P,m} = \frac{7}{2}R$.

PROBLEM 3-39 What are the values of q, w, ΔE, and ΔH for the process in Problem 3-38?

PROBLEM 3-40 Calculate the final pressure if 1 mol of an ideal gas, initially occupying 10.0 L at 5.0×10^5 Pa, expands adiabatically and reversibly to a volume of 35.0 L. $C_{V,m} = \frac{3}{2}R$.

PROBLEM 3-41 What are the values of w, ΔE, and ΔH for the process in Problem 3-40?

PROBLEM 3-42 One mole of an ideal gas ($C_{P,m} = \frac{7}{2}R$), initially at 2.0 atm and 500 K, is allowed to expand adiabatically against an unknown constant pressure. When the pressure of the gas becomes equal to that constant pressure, the temperature is 400 K. What is the constant pressure?

PROBLEM 3-43 What are the values of w, ΔE, and ΔH for the process in Problem 3-42?

PROBLEM 3-44 Calculate the heat required to raise the temperature of 100 g of HCl gas from 300 K to 800 K at constant pressure.

Answer to Supplementary Exercises

3-25 positive, positive; the Zeroth Law; 101 325 Pa; q, w, zero; $\frac{3}{2}R$ **3-26** (**a**) less; (**b**) same **3-27** rise **3-28** 0.110 cal $g^{-1}K^{-1}$ **3-29** 19.6 J **3-30** 2.7×10^2 erg **3-31** 0.158 L atm, 3.83 cal, 16.0 J **3-32** -15 L atm **3-33** -4.03 kJ **3-34** 1.4 kJ **3-35** 3.74 kJ **3-36** (**a**) -16 L atm, (**b**) zero, (**c**) -32 L atm **3-37** $q = 32$ L atm, $\Delta E = \Delta H = 0$ **3-38** 2.6×10^2 K **3-39** $q = 0$; $w = \Delta E = -3.0$ kJ; H $= -4.1$ kJ **3-40** 6.2×10^4 Pa **3-41** $w = \Delta E = -4.3$ kJ; $\Delta H = -7.1$ kJ **3-42** 0.60 atm **3-43** $w = \Delta E = -2.08$ kJ; $\Delta H = -2.91$ kJ **3-44** 40.6 kJ

THERMOCHEMISTRY

THIS CHAPTER IS ABOUT

☑ Calorimetry
☑ Useful Heats of Reaction
☑ Standard Enthalpies
☑ Calculating Enthalpies

Knowledge of the amount of energy required or released by a chemical reaction has a number of possible uses, including the design of reaction vessels, the provision of devices for heating or cooling, the calculation of other thermodynamic properties (e.g., free energies and equilibrium constants), and the prediction (approximate) of spontaneity of a chemical reaction.

4-1. Calorimetry

A. The calorimeter

The usual procedure for measuring the energy released in a chemical process is to generate the reaction in a controlled system, a **calorimeter**, in which energy is exchanged between the process and its immediate surroundings. Total isolation of the system causes the entire energy exchange to appear as a temperature change in the calorimeter. The most common type of reaction studied in this manner is oxidation, in which a substance is ignited in an atmosphere of oxygen. The energy produced is transmitted to the calorimeter system, and the rise of temperature of the water jacket is measured. An essential quantity to be calculated is the **effective heat capacity**, C_{eff}, of the calorimeter; it is determined by igniting a sample of a substance having an accurately known heat of combustion (usually benzoic acid). The value obtained is the effective heat capacity because it incorporates the heat capacities of the several parts of the system when the mass of water is kept constant in successive measurements.

EXAMPLE 4-1: To calibrate a certain constant-volume calorimeter, 0.251 g of benzoic acid ($C_6H_5CO_2H$) was burned in it. The temperature rose from 24.12 °C to 25.46 °C. The heat released by benzoic acid is known accurately to be 770.94 kcal mol^{-1} at 298.15 °C. Calculate the effective heat capacity of the calorimeter.

Solution: We rewrite Eq. (3-26) for the calorimeter, or

$$q = C_{\text{eff}}\,\Delta T \qquad \text{or} \qquad C_{\text{eff}} = \frac{q}{\Delta T} \tag{4-1}$$

The average temperature is about 25 °C, so we can use $q = 770.94$ kcal mol^{-1}. The sign is positive because heat is gained by the calorimeter. Thus

$$C_{\text{eff}} = \frac{(0.251\text{ g})(770.94\text{ kcal mol}^{-1})}{(122.13\text{ g mol}^{-1})(25.46 - 24.12)\text{ °C}} = 1.18\text{ kcal °C}^{-1}$$

EXAMPLE 4-2: In the calorimeter of Example 4-1, the burning of a sample of camphor ($C_{10}H_{16}O$) weighing 0.113 g caused a temperature increase of 1.24 °C from an initial temperature of 24.5 °C. Calculate the experimental value of the heat produced.

Solution: We use Eq. (4-1) to obtain

$$q = C_{eff}\,\Delta t = (1.18 \text{ kcal deg}^{-1})(1.24 \text{ deg}) = 1.46 \text{ kcal}$$

The heat per mole is

$$q' = \left(\frac{1.46 \text{ kcal}}{0.113 \text{ g}}\right)(152. \text{ g mol}^{-1}) = 1.96 \times 10^3 \text{ kcal mol}^{-1}$$

B. Enthalpy change

The usual calorimeter is based on a rigid reaction vessel or "bomb." Thus the volume remains constant, and the energy gained or lost is the internal energy ΔE. Chemists, however, usually work under conditions of constant pressure and thus need ΔH. Remember that $\Delta H = q_P$ [Eq. (3-20)]. To calculate ΔH from the experimentally determined ΔE we need the balanced equation of the reaction and must assume ideal behavior:

$$\Delta H = \Delta E + \Delta(PV)$$

$$\Delta H = \Delta E + RT\,\Delta n \qquad \textbf{[Eq. (3-24)]}$$

This enthalpy change is the **molar heat of combustion** (or **enthalpy of combustion**), i.e., the energy released when one mole of a substance is burned completely in oxygen. Strictly speaking, this is an enthalpy change at constant volume. Its difference from the constant pressure value is zero for ideal gases but could be significant when great precision is required.

EXAMPLE 4-3: Continue Example 4-2 by calculating the enthalpy of combustion of camphor.

Solution: We need Δn in order to calculate ΔH, so we must write the balanced equation

$$C_{10}H_{16}O(s) + \tfrac{27}{2}O_2(g) \longrightarrow 10CO_2(g) + 8H_2O(l)$$

Thus

$$n = 10 - 13.5 = -3.5 \text{ mol}$$

and

$$RT\,\Delta n = (1.987 \text{ cal K}^{-1}\text{mol}^{-1})(298.2 \text{ K})(-3.5 \text{ mol}) = -2.074 \text{ kcal}$$

Using Eq. (3-24), we obtain

$$\Delta H = \Delta E + RT\,\Delta n = (-1.96 \times 10^3 - 2.074) \text{ kcal mol}^{-1} = -1.96 \times 10^3 \text{ kcal mol}^{-1}$$

We see that, within experimental error, $\Delta H = \Delta E$ for camphor. A more precise apparatus would be needed to measure the difference.

4-2. Useful Heats of Reaction

The heat involved in a chemical reaction may be expressed differently and given different names in order to emphasize a certain aspect of the reaction. It is essential to know the balanced chemical equation for any given (numerical) heat of reaction.

A. Heat of combustion

First, let's look again at heat of combustion, that is, the complete burning of one mole of a substance in oxygen.

EXAMPLE 4-4: What is the heat of combustion of aluminum? We can use the balanced chemical equation

$$4Al(s) + 3O_2(g) \longrightarrow 2Al_2O_3(s) + 760 \text{ kcal}$$

Solution: The important point is that 760 kcal is released when 4 mol Al burns completely. Heat of combustion refers to the burning of 1 mol, which in this case would produce one-fourth of 760 kcal. So we divide the equation by 4:

$$Al(s) + \tfrac{3}{4}O_2(g) \longrightarrow \tfrac{1}{2}Al_2O_3(s) + 190 \text{ kcal}$$

We can state the energy loss as an enthalpy change by writing

$$Al(s) + \tfrac{3}{4}O_2(g) \longrightarrow \tfrac{1}{2}Al_2O_3(s) \qquad \Delta H = -190 \text{ kcal}$$

In words, we say, "the heat of combustion of Al is -190 kcal mol^{-1}." Given such a statement for any substance, you should be able to write the proper equation.

B. Heat of formation

The **molar heat of formation,** or **enthalpy of formation,** is the change in enthalpy accompanying the production, or formation, of one mole of a substance from its elements. Often such a reaction is impossible to carry out, and the balanced equation looks strange; nevertheless, the equations are needed and you must be able to write them.

EXAMPLE 4-5: The enthalpy of formation of ammonium sulfate is -281.5 kcal mol^{-1}. Express this information as a balanced chemical equation.

Solution: We need the equation for the reaction in which solid $(NH_4)_2SO_4$ is produced from its elements in their standard states at 298 K. Thus we must use $N_2(g)$, $H_2(g)$, $O_2(g)$, and $S(s)$, or

$$N_2(g) + H_2(g) + S(s) + O_2(g) \longrightarrow (NH_4)_2SO_4(s)$$

In balancing, we want a coefficient of 1 for the product; thus we obtain

$$N_2(g) + 4H_2(g) + S(s) + 2O_2(g) \longrightarrow (NH_4)_2SO_4(s) \qquad \Delta H = -281.5 \text{ kcal}$$

EXAMPLE 4-6: Use the data in Example 4-4 to write a balanced chemical equation for the reaction and calculate the heat of formation of aluminum oxide.

Solution: From Example 4-4, we see that 760 kcal is produced when 2 mol Al_2O_3 is formed from the elements; thus 380 kcal would be produced when one mole is formed. We can write

$$2Al(s) + \tfrac{3}{2}O_2(g) \longrightarrow Al_2O_3(s) \qquad \Delta H = -380 \text{ kcal}$$

In words, we say, "the heat of formation of Al_2O_3 is -380 kcal mol^{-1}."

C. Heats for solutions

There are several names for the enthalpies or *heats* that relate to processes in solution. The definitions of these heats also are based on the change in enthalpy when one mole of substance undergoes the stated process.

1. The **integral enthalpy of solution** is the change in enthalpy accompanying the dissolving of one mole of a substance in a specific quantity of solvent. Its value changes with the amount of solvent, so concentration must be stated. For example,

$$HCl(g) + 100H_2O(l) \longrightarrow HCl(100H_2O, l) \qquad \Delta H = -166 \text{ kJ}$$

2. The **enthalpy of dilution** is the difference between the integral enthalpies of solution of the original and final solutions.
3. The **enthalpy of neutralization** is the change in enthalpy when one mole of an acid (or a base) reacts with a chemically equivalent amount of a base (or an acid) to produce a salt and water. With strong acids and strong bases, when the salt remains completely dissolved and ionized, the heat released per mole of water formed is always the same, i.e., that required for the production of water from its ions:

$$H^+(aq) + OH^-(aq) \longrightarrow H_2O(l) \qquad \Delta H = -55.8 \text{ kJ}$$

When the neutralization involves a weak acid or a weak base, the net energy change will include a contribution arising from the dissociation of the molecules.

4. The **enthalpy of ionization** is the change in enthalpy when one mole of a compound in solution dissociates into ions. For example,

$$CH_3COOH(aq) \longrightarrow H^+(aq) + CH_3COO^-(aq) \qquad \Delta H = -0.25 \text{ kJ}$$

D. Applications

Calculations involving heats of reaction are based on three fundamental principles:

1. The amount of heat produced is directly proportional to the amount of material that reacts.
2. The amount of heat consumed by the reverse reaction is the same as that produced by the forward reaction (with opposite signs, of course). Because enthalpy is a state function, its total change must be zero in a cyclic process.
3. If a given chemical change may be achieved by a series of two or more reactions, the heat of reaction of the change, if it occurred in one step, equals the sum of the heats of reaction of the several steps. This is known as **Hess's law of constant heat summation** (recognized by Hess in 1840). Because enthalpy is a state function, the value of ΔH for a change must be independent of the path.

The utility of these three principles is that heats of reaction may be calculated for reactions that are difficult, dangerous, or impossible to carry out in the laboratory. In general, we find a set of equations that may be combined to give the equation of the reaction of interest.

EXAMPLE 4-7: What is ΔH if 1 mol of $Al_2O_3(s)$ is decomposed into its elements at 298 K?

Solution: The reaction is

$$Al_2O_3(s) \longrightarrow 2Al(s) + \tfrac{3}{2}O_2(g)$$

You can see that this is the reverse of the reaction in Example 4-6; thus ΔH of decomposition is equal in magnitude to ΔH of formation but opposite in sign, i.e.,

$$Al_2O_3(s) \longrightarrow 2Al(s) + \tfrac{3}{2}O_2(g) \qquad \Delta H = +380 \text{ kcal}$$

EXAMPLE 4-8: In the reaction $S(s) + O_2(g) \longrightarrow SO_2(g)$, it is difficult to prevent contamination by SO_3 and to measure a correct ΔH. We can calculate the heat of the reaction, however, from the heats of these reactions, assuming complete combustion to form pure products:

$$2SO_2(g) + O_2(g) \longrightarrow 2SO_3(g) \qquad \Delta H = -46.0 \text{ kcal}$$
$$2S(s) + 3O_2(g) \longrightarrow 2SO_3(g) \qquad \Delta H = -187.8 \text{ kcal}$$

Show how to find ΔH for the given reaction.

Solution: We have to add, subtract, multiply, divide, or reverse these two equations to obtain the desired equation. At the same time, we must make appropriate changes in the ΔH values.

The first equation has SO_2 on the left, but we want it on the right. We reverse this equation, which requires changing the sign of ΔH, and then add it to the second equation.

$$\begin{array}{rcll} 2SO_3 & \longrightarrow & 2SO_2 + O_2 & \Delta H = +46.0 \text{ kcal} \\ 2S + 3O_2 & \longrightarrow & 2SO_3 & \Delta H = -187.8 \text{ kcal} \\ \hline 2S + 3O_2 + 2SO_3 & \longrightarrow & 2SO_2 + O_2 + 2SO_3 & \Delta H = -141.8 \text{ kcal} \end{array}$$

Canceling items that appear on both sides, we get

$$2S + 2O_2 \longrightarrow 2SO_2 \qquad \Delta H = -141.8 \text{ kcal}$$

We want to form only one mole of SO_2, so we divide by 2:

$$S + O_2 \longrightarrow SO_2 \qquad \Delta H = -70.9 \text{ kcal}$$

4-3. Standard Enthalpies

A. Standard enthalpies of formation

Because the enthalpy change of a particular reaction will vary with temperature, concentration, or physical state of the reactants, it is necessary to define a reference point in order to tabulate comparable data concisely. The **standard molar enthalpy of formation** of a substance is the change in enthalpy in the reaction in which one mole of that substance is produced from its elements, if the product and the reactants all are in their standard states. The **standard state** of a substance is the physical state that is stable at 1 atm pressure, designated by a superscript degree sign (e.g., H°). The temperature must be stated; usually it is 298.15 K.

In general, the heat of a reaction may be expressed as

$$\Delta H_r = \Sigma nH(\text{products}) - \Sigma nH(\text{reactants}) \qquad \textbf{(4-2)}$$

where Σ means summation, H is the actual enthalpy of the substance, and n is the amount of substance. For a formation reaction with products and reactants in their standard states, this becomes

$$\Delta H_f^\circ(\text{compound}) = H^\circ(\text{compound}) - \Sigma nH^\circ(\text{elements}) \qquad \textbf{(4-3)}$$

where we set $n = 1$ mol for the product so that ΔH_f° is a molar quantity.

Since we can't determine the actual enthalpy, these expressions appear to be useless. However, if an arbitrary reference point can be established, we can calculate enthalpy values for each substance, relative to the reference point. The reference point is obtained by defining the enthalpy of an element in its standard state at 298.15 K to be zero. Then Eq. (4-3) becomes

$$\Delta H_f^\circ(\text{compound}) = H^\circ(\text{compound}) \qquad \textbf{(4-4)}$$

which means we can use ΔH_f° values as if they were actual enthalpies. Substituting Eq. (4-4) into Eq. (4-2) gives

$$\Delta H_r^\circ = \Sigma n\, \Delta H_f^\circ(\text{products}) - \Sigma n\, \Delta H_f^\circ(\text{reactants}) \qquad \textbf{(4-5)}$$

note: Some authors develop Eq. (4-5) in a different way, obtaining the reference point by defining the standard enthalpy of formation of an element at 298.15 K to be zero.

B. Standard enthalpies for solutions

For reactions of electrolytes in solution, data may be tabulated as the standard enthalpies of formation of ions in solution at infinite dilution. Recall that enthalpies of solution depend on concentration. A first reference point is obtained by defining the **enthalpy of solution at infinite dilution** as the enthalpy change when one mole of a substance dissolves in an infinite amount of solvent. The addition of more solvent would cause no detectable change in temperature. For example, consider the following reaction:

$$CH_3COOH(l) \xrightarrow[\infty H_2O]{} CH_3COOH(aq) \qquad \Delta H = -1.3 \text{ kJ}$$

We can obtain the standard enthalpy of formation of the substance *in solution* by combining this equation with the equation and the ΔH_f° value for the undissolved substance ($\Delta H_f^\circ = -484.5$ kJ):

$$2C(gr) + 2H_2(g) + O_2(g) \xrightarrow[\infty H_2O]{} CH_3COOH(aq) \qquad \Delta H_f^\circ = -485.8 \text{ kJ}$$

Tables of ΔH_f° values for all electrolytes would be huge. A simplified method of calculating these values is based on the assumption that the enthalpy of formation of the dissolved compound is the sum of enthalpies of formation of the dissolved ions. Then the preceding equation becomes

$$2C(gr) + 2H_2(g) + O_2(g) \xrightarrow[\infty H_2O]{} H^+(aq) + CH_3COO^-(aq) \qquad \Delta H_f^\circ = -485.8 \text{ kJ}$$

Now, however, a second reference point is needed. The ΔH_f° for the aqueous hydrogen ion is assigned a value of zero. Thus ΔH_f° for $CH_3COO^-(aq) = -485.8$ kJ mol^{-1}. In this way, we can obtain values of **standard enthalpies of formation of ions in solution at infinite dilution**. We can then calculate the ΔH_f° of any ionic compound in solution by adding the appropriate ionic values, or, in general,

$$\Delta H_f^\circ(X_mY_n, aq) = m\,\Delta H_f^\circ(X^{n+}, aq) + n\,\Delta H_f^\circ(Y^{m-}, aq) \qquad \textbf{(4-6)}$$

C. Standard enthalpy values

Tables of standard enthalpies of formation are included in most chemistry textbooks and handbooks. The values in the tables may have been obtained by direct measurement or by calculation, using Hess's law. Units may be calories, kilocalories, or kilojoules. Those in kilojoules probably were obtained by multiplying the original values in calories by 4.184. Usually the reference temperature is 298.15 K. A short table of such values is included in Appendix A, Table A-1.

EXAMPLE 4-9: Calculate the standard molar enthalpy of formation of ethanol at 298 K from the experimental heats of combustion of ethanol(l), $H_2(g)$, and C(gr) and the balanced equations that correspond to each ΔH:

Reaction	$\Delta H/(\mathrm{kJ\ mol^{-1}})$
$C_2H_5OH(l) + 3O_2(g) \longrightarrow 2CO_2(g) + 3H_2O(l)$	-1367
$H_2(g) + \frac{1}{2}O_2(g) \longrightarrow H_2O(l)$	-286
$C(gr) + O_2(g) \longrightarrow CO_2(g)$	-394

Solution: The equation for the formation of one mole of ethanol is

$$2C(gr) + 3H_2(g) + \tfrac{1}{2}O_2(g) \longrightarrow C_2H_5OH(l) \qquad \Delta H_f^\circ = ?$$

Now we apply the principles stated in Section 4-2D. We could start with any of the three given equations, but a good choice is the one that contains the principal compound, which is the first equation. However, it must be reversed to get ethanol on the product side:

$$2CO_2(g) + 3H_2O(l) \longrightarrow C_2H_5OH(l) + 3O_2(g) \qquad \Delta H = +1367\ \mathrm{kJ\ mol^{-1}}$$

We need 3 mol H_2, so we multiply the second equation by 3:

$$3H_2(g) + \tfrac{3}{2}O_2(g) \longrightarrow 3H_2O(l) \qquad \Delta H = -858\ \mathrm{kJ\ mol^{-1}}$$

Now we multiply the third equation by 2, since we need 2 mol C:

$$2C(gr) + 2O_2(g) \longrightarrow 2CO_2(g) \qquad \Delta H = -788\ \mathrm{kJ\ mol^{-1}}$$

Adding the three equations and canceling like substances on opposite sides of the arrow, we get

$$2C(gr) + 3H_2(g) + \tfrac{1}{2}O_2(g) \longrightarrow C_2H_5OH(l)$$

which is what we want. Adding the ΔH's, we obtain

$$\Delta H_f^\circ = \Delta H = +1367 - 858 - 788 = -279\ \mathrm{kJ\ mol^{-1}}$$

We could have reached the same result by using an equation similar to Eq. (4-5), replacing enthalpies of formation with enthalpies of combustion:

$$\Delta H_r^\circ = \Sigma n\,\Delta H_{\mathrm{cmbn}}(\text{products}) - \Sigma n\,\Delta H_{\mathrm{cmbn}}(\text{reactants}) \qquad \textbf{(4-5a)}$$

D. Applications

Recall that the standard enthalpies of elements are zero, that this permits us to use the standard enthalpies of formation of compounds as if they were the actual enthalpies, and that the change in enthalpy in a process may be calculated using Eq. (4-5).

1. Reactions not in solution

EXAMPLE 4-10: Calculate the heat released when 1 mol FeO reacts with oxygen to produce Fe_2O_3 at 1 atm and 298.15 K. The ΔH_f° values are -64.3 kcal mol^{-1} for FeO and -196.5 kcal mol^{-1} for Fe_2O_3.

Solution: First, we write the balanced equation:

$$2FeO(s) + \tfrac{1}{2}O_2(g) \longrightarrow Fe_2O_3(s)$$

Then we find the enthalpy change. Using Eq. (4-5), we get

$$\begin{aligned}\Delta H_r^\circ &= \Delta H_f^\circ(Fe_2O_3) - [2\Delta H_f^\circ(FeO) + \tfrac{1}{2}\Delta H_f^\circ(O_2)] \\ &= -196.5 \text{ kcal mol}^{-1} - [2(-64.3) + 0] \text{ kcal mol}^{-1} \\ &= -67.9 \text{ kcal mol}^{-1}\end{aligned}$$

note: The ΔH_f° value for oxygen is zero because oxygen is an element.

EXAMPLE 4-11: Calculate the amount of heat released by the reaction

$$Na_2O(s) + H_2O(l) \longrightarrow 2NaOH(s)$$

Solution: Again, using Eq. (4-5), we get

$$\begin{aligned}\Delta H^\circ &= \Sigma n\, \Delta H_f^\circ(\text{product}) - \Sigma n\, \Delta H_f^\circ(\text{reactants}) \\ &= 2\,\Delta H_f^\circ(NaOH) - 1\,\Delta H_f^\circ(Na_2O) - 1\,\Delta H_f^\circ(H_2O)\end{aligned}$$

We can refer to Table A-1 for the required values. Substituting these values directly (and rounding to three significant figures), we obtain

$$\Delta H^\circ = 2(-426) - 1(-414) - 1(-286) = -152 \text{ kJ mol}^{-1}$$

---

2. Phase transitions

EXAMPLE 4-12: Calculate the heat required to evaporate 100 grams of water at 25 °C.

Solution: This transition is not a chemical reaction, and you may wonder how enthalpies of formation are involved. Recall that we have been emphasizing that water must be a liquid in the chemical equations at 298 K. In Example 4-11, we used ΔH_f° for liquid water (-286 kJ mol^{-1}). If you didn't notice it before, look now in Table A-1 for $H_2O(g)$: Its ΔH_f° is -242 kJ mol^{-1}. The difference is the energy required to convert one mole of liquid water to water vapor at 298 K. For

$$H_2O(l) \longrightarrow H_2O(g) \qquad \Delta H = ?$$

Equation (4-5) becomes

$$\Delta H^\circ = 1\Delta H_f^\circ(H_2O, g) - 1\Delta H_f^\circ(H_2O, l) = -242 - (-286) = +44 \text{ kJ mol}^{-1}$$

Note that the mass is given in grams, not moles, so we must calculate the energy for 100 g H_2O:

$$q = (44 \text{ kJ mol}^{-1})\left(\frac{100 \text{ g}}{18 \text{ g mol}^{-1}}\right) = 2.4 \times 10^2 \text{ kJ}$$

This example is a special case. Because the table includes ΔH_f° values for two states of water at the same temperature, we can calculate the energy change for the physical process of a phase transition. (For most compounds, however, the table does not contain both values.) The example emphasizes that you must note carefully the physical state of water given in a balanced equation and then must be careful to use the proper value from the table.

C. Processes in solution

Calculations for processes in solution require careful attention to the states of the reacting species.

EXAMPLE 4-13: Calculate the standard enthalpy of formation of $CaCl_2$ at infinite dilution.

Solution: Using Eq. (4-6), we can write

$$\Delta H_f^\circ(CaCl_2, aq) = \Delta H_f^\circ(Ca^{2+}, aq) + 2\Delta H_f^\circ(Cl^-, aq)$$

Using values from Table A-1, we calculate

$$\Delta H_f^\circ(\text{CaCl}_2, aq) = -542.8 + 2(-167.2) = -877.2 \text{ kJ mol}^{-1}$$

EXAMPLE 4-14: The chemical equation for the process of dissolving H_2SO_4 is

$$H_2SO_4(l) \xrightarrow[\infty H_2O]{} 2H^+(aq) + SO_4^{2-}(aq)$$

Calculate the integral enthalpy of solution at infinite dilution of sulfuric acid at 298 K.

Solution: Using Eq. (4-5), we can write

$$\Delta H_{\text{sol}}^\circ = 2\Delta H_f^\circ(\text{H}^+, aq) + \Delta H_f^\circ(\text{SO}_4^{2-}, aq) - H_f^\circ(\text{H}_2\text{SO}_4, l)$$

Substituting values from Table A-1, we obtain

$$\Delta H_{\text{sol}}^\circ = 0 - 909 - (-814) = -95 \text{ kJ mol}^{-1}$$

4-4. Calculating Enthalpies of Reaction

A. Dependence of enthalpy on temperature

You know how to calculate the difference between the enthalpies of a given substance at two temperatures. Now look back at Table 3-1, which lists the constants in the heat capacity expression for four substances. This should remind you that each substance takes in different amounts of energy as the temperature rises. Thus, if the temperature is raised in a chemical reaction system, the overall enthalpy change at the new temperature will be determined by the heat capacities of all substances present.

There are three approaches for calculating the heat of reaction at a temperature other than 298 K. The second is a simplification of the first, and all require standard enthalpies of formation at 298 K, which can be treated as absolute enthalpies as we have seen. We assume that no phase change occurs within the range of temperature.

1. First approach

We use the procedure in Example 3-27 to calculate ΔH_f° for each reactant and product at the temperature of the reaction and then calculate ΔH from Eq. (4-5). We start with Eq. (3-30): $dH = C_P\, dT$. Integration with limits gives

$$H_{T_2} - H_{T_1} = \int_{T_1}^{T_2} C_P\, dT \tag{4-7}$$

The second step is to substitute the values at T_2 for each substance into Eq. (4-5). Over a small range of temperature, C_P would be essentially constant, even though, generally, it isn't constant.

EXAMPLE 4-15: Obtain an expression for calculating the heat of reaction for $H_2(g) + Cl_2(g) \rightarrow 2HCl(g)$ at 1000 K by first obtaining enthalpies for each substance at 1000 K [apply Eq. (4-7)] and then using Eq. (4-5).

Solution: We write Eq. (4-7) for each substance, choosing 298 K as T_1 because the tabulated values are for that temperature. Recall that $H_{298}^\circ = 0$ for elements in their standard state.

$$\text{H}_2: \quad H_{1000}^\circ = 0 + \int_{298}^{1000} (a + bT + cT^2)\, dT$$

$$\text{Cl}_2: \quad H_{1000}^\circ = 0 + \int_{298}^{1000} (a + bT + cT^2)\, dT$$

$$\text{HCl}: \quad \Delta H_{1000}^\circ = \Delta H_{298}^\circ + \int_{298}^{1000} (a + bT + cT^2)\, dT$$

Substituting the calculated values in Eq. (4-5) gives

$$\Delta H_{1000}^\circ = 2\Delta H_{1000}^\circ(\text{HCl}) - H_{1000}^\circ(\text{H}_2) - H_{1000}^\circ(\text{Cl}_2)$$

You can see that the calculations needed to solve the equations in this example would be lengthy and repetitious.

---

2. Second approach

In the second approach, we can shorten the calculations by simplifying the equations algebraically before substituting numbers. If we substitute into Eq. (4-5) equations for each substance, such as those obtained in Example 4-15, and collect like terms, we get

$$\Delta H^\circ_{T_2} = \Delta H^\circ_{T_1} + \int_{T_1}^{T_2} \Delta C_P \, dT \tag{4-8}$$

where

$$\Delta C_P = \Sigma na(\text{products}) - \Sigma na(\text{reactants}) + [\Sigma nb(\text{products}) - \Sigma nb(\text{reactants})]T + [\Sigma nc(\text{products}) - \Sigma nc(\text{reactants})]T^2$$

or

$$\Delta C_P = \Delta a + \Delta b T + \Delta c T^2 \tag{4-9}$$

Equation (4-8) becomes

$$\Delta H^\circ_{T_2} = \Delta H^\circ_{T_1} + \int_{T_1}^{T_2} \Delta a \, dT + \int_{T_1}^{T_2} \Delta b T \, dT + \int_{T_1}^{T_2} \Delta c T^2 \, dT$$

or

$$\Delta H^\circ_{T_2} = \Delta H^\circ_{T_1} + \Delta a(T_2 - T_1) + \frac{\Delta b}{2}(T_2^2 - T_1^2) + \frac{\Delta c}{3}(T_2^3 - T_1^3) \tag{4-10}$$

EXAMPLE 4-16: Repeat Example 4-15, and use Eq. (4-10) to solve for ΔH°_{1000}.

Solution: First, we need to determine ΔH°_{298} and ΔC_P. For ΔH°_{298}, we use Eq. (4-5):

$$\begin{aligned}\Delta H^\circ_{298} &= 2\Delta H^\circ_f(\text{HCl}) - H^\circ(\text{H}_2) - H^\circ(\text{Cl}_2)\\ &= 2(-92.3) - 0 - 0 = -184.6 \text{ kJ}\end{aligned}$$

For ΔC_P, we first organize the data from Appendix Table A-2:

	a	$b \times 10^3$	$c \times 10^7$
$HCl(g) \times 2$	56.34	3.62	30.94
$H_2(g)$	29.07 } 60.77	−0.83 } 9.31	20.12 } −20.25
$Cl_2(g)$	31.70	10.14	−40.37
Δ values	−4.43	−5.69	51.19

(Recall that the $\int C_P \, dT$ values have units of J mol^{-1}.) Then we substitute in Eq. (4-10):

$$\begin{aligned}\Delta H^\circ_{1000} &= -184.6 \text{ kJ} + (-4.43)(1000 - 298) + \left(\frac{-5.69 \times 10^{-3}}{2}\right)(1000^2 - 298^2)\\ &\quad + \left(\frac{51.19 \times 10^{-7}}{3}\right)(1000^3 - 298^3)\\ &= (-184.6 - 3.11 - 2.59 + 1.66) \times 10^3 \text{ J} = -188.6 \text{ kJ}\end{aligned}$$

---

3. Third approach

The third approach involves an analytic expression in T from which ΔH can be calculated for any T. We obtain a variation of Eq. (4-8) by integrating the equation

$$d(\Delta H) = \Delta C_P \, dT \tag{4-11}$$

Replacing ΔC_P by Eq. (4-9) and integrating without limits, we get

$$\Delta H_T = \Delta a T + \frac{\Delta b}{2} T^2 + \frac{\Delta c}{3} T^3 + C \qquad \textbf{(4-12)}$$

Since we can get ΔH at 298 K from the tabulated ΔH_f° values, we can also evaluate the integration constant. With C known, we can calculate ΔH for any desired T, within the limits for which the C_P coefficients are valid. This method is convenient for calculating the ΔH of a given reaction at several temperatures. The second approach [Eq. (4-8)] is more convenient for only one temperature.

EXAMPLE 4-17: Calculate ΔH° at 400 K and 600 K for the reaction used in Examples 4-15 and 4-16.

Solution: To use Eq. (4-12), we must evaluate C. In Example 4-16, we calculated ΔC_P and ΔH_{298}°. Solving Eq. (4-12) for C gives

$$C = \Delta H_{298}^\circ - \Delta a T - \frac{\Delta b}{2} T^2 - \frac{\Delta c}{3} T^3$$

Substituting the calculated values, we obtain

$$\begin{aligned} C &= -184.6 \times 10^3 - (-4.43)(298) - (-2.85 \times 10^{-3})(298)^2 - (17.06 \times 10^{-7})(298)^3 \text{ J} \\ &= -184.6 + 1.32 + 0.253 - 0.045 \text{ kJ} \\ &= -183.1 \text{ kJ} \end{aligned}$$

For $T = 400$ K, substituting in Eq. (4-12) yields

$$\begin{aligned} \Delta H_{400}^\circ &= -4.43(400) + (-2.85 \times 10^{-3})(400)^2 + (17.06 \times 10^{-7})(400)^3 - 183.1 \times 10^3 \text{ J} \\ &= -1.77 - 0.456 + 0.109 - 183.1 \text{ kJ} \\ &= -185.2 \text{ kJ} \end{aligned}$$

and for $T = 600$ K,

$$\begin{aligned} \Delta H_{600}^\circ &= -4.43(600) + (-2.85 \times 10^{-3})(600)^2 + (17.06 \times 10^{-7})(600)^3 - 183.1 \times 10^3 \text{ J} \\ &= -2.66 - 1.03 + 0.368 - 183.1 \text{ kJ} \\ &= -186.4 \text{ kJ} \end{aligned}$$

B. Estimating heats of reaction with bond enthalpies

The method of bond enthalpies is based on the energy required to break specific bonds in molecules and thus departs from classical thermodynamics. Strictly speaking, the energy involved is a change in internal energy, but the difference from enthalpy is quite small and is neglected, since the method gives only approximate values anyway.

The term **bond dissociation energy**, or **bond energy**, refers to the energy required to break a single bond. For a molecule with multiple bonds, e.g., CCl_4, we would expect that the energies to break the second, third, and fourth C—Cl bonds will differ from the energy required to break the first; the species is different each time, and internal forces will be different. Similarly, in CH_3CH_2Cl the C—Cl bond probably has a slightly different strength than any C—Cl bond in CCl_4. All C—Cl bond energies

TABLE 4-1: Bond Enthalpies in kJ mol^{-1}, at 298.15 K

Single Covalent Bonds					
	H	C	O	Cl	Br
H	435	414	464	431	368
C		347	351	331	276
O			138		
Cl				243	
Br					192

Multiple Covalent Bonds			
C=C	615	C=O	711
C≡C	812	O=O	494

are about the same, however, and the average of such values for a certain bond measured for many compounds is the **mean bond dissociation enthalpy**, or **bond enthalpy**. These empirical numbers have been collected in tables and can be used to estimate heats of reactions (see Table 4-1).

There are two restrictions on the method of estimating the heat of reaction from bond enthalpies:

- The method applies only for covalent bonds.
- All species must be in the gas phase.

If we followed the method rigorously, we would add the bond enthalpy values for each compound to obtain an estimate of each ΔH_f° and then substitute in Eq. (4-5). We can simplify this procedure greatly, however, because any bond not affected by the reaction will appear both in a product and in a reactant, and its bond enthalpy will cancel out. Thus we need to be concerned only with bonds that are broken in the reactants and those formed in the products. If we let D represent bond enthalpies, Eq. (4-5) becomes

$$\Delta H_r = -\Sigma D(\text{bonds formed}) + \Sigma D(\text{bonds broken}) \qquad \textbf{(4-13)}$$

Note that the signs in Eq. (4-13) agree with our usual sign convention.. Since energy is released when bonds are formed, H will decrease, so the signs of the bond enthalpies must be negative in the first term. In the second term, they must be positive, because energy has to be absorbed to break a bond.

EXAMPLE 4-18: Estimate the heat of the reaction when one mole of hydrogen reacts with chlorine to produce hydrogen chloride when all species are gases.

Solution: First, we write the balanced equation, using structural formulas:

$$\text{H—H} + \text{Cl—Cl} \longrightarrow 2\text{H—Cl}$$

The bonds broken and formed are obvious in this case, and Eq. (4-13) becomes

$$\Delta H_r = -2D(\text{H—Cl}) + D(\text{H—H}) + D(\text{Cl—Cl})$$

Substituting values from Table 4-1, we have

$$\Delta H_r = -2(431) + 435 + 243 = -184 \text{ kJ}$$

EXAMPLE 4-19: Estimate the heat released when one mole of ethylene reacts with chlorine to form dichloroethane, when all species are gases.

Solution: When using this method with organic compounds, it is especially important to write the balanced equation using structural formulas:

```
                                    Cl  Cl
                                    |   |
H—C = C—H + Cl—Cl  ——→  H—C—C—H
   |     |                          |   |
   H     H                          H   H
```

Now we have to identify the bonds that are broken and those formed. The broken bonds must be C=C and Cl—Cl, and it is obvious that two C—Cl bonds have formed. Also—perhaps not so obviously to the beginner—one C—C bond has been formed. The bond enthalpy given in the table for C=C (615 kJ) applies to the breaking of both bonds, i.e., the complete separation of the two atoms. (Note that this value is approximately equal to, but less than, twice the bond enthalpy for a single bond between two carbon atoms.) In the product molecule, the two carbon atoms are joined, releasing the C—C bond enthalpy. Equation (4-13) becomes

$$\Delta H_r = -2D(\text{C—Cl}) - D(\text{C—C}) + D(\text{Cl—Cl}) + D(\text{C=C})$$

Substituting values from Table 4-1, we get

$$\Delta H_r = -2(331) - 347 + 243 + 615 = -151 \text{ kJ}$$

SUMMARY

Table S-4 summarizes the major equations of this chapter and shows connections between them. You should determine the conditions that are necessary for each equation to be valid or the conditions that are imposed to make the connections. Other important things you should know are listed below.

1. In calorimetry, the heat capacities of the components of the calorimeter are collected into an effective heat capacity.
2. In the usual "bomb" calorimeter, the calculated result is ΔE, and ideal gas behavior must be assumed to calculate ΔH.
3. The molar enthalpy of combustion, or heat of combustion, is the enthalpy change when one mole of a substance is burned completely in oxygen.
4. The molar enthalpy of formation is the enthalpy change when one mole of a substance is formed from its elements.
5. The molar integral enthalpy of solution is the enthalpy change when one mole of a substance is dissolved in a stated quantity of solvent. Similar terms for solutions are enthalpy of dilution, enthalpy of neutralization, and enthalpy of ionization.
6. Principles governing thermochemical calculations are

   (a) the amount of heat produced is proportional to the amount of material that reacts;
   (b) the heat of the reverse reaction is equal in magnitude but opposite in sign to the heat of the forward reaction;
   (c) the heat of a net reaction is the sum of the heats of each reaction in several steps of the overall change.

7. The standard state of a substance is the physical state which is stable at one atmosphere pressure; it is designated by a superscript degree sign, $H°$. The temperature may vary and must be specified.
8. Tables of standard molar enthalpies of formation are based on an arbitrary reference point for enthalpies. We have used $H^\circ_{298} = 0$ for an element in its standard state at 298.15 K; some texts use $\Delta H^\circ_{\mathrm{f},\,298} = 0$.
9. Tables of standard molar enthalpies of formation of ions in solution at infinite dilution are based on the reference point, $\Delta H^\circ_\mathrm{f}(\mathrm{H}^+) = 0$ at 298.15 K.
10. The enthalpy of reaction at a new temperature can be calculated from an enthalpy of reaction at one temperature and heat capacity data for all species in the reaction.

TABLE S-4: Summary of Important Equations

$\Delta H_\mathrm{r} = \Sigma nH\,(\text{products}) - \Sigma nH\,(\text{reactants})$ (4-2)

→ $\Delta H_\mathrm{r} = \Sigma n\,\Delta H^\circ_\mathrm{f}\,(\text{products}) - \Sigma n\,\Delta H^\circ_\mathrm{f}\,(\text{reactants})$ (4-5)

→ $\Delta H_\mathrm{r} = \Sigma n\,\Delta H_\mathrm{cmbn}\,(\text{products}) - \Sigma n\,\Delta H_\mathrm{cmbn}\,(\text{reactants})$ (4-5a)

→ $\Delta H_\mathrm{r} = -\Sigma D\,(\text{bonds formed}) + \Sigma D\,(\text{bonds broken})$ (4-13)

$dH = C_p\,dT$ (3-31)

→ $H_{T_2} - H_{T_1} = \int_{T_1}^{T_2} C_P\,dT$ (4-7)

↓ $\Delta H^\circ_{T_2} = \Delta H^\circ_{T_1} + \int_{T_1}^{T_2} \Delta C_P\,dT$ (4-8)

$\Delta C_P = \Delta a + \Delta bT + \Delta cT^2$ (4-9)

$\Delta H^\circ_{T_2} = \Delta H^\circ_{T_1} + \Delta a(T_2 - T_1) + \frac{\Delta b}{2}(T_2^2 - T_1^2) + \frac{\Delta c}{3}(T_2^3 - T_1^3)$ (4-10)

→ $d(\Delta H) = \Delta C_P\,dT$ (4-11)

$\Delta H_T = \Delta aT + \frac{\Delta b}{2}T^2 + \frac{\Delta c}{3}T^3 + C$ (4-12)

11. A bond enthalpy is the average of the bond dissociation energy of a bond between two given atoms measured in many species containing the two atoms. The method applies only to covalent molecules and gaseous species.

SOLVED PROBLEMS

Calorimetry and Heats of Reaction

PROBLEM 4-1 Calculate the difference between ΔE and ΔH for the burning of one mole of hexadecane ($C_{16}H_{34}$) at 20 °C.

Solution: Since the problem asks for a difference, you won't need ΔE or ΔH if there is a relationship between them. Recall that $\Delta H = \Delta E + \Delta(PV)$. For this problem, you can obtain $\Delta(PV)$ only by assuming ideal behavior. Then,

$$\Delta H - \Delta E = RT\Delta n \quad \text{[Eq. (3-24)]}$$

To determine Δn, you need the balanced equation for the reaction, written to make $n = 1$ mol for $C_{16}H_{34}$:

$$C_{16}H_{34}(s) + \tfrac{49}{2}O_2(g) \longrightarrow 16CO_2(g) + 17H_2O(l)$$

$$\begin{aligned}\Delta n &= \text{Moles of gaseous products} - \text{Moles of gaseous reactants}\\ &= 16 - 24.5 = -8.5 \text{ mol per mol } C_{16}H_{34}\end{aligned}$$

$$\Delta H - \Delta E = (1.987 \text{ cal K}^{-1} \text{ mol}^{-1})(293.2 \text{ K})(-8.5) = -4.952 \text{ kcal mol}^{-1}$$

The result is approximate because you assumed ideal behavior.

PROBLEM 4-2 If benzoic acid ($C_6H_5CO_2H$) is found to release 770.94 kcal mol^{-1} in a bomb calorimeter at 25 °C, what is the molar heat of combustion?

Solution: Complete combustion will yield carbon dioxide and water. Don't forget that the latter is a liquid at 25 °C. Write the balanced equation so that $n = 1$ mol for the compound:

$$C_6H_5CO_2H(s) + \tfrac{15}{2}O_2(g) \longrightarrow 7CO_2(g) + 3H_2O(l)$$

$$\Delta n = 7 - 7.5 = -0.5 \text{ mol per mol } C_6H_5CO_2H$$

Remember that the energy released is ΔE and that it is given a negative sign.

$$\begin{aligned}\Delta H &= \Delta E + RT\,\Delta n\\ &= -770.94 \times 10^3 \text{ cal mol}^{-1} + (1.987 \text{ cal K}^{-1}\text{mol}^{-1})(298.2 \text{ K})(-0.5)\\ &= -771.24 \text{ kcal mol}^{-1}\end{aligned}$$

Note that again ideal behavior has been assumed.

PROBLEM 4-3 It has been calculated that 11.55 kcal is released when 1.210 g of solid naphthalene, $C_{10}H_8$, is burned in oxygen in a bomb calorimeter, which is in a thermostat at 25.0 °C. What is the enthalpy of combustion per mole of naphthalene?

Solution: The usual calorimeter calculations have been simplified because the calculated energy (ΔE) is supplied. To use Eq. (3-24), you will need the balanced equation for the reaction, must calculate ΔE per mole, and assume ideal behavior.

$$C_{10}H_8(s) + 12O_2(g) \longrightarrow 10CO_2(g) + 4H_2O(l)$$

$$\Delta E = \left(-\frac{11.55 \text{ kcal}}{1.210 \text{ g}}\right)\left(\frac{128.2 \text{ g}}{1 \text{ mol}}\right) = -1.224 \times 10^3 \text{ kcal mol}^{-1}$$

Since only moles of gas are significant, $\Delta n = -2$ mol for one mole of naphthalene.

$$RT\,\Delta n = (1.987 \text{ cal K}^{-1}\text{mol}^{-1})(298.2 \text{ K})(-2) = -1.185 \text{ kcal mol}^{-1}$$

Substituting in Eq. (3-24), you get

$$\Delta H = (-1.224 \times 10^3 - 1.185) \text{ kcal mol}^{-1} = -1225 \text{ kcal mol}^{-1}$$

PROBLEM 4-4 Use the following experimental data to calculate the enthalpy of combustion of phenol, C_6H_5OH:

Run	Compound	Weight of Sample	Temperature, °C	
			Initial	Final
Calibration	benzoic acid	0.3182 g	24.43	25.67
Sample	phenol	0.5118 g	24.61	27.06

Solution: First you must determine the effective specific heat of the calorimeter. Rearranging Eq. (4-1), you get

$$C_{\text{eff}} = \frac{q}{\Delta t}$$

The heat is obtained from the known heat of combustion of benzoic acid: $\Delta E = 770.94$ kcal mol^{-1}. Thus

$$q = \left(\frac{770.94 \text{ kcal mol}^{-1}}{122.13 \text{ g mol}^{-1}}\right)(0.3182 \text{ g}) = 2.009 \text{ kcal}$$

$$C_{\text{eff}} = \frac{2.009}{25.67 - 24.43} = 1.62 \text{ kcal °C}^{-1}$$

Next you must calculate the heat produced by the phenol sample, using Eq. (4-1) again:

$$q = C_{\text{eff}}\,\Delta t = (1.62 \text{ kcal °C}^{-1})(27.06 - 24.61)\text{ °C} = 3.97 \text{ kcal}$$

$$\Delta E = \frac{q}{\text{mass}} M_m = \left(\frac{-3.97 \text{ kcal}}{0.5118 \text{ g}}\right)(94.11 \text{ g mol}^{-1}) = -730 \text{ kcal mol}^{-1}$$

Write the balanced equation so that $n = 1$ mol for phenol:

$$C_6H_5OH(s) + 7O_2(g) \longrightarrow 6CO_2(g) + 3H_2O(l)$$

For moles of gases only, $\Delta n = 6 - 7 = -1$ and

$$\begin{aligned}\Delta H &= \Delta E + RT\,\Delta n \\ &= (-730 \times 10^3 \text{ cal mol}^{-1}) + (1.987 \text{ cal K}^{-1}\text{mol}^{-1})(298.2 \text{ K})(-1) = -731 \text{ kcal mol}^{-1}\end{aligned}$$

PROBLEM 4-5 Calculate the heat of formation of carbon monoxide from the following information:

$$C(gr) + O_2(g) \longrightarrow CO_2(g) \qquad \Delta H = -393 \text{ kJ}$$
$$CO(g) + \tfrac{1}{2}O_2(g) \longrightarrow CO_2(g) \qquad \Delta H = -283 \text{ kJ}$$

Solution: First, write the balanced equation corresponding to the heat you want (see Example 4-5). In this problem, it is

$$C(gr) + \tfrac{1}{2}O_2(g) \longrightarrow CO(g) \qquad \Delta H = ?$$

Looking at the equations given in this problem, you see that the first has C in the desired location, so you should start with it. Now, you need to eliminate the CO_2. You can do this by reversing the second equation, adding it to the first, and canceling like terms:

$$\begin{array}{rcll} C(gr) + O_2(g) & \longrightarrow & CO_2(g) & \Delta H = -393 \text{ kJ} \\ CO_2(g) & \longrightarrow & CO(g) + \tfrac{1}{2}O_2 & \Delta H = +283 \text{ kJ} \\ \hline C(gr) + O_2(g) + \cancel{CO_2}(g) & \longrightarrow & CO(g) + \tfrac{1}{2}O_2(g) + \cancel{CO_2}(g) & \Delta H = -110 \text{ kJ} \\ C(gr) + \tfrac{1}{2}O_2(g) & \longrightarrow & CO(g) & \Delta H = -110 \text{ kJ} \end{array}$$

This is the equation you wanted. Usually you will cancel and add in the same step, making the third equation unnecessary.

Standard Enthalpies

PROBLEM 4-6 If the heat of combustion of gaseous cyclopropane (C_3H_6) is -2907 kJ mol^{-1} at 298 K, calculate its standard enthalpy of formation.

Solution: In this problem, you are not given all of the information you need, and must recognize that it is assumed you will use tables to obtain some of the data. First, you should write the balanced equations for the formation of one mole of cyclopropane and for its combustion.

$$3C(gr) + 3H_2(g) \longrightarrow C_3H_6(g) \qquad \Delta H = \Delta H_f^\circ = ?$$
$$C_3H_6(g) + \tfrac{9}{2}O_2(g) \longrightarrow 3CO_2(g) + 3H_2O(l) \qquad \Delta H = -2907 \text{ kJ mol}^{-1}$$

You need to convert the second equation into the first, reversing it and replacing CO_2 and H_2O with C and H_2. It should be obvious that you need the heats of formation of $CO_2(g)$ and $H_2O(l)$. You get the values from the tables, but have to construct the appropriate balanced equations. They are

$$C(gr) + O_2(g) \longrightarrow CO_2(g) \qquad \Delta H_f^\circ = -393.5 \text{ kJ mol}^{-1}$$
$$H_2(g) + \tfrac{1}{2}O_2(g) \longrightarrow H_2O(l) \qquad \Delta H_f^\circ = -285.8 \text{ kJ mol}^{-1}$$

You need to multiply these two equations by three and add them to the reversed forms of your combustion equation:

$$3CO_2(g) + 3H_2O(l) \longrightarrow C_3H_6(g) + \tfrac{9}{2}O_2(g) \qquad \Delta H = +2907 \text{ kJ mol}^{-1}$$
$$3C(gr) + 3O_2(g) \longrightarrow 3CO_2(g) \qquad \Delta H = -1181 \text{ kJ mol}^{-1}$$
$$3H_2(g) + \tfrac{3}{2}O_2(g) \longrightarrow 3H_2O(l) \qquad \Delta H = -857.4 \text{ kJ mol}^{-1}$$
$$\overline{3C(gr) + 3H_2(g) \longrightarrow C_3H_6(g) \qquad \Delta H = +869 \text{ kJ mol}^{-1}}$$

Thus $\Delta H_f^\circ = +869$ kJ mol^{-1}. Without elaborating, we note that such a large positive value suggests that it would be difficult to make cyclopropane by this reaction.

PROBLEM 4-7 Calculate the heat of the reaction by which ozone is produced in the stratosphere. The balanced equation for the reaction is $O_2(g) + O(g) \rightarrow O_3(g)$.

Solution: Equation (4-5) and ΔH_f° values from Table A-1 will give you the value at 298 K, or

$$\Delta H_r^\circ = n\,\Delta H_f^\circ(O_3) - n\,\Delta H_f^\circ(O_2) - n\,\Delta H_f^\circ(O) = 1(143) - 0 - 1(249) = -106 \text{ kJ mol}^{-1}$$

PROBLEM 4-8 If the integral enthalpy of solution of NaOH(s) in 10 mol H_2O has been found to be -43.5 kJ mol^{-1}, how much heat will be released if a solution of 1 mol NaOH in 10 mol H_2O is infinitely diluted at 298.15 K?

Solution: The desired heat will be the difference between the integral heats of solution of the initial and final solutions. You are given the first of these, which corresponds to the process

$$NaOH(s) \xrightarrow[10H_2O]{} NaOH(10H_2O)$$

The second corresponds to the process

$$NaOH(s) \xrightarrow[\infty H_2O]{} NaOH(aq)$$

To obtain the enthalpy change for the second process, you need to determine the standard enthalpy of formation of the NaOH in solution at infinite dilution by adding the values for the ions and then subtract the standard enthalpy of formation of NaOH(s). To follow what is happening, write the appropriate equations. For the individual ions, you have

$$Na(s) \xrightarrow[\infty H_2O]{} Na^+(aq) + e^- \qquad \Delta H_f^\circ(Na^+, aq)$$
$$\tfrac{1}{2}H_2(g) + \tfrac{1}{2}O_2(g) + e^- \xrightarrow[\infty H_2O]{} OH^-(aq) \qquad \Delta H_f^\circ(OH^-, aq)$$
$$\overline{\tfrac{1}{2}H_2(g) + \tfrac{1}{2}O_2(g) + Na(s) \xrightarrow[\infty H_2O]{} NaOH(aq) \qquad \Delta H_f^\circ(NaOH, aq)}$$

This summation is equivalent to Eq. (4-6) and gives you the standard enthalpy of formation of a solution of NaOH at infinite dilution. Obtain the values from Table A-1. Then

$$\Delta H_f^\circ(NaOH, aq) = \Delta H_f^\circ(Na^+, aq) + \Delta H_f^\circ(OH^-, aq) = -240.1 - 230.0 = -470.1 \text{ kJ mol}^{-1}$$

To obtain the enthalpy of solution, use

$$Na(s) + \tfrac{1}{2}O_2(g) + \tfrac{1}{2}H_2(g) \longrightarrow NaOH(s) \qquad \Delta H_f^\circ = -425.6 \text{ kJ mol}^{-1}$$

If you reverse this equation and add to the equation forming NaOH(aq), you get

$$NaOH(s) \xrightarrow[\infty H_2O]{} NaOH(aq)$$

and

$$\Delta H^\circ_{sol}(\text{NaOH}, aq) = -470.1 + 425.6 = -44.5 \text{ kJ mol}^{-1}$$

Finally, the heat released in the dilution process is

$$q = \Delta H^\circ_{sol}(\text{NaOH}, aq) - \Delta H^\circ_{sol}(\text{NaOH}, 10\text{H}_2\text{O}) = -44.5 - (-43.5) = -1.0 \text{ kJ mol}^{-1}$$

This q is the integral heat of dilution for this process.

PROBLEM 4-9 Calculate the standard heat of neutralization of acetic acid and sodium hydroxide.

Solution: The equation for the process is

$$\text{CH}_3\text{COOH}(aq) + \text{OH}^-(aq) \longrightarrow \text{H}_2\text{O}(l) + \text{CH}_3\text{COO}^-(aq)$$

Using Eq. (4-15) and values from Table A-1, you get

$$\begin{aligned}\Delta H_{neut} &= \Delta H^\circ_f(\text{H}_2\text{O}, l) + \Delta H^\circ_f(\text{CH}_3\text{COO}^-, aq) - \Delta H^\circ_f(\text{CH}_3\text{COOH}, aq) - \Delta H^\circ_f(\text{OH}^-, aq)\\ &= -285.8 - 486.0 - (-485.8) - (-230.0) = -56.0 \text{ kJ mol}^{-1}\end{aligned}$$

Calculating Enthalpies

PROBLEM 4-10 Calculate the heat of reaction at 1200 K for the production of water gas. The balanced equation for the reaction is $\text{C}(gr) + \text{H}_2\text{O}(g) \rightarrow \text{CO}(g) + \text{H}_2(g)$.

Solution: To calculate a ΔH at a temperature other than 298 K, you must use Eq. (4-8) or Eq. (4-11). In either case you need ΔH at 298 K and ΔC_P. Using Eq. (4-5) and obtaining values from Table A-1, you get

$$\begin{aligned}\Delta H^\circ_{298} &= \Delta H^\circ_f(\text{CO}, g) + \Delta H^\circ_f(\text{H}_2, g) - \Delta H^\circ_f(\text{C}, gr) - \Delta H^\circ_f(\text{H}_2\text{O}, g)\\ &= -110.5 + 0 - 0 - (-241.8) = +131.3 \text{ kJ mol}^{-1}\end{aligned}$$

(Note that water is included as a gas because it will be in that state at the higher temperature.)

For ΔC_P:

Substance	a	$b \times 10^3$	$c \times 10^7$
$\text{C}(gr)$	-5.296	58.60	-432.24
$\text{H}_2\text{O}(g)$	30.359	9.61	11.84
Total for products	25.063	68.21	-420.40
$\text{CO}(g)$	26.861	6.97	-8.19
$\text{H}_2(g)$	29.066	-0.83	20.12
Total for reactants	55.927	6.14	11.93
Products − reactants	-30.864	62.07	-432.33

Since the problem calls for only one higher temperature, you should start with Eq. (4-8), integrating with limits, or

$$\Delta H^\circ_{T_2} = \Delta H^\circ_{T_1} + \int_{T_1}^{T_2} \Delta C_P \, dT$$

Now substitute for ΔC_P [Eq. (4-9)] and separate the integral:

$$\begin{aligned}\Delta H^\circ_{T_2} &= \Delta H^\circ_{T_1} + \int_{T_1}^{T_2} \Delta a \, dT + \int_{T_1}^{T_2} \Delta b T \, dT + \int_{T_1}^{T_2} \Delta c T^2 \, dT\\ &= \Delta H^\circ_{T_1} + \Delta a(T_2 - T_1) + \frac{\Delta b}{2}(T_2^2 - T_1^2) + \frac{\Delta c}{3}(T_2^3 - T_1^3) \qquad \text{[Eq. (4-10)]}\end{aligned}$$

Substitute the calculated values and solve:

$$\begin{aligned}\Delta H^\circ_{1200} &= 131.3 \times 10^3 + (-30.864)(1200 - 298.15) + \left(\frac{62.07 \times 10^{-3}}{2}\right)(1200^2 - 298.15^2)\\ &\quad + \left(\frac{-432.33 \times 10^{-7}}{3}\right)(1200^3 - 298.15^3)\\ &= 131.3 \times 10^3 - 27.8 \times 10^3 + 41.9 \times 10^3 - 24.5 \times 10^3 = 120.9 \text{ kJ}\end{aligned}$$

PROBLEM 4-11 Use bond enthalpies to estimate the heat of formation of $H_2O(g)$ at 298 K and compare your answer to the value in the table of standard enthalpies of formation.

Solution: The balanced equation is 2H − H + O═O → 2H—O—H. Note that we treat O_2 as having a double bond. Equation (4-13) becomes

$$\Delta H_r = -2(2)[D(\text{O—H})] + 2[D(\text{H—H})] + D(\text{O═O}) = -4(464) + 2(435) + 494 = -492 \text{ kJ}$$

This value is for 2 mol H_2O. Thus $\Delta H_f^\circ(H_2O, g) = -246$ kJ. The value from the table is -241.8 kJ mol^{-1}, so the error in the estimate is less than 2%.

PROBLEM 4-12 Using bond enthalpies, estimate the heat generated if one mole of ethanol reacts with oxygen to form acetic acid and water at 298 K.

Solution: Recall that all substances must be gases for bond enthalpy calculations, so the ethanol, acetic acid, and water are not in their standard states. First, write the equation with structural formulas:

```
        H                          O
        |                          ||
CH3—C—O—H + O═O  ——→  CH3—C—O—H + H—O—H
        |
        H
```

Using Eq. (4-13) and assuming that the CH_3—C—O—H part of the molecule remains unchanged:

$$\Delta H = -D(\text{C═O}) - 2[D(\text{H—O})] + 2[D(\text{C—H})] + D(\text{O═O})$$
$$= -711 - 2(464) + 2(414) + 494 = -317 \text{ kJ}$$

PROBLEM 4-13 Repeat Problem 4-12, using standard enthalpies of formation.

Solution: There is a complication here because the values in the table for acetic acid and ethanol are for the liquid state. (The values for both the liquid and gaseous states are listed for water). Thus you must include an enthalpy term for converting each liquid to the gaseous state. This quantity is the enthalpy of vaporization (see Section 8-2C). Normally these are determined at the boiling point, but you need them at 298 K. For this problem, use the approximate values of 42 kJ mole^{-1} for acetic acid and 43 kJ mol^{-1} for ethanol and substitute them in Eq. (4-5):

$$\Delta H_r^\circ = \Delta H_f^\circ(CH_3CO_2H, l) + \Delta H_v(CH_3CO_2H) + \Delta H_f^\circ(H_2O, g) - \Delta H_f^\circ(C_2H_5OH, l) - \Delta H_v(C_2H_5OH)$$
$$= -485 + 42 - 242 - (-278) - 43 = -450 \text{ kJ}$$

This result does not agree very well with that of Problem 4-12. Several factors may contribute to the difference, including the approximations of the bond enthalpy method, association and hydrogen bonding in acetic acid vapor, and an abnormally weak C—O bond in alcohols.

Supplementary Exercises

PROBLEM 4-14 The substance used frequently as a standard for determining the effective specific heat of a calorimeter is __________. Based on the reaction $N_2(g) + O_2(g) +$ 180.8 kJ → $2NO(g)$, what is the molar heat of formation of $NO(g)$? How much heat would be produced if 0.5 mol HCl reacts with 0.5 mol NaOH? When the heat of an exothermic reaction is known at one temperature and the reaction is carried out at a higher temperature, the amount of heat produced (**a**) will increase; (**b**) will decrease; (**c**) will stay the same; or (**d**) can't be predicted without more information.

PROBLEM 4-15 A new bomb calorimeter was calibrated by burning 0.218 g of solid benzoic acid in oxygen. The temperature changed from 23.84 °C to 24.97 °C. Calculate the effective heat capacity of the calorimeter. The enthalpy of combustion of benzoic acid is -771.24 kcal mol^{-1}.

PROBLEM 4-16 In the calorimeter in Problem 4-15, 0.326 g of adipic acid, $(CH_2)_4(COOH)_2$, was burned in oxygen and the temperature rose from 24.38 °C to 25.61 °C. Calculate the experimental value of the enthalpy of combustion of adipic acid ($M_m = 146.14$).

PROBLEM 4-17 Calculate Δn for the combustion in oxygen of the following solid compounds: (**a**) menthol ($C_{10}H_{20}O$); (**b**) sucrose ($C_{12}H_{22}O_{11}$); and (**c**) nitroaniline ($NO_2C_6H_4NH_2$).

PROBLEM 4-18 Calculate the heat of combustion of methane gas (CH_4) at 298 K. Assume that the water formed remains a gas.

PROBLEM 4-19 Continuing Problem 4-17, what is the value of ΔE for methane?

PROBLEM 4-20 Calculate the heat produced in the hydrogenation of one mole of gaseous benzene to cyclohexane gas at 298 K.

PROBLEM 4-21 What is the standard enthalpy of solution to infinite dilution of $Ca(OH)_2$?

PROBLEM 4-22 Using the answer to Problem 4-19, calculate the heat of combustion of methane at 600 K.

PROBLEM 4-23 Use bond enthalpies to calculate an estimated heat of reaction for the reaction of one mole of acetone with hydrogen to form isopropyl alcohol.

PROBLEM 4-24 Use the bond enthalpy method to estimate the amount of heat released per mole of acetylene in a torch where acetylene is burned in oxygen to form carbon dioxide and water.

Answers to Supplementary Exercises

4-14 benzoic acid; $+90.4$ kJ mol^{-1}; -27.9 kJ; (d) **4-15** 1.22 kcal deg^{-1} **4-16** -672.7 kcal mol^{-1}
4-17 (**a**) -4.5; (**b**) 0; (**c**) $+0.5$ **4-18** -802.3 kJ mol^{-1} **4-19** -802.3 kJ mol^{-1}
4-20 -206.1 kJ mol^{-1} **4-21** -16.7 kJ mol^{-1} **4-22** -800.4 kJ mol^{-1} **4-23** -83 kJ
4-24 -897 kJ mol^{-1}

5 THE SECOND AND THIRD LAWS OF THERMODYNAMICS

THIS CHAPTER IS ABOUT

- ☑ **Natural Processes, Heat Engines, and the Second Law**
- ☑ **Entropy**
- ☑ **The Fundamental Equation of Thermodynamics and Calculation of Entropy Change**
- ☑ **Entropy Changes during Physical Processes**
- ☑ **The Third Law of Thermodynamics and Third-Law Entropies**
- ☑ **Entropy Changes during Chemical Processes**

5-1. Natural Processes, Heat Engines, and the Second Law

You will find almost as many methods of developing the **second law of thermodynamics** and the concept of entropy as there are textbooks about physical chemistry. The two major approaches may be called *statistical* and *thermodynamic*; the first involves the use of probabilities and the second is often based on the efficiencies of heat engines. In this chapter, we'll consider only the thermodynamic approach, include only fundamental derivations and proofs, and concentrate primarily on terminology and applications.

A. Natural processes

In nature, some processes occur when given the opportunity, whereas the opposite processes don't. If a coiled spring is released, it unwinds. If a piece of zinc is placed in acid, hydrogen appears. These events occur naturally or *spontaneously*. Although the opposite processes don't violate the first law, they can only be carried out by supplying energy or work; they are unnatural or *nonspontaneous*.

- Don't confuse the terms spontaneous and instantaneous. Spontaneous processes tend to take place by themselves, but nothing is implied about the *time* it takes for such processes to occur.

EXAMPLE 5-1: A mixture of hydrogen and oxygen gases will stand for years without producing a measurable amount of water. But, upon the introduction of a catalyst, water is produced in a reaction so rapid as to be explosive. Is this reaction spontaneous?

Solution: Because the catalyst increases the rate of reaction, we conclude that the process is spontaneous, but proceeds extremely slowly in the absence of a catalyst.

Spontaneous processes, when they occur, always lead to a degradation of energy, i.e., to a decrease in the amount of energy available to do work. This fact is the basis for one statement of the second law:

(1) *The available energy of an isolated system decreases in all natural processes (and is conserved in reversible processes).*

Note that natural processes are irreversible.

B. Heat engines

Work can be converted completely into heat, but the opposite isn't true; when heat is converted into work, some of the heat can't be used, and the temperature of the system drops. This observation was generalized by Lord Kelvin into another statement of the second law:

(2) *It is impossible to construct a cyclic engine which will produce no other effect than to extract heat from a reservoir and to convert it into an equivalent amount of work.*

A machine that *would* do what Kelvin said is impossible is called a **perpetual motion machine of the second kind**. Hence another statement of the second law (really the same as Kelvin's) is:

(3) *It is impossible to construct a perpetual motion machine of the second kind.*

A heat engine absorbs thermal energy at some temperature (T_2), converts part of it into work, and releases unused heat at a temperature (T_1) lower than the initial temperature. To repeat the process, the engine must be returned to the starting temperature; thus work must be done on it and heat removed from it. We say that it completes a **cycle**.

The **efficiency** of a heat engine is defined as the ratio of work done to the energy supplied:

$$\text{Eff} = \frac{\text{Energy out}}{\text{Energy in}} = \frac{\text{Work done}}{\text{Energy supplied}} = \frac{\text{Heat supplied} - \text{Heat discharged}}{\text{Heat supplied}}$$

or

$$\text{Eff} = \frac{q_{T_2} - q_{T_1}}{q_{T_2}} \quad \textbf{(5-1)}$$

From a detailed analysis of the reversible Carnot cycle (described later), Lord Kelvin concluded that the quantities of heat for reversible engines are proportional to the absolute temperature. Thus

$$\text{Eff} = \frac{q_{T_2} - q_{T_1}}{q_{T_2}} = \frac{T_2 - T_1}{T_2} \quad \textbf{(5-2)}$$

For convenience, we assume that the reservoirs at T_1 and T_2 are large enough that operation of the cycle doesn't change the temperature.

EXAMPLE 5-2: What is the efficiency of a reversible cyclic engine that operates between reservoirs at 10 °C and 300 °C?

Solution: Substituting the temperatures in Eq. (5-2), we get

$$\text{Eff} = \frac{T_2 - T_1}{T_2} = \frac{573 - 283}{573} = 0.506$$

Further analysis of the reversible Carnot cycle leads to further conclusions about efficiencies:

1. The efficiency is the same for all reversible cyclic heat engines operating between two temperatures (Carnot's theorem). In other words, the efficiencies of all reversible cyclic heat engines are the same as that of the ideal gas Carnot reversible cyclic heat engine (described later).
2. The efficiencies of all cyclic heat engines employing any irreversible process are less than that of the Carnot reversible cyclic heat engine.

C. Direction of heat flow

Heat does not flow from cold to hot, which leads to Clausius' statement of the second law:

(4) *Heat cannot pass from a colder to a hotter body without some other change occurring at the same time.*

EXAMPLE 5-3: The direction of heat flow in a kitchen refrigerator is from cold to hot. Explain why refrigerators do not violate the second law.

Solution: A kitchen refrigerator takes heat from its cold interior and discharges it to the warmer room but *only* because a compressor provides work to keep the process going.

5-2. Entropy

A. A new state function

Chemists want to know which way to draw the arrow in a possible chemical reaction. Which direction is the spontaneous one?

The second law provides a way of determining the spontaneity of a potential change: It yields a state function (thermodynamic property) that changes in a particular direction when a process is spontaneous. One method of defining the new function employs the properties of exact differentials and Eq. (5-2). First, we rearrange Eq. (5-2):

$$\frac{q_{T_2}}{q_{T_2}} - \frac{q_{T_1}}{q_{T_2}} = \frac{T_2}{T_2} - \frac{T_1}{T_2} \quad \text{or} \quad 1 - \frac{q_1}{q_2} = 1 - \frac{T_1}{T_2}$$

Then

$$\frac{q_1}{q_2} = \frac{T_1}{T_2}, \qquad \frac{q_1}{T_1} = \frac{q_2}{T_2}, \qquad \text{and} \qquad \frac{q_2}{T_2} - \frac{q_1}{T_1} = 0$$

or

$$\sum \frac{Q}{T} = 0 = \oint \frac{dQ}{T}$$

The term $\frac{đQ}{T}$ must be an exact differential because its cyclic integral equals zero. We give it its own symbol, S, where $dS = \frac{đQ}{T}$, and its own name, **entropy**.

One important restriction is that the heat term must be for a *reversible* process. If we want to determine the entropy change for an irreversible process, we have to devise a reversible path—between the same initial and final states—for which we can obtain the heat quantities and corresponding temperatures. Thus the proper definition is

ENTROPY CHANGE
$$dS = \frac{đq_{\text{rev}}}{T} \tag{5-3}$$

For macroscopic processes, we use the integrated form:

ENTROPY CHANGE (integrated form)
$$\Delta S = \frac{q_{\text{rev}}}{T} \tag{5-4}$$

B. The method of integrating multipliers

If a gas expands, the first law tells us that

$$dE = đq + đw \qquad \text{[Eq. (3-1a)]}$$

Rearranging and substituting Eqs. (3-11) and (3-28), we obtain

$$đq = nC_{V,m}\,dT + P_{\text{ext}}\,dV \tag{5-5}$$

If the process is reversible, $P_{\text{ext}} = P_{\text{int}} = P_{\text{gas}}$. Assuming that the gas is ideal,

$$đq_{\text{rev}} = nC_{V,m}\,dT + \frac{nRT}{V}\,dV$$

Since an inexact differential has a different value for each path, the value of its integral also depends on the path. If we specified a number of paths, we could obtain a number of integrals.

If we multiply by $1/T$, however, we obtain an expression with an integral that has only one value:

$$\frac{đq_{\text{rev}}}{T} = nC_{V,m}\frac{dT}{T} + nR\frac{dV}{V} \tag{5-6}$$

Integrating without limits, we get

$$\int \frac{đq_{\text{rev}}}{T} = nC_{V,m} \ln T + nR \ln V + C \tag{5-7 Ideal}$$

Thus $\frac{đq}{T}$ is exact; i.e., there is an explicit function that can be differentiated to give $\frac{đq}{T}$. Again, we can say that $\frac{đq}{T}$ is the differential of a state function and give it a new symbol, dS.

note: We have illustrated a general technique in mathematics for dealing with inexact differentials. The $1/T$ term is called an *integrating multiplier*, or *integrating factor*, which converts an inexact differential into an exact differential.

EXAMPLE 5-4: Consider the expression $du = \frac{x^2}{y^2}\,dx$. The right-hand side of the equation can't be integrated, so what can we do?

Solution: If we multiply by y^2, we get

$$y^2\,du = x^2\,dx$$

Then we apply Eq. (1-21) to get

$$\int y^2\,du = \frac{x^3}{3} + C$$

C. Applying the second law

You may have noted that the statements (2), (3), and (4) of the second law were expressed negatively, i.e., as the impossibility of doing something. Such *principles of impotence* are useful ways of stating basic laws of nature, but they aren't helpful in predicting whether a certain process is spontaneous. However, Clausius provided another statement of the second law:

(5) *The entropy of the universe strives towards a maximum.*

In other words,

(6) *All natural processes occur with an increase in the entropy of the universe.*

Reducing the scope of the statement from the universal to the human scale, let's consider an **isolated system**, i.e., a particular process and a heat reservoir that are insulated adiabatically from the rest of the universe. Now statement (6) becomes

(7) *All natural (irreversible) processes in an isolated system occur with an increase in the entropy of the system.*

Note that the entropy of the isolated system would not change if the process were carried out reversibly.

EXAMPLE 5-5: In an isolated system at 300 K, an ideal gas expands reversibly. This requires 900 J of heat from the reservoir. Calculate the entropy changes of the gas, the reservoir, and the entire system.

Solution: We can apply Eq. (5-4) in order to calculate the entropy changes of the gas and the reservoir and then add the two values to get the total for the system:

For the gas: $\Delta S_g = \frac{900}{300} = 3.0 \text{ J K}^{-1}$

For the reservoir: $\Delta S_r = -\frac{900}{300} = -3.0 \text{ J K}^{-1}$

For the system: $\Delta S_{sys} = \Delta S_g + \Delta S_r = 3.0 - 3.0 = 0$

EXAMPLE 5-6: Let's continue with Example 5-5 to find the three entropy changes if the expansion (between the same initial and final volumes) is carried out irreversibly by a process that requires 600 J of heat from the reservoir. Note that the amount of heat required is less than in the reversible case. A reversible process uses the maximum amount of heat, just as it performs the maximum amount of work (see Section 3-3).

Solution: For the gas, the entropy change is the same as that calculated for the reversible case, 3.0 J K^{-1}. Recall that ΔS is defined as the heat in a *reversible* process divided by T [Eqs. (5-3) and (5-4)]. Also, since entropy is a state function, the change in entropy between specific initial and final states must be the same for every possible path.

The problem doesn't identify the process used to produce the heat lost by the reservoir, so we assume that some reversible process is possible and calculate

$$\Delta S_r = \frac{-600}{300} = -2.0 \text{ J K}^{-1}$$

The entropy change for the system is

$$\Delta S_{sys} = \Delta S_g + \Delta S_r = 3.0 - 2.0 = 1.0 \text{ J K}^{-1}$$

Thus entropy of the entire system has increased, as predicted by statement (7).

Let's now consider an adiabatic expansion of a gas. Imagine that the gas in Examples 5-5 and 5-6 has no connecting reservoir; the gas and its container are isolated from the rest of the universe so that no heat can enter or leave. We've seen that $\Delta S = q_{rev}/T$ [Eq. (5-4)] and that $q_{rev} > q_{irrev}$ (Example 5-6); thus we have

$$\Delta S > \frac{q_{irrev}}{T} \tag{5-8}$$

For any adiabatic process, $q = 0$ by definition. For a reversible adiabatic process, $q_{rev} = 0$, and $\Delta S_{rev} = 0$; for an irreversible adiabatic process, $q_{irrev} = 0$, but $\Delta S_{irrev} > 0$. Thus entropy is a property that is conserved in a reversible adiabatic process but increases in a "real" adiabatic process. Recall, however, that spontaneous processes are always irreversible. Frequently Eqs. (5-4) and (5-8) are combined into an expression known as the *Clausius inequality*:

CLAUSIUS INEQUALITY

$$\Delta S \geq \frac{q}{T} \tag{5-9}$$

This equation is valid for an isolated system. Within the system, a process having a negative entropy change may occur, but this will be compensated by an equal or larger positive change somewhere in the system.

warning: This seems to be an appropriate point to comment on the sometimes confusing use of the word *system.* It may refer to the process or substance involved (e.g., the gas in Examples 5-5 and 5-6), or it may refer to the whole of the isolated region of space (i.e., the gas and the reservoir). To illustrate this difference, one textbook speaks of treating "the system and its surroundings as an isolated system"; often, however, "the surroundings" mean the rest of the universe, treated as one vast reservoir. In short, make sure you know what the *system* is when you read about one in your text.

D. Summary of statements about entropy and entropy change

1. ΔS is a measure of the degradation of energy in a system when a process occurs.
2. ΔS is defined as $\int \frac{dq_{\text{rev}}}{T}$, or simply $\frac{q_{\text{rev}}}{T}$.
3. ΔS is greater than $\frac{q_{\text{irrev}}}{T}$ [see Examples 5-5 and 5-6 and Eq. (5-7)].
4. ΔS can be calculated only from reversible paths.
5. S is the property of a system that is conserved in an *adiabatic* reversible process.
6. S of the universe will increase in any process that isn't carried out reversibly.

EXAMPLE 5-7: Which property of a system is conserved in an *isothermal* reversible process?

Solution: First, let's be sure that the meaning of *conserved* is clear. It is used more by physicists than by chemists and means that the property in question does not change; e.g., the law of conservation of mass.

That should give us a clue to the answer. Remember the law of conservation of energy, the first law of thermodynamics? It tells us that the energy of the universe doesn't change. Thus, for an ideal gas, the internal energy, E, is conserved in an isothermal reversible process (see Chapter 3).

5-3. The Fundamental Equation of Thermodynamics and Calculation of Entropy Change

A. Combining the first and second laws

If we substitute the equation of the second law, Eq. (5-3), into that of the first law, Eq. (3-1a), and assume only PV work, we have the **fundamental equation of thermodynamics**:

$$dE = T\,dS - P\,dV \qquad \text{(5-10, } PV \text{ work)}$$

which can be rearranged to

$$dS = \frac{dE}{T} + \frac{P\,dV}{T}$$

or

$$dS = \frac{C_V\,dT}{T} + \frac{P\,dV}{T} \qquad \text{[see Eq. (3-28)]} \qquad \textbf{(5-11)}$$

The completely general forms of Eq. (5-11) allow for the variation of P or V with temperature and are the forms you should learn and use. If S and E are taken to be functions of T and V, then

$$dS = \frac{C_V}{T}\,dT + \left(\frac{\partial P}{\partial T}\right)_V dV \qquad \text{(5-12, } PV \text{ work)}$$

If S and E are taken to be functions of T and P, we obtain

$$dS = \frac{C_P}{T}\,dT - \left(\frac{\partial V}{\partial T}\right)_P dV \qquad \text{(5-13, } PV \text{ work)}$$

B. Expressions for ΔS

In order to integrate the preceding equations, we must know more about the coefficients of dT and dV or dP. The partial derivative terms could be evaluated from experimental data or from an equation of state. If we assume ideal gas behavior, C_V and C_P are constant. Then we can use $PV = nRT$. Thus for Eq. (5-12),

$$\left(\frac{\partial P}{\partial T}\right)_V = \left(\frac{\partial nRT/V}{\partial T}\right)_V = \frac{nR}{V}$$

Integration of Eq. (5-12) with limits yields

$$\Delta S = nC_{V,m} \ln \frac{T_2}{T_1} + nR \ln \frac{V_2}{V_1} \qquad \begin{pmatrix} \textbf{5-14} \\ \textbf{Ideal, } \boldsymbol{PV} \textbf{ work} \end{pmatrix}$$

Compare this result to Eq. (5-7), which resulted from integration without limits.

Similarly, integration of Eq. (5-13) yields

$$\Delta S = nC_{P,m} \ln \frac{T_2}{T_1} - nR \ln \frac{P_2}{P_1} \qquad \begin{pmatrix} \textbf{5-15} \\ \textbf{Ideal, } \boldsymbol{PV} \textbf{ work} \end{pmatrix}$$

EXAMPLE 5-8: Derive Eq. (5-15) from Eq. (5-14).

Solution: Both equations are restricted to ideal gases, so we start with the volume term in Eq. (5-14), substituting from the ideal gas equation. We can't use the simpler Boyle's law because temperature is not constant.

$$V_2 = \frac{nRT_2}{P_2} \qquad \text{and} \qquad V_1 = \frac{nRT_1}{P_1}$$

Canceling yields

$$\frac{V_2}{V_1} = \left(\frac{T_2}{P_2}\right)\left(\frac{P_1}{T_1}\right) = \left(\frac{P_1}{P_2}\right)\left(\frac{T_2}{T_1}\right) \qquad \text{or} \qquad \ln \frac{V_2}{V_1} = \ln \frac{P_1}{P_2} + \ln \frac{T_2}{T_1}$$

Substituting in Eq. (5-14), we get

$$\begin{aligned} \Delta S &= nC_{V,m} \ln \frac{T_2}{T_1} + nR \ln \frac{P_1}{P_2} + nR \ln \frac{T_2}{T_1} \\ &= n(C_{V,m} + R) \ln \frac{T_2}{T_1} + nR \ln \frac{P_1}{P_2} = nC_{P,m} \ln \frac{T_2}{T_1} - nR \ln \frac{P_2}{P_1} \end{aligned}$$

5-4. Entropy Changes during Physical Processes

A. Volume, pressure, and temperature changes

Equations (5-14) and (5-15) are integrated forms of Eq. (5-11) for the ideal case. If the gas is not ideal, either Eq. (5-12) or (5-13) must be integrated after substituting appropriate expressions for C and the partial derivative. If only one variable changes, the equations reduce to one term.

EXAMPLE 5-9: Calculate the entropy change in 5.00 mol of an ideal gas when it expands isothermally and reversibly from 100 L to 200 L.

Solution: We start with Eq. (5-12) because the process is a volume change. Since the process is isothermal, we can omit the first term and thus don't need the heat capacity, so

dT = Ø

$$dS = \left(\frac{\partial P}{\partial T}\right)_V dV$$

For an ideal gas,

$$\left(\frac{\partial P}{\partial T}\right)_V = \left(\frac{\partial nRT/V}{\partial T}\right)_V = \frac{nR}{V}$$

Thus

$$\int_{S_1}^{S_2} dS = \int_{V_1}^{V_2} \frac{nR}{V} dV$$

and

$$\Delta S = nR \ln \frac{V_2}{V_1} = 5.00(8.314) \ln \frac{200}{100} = 28.8 \text{ J K}^{-1}$$

EXAMPLE 5-10: Calculate the entropy change if 2.0 mol of a monatomic gas ($C_{P,m} = \frac{5}{2}R$) is heated from 300 K to 500 K and the pressure increases from 2.5 atm to 4.0 atm.

Solution: Since the pressure changes, we use Eq. (5-13), and we'll need to evaluate $(\partial V/\partial T)_P$. Again using $PV = nRT$, we have

$$\left(\frac{\partial V}{\partial T}\right)_P = \left(\frac{\partial nRT/P}{\partial T}\right)_P = \frac{nR}{P}\frac{dT}{dT} = \frac{nR}{P}$$

Since $C_{P,m}$ is constant, the integration of Eq. (5-13) becomes

$$\int_{S_1}^{S_2} dS = nC_{P,m} \int_{T_1}^{T_2} \frac{dT}{T} - nR \int_{P_1}^{P_2} \frac{dP}{P}$$

$$\Delta S = nC_{P,m} \ln \frac{T_2}{T_1} - nR \ln \frac{P_2}{P_1} \qquad \text{[Eq. (5-15)]}$$

Substituting the values given, we obtain

$$\Delta S = 2.0\left(\frac{5}{2}\right)(8.314)\left(\ln \frac{500}{300}\right) - 2.0(8.314)\left(\ln \frac{4.0}{2.5}\right) = 21.2 - 7.8 = 13.4 \text{ J K}^{-1}$$

B. Phase transitions

1. A phase transition at the transition temperature under 1 atm pressure is a reversible isothermal process.

The temperature cannot change so long as both phases are present, and the relative amounts of the phases can be changed by adding or removing heat. The heat required to change one mole of the substance from one phase to the other is called the **enthalpy of transition**. The entropy change for the process may be calculated using Eq. (5-4), where q_{rev} is the appropriate enthalpy change, or

$$\Delta S_{\text{trans}} = \frac{\Delta H_{\text{trans}}}{T_{\text{trans}}} \qquad \textbf{(5-16)}$$

EXAMPLE 5-11: Calculate the change in entropy when one mole of liquid ethanol at its normal boiling point (pressure = 1 atm) of 78.3 °C changes to vapor at the same temperature. The heat of vaporization is 9380 cal mol^{-1} at this temperature.

Solution: We need merely to substitute in Eq. (5-16):

$$\Delta S = \frac{9380}{351.5} = 26.69 \text{ cal K}^{-1}$$

---

2. A phase transition often occurs under irreversible conditions.

The usual illustration is the freezing of a liquid at a constant temperature below its normal freezing point; e.g., liquid water, supercooled to −5 °C at 1 atm pressure, changes to ice if disturbed, but ice at −5 °C and 1 atm doesn't melt spontaneously. In order to calculate ΔS for such a case, we have to devise a means of carrying out the same overall change by a reversible path, which usually involves two or more reversible steps. Since entropy is a state function, ΔS for the irreversible path must equal that for the reversible path.

EXAMPLE 5-12: Calculate the entropy change when 1 mole of liquid water at -5 °C and 1 atm pressure changes to ice at the same temperature and pressure.

Solution: We know that the freezing of water at 0 °C and 1 atm is a reversible process, as is the raising and lowering of the temperature at constant pressure. Thus the simplest series of reversible steps to achieve the desired change is to

(a) heat the liquid to 0 °C,
(b) change the liquid to a solid at 0 °C, and
(c) cool the solid to -5 °C.

We can calculate ΔS for step (b) as we did in Example 5-11, using Eq. (5-16). The enthalpy of crystallization (negative of enthalpy of fusion) for water is -1436 cal mol^{-1}. For steps (a) and (c), we apply the definition of entropy, Eq. (5-3), but now q_{rev} is given by $C_P\,dT$ [Eq. (3-30)]:

$$dS = \frac{C_P\,dT}{T} \tag{5-17}$$

$$\int dS = \int C_P \frac{dT}{T}$$

Assuming that C_P is constant over the short range of temperature and integrating with limits, we have

$$\Delta S = C_P \ln \frac{T_2}{T_1} = nC_{P,m} \ln \frac{T_2}{T_1} \tag{5-18}$$

which is merely Eq. (15-15) for the case of constant pressure.

Don't forget that the heat capacities for the solid and liquid forms of a substance are different. For liquid water, $C_{P,m} = 18.0$ cal K^{-1}mol^{-1}, and, for ice, $C_{P,m} = 8.79$ cal K^{-1}mol^{-1}.

Collecting the calculations for the three steps, we have, for $n = 1$:

(a) $\Delta S(-5° \text{ to } 0°) = C_{P,m}(\text{liq}) \ln \dfrac{T_2}{T_1} = 18.0 \ln \dfrac{273.2}{268.2}$

(b) $\Delta S(\text{liquid to solid at } 0°) = \dfrac{\Delta H}{T} = -\dfrac{1436}{273.2}$

(c) $\Delta S(0° \text{ to } -5°) = C_{P,m}(\text{sol}) \ln \dfrac{T_2}{T_1} = 8.79 \ln \dfrac{268.2}{273.2}$

Completing the calculations and adding, we get

$$\Delta S(\text{total}) = 0.332 - 5.256 - 0.162 = -5.09 \text{ cal K}^{-1}$$

C. Mixing of gases

In this case, the total entropy change is the sum of the changes for each individual gas. An expression for ΔS_{mix} may be obtained by starting with either Eq. (5-14) (volume) or Eq. (5-15) (pressure), which were obtained from the differential forms, Eq. (5-12) and (5-13), respectively. We assume that ideal gases are mixed isothermally. If n_A moles of gas A at pressure P_A mix with one or more gases, so that the new pressure of A is P'_A, from Eq. (5-15) we have

$$\Delta S_A = -n_A R \ln \frac{P'_A}{P_A}$$

With similar expressions for all the gases involved, we can write their sum as

$$\Delta S(\text{for } n_i) = \sum_i -n_i R \ln \frac{P'_i}{P_i} \tag{5-19}$$

In a mixture of gases, $P_i' = X_i P_t$. Thus,

$$\ln \frac{X_i P_t}{P_i}$$

The usual procedure is to take the special case where all the gases are at the same initial pressure. If the mixing is done so that the initial volumes add, the final pressure would equal the initial pressure, or

$$P_2 = P_1 = P_A = P_B = \cdots$$

The logarithm term becomes $\ln X_i$, and Eq. (5-19) becomes

$$\Delta S(\text{mix of } n_i) = -R \sum_i n_i \ln X_i \qquad \text{(Ideal, } PV \text{ work)} \qquad \mathbf{(5\text{-}20)}$$

EXAMPLE 5-13: If 2 mol N_2 and 3 mol Ar, initially at the same pressure and temperature, are allowed to mix by removing a partition that had kept them apart, what is the entropy change? Assume ideal behavior.

Solution: The important conditions are that the gases were at the same initial pressure and that the two original volumes have been added. Since behavior is ideal, Eq. (5-20) is applicable. The two mole fractions are $X(N_2) = \frac{2}{5}$ and $X(\text{Ar}) = \frac{3}{5}$. Thus

$$\Delta S = -8.314(2 \ln \tfrac{2}{5} + 3 \ln \tfrac{3}{5}) = 28.0 \text{ J K}^{-1}$$

D. Carnot cycle

Frequently, the **Carnot cycle** is used to introduce the workings of a heat engine, to develop the definition of entropy, and to show that the thermodynamic temperature scale is equivalent to the ideal gas temperature scale. We used a different approach in developing these concepts, but you should know the essential features of a Carnot cycle and how to make calculations when the working substance is an ideal gas. In addition, it serves as an illustration of the entropy change in a cyclic process.

Carnot proposed an ideal heat engine, having a weightless, frictionless piston and a working substance (usually taken to be a gas) that expands and contracts when heat is added or removed. The four steps in the cycle are reversible:

Step 1: Isothermal expansion at higher temperature, T_h; $V_1 \rightarrow V_2$.
Step 2: Adiabatic expansion; $T_h \rightarrow T_l$, $V_2 \rightarrow V_3$.
Step 3: Isothermal compression at lower temperature, T_l; $V_3 \rightarrow V_4$.
Step 4: Adiabatic compression; $T_l \rightarrow T_h$, $V_4 \rightarrow V_1$.

In step 1, heat q_2 is removed from a reservoir at T_h; in step 3, heat q_1 is discharged to a reservoir at T_l. In steps 2 and 4, of course, $q = 0$. Work is done in each step, and the total work is $w_{\text{net}} = w_1 + w_2 + w_3 + w_4$.

Calculations of heat and work require the methods and equations presented in Chapter 3. We will also consider the entropy changes in each step and for the complete cycle.

EXAMPLE 5-14: If the working substance of a Carnot engine is one mole of an ideal gas, write the expressions for calculating the work done in each step and the net work.

Solution: From Chapter 3, we need the equations for the work in reversible isothermal and adiabatic changes, or Eqs. (3-13) and (3-45). For an ideal gas undergoing the four steps in the Carnot cycle,

Step 1: Isothermal expansion $\qquad w_1 = -RT_h \ln \dfrac{V_2}{V_1}$

Step 2: Adiabatic expansion $w_2 = C_V(T_l - T_h)$

Step 3: Isothermal compression $w_3 = -RT_l \ln \frac{V_4}{V_3}$

Step 4: Adiabatic compression $w_4 = C_V(T_h - T_l)$

Note that w_4 is the negative of w_2, so they cancel. Inverting the ln term in w_1 to make it positive, we have

$$w_{\text{net}} = RT_h \ln \frac{V_1}{V_2} - RT_l \ln \frac{V_4}{V_3}$$

This expression can be simplified by using Eq. (3-52) to show that

$$\frac{V_4}{V_3} = \frac{V_1}{V_2}$$

Thus

$$w_{\text{net}} = R(T_h - T_l) \ln \frac{V_1}{V_2}$$

EXAMPLE 5-15: Let's continue Example 5-14 by writing the expressions for the heat involved and the entropy change in each step of the Carnot cycle and their totals.

Solution: In the isothermal steps, $q = -w$ because $\Delta E = 0$. The entropy change can be calculated from q, using Eq. (5-4), or directly, using Eq. (5-14).

	q	S
Step 1: Isothermal expansion	$RT_h \ln \frac{V_2}{V_1}$	$R \ln \frac{V_2}{V_1}$
Step 2: Adiabatic expansion	0	0
Step 3: Isothermal compression	$RT_l \ln \frac{V_4}{V_3}$	$R \ln \frac{V_4}{V_3}$
Step 4: Adiabatic compression	0	0

Substituting $V_4/V_3 = V_1/V_2$ (see Example 5-14) and inverting the ln term to change the sign, we get the sums

$$q_{\text{net}} = R(T_h - T_l) \ln \frac{V_2}{V_1}$$

$$\Delta S_{\text{net}} = R \ln \frac{V_2}{V_1} - R \ln \frac{V_2}{V_1} = 0$$

Note that q(cycle) $= -w$(cycle), as expressed by Eq. (3-16). Also note that ΔS(cycle) $= 0$, as expected for a state function.

5-5. The Third Law of Thermodynamics and Third-Law Entropies

A. Third law of thermodynamics

The **third law of thermodynamics** has been stated as a principle of impotence:

It is impossible to attain the absolute zero of temperature.

This statement is not satisfactory, however, because it isn't equivalent to and doesn't imply the statements based on entropy.

Measurements of ΔG as a function of temperature have led to statements such as

In an isothermal process involving condensed pure substances in equilibrium, the entropy change approaches zero as the absolute temperature approaches zero and equals zero when the temperature equals zero.

This statement of the third law gives us a basis for calculating the entropy change during a chemical process by providing a reference point from which we can determine a value for the entropy of a substance; this approach is similar to that of using standard enthalpies of formation for the enthalpy of a substance. First, we assign a value of zero to the entropy of a pure crystalline element at $T = 0$ K and 1 atm. It follows that the entropy of a pure compound also is zero at $T = 0$, which leads to a third statement of the third law:

> *At absolute zero, the entropy of all pure perfect crystalline substances may be taken to be zero.*

Factors that prevent the entropy from actually equaling zero will cancel out in calculations.

The second step is to determine an entropy for each substance at the standard-state temperature of 298.15 K, as described in Section 5-5B. Then we can calculate entropy changes from

$$\Delta S_r^\circ = \Sigma n S^\circ(\text{products}) - \Sigma n S^\circ(\text{reactants}) \tag{5-21}$$

B. Third-law entropies

The entropy of a substance at 298.15 K is obtained by calculating the changes in entropy that occur as the temperature of the substance is raised from absolute zero to 298.15 K, and is called a **third-law entropy**. Phase transitions will occur; thus we need to know transition temperatures and enthalpies of transition as well as heat capacities of each state. The entropy changes are given by Eqs. (5-12) and (5-13).

One minor complication is that heat capacity is difficult to measure at very low temperatures and its dependence on temperature is different from that at high temperatures. Fortunately, an expression obtained by Debye may be used to calculate a value of C_V for use below the lowest measured value. For monatomic substances,

$$C_V \text{ (or } C_P) = 464.5\left(\frac{T^3}{\Theta^3}\right) \tag{5-22}$$

Here, Θ represents the **Debye characteristic temperature**, which must be evaluated from one measurement of C. The calculated Θ will vary slightly with T, so it's desirable to use the value of C_V at the lowest available T. The range of applicability of the expression varies widely for different substances. One rule of thumb is that Eq. (5-22) may be used if Θ is less than $10T$; if it isn't, the complete Debye expression should be used.

Since C is so small at low T, little error is introduced by the Debye approximation or the assumption that $C_V = C_P$. Substituting Eq. (5-22) into Eq. (5-12) and holding the volume constant gives

$$dS = \frac{464.5T^2}{\Theta^3}\,dT \tag{5-23}$$

Integrating between $T = 0$ and T yields

$$S_T - S_0 = \frac{464.5}{\Theta^3}\int_0^T T^2\,dT \tag{5-24}$$

By the third law, $S_0 = 0$, so

$$S_T = \left(\frac{464.5}{\Theta^3}\right)\left(\frac{T^3}{3}\right) \tag{5-25}$$

EXAMPLE 5-16: Calculate the entropy at 15 K of a substance if its C_P at 15 K has been measured to be 3.0 J K^{-1} mol^{-1}.

Solution: First, we evaluate Θ^3 from Eq. (5-22).

$$\Theta^3 = \frac{464.5T^3}{C_P} = \left(\frac{464.5}{3.0}\right)(15)^3 = 5.2 \times 10^5$$

From the third law, we know that $S_0 = 0$ if the substance is pure and perfectly crystalline.

Assuming that it is and substituting Θ^3 in Eq. (5-24), we get

$$S_{15} - 0 = \frac{464.5}{5.2 \times 10^5} \int_0^{15} T^2\, dT$$

$$S_{15} = (8.9 \times 10^{-4})\left(\frac{T^3}{3} - 0\right)$$

$$= (8.9 \times 10^{-4})\left(\frac{3375}{3}\right) \qquad \text{[Eq. (5-25)]}$$

$$= 1.00\ \text{J K}^{-1}\,\text{mol}^{-1}$$

The remaining steps in calculating S° at 298 K, for the substance in Example 5-16, if the substance is a liquid at that temperature and if it has only one solid form, are

(b) $\Delta S(15\ \text{K} \rightarrow T_f) = \int_{15}^{T_f} \frac{C_P(\text{sol})}{T}\, dT$

(c) $\Delta S(\text{sol} \rightarrow \text{liq}) = \frac{\Delta H_f}{T_f}$

(d) $\Delta S(T_f \rightarrow 298\ \text{K}) = \int_{T_f}^{298} \frac{C_P(\text{liq})}{T}\, dT$

Finally

$$S^\circ_{298} = S_{15} + \Delta S_b + \Delta S_c + \Delta S_d$$

The integrals for (b) and (d) are often integrated graphically from a plot of experimental values of C_P at various T's.

C. Tables of third-law entropies

Similar calculations (but usually involving several more steps) have been made for many substances. A table of third-law entropies is contained in every textbook. A short list of these values is included in Table A-1.

5-6. Entropy Changes during Chemical Processes

A. Entropy of reaction

From Table A-1, we obtain data for calculating the change in entropy for chemical reactions. We use Eq. (5-21).

EXAMPLE 5-17: Determine the change in entropy at 298 K for the reaction

$$H_2(g) + Cl_2(g) \longrightarrow 2HCl(g)$$

Solution: Applying Eq. (5-21), we get

$$\Delta S^\circ_r = \Sigma n S^\circ(\text{products}) - \Sigma n S^\circ(\text{reactants})$$

$$\Delta S^\circ_{298} = 2S^\circ(\text{HCl}) - S^\circ(\text{H}_2) - S^\circ(\text{Cl}_2) = 2(186.9) - 130.7 - 223.1 = 20.0\ \text{J K}^{-1}$$

B. Temperature dependence of entropy of reaction

As with enthalpies of reaction, ΔS at any temperature can be calculated from a known ΔS using heat capacities. The equation is analogous to that for ΔH [Eq. (4-17)].

$$\Delta S_{T_2} = \Delta S_{T_1} + \int_{T_1}^{T_2} \frac{\Delta C_P}{T}\, dT \qquad \textbf{(5-26)}$$

And, as before,

$$\Delta C_P = \Delta a + \Delta b T + \Delta c T^2 \qquad \text{[Eq. (4-9)]}$$

Usually T_1 is 298 K, because the known ΔS was calculated from S° values at 298 K.

EXAMPLE 5-18: Calculate ΔS° at 1000 K for the reaction in Example 5-17.

Solution: We would start by setting up a table of the C_P coefficients as before, but, since this reaction is the one used in Example 4-16, the three coefficients of ΔC_P are the same:

$$\Delta a = -4.43 \qquad \Delta b = -5.69 \times 10^{-3} \qquad \Delta c = 51.19 \times 10^{-7}$$

Substituting Eq. (4-9) in Eq. (5-26) and integrating, we obtain

$$\Delta S^\circ_{T_2} = \Delta S^\circ_{T_1} + \Delta a \ln \frac{T_2}{T_1} + \Delta b(T_2 - T_1) + \frac{\Delta c}{2}(T_2^2 - T_1^2)$$

Substituting ΔS°_{298} from Example 5-17, the given temperatures, and the coefficients of ΔC_P, we get

$$\Delta S^\circ_{1000} = 20.0 - \left(4.43 \ln \frac{1000}{298}\right) - (5.69 \times 10^{-3})(1000 - 298) + \left(\frac{51.19 \times 10^{-7}}{2}\right)(1000^2 - 298^2)$$

$$= 20.0 - 5.36 - 3.99 + 2.33 = 13.0 \text{ J K}^{-1}$$

SUMMARY

Table S-5 summarizes the major equations of this chapter and shows connections between them. You should determine the conditions that are necessary for each equation to be valid or the conditions that are imposed to make the connections. Other important things you should know are

1. A process is spontaneous if it tends to occur by itself regardless of rate.
2. Spontaneous processes are irreversible and lead to degradation of energy.
3. When heat is converted into work, complete conversion is impossible.
4. The efficiency of a reversible heat engine is determined by the two operating temperatures.
5. Heat can be transferred from a certain temperature to a higher temperature if work is done on the system.

TABLE S-5: Summary of Important Equations

$$\text{Eff} = \frac{q_{T_2} - q_{T_1}}{q_{T_2}} \qquad (5\text{-}1)$$

→ $$\text{Eff} = \frac{T_2 - T_1}{T_2} \qquad (5\text{-}2)$$

$$dS = \frac{đq_{\text{rev}}}{T} \qquad (5\text{-}3)$$

→ → $$\Delta S_{\text{trans}} = \frac{\Delta H_{\text{trans}}}{T_{\text{trans}}} \qquad (5\text{-}16)$$

→ $$\Delta S = \frac{q_{\text{rev}}}{T} \qquad (5\text{-}4)$$

→ $$dE = T\,dS - P\,dV \qquad (5\text{-}10)$$

→ → $$dS = \frac{C_V}{T}\,dT + \left(\frac{\partial P}{\partial T}\right)_V dV \qquad (5\text{-}12)$$

→ → $$dS = \frac{C_P}{T}\,dT - \left(\frac{\partial V}{\partial T}\right)_P dP \qquad (5\text{-}13)$$

$$\Delta S^\circ_r = \Sigma n S^\circ(\text{products}) - \Sigma n S^\circ(\text{reactants}) \qquad (5\text{-}21)$$

$$\Delta S_{T_2} = \Delta S_{T_1} + \int_{T_1}^{T_2} \frac{\Delta C_P}{T}\,dT \qquad (5\text{-}26)$$

6. Entropy is a state function, defined by $dS = \frac{đq_{\text{rev}}}{T}$.
7. The entropy of the universe (or of an isolated system) always increases when a spontaneous process occurs, but remains unchanged when a reversible process occurs.
8. The entropy change for an irreversible process can be determined only if the entropy change can be calculated for an alternate reversible path between the same initial and final states.
9. The four steps of the Carnot cycle are isothermal expansion, adiabatic expansion, isothermal compression, and adiabatic compression.
10. At very low temperatures, the heat capacity may be represented by the Debye relationship: $C = \frac{464.5T^3}{\Theta^3}$.
11. The entropy change for a reaction may be calculated using third-law entropies.

SOLVED PROBLEMS

Efficiency of Heat Engines

PROBLEM 5-1 What is the efficiency of a cyclic heat engine operating with its higher heat reservoir at 600 K and its lower reservoir at 200 K?

Solution: This is a straightforward application of Eq. (5-2):

$$\text{Eff} = \frac{T_2 - T_1}{T_2} = \frac{600 - 200}{600} = 0.667$$

PROBLEM 5-2 What condition would be necessary for a cyclic heat engine to be 100% efficient?

Solution: Referring to Eq. (5-2),

$$\text{Eff} = \frac{T_2 - T_1}{T_2}$$

we see that Eff = 1 only if $T_1 = 0$. In other words, the lower reservoir would have to operate at absolute zero. What does the third law tell you about this? That it is impossible to achieve this condition. Thus an engine of 100% efficiency is impossible.

PROBLEM 5-3 An engine operates with its low temperature reservoir at 15.0 °C, but the temperature in the high temperature reservoir can be varied. (**a**) What is the maximum work per 1000 J of heat used that can be obtained if the high temperature is 100 °C? (**b**) What is it at 200 °C?

Solution: This problem involves the efficiency expression, Eq. (5-2), and the definition of efficiency, Eq. (5-1). Combining them gives

$$\text{Eff} = \frac{\text{Work done}}{\text{Energy supplied}} = \frac{T_2 - T_1}{T_2} \qquad \text{or} \qquad w = \left(\frac{T_2 - T_1}{T_2}\right)q$$

(**a**) $w = \left(\frac{373.2 - 288.2}{373.2}\right)(1000\text{ J}) = 227.8\text{ J}$ (**b**) $w = \left(\frac{473.2 - 288.2}{473.2}\right)(1000\text{ J}) = 391.0\text{ J}$

Entropy Change: Physical Processes

PROBLEM 5-4 For an ideal gas having $C_P = \frac{7}{2}R$, calculate the entropy changes of 5 mol of the gas when it is heated from room temperature (298.2 K) to 500 K (**a**) at constant volume and (**b**) at constant pressure.

Solution: The general equations for entropy changes are Eqs. (5-12) and (5-13). Since (a) involves volume, you choose Eq. (5-12) because it contains C_V and dV. For (b), you use Eq. (5-13).

(a) $dS = \frac{C_V}{T}\,dT + \left(\frac{\partial P}{\partial T}\right)_V dV$ [Eq. (5-12)]

Since V is constant, $dV = 0$. Changing C_V to $nC_{V,m}$, you have

$$dS = nC_{V,m}\frac{dT}{T}$$

Assuming that C_V is constant and integrating with limits, you get

$$\Delta S = nC_{V,m}\ln\frac{T_2}{T_1} \qquad \text{[Eq. (5-14) at constant } V\text{]}$$

Recall that $C_{P,m} - C_{V,m} = R$, or $C_{V,m} = C_{P,m} - R$. Thus, in this case, $C_{V,m} = \frac{7}{2}R - R = \frac{5}{2}R$, and

$$\Delta S = 5\left(\frac{5}{2}R\right)\ln\frac{500}{298.2} = 53.7\ \text{J K}^{-1}$$

(b) $dS = \frac{C_P}{T}\,dT - \left(\frac{\partial V}{\partial T}\right)_P dV$ [Eq. (5-13)]

In a similar manner, imposing constant pressure on Eq. (5-13), assuming C_P constant, and integrating with limits gives

$$\Delta S = nC_{P,m}\ln\frac{T_2}{T_1} \qquad \text{[Eq. (5-15) at constant } P\text{]}$$

$$= 5\left(\frac{7}{2}R\right)\ln\frac{500}{298.2} = 75.2\ \text{J K}^{-1}$$

PROBLEM 5-5 Suppose that Problem 5-4 had asked merely for the difference between the ΔS's for the two processes. Derive an expression for calculating the difference directly, rather than by subtracting the two answers in Problem 5-4.

Solution: The only difference between the two calculations is the two heat capacities. They differ by R; thus the two answers must differ by R times the identical factors, or

$$\text{Diff} = R\left(5\ln\frac{T_2}{T_1}\right) = 5(8.314)\ln\frac{500}{298.2} = 21.5\ \text{J K}^{-1}$$

PROBLEM 5-6 Calculate the entropy change undergone by 1 mol $H_2(g)$ when it is heated from 300 K to 1200 K at constant volume. The heat capacity is not constant over this range of temperature.

Solution: You have to integrate Eq. (5-11), substituting an analytical expression for C_V. Since the volume is constant, the equation reduces to

$$\int dS = \int\frac{C_V\,dT}{T} + 0$$

From Eq. (3-58),

$$C_P = a + bT + cT^2$$

To change to C_V, assume ideal behavior so that

$$C_{V,m} = C_{P,m} - R = a - R + bT + cT^2$$

Then

$$\Delta S = \int(a-R)\frac{dT}{T} + \int bT\frac{dT}{T} + \int cT^2\frac{dT}{T} = (a-R)\ln\frac{T_2}{T_1} + b(T_2 - T_1) + \frac{c}{2}(T_2^2 - T_1^2)$$

From Table A-2,

$$a = 29.07, \qquad b = -0.83\times10^{-3}, \qquad c = 20.12\times10^{-7}$$

Thus

$$\Delta S = (29.07 - 8.314) \ln \frac{1200}{300} - (0.83 \times 10^{-3})(1200 - 300) + \left(\frac{20.12 \times 10^{-7}}{2}\right)(1200^2 - 300^2)$$

$$= 28.78 - 0.75 + 1.36 = 29.39 \text{ J K}^{-1}$$

PROBLEM 5-7 Calculate the entropy change of 10 g $H_2O(l)$ which changes to vapor at the *normal* boiling point.

Solution: The entropy of transition is given by Eq. (5-16). You need the heat of vaporization of water at 100 °C, which is 40.7 kJ mol^{-1}. Then

$$\Delta S = \frac{40.7 \times 10^3 \text{ J mol}^{-1}}{373.2 \text{ K}} = 109 \text{ J K}^{-1}\text{mol}^{-1}$$

For 10 g,

$$\Delta S = \left(\frac{10 \text{ g}}{18 \text{ g mol}^{-1}}\right)(109 \text{ J K mol}^{-1}) = 60.1 \text{ J K}^{-1}$$

PROBLEM 5-8 What is the entropy change in one mole of liquid glycol when it freezes at -25.0 °C? At the normal melting point (-15.0 °C), the heat of fusion of glycol is 2688 cal mol^{-1}, and the specific heats of liquid and solid glycol are 33.8 and 20.1 cal K^{-1}mol^{-1}, respectively.

Solution: Since this is not a reversible process, you must find a reversible path for attaining the same overall change. This can be done in three steps, all of which are reversible:

(a) Heat the liquid from $-25°$ to $-15°$.
(b) Freeze the liquid at $-15°$.
(c) Cool the solid from $-15°$ to $-25°$.

The corresponding expressions, from Eqs. (5-16) and (5-18), are

(a) $\Delta S_a = C_P(\text{liq}) \ln \dfrac{T_2}{T_1}$ **(b)** $\Delta S_b = \dfrac{\Delta H(\text{crys})}{T_2}$ **(c)** $\Delta S_c = C_P(\text{sol}) \ln \dfrac{T_2}{T_1}$

Then

$$\Delta S = \Delta S_a + \Delta S_b + \Delta S_c = 33.8 \ln \frac{258.2}{248.2} - \frac{2688}{258.2} + 20.1 \ln \frac{248.2}{258.2}$$

Note two points here: (1) the heat of crystallization is the negative of the heat of fusion (energy is lost); and (2) in a problem of this type, one logarithmic term necessarily is the reciprocal of the other. By inverting one (with a sign change), you can simplify the calculation to

$$\Delta S = 33.8 \ln \frac{258.2}{248.2} - \frac{2688}{258.2} - 20.1 \ln \frac{258.2}{248.2} = 13.7 \ln 1.040 - 10.411 = 0.541 - 10.411 = -9.870 \text{ cal K}^{-1}$$

If the negative result seems to violate the second law, it is because this ΔS is for the glycol only; go on to Problem 5-9.

PROBLEM 5-9 Continue Problem 5-8 by considering the glycol to be connected to a heat reservoir and this whole system to be isolated. Calculate ΔS for the reservoir and for the isolated system.

Solution: You need to find the total heat taken by the reservoir from the glycol; then you can assume that it would absorbed into the reservoir by some reversible process and calculate ΔS_{res}, using Eq. (5-4). You don't have the heat of crystallization at $-25°$, but you can calculate it using Hess's law and the three steps in Problem 5-8.

$$\Delta H_f(-25°) = \Delta H_f(-15°) + q_a + q_c$$
$$q_a = C_P(\text{liq})\,\Delta T = 33.8(10) = 338 \text{ cal}$$
$$q_c = C_P(\text{sol})\,\Delta T = 20.1 - (10) = -201 \text{ cal}$$
$$\Delta H_f(-25°) = -2688 + 338 - 201 = -2551 \text{ cal}$$

The reservoir gains +2551 cal; thus

$$\Delta S_{res} = \frac{2551}{248.2} = +10.28 \text{ cal K}^{-1}$$

$$\Delta S_{sys} = \Delta S_{glycol} + \Delta S_{res} = -9.87 + 10.28 = 0.41 \text{ cal K}^{-1}$$

The value is positive, as required by the second law.

Entropy Change: Chemical Processes

PROBLEM 5-10 Calculate the entropy changes for the system and the surroundings when one mole of liquid water is formed by direct reaction of hydrogen and oxygen at 298 K and 1 atm (ΔH_f° for $H_2O(l) = -286$ kJ mol^{-1}). (Don't use any tables.) With the proper setup, electrochemical reactions can be made essentially reversible. In one such apparatus, the heat equivalent of the electric current required to electrolyze water at 298 K was 48.5 kJ mol^{-1}.

Solution: The need to supply the information about electrolysis gives away the crucial point: The formation of water from its elements at constant pressure is not a reversible process. In order to calculate ΔS for the system, you need a reversible path. The reaction that you use is $H_2(g) + \frac{1}{2}O_2(g) \rightarrow H_2O(l)$, which is the opposite of the electrolytic reaction. Thus you must change the sign to get $\Delta H_{rev} = -48.5$ kJ mol^{-1} for the formation reaction. The temperature is constant, so for 1 mol,

$$\Delta S_{rev} = \frac{q}{T} = \frac{-48.5 \text{ kJ}}{298 \text{ K}} = -163 \text{ J K}^{-1}$$

The actual enthalpy change was -286 kJ (ΔH_f°) for 1 mol, thus the heat absorbed by the surroundings was +286 kJ. Assuming a reversible process in the surroundings, you have

$$\Delta S_{surr} = \frac{286}{298} = +960 \text{ J K}^{-1}$$

Again the net ΔS is positive, as required by the second law.

PROBLEM 5-11 Assuming ideal behavior and no chemical reaction, calculate the entropy change when 5 mol O_2 is mixed with 10 mol H.

Solution: This is a straightforward mixing problem, calling for Eq. (5-20). Writing the equation for two components, you obtain

$$\Delta S = -R(n_{O_2} \ln X_{O_2} + n_{H_2} \ln X_{H_2}) = (-8.314 \text{ J K}^{-1}\text{mol}^{-1})\left(5 \text{ mol} \ln \frac{5}{15} + 10 \text{ mol} \ln \frac{10}{15}\right)$$
$$= -8.314(-5.493 - 4.055) \text{ J K}^{-1} = 79.38 \text{ J K}^{-1}$$

PROBLEM 5-12 A certain Carnot cycle engine uses one mole of an ideal gas ($C_V = \frac{5}{2}R$) and operates between 800 K and 300 K. If 200 kJ is taken from the reservoir in step 1, calculate **(a)** the entropy change in step 1; **(b)** the heat discharged in step 3; **(c)** the net work done in the cycle; and **(d)** the net entropy change in the cycle.

Solution:

(a) The entropy change in a reversible isothermal process is given by Eq. (5-4), or

$$\Delta S_1 = \frac{q}{T} = \frac{200}{800} = 0.250 \text{ kJ K}^{-1}$$

(b) One way to find q_3 is to recognize that $\Delta S_3 = -\Delta S_1$ [see part (d)] and to use Eq. (5-4) again, but rearranged to yield

$$q_3 = T_1 \Delta S_3 = (300 \text{ K})(-0.250 \text{ kJ K}^{-1}) = -75.0 \text{ kJ}$$

(c) The net work in a cyclic process is the negative of the net heat, so

$$w_{net} = -q_{net} = -(200 - 75.0) \text{ kJ} = -125 \text{ kJ}$$

(d) Since S is a state function, the net change for a cycle is zero. Actually you recognized this in (b), along with the fact that the sum of ΔS for the two adiabatic steps equals zero.

PROBLEM 5-13 A certain monatomic substance has only one solid form, which melts at 190.2 K and then remains liquid to above 298.2 K. The heat of fusion is 5891 J mol^{-1}, and the average heat capacities of the solid and the liquid may be taken to be 35.6 and 53.1 J $K^{-1}mol^{-1}$, respectively. At 18 K, the heat capacity was found to be 2.0 J $K^{-1}mol^{-1}$. Calculate the approximate third-law entropy of this compound.

Solution: You have to calculate ΔS for each step involved in raising the temperature of the compound from 0 K to 298.2 K and then add them. Starting at the bottom, evaluate ΔS from 0 to 18 K using the Debye relationship, Eq. (5-22), and then Eq. (5-23). Rearranging Eq. (5-22), you get

$$\Theta^3 = \frac{464.5}{C_P} T^3 = \frac{464.5}{2.0}(18)^3$$

You can reduce the number of calculations by noting that the constant term (464.5) will cancel out when you substitute this expression for Θ^3 in Eq. (5-23). Thus you should not calculate the value of Θ unless you have some other need for it. Another simplification is possible because 18 K appears twice. Don't calculate yet, but integrate Eq. (5-23) with limits of 18 and 0 to obtain

$$\int_0^{18} dS = \int_0^{18} \frac{464.5}{\Theta^3} T^2\, dT$$

$$S_{18} - S_0 = \left(\frac{464.5}{\Theta^3}\right)\left(\frac{T^3}{3}\right)\Bigg|_0^{18}$$

Now, substitute $S_0 = 0$ and the value of Θ^3 to get the value of S_{18} and then calculate the other required enthalpies.

(a) $$S_{18} - 0 = \frac{2.0}{(18)^3}\left(\frac{T^3}{3} - 0\right) = \left[\frac{2.0}{(18)^3}\right]\left[\frac{(18)^3}{3}\right]$$

$$S_{18} = 0.67$$

(b) $$S_{190.2} - S_{18} = \int_{18}^{190.2} \frac{35.6}{T}\, dT \qquad \text{[Eq. (5-13), constant } P\text{]}$$

$$\Delta S(18\text{ K to }190.2\text{ K}) = 35.6 \ln\frac{190.2}{18} = 83.9$$

(c) $$\Delta S(\text{sol} \rightarrow \text{liq}) = \frac{\Delta H_f}{T_f} = \frac{5891}{190.2} = 30.97 \qquad \text{[Eq. (5-16)]}$$

(d) $$\Delta S(190.2\text{ K to }298.2\text{ K}) = \int_{190.2}^{298.2} \frac{53.1}{T}\, dT = 53.1 \ln\frac{298.2}{190.2} = 23.9 \qquad \text{[Eq. (5-13), constant } P\text{]}$$

And, finally,

$$S° \text{ (at 298.2 K)} = \Delta S_a + \Delta S_b + \Delta S_c + \Delta S_d = 0.67 + 83.9 + 30.97 + 23.9 = 139.4 \text{ J K}^{-1}\text{mol}^{-1}$$

PROBLEM 5-14 What is the change in entropy when one mole of calcite ($CaCO_3$) decomposes at 298.2 K to $CaO(s)$ and $CO_2(g)$?

Solution: Look up the $S°$ values in Table A-1 and substitute them into Eq. (5-21):

$$\Delta S_r = \Sigma nS° \text{ (products)} - \Sigma nS° \text{ (reactants)} = S°(\text{CaO}) + S°(\text{CO}_2) - S°(\text{CaCO}_3)$$
$$= 39.8 + 213.7 - 92.9 = 160.6 \text{ J K}^{-1}$$

Note that this relatively large value results when one product is a gas.

PROBLEM 5-15 Use Table A-1 and calculate the entropy change for the formation of liquid water from its elements at 298 K and 1 atm. Compare your result to the value in Problem 5-10.

Solution: The equation of the reaction is $H_2(g) + \frac{1}{2}O_2(g) \rightarrow H_2O(l)$. You use Eq. (5-21) again to get

$$\Delta S_r^\circ = S°(\text{H}_2\text{O}) - S°(\text{H}_2) - \tfrac{1}{2}S°(\text{O}_2) = 69.9 - 130.7 - \tfrac{1}{2}(205.1) = -163.4 \text{ J K}^{-1}$$

Agreement of this value with that obtained for the electrolysis of water is good.

PROBLEM 5-16 What would ΔS be at 1000 K for the reaction in Problem 5-15?

Solution: By now you should see that a calculation involving ΔC_P is required to obtain a value at a new temperature, and you might write down Eq. (5-26) almost without thinking. However, there is a complication in this problem: In Problem 5-15 the water is liquid, and, of course, will change to gas as the temperature rises. So, in addition to substituting in Eq. (5-26), you must account for the phase change. The simplest approach is to convert the equation to

$$H_2(g) + \tfrac{1}{2}O_2(g) \longrightarrow H_2O(g)$$

and add to ΔS°_{298} the ΔS for $H_2O(l) \rightarrow H_2O(g)$, or

$$\Delta S(\text{liq} \rightarrow \text{gas}) = S^\circ(\text{gas}) - S^\circ(\text{liq})$$

From Table A-1,

$$\Delta S = 188.8 - 69.9 = 118.9 \text{ J K}^{-1}$$

Then

$$\Delta S^\circ_r(\text{at } 298 \text{ K}) = -163.4 + 118.9 = -44.5 \text{ J K}^{-1}$$

Now you can change to the new temperature with Eq. (5-26):

$$\Delta S^\circ_{1000} = \Delta S^\circ_{298} + \int_{298}^{1000} \left(\frac{\Delta a}{T} + \Delta b + \Delta c T^2 \right) dT$$

For ΔC_P:

Component	a	$b \times 10^3$	$c \times 10^7$
$H_2O(g)$	30.36	9.61	11.84
$H_2(g)$	29.07	−0.83	20.12
$O_2[\times \frac{1}{2}]$	12.86	6.49	−19.31
Sum of reactants	41.93	5.66	0.81
Product − Reactants	−11.57	3.95	11.03

Substitute and solve:

$$\begin{aligned}
\Delta S^\circ_{1000} &= -44.5 + \Delta a \ln \frac{1000}{298} + \Delta b(1000 - 298) + \frac{\Delta c}{2}(1000^2 - 298^2) \\
&= -44.5 - 11.57 \ln \frac{1000}{298} + (3.95 \times 10^{-3})(1000 - 298) - \left(\frac{11.03 \times 10^{-7}}{2} \right)(1000^2 - 298^2) \\
&= -44.5 - 14.01 + 2.77 - 0.503 = -56.2 \text{ J K}^{-1}
\end{aligned}$$

Note that the simple approach, i.e., calculating ΔS° for the reaction at 298 K when the water is in the gaseous state, is possible because S° for $H_2O(g)$ is given in the table. If it hadn't been, you would have had to calculate ΔS(298 K to 373 K) using C_P for liquid water, ΔS(liq → gas at 373 K), and ΔS(373 K to 1000 K) using C_P for gaseous water, and then add the three values.

Supplementary Exercises

PROBLEM 5-17 If a natural process occurs, the available energy of a system ______________. Can a spontaneous process be a reversible process? Which cyclic heat engine would be more efficient: No. 1, operating between 300 K and 500 K, or No. 2, operating between 200 K and 500 K? All natural processes occur with an increase in the entropy of the substance undergoing the process (true or false?). For a process in which entropy is conserved, ______________ and ______________ conditions must hold.

PROBLEM 5-18 For a certain reversible cyclic heat engine, the maximum safe high temperature is 500 K. What must the low temperature be to achieve an efficiency of 45%?

PROBLEM 5-19 If you want to obtain 1.0 kJ of work from the engine in Problem 5-8, how much heat must be taken from the high temperature reservoir?

PROBLEM 5-20 Calculate the entropy changes (**a**) for the gas and (**b**) for the surroundings if one mole of an ideal gas expands isothermally and reversibly from 4.0 L to 9.0 L at 300 K.

PROBLEM 5-21 If the same volume change as in Problem 5-20 is achieved isothermally and irreversibly, with the gas absorbing 1.2 kJ of heat, what are the entropy changes for (**a**) the gas and (**b**) the surroundings?

PROBLEM 5-22 Calculate ΔS for 60.0 g of SO_2 gas if it is heated from 300 K to 700 K in a rigid container. Assume ideal behavior and $C_V = 4R$.

PROBLEM 5-23 One mole of an ideal gas ($C_V = \frac{5}{2}R$), initially at 300 K and 2 atm, was compressed until the pressure was 5 atm. The temperature rose to 550 K. Calculate ΔS.

PROBLEM 5-24 If the same volume change as in Problem 5-20 is achieved adiabatically and irreversibly, with the performance of 770 J of work, the final pressure is 2.4 atm. What are the entropy changes for (**a**) the gas and (**b**) the surroundings? Assume $C_{V,m} = 20.8\ \text{J K}^{-1}\text{mol}^{-1}$ for an ideal gas.

PROBLEM 5-25 Calculate the entropy change if 1 mol $HCl(g)$ is heated from 400 K to 800 K at constant volume (see Table 3-1).

PROBLEM 5-26 The normal boiling point of bromine is 58.8 °C, and the heat of vaporization is 44.8 cal g^{-1}. Calculate the molar entropy of vaporization.

PROBLEM 5-27 At the normal melting point of 47.0 °C, the heat of fusion of lauric acid is 43.7 cal g^{-1}. When 100.0 g of the solid compound is brought to a constant temperature of 60 °C, the solid melts. Calculate the entropy change in the compound. The specific heats are: solid = 0.430 cal $g^{-1}deg^{-1}$ and liquid = 0.515 cal $g^{-1}deg^{-1}$.

PROBLEM 5-28 Calculate the entropy change when 4 mol Ne is mixed with 7 mol Ar, all at the same pressure.

PROBLEM 5-29 Calculate the third-law entropy at 20.0 K of a substance having a measured C_P at 12.0 K of 1.78 J $K^{-1}mol^{-1}$.

PROBLEM 5-30 Calculate ΔS if 1 mol H_2S is burned in oxygen to produce $SO_2(g)$ and $H_2O(g)$ at 298.2 K.

PROBLEM 5-31 Calculate ΔS at 800 K for the reaction in Problem 5-30.

Answers to Supplementary Exercises

5-17 decreases; no; No. 2; false; adiabatic, reversible **5-18** 275 K **5-19** 2.22 kJ
5-20 (**a**) 6.7 J K^{-1}; (**b**) −6.7 J K^{-1} **5-21** (**a**) 6.7 J K^{-1}; (**b**) −4.0 J K^{-1} **5-22** 26.4 J K^{-1}
5-23 10.0 J K^{-1} **5-24** (**a**) 4.0 J K^{-1}; (**b**) 0 **5-25** 14.9 J K^{-1} **5-26** 21.6 cal $K^{-1}mol^{-1}$
5-27 14.0 cal K^{-1} **5-28** 59.9 J K^{-1} **5-29** 2.75 J $K^{-1}mol^{-1}$ **5-30** −76.5 J K^{-1}
5-31 −83.0 J K^{-1}

MID-TERM EXAM (Chapters 1–5)

Time: 50 minutes Total Points: 100

A. Short Answers (2 points per answer = 18)

1. (**a**) What is the SI unit for energy? ________________
(**b**) What is its definition in basic units? ________________

2. What is the derivative of x^{-2}? ________________

3. Write the partial derivative expressions that have these names:
(**a**) internal pressure ________________
(**b**) Joule–Thomson coefficient ________________
(**c**) Heat capacity at constant volume ________________

4. In an experiment to measure ΔE for a chemical reaction, what variable in addition to temperature must be held constant? ________________

5. If one wishes to liquefy a gas by a series of Joule–Thomson expansions, what must be the sign of the Joule–Thomson coefficient? ________________

6. What happens to the entropy of an isolated system during an irreversible process?
____

B. Long Answers (Total points = 29)

1. Is $dL = y^3\,dx + x^3\,dy$ an exact differential? Show how you arrive at your answer. (5)

2. List the three properties of exact differentials. (6)

3. Describe briefly the essential difference between a reversible and an irreversible expansion of a gas contained in a cylinder by a piston. (5)

4. Outline, using brief statements, the steps you would take to calculate the heat of a reaction at 1000 K. Write out any equations you would use and name any data you would need from tables. (7)

5. List the steps of the Carnot cycle by giving the essential features of each. Beside each, write the expression for calculating the work associated with that step if one mole of an ideal gas is used. (6)

C. Problems (Total points = 53). To save time, do not do the arithmetic unless the problem is labeled CALCULATE; otherwise just *set up* the final calculation. Solve all equations for the unknown, and substitute numerical values for all symbols. Put the proper units on the final answer. For tables of data you might need, see the Appendix.

1. Use the van der Waals equation to calculate the pressure exerted by 1 mol of nitrogen gas contained in a 2-L vessel at 500 K if $a = 1.41 \times 10^5$ Pa L^2 mol^{-2} and $b = 3.91 \times 10^{-2}$ L mol^{-1}. (4)

2. Find (**a**) the collision frequency and (**b**) the mean free path of the chlorine gas molecules in a 5-L flask containing 1 mol of Cl_2 at 298 K. The average molecular speed at this temperature is 300 m s^{-1}, and the effective molecular diameter is 5.4×10^{-10} m. (8)

3. Find w, q, ΔE, and ΔH if one mole of an ideal gas, initially at 200 K and 10.0 atm, is allowed to expand at constant temperature against a constant pressure of 4.0 atm until the gas pressure is 4.0 atm. (6)

4. What is the final temperature if one mole of an ideal gas ($C_P = \frac{7}{2}R$) expands adiabatically and reversibly from STP to a pressure of 0.50 atm? (8)

5. CALCULATE. If the heat of combustion of gaseous cyclopropane (C_3H_6) is -2095.7 kJ mol^{-1} at 298 K, calculate the standard enthalpy of formation of C_3H_6. (14)

6. What is the entropy change of 3 mol of an ideal gas, initially at 4 atm and 40 °C, heated until the pressure is 12 atm and the temperature is 200 °C? $C_P = 9.88$ cal K^{-1} mol^{-1}. (6)

7. CALCULATE. Determine the entropy change at 25 °C for the reaction

$$C_3H_8\,(g) + 5O_2\,(g) \longrightarrow 3CO_2\,(g) + 4H_2O\,(l) \qquad (7)$$

Solutions to Mid-Term Exam

A. Short Answers

1. **(a)** joule **(b)** kg m^2 s^{-2} [Table 1-1]
2. Apply Eq. (1-12), $du^n = nu^{n-1}\,du$: $dx^{-2} = -2x^{-3}$.
3. **(a)** $\left(\frac{\partial E}{\partial V}\right)_T$ [Section 3-5B]

   (b) $\left(\frac{\partial T}{\partial P}\right)_H$ [Eq. (3-59)]

   (c) $\left(\frac{\partial E}{\partial T}\right)_V = C_V$ [Eq. (3-27)]
4. volume [Section 4-1B]
5. Positive. If μ_{JT} is positive, T and P change in the same direction, so reducing the pressure (expansion) reduces the temperature. [Section 3-7B]
6. increases [Section 5-2C]

B. Long Answers

1. The test for exactness is Euler's reciprocity theorem, Eq. (1-26). If dL is exact, $\frac{dM}{dy} = \frac{dN}{dx}$. Here $M = y^3$ and $N = x^3$:

$$\frac{dM}{dy} = \frac{d}{dy}\,y^3 = 3y^2$$
$$\frac{dN}{dx} = \frac{d}{dx}\,x^3 = 3x^2$$

These are not equal; therefore this dL is not exact.

2. **(a)** The line integral is independent of the path. **(b)** The cyclic integral is zero. **(c)** There exists a function that can be differentiated to give the differential. [Section 1-4E]
3. In a reversible expansion, the pressure must be changed by infinitesimal increments so that the internal pressure (the gas pressure) always is equal to the opposing pressure of the piston. In an irreversible expansion, the pressure is released, then held constant at some lower value. The pressure of the gas cannot be known until the system has reached a steady state at the new pressure. [Section 3-3E]
4. The best choice for a single calculation is the second of the three approaches described in Section 4-4A.
   Step 1: Calculate ΔH° at 298 K by using ΔH_f° values from tables and the equation $\Delta H_r^\circ = \Sigma n\,\Delta H_f^\circ$ (products) $- \Sigma n\,\Delta H_f^\circ$ (reactants), Eq. (4-5).
   Step 2: Calculate values of Δa, Δb, and Δc, getting the values of a, b, and c from tables: $\Delta a = \Sigma na$ (products) $- \Sigma na$ (reactants).
   Step 3: Calculate ΔH° at 1000 K by using Eq. (4-10):

$$\Delta H^\circ(1000) = \Delta H^\circ(298) + \Delta a(T_2 - T_1) + \frac{\Delta b}{2}(T_2^2 - T_1^2) + \frac{\Delta c}{3}(T_2^3 - T_1^3)$$

5. All steps are reversible. [Section 5-4D]

   (a) Isothermal expansion at T_h $\quad w = -RT_h \ln\frac{V_2}{V_1}$

(**b**) Adiabatic expansion $\quad w = C_V(T_l - T_h)$

(**c**) Isothermal compression at T_l $\quad w = -nRT_l \ln \dfrac{V_4}{V_3}$

(**d**) Adiabatic compression $\quad w = C_V(T_h - T_l)$

C. Problems

1. Solve the van der Waals equation, Eq. (2-13), for P. In this problem, $n = 1$ mol, so the equation reduces to

$$\left(P + \frac{a}{V^2}\right)(V - b) = RT \qquad P = \frac{RT}{V - b} - \frac{a}{V^2}$$

Substituting the values gives

$$P = \left[\frac{8.314(500)}{2 - 0.0391} - \frac{1.41 \times 10^5}{(2)^2}\right] \text{Pa}$$

2. (**a**) Since only one type of molecule is present, you should use Eq. (2-42) for the collision frequency:

$$Z_{AA} = \frac{\bar{u}\sigma M^2}{\sqrt{2}V^2} \qquad \text{where} \qquad \sigma = \pi d^2$$

$$= \left[\frac{(300 \text{ m s}^{-1})(\pi)(5.4 \times 10^{-10} \text{ m})^2(6.02 \times 10^{23})^2}{\sqrt{2}(5 \times 10^{-3} \text{ m}^3)^2}\right] \text{s}^{-1} \text{ m}^{-3}$$

(**b**) The expression for the mean free path is Eq. (2-45):

$$\text{mfp} = \frac{V}{\sqrt{2}\sigma M} = \left[\frac{5 \times 10^{-3} \text{ m}^3}{\sqrt{2}(\pi)(5.4 \times 10^{-10})^2(6.02 \times 10^{23})}\right] \text{m}$$

3. First notice the conditions: ideal, isothermal, and constant pressure. Since the gas is ideal and the process is isothermal, ΔE and $\Delta H = 0$. Since $\Delta E = 0$, $q = -w$. From Eq. (3-11) you have

$$w = -\int_{V_1}^{V_2} P\, dV = -P_2(V_2 - V_1)$$

Since the volumes are not given, you should substitute from the ideal gas equation:

$$w = -P_2\left(\frac{RT_2}{P_2} - \frac{RT_1}{P_1}\right)$$

Since T is constant, $w = -RT\left(1 - \dfrac{P_2}{P_1}\right)$

$$w = (-8.314)(200)\left(1 - \frac{4.0}{10.0}\right) \text{J}$$

$$q = -w = (+8.314)(200)(0.60) \text{ J}$$

4. Of the several equations involving reversible adiabatic expansions, Eqs. (3-51) to (3-54), Eq. (3-54) is convenient for this problem.

$$TP^{(1-\gamma)/\gamma} = \text{constant} \qquad T_1 P_1^{(1-\gamma)/\gamma} = T_2 P_2^{(1-\gamma)/\gamma} \qquad \frac{T_2}{T_1} = \left(\frac{P_1}{P_2}\right)^{(1-\gamma)/\gamma}$$

Since

$$\gamma = \frac{C_{P,m}}{C_{V,m}} = \frac{\frac{7}{2}R}{\frac{5}{2}R} = \frac{7}{5} = 1.4$$

$$\frac{1-\gamma}{\gamma} = \frac{1 - 1.4}{1.4} = -0.286$$

So

$$T_2 = T_1\left(\frac{P_1}{P_2}\right)^{-0.286} = T_1\left(\frac{P_2}{P_1}\right)^{0.286}$$

$$T_2 = 298\left(\frac{0.5}{1}\right)^{0.286} \text{K}$$

5. Apply Hess's law to convert the equation for the combustion of cyclopropane into the equation for the formation of cyclohexane. This conversion requires that you (1) write the reverse of the combustion reaction, changing the sign of -2095.7, (2) write the equations for formation of CO_2 and H_2O and find the standard enthalpies of each reaction from Table A-1, (3) multiply both those equations and their

ΔH_f° values by 3, and (4) add all three equations and their ΔH values. [Section 4-2D]

$$\begin{array}{rcll} 3CO_2 + 3H_2O & \longrightarrow & C_3H_6 + \frac{9}{2}O_2 & +2095.7 \\ 3C + 3O_2 & \longrightarrow & 3CO_2 & -1180.5 \\ 3H_2 + \frac{3}{2}O_2 & \longrightarrow & 3H_2O & -857.4 \\ \hline 3C + 3H_2 & \longrightarrow & C_3H_6 & \Delta H_f^\circ = \quad +57.8 \text{ kJ mol}^{-1} \end{array}$$

6. To calculate ΔS when both temperature and pressure change, use Eq. (5-15):

$$\Delta S = nC_P \ln\left(\frac{T_2}{T_1}\right) - nR \ln\left(\frac{P_2}{P_1}\right)$$

$$= 3 \text{ mol}(9.88 \text{ cal K}^{-1}\text{mol}^{-1}) \ln\left(\frac{473}{313}\right) - 3 \text{ mol}(1.987 \text{ cal K}^{-1}\text{mol}^{-1}) \ln\left(\frac{12}{4}\right)$$

7. Look up the third-law entropies of each substance in Table A-1 and substitute them in Eq. (5-21):

$$\begin{aligned} \Delta S^\circ(\text{reaction}) &= \Sigma nS^\circ(\text{products}) - \Sigma nS^\circ(\text{reactants}) \\ &= 3S^\circ(CO_2) + 4S^\circ(H_2O, l) - S^\circ(C_3H_8) - 5S^\circ(O_2) \\ &= 3(213.7) + 4(69.9) - 270.0 - 5(205.1) = 374.8 \text{ J} \end{aligned}$$

6 SPONTANEITY, EQUILIBRIUM, AND FREE ENERGY

THIS CHAPTER IS ABOUT

- ☑ Enthalpy, Internal Energy, and Entropy: Unsatisfactory Criteria of Spontaneity and Equilibrium
- ☑ The Free Energy State Functions: Satisfactory Criteria for Spontaneity and Equilibrium
- ☑ Relationships between State Functions
- ☑ Free Energy Changes in Physical Processes
- ☑ Free Energy Changes in Chemical Processes
- ☑ Temperature Dependence of Free Energy Change of a Reaction
- ☑ Fugacity, Activity, and Fugacity Coefficient
- ☑ Chemical Potential

6-1. Enthalpy, Internal Energy, and Entropy: Unsatisfactory Criteria of Spontaneity and Equilibrium

The practicing chemist needs ways to predict (1) whether a reaction is spontaneous and (2) what the ratio of products to reactants will be when the reaction mixture reaches equilibrium. Thermodynamic measures that serve as predictors are called a **criteria of spontaneity and equilibrium**.

A. Enthalpy and internal energy are not satisfactory criteria of spontaneity.

In the past it was thought that a negative sign for ΔE or ΔH for a given process meant that the process was spontaneous. Now we know that a number of endothermic (positive ΔH) processes occur spontaneously. For example, many salts (especially ammonium compounds such as the bromides, chlorides, and nitrates) have negative heats of solution. Ice at 25 °C certainly melts spontaneously and takes heat from its surroundings.

If a process occurs under conditions of constant entropy and constant pressure, ΔH is the criterion of spontaneity and equilibrium. If $\Delta H_{S,P}$ is negative, i.e.,

$$\Delta H_{S,P} \leq 0 \qquad \textbf{(6-1)}$$

the process is spontaneous. Similarly, for conditions of constant entropy and constant volume,

$$\Delta E_{S,V} \leq 0 \qquad \textbf{(6-2)}$$

If $\Delta H_{S,P}$ is zero, the system is at equilibrium. Unfortunately, none of these conditions is found in typical chemical reactions.

B. Entropy is not a satisfactory criterion of spontaneity.

As we have already noted for an isolated system, entropy always increases in a spontaneous process and is zero in a reversible (equilibrium) process. An isolated system, however, involves constant energy and volume, for which the Clausius inequality, Eq. (5-9), can be written:

$$\Delta S_{E,V} \geq 0 \qquad \textbf{(6-3)}$$

This expression is inadequate as a criterion of spontaneity because evaluation of ΔS requires knowledge of changes in the surroundings as well as in the process itself (for which ΔS may be negative). A more useful criterion would depend only on the process or system and would apply under the usual conditions for chemical reactions: constant T and P or constant T and V.

6-2. The Free Energy State Functions: Satisfactory Criteria for Spontaneity and Equilibrium

A. Helmholtz free energy

The fundamental equation of thermodynamics, Eq. (5-10), can be made more general by combining it with the Clausius inequality, Eq. (5-9), and removing the restriction to PV work. We start with the first law in its differential form, or

$$dE = đq + đw \qquad \text{[Eq. (3-1a)]}$$

We solve the differential form of Eq. (5-9) for $đq$ and substitute to obtain

$$dS \geq đq/T \qquad \text{[See Eq. (5-3)]}$$

$$T\,dS \geq đq$$

$$dE \leq T\,dS + đw \tag{6-4}$$

If we restrict the work to PV work, this expression becomes

$$E \leq T\,dS - P\,dV \qquad (PV \text{ work only}) \tag{6-5}$$

Rearranging Eq. (6-4), we get

$$dE - T\,dS \leq đw \tag{6-6}$$

If we add the restriction of reversibility, $đw = đw_{max}$. Then we substitute in Eq. (6-5) to get

$$dE - T\,dS = dw_{max} \qquad \text{(reversible)} \tag{6-7}$$

The work differential is now exact because it is expressed by two exact differentials.

We can establish a new state function by subtracting $d(TS)$ from both sides of Eq. (6-4). As before, we choose this quantity because we know the result we want.

$$d(TS) = S\,dT + T\,dS$$

$$dE - d(TS) \leq T\,dS + đw - S\,dT - T\,dS$$

or

$$d(E - TS) \leq đw - S\,dT \tag{6-8}$$

Since E, T, and S are state functions, the term $(E - TS)$ also must be a state function. It has its own symbol, A, and name, **Helmholtz free energy**, and is defined as

HELMHOLTZ FREE ENERGY

$$A = E - TS \tag{6-9}$$

note: Current texts often refer to it as *Helmholtz energy*, or *Helmholtz function*; earlier texts used the terms *work function* and *work content*. The symbol A comes from *Arbeit*, the German word for work.

If we substitute A for $E - TS$ in Eq. (6-8), impose a constant temperature restriction (so that $-S\,dT = 0$), and reverse the sign (since we're interested in the work done *by* the system), we get

$$-dA \geq -đw + S\,dT \tag{6-10}$$

$$-dA \geq -đw \qquad (\text{constant } T) \tag{6-11}$$

For a reversible process, $đw = đw_{max}$, so

$$-dA = -dw_{max} \qquad (\text{reversible, constant } T) \tag{6-12}$$

Integration of Eqs. (6-11) and (6-12) gives

$$-\Delta A \geq -w \qquad (\text{constant } T) \tag{6-13}$$

$$-\Delta A = -w_{max} \qquad (\text{reversible, constant } T) \tag{6-14}$$

Thus the function A represents the amount of energy available to do work. In an isothermal reversible change, this energy is converted completely into work; in an irreversible change (between the same two states), the work will be less, although the change in A is the same for both paths. Because of these relationships with work, A is also called the *maximum-work function.*

B. Gibbs free energy

Until now, we've considered only work of expansion, often called *pressure-volume* or PV work. We now extend our consideration to include other types of work, such as electrical and magnetic work. We designate these forms of work $w_{\text{non}PV}$. Part of any w_{PV} is usually expended to push back the surrounding environment as a system expands; it can't be harnessed to run an engine, so it is wasted for practical purposes. The work term must include all possible components, i.e.,

$$đw = đw_{PV} + dw_{\text{non}PV} \tag{6-15}$$

We are led to another, very important, state function by substituting Eq. (6-15) into Eq. (6-6) and adding $d(PV)$ to both sides:

$$dE - T\,dS \leq đw_{PV} + đw_{\text{non}PV}$$
$$d(PV) = P\,dV + V\,dP$$

We also substitute $-P\,dV$ for $đw_{PV}$ to obtain

$$dE - T\,dS + d(PV) \leq -P\,dV + đw_{\text{non}PV} + P\,dV + V\,dP$$

Now we impose constant temperature, so that $T\,dS = d(TS)$ and have

$$dE - d(TS) + d(PV) \leq đw_{\text{non}PV} + V\,dP$$

or

$$d(E - TS + PV) \leq đw_{\text{non}PV} + V\,dP \qquad \text{(constant } T\text{)} \tag{6-16}$$

Since E, T, S, P, and V are state functions, the term $(E - TS + PV)$ must be a state function. It has its own symbol, G (older books and tables use F), and name, **Gibbs free energy**, and is defined as

GIBBS FREE ENERGY
$$G = E - TS + PV \tag{6-17}$$

Current texts tend to use the names *Gibbs energy* and *Gibbs function.*

Putting G in Eq. (6-16), imposing constant pressure, and changing the signs, we have

$$-dG \geq -đw_{\text{non}PV} - V\,dP \qquad \text{(constant } T\text{)} \tag{6-18}$$
$$-dG \geq -dw_{\text{non}PV} \qquad \text{(constant } T \text{ and } P\text{)} \tag{6-19}$$

For a reversible process, $dw = dw_{\text{max}}$, so

$$-dG \geq -dw_{\text{max, non}PV} \qquad \text{(reversible, constant } T \text{ and } P\text{)} \tag{6-20}$$

Integration of Eqs. (6-19) and (6-20) gives

$$-\Delta G \geq -w_{\text{non}PV} \qquad \text{(constant } T \text{ and } P\text{)} \tag{6-21}$$
$$-\Delta G = -w_{\text{max, non}PV} \qquad \text{(reversible, constant } T \text{ and } P\text{)} \tag{6-22}$$

In summary, the function G represents the amount of energy available to do nonPV work at constant temperature and pressure. An important application is in electrochemical cells: Conditions can be controlled to carry out chemical reactions very close to reversibly. Under irreversible conditions, for the same ΔG, the amount of work done will be less than that done under reversible conditions. Recalling that $E + PV = H$, we can convert Eq. (6-17) to

$$G = H - TS \tag{6-23}$$

Similarly $E - TS = A$, and we have

$$G = A + PV \tag{6-24}$$

Note that the energies A and G are *free* in the sense that they are unrestricted or available to do work. In any real process, that part of a system's energy available to do work must decrease.

C. Macroscopic processes

We can integrate and manipulate the various definitions and differential expressions in Sections 6-2A and 6-2B to get corresponding expressions for actual changes, i.e.,

$$\Delta A = \Delta E - \Delta(TS) \qquad \textbf{(6-25)}$$
$$\Delta A = \Delta E - T\,\Delta S \quad \text{(constant } T\text{)} \qquad \textbf{(6-26)}$$
$$\Delta G = \Delta E - \Delta(TS) + \Delta(PV) \qquad \textbf{(6-27)}$$
$$\Delta G = \Delta H - \Delta(TS) \qquad \textbf{(6-28)}$$
$$\Delta G = \Delta H - T\,\Delta S \quad \text{(constant } T\text{)} \qquad \textbf{(6-29)}$$
$$\Delta G = \Delta A + \Delta(PV) \qquad \textbf{(6-30)}$$

EXAMPLE 6-1: Show how to obtain Eq. (6-30) from any of the Eqs. (6-25) through (6-29).

Solution: We can substitute Eq. (6-26) into Eq. (6-27):

$$\Delta G = \Delta E - \Delta(TS) + \Delta(PV)$$
$$\Delta E - \Delta(TS) = \Delta A$$
$$\Delta G = \Delta A + \Delta(PV)$$

D. Satisfactory criteria of spontaneity and equilibrium

The Helmholtz and Gibbs free energy functions provide satisfactory criteria of spontaneity and equilibrium. We've seen that free energies decrease for spontaneous processes. When a system is at equilibrium, the process is reversible, no work is done, and there is no change in free energy. We can summarize these statements with

$$\Delta A \le 0 \quad \text{(constant } T \text{ and } V\text{)} \qquad \textbf{(6-31)}$$
$$\Delta G \le 0 \quad \text{(constant } T \text{ and } P\text{)} \qquad \textbf{(6-32)}$$

6-3. Relationships between State Functions

A. Defining expressions

The functions E and S are defined by the first and second laws, but H, A, and G are defined using E and S. The four relationships are

$$H = E + PV \qquad \textbf{[Eq. (3-18)]}$$
$$A = E - TS \qquad \textbf{[Eq. (6-9)]}$$
$$G = H - TS \qquad \textbf{[Eq. (6-23)]}$$
$$G = A + PV \qquad \textbf{[Eq. (6-24)]}$$

B. Gibbs equations

Recall the fundamental equation, $dE = T\,dS - P\,dV$ [Eq. (5-10)]. A set of equations analogous to the fundamental equation is known as the Gibbs equations. Expressions for H, A, and G similar to Eq. (5-10) for E can also be obtained.

EXAMPLE 6-2: Show that $dH = T\,dS + V\,dP$.

Solution: First, we write the differential of H from its definition:

$$dH = dE + d(PV) = dE + P\,dV + V\,dP$$

Substituting Eq. (5-10) gives us

$$dH = T\,dS - P\,dV + P\,dV + V\,dP$$

so

$$dH = T\,dS + V\,dP \qquad \textbf{(6-33)}$$

Similarly, we can obtain comparable expressions for dA and dG. Again collecting all expressions, we have

GIBBS EQUATIONS

$$dE = T\,dS - P\,dV \qquad \textbf{[Eq. (5-10)]}$$
$$dH = T\,dS + V\,dP \qquad \textbf{[Eq. (6-33)]}$$
$$dA = -S\,dT - P\,dV \qquad \textbf{(6-34)}$$
$$dG = -S\,dT + V\,dP \qquad \textbf{(6-35)}$$

These *Gibbs equations* apply for reversible processes in which all work done is PV work. They have three important features, which will help you to remember them:

1. The equations for the two free energies include $S\,dT$, whereas those for the other energies include $T\,dS$.
2. The sign of a particular term is always the same.
3. The equations for the constant-pressure energies (H and G) include $V\,dP$, whereas those for the constant-volume energies (E and A) include $P\,dV$.

C. Partial derivatives

Partial derivatives can be read as a rate of change in one variable when a second variable changes, holding all other variables constant. For example, the speed of an automobile is 50 miles per hour. This could be expressed as the change in distance for a certain change in time, or

$$\left(\frac{\partial d}{\partial t}\right) = 50 \text{ mi h}^{-1}$$

At constant pressure, the change of volume when temperature changes may be expressed as a coefficient, $\left(\frac{\partial V}{\partial T}\right)_P$. When multiplied by $1/V$, this quantity is called the **coefficient of thermal expansion**, or the **thermal expansivity**, and is designated α (alpha). Thus

COEFFICIENT OF THERMAL EXPANSION

$$\alpha = \left(\frac{1}{V}\right)\left(\frac{\partial V}{\partial T}\right)_P \tag{6-36}$$

Another commonly encountered coefficient expresses the change of volume with pressure at constant temperature, i.e., $\left(\frac{\partial V}{\partial P}\right)_T$. Multiplying by $-1/V$ gives the quantity known as the **coefficient of compressibility**, which is represented by κ (kappa). Thus

COEFFICIENT OF COMPRESSIBILITY

$$\kappa = -\left(\frac{1}{V}\right)\left(\frac{\partial V}{\partial P}\right)_T \tag{6-37}$$

These two quantities, α and κ, are useful because they can be measured directly.

Partial derivatives arise when we write the total differential of a function of two or more variables (see Section 1-4D). We know that G is a function of T and P. Thus the total differential of G is

$$dG = \left(\frac{\partial G}{\partial T}\right)_P dT + \left(\frac{\partial G}{\partial P}\right)_T dP \tag{6-38}$$

Equation (6-35) also is an expression of dG, so it must be equivalent to Eq. (6-38), i.e., the coefficients of dT and dP must be the same. Thus

$$\left(\frac{\partial G}{\partial T}\right)_P = -S \tag{6-39}$$

and

$$\left(\frac{\partial G}{\partial P}\right)_T = V \tag{6-40}$$

EXAMPLE 6-3: The expression for A similar to Eq. (6-39) is $\left(\frac{\partial A}{\partial T}\right)_V = -S$. Show how to derive it.

Solution: Since the partial derivative calls for constant volume, we apply this condition to Eq. (6-34) and get $dA = -S\,dT$. Dividing by dT and showing that V is constant gives the partial derivative,

$$\left(\frac{\partial A}{\partial T}\right)_V = -S$$

Proceeding similarly for E, H, and A, we get comparable expressions. It's not worth listing them or memorizing them because they can be derived easily from the Gibbs equations. Qualitative conclusions may also be drawn from these relationships.

EXAMPLE 6-4: What can we say about the direction of change of G of a substance with change in temperature? with change in pressure?

Solution: Since the S of a substance always is positive, we can predict from Eq. (6-39) that G must decrease with an increase in T at constant P. For a change in P, we use Eq. (6-40). Since V is always positive, G must increase with an increase in P at constant T. Alternatively, we could reach the same conclusions from the Gibbs equation, [Eq. (6-35)], $dG = -S\,dT + V\,dP$.

EXAMPLE 6-5: Compare the temperature effects on G for the gaseous and the solid states of the same substance.

Solution: They will be different because the S of a gas is greater than that of the solid state of the same substance. The change in G of the gas would be greater than that of the solid for the same temperature change [see Eq. (6-39)].

D. Maxwell relations

In Section 1-4F, we noted that the reciprocity theorem is a test for the exactness of a differential. The theorem tells us that the coefficients in the expression of an exact differential, such as Eq. (6-35), $dG = -S\,dT + V\,dP$, are related in the following way:

$$\left(\frac{\partial(-S)}{\partial P}\right)_T = \left(\frac{\partial V}{\partial T}\right)_P \quad \text{or} \quad \left(\frac{\partial S}{\partial P}\right)_T = -\left(\frac{\partial V}{\partial T}\right)_P \qquad \textbf{(6-41)}$$

The second term involves P, V, and T and can be evaluated experimentally or by calculation from an equation of state. An alternative equation for the dependence of S on P can be obtained by substituting Eq. (6-36) to obtain

$$\left(\frac{\partial S}{\partial P}\right)_T = -\alpha V \qquad \textbf{(6-42)}$$

Application of the Euler theorem to each of the Gibbs equations yields four equations, including Eq. (6-41), which are called the **Maxwell relations**:

MAXWELL RELATIONS

$$\left(\frac{\partial T}{\partial V}\right)_S = -\left(\frac{\partial P}{\partial S}\right)_V \qquad \text{[from Eq. (5-10)]} \qquad \textbf{(6-43)}$$

$$\left(\frac{\partial T}{\partial P}\right)_S = \left(\frac{\partial V}{\partial S}\right)_P \qquad \text{[from Eq. (6-33)]} \qquad \textbf{(6-44)}$$

$$\left(\frac{\partial S}{\partial V}\right)_T = \left(\frac{\partial P}{\partial T}\right)_V \qquad \text{[from Eq. (6-34)]} \qquad \textbf{(6-45)}$$

$$\left(\frac{\partial S}{\partial P}\right)_T = -\left(\frac{\partial V}{\partial T}\right)_P \qquad \text{[from Eq. (6-35)]} \qquad \textbf{[Eq. (6-41)]}$$

These expressions may be used to obtain other useful expressions in P, V, and T.

EXAMPLE 6-6: Obtain an expression involving only P, V, and T for the isothermal change of enthalpy with pressure.

Solution: We start with the Gibbs equation that contains H and P:

$$dH = T\,dS + V\,dP \qquad \text{[Eq. (6-33)]}$$

Taking the derivative with respect to pressure and showing that T is constant, we have

$$\left(\frac{\partial H}{\partial P}\right)_T = T\left(\frac{\partial S}{\partial P}\right)_T + V$$

Substituting Eq. (6-41) gives

$$\left(\frac{\partial H}{\partial P}\right)_T = -T\left(\frac{\partial V}{\partial T}\right)_P + V \qquad \textbf{(6-46)}$$

We can simplify the equation by substituting from Eq. (6-36):

$$\left(\frac{\partial V}{\partial T}\right)_P = \alpha V \qquad \textbf{[Eq. (6-36)]}$$

$$\left(\frac{\partial H}{\partial P}\right)_T = -T\alpha V + V$$

$$\left(\frac{\partial H}{\partial P}\right)_T = V(1 - \alpha T) \qquad \textbf{(6-47)}$$

EXAMPLE 6-7: Calculate the change in enthalpy if the pressure on one mole of liquid water at 298 K is increased from 1 atm to 11 atm, assuming that V and α are independent of pressure. At room temperature, α for water is approximately $3.0 \times 10^{-4}\ \mathrm{K}^{-1}$.

Solution: We rearrange Eq. (6-47) to get

$$\int_{H_1}^{H_2} dH = V(1 - \alpha T)\int_{P_1}^{P_2} dP$$

Integrating, we obtain

$$\Delta H = V(1 - \alpha T)\,\Delta P \qquad \textbf{(6-48)}$$

The volume of one mole of water is about 18 mL, or 0.018 L; thus

$$\Delta H = 0.018[1 - (3.0 \times 10^{-4})(298)](11 - 1) = 0.018(1 - 0.089)(10) = 0.16\ \mathrm{L\ atm}$$

6-4. Free Energy Changes in Physical Processes

A. Dependence of G on P

The change in G for a change in pressure may be calculated easily. From Eq. (6-35) at constant T, or from Eq. (6-40), $dG = V\,dP$. In order to integrate, we must have an expression for V as a function of P, which we can get from an equation of state. For an ideal gas, $V = nRT/P$, so Eq. (6-40) becomes

$$\int_{G_1}^{G_2} dG = nRT\int_{P_1}^{P_2} \frac{dP}{P}$$

$$\Delta G = nRT \ln \frac{P_2}{P_1} \qquad \textbf{(6-49)}$$

EXAMPLE 6-8: Calculate the change in G if the pressure on two moles of an ideal gas is increased from 1 atm to 10 atm at 300 K.

Solution: We can substitute directly in Eq. (6-49):

$$\Delta G = 2(8.314\ \mathrm{J\ K^{-1}mol^{-1}})(300) \ln \frac{10}{1} = 11.5\ \mathrm{kJ}$$

B. Dependence of G on T

For a single substance, we can make qualitative predictions of the change of G with T from Eq. (6-39), as in Example 6-8. If we try to integrate this equation in order to make quantitative calculations, we have

$$\int dG = -\int S\,dT$$

but usually we won't know S as a function of T. Because S changes rapidly with T, it isn't valid to assume that S is constant. We can get around this difficulty by using a different expression for the dependence of G on T, which is called the **Gibbs–Helmholtz equation** and includes enthalpy rather than entropy:

GIBBS–HELMHOLTZ EQUATION (single substance)

$$\left(\frac{\partial(G/T)}{\partial T}\right)_P = -\frac{H}{T^2} \tag{6-50}$$

We can derive this expression by differentiating G/T and using the definition of G, Eq. (6-23). Perhaps a better choice is to start with Eq. (6-39).

EXAMPLE 6-9: Derive the Gibbs–Helmholtz equation.

Solution: We start with Eq. (6-39):

$$\left(\frac{\partial G}{\partial T}\right)_P = -S$$

Next, we rearrange Eq. (6-23) to get

$$-S = \frac{G-H}{T}$$

Then

$$\left(\frac{\partial G}{\partial T}\right)_P = \frac{G-H}{T} = \frac{G}{T} - \frac{H}{T}$$

$$\left(\frac{\partial G}{\partial T}\right)_P - \frac{G}{T} = -\frac{H}{T} \tag{a}$$

Now we have another case where knowing what we want helps, but, even if we didn't know, the form of the left-hand side of the equation suggests the derivative of G/T. We differentiate this fraction as a product, $G(1/T)$,using Eq. (1-12):

$$\frac{d(G/T)}{dT} = G\left(\frac{d(1/T)}{dT}\right) + \left(\frac{1}{T}\right)\left(\frac{dG}{dT}\right) = G\left(-\frac{1}{T^2}\right) + \left(\frac{1}{T}\right)\left(\frac{dG}{dT}\right) = \frac{1}{T}\left(-\frac{G}{T} + \frac{dG}{dT}\right)$$

Now we show explicitly that pressure is constant:

$$\left(\frac{\partial(G/T)}{\partial T}\right)_P = \frac{1}{T}\left[-\frac{G}{T} + \left(\frac{\partial G}{\partial T}\right)_P\right]$$

From eq. (a), the term in the brackets equals $-H/T$, so

$$\left(\frac{\partial(G/T)}{\partial T}\right)_P = \frac{1}{T}\left(-\frac{H}{T}\right) = -\frac{H}{T^2}$$

which is Eq. (6-50), the Gibbs–Helmholtz equation.

C. Dependence of A on T

To make quantitative calculations of changes in A with T for a single substance, we can't use the integrated form of the equation analogous to Eq. (6-39), i.e., $\left(\frac{\partial A}{\partial T}\right)_V = -S$ (obtained in Example 6-3). Again, S varies rapidly with T, but we could derive an expression analogous to Eq. (6.50).

D. Change in G at constant temperature and pressure

We have to consider both reversible and irreversible processes.

1. A reversible process always is at equilibrium; thus $\Delta G = 0$. The typical illustration of this is a phase transition for which the two phases are in equilibrium.

2. In an irreversible process, $\Delta G \neq 0$. This is illustrated by a phase transition for which the two phases are not in equilibrium.

EXAMPLE 6-10: At -5.0 °C, the vapor pressure of ice is 3.013 mm and that of liquid water is 3.163 mm. Calculate ΔG for the change of one mole of liquid water at $-5.0°$ to solid at $-5.0°$.

Solution: The system will not be in equilibrium, so ΔG is not zero. The situation is like the one presented in Section 5-4B for ΔS when a reversible path is required, and we have to find a series of equilibrium steps between the initial and final states. For ΔS, we raised the temperature to 0° but won't do that now because of the difficulty of calculating ΔG for a change in T (see Section 6-4B).

The clue to the solution is given by the inclusion of vapor pressures in the problem. We can calculate ΔG for a change in pressure at constant temperature. The alternative path for the process has three steps, all at $-5°$:

Step 1: Liquid water at 3.163 mm → vapor at 3.163 mm.
Step 2: Vapor at 3.163 mm → vapor at 3.013 mm.
Step 3: Vapor at 3.013 mm → solid water at 3.013 mm.

Steps (1) and (3) are equilibrium processes ($\Delta G = 0$), so the total ΔG is the ΔG of step (2). We use Eq. (6-49) to obtain

$$\Delta G = nRT \ln \frac{P_2}{P_1} = 1(8.314)(268.2) \ln \frac{3.013}{3.163} = -108.3 \text{ J}$$

6-5. Free Energy Changes in Chemical Processes

A. Calculation using enthalpy and entropy changes

Using the methods presented in Chapters 3–5 to determine ΔH and ΔS, we can calculate ΔG for a chemical reaction by using Eq. (6-29), $\Delta G = \Delta H - T\,\Delta S$.

EXAMPLE 6-11: Calculate $\Delta G°$ at 298 K for the reaction $H_2(g) + Br_2(g) \rightarrow 2HBr(g)$.

Solution: We want to use Eq. (6-29), so we must calculate $\Delta H°$ and $\Delta S°$ for the reaction. For $\Delta H°$, we need Eq. (4-5):

$$\Delta H_r^\circ = \Sigma n\, \Delta H_f^\circ(\text{products}) - \Sigma n\, \Delta H_f^\circ(\text{reactants}) = 2\,\Delta H_f^\circ(\text{HBr}) - \Delta H_f^\circ(\text{H}_2) - \Delta H_f^\circ(\text{Br}_2, g)$$

Then we substitute values from Table A-1 to get

$$\Delta H° = 2(-36.4) - 0 - 30.9 = -103.7 \text{ kJ}$$

Why is the ΔH_f° for bromine not zero as it is for hydrogen? Because bromine is in the gaseous state here, and the standard state for bromine at 298 K is the liquid state.

For ΔS_r°, we need Eq. (5-21):

$$\begin{aligned}\Delta S_r^\circ &= \Sigma n S°(\text{products}) - \Sigma n S°(\text{reactants}) \\ &= 2S°(\text{HBr}) - S°(\text{H}_2) - S°(\text{Br}_2, g) = 2(198.7) - 130.7 - 245.5 = 21.2 \text{ J K}^{-1}\end{aligned}$$

When calculating $\Delta G°$, watch the units; i.e., $\Delta H°$ is in kJ but $\Delta S°$ is in J.

$$\Delta G° = -103.7 \text{ kJ} - 298.2(21.2 \text{ J})(0.001 \text{ kJ J}^{-1}) = -110.0 \text{ kJ}$$

B. Calculation using free energies of formation

The approach used for enthalpy changes can be applied to free energy changes. The ΔG for a chemical reaction means

$$\Delta G_r = \Sigma n G(\text{products}) - \Sigma n G(\text{reactants}) \tag{6-51}$$

As with enthalpy, we can't obtain absolute values of free energy, so we again resort to defining a reference point from which arbitrary values for the G of compounds can be assigned.

The reference chosen is similar to that for H. First, we define the **standard free energy of formation** of a substance as the free energy change in the reaction in which the substance is formed from its elements, if all the substances are in their standard states at the temperature of the reaction. Then we let the standard free energy of formation of elements be zero at 298 K. Tables of ΔG_f° values have been compiled from calculations like that in Example 6-11. When we assume that ΔG_f° is the G of a substance, Eq. (6-51) becomes

$$\Delta G_r^\circ = \Sigma n\, \Delta G_f^\circ(\text{products}) - \Sigma n\, \Delta G_f^\circ(\text{reactants}) \qquad \textbf{(6-52)}$$

EXAMPLE 6-12: Repeat Example 6-11 using values of ΔG_f°.

Solution: For the HBr reaction, Eq. (6-52) becomes

$$\Delta G^\circ = 2\,\Delta G_f^\circ(\text{HBr}) - \Delta G_f^\circ(\text{H}_2) - \Delta G_f^\circ(\text{Br}_2, g) = 2(-53.5) - 0 - 3.1 = -110.1 \text{ kJ}$$

6-6. Temperature Dependence of Free Energy Change of a Reaction

A. General expression for dependence of ΔG on T and P

The Gibbs equations express the dependence of the various functions on P, V, T, and S. For the Gibbs energy, the Gibbs equation is

$$dG = -S\,dT + V\,dP \qquad \textbf{[Eq. (6-35)]}$$

For a chemical reaction, ΔG is given by Eq. (6-51). We can rewrite that equation slightly, letting G represent the total free energy, i.e., the summation term:

$$\Delta G_r = G(\text{products}) - G(\text{reactants})$$

Then

$$d(\Delta G) = dG(\text{products}) - dG(\text{reactants})$$

Substituting Eq. (6-35) gives

$$d(\Delta G) = (-S\,dT + V\,dP)(\text{products}) - (-S\,dT + V\,dP)(\text{reactants})$$

Since dT and dP must be equal for products and reactants, we combine terms, and

$$d(\Delta G) = -\Delta S\,dT + \Delta V\,dP \qquad \textbf{(6-53)}$$

At constant pressure, $d(\Delta G) = -\Delta S\,dT$, or

$$\left(\frac{\partial(\Delta G)}{\partial T}\right)_P = -\Delta S \qquad \textbf{(6-54)}$$

At constant temperature, $d(\Delta G) = \Delta V\,dP$, or

$$\left(\frac{\partial(\Delta G)}{\partial P}\right)_T = \Delta V \qquad \textbf{(6-55)}$$

The last two expressions are analogous to Eqs. (6-39) and (6-40) for a single substance.

B. Gibbs–Helmholtz equation for a reaction

In Example 6-9, we derived the Gibbs–Helmholtz equation for a single substance [Eq. (6-50)]. Starting with Eq. (6-54) instead of Eq. (6-39) yields the Gibbs–Helmholtz equation for a chemical reaction:

GIBBS–HELMHOLTZ EQUATION (chemical reaction)

$$\left(\frac{\partial(\Delta G/T)}{\partial T}\right)_P = -\frac{\Delta H}{T^2} \qquad \textbf{(6-56)}$$

Some authors apply the name Gibbs–Helmholtz to this form rather than to Eq. (6-50). Qualitatively, we can see that, if ΔH is negative, $\Delta(G/T)$ must increase when T increases.

C. Integration of the Gibbs–Helmholtz equation

In order to integrate, we need to express ΔH as a function of T, which we did in Chapter 4, obtaining Eq. (4-12):

$$\Delta H_T = \Delta a T + \frac{\Delta b}{2} T^2 + \frac{\Delta c}{3} T^3 + C$$

We substitute in Eq. (6-56) to get

$$\left(\frac{\partial(\Delta G/T)}{\partial T}\right)_P = -\frac{\Delta a}{T} - \frac{\Delta b}{2} - \frac{\Delta c}{3} T - \frac{C}{T^2} \qquad \textbf{(6-57)}$$

1. **Integration without limits gives an analytic expression with an additional integration constant.** Integration of Eq. (6-57) without limits yields

$$\frac{\Delta G}{T} = -\Delta a \ln T - \frac{\Delta b}{2} T - \frac{\Delta c}{6} T^2 + \frac{C}{T} + M \qquad \textbf{(6-58)}$$

EXAMPLE 6-13: Extend Example 6-11 by calculating ΔG° at 600 K for the same reaction.

Solution: The problem has become complicated because of the two integration constants, but the mathematics is no different from that we've used before. We need to decide which calculations to do and their sequence. To calculate C, we need ΔH at some T and the heat capacity coefficients; to evaluate M, we need C, ΔG at some T, and the heat capacity coefficients.

From Example 6-11, $\Delta H^\circ(298) = -103.7$ kJ, and $\Delta G^\circ(298) = -110.0$ kJ. To obtain the heat capacity coefficients, we use a familiar setup. For ΔC_P:

Substance	a	$b \times 10^3$	$c \times 10^7$
HBr (×2)	55.04	8.00	13.22
H_2	29.07	−0.83	20.12
$Br_2(g)$	35.24	4.07	−14.86
Total for reactants	64.31	3.24	5.26
Products − reactants	−9.27	4.76	7.96

From Eq. (4-12),

$$C = \Delta H_{298} - \Delta a T - \frac{\Delta b}{2} T^2 - \frac{\Delta c}{3} T^3$$
$$= -103.7 \times 10^3 + 9.27(298) - (2.38 \times 10^{-3})(298)^2 - (2.65 \times 10^{-7})(298)^3 \text{ J} = -101.2 \text{ kJ}$$

From Eq. (6-58),

$$M = \frac{\Delta G^\circ}{T} + \Delta a \ln T + \frac{\Delta b}{2} T + \frac{\Delta c}{6} T^2 - \frac{C}{T}$$
$$= \frac{-110.0 \times 10^3}{298} - 9.27 \ln 298 + (2.38 \times 10^{-3})(298) + (1.33 \times 10^{-7})(298)^2 + \frac{101.2 \times 10^3}{298}$$
$$= -81.6 \text{ J K}^{-1}$$

Before using Eq. (6-58), we could multiply through by T, but that would complicate the arithmetic. We solve for $\Delta G/T$ by substituting into Eq. (6-58) at 600 K.

$$\frac{\Delta G^\circ}{600} = 9.27 \ln 600 - (2.38 \times 10^{-3})(600) - (1.33 \times 10^{-7})(600)^2 - \frac{101.2 \times 10^3}{600} - 81.6 \text{ J K}^{-1}$$
$$= -192 \text{ J K}^{-1}$$
$$\Delta G^\circ = -115 \text{ kJ}$$

EXAMPLE 6-14: Use an approach similar to that of Example 6-13 to find ΔS at 600 K for the HBr reaction of Example 6-11.

Solution: At constant pressure, Eq. (6-35) becomes $dG = -S\,dT$. Also, $d(\Delta G) = -\Delta S\,dT$, or

$$\left(\frac{\partial(\Delta G)}{\partial T}\right)_P = -\Delta S$$

We can get an expression for ΔG as a function of T by multiplying both sides of Eq. (6-58) by T:

$$\Delta G = -\Delta a\,T \ln T - \frac{\Delta b}{2}T^2 - \frac{\Delta c}{6}T^3 - C + MT$$

Differentiating at constant pressure gives

$$\left(\frac{\partial(\Delta G)}{\partial T}\right)_P = -\Delta a - \Delta a \ln T - \Delta b\,T - \frac{\Delta c}{2}T^2 - 0 + M$$

The differentiation of the first term may not be clear. If we write it $-\Delta a\,d(T \ln T)$, it is of the type $d(uv) = u\,dv + v\,du$. Then

$$d(T \ln T) = T\,d(\ln T) + \ln T\,dT \qquad \text{and} \qquad T\,d(\ln T) = T\frac{dT}{T} = dT$$

Thus

$$\frac{d}{dT}(T \ln T) = \frac{dT}{dT} + \ln T\frac{dT}{dT} = 1 + \ln T$$

We change the sign of ΔS to positive to get

$$\Delta S = \Delta a + \Delta a \ln T + \Delta b T + \frac{\Delta c}{2}T^2 - M \tag{6-59}$$

and substitute values from Example 6-13, which results in

$$\begin{aligned}\Delta S &= -9.27 - 9.27 \ln 600 + (4.76 \times 10^{-3})(600) + (3.98 \times 10^{-7})(600)^2 - (-81.6)\ \text{J K}^{-1}\\ &= 16.1\ \text{J K}^{-1}\end{aligned}$$

2. **Integration with limits avoids the integration constant M but not C.** Integrating Eq. (6-57) between T_1 and T_2 gives

$$\left(\frac{\Delta G}{T}\right)_{T_2} - \left(\frac{\Delta G}{T}\right)_{T_1} = -\Delta a \ln\frac{T_2}{T_1} - \frac{\Delta b}{2}(T_2 - T_1) - \frac{\Delta c}{6}(T_2^2 - T_1^2) + C\left(\frac{1}{T_2} - \frac{1}{T_1}\right) \tag{6-60}$$

As with the methods of calculating the dependence of enthalpy change on temperature (see Section 4-4), Eq. (6-60) is more convenient if we want ΔG at only one high temperature. If we want ΔG at several temperatures, Eq. (6-58) is quicker and easier to use.

EXAMPLE 6-15: Repeat Example 6-13, using Eq. (6-60). Most of the numbers are the same; we merely wish to illustrate the use of Eq. (6-60).

Solution: We rewrite the equation slightly as we substitute to obtain

$$\begin{aligned}\frac{\Delta G_{600}}{600} &= \frac{-110.0 \times 10^3}{298} + 9.27 \ln\frac{600}{298} - (2.38 \times 10^{-3})(600 - 298)\\ &\quad -(1.33 \times 10^{-7})(600^2 - 298^2) - (101.2 \times 10^3)\left(\frac{1}{600} - \frac{1}{298}\right) = -192\ \text{J K}^{-1}\end{aligned}$$

$$\Delta G = -115\ \text{kJ}$$

3. **A different approach shows that the apparently distinct methods of integrating the Gibbs–Helmholtz equation with and without limits amount to the same thing.**

If T_1 is the reference temperature, all terms in T_1 are constant. If they are collected and designated M, we have a rearranged form of Eq. (6-58):

$$M = \frac{\Delta G_{T_1}}{T_1} + \Delta a \ln T_1 + \frac{\Delta b}{2} T_1 + \frac{\Delta c}{6} T_1^2 - \frac{C}{T_1}$$

The terms for the lower limit always constitute the integration constant for integration without limits. Thus we could write C immediately as

$$C = \Delta H_{T_1} - \Delta a\, T_1 - \frac{\Delta b}{2} T_1^2 - \frac{\Delta c}{3} T_1^3$$

A convenient technique to use in these cases is to write the terms for the lower limit on a separate line aligned below the upper limit term, thus identifying clearly the integration constant. Instead of using the form of Eq. (6-60), we would write

$$\begin{aligned} \frac{\Delta G_{T_2}}{T_2} = & \qquad\quad - \Delta a \ln T_2 - \frac{\Delta b}{2} T_2 - \frac{\Delta c}{6} T_2^2 + \frac{C}{T_2} \\ & + \frac{\Delta G_{T_1}}{T_1} + \Delta a \ln T_1 + \frac{\Delta b}{2} T_1 + \frac{\Delta c}{6} T_1^3 - \frac{C}{T_1} \end{aligned}$$

The second line is M, as given above. After evaluating it, we could calculate ΔG conveniently for several temperatures.

6-7. Fugacity, Activity, and Fugacity Coefficient

A. Fugacity and activity

Fugacity and activity account for deviations from ideal behavior. When a gas is not ideal, we could obtain an expression like Eq. (6-49) by using one of the more complex equations of state. It is more convenient, however, to maintain the simple form of the equation by using the concept called **fugacity**, and write for one mole of a real gas

$$\Delta G = RT \ln \frac{f_2}{f_1} \qquad \textbf{(6-61)}$$

If we refer to a standard state,

$$\Delta G = G - G^\circ = RT \ln \frac{f}{f^\circ} \qquad \textbf{(6-62)}$$

The standard state for fugacity merits explanation, but we will only state its definition, as used by Laidler and Meiser. It is the *state at which the fugacity would be equal to 1 atm if the gas remained ideal from low pressures to 1 atm pressure.* Thus $f^\circ = 1$, and we can write

$$G = G^\circ + RT \ln f \qquad \textbf{(6-63)}$$

If we define the ratio f/f° as the **activity**, a, we have

$$G = G^\circ + RT \ln a \qquad \textbf{(6-64)}$$

Although *fugacity* is limited to gases, *activity* may be used for all states, including those of solution components. Both terms give exact values of ΔG because they incorporate all of the factors that are responsible for nonideal behavior.

For a change of state at constant temperature,

$$\Delta G = G_2 - G_1 = G^\circ + RT \ln a_2 - G^\circ - RT \ln a_1$$

$$\Delta G = RT \ln \frac{a_2}{a_1} \qquad \textbf{(6-65)}$$

B. Fugacity coefficient

The fugacity coefficient relates fugacity to pressure and provides a way to determine the fugacity. Gases behave more ideally at low pressures; thus f approaches P in value and at $P = 0$, $f = P$. A better way to express this is to say that the limit of f/P equals 1 as P approaches 0. The ratio f/P is called the **fugacity coefficient**, or **activity coefficient**, and is represented by γ:

FUGACITY COEFFICIENT $$\gamma = f/P \tag{6-66}$$

This quantity can be related to measurable properties by differentiating Eq. (6-63) and comparing the result to Eq. (6-40), written as $dG = V\,dP$; we get

$$d(\ln f) = \frac{V}{RT}\,dP$$

If we express the actual volume as a simple deviation from the ideal volume $V = (RT)/P + A$, substitute, and rearrange, we have

$$d(\ln \gamma) = \frac{A}{RT}\,dP \qquad \text{or} \qquad \ln \gamma = \frac{1}{RT}\int_0^P A\,dP \tag{6-67}$$

The complete derivation is given in Solved Problem 6-21.

If experimental P–V data are plotted as A versus P, graphical integration will give the value of the integral at the desired P. If we know the proper equation of state, we can find A as a function of P and integrate.

Plots of γ versus reduced pressure at various reduced temperatures are available. Thus, for any gas, we can calculate P_r and T_r for the gas and the given conditions, read the corresponding γ, and calculate f from Eq. (6-66).

EXAMPLE 6-16: Compare the free energy changes when 1 mol NH_3, initially at 200 °C and 50 atm, is compressed isothermally to 300 atm (**a**) assuming ideal behavior, and (**b**) assuming nonideal behavior. The fugacity coefficients are 0.91 at 50 atm and 0.642 at 300 atm.

Solution:

(**a**) This part is like Example 6-8, and we merely substitute in Eq. (6-49) to obtain

$$\Delta G = nRT \ln \frac{P_2}{P_1} = 1(8.314)(473.2) \ln \frac{300}{50} = 7.05 \text{ kJ}$$

(**b**) For nonideal behavior, we have to use Eq. (6-61), which requires calculation of the fugacities using Eq. (6-66):

$$f_{50} = \gamma P = 0.91(50) = 45.5 \qquad f_{300} = 0.642(300) = 192.6$$

Substituting these values, we get

$$\Delta G = nRT \ln \frac{f_2}{f_1} = 1(8.314)(473.2) \ln \frac{192.6}{45.5} = 5.68 \text{ kJ}$$

6-8. Chemical Potential

A. Dependence of G on concentration

So far we've dealt with a fixed amount of a single substance or assumed that the amounts of substances in a system didn't change. Clearly, neither of these conditions holds in an actual chemical reaction. To be completely general, we take G to be a function of the amount of each component, represented by n_i, as well as of T and P; i.e., $G = G(T, P, n_i)$. Then

$$dG = \left(\frac{\partial G}{\partial T}\right)_{P,n_i} dT + \left(\frac{\partial G}{\partial P}\right)_{T,n_i} dP + \left(\frac{\partial G}{\partial n_1}\right)_{T,P,n_j} dn_1 + \left(\frac{\partial G}{\partial n_1}\right)_{T,P,n_j} dn_2 + \cdots \tag{6-68}$$

Here, n_j represents all n except the one involved in the differentiation. The partial derivative with respect to n is called the **chemical potential**, μ, where

$$\left(\frac{\partial G}{\partial n_1}\right)_{T,P,n_j} = \mu_1, \ldots \tag{6-69}$$

This is a coefficient, or rate of change, representing the change in G when some quantity of a component is added to or removed from the system.

Now we can write the general form of the Gibbs equation, Eq. (6-35):

$$dG = -S\,dT + V\,dP + \Sigma\,\mu_i\,dn_i \tag{6-70}$$

This is the *true* fundamental equation of thermodynamics.

B. Other state functions related to energy

For each state function, if we hold constant the two variables usually associated with it (Gibbs equations), we can write

$$\left(\frac{\partial A}{\partial n_i}\right)_{T,V,n_j} = \mu_i = \left(\frac{\partial E}{\partial n_i}\right)_{S,V,n_j} = \left(\frac{\partial H}{\partial n_i}\right)_{S,P,n_j} = \left(\frac{\partial S}{\partial n_i}\right)_{E,V,n_j} \tag{6-71}$$

EXAMPLE 6-17: Derive the *chemical potential* term for enthalpy as shown in Eq. (6-68).

Solution: Recall that the Gibbs equation for H is $dH = T\,dS + V\,dP$, Eq. (6-33). This shows us that H depends on S and P for a fixed amount of a substance. In general, then, $H = H_{S,P,n_i}$. Taking the derivative, we get

$$dH = \left(\frac{\partial H}{\partial S}\right)_{P,n_i} dS + \left(\frac{\partial H}{\partial P}\right)_{S,n_i} dP + \left(\frac{\partial H}{\partial n_1}\right)_{S,P,n_j} dn_1 + \cdots$$

Writing the last term for all n_i, we have $\left(\frac{\partial H}{\partial n_i}\right)_{S,P,n_j}$, which is the chemical potential.

Usually, however, **chemical potential** is understood to mean the potential of the Gibbs free energy, since it applies to the usual working conditions of chemistry. The various equations using G can be rewritten using μ and other partial molar quantities. Some of these modified equations are

$$\mu = \left(\frac{\partial H}{\partial n_i}\right)_{P,T,n_j} - T\left(\frac{\partial S}{\partial n_i}\right)_{P,T,n_j} \tag{6-72}$$

$$\left(\frac{\partial \mu}{\partial T}\right)_P = -\left(\frac{\partial S}{\partial n_i}\right)_{P,T,n_j} \tag{6-73}$$

$$\mu_2 - \mu_1 = RT \ln \frac{P_2}{P_1} \tag{6-74}$$

C. Another satisfactory criterion of equilibrium

At constant T and P, from Eq. (6-70),

$$dG = \Sigma\,\mu_i\,dn_i \tag{6-75}$$

At equilibrium, $dG = 0$ and

$$\mu_1\,dn_1 + \mu_2\,dn_2 = 0 \tag{6-76}$$

For two phases, α and β, and one substance, if a quantity of substance passes from one phase to the other,

$$\mu^\alpha\,dn^\alpha = -\mu^\beta\,dn^\beta \tag{6-77}$$

And since the loss from one phase equals the gain of the other, $dn^\alpha = -dn^\beta$. Then

$$\mu^\alpha = \mu^\beta \tag{6-78}$$

In other words, in order for equilibrium to exist between the two phases, the chemical potentials for a substance in each phase must be equal. Conversely, if the μ's are not equal, the potential exists for matter to pass from one phase to the other, spontaneously from the phase of high μ to that of low μ. Thus μ serves as a satisfactory criterion of spontaneity and equilibrium.

D. Partial molar quantities

In a system that contains at least two substances, the total value of any **extensive property** of the system, i.e., any property proportional to the mass of the system (e.g., n and V), is the sum of the contribution of each substance to that property. The contribution of one mole of a substance is called the **partial molar property** of that component.

EXAMPLE 6-18: If n_A moles of liquid A are mixed with n_B moles of liquid B, what is the final volume of the solution?

Solution: The final volume would equal the volume of A plus the volume of B only for ideal liquids. In real solutions, the final volume is determined by the partial molar volumes of each liquid $\left(\frac{\partial V}{\partial n_i}\right)_{P,T,n_j}$. Thus the volume of the solution can be obtained by integrating the equation

$$dV(\text{soln}) = \left(\frac{\partial V}{\partial n_A}\right) dn_A + \left(\frac{\partial V}{\partial n_B}\right) dn_B \tag{6-79}$$

The partial molar Gibbs free energy would be written $\left(\frac{\partial G}{\partial n_i}\right)_{P,T,n_j}$, which is the term we called the chemical potential, Eq. (6-69). Analogous to Eq. (6-79), we can write

$$dG(\text{mixture}) = \left(\frac{\partial G}{\partial n_1}\right)_{P,T,n_j} dn_1 + \left(\frac{\partial G}{\partial n_2}\right)_{P,T,n_j} dn_2 + \cdots \tag{6-80}$$

or

$$G = \mu_1 n_1 + \mu_2 n_2 + \cdots = \Sigma\, \mu_i n_i \tag{6-81}$$

A useful expression is obtained by differentiating and imposing the condition of equilibrium. Differentiation gives

$$dG = \mu_1\, dn_1 + n_1\, d\mu_1 + \mu_2\, dn_2 + n_2\, d\mu_2 \tag{a}$$

At equilibrium, $dG = 0$, and Eq. (6-76) applies, i.e., $\mu_1\, dn_1 + \mu_2\, dn_2 = 0$. Then eq. (**a**) becomes

$$0 = n_1\, d\mu_1 + n_2\, d\mu_2$$

or

GIBBS–DUHEM EQUATION

$$d\mu_1 = -\frac{n_2}{n_1}\, d\mu_2 \tag{6-82}$$

This expression is known as the **Gibbs–Duhem equation,** which is useful in dealing with solutions. A similar equation may be written for any extensive property.

SUMMARY

Table S-6 summarizes the major equations of this chapter and shows connections between them. Some analogous equations, which haven't been derived, are included. You should determine the conditions that are necessary for each equation to be valid or the conditions that are imposed to make the connections; you should also be able to derive all of the unnumbered equations. Note the similarities between the groups of equations—these should help you in remembering the primary equations and the forms you should expect when deriving the secondary equations. Other important things you should know are

1. The second law of thermodynamics provides a criterion of spontaneity, entropy, but ΔS is inconvenient for chemical applications.

TABLE S-6: Summary of Important Equations

Definitions

Substance or system

$$dE = đq - đw \quad \text{(3-1a)}$$
$$dS = \frac{đq(\text{rev})}{T} \quad \text{(5-3)}$$
$$H = E + PV \quad \text{(3-18)}$$
$$A = E - TS \quad \text{(6-9)}$$
$$G = H - TS \quad \text{(6-23)}$$

Gibbs

$$dE = T\,dS - P\,dV \quad \text{(5-10)}$$
$$dH = T\,dS + V\,dP \quad \text{(6-33)}$$
$$dA = -S\,dT - P\,dV \quad \text{(6-34)}$$
$$dG = -S\,dT + V\,dP \quad \text{(6-35)}$$

Maxwell

$$\left(\frac{\partial T}{\partial V}\right)_S = -\left(\frac{\partial P}{\partial S}\right)_V \quad \text{(6-43)}$$
$$\left(\frac{\partial T}{\partial P}\right)_S = \left(\frac{\partial V}{\partial S}\right)_P \quad \text{(6-44)}$$
$$\left(\frac{\partial S}{\partial V}\right)_T = \left(\frac{\partial P}{\partial T}\right)_V \quad \text{(6-45)}$$
$$\left(\frac{\partial S}{\partial P}\right)_T = -\left(\frac{\partial V}{\partial T}\right)_P \quad \text{(6-41)}$$

$$\left(\frac{\partial E}{\partial V}\right)_S = -P \qquad \left(\frac{\partial E}{\partial S}\right)_V = T$$
$$\left(\frac{\partial H}{\partial P}\right)_S = V \qquad \left(\frac{\partial H}{\partial S}\right)_P = T$$
$$\left(\frac{\partial A}{\partial V}\right)_T = -P \qquad \left(\frac{\partial A}{\partial T}\right)_V = -S$$
$$\left(\frac{\partial G}{\partial P}\right)_T = V \qquad \left(\frac{\partial G}{\partial T}\right)_P = -S \quad \text{(6-40; 6-39)}$$

Gibbs–Helmholtz

$$\longrightarrow \left[\frac{\partial(G/T)}{\partial T}\right] = -\frac{H}{T^2} \quad \text{(6-50)}$$

Process

$$\Delta X_r = \Sigma X(\text{products}) - \Sigma X(\text{reactants})$$

$$d(\Delta E) = \Delta T\,dS - \Delta P\,dV$$
$$d(\Delta H) = \Delta T\,dS + \Delta V\,dP$$
$$d(\Delta A) = -\Delta S\,dT - \Delta P\,dV$$
$$d(\Delta G) = -\Delta S\,dT + \Delta V\,dP \quad \text{(6-53)}$$

$$\left(\frac{\partial \Delta E}{\partial V}\right)_S = -\Delta P \qquad \left(\frac{\partial \Delta E}{\partial S}\right)_V = \Delta T$$
$$\left(\frac{\partial \Delta H}{\partial P}\right)_S = \Delta V \qquad \left(\frac{\partial \Delta H}{\partial S}\right)_P = \Delta T$$
$$\left(\frac{\partial \Delta A}{\partial V}\right)_T = -P \qquad \left(\frac{\partial \Delta A}{\partial T}\right)_V = -\Delta S$$
$$\left(\frac{\partial \Delta G}{\partial P}\right)_T = \Delta V \qquad \left(\frac{\partial \Delta G}{\partial T}\right)_P = -\Delta S \quad \text{(6-55; 6-54)}$$

$$\longrightarrow \left[\frac{\partial(\Delta G/T)}{\partial T}\right]_P = -\frac{\Delta H}{T^2} \quad \text{(6-56)}$$

Molar change

$$dE = T\,dS - P\,dV + \sum\left(\frac{\partial E}{\partial n_i}\right)_{S,V,n_j} dn_i$$
$$dH = T\,dS + V\,dP + \sum\left(\frac{\partial H}{\partial n_i}\right)_{S,P,n_j} dn_i$$
$$dA = -S\,dT - P\,dV + \sum\left(\frac{\partial A}{\partial n_i}\right)_{T,V,n_j} dn_i$$
$$\left(\frac{\partial G}{\partial n_i}\right)_{T,P,n_j} = \mu_i \quad \text{(6-69)} \longrightarrow dG = -S\,dT + V\,dP + \sum\left(\frac{\partial G}{\partial n_i}\right)_{T,P,n_j} dn_i \quad \text{(6-70)}$$

Gibbs–Duhem

$$\longrightarrow d\mu_i = -\frac{n_2}{n_1}\,d\mu_2 \quad \text{(6-82)}$$

2. The new state functions A and G provide the desired criteria of spontaneity: at constant V and T, $\Delta A \le 0$; at constant P and T, $\Delta G \le 0$.
3. $\Delta A \le w_{\text{max, all work}}$
4. $\Delta G \le w_{\text{max, non}PV}$
5. Two partial derivatives involving P, V, and T are given special names and symbols, due to their usefulness:

   (a) coefficient of thermal expansivity: $\alpha = \dfrac{1}{V}\left(\dfrac{\partial V}{\partial T}\right)_P$

   (b) coefficient of compressibility: $\kappa = -\dfrac{1}{V}\left(\dfrac{\partial V}{\partial P}\right)_T$
6. For an irreversible (nonequilibrium) process, ΔG can be calculated only by an alternative path of reversible (equilibrium) steps.
7. For a chemical reaction, ΔG can be calculated from

   (a) $\Delta G = \Delta H - T\,\Delta S$

   (b) standard free energies of formation
8. ΔG of a reaction may be calculated at a second temperature if ΔG is known at one temperature, if ΔH is known at one temperature, and if the coefficients in the heat capacity equation are available for all substances involved.
9. Fugacity is an idealized pressure, designed to allow for nonideal behavior in gases.
10. Activity is the ratio of the actual fugacity to the fugacity in the standard state: $a = f/f^\circ$.
11. The fugacity coefficient is the ratio of the fugacity to the pressure: $\gamma = f/P$.
12. Chemical potential expresses the dependence of a state function on amount of substance.
13. For two phases of a substance to be in equilibrium, the chemical potentials in each phase must be equal.
14. For two substances in a mixture, the chemical potentials change as concentration changes (Gibbs–Duhem equation).

SOLVED PROBLEMS

Free Energy State Functions

PROBLEM 6-1 Various differential and "Δ" forms of equations for a particular thermodynamic property are needed in different situations. They may be considered alternative forms, and you should be able to write them almost automatically, starting from the fundamental relationships. From $G = H - TS$, write a series of expressions for dG and ΔG.

Solution: You should write the total differential first; the other expressions will follow from it.

$dG = dH - d(TS)$			
$dG = dH - T\,dS - S\,dT$		$\Delta G = \Delta H - \Delta(TS)$	
$dG = dH - T\,dS$	(at constant T)	$\Delta G = \Delta H - T\,\Delta S$	(at constant T)
$dG = dH - S\,dT$	(at constant S)	$\Delta G = \Delta H - S\,\Delta T$	(at constant S)
$dG = -T\,dS - S\,dT$	(at constant H)	$\Delta G = -\Delta(TS)$	(at constant H)

PROBLEM 6-2 Derive the Gibbs equation, $dG = -S\,dT + V\,dP$ [Eq. (6-35)].

Solution: You should write the differential of G from its definition, then substitute the fundamental equation, Eq. (6-5). From Eq. (6-17),

$$dG = dE - T\,dS - S\,dT + P\,dV + V\,dP$$

Substituting $dE = T\,dS - P\,dV$ and canceling like terms, you get

$$dG = -S\,dT + V\,dP$$

PROBLEM 6-3 Show that $\left(\frac{\partial H}{\partial P}\right)_S = V$.

Solution: Since an energy function appears in the partial derivative, you may use the Gibbs equation. For H, this is Eq. (6-33):

$$dH = T\,dS + V\,dP$$

Now set S = constant, which means that $dS = 0$. Dividing by dP gives you

$$\left(\frac{\partial H}{\partial P}\right)_S = V$$

PROBLEM 6-4 Will the Helmholtz energy of a substance increase or decrease when the volume increases isothermally?

Solution: First, you need a relationship between the Helmholtz energy A and V at constant T. From the Gibbs equation for dA, Eq. (6-34), when T is constant

$$dA = -P\,dV \qquad \text{or} \qquad \left(\frac{\partial A}{\partial V}\right)_T = -P$$

Since P must be positive, the coefficient has to be negative, and A must decrease when V increases.

PROBLEM 6-5 Derive the expression $\left(\frac{\partial T}{\partial P}\right)_S = \left(\frac{\partial V}{\partial S}\right)_P$.

Solution: You should notice that this looks like one of the Maxwell relations and suspect that you should start with one of the Gibbs equations. Which one? The variables held constant in the desired expression appear as differentials in the parent Gibbs equation. In this case, they are S and P, thus you want Eq. (6-33), or

$$dH = T\,dS + V\,dP$$

Apply the Euler reciprocity theorem to get

$$\left(\frac{\partial T}{\partial P}\right)_S = \left(\frac{\partial V}{\partial S}\right)_P$$

PROBLEM 6-6 Evaluate $\left(\frac{\partial H}{\partial P}\right)_T$ for one mole of an ideal gas.

Solution: You could start from the Gibbs equation for H, but we already did this in Example 6-6, where we obtained Eq. (6-47):

$$\left(\frac{\partial H}{\partial P}\right)_T = V(1 - \alpha T) \qquad \left[\alpha = \left(\frac{1}{V}\right)\left(\frac{\partial V}{\partial T}\right)_P\right]$$

or

$$\left(\frac{\partial H}{\partial P}\right)_T = V - VT\left[\left(\frac{1}{V}\right)\left(\frac{\partial V}{\partial T}\right)_P\right]$$

For one mole of an ideal gas, $V = RT/P$ and

$$\left(\frac{\partial V}{\partial T}\right)_P = \frac{R}{P}$$

Now

$$\left(\frac{\partial H}{\partial P}\right)_T = \frac{RT}{P} - T\frac{R}{P} = 0$$

PROBLEM 6-7 At 20 °C, the value of α for acetone is $1.487 \times 10^{-3}\ \text{K}^{-1}$. Calculate the change in enthalpy if the pressure on 1 L of acetone is increased from 1 atm to 5 atm.

Solution: This is a numerical application of Eq. (6-47):

$$\left(\frac{\partial H}{\partial P}\right)_T = V(1 - \alpha T) = (1\ \text{L})[1 - (1.487 \times 10^{-3}\ \text{K}^{-1})(293.2\ \text{K})] = 0.564\ \text{L}$$

Then $\qquad dH = 0.564\,dP$

and integration gives

$$\Delta H = 0.564\ \Delta P = 0.564(5 - 1) = 2.26\ \text{L atm} = 229\ \text{J}$$

Free Energy Changes: Physical Processes

PROBLEM 6-8 Derive an expression for calculating ΔA for an isothermal change in volume of an ideal gas.

Solution: The Gibbs equations give the dependence of the energy functions on other variables. For A, you need Eq. (6-34), or

$$dA = -S\,dT - P\,dV$$

At constant T,

$$dA = -P\,dV$$

For an ideal gas,

$$\int_{A_1}^{A_2} dA = -\int_{V_1}^{V_2} \frac{nRT}{V}\,dV = -nRT\int_{V_1}^{V_2} d\ln V$$

$$\Delta A = -nRT\ln\frac{V_2}{V_1} = nRT\ln\frac{V_1}{V_2}$$

PROBLEM 6-9 If the pressure on 5.0 moles of an ideal gas at 300 K is reduced to 1/4 of its original value isothermally and reversibly, what is the change in G?

Solution: Again you need a Gibbs equation, this time for G, i.e., Eq. (6-35),

$$dG = -S\,dT + V\,dP$$

At constant T and for an ideal gas,

$$dG = \frac{nRT}{P}\,dP$$

Integrating, you get

$$\Delta G = nRT\ln\frac{P_2}{P_1} \qquad \text{[Eq. (6-49)]}$$

Since $$P_2 = \frac{P_1}{4} \quad \text{or} \quad \frac{P_2}{P_1} = \frac{1}{4}$$

$$\Delta G = 5.0(8.314)(300)\ln\frac{1}{4} = -17\ \text{kJ}$$

PROBLEM 6-10 Calculate q, w, ΔE, ΔH, ΔS, ΔA, and ΔG for the isothermal, reversible compression of one mole of an ideal gas at 273.2 K from 1 atm to 10 atm.

Solution: In all-inclusive problems of this type, usually some quantities are zero and others are equal to each other or have the same absolute values. Don't rush in, calculating each quantity as listed; look for the obvious ones first. Since you have an ideal gas and isothermal conditions, $\Delta E = \Delta H = 0$. Thus $q = -w$. Now use Eq. (3-13) to obtain

$$w = -nRT\ln\frac{V_2}{V_1} = -nRT\ln\frac{P_1}{P_2} = -1(8.314)(273.2)\ln\frac{1}{10} = +5.230\ \text{kJ}$$

So $$q = -5.230\ \text{kJ}$$

Using Eqs. (5-14) and (5-4) gives you

$$\Delta S = nR\ln\frac{V_2}{V_1} = \frac{q}{T} = -\frac{5.230}{273.2} = -19.14\ \text{J K}^{-1}$$

You can calculate the free energy changes in several ways. Since the process involves only PV work, and is reversible and isothermal, $dA = dw_{max}$ [Eq. (6-12)]. Integrating yields

$$\Delta A = w_{max} = 5.230\ \text{kJ} \qquad \textbf{[Eq. (6-14)]}$$

Another option is to start with Eq. (6-26):

$$\Delta A = \Delta E - T\Delta S = 0 - q = w = 5.230\ \text{kJ}$$

For ΔG, you have three options. (a) Use Eq. (6-29),

$$\Delta G = \Delta H - T\Delta S = 0 - q = w = 5.230 \text{ kJ}$$

(b) Use Eq. (6-30),

$$\Delta G = \Delta A + \Delta(PV) = \Delta A + 0 = 5.230 \text{ kJ}$$

You should recognize that $\Delta(PV) = 0$, because the isothermal condition is the requirement for Boyle's law, $PV = \text{constant}$.
Or (c) use Eq. (6-49):

$$\Delta G = nRT \ln \frac{P_2}{P_1} = 1(8.314)(273.2) \ln \frac{10}{1} = 5.230 \text{ kJ}$$

Or you could have compared the last expression to the first, for w from Eq. (3-13), noting that $\Delta G = w = 5.230$ kJ.

PROBLEM 6-11 If the gas in Problem 6-10 is allowed to return to its original state by expanding isothermally against a constant pressure of 1 atm until the gas pressure is 1 atm, what are the values of q, w, ΔE, ΔH, ΔS, ΔA, and ΔG?

Solution: Again, look first for the obvious values. The process is isothermal and the gas is ideal, so $\Delta E = \Delta H = 0$. Since the system has returned to its initial state, the total change in the state functions must be zero; therefore ΔE, ΔH, ΔS, ΔA, and ΔG must be equal but opposite in sign to the values in Problem 6-10:

$$\Delta E = \Delta H = 0 \qquad \Delta S = +19.14 \text{ J K}^{-1} \qquad \Delta A = -5.230 \text{ kJ} \qquad \Delta G = -5.230 \text{ kJ}$$

What about q and w? Since $\Delta E = 0$, you still have $q = -w$, but w is different because the process is not reversible. For constant pressure,

$$w = -P\,\Delta V = -P(V_2 - V_1) \qquad \text{[Eq. (3-12)]}$$

$$= -1\left(\frac{RT_2}{P_2} - \frac{RT_1}{P_1}\right) = -RT\left(\frac{1}{P_2} - \frac{1}{P_1}\right) = -8.314(273.2)\left(\frac{1}{1} - \frac{1}{10}\right) = -2.044 \text{ kJ}$$

So $\quad q = +2.044$ kJ

PROBLEM 6-12 For the state function A, derive an expression similar to the Gibbs–Helmholtz equation.

Solution: Knowing that the expected expression will contain the term A/T, you could differentiate this quantity. Instead, let's start with $A = E - TS$ [Eq. (6-9)] and $\left(\frac{\partial A}{\partial T}\right)_V = -S$ (obtained in Example 6-3). From Eq. (6-9),

$$-S = \frac{A - E}{T}$$

Thus

$$\left(\frac{\partial A}{\partial T}\right)_V = \frac{A - E}{T} = \frac{A}{T} - \frac{E}{T}$$

or

$$\left(\frac{\partial A}{\partial T}\right)_V - \frac{A}{T} = -\frac{E}{T} \qquad \textbf{(a)}$$

The left-hand side of eq. (a) looks as though it may have come from the derivative of A/T, so we'll try that:

$$\frac{d(A/T)}{dT} = A\frac{d(1/T)}{dT} + \frac{1}{T}\frac{dA}{dT} = A(-T^{-2}) + \frac{1}{T}\frac{dA}{dT} = \frac{1}{T}\left(-\frac{A}{T} + \frac{dA}{dT}\right)$$

Now show that V is constant:

$$\left(\frac{\partial A/T}{\partial T}\right)_V = \frac{1}{T}\left(\frac{\partial A}{\partial T}\right)_V - \frac{A}{T}$$

Substitute eq. (a) and multiply:

$$\left(\frac{\partial (A/T)}{\partial T}\right)_V = -\frac{E}{T^2}$$

PROBLEM 6-13 If $\Delta G = -118.1$ J when 10.0 g $H_2O(l)$ at -10 °C changes to a solid, what is the vapor pressure of ice at -10 °C? The vapor pressure of $H_2O(l)$ at -10 °C is 2.149 mm Hg.

Solution: This problem is similar to Example 6-10. The two phases of water can't be at equilibrium at -10 °, so the transition described must be expressed as a series of reversible steps for which you can calculate ΔG. As in Example 6-10, you can imagine the 10.0 g H_2O passing through the vapor state when going from the liquid to the solid state isothermally.

Step 1: Liquid water at 2.149 mm → vapor at 2.149 mm.
Step 2: Vapor at 2.149 mm → vapor at P(sol).
Step 3: Vapor at P(sol) → solid at P(sol).

Steps (1) and (3) are equilibrium processes; thus $\Delta G = 0$, and the net ΔG (-118.1 J) must be ΔG for step (2). Now you can apply Eq. (6-49):

$$\Delta G = nRT \ln \frac{P_2}{P_1} = nRT \ln \frac{P(\text{sol})}{P(\text{liq})}$$

Before substituting numbers, you should solve for ln P(sol):

$$\frac{\Delta G}{nRT} = \ln P(\text{sol}) - \ln P(\text{liq})$$

$$\ln P(\text{sol}) = \frac{\Delta G}{nRT} + \ln P(\text{liq})$$

Substituting numbers gives you

$$\ln P(\text{sol}) = \frac{-118.1}{(10.0/18.0)(8.314)(263.2)} + \ln 2.149 = -0.0971 + 0.7650 = 0.6679$$

$$P(\text{sol}) = 1.95 \text{ mm Hg}$$

Free Energy Changes: Chemical Processes

PROBLEM 6-14 Use ΔG_f° and ΔH_f° values to determine ΔS° at 298 K for the reaction $2SO_2(g) + O_2(g) \rightarrow 2SO_3(g)$.

Solution: When ΔG and ΔH are mentioned, you should automatically think of $\Delta G = \Delta H - T\,\Delta S$ [Eq. (6-29)]. You can rearrange it to $T\,\Delta S = \Delta H - \Delta G$ and then calculate ΔH and ΔG, using Eqs. (4-5) and (6-52). From Eq. (4-5),

$$\begin{aligned}\Delta H_r^0 &= 2\,\Delta H_f^0(SO_3) - 2\,\Delta H_f^0(SO_2) - \Delta H_f^0(O_2)\\ &= (2 \text{ mol})(-395.7 \text{ kJ mol}^{-1}) - (2 \text{ mol})(-296.8 \text{ kJ mol}^{-1}) - 0 = -791.4 + 593.6 = -197.8 \text{ kJ}\end{aligned}$$

and from Eq. (6-29),

$$\begin{aligned}\Delta G^\circ &= 2\,\Delta G_f^\circ(SO_3) - 2\,\Delta G_f^\circ(SO_2) - \Delta G_f^\circ(O_2)\\ &= 2(-371.1) - 2(-300.2) = -742.2 + 600.4 = -141.8 \text{ kJ}\end{aligned}$$

Now,

$$T\,\Delta S^\circ = \Delta H^\circ - \Delta G^\circ = -197.8 - (-141.8) \text{ kJ}$$

$$\Delta S^\circ = \frac{-56.0 \times 10^3 \text{ J}}{298 \text{ K}} = -188 \text{ J K}^{-1}$$

PROBLEM 6-15 Nitrogen dioxide (NO_2) will form a dimer (N_2O_4). Which form is more stable at 298 K?

Solution: The equation for the reaction is $2NO_2 \rightleftharpoons N_2O_4$. The question really is, "Which way does the reaction go at 298 K?" You can answer it if you find the direction that has the negative ΔG. For the forward reaction, Eq. (6-52) gives

$$\Delta G_r^\circ = \Delta G_f^\circ(N_2O_4) - 2\,\Delta G_f^\circ(NO_2) = 97.9 - 2(51.3) = -4.7 \text{ kJ}$$

Thus, at room temperature, the spontaneous reaction is in the direction of N_2O_4.

PROBLEM 6-16 A beginning student in organic chemistry has the idea that, if he can find the proper catalyst, he can make butane from propane and methane by the reaction

$$C_3H_8 + CH_4 \rightleftharpoons C_4H_{10} + H_2$$

Based on thermodynamic calculations, what would you recommend to this student?

Solution: In order to determine whether a reaction can occur, you have to know the sign of ΔG. Using Eq. (6-52) and omitting n because all $n = 1$ mol, you would get

$$\Delta G^\circ_r = \Delta G^\circ_f(C_4H_{10}) + \Delta G^\circ_f(H_2) - \Delta G^\circ_f(C_3H_8) - \Delta G^\circ_f(CH_4)$$
$$= -17.0 + 0 - (-23.4) - (-50.7) = +57.1 \text{ kJ}$$

Since the sign is positive, your recommendation should be: "Forget it!"

PROBLEM 6-17 The student mentioned in Problem 6-16 is persistent and suggests that the reaction might go if he increases the temperature to 1000 or 1200 K. Should he try it?

Solution: There are at least two calculations that would lead to an answer. The first would be to calculate ΔG at a high temperature by one of the methods in Section 6-6C. This would be a lengthy calculation, even if we had the C_P coefficients, so let's consider a second approach, which requires less calculation.

Useful predictions can be made using Eq. (6-29), $\Delta G = \Delta H - T\,\Delta S$. In this case, since ΔG is large at 298 K, ΔH almost certainly will have the same sign (positive). The only way to get a negative ΔG is to make the term $-T\,\Delta S$ numerically larger than a positive ΔH, so that their sum becomes negative. This can be done by increasing T, but only if ΔS is positive; if ΔS is negative, $-T\,\Delta S$ becomes positive and would make ΔG positive regardless of T. Thus a calculation of ΔS may give you a clue. Recall Eq. (5-21):

$$\Delta S^\circ_r = \Sigma nS^\circ(\text{products}) - \Sigma nS^\circ(\text{reactants})$$

For the reaction given in Problem 6-16,

$$\Delta S^\circ_r = S^\circ(C_4H_{10}) + S^\circ(H_2) - S^\circ(C_3H_8) - S^\circ(CH_4) = 310.2 + 130.7 - 270.0 - 186.3 = -15.4 \text{ J K}^{-1}$$

Since the sign is negative, the $T\,\Delta S$ term would be positive and you would predict that ΔG would remain positive at all temperatures. One other slight possibility is that the heat capacities would be such that a calculation of ΔS at a high T, using Eq. (5-26), would change the sign of ΔS. However, this is very unlikely; in Example 5-18, the change resulting from the $\int \Delta C_P\, dT$ was only 7 J K^{-1}.

You can feel quite safe in recommending to the beginner that he not waste his time on this proposal. The reaction is nonspontaneous at all temperatures.

PROBLEM 6-18 **(a)** Show by calculation that the transformation of water from a liquid to a gas at 1.00 atm is spontaneous at 110 °C and nonspontaneous at 90 °C. **(b)** Find the temperature at which liquid and gaseous water are in equilibrium. Assume that the values of ΔH and ΔS at 298 K are valid at these temperatures.

Solution: You have seen from previous calculations that ΔH and ΔS change only slightly with temperature. Thus the values calculated from the tables (at 298 K) will introduce little error when used for about 100 °C.

(a) You need ΔH and ΔS in order to calculate ΔG from Eq. (6-29), $\Delta G = \Delta H - T\,\Delta S$. The sign of ΔG will tell you whether the process is spontaneous. The equation for the process is

$$H_2O(l, 1 \text{ atm}) \rightleftharpoons H_2O(g, 1 \text{ atm})$$

From Eq. (4-5),

$$\Delta H^\circ_r = \Delta H^\circ_f(H_2O, g) - \Delta H^\circ_f(H_2O, l) = -241.8 - (-285.8) = +44.0 \text{ kJ}$$

and, from Eq. (5-21),

$$\Delta S^\circ_r = S^\circ(H_2O, g) - S^\circ(H_2O, l) = 188.8 - 69.9 = 118.9 \text{ J K}^{-1}$$

Thus, at 90 °C,

$$\Delta G^\circ = \Delta H^\circ - T\,\Delta S^\circ$$
$$= 44.0 \text{ kJ} - (363.2 \text{ K})(118.9 \text{ J K}^{-1}) = 44.0 - 43.2 \text{ kJ} = +0.8 \text{ kJ}$$

and, at 110 °C,

$$\Delta G^\circ = 44.0 \text{ kJ} - (383.2 \text{ K})(118.9 \text{ J K}^{-1}) = 44.0 - 45.6 \text{ kJ} = -1.6 \text{ kJ}$$

The signs of ΔG tell you that the process is spontaneous at 110° but isn't spontaneous at 90°.

(b) The two phases will be in equilibrium when ΔG for the process equals zero. You need to calculate the T at which this occurs, still using Eq. (6-29):

$$0 = \Delta H - T\,\Delta S \qquad T\,\Delta S = \Delta H$$

$$T = \frac{\Delta H}{\Delta S} = \frac{44.0\ \text{kJ}}{118.9\ \text{J K}^{-1}} = 370.1\ \text{K}$$

You know, of course, that the answer should be 373.2 K, the normal boiling point of water. The discrepancy probably is caused by the use of values for 298 K.

PROBLEM 6-19 Calculate the normal boiling point of Br_2, using values of ΔH_f° and S° from Table A-1.

Solution: This problem is like Problem 6-18b. You want the temperature at which $\Delta G = 0$ under 1 atm pressure. You'll need ΔH° and ΔS° for the process

$$Br_2(l,\ 1\ \text{atm}) \longrightarrow Br_2(g,\ 1\ \text{atm})$$

From Eq. (4-5), assuming $n = 1$, you get

$$\Delta H_r^\circ = \Delta H_f^\circ(Br_2, g) - \Delta H_f^\circ(Br_2, l) = 30.9 - 0 = 30.9\ \text{kJ}$$

and, from Eq. (5-21),

$$\Delta S_r^\circ = S^\circ(Br_2, g) - S^\circ(Br_2, l) = 245.5 - 152.2 = 93.3\ \text{J K}^{-1}$$

Then, from Eq. (6-29), when $\Delta G = 0$,

$$T = \frac{\Delta H}{\Delta S} = \frac{30.9\ \text{kJ}}{93.3\ \text{J K}^{-1}} = 331\ \text{K} = 58\ ^\circ\text{C}$$

The accepted value is 58.8 °C, so the agreement is better than in Problem 6-18, perhaps because the temperature is closer to 298 K, or because the C_P's of gas and liquid are closer in value for bromine.

PROBLEM 6-20 Will the "water gas" reaction $C(gr) + H_2O(g) \rightarrow CO(g) + H_2(g)$ be spontaneous at **(a)** 298 K? **(b)** 1500 K? Base your answer on calculations of ΔG.

Solution:

(a) At 298 K, the calculation is a straightforward application of Eq. (6-52), with values of ΔG_f° obtained from Table A-1:

$$\begin{aligned}\Delta G_r^\circ &= \Delta G_f^\circ(CO) + \Delta G_f^\circ(H_2) - \Delta G_f^\circ(C, gr) - \Delta G_f^\circ(H_2O, g)\\ &= -137.2 + 0 - 0 - (-228.6) = +91.4\ \text{kJ}\end{aligned}$$

The reaction is not spontaneous at 298 K.

(b) To calculate ΔG at another temperature, you must use the Gibbs–Helmholtz equation, Eq. (6-56), as modified to give Eq. (6-57). This equation may be integrated with or without limits to give either Eq. (6-58) or Eq. (6-60). For either case, you must evaluate the integration constant of Eq. (4-12). To do this, you need ΔH and heat capacity coefficients at some temperature.

To save time, refer to Problem 4-10, where the same reaction was used and $\Delta H^\circ(298)$, Δa, Δb, and Δc are given. Solving Eq. (4-12) for C gives

$$C = \Delta H - \Delta a T - \frac{\Delta b}{2}T^2 - \frac{\Delta c}{3}T^3$$

Now substitute the numbers from Problem 4-10 to obtain

$$C = 120.9 \times 10^3 - (-30.864)T - \left(\frac{62.07 \times 10^{-3}}{2}\right)T^2 - \left(\frac{-432.33 \times 10^{-7}}{3}\right)T^3$$

At 298 K,

$$C = 120.9 \times 10^3 + 30.864(298) - (31.04 \times 10^{-3})(298)^2 + (144.11 \times 10^{-7})(298)^3 = 127.7\ \text{kJ}$$

Because only one higher temperature is involved, you should use the method of integrating with limits, which yields Eq. (6-60), or

$$\left(\frac{\Delta G}{T}\right)_{T_2} - \left(\frac{\Delta G}{T}\right)_{T_1} = -\Delta a \ln\frac{T_2}{T_1} - \frac{\Delta b}{2}(T_2 - T_1) - \frac{\Delta c}{6}(T_2^2 - T_1^2) + C\left(\frac{1}{T_2} - \frac{1}{T_1}\right)$$

In part (a), you determined ΔG at 298 K ($= T_1$). Transposing the $(\Delta G/T)_{T_1}$ term and substituting numbers, you get

$$\frac{\Delta G}{1500} = \frac{91.4 \times 10^3}{298} - (-30.864) \ln \frac{1500}{298} - \left(\frac{62.07 \times 10^{-3}}{2}\right)(1500 - 298)$$
$$- \left(\frac{-432.33 \times 10^{-7}}{6}\right)(1500^2 - 298^2) + (127.7 \times 10^3)\left(\frac{1}{1500} - \frac{1}{298}\right) = -8.5 \text{ J}$$
$$\Delta G = -13 \text{ kJ}$$

The negative sign tells you that, at 1500 K, the water gas reaction is spontaneous.

PROBLEM 6-21 Complete the derivation of Eq. (6-67): $\ln \gamma = \frac{1}{RT} \int_0^P A\, dP$.

Solution: You need to differentiate Eq. (6-63) and compare the result to Eq. (6-40), as indicated in Section 6-7B. Therefore you begin with

$$G = G^\circ + RT \ln f \qquad \textbf{[Eq. (6-63)]}$$

and differentiate (remembering that G° is a constant) to obtain

$$dG = 0 + RT\, d(\ln f)$$

From Eq. (6-40),

$$dG = V\, dP$$

Thus

$$RT\, d(\ln f) = V\, dP \qquad \text{and} \qquad d(\ln f) = \frac{V}{RT}\, dP$$

Now you express the actual volume, V, as the ideal volume plus a factor A, which represents the deviation from ideality, i.e.,

$$V = \frac{RT}{P} + A$$

Substitute to get

$$d(\ln f) = \frac{1}{RT}\left(\frac{RT}{P} + A\right) dP = \frac{dP}{P} + \frac{A}{RT}\, dP$$

Since

$$\frac{dP}{P} = d(\ln P)$$

$$d(\ln f) - d(\ln P) = \frac{A}{RT}\, dP$$

And since $\gamma = f/P$ [Eq. (6-66)],

$$d\left(\ln \frac{f}{P}\right) = d(\ln \gamma) = \frac{A}{RT}\, dP$$

Integrating between P and $P = 0$ gives

$$\ln \gamma(P) - \ln \gamma(P = 0) = \frac{1}{RT} \int_0^P A\, dP$$

Since $\gamma(P = 0) = 1$ and $\ln(1) = 0$,

$$\ln \gamma = \frac{1}{RT} \int_0^P A\, dP \qquad \textbf{[Eq. (6-67)]}$$

PROBLEM 6-22 Continue Problem 6-21 by deriving another form of Eq. (6-67) that contains the compressibility factor ($Z = PV/RT$).

Solution: This involves manipulation of the factor A/RT. You can rearrange the expression (in Problem 6-21) that defines A to

$$A = V - \frac{RT}{P}$$

Divide both sides by RT to get

$$\frac{A}{RT} = \left(\frac{1}{RT}\right)\left(V - \frac{RT}{P}\right)$$

Since

$$Z = \frac{PV}{RT}, \qquad V = \frac{ZRT}{P}$$

Then

$$\frac{A}{RT} = \left(\frac{1}{RT}\right)\left(\frac{ZRT}{P} - \frac{RT}{P}\right) = \frac{Z}{P} - \frac{1}{P}$$

Thus Eq. (6-67) becomes

$$\ln \gamma = \int_0^P \frac{Z-1}{P}\, dP$$

PROBLEM 6-23 Estimate the fugacity of water vapor at 700 K and 500 atm using a typical plot of fugacity coefficient versus reduced pressure.

Solution: First, you must calculate the reduced temperature and reduced pressure, using Eq. (2-17). The critical constants for water are given in Table 2-1, i.e., $T_c = 374\ °C$ and $P_c = 218$ atm.

$$T_r = \frac{T}{T_c} = \frac{700\text{ K}}{647\text{ K}} = 1.08; \qquad P_r = \frac{P}{P_c} = \frac{500\text{ atm}}{218\text{ atm}} = 2.29$$

A plot of fugacity coefficient versus reduced pressure is not included in this book, but there should be one in your textbook. Estimating between the curves of T_r for 1.0 and 1.1 at P_r of 2.29, you should get a value for γ of about 0.49. From Eq. (6-66),

$$f = \gamma P = 0.49(500\text{ atm}) = 245\text{ atm}$$

Supplementary Exercises

PROBLEM 6-24 Are all spontaneous reactions exothermic? A substance undergoing a spontaneous process always increases in entropy (true or false?). Free energy (increases) (decreases) (stays the same) in a spontaneous change, but ______________ when action proceeds at equilibrium. The Gibbs free energy of a gas ______________ when the gas is compressed. If a process is occurring at equilibrium and at constant volume and temperature, which thermodynamic state function remains constant? When 2.0 mol $H_2O(l)$ evaporates at 1 atm pressure and 100 °C, $\Delta G =$ _____.

PROBLEM 6-25 The pressure on 2.00 L of choloroform is raised from 1.00 atm to 10.0 atm at 20 °C. What is the change in the enthalpy of the compound? The coefficient of thermal expansion is $1.273 \times 10^{-3}\ K^{-1}$.

PROBLEM 6-26 If the pressure on 3 mol of an ideal gas initially at STP is increased isothermally to 5 atm, what is the change in the Gibbs free energy?

PROBLEM 6-27 One mole of a certain gas occupies 26.0 L at 300 K. If it undergoes an isothermal expansion with $\Delta A = 2.0$ kJ, what is the new volume?

PROBLEM 6-28 Calculate ΔG for the isothermal solidification of 1.0 mol $H_2O(l)$ at -8 °C. At this temperature, the vapor pressures are $H_2O(s) = 2.326$ mm Hg and $H_2O(l) = 2.514$ mm Hg.

PROBLEM 6-29 An ideal gas ($C_P = \frac{7}{2}R$; $n = 2$ mol) at 300 K and 3 atm is allowed to expand isothermally and reversibly until the pressure is 0.50 atm. Calculate q, w, ΔE, ΔH, ΔS, ΔA, and ΔG for this process.

PROBLEM 6-30 Calculate ΔH°, ΔS°, and ΔG° at 25 °C for the burning of one mole of gaseous benzene.

PROBLEM 6-31 For the reaction in Problem 6-30, what is the numerical difference between ΔG° and ΔA°?

PROBLEM 6-32 For the reaction in Problem 6-30, what is the numerical value of the partial derivative $\left(\frac{\partial(\Delta G/T)}{\partial T}\right)_P$?

PROBLEM 6-33 Using a table of standard free energies of formation, calculate ΔG° at 298.2 K for the complete combustion of propane.

PROBLEM 6-34 Calculate ΔG° at 700 K for the Haber process

$$N_2(g) + 3H_2(g) \longrightarrow 2NH_3(g)$$

Use the appropriate tables to obtain the necessary values. What does the answer tell you?

Answers to Supplementary Exercises

6-24 No; false; decreases, stays the same; increases; Helmholtz energy; zero **6-25** 11.3 L atm **6-26** 11.0 kJ **6-27** 11.7 L **6-28** -171.4 J **6-29** $w = \Delta A = \Delta G = -8.94$ kJ; $q = 8.94$ kJ; $\Delta E = \Delta H = 0$; $\Delta S = 29.8$ J K^{-1} **6-30** $\Delta H^\circ = -3.30 \times 10^3$ kJ; $\Delta S^\circ = -316$ J; $\Delta G^\circ = -3.21 \times 10^3$ kJ **6-31** 6.19 kJ **6-32** 37.1 K^{-2} **6-33** -2.11×10^3 kJ **6-34** 53.3 kJ; The reaction isn't spontaneous.

CHEMICAL EQUILIBRIUM

THIS CHAPTER IS ABOUT

☑ Equilibrium State
☑ Equilibrium Constants
☑ Applications
☑ The Thermodynamic Equilibrium Constant

7-1. Equilibrium State

A. Physical and chemical change

A process in which the nature of the substance(s) involved doesn't change is called a *physical change* or *physical process*. But when one or more of the substances involved in a process are changed into one or more other substances, the process is called a **chemical change** or **chemical process**.

B. Static and dynamic equilibrium

A system in which no activity is occurring is said to be in **static equilibrium**. A book lying on a desk is an example of the state of static equilibrium.

Most systems of interest to chemists involve some activity—usually the motion of molecules. Molecules are in motion, for example, during the evaporation of liquid H_2O (physical change) and the dissociation of gaseous PCl_5 into PCl_3 and Cl_2 (chemical change). If the products of these changes escape, both processes will continue until the starting material is gone. But, if the system is enclosed, the products eventually will begin to form the original material: Some water vapor will condense to liquid; some PCl_3 will react with Cl_2 to form PCl_5. When the forward and reverse processes proceed at the same rate, we say that the system is at *equilibrium*; we can't observe any change, and the relative concentrations of all components remain constant. To make it clear that activity continues, we call this a state of **dynamic equilibrium**. We identify a system that reaches chemical dynamic equilibrium by a double arrow in the equation for the reaction and call it a *reversible reaction.*

warning: A spontaneous reaction, which is *thermodynamically irreversible*, may be *chemically reversible*.

EXAMPLE 7-1: Write balanced equations for some reversible chemical reactions that come to a state of dynamic equilibrium.

Solution: There are many chemical reactions that come to a state of dynamic equilibrium. For example,

$$PCl_5(g) \rightleftharpoons PCl_3(g) + Cl_2(g)$$
$$N_2(g) + 3H_2(g) \rightleftharpoons 2NH_3(g)$$
$$2NO_2(g) \rightleftharpoons N_2O_4(g)$$

C. Homogeneous and heterogeneous equilibrium

If all components of a system in dynamic equilibrium are in the same phase, we call the equilibrium *homogeneous*. The three systems in Example 7-1 are of this type: All components are in the gas phase. If two or more phases are present in a system in dynamic equilibrium, we call the equilibrium *heterogeneous*.

EXAMPLE 7-2: Write balanced equations for some systems in heterogeneous equilibrium.

Solution: Again, there are many possible examples.

Physical equilibrium: $H_2O(l) \rightleftharpoons H_2O(g)$ at 100 °C

$S_{\text{rhombic}} \rightleftharpoons S_{\text{monoclinic}}$ at 95.6 °C

Chemical equilibrium: $NH_4CO_2NH_2(s) \rightleftharpoons CO_2(g) + 2NH_3(g)$

D. Extent of reaction

Although this concept was introduced as the *degree of advancement* by De Donder in 1922, it didn't begin appearing in most introductory physical chemistry texts until the IUPAC recommended its use. (It is quite similar to the *degree of dissociation*, which we will describe later.)

We may write a general expression for any chemical reaction as

$$aA + bB = cC + dD$$

or as

$$\nu_1 A_1 + \nu_2 A_2 = \nu_3 A_3 + \nu_4 A_4 \tag{7-1}$$

Then we can rearrange

$$\nu_3 A_3 + \nu_4 A_4 - \nu_1 A_1 - \nu_2 A_2 = 0$$

and consolidate to get

$$\sum_i \nu_i A_i = 0 \tag{7-2}$$

This expression is a shorthand way of writing a chemical equation in which each A represents a chemical formula. The ν's are called the **stoichiometric coefficients** and are negative for reactants and positive for products. They are simply numbers, representing the relative amounts of each participant in the reaction. We could read Eq. (7-1) as "ν_2 molecules of A_2 will react with ν_1 molecules of A_1" or as "ν_2/ν_1 moles of A_2 react with one mole of A_1."

EXAMPLE 7-3: What are the values of ν in the equation

$$C_3H_8 + 5O_2 \longrightarrow 3CO_2 + 4H_2O$$

Solution: $\nu_{C_3H_8} = -1$ $\quad \nu_{O_2} = -5$ $\quad \nu_{CO_2} = 3$ $\quad \nu_{H_2O} = 4$

Recall that n refers to the amount of substance and has the unit of mole. We represent the initial amount of substance i by $n_{i,0}$ and the amount at any other time by n_i. The amount that has reacted can be expressed using the **extent of reaction**, ξ, defined as

$$-\nu_i \xi = n_{i,0} - n_i$$

or

EXTENT OF REACTION

$$\xi = \frac{n_i - n_{i,0}}{\nu_i} \tag{7-3}$$

The extent of reaction is the *same* for every reactant and product at any given time.

EXAMPLE 7-4: If 2.0 mol O_2 is used in the reaction in Example 7-3, what is the extent of reaction?

Solution: Since 2 mol is consumed, we have $n_i - n_{i,0} = -2$ mol. For oxygen, $\nu = -5$. Substituting in Eq. (7-3), we obtain

$$\xi = \frac{-2\text{ mol}}{-5} = 0.4\text{ mol}$$

EXAMPLE 7-5: What is the extent of reaction if we choose carbon dioxide as the basis for the calculation in Example 7-4?

Solution: From the balanced equation in Example 7-3, 5 mol O_2 would produce 3 mol CO_2, or the 2 moles would produce $(\frac{2}{5})(3) = 1.2$ mol CO_2. Thus

$$n_{CO_2} - n_{CO_2,0} = 1.2\text{ mol}$$

and

$$\xi = \frac{1.2\text{ mol}}{3} = 0.4\text{ mol}$$

Examples 7-4 and 7-5 show that the extent of reaction is the same for each participant in the reaction. Consequently, we can use the term *mole of reaction*; i.e., we can express the results of those Examples as "0.4 mol of C_3H_8 reacted," "0.4 mol of $5O_2$ reacted," or "0.4 mol of $4H_2O$ was produced." Of course, 0.4 mol $4H_2O$ is the same as 1.6 mol H_2O.

This odd-sounding terminology is helpful when we use heats of reaction. The ΔH° for the reaction is 2×10^3 kJ mol^{-1}, which now means 2×10^3 kJ per mole of $5O_2$, 2×10^3 kJ per mole of $4H_2O$, etc. Thus it becomes essential to state the stoichiometric equation that accompanies any ΔH or ΔG.

E. Criteria for equilibrium

In Chapter 6 we saw that the conditions for a system to be at equilibrium are

$$\text{at constant } T \text{ and } V, \quad \Delta A = 0 \qquad \textbf{[Eq. (6-31)]}$$

and

$$\text{at constant } T \text{ and } P, \quad \Delta G = 0 \qquad \textbf{[Eq. (6-32)]}$$

We also saw that

$$\text{at constant } T \text{ and } P, \quad dG = \Sigma \mu_i \, dn_i \qquad \textbf{[Eq. (6-75)]}$$

For a system in physical equilibrium, each component's chemical potential in every phase must be the same:

$$\begin{aligned} \mu_1(\alpha) &= \mu_1(\beta) = \cdots \\ \mu_2(\alpha) &= \mu_2(\beta) = \cdots \end{aligned} \qquad \textbf{[Eq. (6-78)]}$$

For a chemical reaction at equilibrium, the sum of the chemical potential multiplied by the change in amount for each product must equal the sum of the chemical potential multiplied by the change in amount for each reactant. Overall, since $dG = 0$ at equilibrium, Eq. (6-75) becomes

$$\Sigma \mu_i \, dn_i = 0 \qquad \textbf{[Eq. (6-76)]}$$

Let's rewrite Eq. (7-3) and differentiate to get

$$n_i = n_{i,0} + \nu_i \xi$$

Since $n_{i,0}$ and ν_i are constant,

$$dn_i = \nu_i \, d\xi$$

Substituting in Eq. (6-75), we obtain

$$dG = \Sigma \mu_i \nu_i \, d\xi$$

or

$$\frac{dG}{d\xi} = \left(\frac{\partial G}{\partial \xi}\right)_{T,P} = \sum_i \nu_i \mu_i \tag{7-4}$$

At equilibrium, G for the reaction mixture is at a minimum; thus

$$\left(\frac{\partial G}{\partial \xi}\right)_{T,P} = 0$$

and

$$\left(\sum_i \nu_i \mu_i\right) = 0 \tag{7-5}$$

Equation (7-5) applies to all chemical equilibria in a closed system.

7-2. Equilibrium Constants

A. Ideal gases; homogeneous systems

In your first-year chemistry course, you worked many problems using equilibrium constants that were stated in concentrations, particularly for the special cases of ionization constants and solubility products. You probably learned a definition something like:

The equilibrium constant of a reaction equals the product of the equilibrium concentrations of the reaction products divided by the product of the equilibrium concentrations of the reactants, each concentration being raised to the power equal to the coefficient of its compound in the balanced equation.

Thus, for the reaction $a\text{A} + b\text{B} \rightleftharpoons c\text{C} + d\text{D}$,

$$K_c = \frac{[\text{C}]^c[\text{D}]^d}{[\text{A}]^a[\text{B}]^b} \tag{7-6}$$

where the brackets represent molar concentrations (mol L^{-1}).

EXAMPLE 7-6: Write the equilibrium constant expression K_c for the reaction

$$2\text{SO}_2(g) + \text{O}_2(g) \rightleftharpoons 2\text{SO}_3(g)$$

Solution: We need to write an expression like Eq. (7-6). Putting the products on top, we have

$$K_c = \frac{[\text{SO}_3]^2}{[\text{SO}_2]^2[\text{O}_2]}$$

Expressions similar to Eq. (7-6) can be written using mole fractions, pressures, fugacities, and activities. The last two, of course, apply to both nonideal and ideal behavior.

EXAMPLE 7-7: Write the equilibrium constant expression in pressures for the reaction

$$2\text{NO}_2(g) \rightleftharpoons 2\text{NO}(g) + \text{O}_2(g)$$

Solution: We first write a general expression like Eq. (7-6), using the partial pressures of the gases at equilibrium:

$$K_P = \frac{P_\text{C}^c P_\text{D}^d}{P_\text{A}^a P_\text{B}^b} \tag{7-7}$$

For this reaction

$$K_P = \frac{P_{NO}^2 P_{O_2}}{P_{NO_2}^2}$$

EXAMPLE 7-8: Assume ideal behavior and derive an expression relating K_c and K_P for a given reaction.

Solution: From the ideal gas equation, for a given gas A,

$$P_A = \frac{n_A RT}{V}$$

but

$$\frac{n_A}{V} = \text{mol}_A\ \text{L}^{-1} = M_A = [\text{A}]$$

Thus we can write

$$P_A = [\text{A}]RT$$

Substituting similar expressions for the other partial pressure terms in Eq. (7-7) gives

$$K_P = \frac{([\text{C}]RT)^c([\text{D}]RT)^d}{([\text{A}]RT)^a([\text{B}]RT)^b}$$

Separating the concentration terms from the RT terms yields

$$K_P = \frac{[\text{C}]^c[\text{D}]^d}{[\text{A}]^a[\text{B}]^b}\frac{(RT)^c(RT)^d}{(RT)^a(RT)^b} = K_c(RT)^{c+d-a-b}$$

$$K_P = K_c(RT)^{\Sigma\nu} \tag{7-8}$$

where $\Sigma\nu$ has its usual meaning but refers to gases only. Note that when we state volume in liters, we have to express pressure in atmospheres. In general, the pressures in Eq. (7-7) could be expressed in any units.

B. Real gases; homogeneous systems

If pressures are replaced by fugacities, Eq. (7-7) becomes

$$K_f = \frac{f_C^c f_D^d}{f_A^a f_B^b} \tag{7-9}$$

We can relate K_f to K_P by using the fugacity coefficient, $\gamma = f/P$ [Eq. (6-66)]. Substituting $f = \gamma P$ into Eq. (7-9) and rearranging gives

$$K_f = \frac{\gamma_C^c \gamma_D^d}{\gamma_A^a \gamma_B^b}\frac{P_C^c P_D^d}{P_A^a P_B^b}$$

or

$$K_f = K_\gamma K_P \tag{7-10}$$

The most general equilibrium constant is expressed in activities; i.e.,

$$K_a = \frac{a_C^c a_D^d}{a_A^a a_B^b} \tag{7-11}$$

C. Characteristics of constants

Three characteristics of equilibrium constants are important in calculations:

1. The value of the constant depends on how the stoichiometric equation is written.
2. The equilibrium constant for the reverse reaction is the reciprocal of the constant for the forward reaction.

3. If a given system consists of the sum of a series of equilibrium reactions, the overall equilibrium constant is the product of the constants of the several reactions, or

$$K_{\text{overall}} = K_1 K_2 \cdots K_n \qquad \textbf{(7-12)}$$

EXAMPLE 7-9: Write the equilibrium constant expression in pressures for the reaction of Example 7-7, balanced as

$$NO_2(g) \rightleftharpoons NO(g) + \tfrac{1}{2}O_2(g)$$

and compare it to the expression obtained in Example 7-7.

Solution: Substituting in Eq. (7-7), we get

$$K'_P = \frac{P_{NO}P_{O_2}^{1/2}}{P_{NO_2}}$$

In Example 7-7 we had

$$K_P = \frac{P_{NO}^2 P_{O_2}}{P_{NO_2}^2}$$

Clearly, $K_P = (K'_P)^2$. In general, if a balanced equation is multiplied by a factor, the corresponding equilibrium constant is changed by the power equal to that factor.

EXAMPLE 7-10: Write the equilibrium constant expressions for the forward and reverse reactions of

$$H^+(aq) + CN^-(aq) \rightleftharpoons HCN(aq)$$

and compare the expressions.

Solution: Substituting in Eq. (7-6), we have

$$K_{c_{\text{fwd}}} = \frac{[HCN]}{[H^+][CN^-]} \qquad \text{and} \qquad K_{c_{\text{rev}}} = \frac{[H^+][CN^-]}{[HCN]}$$

Clearly,

$$K_{c_{\text{rev}}} = (K_{c_{\text{fwd}}})^{-1} \qquad \textbf{(7-13)}$$

EXAMPLE 7-11:

(a) Write the equilibrium constant expressions for the following reactions:

1. $H_2SO_4 \rightleftharpoons 2H^+ + SO_4^{2-}$ 2. $H_2SO_4 \rightleftharpoons H^+ + HSO_4^-$ 3. $HSO_4^- \rightleftharpoons H^+ + SO_4^{2-}$

(b) Multiply K_{c_2} and K_{c_3} and compare to K_{c_1}.

Solution:

(a) Substitute in Eq. (7-6):

$$K_{c_1} = \frac{[H^+]^2[SO_4^{2-}]}{[H_2SO_4]} \qquad K_{c_2} = \frac{[H^+][HSO_4^-]}{[H_2SO_4]} \qquad K_{c_3} = \frac{[H^+][SO_4^{2-}]}{[HSO_4^-]}$$

(b) $K_{c_2}K_{c_3} = \dfrac{[H^+][HSO_4^-]}{[H_2SO_4]}\dfrac{[H^+][SO_4^{2-}]}{[HSO_4^-]} = \dfrac{[H^+]^2[SO_4^{2-}]}{[H_2SO_4]}$

This product is identical to K_{c_1}. Note that we have illustrated Eq. (7-12).

D. Heterogeneous systems

When substances in condensed phases (solid or liquid) are part of the equilibrium mixture, they are omitted from the equilibrium constant (except for the constant expressed in activities, since condensed phases can have activities). Essentially, we assume that the concentrations don't change and are incorporated into K_c. More precisely, we assume that the condensed pure substance is in its standard state for the reaction temperature.

EXAMPLE 7-12: Write the equilibrium constant expression K_P for

(a) $SiF_4(g) + 2H_2O(g) \rightleftharpoons SiO_2(s) + 4HF(g)$ **(b)** $CaCO_3(s) \rightleftharpoons CaO(s) + CO_2(g)$

Solution: Use Eq. (7-7) with the partial pressures of the gases.

(a) $K_P = \dfrac{P_{HF}^4}{P_{SiF_4}P_{H_2O}^2}$ **(b)** $K_P = P_{CO_2}$

7-3. Applications

A. Le Chatelier's Principle

One statement of Le Chatelier's principle is

If a stress is applied to a system that is in a state of dynamic equilibrium, the position of equilibrium *will shift in the direction that will relieve the stress.*

By "position of equilibrium" we mean an imagined point between no reaction and complete reaction at which the concentrations of the components are such that the system is in equilibrium. Note that we didn't say, "the system will shift." The system doesn't move; rather, a change occurs *within* the system. In the case of chemical equilibrium, the rates of the forward and the reverse reactions will change, causing a change in the relative amounts of the components of the system. Eventually, the two rates will become equal again, and a new state of equilibrium will be established. If the temperature is constant, the equilibrium constant doesn't change.

- An increase in the concentration of one component increases the rate in the direction that uses up some of the added material.

EXAMPLE 7-13: A solution of acetic acid is at equilibrium at constant pressure. If some hydrogen ion is added, **(a)** what will happen? **(b)** What can we predict about the concentration of each component, relative to its initial concentration at the new position of equilibrium?

Solution:

(a) First, we need the balanced equation for the overall reaction. We write it so that the forward reaction is the ionization of acetic acid, or

$$HAc(aq) \rightleftharpoons H^+(aq) + Ac^-(aq)$$

If H^+ is added, the rate of the reverse reaction will increase and some of the added H^+ will combine with some of the Ac^- to form HAc. A new state of equilibrium will be established with the adjustment of the three concentrations to again satisfy the equilibrium constant expression.

(b) At the new state of equilibrium, the new $[H^+]$ will be greater than it was at the first state of equilibrium because all of the added amount isn't used up. The new [HAc] will be greater than it was because some HAc is produced by the reverse reaction in consuming some of the added H^+. The new $[Ac^-]$ will be less than it was because some Ac^- is consumed to produce the new HAc.

- An increase in the pressure on the system increases the rate in the direction that reduces the total pressure exerted by the gases in the system.

Since the pressure depends on the number of molecules, the change will occur in the direction that reduces the number of molecules. We determine that direction by examining the balanced equation for the reaction.

EXAMPLE 7-14: Predict the effect of an increase in pressure on the concentrations of the various components of the equilibrium system

$$PCl_5(g) \rightleftharpoons PCl_3(g) + Cl_2(g)$$

Solution: By inspection, we see that there are two molecules on the right-hand side and one molecule on the left-hand side of the equation. Thus the total number of molecules will be reduced when the reverse reaction occurs. Therefore, an increase in pressure will reduce the amounts of PCl_3 and Cl_2 by causing those gases to react, increasing the amount of PCl_5.

- The presence or absence of a catalyst has no effect on the position of equilibrium. However, it would affect the rate at which equilibrium is reached.
- Equilibrium constants change with temperature.

We can predict the direction of the change and the relative concentrations of the components at the new state of equilibrium if we know the heat of reaction. We can avoid having to learn rules for endothermic and exothermic cases, or for positive and negative enthalpies of reaction, by adding heat to the balanced equation as though it were a reactant or product. Raising the temperature makes more heat available, and the effect is the same as adding more of one of the chemical reactants or products: Some of the excess will be consumed. Actually, raising the temperature will increase the rates of both the forward and the reverse reactions; however, the increase will be greater for the direction in which heat is consumed. Thus the position of equilibrium will be shifted in the direction that uses heat.

EXAMPLE 7-15: The enthalpy of reaction of $2NO_2(g) \rightleftharpoons N_2O_4(g)$ is about -55 kJ at 298 K. If the system is at equilibrium at 298 K and the temperature is raised, **(a)** which component will increase in concentration? **(b)** Will K_c increase or decrease?

Solution: We don't need the value of ΔH; we merely need to know that heat is given off, which tells us to write the equation

$$2NO_2(g) \rightleftharpoons N_2O_4(g) + \text{Heat}$$

(a) If we add heat (i.e., raise the temperature), the rate of the reverse reaction will increase. Some of the N_2O_4 will break down, and the concentration of NO_2 will increase.

(b) The expression for K_c is

$$K_c = \frac{[N_2O_4]}{[NO_2]^2}$$

If the concentration of NO_2 is larger and that of N_2O_4 is smaller at a higher temperature, then K_c at the higher temperature must be smaller than that at the lower temperature.

B. Calculations using concentrations

The calculation of K_c when all equilibrium concentrations are known involves only a straightforward substitution. The reverse calculation is that of equilibrium concentrations when K_c and the starting concentrations are known.

EXAMPLE 7-16: A solution of hydrocyanic acid is prepared by dissolving 0.10 mol HCN in enough water to make 1.0 L of solution. After some time, the concentration of hydrogen ion is found to be 6.3×10^{-6} *M*. Calculate the equilibrium constant.

Solution: We assume that a state of dynamic equilibrium has been reached and write the balanced equation

$$HCN(aq) \rightleftharpoons H^+(aq) + CN^-(aq)$$

Since all the coefficients are 1, we know that

$$[CN^-] = [H^+] = 6.3 \times 10^{-6}\ M$$
$$[HCN] = 0.10 - [H^+] \approx 0.10\ M$$

So

$$K_c = \frac{[H^+][CN^-]}{[HCN]} = \frac{(6.3 \times 10^{-6})(6.3 \times 10^{-6})}{0.10} = 4.0 \times 10^{-10}\ \text{mol L}^{-1}$$

EXAMPLE 7-17: For the reaction $PCl_5(g) \rightleftharpoons PCl_3(g) + Cl_2(g)$, $K_c = 4.0 \times 10^{-2}$. Calculate the equilibrium concentrations of the three gases if 5.0 mol PCl_5 is placed in a 50-L flask.

Solution: We need to use Eq. (7-6) and need algebraic quantities to represent the equilibrium concentrations. The best way to obtain these quantities is to set up a table showing the initial concentrations, the changes that occur, and the final concentrations. We fill in the table by using any numbers given, letting x represent a selected unknown, and referring to the balanced equation. For convenience we arrange the table so that entries for each reactant are in a column under that reactant in the balanced equation. We start with

	PCl_5	$\rightleftharpoons PCl_3$	$+ Cl_2$
Initial concentration	$\frac{5.0\ \text{mol}}{50\ \text{L}}$	0	0
Change			
Equilibrium concentration			

We let x represent the change in concentration of PCl_5, so a decrease is $-x$. Since all the coefficients in the balanced equation are 1, the amount of PCl_3 produced equals the amount of PCl_5 lost. Thus the change in PCl_3 is $+x$, and, similarly, the change in Cl_2 is $+x$. Entering these quantities in the table and adding the first and second rows to get the third, we have

	PCl_5	$\rightleftharpoons PCl_3$	$+ Cl_2$
Initial concentration	$0.10\ M$	0	0
Change	$-x$	$+x$	$+x$
Equilibrium concentration	$0.10 - x$	x	x

Now, we substitute the expressions for the equilibrium concentrations into Eq. (7-6) along with the value of K_c to obtain

$$4.0 \times 10^{-2} = \frac{(x)(x)}{0.10 - x}$$

Next, we cross-multiply and simplify by removing the powers of 10, so that

$$x^2 = 0.0040 - 0.040x$$

Rearranging, we get

$$x^2 + 0.040x - 0.0040 = 0$$

To solve for x, we must use the quadratic formula,

$$x = \frac{-b \pm \sqrt{b^2 - 4ac}}{2a}$$

For this case, $a = 1$, $b = 0.040$, and $c = -0.0040$, so

$$x = 0.046\ M$$

We chose the plus sign for the evaluation of the square root because the minus sign makes x negative, which is impossible. Thus the equilibrium concentrations are

$$[PCl_3] = [Cl_2] = x = 0.046\ M$$
$$[PCl_5] = 1 - x = 0.054\ M$$

A more general approach is to choose moles (y) rather than concentration for the unknown quantity. (In fact, it is the essential approach if volumes change.) For this problem, we would have

	$PCl_5 \rightleftharpoons$	PCl_3 +	Cl_2
Initial amount	5 mol	0	0
Change	$-y$	$+y$	$+y$
Equilibrium amount	$5 - y$	y	y

Equation (7-6) requires molar concentrations, so we must divide each term in the table by the volume (50 L), or

$$4.0 \times 10^{-2} = \frac{(y/50)(y/50)}{(5 - y)/50}$$

$$y = 2.3 \text{ mol}$$

The equilibrium concentration of Cl_2 is 2.3 mol/50 L = 0.046 M, as in the first calculation.

In Example 7-13, we saw the qualitative effect of adding more of one participant to a system in equilibrium. In the chemistry of solutions, the disturbance of the equilibrium by the addition of a compound containing one of the ions of the original compound is known as the **common ion effect**. The resulting depression of ionization is the basis of buffered solutions.

EXAMPLE 7-18: Let's extend Example 7-16 by adding a common ion and comparing the two hydrogen-ion concentrations. If 0.050 mol NaCN is added to the solution with no increase in volume, what is the concentration of H^+ when equilibrium is reestablished?

Solution: We proceed as before, setting up a table to help us arrive at algebraic expressions for the equilibrium concentrations. We assume that both acid and salt were added simultaneously and that the salt dissolves instantly to release its ions. Thus, before any acid molecules dissociate, the concentration of CN^- ions is 0.050 M.

	$HCN(aq) \rightleftharpoons$	$H^+(aq)$ +	$CN^-(aq)$
Initial concentration	0.10	0	0.050
Change	$-x$	$+x$	$+x$
Equilibrium concentration	$0.10 - x$	x	$0.050 + x$

Substituting into Eq. (7-6),

$$K_c = \frac{[H^+][CN^-]}{[HCN]}$$

we obtain

$$4.0 \times 10^{-10} = \frac{x(0.050 + x)}{0.10 - x}$$

When K_c is less than about 10^{-4}, x is usually small enough in comparison to the initial concentration to be neglected. This is even truer in common-ion problems, so we are justified in simplifying the preceding equation to

$$4.0 \times 10^{-10} = \frac{x(0.050)}{0.10}$$

$$x = 8.0 \times 10^{-10} = [H^+]$$

Comparing this concentration to that of 6.3×10^{-6} from Example 7-16, we see that the $[H^+]$ is smaller by a factor of 10^4.

C. Calculations using pressure

When all the components of a system are gases, pressure is an appropriate and convenient way to express amounts. Since pressure is proportional to concentration, there isn't much difference in the use of the two properties. With pressure, however, you may need to use the gas laws or work with mole fractions.

EXAMPLE 7-19: Calculate K_P for the reaction $N_2O_4(g) \rightleftharpoons 2NO_2(g)$ if, at equilibrium and at a total pressure of 0.50 atm, there are 3.0 mol N_2O_4 and 9.0 mol NO_2.

Solution: Before substituting in Eq. (7-7), we have to calculate the partial pressures of the two gases. We do this by multiplying the mole fractions by the total pressure:

$$P_{N_2O_4} = \left(\frac{3.0}{12.0}\right)(0.50) = 0.125 \qquad \text{and} \qquad P_{NO_2} = \left(\frac{9.0}{12.0}\right)(0.50) = 0.375$$

Substituting these values in Eq. (7-7), we get

$$K_P = \frac{P_{NO_2}^2}{P_{N_2O_4}} = \frac{(0.375)^2}{(0.125)} = 1.13 \text{ atm}$$

EXAMPLE 7-20: At 723 K, $K_P = 2.0 \times 10^{-2}$ for the reaction $2HI(g) \rightleftharpoons H_2(g) + I_2(g)$. If 5.00 mol HI is placed in a 100-L container and heated to 723 K, how many moles of H_2 will be present at equilibrium?

Solution: In order to use K_P, we need to determine the equilibrium partial pressure of each gas. Thus we need the total pressure and the mole fractions. First, we set up a table to get expressions for the equilibrium concentrations. In this case, we need only the number of moles, not the actual concentrations; we'll let x represent the moles of H_2.

	$2HI(g)$	$\rightleftharpoons H_2(g)$	$+ I_2(g)$
Initial amount	5 mol	0	0
Change	$-2x$	$+x$	$+x$
Equilibrium amount	$5 - 2x$	x	x

$$\text{Total moles} = 5 - 2x + x + x = 5$$

Note that the number of moles at equilibrium is equal to the starting number whenever the sums of the coefficients of gases on both sides of the balanced equation are equal; i.e., $\Sigma\nu = 0$.

Second, we need the total pressure. Using the ideal gas equation, we obtain

$$P = \frac{nRT}{V} = \frac{(5.00)(0.0821)(723)}{100} = 2.97 \text{ atm}$$

Then, to get the partial pressure of each gas, we multiply the total pressure by each mole fraction, or

$$P_{H_2} = P_{I_2} = \left(\frac{x}{5}\right)(2.97)$$

$$P_{HI} = \left(\frac{5 - 2x}{5}\right)(2.97)$$

Sometimes, it is convenient to include the expressions for mole fractions in the table. Instead of calculating them, we would usually substitute these expressions directly into the K_P expression. Often, we can use cancellation to simplify calculations; i.e.,

$$K_P = \frac{P_{H_2}P_{I_2}}{P_{HI}^2} = \frac{(X_{H_2}P)(X_{I_2}P)}{(X_{HI}P)^2}$$

We see that the total pressure cancels out, so it was not necessary to calculate it for this case. This will occur whenever the coefficients of gases on both sides of the equation are equal. (If $\Sigma\nu \neq 0$, some pressures won't cancel.) Now

$$K_P = \frac{X_{H_2}X_{I_2}}{X_{HI}^2} = \frac{\left(\frac{x}{5}\right)\left(\frac{x}{5}\right)}{\left(\frac{5-2x}{5}\right)^2} = \frac{x^2}{(5-2x)^2} = 2.0 \times 10^{-2}$$

Another simplification is possible, i.e., taking the square root of both sides of the equation:

$$\frac{x}{5-2x} = 1.4 \times 10^{-1} = 0.14$$

$$x = 0.70 - 0.28x = \frac{0.70}{1.28} = 0.55 \text{ mol } H_2$$

In Example 7-20, we effectively derived another equilibrium constant. In addition to those of Eqs. (7-6), (7-7), (7-9), and (7-11), we can write one in mole fractions:

$$K_X = \frac{X_C^c X_D^d}{X_A^a X_B^b} \tag{7-14}$$

Also from Example 7-20, it should be clear that

$$K_P = K_X P^{\Sigma\nu} \tag{7-15}$$

Note that the simplifications in Example 7-20 were made possible by two facts: $\Sigma\nu(\text{gas}) = 0$, and there was only one initial concentration. In other cases, the calculations may be more complex. Always set up the algebraic expression and look for ways to simplify it before doing any algebra or arithmetic; you may be able to cancel or take a root.

We could have set up the solution to Example 7-20 using the extent of reaction. From Eq. (7-3), we have the amount of substance that has reacted, or

$$n_{i,0} - n_i = -\nu\xi$$

The table of values would be

	$2HI(g) \rightleftharpoons$	$H_2(g)$	$+ I_2(g)$
Initial amount	5 mol	0	0
Change	$-\nu\xi$	$+\nu\xi$	$+\nu\xi$
Equilibrium amount	$5 - 2\xi$	ξ	ξ

The last line is the same as that obtained using x. The rest of the calculation is identical, giving $\xi = 0.55$ mol. An additional step is necessary to get the answer:

$$\text{Moles of } H_2 = \nu\xi = (1)(0.55) = 0.55 \text{ mol}$$

D. Degree of dissociation

Many equilibrium processes involve a single substance that is decomposing, or *dissociating*, into two or more substances. We already have used $PCl_5 \rightleftharpoons PCl_3 + Cl_2$ and $2HI \rightleftharpoons H_2 + I$. The amount of decomposition often is described by the **degree of dissociation** α, defined as the fraction of the initial amount a (or initial concentration, C_0) decomposed:

$$\alpha = \frac{\text{Amount dissociated}}{\text{Initial amount}} = \frac{x}{a} = \frac{x}{C_0} = \frac{\xi}{n_0} \tag{7-16}$$

If we convert the fraction to a percentage, we have the **percent dissociation**. In the special case of dissociation in solution to form ions, we use **degree of ionization** and **percent ionization**. For dissociation systems, the equilibrium constant may be expressed in terms of α.

EXAMPLE 7-21: Derive expressions for K_c using α and C_0 for the following general reactions:

(a) $AB \rightleftharpoons A + B$ **(b)** $AB_2 \rightleftharpoons A + 2B$

Solution:

(a) We set up the usual table, with C_0 as the initial concentration of AB. The amount of AB that breaks down is x, but, by Eq. (7-16), $x = \alpha C_0$.

	AB	$\rightleftharpoons$	A	+	B
Initial concentration	C_0		0		0
Change	$-\alpha C_0$		$+\alpha C_0$		$+\alpha C_0$
Equilibrium concentration	$C_0 - \alpha C_0$		αC_0		αC_0

From Eq. (7-6)

$$K_c = \frac{[A][B]}{[AB]} = \frac{(\alpha C_0)(\alpha C_0)}{C_0(1-\alpha)} = \frac{\alpha^2}{1-\alpha} C_0$$

(b) Again, we set up a table:

	AB_2	$\rightleftharpoons$	A	+	2B
Initial concentration	C_0		0		0
Change	$-\alpha C_0$		$+\alpha C_0$		$+2\alpha C_0$
Equilibrium concentration	$C_0(1-\alpha)$		αC_0		$2\alpha C_0$

Substituting in Eq. (7-6), we obtain

$$K_c = \frac{[A][B]^2}{[AB_2]} = \frac{\alpha C_0(2\alpha C_0)^2}{C_0(1-\alpha)} = \frac{4\alpha^3}{1-\alpha} C_0^2$$

We see that the actual expression depends on the coefficients in the balanced equation.

E. Fundamental concept of equilibrium: A self-test

To conclude this section on applications, we ask you to take a brief test without further explanation and without looking at the answer. This test, presented as Example 7-22, was adapted from Johnstone, McDonald, and Webb, *Education in Chemistry*, *14*:169 (1977).

EXAMPLE 7-22: In the familiar Haber industrial process, a mixture of three gases exists at equilibrium:

$$N_2(g) + 3H_2(g) \rightleftharpoons 2NH_3(g)$$

From the list below, select the best response to the following statement: Application of increased pressure to *only* the right-hand side of the system will drive the position of equilibrium to the left.

(A) The statement is correct.
(B) The statement is incorrect; the position of equilibrium would in fact be driven to the right.
(C) The process described is impossible.
(D) The statement is correct so long as the nitrogen and hydrogen are continuously removed.
(E) Insufficient information is given to assess the accuracy of the statement.

Solution: The correct response is answer C. If you chose incorrectly, think carefully about the nature of a system of gases.

The crucial point in the question is the phrase "to *only* the right-hand side of the system." Since the system is a mixture of completely intermingled gases, it is impossible to apply pressure to two of the gases and not to the third. The gases aren't separated into two compartments, one on the left holding reactants and the other on the right holding products. (Such a system *could* be devised, using semipermeable membranes.) Confusion apparently arises from statements such as "an increase in temperature shifts the equilibrium to the right." Remember that right and left refer only to the chemical equation that represents the process. Perhaps we should speak of shifting the position of equilibrium "so as to form more products" and avoid right and left completely.

7-4. The Thermodynamic Equilibrium Constant

A. The reaction isotherm

Every physical chemistry text gives a derivation of the expression for ΔG that includes an equilibrium constant of the form in Eqs. (7-6) and (7-7). The derivation may be based on pressure (assuming ideal gases), fugacities, or activities. In summary, it starts with a general reaction, such as we used for Eqs. (7-6) and (6-49) for pressures (or Eq. (6-65) for activities). For 1 mol, Eq. (6-49) becomes

$$\Delta G = G_2 - G_1 = RT \ln \frac{P_2}{P_1}$$

If state 1 is the standard state, $G_1 = G^\circ$, $P_1 = 1$ atm, and Eq. (6-49) becomes

$$G = G^\circ + RT \ln \frac{P}{\text{atm}} \tag{7-17}$$

Writing Eq. (7-17) for each reactant and product, substituting into Eq. (6-51), and rearranging gives

$$\Delta G = \Delta G^\circ + RT \ln \frac{P_C^c P_D^d}{P_A^a P_B^b} \tag{7-18}$$

If we designate the logarithmic term as Q_P, the quotient of the pressures, we have

REACTION ISOTHERM

$$\Delta G = \Delta G^\circ + RT \ln Q_P \tag{7-19}$$

This relationship is called the **reaction isotherm**. At a specified temperature, ΔG° is constant, and the sign of ΔG is determined by Q_P.

EXAMPLE 7-23: From an analysis of Eq. (7-18), what could we do to obtain more products from a given reaction?

Solution: The tendency for products to form increases as ΔG becomes more negative. If Q_P is less than 1, $RT \ln Q_P$ will be negative, making ΔG more negative than the constant ΔG°. The smaller we make Q_P, the greater is the effect. Thus we should increase the partial pressures of the reactants to provide more favorable conditions for the formation of products.

B. The equilibrium constant

Now we consider what the reaction isotherm tells us when the reaction reaches a state of equilibrium. In Eq. (7-18), P_A and P_B represent partial pressures of reactants at the start of the reaction, and P_C and P_D represent partial pressures of products at the end of the reaction. At equilibrium at constant pressure and temperature, we know that $\Delta G = 0$. Now the partial pressures must be the equilibrium partial pressures: $P_A = P_{A,\text{eq}}$, etc., and Eq. (7-19) becomes

$$\Delta G^\circ = -RT \ln Q_{P,\text{eq}} = -RT \ln \frac{P_{C,\text{eq}}^c P_{D,\text{eq}}^d}{P_{A,\text{eq}}^a P_{B,\text{eq}}^b} \tag{7-20}$$

Since ΔG° is constant for a given T, $Q_{P,\text{eq}}$ must be constant. We now call it the **equilibrium constant** and give it a new symbol, K_P. So,

$$Q_{P,\text{eq}} = K_P = \frac{P_{C,\text{eq}}^c P_{D,\text{eq}}^d}{P_{A,\text{eq}}^a P_{B,\text{eq}}^b} \tag{7-21}$$

We use the subscript eq for emphasis here but normally omit it, as in Eq. (7-6). It is understood that the pressures are equilibrium partial pressures.

Now, Eq. (7-20) becomes

$$\Delta G^\circ = -RT \ln K_P \tag{7-22}$$

This expression is an important link between thermodynamic properties, which sometimes seem a little esoteric or impractical, and the practical chemist's concern with how far a reaction will proceed toward completion. It provides a method of obtaining values of ΔG° that does not involve calorimetric measurements and a method of calculating K_P for experimentally difficult cases, if the ΔG_f° values are known.

EXAMPLE 7-24: In Example 7-20 we used the dissociation reaction of HI at 723 K for which $K_P = 2.0 \times 10^{-2}$. **(a)** Calculate ΔG° for this reaction; **(b)** interpret the result.

Solution:

(a) This solution involves a simple substitution into Eq. (7-22):

$$\Delta G^\circ = -RT \ln K_P = -(8.314\ \mathrm{J\,K^{-1}mol^{-1}})(723\ \mathrm{K})(\ln 0.02) = +23.5\ \mathrm{kJ}$$

(b) Recall that ΔG° refers to a system where reactants and products are in their standard states. In this case, the system would be a mixture of HI, H_2, and I_2 at 723 K in which each is at 1 atm pressure (better: unit activity). The positive sign of ΔG° tells us that the dissociation of HI is not a spontaneous process. Thus the reverse should occur, i.e., H_2 and I_2 combining to produce HI. When equilibrium is reached, the pressure of HI should be greater than 1 and the pressures of H_2 and I_2 less than 1. Since H_2 and I_2 are in the numerator of K_P, we would predict that $K_P < 1$, which in fact it is.

The numerical value of ΔG° suggests how far the standard state system is from the equilibrium system. In this case, it is not far: $K_P \approx 10^{-2}$. Compare this to typical K_a and K_{sp} values.

EXAMPLE 7-25: Using tables of ΔG_f°, calculate K_P at 298 K for the reaction

$$2NO_2(g) \rightleftharpoons N_2O_4(g)$$

Solution: To calculate ΔG° from ΔG_f° values, we have to use Eq. (6-52):

$$\Delta G^\circ = \Sigma n\, \Delta G_f^\circ\,(\text{products}) - \Sigma n\, \Delta G_f^\circ\,(\text{reactants})$$

For this reaction, $\Delta G^\circ = \Delta G_f^\circ(N_2O_4) - 2\,\Delta G_f^\circ(NO_2)$. Obtaining values from Appendix Table A-1 and substituting, we get

$$\Delta G^\circ = 97.9 - 2(51.3) = -4.7\ \mathrm{kJ}$$

Rearranging Eq. (7-22), we obtain

$$\ln K_P = -\frac{\Delta G^\circ}{RT} = -\frac{-4.7 \times 10^3\ \mathrm{J\,mol^{-1}}}{(8.314\ \mathrm{J\,K^{-1}mol^{-1}})(298\ \mathrm{K})} = 1.9$$

$$K_P = 6.7$$

C. General expression for ΔG

If we combine Eqs. (7-19) and (7-22), we obtain

$$\Delta G = -RT \ln K_P + RT \ln Q_P \tag{7-23}$$

or

$$\Delta G = RT \ln \frac{Q_P}{K_P} \tag{7-24}$$

These equations allow us to make qualitative analyses or quantitative calculations for systems not at equilibrium and to predict the directions in which change might occur.

EXAMPLE 7-26: Consider again the reaction in Example 7-25, when we mix 3 mol NO_2 and 4 mol N_2O_4 under a total pressure of 2 atm at 298 K. (**a**) What is ΔG for the system immediately after mixing? (**b**) Describe what will happen as the system changes toward and reaches equilibrium. (**c**) Calculate the amount of each gas at equilibrium.

Solution:

(**a**) The system probably is not at equilibrium initially, but we can make sure by calculating Q_P. If it equals K_P, equilibrium does exist.

$$Q_P = \frac{P_{N_2O_4}}{P^2_{NO_2}} = \frac{X_{N_2O_4}P}{(X_{NO_2}P)^2} = \frac{(4/7)(2)}{(3/7)^2(2)^2} = 1.6$$

This value is less than K_P, which, from Example 7-25, is 6.7, and verifies our assumption. To calculate ΔG, we substitute into Eq. (7-24):

$$\Delta G = (8.314\ \mathrm{J\,K^{-1}mol^{-1}})(298\ \mathrm{K})\left(\ln\frac{1.6}{6.7}\right) = -3.5\ \mathrm{kJ}$$

(**b**) As soon as we saw that the value of Q_P is less than that of K_P, we could predict that a change must occur to increase the amount of N_2O_4, since this would make the numerator of our expression for Q_P, and Q_P itself, larger. Simultaneously, of course, the amount of NO_2 will decrease.

From the negative sign of ΔG, we predict that the reaction is favored in the direction it is written. Again this would produce more N_2O_4.

Thus some NO_2 will change into N_2O_4, and Q_P will increase. When $Q_P = K_P, \Delta G = 0$, as it should at equilibrium.

(**c**) Let $2x$ be the amount of NO_2 that disappears; then x is the amount of N_2O_4 that appears. The table is

	$2NO_2$	N_2O_4
Initial moles	3	4
Change	$-2x$	$+x$
Equilibrium moles	$3-2x$	$4+x$
Mole fractions	$\frac{3-2x}{7-x}$	$\frac{4+x}{7-x}$

Thus

$$K_P = 6.7 = \frac{\left(\frac{4+x}{7-x}\right)(2)}{\left[\left(\frac{3-2x}{7-x}\right)(2)\right]^2}$$

Canceling $(7-x)$ and 2 yields

$$6.7 = \frac{(4+x)(7-x)}{(3-2x)^2(2)}$$

$$28 + 3x - x^2 = 13.4(9 - 12x + 4x^2) = 121 - 161x + 53.6x^2$$

or

$$54.6x^2 - 164x + 93 = 0$$

Using the quadratic formula to solve for x, we get $x = 0.76$. At equilibrium,

$$\text{moles } NO_2 = 3 - 2x = 1.48 \qquad \text{moles } N_2O_4 = 4 + x = 4.76$$

D. Dependence on temperature

We've seen that ΔG changes with temperature and that the calculation of ΔG at a new temperature is rather complicated: It involves either one or two integration constants, depending on the method. For K_P, the situation is similar.

We obtain the differential relationship by differentiating Eq. (7-22) and combining the result with expressions from Chapter 6. In Example 6-14, we used

$$\left(\frac{\partial \Delta G^\circ}{\partial T}\right)_P = -\Delta S^\circ \qquad \text{[Eq. (6-54)]}$$

Then, rearranging Eq. (6-29), we have

$$-\Delta S^\circ = \frac{\Delta G^\circ - \Delta H^\circ}{T}$$

Thus

$$\left(\frac{\partial \Delta G^\circ}{\partial T}\right)_P = \frac{\Delta G^\circ - \Delta H^\circ}{T} = \frac{d(\Delta G^\circ)}{dT} \tag{7-25}$$

Differentiating Eq. (7-22) gives

$$-\frac{d\Delta G^\circ}{dT} = R \ln K_P + RT \frac{d \ln K_P}{dT}$$

Substituting from Eq. (7-25) yields

$$-\frac{\Delta G^\circ - \Delta H^\circ}{T} = R \ln K_p + RT \frac{d \ln K_P}{dT}$$

Multiplying by T and canceling $-\Delta G^\circ$ and $RT \ln K_P$ [from Eq. (7-22)], we obtain

$$\Delta H^\circ = RT^2 \frac{d \ln K_P}{dT} \qquad \text{or} \qquad \frac{d \ln K_P}{dT} = \frac{\Delta H^\circ}{RT^2} \tag{7-26}$$

(For a derivation based on the Gibbs–Helmholtz equation, see Problem 7-21.)

Qualitatively we can predict that, for most spontaneous chemical changes (usually ΔH° is negative), K_P will decrease when T increases.

EXAMPLE 7-27: Consider a dissociation reaction such as

$$PCl_5(g) \rightleftharpoons PCl_3(g) + Cl_2(g)$$

Predict whether K_P will increase or decrease if the temperature is raised from 300 K.

Solution: First we must predict the sign of ΔH. At 300 K, PCl_5 probably would have little tendency to dissociate; i.e., heating probably is required or ΔH is positive. From Eq. (7-26), the coefficient of ln K_P to T would be positive; thus K_P would increase when T increases. This prediction could be verified by calculating ΔH using ΔH_f° values.

We can integrate Eq. (7-26) four ways. We could treat ΔH° as constant or as a function of temperature and, for either of these, integrate with or without limits. In all cases, we start by rearranging Eq. (7-26) to

$$\int d \ln K_P = \int \frac{\Delta H^\circ}{R} \frac{1}{T^2} \, dT \tag{7-27}$$

For constant ΔH°, Eq. (7-27) becomes

$$\int d \ln K_P = \frac{\Delta H^\circ}{R} \int \frac{1}{T^2} \, dT \tag{7-28}$$

Integrating without limits gives

$$\ln K_P = -\frac{\Delta H^\circ}{R}\left(\frac{1}{T}\right) + C \tag{7-29}$$

Note that Eq. (7-29) is the equation of a straight line; if $\ln K_P = y$ and $1/T = x$, the slope is $-\Delta H^\circ/R$ and the intercept is C. The slope and the intercept can be determined experimentally by measuring K_P at several T's and plotting the readings.

Integrating with limits gives

$$\ln \frac{K_{P_2}}{K_{P_1}} = \left(\frac{\Delta H^\circ}{R}\right)\left(\frac{1}{T_1} - \frac{1}{T_2}\right) \tag{7-30}$$

or

$$\ln \frac{K_{P_2}}{K_{P_1}} = \left(\frac{\Delta H^\circ}{R}\right)\left(\frac{T_2 - T_1}{T_1 T_2}\right) \tag{7-31}$$

As with the similar equations for the dependence of ΔH° on T, you would use Eq. (7-29) if you want to calculate K_P at several temperatures from the values of K_P and ΔH° at one temperature.

EXAMPLE 7-28: Calculate K_P at 500 K for the system $2HI(g) \rightleftharpoons H_2(g) + I_2(g)$ if K_P at 300 K is 2.16×10^{-3}, assuming that ΔH° is constant.

Solution: We have to calculate only one K_P, so we integrate Eq. (7-27) with limits, obtaining Eq. (7-31). We'll need ΔH°, which we calculate from ΔH_f° values:

$$\Delta H^\circ = \Delta H_f^\circ(H_2) + \Delta H_f^\circ(I_2, g) - 2\,\Delta H_f^\circ(HI)$$

Taking values from Table A-1 and substituting, we get

$$\Delta H^\circ = 0 + 62.44 - 2(26.48) = 9.48 \text{ kJ}$$

Rearranging Eq. (7-31) gives

$$\ln K_{P_2} = \ln K_{P_1} + \frac{\Delta H^\circ}{R}\left(\frac{T_2 - T_1}{T_2 T_1}\right)$$

Then we substitute and evaluate to obtain

$$\ln K_{P_2} = \ln 0.00216 + \frac{9.48 \times 10^3}{8.314}\left(\frac{500 - 300}{500(300)}\right) = -6.1376 + 1.5203 = -4.6173$$

so

$$K_{P_2} = 9.88 \times 10^{-3}$$

When ΔH° is a function of T, we must include heat capacities in order to express this dependence. In Chapter 4, we obtained an analytic expression for ΔH at any T, i.e.,

$$\Delta H = \Delta a T + \frac{\Delta b}{2} T^2 + \frac{\Delta c}{3} T^3 + C \qquad \text{[Eq. (4-12)]}$$

Now we substitute this expression into Eq. (7-27), dividing by T^2 as we go:

$$\int d \ln K_P = \frac{1}{R}\left[\int \frac{\Delta a}{T}\, dT + \int \frac{\Delta b}{2}\, dT + \int \frac{\Delta c}{3}\, T\, dT + \int \frac{C}{T^2}\, dT\right] \tag{7-32}$$

Integrating without limits gives

$$\ln K_P = \frac{1}{R}\left[\Delta a \ln T + \frac{\Delta b}{2} T + \frac{\Delta c}{6} T^2 - \frac{C}{T}\right] + M \tag{7-33}$$

where M is a second integration constant. We use this expression in the same way that we use the similar one for ΔG, Eq. (6-58). When we have evaluated C, M, and the heat capacity coefficients, we can conveniently use Eq. (4-14) for calculating K_P's at a series of temperatures.

Integrating with limits gives

$$\ln \frac{K_{P_2}}{K_{P_1}} = \frac{1}{R}\left[\Delta a \ln \frac{T_2}{T_1} + \frac{\Delta b}{2}(T_2 - T_1) + \frac{\Delta c}{6}(T_2^2 - T_1^2) - C\left(\frac{1}{T_2} - \frac{1}{T_1}\right)\right] \tag{7-34}$$

As before, this expression is the one to choose if you need to make only one calculation.

EXAMPLE 7-29: The dissociation of water vapor is negligible at 298 K, as shown by the small K_P of 8.33×10^{-81}. Would the dissociation be much greater at 1500 K? Calculate K_P at 1500 K for

$$2H_2O(g) \rightleftharpoons 2H_2(g) + O_2(g)$$

Solution: This calculation is a long one and must be taken step-by-step. We need ΔH°_{298}, the heat capacity coefficients, and C in order to make the final calculation.

Step 1: To calculate ΔH°, we use ΔH°_f values from Table A-1 but note that the reaction is the reverse of the formation of water, so its ΔH is the negative of ΔH°_f for water, i.e.,

$$\Delta H^\circ = -(-241.8) = 241.8 \text{ kJ mol}^{-1}$$

For the two moles of the balanced equation, $\Delta H^\circ = 483.6$ kJ

Step 2: For the heat capacity terms, we obtain

	a	$b \times 10^3$	$c \times 10^7$
H_2 (×2)	58.13	− 1.66	40.24
O_2	25.72	12.98	−38.61
Sum for products	83.85	11.32	1.63
H_2O (×2)	60.72	19.22	23.68
Products minus reactants	23.13	− 7.90	−22.05

Step 3: To calculate C, we rearrange Eq. (4-12) and substitute:

$$C = \Delta H^\circ - \Delta aT - \frac{\Delta b}{2}T^2 - \frac{\Delta c}{3}T^3$$

$$= 483.6 \times 10^3 - 23.13(298) - \left(\frac{-7.90 \times 10^{-3}}{2}\right)(298)^2 - \left(\frac{-22.05 \times 10^{-7}}{3}\right)(298)^3$$

$$= 477.1 \text{ kJ}$$

Step 4: Now we can substitute into Eq. (7-34) and evaluate $K_{P,1500}$:

$$\ln\frac{K_{P,1500}}{K_{P,298}} = \frac{1}{8.314}\left[23.13 \ln\frac{1500}{298} + \left(\frac{-7.90 \times 10^{-3}}{2}\right)(1500 - 298) + \left(\frac{-22.05 \times 10^{-7}}{6}\right)(1500^2 - 298^2) - (477.1 \times 10^3)\left(\frac{1}{1500} - \frac{1}{298}\right)\right]$$

$$= \left(\frac{1}{8.314}\right)(37.38 - 4.75 - 0.79 + 1282.94) = 158.14$$

$$\ln K_{P,1500} = 158.14 + \ln(8.33 \times 10^{-81}) = 158.14 - 184.39 = -26.25$$

$$K_{P,1500} = 3.98 \times 10^{-12}$$

We see that the dissociation of water is much greater at 1500 K, although it still is essentially negligible. We predict that, at higher temperatures, it would become more significant.

EXAMPLE 7-30: In Example 7-29, the change in K_P was quite large. **(a)** Which factor was primarily responsible for this? **(b)** What general rule for qualitative predictions can be stated concerning the change of K_P with temperature?

Solution:

(a) Looking at Step 4 of the solution in Example 7-29, we see that the large contributor to the difference between the logarithms is the value 1282, which comes from C, which in turn comes from ΔH°.

(b) The general rule is that, if ΔH at the lower temperature is large, we can expect a larger variation in K_P with temperature than if ΔH is small.

SUMMARY

The major equations of this chapter and connections between them are shown in Table S-7. You should determine the conditions that are necessary for each equation to be valid or the conditions that are imposed to make the connections. Other important things you should know are

1. Dynamic equilibrium may occur for both chemical and physical processes.
2. Equilibria may be homogeneous or heterogeneous, with appropriate differences in the equilibrium constants.
3. The criterion for physical equilibrium is that the chemical potential of each component must be the same in each phase: $\mu_i(\alpha) = \mu_i(\beta) = \cdots$.
4. The criterion for chemical equilibrium is $\Sigma\mu_i\, dn_i = 0$ or $\Sigma\nu_i\mu_i = 0$.
5. The equilibrium constant of a reaction is the ratio of the product of the equilibrium concentrations of the products to the product of the equilibrium concentrations of the reactants, all raised to the power of the appropriate coefficients in the balanced equation.
6. The most general equilibrium constant is that expressed in activities, but pressure or fugacity may also be used.
7. Major properties of equilibrium constants are that they

   (a) are valid only at equilibrium;
   (b) are dependent on temperature and the way the equation is balanced;
   (c) are independent of total pressure, initial concentration, volume of container, or direction of approach to equilibrium (at a given temperature);
   (d) indicate qualitatively the extent of the reaction; and
   (e) express quantitatively the dependence of the extent of the reaction on concentration.

8. Le Chatelier's principle states that if a stress is applied to a system in a state of dynamic equilibrium, the position of equilibrium will shift in the direction that will relieve the stress.
9. The degree of dissociation is another way of expressing the extent of the reaction when a single reactant forms two or more products.

TABLE S-7: Summary of Important Equations

$$\sum_i \nu_i \mathrm{A}_i = 0 \qquad \text{(7-2)}$$

$$\xi = \frac{n_i - n_{i,0}}{\nu_i} \qquad \text{(7-3)}$$

$$\rightarrow \sum_i \nu_i \mu_i = 0 \qquad \text{(7-5)}$$

$$K_c = \frac{[\mathrm{C}]^c[\mathrm{D}]^d}{[\mathrm{A}]^a[\mathrm{B}]^b} \qquad \text{(7-6)}$$

$$\rightarrow K_P = K_c(RT)^{\Sigma\nu} \qquad \text{(7-8)}$$

$$K_P = \frac{P_{\mathrm{C}}^c P_{\mathrm{D}}^d}{P_{\mathrm{A}}^a P_{\mathrm{B}}^b} \qquad \text{(7-7)}$$

→ (See also Eqs. 7-9, 7-10, 7-11, and 7-14 for expressions for K_f, K_γ, K_a, and K_X)

$$\Delta G = RT \ln \frac{P_2}{P_1} \qquad \text{(6-49, for 1 mol)}$$

$$\rightarrow \Delta G = \Delta G^\circ + RT \ln Q_P \qquad \text{(7-19)}$$

$$\rightarrow \Delta G^\circ = -RT \ln K_P \qquad \text{(7-22)}$$

$$\rightarrow \frac{d \ln K_P}{dT} = \frac{\Delta H^\circ}{RT^2} \qquad \text{(7-26)}$$

10. The reaction quotient is defined in the same manner as the equilibrium constant, except that the quantities involved are not equilibrium quantities.
11. The thermodynamic equilibrium constant is defined by $\Delta G^\circ = -RT \ln K_P$.
12. The equilibrium constant varies with temperature and is governed primarily by the magnitude of ΔH.

SOLVED PROBLEMS

Equilibrium States

PROBLEM 7-1 Calculate the extent of reaction if equilibrium is reached after 50.0 g of SO_3 has dissociated according to

$$2SO_3(g) \rightleftharpoons 2SO_2(g) + O_2(g)$$

Solution: You must convert grams to moles and then apply Eq. (7-3):

$$\frac{50.0 \text{ g}}{96.0 \text{ g mol}^{-1}} = 0.521 \text{ mol}^{-1}$$

$$\xi = \frac{n_i - n_{i,0}}{\nu} = \frac{\text{Amount reacted}}{\nu} = \frac{-0.521}{-2} = 0.261 \text{ mol} \qquad \text{[Eq. (7-3)]}$$

Equilibrium Constants and Applications

PROBLEM 7-2 Write the equilibrium constant expression, K_c, for the equilibrium system

$$4NH_3(g) + 5O_2(g) \rightleftharpoons 4NO(g) + 6H_2O(g)$$

Solution: You apply Eq. (7-6), representing the molar concentrations of each component by brackets:

$$K_c = \frac{[NO]^4[H_2O]^6}{[NH_3]^4[O_2]^5}$$

PROBLEM 7-3 For an all-gas reaction like that in Problem 7-2, you probably would use K_P. If you knew the value of K_c at 300 K, what relationship and numbers would you use to calculate K_P?

Solution: You would have to assume ideal behavior and use Eq. (7-8). The energy units of R must be L atm because K_c is in mol L^{-1}.

$$\Delta n = 10 - 9 = 1$$

$$K_P = K_c[0.0821(300)]^1 = 24.6 \text{ L atm mol}^{-1}(K_c)$$

PROBLEM 7-4 Calculate K_c for the reaction $A + 2B \rightleftharpoons 3C + D$ if the concentrations (in mol L^{-1}) in the equilibrium mixture are A, 2.0; B, 3.0; C, 2.0; and D, 1.5.

Solution: You should write the expression for K_c using Eq. (7-6) and substitute the given numbers:

$$K_c = \frac{[C]^3[D]}{[A][B]^2} = \frac{2^3(1.5)}{2(3)^2} = 0.67$$

PROBLEM 7-5 Write and compare the equilibrium constant expressions, K_c, for the HI system at 298 K and 450 K. The equations are

$$298 \text{ K:} \quad H_2(g) + I_2(s) \rightleftharpoons 2HI(g) \qquad 450 \text{ K:} \quad H_2(g) + I_2(g) \rightleftharpoons 2HI(g)$$

Solution: You set up the expressions using Eq. (7-6):

$$K_{c,298} = \frac{[HI]^2}{[H_2]} \qquad K_{c,450} = \frac{[HI]^2}{[H_2][I_2]}$$

The purpose of this problem is to emphasize that condensed phases are not included in the expression for K_c or K_P.

PROBLEM 7-6 Calculate the ionization constant for the complete ionization of carbonic acid

$$H_2CO_3 \rightleftharpoons 2H^+ + CO_3^{2-}$$

from this information:

$$H_2CO_3 \rightleftharpoons H^+ + HCO_3^- \qquad K_{a_1} = 4.2 \times 10^{-7}$$
$$HCO_3^- \rightleftharpoons H^+ + CO_3^{2-} \qquad K_{a_2} = 4.8 \times 10^{-11}$$

Solution: You have the two steps in the ionization of H_2CO_3. You can obtain the overall reaction by adding the two steps, but the overall K is the product of the K's of the steps. Apply Eq. (7-12) and evaluate:

$$K_i(H_2CO_3) = K_{a_1}K_{a_2} = (4.2 \times 10^{-7})(4.8 \times 10^{-11}) = 2.0 \times 10^{-17}$$

PROBLEM 7-7 When a system of gases, $CO + Cl_2 \rightleftharpoons COCl_2$, reached equilibrium in a 2.0-L vessel at 100 °C, the equilibrium mixture was found to contain 0.30 mol CO, 0.20 mol Cl_2, and 0.80 mol $COCl_2$. Calculate K_c.

Solution: Knowing the equilibrium concentrations, you merely substitute in Eq. (7-6), or

$$K_c = \frac{[COCl_2]}{[CO][Cl_2]} = \frac{0.80/2.0}{(0.30/2.0)(0.20/2.0)} = 26.7$$

PROBLEM 7-8 For the system in Problem 7-7, calculate the concentration of the three gases at equilibrium if 1.0 mol $COCl_2$ is placed in a 10-L vessel at 100 °C.

Solution: The equation for the reaction is the reverse of that in Problem 7-7, so K will be the reciprocal of 26.7. Set up the usual table:

	$COCl_2$	$\rightleftharpoons CO$	$+ Cl_2$
Initial concentration	0.10	0	0
Change	$-x$	$+x$	$+x$
Equilibrium concentration	$0.10 - x$	x	x

Then use Eq. (7-6) to obtain

$$K_c = \frac{[CO][Cl_2]}{[COCl_2]} = \frac{x(x)}{0.10 - x} = \frac{1}{26.7}$$
$$26.7x^2 = 0.10 - x$$
$$26.7x^2 + x - 0.10 = 0$$

Finally, using the quadratic formula gives

$$x = 0.045$$

so $\quad [CO] = [Cl_2] = 4.5 \times 10^{-2} \quad$ and $\quad [COCl_2] = 0.10 - x = 5.5 \times 10^{-2}$

PROBLEM 7-9 The reaction known as the "water gas reaction" has a K_c of 1.60 at 986 °C. Calculate the moles present at equilibrium if 1 mol H_2 is mixed with 1 mol CO_2 at that temperature.

$$H_2(g) + CO_2(g) \rightleftharpoons H_2O(g) + CO(g)$$

Solution: This problem appears very similar to the last, but, as you start to solve it in the same way, you'll find that the volume isn't given. Actually, the volume isn't needed because this is a case for which $\Delta n_{gas} = 0$, which means the volume will cancel. To show that the volume will cancel, you can set up the equation for K_c using V for the unknown volume.

$$K_c = \frac{\left(\frac{\text{Moles } H_2O}{V}\right)\left(\frac{\text{Moles CO}}{V}\right)}{\left(\frac{\text{Moles } H_2}{V}\right)\left(\frac{\text{Moles } CO_2}{V}\right)} = \frac{(n_{H_2O})(n_{CO})}{(n_{H_2})(n_{CO_2})}$$

You can fill in the table with moles instead of molarity. You should be proficient enough now to omit the line for change, at least in this simple system.

	H_2	$+ CO_2 \rightleftharpoons$	$H_2O +$	CO
Initial concentration	1	1	0	0
Equilibrium concentration	$1-x$	$1-x$	x	x

Now

$$K_c = \frac{(x)(x)}{(1-x)(1-x)} = \frac{x^2}{(1-x)^2} = 1.60$$

You can take the square root, giving

$$\frac{x}{1-x} = 1.26 \qquad x = 1.26 - 1.26x \qquad x = 0.56$$

Now

$$\text{Moles } H_2O = \text{Moles CO} = 0.56$$
$$\text{Moles } H_2 = \text{Moles } CO_2 = 0.44$$

PROBLEM 7-10 Calculate the degree of dissociation of PCl_5 at 300 °C (**a**) starting with 1.0 mol L^{-1}; and (**b**) starting with 0.010 mol L^{-1}. Use $K_c = 4.6 \times 10^{-2}$.

$$PCl_5(g) \rightleftharpoons PCl_3(g) + Cl_2(g)$$

Solution: You can set up the calculation in either of two ways: (1) solve for x as in Problems 7-8 and 7-9; then calculate α using Eq. (7-16); or (2) derive an expression for K_c containing α, as in Example 7-21. Let's use method (1) for part (a) and method (2) for part (b).

(**a**) Set up the table:

	$PCl_5 \rightleftharpoons$	$PCl_3 +$	Cl_2
Initial concentration	1	0	0
Equilibrium concentration	$1-x$	x	x

$$K_c = \frac{[PCl_3][Cl_2]}{[PCl_5]} = \frac{x^2}{1-x} = 0.046$$
$$x^2 = 0.046 - 0.046x$$
$$x^2 + 0.046x - 0.046 = 0$$
$$x = 0.19$$

Using Eq. (7-16), you get

$$\alpha = \frac{x}{C_0} = \frac{0.19}{1} = 0.19$$

(**b**) The expression for K_c in terms of α was derived in Example 7-21a. The PCl_5 dissociation is of the general form $AB \rightleftharpoons A + B$. You can use

$$K_c = \frac{\alpha^2}{1-\alpha} C_0$$

and rearrange it to

$$\frac{\alpha^2}{1-\alpha} = \frac{K_c}{C_0} = \frac{0.046}{0.010} = 4.6$$
$$\alpha^2 = 4.6 - 4.6\alpha$$
$$\alpha^2 + 4.6\alpha - 4.6 = 0$$
$$\alpha = 0.85$$

Comparison of the two results shows the large increase in dissociation when the concentration is reduced by a factor of 100.

PROBLEM 7-11 Suppose that you mix 2.00 moles of butane and 3.00 moles of isobutane in a 6.00-L container. What is the equilibrium concentration of butane if $K_c = 0.400$?

$$\text{butane}(g) \rightleftharpoons \text{isobutane}(g)$$

Solution: As you set up the usual table, you must decide which side gets $-x$. In other words, which way will the reaction go? Qualitatively, the clue comes from the value of K_c. Since it is less than one, the numerator is smaller than the denominator. Except for very special cases, this means that, at equilibrium, there must be less product than reactant—in this case, less isobutane than butane. At the start there is more isobutane; therefore some of it must change to butane. Note that the volume will cancel, so you can use moles.

The table becomes

	butane	$\rightleftharpoons$ isobutane
Initial moles	2	3
Equilibrium moles	$2 + x$	$3 - x$

$$K_c = \frac{\text{Moles isobutane}}{\text{Moles butane}} = \frac{3 - x}{2 + x} = 0.4$$

$$3 - x = 0.8 + 0.4x$$

$$2.2 = 1.4x$$

$$x = 1.57 \text{ mol}$$

$$\text{Concentration of butane} = \frac{1.57}{6.00} = 0.262 \text{ mol L}^{-1}$$

Quantitatively, you could determine which way the reaction will go to reach equilibrium by calculating Q_c, analogous to the Q_P of Eq. (7-19). At equilibrium, $Q_c = K_c$, as in Eq. (7-21). If Q_c is less than K_c, then, to reach equilibrium, products must change to reactants [see Example 7-26b].

PROBLEM 7-12 When 8.0 mol PCl_5 is placed in a container and allowed to come to equilibrium, we find that the compound is 20 percent dissociated and that the total pressure is 0.80 atm. Calculate K_P for the reaction

$$PCl_5(g) \rightleftharpoons PCl_3(g) + Cl_2(g)$$

Solution: When using K_P, you usually must find mole fractions in order to get partial pressures. In this problem, the equilibrium amounts are disguised slightly. If PCl_5 is 20 percent dissociated, the number of moles that dissociated is $0.20 \times 8.0 = 1.6$. To be more formal, you could obtain this number by solving Eq. (7-16) and using $\alpha = 0.20$ to obtain

$$\text{Amount dissociated} = (\alpha)(\text{Initial amount}) = 0.20(8.0) = 1.6$$

Now set up the table:

	PCl_5	$\rightleftharpoons PCl_3$ +	Cl_2
Initial moles	8.0	0	0
Change	−1.6	+1.6	+1.6
Equilibrium moles	6.4	1.6	1.6
Mole fractions	$\frac{6.4}{9.6}$	$\frac{1.6}{9.6}$	$\frac{1.6}{9.6}$

Using Eq. (7-7) gives

$$K_P = \frac{P_{PCl_3}P_{Cl_2}}{P_{PCl_5}} = \frac{(1.6/9.6)(0.80)(1.6/9.6)(0.80)}{(6.4/9.6)(0.80)} = \frac{(1.6)^2(0.80)}{(9.6)(6.4)} = 3.3 \times 10^{-2}$$

PROBLEM 7-13 Three moles of SO_3 enters an evacuated 4.00-L flask. When equilibrium is reached, the pressure is 1.50 atm and 2.00 mol SO_3 is present. Calculate K_P for the reaction

$$2SO_3(g) \rightleftharpoons 2SO_2(g) + O_2(g)$$

Solution: In this problem, we introduce another twist for getting the equilibrium amounts. First, set up the table with what you know:

	$2SO_3$	$\rightleftharpoons 2SO_2$ +	O_2
Initial moles	3 mol	0	0
Change			
Equilibrium moles	2 mol		

It is clear that the change for SO_3 is -1 mol. From the balanced equation, you can see that 1 mol SO_3 turns into 1 mol SO_2 and 0.5 mol O_2. The table becomes

	$2SO_3$	$\rightleftharpoons 2SO_2$	$+ \; O_2$
Initial moles	3	0	0
Change	-1	$+1$	$+0.5$
Equilibrium moles	2	1	0.5
Mole fractions	$\frac{2}{3.5}$	$\frac{1}{3.5}$	$\frac{0.5}{3.5}$

Using Eq. (7-7), you get

$$K_P = \frac{P_{SO_2}^2 P_{O_2}}{P_{SO_3}^2} = \frac{[(1/3.5)(1.5)]^2(0.5/3.5)(1.5)}{[(2/3.5)(1.5)]^2} = \frac{(0.5)(1.5)}{(3.5)(2)^2} = 0.54$$

PROBLEM 7-14 The K_P for the production of phosgene is 18.0. If one mole each of CO and Cl_2 are mixed at a pressure of 1 atm, what will be the partial pressure of $COCl_2$ after equilibrium is reached?

$$CO(g) + Cl_2(g) \rightleftharpoons COCl_2(g)$$

Solution: The pressure has been made 1 atm to simplify the calculations. You need the table again:

	CO	$+ \; Cl_2$	$\rightleftharpoons COCl_2$
Initial moles	1	1	0
Change	$-x$	$-x$	$+x$
Equilibrium moles	$1-x$	$1-x$	x
Mole fractions	$\frac{1-x}{2-x}$	$\frac{1-x}{2-x}$	$\frac{x}{2-x}$

The pressure is 1, so

$$K_P = \frac{X_{COCl_2}}{X_{CO}X_{Cl_2}} = \frac{\dfrac{x}{2-x}}{\left(\dfrac{1-x}{2-x}\right)^2} = \frac{x(2-x)}{(1-x)^2} = 18$$

$$2x - x^2 = 18(1 - 2x + x^2) = 18 - 36x + 18x^2$$

$$19x^2 - 38x + 18 = 0$$

From the quadratic formula, $x = 0.77$. The problem asks for the partial pressure of $COCl_2$, which is given by

$$P_{COCl} = X_{COCl}P = \left(\frac{x}{2-x}\right)(1) = \frac{0.77}{1.23} = 0.63 \text{ atm}$$

PROBLEM 7-15 Obtain an expression for K_P in terms of the degree of dissociation for the dissociation of sulfur trioxide [see Problem 7-14].

Solution: You should set up the usual table. Use α, P, and n to represent the degree of dissociation, the total pressure, and the initial moles of SO_3, respectively.

	$2SO_3$	$\rightleftharpoons \; 2SO_2$	$+ \; O_2$
Initial moles	n	0	0
Change	$-\alpha n$	$+\alpha n$	$+0.5\alpha n$
Equilibrium moles	$n - \alpha n$	αn	$0.5\alpha n$
Mole fractions	$\frac{n(1-\alpha)}{n(1+0.5\alpha)}$	$\frac{\alpha n}{n(1+0.5\alpha)}$	$\frac{0.5\alpha n}{n(1+0.5\alpha)}$
or	$\frac{(1-\alpha)}{(1+0.5\alpha)}$	$\frac{\alpha}{(1+0.5\alpha)}$	$\frac{0.5\alpha}{(1+0.5\alpha)}$

Thus

$$K_P = \frac{\left(\dfrac{\alpha}{1+0.5\alpha}P\right)^2\left(\dfrac{0.5\alpha}{1+0.5\alpha}P\right)}{\left(\dfrac{1-\alpha}{1+0.5\alpha}P\right)^2} = \frac{\alpha^2(0.5\alpha P)}{(1-\alpha)^2(1+0.5\alpha)} = \frac{0.5\alpha^3}{(1-\alpha)^2(1+0.5\alpha)}P$$

You can obtain a slightly neater form by changing 0.5 to 1/2, or

$$K_P = \frac{\alpha^3/2}{(1-\alpha)^2\left(\dfrac{2+\alpha}{2}\right)}P = \frac{\alpha^3}{(1-\alpha)^2(2+\alpha)}P$$

PROBLEM 7-16 Calculate the degree of dissociation of 1 mol $SbCl_5$ under a total pressure of 1 atm if $K_P = 1.5$. The balanced equation is

$$SbCl_5(g) \rightleftharpoons SbCl_3(g) + Cl_2(g)$$

Solution: First, derive an expression for K_P in terms of α, using n for the initial moles. Set up the usual table:

	$SbCl_5 \rightleftharpoons$	$SbCl_3$ +	Cl_2
Initial moles	n	0	0
Change	$-\alpha n$	$+\alpha n$	$+\alpha n$
Equilibrium moles	$n - \alpha n$	αn	αn
Mole fractions	$\dfrac{(1-\alpha)}{(1+\alpha)}$	$\dfrac{\alpha}{(1+\alpha)}$	$\dfrac{\alpha}{(1+\alpha)}$

Note that here, as in Problem 7-15, the n's cancel in the mole fractions. Now substitute in Eq. (7-6):

$$K_P = \frac{P_{SbCl_3}P_{Cl_2}}{P_{SbCl_5}} = \frac{\left(\dfrac{\alpha}{1+\alpha}P\right)\left(\dfrac{\alpha}{1+\alpha}P\right)}{\dfrac{1-\alpha}{1+\alpha}P} = \frac{\alpha^2 P}{(1+\alpha)(1-\alpha)} = \frac{\alpha^2}{1-\alpha^2}P$$

Note that the α terms are different from those in the expression for K_c for the same reaction, derived in Example 7-21a. Rearrange and substitute the given values to get

$$\frac{\alpha^2}{1-\alpha^2} = \frac{K_P}{P} = \frac{1.5}{1} = 1.5$$

$$\alpha^2 = 1.5 - 1.5\alpha^2$$

$$2.5\alpha^2 = 1.5$$

$$\alpha^2 = \frac{1.5}{2.5} = 0.60$$

$$\alpha = 0.77$$

note: You could have solved for the amount dissociated and then calculated α from Eq. (7-16).

PROBLEM 7-17 Suppose that the vessel containing the $SbCl_5$ in Problem 7-16 had also contained enough nitrogen that, at equilibrium, the nitrogen contributed 0.30 atm to the total pressure (1 atm). **(a)** What is the degree of dissociation of $SbCl_5$ in this situation? **(b)** Compare your result to that of Problem 7-16. **(c)** Are the results what you would expect from applying Le Chatelier's principle? Why?

Solution:

(a) In situations like this, the gas that doesn't participate in the chemical reaction is called an *inert gas*. Its presence will reduce the partial pressures of all of the gases. In this case, the total pressure exerted by the other gases is

$$P' = 1 - P_{N_2} = 1 - 0.3 = 0.7 \text{ atm}$$

You can use this pressure as the total equilibrium pressure in the expression for α obtained in Problem 7-16 to calculate

$$\frac{\alpha^2}{1-\alpha^2} = \frac{K_P}{P} = \frac{1.5}{0.7} = 2.14$$

$$\alpha^2 = 2.14 - 2.14\alpha^2$$

$$3.14\alpha^2 = 2.14$$

$$\alpha^2 = \frac{2.14}{3.14} = 0.682$$

$$\alpha = 0.83$$

(b) In Problem 7-16, α was 0.77, so the result of part (a) shows that decreasing the pressure increases α. Further, you can see that the presence of an inert gas is equivalent to a reduction in pressure. The effect can also be viewed as a dilution, reducing all concentrations.

(c) According to Le Chatelier's principle, a reduction in pressure will cause a change in the direction that will tend to restore the pressure. The pressure will increase if there are more moles of gas, so some of the $SbCl_5$ should dissociate. Your calculation shows that this does indeed happen.

PROBLEM 7-18 At 1080 °C, the pressure of oxygen from the decomposition of cupric oxide is 0.51 atm for

$$4CuO(s) \rightleftharpoons 2Cu_2O(s) + O_2(g)$$

(a) Calculate K_P. **(b)** What fraction of the CuO will dissociate if 0.20 mol is placed in a 1-L flask at 1080 °C? **(c)** What would the fraction be if a 1.0-mol sample were used? **(d)** What is the smallest amount of CuO that would establish equilibrium?

Solution:

(a) From Eq. (7-7), $K_P = P_{O_2}$; the two solids are omitted. Thus $K_P = 0.51$.

(b) You cannot handle this problem like previous problems because the amount of reactant does not appear in K_P. The quantity of oxygen produced depends only on the pressure, volume, and temperature. Assuming ideal behavior, you have

$$n_{O_2} = \frac{PV}{RT} = \frac{0.51(1)}{0.0821(1353)} = 0.0046 \text{ mol}$$

In the balanced equation for the reaction, $4CuO \rightarrow 1O_2$; thus $n_{CuO} = 4 \times n_{O_2} = 0.0184$ mol.

$$\text{Fraction dissociated} = \frac{0.0184}{0.20} = 0.092$$

(c) If 1.0 mol CuO were used, the amount of oxygen would be exactly the same, but

$$\text{Fraction dissociated} = \frac{0.0184}{1.0} = 0.0184$$

(d) If the amount of CuO were less than 0.0184 mol, all of it would dissociate. There would not be enough oxygen to achieve the pressure of 0.51 atm. Equilibrium cannot be established unless some CuO(s) is present; thus you would need slightly more than 0.0184 mol CuO.

The Thermodynamic Equilibrium Constant

PROBLEM 7-19 Calculate ΔG° for the production of phosgene from carbon monoxide and chlorine if $K_P = 4.35$ at 724 K.

Solution: The relationship between K_P and ΔG° is $\Delta G^\circ = -RT \ln K_P$ [Eq. (7-22)]. So

$$\Delta G^\circ = -8.314(724) \ln 4.35 = -8.85 \text{ kJ}$$

PROBLEM 7-20 Calculate K_P at 298 K for the reaction

$$4HCl(g) + O_2(g) \rightleftharpoons 2H_2O(g) + 2Cl_2(g)$$

Solution: Since the method is not specified, you should recognize that the reasonable approach is to use ΔG_f° values to calculate ΔG° and then to calculate K_P using Eq. (7-22). Writing Eq. (6-52) for this reaction, you have

$$\Delta G^\circ = 2\Delta G_f^\circ(H_2O) + 2\Delta G_f^\circ(Cl_2) - 4\Delta G_f^\circ(HCl) - \Delta G_f^\circ(O_2)$$
$$= 2(-228.6) + 0 - 4(-95.3) - 0 = -76.0 \text{ kJ}$$

Rearranging Eq. (7-22) and substituting values, you obtain

$$\ln K_P = -\frac{-76.0 \times 10^3}{8.314(298)} = 30.68$$
$$K_P = 2.1 \times 10^{13}$$

PROBLEM 7-21 Derive the expression for the temperature dependence of K_P [Eq. (7-26)] using the second form of the Gibbs–Helmholtz equation, Eq. (6-56).

Solution: The difference from the derivation outlined in Section 7-4D is that the second form of the Gibbs–Helmholtz equation includes $d(\Delta G^\circ/T)$. To get this quantity, you must rearrange Eq. (7-22) before differentiating to get

$$\ln K_P = -\frac{\Delta G^\circ}{RT} = -\left(\frac{1}{R}\right)\left(\frac{\Delta G^\circ}{T}\right)$$

Now

$$\frac{d \ln K_P}{dT} = -\left(\frac{1}{R}\right)\left[\frac{d(\Delta G^\circ/T)}{dT}\right]$$

But

$$\frac{d(\Delta G^\circ/T)}{dT} = -\frac{\Delta H}{T^2} \qquad \text{[Eq. (6-56)]}$$

Thus

$$\frac{d \ln K_P}{dT} = \frac{\Delta H}{RT^2} \qquad \text{[Eq. (7-26)]}$$

PROBLEM 7-22 Repeat Example 7-29, assuming that ΔH doesn't depend on temperature. Compare the two answers.

Solution: When ΔH is constant and only one calculation is to be made, you should choose either Eq. (7-30) or Eq. (7-31). Let's rewrite and use Eq. (7-31):

$$\ln K_{P_2} = \ln K_{P_1} + \left(\frac{\Delta H^\circ}{R}\right)\left(\frac{T_2 - T_1}{T_2 T_1}\right)$$

Now substitute values from Example 7-29 to get

$$\ln K_{P_2} = \ln 8.33 \times 10^{-81} + \left(\frac{483.6 \times 10^3}{8.314}\right)\left[\frac{1500 - 298}{1500(298)}\right] = -184.39 + 156.41 = -27.98$$
$$K_{P,1500} = 7.05 \times 10^{-13}$$

In Example 7-29, we found that $K_P = 39.4 \times 10^{-13}$. Thus the approximate answer is about 1/5 of the more accurate answer.

Supplementary Exercises

PROBLEM 7-23 The expression for K_P for the reaction $CaCO_3(s) \rightleftharpoons CaO(s) + CO_2(g)$ is ________. If K_c for the reaction $H_2(g) + I_2(g) \rightleftharpoons 2HI(g)$ is 50, K_P is ________. For the same data, K_c for the dissociation of 2HI is ________. Use the preceding answer to determine K_c for the dissociation of 1HI: ________. For the equilibrium system $4NH_3(g) + 5O_2(g) \rightleftharpoons 4NO(g) + 6H_2O(g)$ + heat, at 200 °C and a fixed pressure of 1 atm, predict whether the position of equilibrium will be shifted in favor of products or reactants or will not be changed by (**a**) adding more NO ________; (**b**) raising the temperature to 300 °C ________; (**c**) increasing the pressure ________; (**d**) adding more O_2 ________; (**e**) adding a catalyst ________; and (**f**) introducing an inert gas ________. When $\Delta G = 0$, Q_P = ________.

PROBLEM 7-24 A 20.0-L vessel contains 1.50 mol O_2, 1.50 mol N_2, and 0.10 mol NO after equilibrium has been reached. Calculate K_c for the reaction $2NO \rightleftharpoons N_2 + O_2$.

PROBLEM 7-25 For the NO system of Problem 7-24, what is the equilibrium concentration of the three gases if 2.0 mol NO is placed in a 40.0-L flask?

PROBLEM 7-26 Using information from Problem 7-25, calculate the degree of dissociation of NO.

PROBLEM 7-27 For the system $CO_2(g) + H_2(g) \rightleftharpoons CO(g) + H_2O(g)$, at 700 °C, $K_c = 0.534$. Calculate the number of moles of CO present at equilibrium if 2 mol CO_2 and 1 mol H_2 were mixed initially.

PROBLEM 7-28 For the reaction $CO(g) + 2H_2(g) \rightleftharpoons CH_3OH(l)$, the value of K_c at 25 °C is 45.0. Calculate K_P.

PROBLEM 7-29 When 4.0 mol H_2O was heated to 1500 °C in a 6.0-L vessel, the water was found to be 1.6 percent dissociated. Calculate K_c.

PROBLEM 7-30 The system $SbCl_3(g) + Cl_2(g) \rightleftharpoons SbCl_5(g)$, under a pressure of 4.0 atm and at equilibrium, is found to contain 3.0 mol $SbCl_3$, 4.0 mol Cl_2, and 9.0 mol $SbCl_5$. Calculate K_P.

PROBLEM 7-31 For the system $2H_2S(g) \rightleftharpoons 2H_2(g) + S_2(g)$, at 1065 °C and 1 atm, H_2S is 24.7 percent dissociated. Calculate K_P.

PROBLEM 7-32 At 25 °C, K_P is 4.0 for the reaction $2NOBr \rightleftharpoons 2NO + Br_2$. Calculate $\Delta G°$.

PROBLEM 7-33 Calculate K_P at 298 K for the reaction $C_2H_4(g) + H_2O(g) \rightleftharpoons C_2H_5OH(g)$.

PROBLEM 7-34 Continue Problem 7-33 by calculating K_P at 600 K. Assume that ΔH is constant.

PROBLEM 7-35 Compare the answers for Problems 7-33 and 7-34. What do they tell you about ΔS for the reaction?

PROBLEM 7-36 At 0 °C, $K_P = 4.1 \times 10^{-25}$ for the reaction $Na_2SO_4 \cdot 10H_2O(s) \rightleftharpoons Na_2SO_4(s) + 10H_2O(g)$. Calculate the equilibrium pressure of H_2O vapor **(a)** at 0 °C and **(b)** at 25 °C.

Answers to Supplementary Exercises

7-23 P_{CO_2}; 50; 2.0×10^{-2}; 0.14; **(a)** reactants, **(b)** reactants, **(c)** reactants, **(d)** products, **(e)** not changed, and **(f)** products; K_P **7-24** 2.25×10^2
7-25 1.6×10^{-3}; 2.42×10^{-2} **7-26** 0.968 **7-27** 0.572 **7-28** 3.1×10^{-3}
7-29 1.41×10^{-6} **7-30** 3.0 **7-31** 1.18×10^{-2} **7-32** 3.4 kJ **7-33** 25.9
7-34 2.49×10^{-3} **7-35** It is negative. **7-36** **(a)** 3.6×10^{-3} atm; **(b)** 2.51×10^{-2} atm

8 PHYSICAL EQUILIBRIUM: A SINGLE SUBSTANCE

THIS CHAPTER IS ABOUT

☑ **Phase Transitions**
☑ **Phase Diagrams**
☑ **Clapeyron Equation**
☑ **Clausius–Clapeyron Equation**

In Chapters 6 and 7, we showed that, for a system to be in physical equilibrium, the chemical potential of each substance in each phase must be the same, i.e., $\mu_i^{\alpha} = \mu_i^{\beta} = \ldots$ [Eq. (6-78)]. We must now deal with several aspects of equilibrium in systems that meet this criterion. There are three types of system in which physical equilibrium can occur, and these are presented in Chapters 8–10:

1. A single substance (all phases)—Chapter 8.
2. Two substances (liquid and vapor phases)—Chapter 9.
3. Two or more substances (liquid–liquid, liquid–solid, and solid–solid phases)—Chapter 10.

8-1. Phase Transitions

A. Phases

A **phase** is a homogeneous, physically distinct, and mechanically separable portion of a system. For example, in a mixture of powdered sulfur and iron filings, the two solid phases are clearly distinguishable by their colors, and they could be sorted by hand. Every yellow particle would be exactly like every other yellow particle in composition. Although separated from each other, the yellow particles comprise one solid phase, homogeneous in composition. The mixture, of course, is heterogeneous.

The three phases of substances are solid, liquid, and gas. At a given pressure, the phase of a substance depends on its temperature or, more fundamentally, on the kinetic energy of its molecules. A substance may exist in more than one solid phase, a phenomenon called **polymorphism**. (With reference to *elements*, the term **allotropy** usually is used.) The terms mean that a substance exists in two or more forms, which have significantly different physical or chemical properties. For each of the three basic phases, the reason for the occurrence of allotropic forms is different:

Phase	Difference in	Example
Solid	Crystal structure	Graphite and diamond
Liquid	Liquid structure	Helium I and helium II
Gas	Number of atoms in molecule	O_2 and O_3

EXAMPLE 8-1: List some substances that exist in more than one solid form.

Solution:

Some allotropes of elements		Some polymorphs of compounds	
Sulfur:	rhombic	$CaCO_3$:	rhombohedral (calcite)
	monoclinic		orthorhombic (aragonite)
Tin:	tetragonal (metallic)	AgI:	hexagonal
	cubic (gray)		cubic

B. Changes of phase

A **phase transition** is the change of a substance from one phase to another. It occurs at constant temperature and is an equilibrium process. Thus a phase transition is reversible in the thermodynamic sense, and $\Delta G = 0$.

EXAMPLE 8-2: Give the specific names for the six common phase transitions.

Solution:

Liquid to vapor:	**vaporization;**	reverse: **condensation**
Solid to vapor:	**sublimation;**	reverse: **condensation**
Solid to liquid:	**fusion;**	reverse: **crystallization**

Note that solid-to-solid transitions do not have names.

C. Heats of transition

Any phase change requires addition or removal of energy in order to change the average kinetic energy of the molecules to the value appropriate to the new phase. The temperature of water vapor at 100 °C and the temperature of liquid water at 100 °C are the same, but the average kinetic energy of the molecules obviously is different. The amount of energy required to raise the average kinetic energy enough to overcome the attractive forces of the liquid state is

$$\Delta E_{\mathrm{v}} = E_{\mathrm{vapor}} - E_{\mathrm{liquid}} = E_v - E_l \tag{8-1}$$

Because transitions normally are at constant pressure, we use enthalpy to express the magnitude of the changes. We write the **heat of vaporization** per mole at the boiling point as ΔH_{v}, representing the difference between the enthalpies of the liquid and the vapor. Thus

$$\Delta H_{\mathrm{v}} = H_v - H_l \tag{8-2}$$

This difference includes ΔE and the work of pushing back the atmosphere by the vapor as it is formed, so

$$\Delta H_{\mathrm{v}} = \Delta E_{\mathrm{v}} + P(V_v - V_l) \tag{8-3}$$

Similarly, we can define the **heat of fusion,** ΔH_{f}, and the **heat of sublimation,** ΔH_{s}. Some typical phase transition values are listed in Table 8-1.

EXAMPLE 8-3: Calculate the heat required to melt 100 g of benzene at its normal melting point.

Solution: From Table 8-1, we find that $\Delta H_{\mathrm{f}} = 9.821\ \mathrm{kJ\ mol^{-1}}$. Thus

$$\text{Heat} = n(\Delta H_{\mathrm{f}}) = (100\ \mathrm{g})\left(\frac{1\ \mathrm{mol}}{78.1\ \mathrm{g}}\right)(9.821\ \mathrm{kJ\,mol^{-1}}) = 12.6\ \mathrm{kJ}$$

EXAMPLE 8-4: A student who needed the heat of sublimation of water for a problem reasoned that sublimation was the overall result of fusion plus evaporization, so she added ΔH_{f} and ΔH_{v} from Table 8-1 to get ΔH_{s}. Is this a valid procedure? Why?

Solution: No. The basic reasoning is valid; the error is in using the values from the table. If all values are for the *same temperature,*

$$\Delta H_{\mathrm{s}} = \Delta H_{\mathrm{v}} + \Delta H_{\mathrm{f}} \tag{8-4}$$

The values in the table, however, refer to the normal melting and boiling points. Remember that ΔH depends on temperature. By appropriate calculations, the student could convert the tabulated values to their values at the sublimation temperature.

TABLE 8-1: Phase Transition Data

	T_f (K)	ΔH_f (kJ mol^{-1})	T_b (K)	ΔH_v (kJ mol^{-1})
Benzene	278.7	9.821	353.4	30.768
Carbon tetrachloride	250.2	2.680	350.0	29.890
Ethanol	161.2	4.791	351.5	39.253
Iron	1808.2	11.204		
Mercury	234.2	2.327		
Water	273.2	5.999	373.2	40.625

D. Trouton's rule

The heat of vaporization may be estimated from a rough relationship known as **Trouton's rule**:

TROUTON'S RULE

$$\frac{\Delta H_v}{T_b} = \text{Constant} \qquad \textbf{(8-5)}$$

The accepted constant is approximately 21 cal K^{-1}mol^{-1}, or 88 J K^{-1}mol^{-1}.

EXAMPLE 8-5: Test Trouton's rule on benzene, carbon tetrachloride, ethanol, and water.

Solution: We obtain the necessary data from Table 8-1 and apply Eq. (8-5) to each compound:

	$\Delta H_v T_b^{-1}$ (J K^{-1}mol^{-1})
Benzene	$(30.768 \times 10^3)/353.4 = 87.1$
Carbon tetrachloride	$(29.890 \times 10^3)/350.0 = 85.4$
Ethanol	$(39.253 \times 10^3)/351.5 = 111.7$
Water	$(40.625 \times 10^3)/373.2 = 108.9$

The first two values fall close to 88 J K^{-1}mol^{-1}, but the last two don't. In general, the rule doesn't work well for compounds with hydrogen bonding or large polarity or those with very low boiling points (i.e., compounds that are gases at room temperature).

E. Entropies of transition

Recall Eq. (5-16):

$$\Delta S_{trans} = \frac{\Delta H_{trans}}{T_{trans}}$$

For evaporation, it is written

$$\Delta S_v = \frac{\Delta H_v}{T_b} \qquad \textbf{(8-6)}$$

Comparing Eq. (8-6) to Eq. (8-5), we see that Trouton's rule says that the entropies of vaporization of many liquids are approximately the same, with values close to 88 J K^{-1}mol^{-1}.

EXAMPLE 8-6: Calculate the entropy of vaporization of liquid bromine using values of $S°$ from Table A-1 and compare the result to the prediction from Trouton's rule.

Solution: From Eq. (5-21),

$$\Delta S_v = S°(Br_2, g) - S°(Br_2, l) = 245.5 - 152.2 = 93.3 \text{ J K}^{-1}\text{mol}^{-1}$$

This result agrees fairly well with the Trouton's rule constant.

8-2. Phase Diagrams

A. Vapor-pressure curves

When a substance evaporates or sublimes in a closed container, pressure develops because of the buildup of vapor. When dynamic equilibrium has been established between liquid and vapor, the pressure exerted by the vapor is called the **vapor pressure** of the liquid. Vapor pressure increases with temperature, as shown by the typical *vapor-pressure curves* in Figure 8-1. In these plots of equilibrium pressure versus the corresponding temperature, each point on the curve represents a state of equilibrium between liquid and vapor; thus each point is a **boiling point**, i.e., the temperature at which the liquid boils under the confining pressure. So the vapor pressure curve also is a *boiling-point curve*.

FIGURE 8-1. Vapor-pressure curves for several liquids.

B. Phase diagram

If the vapor-pressure curve of a substance is combined with similar curves for the solid–vapor and the solid–liquid equilibria, the result is a **phase diagram**. A typical diagram for a one-component (single substance) system is shown in Figure 8-2.

The significant features of a phase diagram are

Curves: Each point on a curve represents an equilibrium state where two phases must be present. In Figure 8-2,

TC = boiling-point curve; liquid $\rightleftharpoons$ vapor
TB = melting-point curve; solid $\rightleftharpoons$ liquid
TA = sublimation-point curve; solid $\rightleftharpoons$ vapor

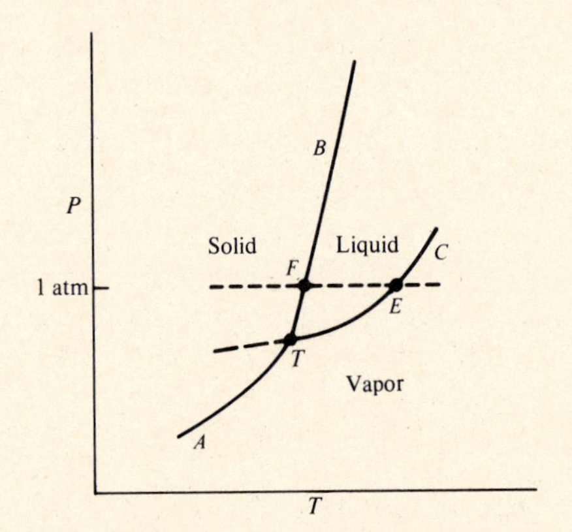

FIGURE 8-2. General phase diagram.

Triple point: The point where the three curves meet is the triple point T. Here, all the three phases, solid $\rightleftharpoons$ liquid $\rightleftharpoons$ vapor, are present. This point is an **invariant point**. The system can be in this state only at that specific temperature and pressure and if the three phases are present and in equilibrium.

Areas: In the areas between the curves, a point represents a single phase. Two phases can't be in equilibrium at a pressure and temperature that correspond to a point between the curves.

Critical point: The point C is the critical point, which we already described in Chapter 2.

Normal boiling and melting points: The boiling point (point E) and melting point (point F) at 1 atm pressure are called the **normal boiling point** and **normal melting point**, respectively.

C. Two special cases

1. Water

Water is one of very few substances that has a diagram unlike Figure 8-2. The difference is that the solid–liquid curve, TB, slopes to the left; in other words, the melting point of water decreases as the pressure increases. In Figure 8-3, the effect is exaggerated to show it clearly. The horizontal difference between points F and T is only 0.01 K: the normal freezing point is 273.15 K and the triple point is 273.16 K. The curve slopes to the left because the density of the solid is less than that of the liquid.

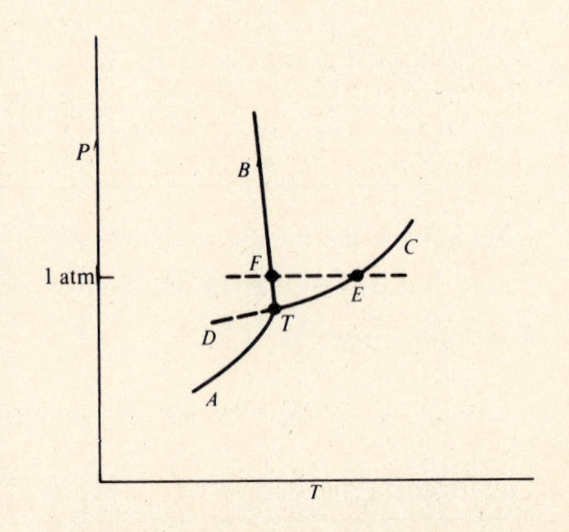

FIGURE 8-3. Phase diagram of water.

EXAMPLE 8-7: Use the dependence of the Gibbs energy on pressure to explain why the change in the melting point of water with pressure is opposite that of most substances.

Solution: The dependence of G on P is given by Eq. (6-40):

$$\left(\frac{\partial G}{\partial P}\right)_T = V \quad \text{or} \quad dG = V\,dP$$

Since V is always positive, if P increases, G also increases.

For most substances, the *density* of the liquid is less than that of the solid; thus, for a given mass, the *volume* of the liquid is greater than that of the solid. Therefore, if the pressure increases by dP, dG_l is greater than dG_s. Momentarily, G_l will be greater than G_s. To maintain equilibrium, G_l must become equal to G_s. Thus some liquid must change to solid, which is achieved by a rise in the melting point, dT. This phenomenon is an application of Le Chatelier's principle: A change that removed the stress has occurred. When equilibrium exists again, overall

$$dG_l = dG_s \tag{8-7}$$

For water, the opposite occurs because the density of liquid water is greater than that of ice. Thus, for a given mass, the volume of the liquid is less than that of the solid, and, for a positive dP, $dG_l < dG_s$. Now the melting point must decrease to hold $G_l = G_s$.

2. Carbon dioxide

Carbon dioxide is different from many substances in that its triple point is at a pressure above 1 atm, as shown in Figure 8-4. The result is the practical phenomenon of *dry ice.*

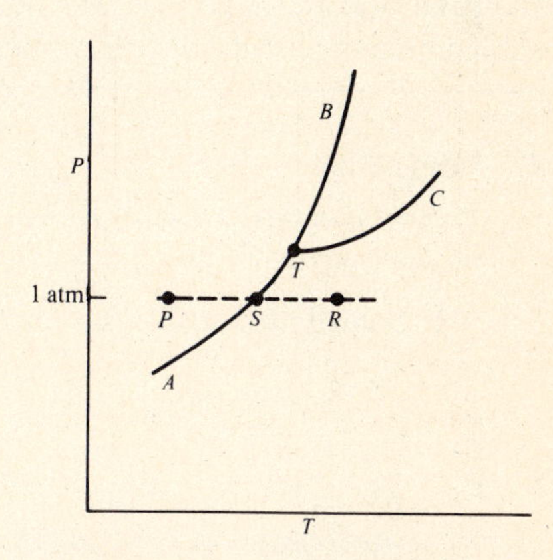

FIGURE 8-4. Phase diagram of carbon dioxide.

EXAMPLE 8-8: Explain why dry ice is dry (refer to Figure 8-4).

Solution: We start with solid CO_2 at a pressure of 1 atm and a temperature to the left of curve TA, at point P. If we warm the solid at constant pressure, the state of the system is described by points along line PR. When the temperature corresponds to point S on curve TA, the solid will sublime. When the system is in the vapor state, the temperature will rise again toward point R. No liquid will appear.

8-3. Clapeyron Equation

A. Derivation

In Example 8-7, we saw that a change in pressure on a system of two phases in equilibrium requires a change in temperature in order to maintain equilibrium. The Gibbs free energy of each phase must change by the same amount, i.e.,

$$dG^\alpha = dG^\beta \tag{8-8}$$

Previously, in Eq. (6-35), we had

$$dG = V\,dP - S\,dT$$

Writing this expression for both phases, substituting in Eq. (8-8), and rearranging gives us

$$\frac{dP}{dT} = \frac{\Delta s}{\Delta V} \tag{8-9}$$

where $\Delta S = S^{\beta} - S^{\alpha}$ and $\Delta V = V^{\beta} - V^{\alpha}$. Then, for a phase transition, we have Eq. (5-16):

$$\Delta S_{\text{trans}} = \frac{\Delta H_{\text{trans}}}{T_{\text{trans}}}$$

Substituting in Eq. (8-9) gives us an expression known as the **Clapeyron equation**:

CLAPEYRON EQUATION: $$\frac{dP}{dT} = \frac{\Delta H_{\text{trans}}}{T_{\text{trans}} \Delta V} \quad \textbf{(8-10)}$$

B. Interpretation

The Clapeyron equation expresses a rate, or coefficient, of change. In fact, it gives the slopes of the various curves in Figures 8-2, 8-3, and 8-4. The plots are curved rather than straight because the rate depends on one of the variables, temperature. For the solid → liquid curve, the slope may be positive or negative because ΔV can be positive or negative. For the solid → gas and the liquid → gas cases, however, ΔV is always positive.

EXAMPLE 8-9: What can we say about the relative slopes of the sublimation and evaporation curves of phase diagrams?

Solution: From Eq. (8-10), we can see that the slope is inversely proportional to the volume. For the solid → gas change, ΔV will be greater than ΔV for the liquid → gas change. Thus the sublimation curve will be steeper than the evaporation curve.

C. Integration

Proper integration usually isn't possible because we don't have expressions for ΔH_{trans} and ΔV as functions of T. Two approximations may be used.

1. Sometimes Eq. (8-10) is rewritten as

$$\frac{P_2 - P_1}{T_2 - T_1} = \frac{\Delta H}{T \Delta V} \quad \textbf{(8-11)}$$

where T is the average of T_2 and T_1. This approximation would be reliable only over small ranges of temperature.

2. If we assume that ΔH and ΔV are constant, which is not a bad assumption for solid–liquid transitions, we can integrate Eq. (8-10):

$$\int_{P1}^{P_2} dP = \frac{\Delta H}{\Delta V} \int_{T_1}^{T_2} \frac{dT}{T}$$

or

$$P_2 - P_1 = \frac{\Delta H}{\Delta V} \ln \frac{T_2}{T_1} \quad \textbf{(8-12)}$$

This expression, although still an approximation, is better than Eq. (8-11) when the temperature change is large. For the melting-point curve, dT/dP is so small that we usually treat melting points as constants.

EXAMPLE 8-10: What pressure would be required to lower the melting point of 1 mol H_2O to -0.5 °C? The densities of ice and water at 0 °C and 1 atm are 0.9168 and 0.999 84 g mL^{-1}.

Solution: We will use Eq. (8-12) but will calculate ΔV first. The volumes of one mole are given by $V = M_m/d$, so

$$V = V_l - V_s = \left(\frac{18.016}{0.999\,84} - \frac{18.016}{0.9168}\right)\left(\frac{\text{g mol}^{-1}}{\text{g mL}^{-1}}\right) = 18.019 - 19.651 = -1.632 \text{ mL mol}^{-1}$$

From Table 8-1, $\Delta H_f = 5999$ J mol^{-1}. Substituting in Eq. (8-12), we obtain

$$P_2 - 1 = \frac{5999 \text{ J mol}^{-1}}{-1.632 \text{ mL mol}^{-1}} \ln \frac{272.65}{273.15}$$

We have to do something about the units. When we don't know an energy conversion factor, often we can use the values of R. To get the pressure in atm, we should convert from J to L atm. Canceling the mol^{-1} units in the previous equation, we now have

$$P_2 - 1 = \left(-\frac{5999 \text{ J}}{1.632 \times 10^{-3} \text{ L}}\right)\left(\frac{0.082\,06 \text{ L atm K}^{-1}\text{mol}^{-1}}{8.314 \text{ J K}^{-1}\text{mol}^{-1}}\right)(-0.001\,832\,2) = 66.47 \text{ atm}$$

$$P_2 = 67.47 \text{ atm}$$

8-4. Clausius–Clapeyron Equation

The Clapeyron equation usually is applied only to solid–liquid or solid–solid equilibria. When one phase is a gas, a very useful extension, developed by Clausius, is used.

A. Derivation

We make two assumptions: 1. $V_g \gg V_l$; therefore $\Delta V = V_g$.
2. Ideal behavior; therefore $V = RT/P$ for one mole.

Substituting in Eq. (8-10) yields

$$\frac{dP}{dT} = \frac{\Delta H}{TV_g} = \left(\frac{\Delta H}{T}\right)\left(\frac{P}{RT}\right)$$

Dividing by P gives $\frac{dP}{P}$, which is $d \ln P$, and the **Clausius–Clapeyron equation:**

CLAUSIUS–CLAPEYRON EQUATION

$$\frac{d \ln P}{dT} = \frac{\Delta H}{RT^2} \qquad \textbf{(8-13)}$$

Here, ΔH may be either the heat of vaporization or the heat of sublimation, and T is the corresponding temperature. Results calculated from this equation are valid only to the extent of the validity of the two assumptions. Near the critical state, V_l will be much closer to V_g, so the equation is less reliable.

B. Integration

Again we have the difficulty that ΔH is a function of T, but again we assume that it's constant and integrate Eq. (8-13):

$$\int d \ln P = \frac{\Delta H}{R} \int \frac{1}{T^2} dT$$

1. Integration without limits yields

$$\ln P = \left(-\frac{\Delta H}{R}\right)\left(\frac{1}{T}\right) + C \qquad \textbf{(8-14)}$$

2. Integration with the limits P_2, T_2 and P_1, T_1 yields

$$\ln \frac{P_2}{P_1} = \left(-\frac{\Delta H}{R}\right)\left(\frac{1}{T_2} - \frac{1}{T_1}\right) \qquad \textbf{(8-15)}$$

or

$$\ln \frac{P_2}{P_1} = \frac{\Delta H}{R}\left(\frac{T_2 - T_1}{T_2 T_1}\right) \qquad \textbf{(8-16)}$$

As with previous equations obtained from the two types of integration, Eq. (8-14) is useful if a series of calculations is required, whereas Eq. (8-15) or Eq. (8-16) is more convenient if only one calculation is required. With Eq. (8-16), the mathematics is more convenient, and we are less likely to lose significant figures.

C. Interpretation

Equation (8-14) is the equation of a straight line; a plot of ln P versus $1/T$ would have the slope $-\Delta H/R$ and the intercept C. Experimental data do give straight lines, so this is a valid method of determining values of ΔH.

EXAMPLE 8-11: A liquid is placed in an apparatus for measuring vapor pressure over a liquid. The results of two pressure–temperature measurements are $P = 239.9$ mm Hg at 357.1 K and $P = 13.80$ mm Hg at 294.1 K. Assuming that these points fall on the straight line of a plot of ln P versus $1/T$, calculate the heat of vaporization of the substance.

Solution: First we convert the data to the form needed for plotting:

$$P_2:\ \ln(239.9/760) = -1.153 \qquad P_1:\ \ln(13.80/760) = -4.009$$
$$T_2:\ 1/357.1 = 2.800 \times 10^{-3} \qquad T_1:\ 1/294.1 = 3.400 \times 10^{-3}$$

The slope is calculated from

$$m = \frac{y_2 - y_1}{x_2 - x_1} = \frac{\ln P_2 - \ln P_1}{1/T_2 - 1/T_1} = \frac{-1.153 - (-4.009)}{2.800 \times 10^{-3} - 3.400 \times 10^{-3}} = -4.760 \times 10^3$$

From Eq. (8-14), the slope is $-\Delta H_v/R$, so

$$-\Delta H_v = -4.760 \times 10^3 \mathrm{R}$$
$$\Delta H_v = 39.57 \text{ kJ mol}^{-1}$$

EXAMPLE 8-12: The vapor pressure of water at 80.0 °C is 355.1 mm Hg. Calculate the vapor pressure at 90.0 °C, assuming that the ΔH_v from Table 8-1 is valid at these temperatures.

Solution: We use Eq. (8-16):

$$\ln \frac{P_2}{P_1} = \frac{\Delta H_v}{R}\left(\frac{T_2 - T_1}{T_2 T_1}\right)$$
$$\ln P_2 - \ln 355.1 = \left(\frac{40\,625 \text{ J mol}^{-1}}{8.314 \text{ J K}^{-1}\text{mol}^{-1}}\right)\left[\frac{10}{363.2(353.2)}\right]$$
$$\ln P_2 = 5.872 + 0.3809$$
$$P_2 = 519.6 \text{ mm Hg}$$

SUMMARY

The major equations of this chapter and connections between them are shown in Table S-8. You should determine the conditions that are necessary for each equation to be valid or the conditions that are imposed to make the connections. Other important things you should know are

1. The criterion for physical equilibrium is $\mu_i^\alpha = \mu_i^\beta = \ldots$, which also can be expressed as $\Delta G = 0$ for any change at equilibrium.
2. Polymorphism and allotropism describe the phenomenon where one substance exists in two or more forms, which have significantly different physical or chemical properties.
3. Phase transitions occur at constant temperature, are equilibrium processes, and are accompanied by a gain or loss of energy.
4. Phase diagrams describe systems under all pressure and temperature combinations; lines represent states of equilibrium between two phases, areas represent one phase, and intersections of lines represent three phases.
5. The solid–liquid line slopes to the right for almost all substances; in other words, an increase in pressure raises the melting point. Water is one of the few exceptions.
6. The Clapeyron equation gives the slope of any phase transition curve, whereas the Clausius–Clapeyron equation applies only to vaporization and sublimation.

TABLE S-8: Summary of Important Equations

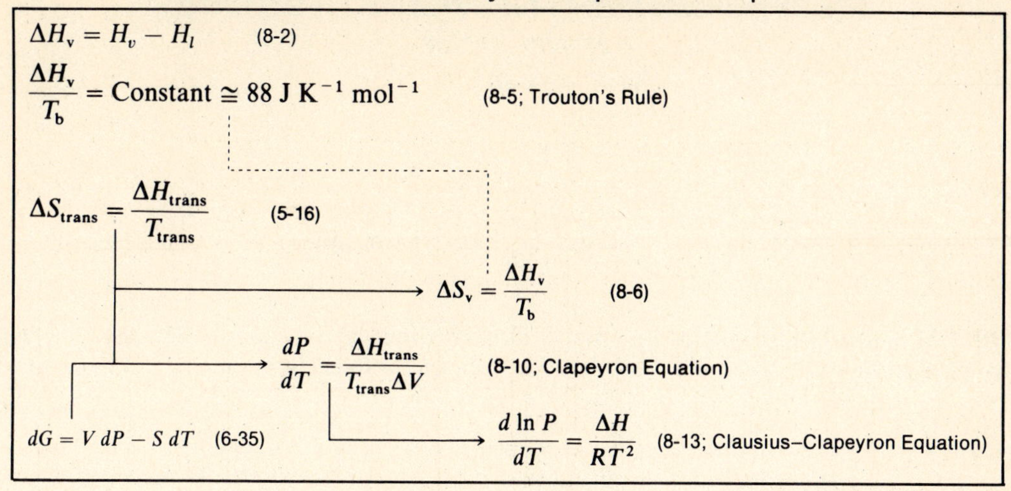

SOLVED PROBLEMS

Phase Transitions and Diagrams

PROBLEM 8-1 Isopropanol boils at 82.5 °C and 1 atm. Estimate the heat of vaporization.

Solution: For an estimate, you can use Trouton's rule. Rearrange Eq. (8-5) and evaluate:

$$\Delta H_v = 88 \text{ J K}^{-1}\text{mol}^{-1}(T_b) = 88(355.7) = 31.3 \text{ kJ mol}^{-1}$$

PROBLEM 8-2 Refer to Figure 8-5 and describe what you would see if 10 g of ice at 1 atm and -10 °C were warmed to 110 °C by a slow application of heat. The temperature is recorded continuously.

Solution: Note that you are not asked to explain what happens but merely to state what you would *see*. You would see the recorder show a rise in temperature with time until $T = 273.15$ K; then T would remain constant. You would see the ice slowly melt to liquid. When the ice is gone, the temperature would rise again. At 373.15 K, the temperature again would remain constant while the liquid disappears, turning into vapor which, of course, you could not see. When all the liquid is gone, the temperature would rise again.

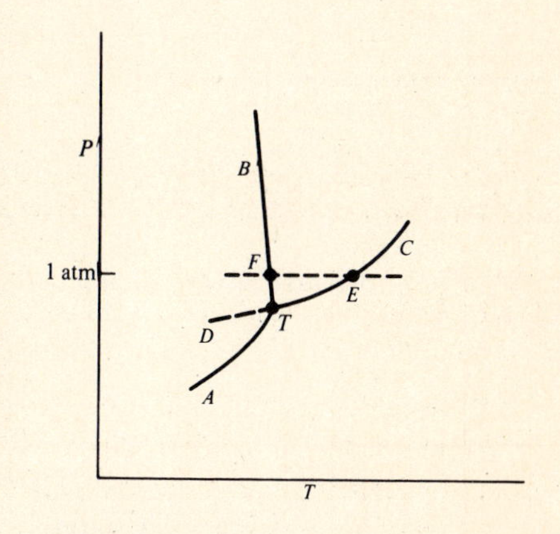

FIGURE 8-5. Phase diagram of water.

Clapeyron Equation

PROBLEM 8-3 Give a complete derivation of the Clapeyron equation; that is, fill in the steps omitted in getting from Eq. (8-8) to Eq. (8-10).

Solution: You should write Eq. (6-35) for each phase and substitute in Eq. (8-8):

$$dG^\alpha = V^\alpha\,dP - S^\alpha\,dT$$

$$dG^\beta = V^\beta\,dP - S^\beta\,dT$$

$$V^\alpha\,dP - S^\alpha\,dT = V^\beta\,dP - S^\beta\,dT$$

Rearrange terms:

$$(S^\beta - S^\alpha)\,dT = (V^\beta - V^\alpha)\,dP$$

$$\Delta S\,dT = \Delta V\,dP$$

$$\frac{dP}{dT} = \frac{\Delta S}{\Delta V} \qquad \text{[Eq. (8-9)]}$$

Now substitute $\Delta S = \Delta H/T$ to get

$$\frac{dP}{dT} = \frac{\Delta H}{T\,\Delta V} \qquad \text{[Eq. (8-10)]}$$

PROBLEM 8-4 At its freezing point, 1 mol of liquid benzene has a volume of 88.7 mL, whereas the volume of the solid is 87.5 mL. Calculate the slope of the freezing-point curve of benzene at 7 °C.

Solution: From Table 8-1, the freezing point of benzene is 278.7 K, or 5.5 °C, and $\Delta H_f = 9821$ J mol^{-1}. Since 7 °C is close to the freezing point, you can assume that ΔH_f and ΔV_f are constant. The slope of the curve is given by Eq. (8-10), so

$$\frac{dP}{dT} = \frac{\Delta H_f}{T\Delta V_f} = \frac{\Delta H_f}{T(V_l - V_s)} = \frac{9821\ \text{J mol}^{-1}}{(280.2\ \text{K})(88.7 - 87.5)\ \text{mL}}$$

You must now convert J to L atm and mL to L to obtain

$$\frac{dP}{dT} = \frac{9821\ \text{J mol}^{-1}}{(280.2\ \text{K})(1.2 \times 10^{-3}\ \text{L})}\left(\frac{0.08205\ \text{L atm K}^{-1}\text{mol}^{-1}}{8.314\ \text{J K}^{-1}\text{mol}^{-1}}\right) = 288.3\ \text{atm K}^{-1}$$

PROBLEM 8-5 Calculate the change in the boiling point of water per atmosphere change in pressure. The density of liquid water at 100 °C is 0.958 35 g mL^{-1}. Assume that water vapor behaves ideally at 100 °C.

Solution: You can represent the change in boiling point per unit change in pressure by $\frac{dT}{dP}$, which is the reciprocal of Eq. (8-10). First, calculate the molar volumes. Since V_g will be much greater than V_l, you should use more significant figures than usual.

$$V_l = \frac{\text{Mass}}{\text{Density}} = \frac{18.016\ \text{g mol}^{-1}}{0.958\,35\ \text{g mL}^{-1}} = 18.799\ \text{mL mol}^{-1}$$

For the volume of the gas, convert the 22.4 L for an ideal gas at 0 °C to 100 °C:

$$V_g = 22.412\left(\frac{373.15}{273.15}\right) = 30.6170\ \text{L mol}^{-1}$$

$$\Delta V = V_g - V_l = 30.6170 - 0.0188 = 30.5982\ \text{L mol}^{-1}$$

Write Eq. (8-10) for the vaporization process and invert it to get

$$\frac{dT}{dP} = \frac{T_b(V_g - V_l)}{\Delta H_v}$$

Obtain ΔH_v from Table 8-1 and convert J to L atm to obtain

$$\frac{dT}{dP} = \frac{373.15\ \text{K}\ (30.598\ \text{L mol}^{-1})}{40\,625\ \text{J mol}^{-1}}\left(\frac{8.314\ \text{J K}^{-1}\text{mol}^{-1}}{0.082\,05\ \text{L atm K}^{-1}\text{mol}^{-1}}\right) = 28.48\ \text{K atm}^{-1}$$

note: This value is an instantaneous slope, valid only at 100 °C and 1 atm. The slope changes rapidly with changes in T or P because V_g changes rapidly.

PROBLEM 8-6 The normal melting point of a certain compound is 340 K. Under 480 atm pressure, it melts at 353 K. If the volume change of one mole upon melting is 21.0 mL, what is the heat of fusion?

Solution: You will have to assume that the volume change and heat of fusion are constant and use either Eq. (8-11) or Eq. (8-12). Try it both ways and compare the results.

(a) In order to use Eq. (8-11), you need the average T:

$$T = \frac{353 + 340}{2} = 346.5 \text{ K}$$

Now solve Eq. (8-11) for ΔH and evaluate:

$$\Delta H = \frac{T\,\Delta V(P_2 - P_1)}{T_2 - T_1} = \frac{346.5 \text{ K } (0.021 \text{ L})(480 - 1)}{(353 - 340) \text{ K}} = 268.1 \text{ L atm mol}^{-1} = 27.17 \text{ kJ mol}^{-1}$$

(b) Solving Eq. (8-12) for ΔH gives you

$$\Delta H = \frac{\Delta V(P_2 - P_1)}{\ln(T_2/T_1)} = \frac{0.021 \text{ L } (480 - 1) \text{ atm}}{\ln(353/340)} = 268.1 \text{ L atm mol}^{-1}$$

For this range of temperature, Eq. (8-11) seems to be adequate.

Clausius–Clapeyron Equation

PROBLEM 8-7 Calculate the vapor pressure of benzene at 50 °C if it is 548 mm Hg at 70 °C.

Solution: This problem is different from the preceding ones in that neither the volume nor the density is given. You should recognize that it deals with liquid–vapor equilibrium and that you can use the Clausius simplification of the Clapeyron equation. Use Eq. (8-16) and let P_1 represent the unknown pressure, so that T_1 will be smaller than T_2. Use ΔH from Table 8-1 and assume that it is constant.

$$\ln P_2 - \ln P_1 = \left(\frac{\Delta H}{R}\right)\left(\frac{T_2 - T_1}{T_2 T}\right)$$

$$\ln P_1 = \ln 548 - \left(\frac{29\,890 \text{ J mol}^{-1}}{8.314 \text{ J K}^{-1}\text{mol}^{-1}}\right)\left[\frac{(343 - 323) \text{ K}}{343(323) \text{ K}^2}\right] = 6.306 - 0.6490 = 5.657$$

$$P_1 = 286 \text{ mm Hg}$$

PROBLEM 8-8 In handbooks, the vapor pressures of substances often are listed by the expression

$$\log P(\text{mm}) = \frac{-52.23A}{T} + B$$

For acetaldehyde, the values of A and B are 27.707 and 7.8206, respectively, between −24 °C and +27 °C. Calculate ΔH_v at the normal boiling point of 20.2 °C.

Solution: At first glance, you may wonder how you will get ΔH from $\log P$. The expression for $\log P$, however, looks like Eq. (8-14), which was obtained by integration without limits of the Clausius–Clapeyron equation. Of course, this equation was used to set up Table 8-1. For comparison, you need the same logarithmic base. Convert Eq. (8-14) to base 10, so that

$$\log P = \left(-\frac{\Delta H_v}{2.303R}\right)\left(\frac{1}{T}\right) + \frac{C}{2.303}$$

Thus

$$B = \frac{C}{2.303} \quad \text{and} \quad -52.23A = -\frac{\Delta H_v}{2.303R}$$

or

$$\Delta H_v = 52.23(A)(2.303)(R) = 52.23(27.707)(2.303)(8.314) = 27.71 \text{ kJ mol}^{-1}$$

PROBLEM 8-9 Calculate the heat of vaporization of toluene if its vapor pressure has been measured to be 413.4 mm Hg at 90 °C and 500.1 mm Hg at 96 °C.

Solution: When you're given two sets of pressure/temperature data, you can apply the Clausius–Clapeyron equation directly. Solving Eq. (8-16) for ΔH_v and evaluating, you have

$$\Delta H_v = R\left(\ln\frac{P_2}{P_1}\right)\left(\frac{T_2 T_1}{T_2 - T_1}\right) = 8.314\left(\ln\frac{500.1}{413.4}\right)\left[\frac{369(363)}{369 - 363}\right] = 35.3 \text{ kJ mol}^{-1}$$

PROBLEM 8-10 The normal boiling point of nitrobenzene is 210.9 °C. If the vapor pressure at 180 °C is 353.2 mm Hg, what is the vapor pressure at 200 °C?

Solution: This problem has a little subproblem. Since ΔH_v isn't given, you must calculate it. This process was illustrated in Problem 8-9, so use the rearranged Clausius–Clapeyron equation obtained there. Remember that the normal boiling point is at 1 atm. You again have two pressures with corresponding temperatures, and

$$\Delta H_v = 8.314\left(\ln\frac{760}{353.2}\right)\left[\frac{484.1(453.2)}{484.1 - 453.2}\right] = 45.23 \text{ kJ mol}^{-1}$$

In order to obtain the vapor pressure at a given temperature, you must use the Clausius–Clapeyron equation again and solve for the desired quantity. Rearrange Eq. (8-16) to obtain $\ln P_2$:

$$\ln P_2 = \ln P_1 + \left(\frac{\Delta H_v}{R}\right)\left[\frac{T_2 - T_1}{T_2 T_1}\right)$$

Choosing 353.2 mm as P_1, you get

$$\ln P_2 = \ln(353.2) + \left(\frac{45.23 \times 10^3}{8.314}\right)\left[\frac{473.2 - 453.2}{473.2(453.2)}\right] = 5.8670 + 0.5074$$

$$P_2 = 586.6 \text{ mm Hg}$$

PROBLEM 8-11 The normal boiling point of methyl alcohol is 64.7 °C. What is its vapor pressure at 25 °C? Do not use tables of data.

Solution: This problem is much like the previous problem, but less information is given. The Clausius–Clapeyron equation seems to be needed, but it requires ΔH_v. How can you obtain it? Your only resort is to estimate ΔH_v from Trouton's rule, Eq. (8-5). Rearranging that expression, you get

$$\Delta H_v = 88T_b = 88(337.9)$$

The rest of the solution is like Problem 8-10. The vapor pressure is given by

$$\ln P_2 = \ln P_1 + \left(\frac{\Delta H_v}{R}\right)\left(\frac{T_2 - T_1}{T_2 T_1}\right)$$

At 298.2 K,

$$\ln P_2 = \ln(760) + \left[\frac{88(337.9)}{8.314}\right]\left[\frac{298.2 - 337.9}{298.2(337.9)}\right] = 6.6333 - 1.4091 = 5.2242$$

$$P_2 = 185.7 \text{ mm Hg}$$

Supplementary Exercises

PROBLEM 8-12 We call two solid forms of the same compound ________. The two common allotropes of carbon are ________ and ________. At a point of intersection of lines in a phase diagram ________ phases are in equilibrium. We call such a point the ________ point. Another name for a vapor-pressure curve is ________ ________ ________.

PROBLEM 8-13 How many calories of heat would be needed to convert 100 g H_2O at 100 °C to steam at 100 °C?

PROBLEM 8-14 Estimate the heat of vaporization of pentane if its normal boiling point is 36.3 °C.

PROBLEM 8-15 The heat of vaporization of *n*-heptane is 76.3 cal g^{-1}. Estimate the normal boiling point of *n*-heptane.

PROBLEM 8-16 Calculate the boiling point of water on a mountain where the barometric pressure is 600 mm Hg.

PROBLEM 8-17 Calculate the slope of the boiling-point curve of carbon tetrachloride at 100 °C if the densities of the liquid and vapor at that temperature are 1.434 g mL^{-1} and 0.0103 g mL^{-1}, respectively.

PROBLEM 8-18 The vapor pressure of camphor is 400.8 mm Hg at 182.0 °C and 593.2 mm Hg at 194.8 °C. Calculate the heat of vaporization.

Answers to Supplementary Exercises

8-12 polymorphs; diamond, graphite; three, triple point; boiling-point curve
8-13 53.9 kcal (225.4 kJ) **8-14** 27.2 kJ **8-15** 363.4 K **8-16** 93.4 °C
8-17 5.33×10^{-2} atm K^{-1} **8-18** 54.3 kJ mol^{-1}

9 PHYSICAL EQUILIBRIUM: LIQUID AND VAPOR MIXTURES

THIS CHAPTER IS ABOUT

- ☑ Mixtures and Solutions
- ☑ Solutions of Gases in Liquids: Henry's Law
- ☑ Liquid Solutions with One Volatile Component: Raoult's Law
- ☑ Colligative Properties
- ☑ Liquid Solutions with Two Volatile Components
- ☑ Immiscible Volatile Liquids

9-1. Mixtures and Solutions

A. Homogeneous and heterogeneous mixtures

A **mixture** is any system containing two or more components. In general, a mixture will be heterogeneous and have two or more phases. But a mixture can also be homogeneous; when it is, it's called a **solution**. Strictly speaking, nine types of solution are possible: Each of the three phases may be dissolved in each of the three phases. The component present in the largest amount normally is designated the **solvent**; all other components are **solutes**.

EXAMPLE 9-1: List the nine possible types of binary solutions.

Solution:

1. gas in gas	2. gas in liquid	3. gas in solid
4. liquid in gas	5. liquid in liquid	6. liquid in solid
7. solid in gas	8. solid in liquid	9. solid in solid

Some of the nine types aren't common, and others aren't customarily called solutions. An example of a gas/gas solution is air, which could be called a solution of oxygen, carbon dioxide, etc., in nitrogen. Usually we simply call it a mixture of gases.

In this chapter, we will deal primarily with gas/gas and liquid/liquid solutions and with heterogeneous systems in which these two types are in equilibrium. Solid/liquid solutions will appear in illustrations of colligative properties.

B. Thermodynamic relationships

1. Ideal solution of gases

We want an expression for ΔG when a mixture is formed from pure ideal gases. We will develop the case of constant-pressure mixing, in which gases at the same specified pressure are mixed in such a way that the final total pressure equals the specified initial pressure. This is the same special case we used to obtain Eq. (5-20).

From Eq. (6-81) we know that

$$G\text{ (mixture)} = \mu_A n_A + \mu_B n_B + \cdots + \Sigma \mu_i n_i$$

and, for an ideal gas, Eq. (6-74) gives us

$$\mu_2 - \mu_1 = RT \ln \frac{P_2}{P_1}$$

If we let state 1 be the pure gas and state 2 be the gas after mixing, Eq. (6-74) becomes

$$\mu\,(\text{mixture}) = \mu^0 + RT \ln \frac{P\,(\text{mixture})}{P^0} \qquad \textbf{(9-1)}$$

where the superscript zero indicates the pure state. Since $\Delta G = G\,(\text{final}) - G\,(\text{initial})$,

$$\Delta G\,(\text{mixing}) = G\,(\text{mixture}) - G^0\,(\text{unmixed components}) \qquad \textbf{(9-2)}$$

Substituting Eq. (9-1) into Eq. (6-81) and the result into Eq. (9-2), we obtain for the mixing of two ideal gases

$$\Delta G\,(\text{mixing}) = RT\left(n_A \ln \frac{P_A\,(\text{mixture})}{P^0} + n_B \ln \frac{P_B\,(\text{mixture})}{P^0}\right) \qquad \textbf{(9-3)}$$

where P^0 is the specified initial pressure, $P^0 = P$, *and* is the sum of the partial pressures (see Problem 9-2). In general, then,

$$\Delta G\,(\text{mixing}) = RT\,\Sigma\, n_i \ln \frac{P_i}{P} \qquad \textbf{(9-4)}$$

Substituting $\Delta G = -T\,\Delta S$ in Eq. (9-4) yields,

$$\Delta S\,(\text{mixing}) = -R\,\Sigma\, n_i \ln \frac{P_i}{P} \qquad \text{[Eq. (5-19)]}$$

These expressions may be stated in terms of mole fraction, since $P_i = X_i P$. Substituting this and $n_i = nX_i$, we get

$$\Delta G\,(\text{mixing}) = nRT\,\Sigma\, X_i \ln X_i \qquad \textbf{(9-5)}$$

and

$$\Delta S\,(\text{mixing}) = -nR\,\Sigma\, X_i \ln X_i \qquad \textbf{(9-6)}$$

These expressions show that ΔG (mixing) is independent of P but changes with T, whereas ΔS (mixing) is constant at all P and T. Other properties of an ideal solution are

$$\Delta V\,(\text{mixing}) = 0 \qquad \text{and} \qquad \Delta H\,(\text{mixing}) = 0$$

EXAMPLE 9-2: If 0.50 mol of ideal gas A at 2.0 atm and 0.80 mol of ideal gas B at 2.0 atm are allowed to mix to form an ideal mixture at 2.0 atm at 300 K, what is the Gibbs free energy of mixing?

Solution: This clearly is a case of mixing at constant pressure, so Eq. (9-3) is applicable. In order to use it, however, we need the partial pressures of the gases in the final mixture. Remember that $P_i = X_i P$ and $X_i = n_i/n$. Combining these expressions for gas A, we have

$$P_A = \frac{n_A}{n_A + n_B} P = \left(\frac{0.50}{0.50 + 0.80}\right)(2.0) = 0.77 \text{ atm}$$

$$P_B = 2 - P_A = 1.23 \text{ atm}$$

Now, we substitute in Eq. (9-3) to obtain

$$\Delta G\,(\text{mixing}) = RT\left(n_A \ln \frac{P_A}{P} + n_B \ln \frac{P_B}{P}\right)$$

$$= (8.314 \text{ J K}^{-1}\text{mol}^{-1})(300 \text{ K})\left[(0.50 \text{ mol})\left(\ln \frac{0.77}{2.0}\right) + (0.80 \text{ mol})\left(\ln \frac{1.23}{2.0}\right)\right]$$

$$= -2.2 \text{ kJ}$$

---

2. Other ideal solutions

The preceding expressions were derived for gases, but they apply to all ideal solutions. Thus an **ideal solution** is a solution that obeys Eq. (9-5). An alternative definition focuses on each component rather than the whole mixture. Letting $P^0 = 1$ atm and substituting $P_i = X_i P$ in Eq. (9-1), then separating the ln term, we get

$$\mu_i = \mu_i(T, P) + RT \ln X_i \tag{9-7}$$

where $\mu_i(T, P) = \mu_i^0 + RT \ln P$ represents μ of component i at P. At this pressure, i may not be in the same physical state as the solution. Thus we let $\mu_i^0(T, P)$ represent the μ of pure i in the same state as the solution and define an ideal solution as a solution in which every component obeys the expression

$$\mu_i = \mu_i^0(T, P) + RT \ln X_i \tag{9-8}$$

3. Real solutions

In most cases, behavior is not ideal. Only if activities are used will actual systems obey the above relationships. Deviations from ideal behavior are caused mainly by differences in the attractive forces between molecules. Thus ideal solutions usually have components with very similar molecular structures, such as benzene and toluene.

9-2. Solutions of Gases in Liquids: Henry's Law

A. Qualitative considerations

The amount of gas that dissolves depends on the pressure, the temperature, and the natures of the gas and the solvent. The nature of the components determines the magnitude of the attractive forces between their molecules. If large enough, these forces may lead to the formation of new compounds, such as the ammine complexes of BF_3, or new species, such as NH_4^+ ions. In such cases, solubility is high. If the attractive forces are weak, however, solubility of gases usually is low. Solubility of a gas always increases with an increase in pressure and usually decreases with an increase in temperature.

B. Quantitative considerations

The direct proportionality between solubility of a gas in a liquid and pressure is expressed by **Henry's law**:

HENRY'S LAW

$$X_2 = kP_2 \tag{9-9}$$

The subscript 2 is used to designate solute, while a subscript 1 will be used for the solvent. The proportionality constant depends on temperature and will be different for each combination of gas and solvent. The pressure is that exerted by the gas above the solution. If several gases are in the solution, P_2 is the partial pressure of each, as X_2 is the mole fraction of each. At very low concentrations, all solutions follow Henry's law; i.e., it is not limited to solutions of gases in liquids. It is followed less well, however, as concentration increases. Thus agreement with experiment is better at high temperatures and low pressures (which give dilute solutions).

Other units of concentration are used, especially molarity and molality. If the constant is given, its units will tell you the units to use for concentration and pressure.

EXAMPLE 9-3: The Henry's law constant for gaseous oxygen in liquid water at 20 °C is 1.38×10^{-3} mol L^{-1}atm^{-1}. Calculate the grams of O_2 in 5.0 L H_2O at that temperature and a partial pressure of 0.50 atm.

Solution: Because the unit of the constant includes molarity, we should write Henry's law as

$$C = kP_2$$

Substitution yields

$$C = (1.38 \times 10^{-3} \text{ mol L}^{-1}\text{atm}^{-1})(0.50 \text{ atm}) = 6.9 \times 10^{-4} \text{ mol L}^{-1}$$

Multiplying by 5.0 L and converting to grams, we get

$$\text{Grams} = (6.9 \times 10^{-4} \text{ mol L}^{-1})(5.0 \text{ L})(32.0 \text{ g mol}^{-1}) = 0.11 \text{ g}$$

EXAMPLE 9-4: A solution of oxygen in water at 25 °C has a mole fraction of 8.0×10^{-6} at a partial pressure of 0.35 atm. Calculate the Henry's law constant.

Solution: Rearranging Eq. (9-9) and substituting, we obtain

$$k = \frac{X_2}{P_2} = \frac{8.0 \times 10^{-6}}{0.35 \text{ atm}} = 2.3 \times 10^{-5} \text{ atm}^{-1}$$

A variation of Henry's law is its expression in reverse, i.e., as an expression of the dependence of vapor pressure on concentration, or

$$P_2 = k'X_2 \qquad (\text{where } k' = 1/k) \qquad \textbf{(9-10)}$$

EXAMPLE 9-5: The partial pressure of a gas is 0.30 atm over a solution of the gas in water, having a mole fraction 1.34×10^{-2}. What would the pressure be if the mole fraction were 2.00×10^{-2}?

Solution: Since we are asked to find the pressure, we can use Eq. (9-10) more conveniently than Eq. (9-9). Solving for k' and substituting, we have

$$k' = \frac{P_2}{X_2} = \frac{0.30 \text{ atm}}{1.34 \times 10^{-2}} = 22.4 \text{ atm}$$

For the new solution,

$$P_2 = k'X_2 = 22.4 \text{ atm } (2.00 \times 10^{-2}) = 0.448 \text{ atm}$$

9-3. Liquid Solutions with One Volatile Component: Raoult's Law

A. Solutions of liquids in general

Gases are miscible in all proportions, whereas liquids aren't. Pairs of liquids may be almost totally immiscible, miscible at certain concentrations but not others, or completely miscible. Miscibility also depends on temperature.

Usually one or more of the components of a liquid solution is volatile; thus there is a vapor phase in equilibrium with the liquid phase, and vapor pressure becomes a significant property of a solution. Changes in vapor pressure with concentration are closely related to changes in melting points and boiling points, which has practical application in antifreeze for car radiators. An ideal solution of liquids is completely miscible in all proportions and obeys Eqs. (9-5) and (9-8).

B. Raoult's law

We will assume an ideal liquid solution in which only the solvent is volatile; i.e., the vapor phase will contain only solvent molecules. Experiments show that the vapor pressure decreases as more solute is added. In dilute solutions, the vapor pressure is directly proportional to the concentration of the solvent, and the proportionality constant is the vapor pressure of the pure solvent. This relationship is known as **Raoult's law**:

RAOULT'S LAW $$P_1 = X_1 P_1^0 \qquad \textbf{(9-11)}$$

Real solutions deviate from this law as X_1 decreases, i.e., as the solution becomes more concentrated. The law provides a limit, which real solutions approach at low concentrations or "infinite dilution." The law can also serve as another definition of *ideal solution:* a solution that obeys Raoult's law at all concentrations.

Raoult's law can be considered to be a special case of Henry's law, although some authors introduce Raoult's law first. If we rearrange Eq. (9-11) to $X_1 = (1/P_1^0)P_1$ and compare it to Eq. (9-9), we see that $k = 1/P_1^0$. It may seem that we have been careless with subscripts, but we really haven't: the subscripts merely identify components. Both P_1 and P_2 refer to pressures exerted by molecules in the gaseous state that are in equilibrium with the corresponding molecules in the solution.

Most textbooks contain diagrams of plots of pressure versus mole fraction for nonideal solutions, showing that Henry's law matches the curve when X_A is less than about 0.2 and that Raoult's law matches the curve when X_A is greater than about 0.9 (Remember that, at low X, component A is the solute, but, at high X, it is the solvent.)

EXAMPLE 9-6: The vapor pressure of pure water at 25 °C is 23.8 mm Hg. What is the pressure over a solution that consists of 50.0 g of glycol, $C_2H_4(OH)_2$, and 50.0 g of water? Assume that behavior is ideal and that glycol isn't volatile.

Solution: In order to apply Raoult's law, we need the mole fraction of the solute, so we must calculate the moles of each component:

$$n_{gly} = \frac{50.0 \text{ g}}{62.0 \text{ g mol}^{-1}} = 0.806 \text{ mol} \qquad n_w = \frac{50.0 \text{ g}}{18.0 \text{ g mol}^{-1}} = 2.78 \text{ mol}$$

Substituting in Eq. (9-11), we get

$$P_1 = P_w = X_w P_w^0 = \left[\frac{2.78 \text{ mol}}{(2.78 + 0.806) \text{ mol}}\right](23.8 \text{ mm Hg}) = 18.4 \text{ mm Hg}$$

We can obtain a useful variation of Raoult's law by using the concentration of the solute, which is the usual way of describing a solution. If there are only two components, $X_1 = 1 - X_2$, and Eq. (9-11) becomes

$$P_1 = (1 - X_2)P_1^0$$

Rearranging, we have

$$P_1 = P_1^0 - X_2 P_1^0$$

$$P_1^0 - P_1 = X_2 P_1^0 = \Delta P \qquad \textbf{(9-12)}$$

Thus $X_2 P_1^0$ expresses the decrease in the vapor pressure of the solvent when a certain amount of nonvolatile solute is added. This effect is called the *lowering of the vapor pressure*.

EXAMPLE 9-7: Repeat Example 9-6, using Eq. (9-12).

Solution: Now we need the mole fraction of glycol, and we will have to subtract ΔP from P^0 to get the answer. First, we substitute in Eq. (9-12) to get

$$\Delta P = \left[\frac{0.806 \text{ mol}}{(2.78 + 0.806) \text{ mol}}\right](23.8 \text{ mm Hg}) = 5.34 \text{ mm Hg}$$

and then rearrange Eq. (9-12) and evaluate:

$$P_1 = P_1^0 - \Delta P = 23.8 - 5.34 = 18.5 \text{ mm Hg}$$

What happens when several nonvolatile solutes are in a solution? According to Eq. (9-11), the pressure is determined solely by the mole fraction of the solvent. To use Eq. (9-12), we interpret X_2 as the sum of the mole fractions of all the solutes. The simplest calculation would be to replace X_2 by $1 - X_1$.

EXAMPLE 9-8: What is the vapor pressure of an aqueous solution at 25 °C that is 4.00 molal in sugar and 3.00 molal in ethylene glycol? (The vapor pressure of water at 25 °C = 23.8 mm Hg.)

Solution: An apparent complication was introduced by expressing the concentrations in molality. Actually, this is a simplification because we don't need to calculate the number of moles of the solutes. Since molality is moles of solute per kilogram of solvent, we have 4.00 moles, 3.00 moles, and 55.56 moles (for 1 kg of water). Then

$$X_1 = \frac{55.56}{4.00 + 3.00 + 55.56} = 0.888$$

Using Eq. (9-11), we have

$$P_1 = X_1 P_1^0 = 0.888(23.8) = 21.1 \text{ mm Hg}$$

9-4. Colligative Properties

A. Common origin

We described the lowering of the vapor pressure of a solvent resulting from the addition of a nonvolatile solute in Section 9-3B. Three other significant characteristics of solutions are related to that property: (1) a freezing point lower than that of pure solvent; (2) a boiling point higher than that of pure solvent; and (3) the phenomenon of osmotic pressure. The four characteristics are called **colligative properties** because they are the result of the same fundamental property of solutions (*colligative* means "tied together").

All these characteristics are explained thermodynamically by the fact that a solvent has a lower chemical potential in solution than when pure, as shown by Eq. (9-8). Since a mole fraction always is less than 1, the ln term is always negative and $\mu_i < \mu_i^0$. The practical result is shown conveniently on the phase diagram, where the vapor-pressure curve of the solution will always be lower than that of the pure solvent. Since most texts have a diagram like Figure 8-2, with typical vapor-pressure and freezing-point curves of a solution added, we won't show one here.

An essential point about all these properties is that they depend only on the number of molecules or other basic particles in the solution. This is shown mathematically by Eq. (9-12), in which P_1^0 is a constant. It, of course, depends on the solvent, but, when the solvent has been designated, ΔP depends only on X_2, which is independent of the nature of the solvent or the solute.

B. Boiling-point elevation

The boiling-point–elevation effect can be explained from either of two viewpoints:

1. Since μ (solution) is less than μ (liquid solvent) but μ (vapor) has remained the same, the temperature must be raised to the point that μ (solution) equals μ (vapor), the condition for boiling to occur.
2. Since the vapor-pressure curve of the solution is lower than that of the pure solvent at any temperature, it is farther to the right at any pressure. Thus, for a given confining pressure, the temperature at which the vapor pressure equals the confining pressure is higher for the solution than for the pure solvent.

Empirically, we express the change of boiling point with concentration by

$$\Delta T_b = K_b m \qquad \textbf{(9-13)}$$

using molality m, as is customary with these properties. The proportionality constant can be determined experimentally.

A lengthy mathematical derivation found in most textbooks yields Eq. (9-13) and gives an expression for K_b in terms of known quantities. More than one approach is possible, but the usual starting point is the equilibrium condition: μ (solution) = μ (vapor).

From Eq. (9-8) for the solvent,

$$\mu \text{ (solution)} = \mu^0 \text{ (liquid)} + RT \ln X_1$$

so

$$\mu \text{ (vapor)} = \mu^0 \text{ (liquid)} + RT \ln X_1$$

At the boiling point, $P = 1$ atm, so μ (vapor) = μ^0 (vapor), and we have

$$\mu^0 \text{ (vapor)} = \mu^0 \text{ (liquid)} + RT \ln X_1$$

For one mole of a pure substance, $\mu^0 = G^0$. Substituting and dividing by T, we get

$$\frac{G^0\,(\text{vapor})}{T} - \frac{G^0\,(\text{liquid})}{T} = R \ln X_1$$

Next, we differentiate with respect to T and substitute from the Gibbs–Helmholtz equation, Eq. (6-50), to obtain

$$\left(\frac{\partial(G^0/T)}{\partial T}\right)_P = -\frac{H^0}{T^2}$$

which yields

$$-\frac{H^0\,(\text{vapor})}{T^2} + \frac{H^0\,(\text{liquid})}{T^2} = R\frac{\partial \ln X_1}{\partial T}$$

Since H^0 (vapor) $-$ H^0 (liquid) $= \Delta H_v$, we arrange to

$$-\frac{\Delta H_v}{RT^2} = \frac{\partial \ln X_1}{\partial T}$$

For the small temperature range, we can take ΔH_v to be constant. We let T_b^0 represent the normal boiling point of the pure solvent and recognize that $X_1 = 1$ at T_b^0. Substituting and integrating

$$\int_1^{X_1} d\ln X_1 = -\frac{\Delta H_v}{R}\int_{T_b^0}^{T_b}\frac{1}{T^2}\,dT$$

we obtain

$$\ln X_1 = -\frac{\Delta H_v}{R}\left(-\frac{1}{T_b} + \frac{1}{T_b^0}\right) = -\frac{\Delta H_v}{R}\left(\frac{T_b - T_b^0}{T_b T_b^0}\right)$$

We let $\Delta T_b = T_b - T_b^0$ and assume that $T_b T_b^0 = (T_b^0)^2$ for the small temperature range. Also, replacing X_1 with $1 - X_2$, we have

$$\ln(1 - X_2) = -\frac{\Delta H_v \Delta T_b}{R(T_b^0)^2}$$

Expanding the logarithm in a power series gives

$$\ln(1 - X_2) = -X_2 - \frac{X_2^2}{2} - \frac{X_3^3}{3}$$

Since X_2 is small in a dilute solution, we neglect the higher terms and have

$$-X_2 = -\frac{\Delta H_v \Delta T_b}{R(T_b^0)^2}$$

Rearranging yields

$$\Delta T_b = \frac{R(T_b^0)^2}{\Delta H_v} X_2$$

It is customary to use molality rather than mole fraction. If we let n_2 be the moles of solute in 1 kg of solvent, then $n_2 = m$ and $n_1 = 1/M_m$, where M_m is the molar mass of the solvent in kg. Now

$$X_2 = \frac{m}{m + n_1} \simeq \frac{m}{1/M_m}$$

but we neglect m in the denominator because, in a very dilute solution, $n_2 \ll n_1$. Our expression becomes

$$\Delta T_b = \frac{R(T_b^0)^2}{\Delta H_v} m M_m = \frac{M_m R(T_b^0)^2}{\Delta H_v} m$$

The first term consists of constants for a given solvent and is designated K_b, so

MOLAL BOILING-POINT–ELEVATION CONSTANT

$$K_b = \frac{MR(T_b^0)^2}{\Delta H_v} \tag{9-14}$$

The term K_b is called the **molal boiling-point–elevation constant** (or the *ebullioscopic constant*). If the molar mass were expressed in grams, we would include a 1000 in the denominator.

Calculated values of K_b for various types of solvents agree quite well with experimentally determined values. Although it is expressed as degrees per 1-molal solution, K_b generally is not valid for a solution as concentrated as 1 *m*, because of the assumptions in the derivation. The range of validity depends on the degree of ideality of the solution and the molar mass of the solvent.

EXAMPLE 9-9: Calculate the theoretical value of K_b for benzene by using Eq. (9-14) and $\Delta H_v = 30.768$ kJ mol^{-1}. Compare your result to the experimentally determined value in Table 9-1.

Solution: From Table 9-1, the boiling point of benzene is 353.4 K, so

$$K_b = \frac{0.0781 \text{ kg mol}^{-1}(8.314 \text{ J K}^{-1}\text{mol}^{-1})(353.4 \text{ K})^2}{30\,768 \text{ J mol}^{-1}} = 2.64 \text{ K kg mol}^{-1}$$

The value of K_b from Table 9-1 is 2.53 K kg mol^{-1}. The agreement is all right:

$$\text{Error} = \left(\frac{0.11}{2.5}\right)(100) = 4.4\%$$

EXAMPLE 9-10: Calculate the boiling point of a 0.050-molal solution of a nonvolatile compound that doesn't associate or dissociate in cyclohexane (see Table 9-1).

Solution: The change in boiling point is given by Eq. (9-13):

$$\Delta T_b = K_b m$$

Note that it isn't necessary to know what the solute is, only that it must yield one mole of entities per mole of compound. From Table 9-1, $T_b^0 = 80.9$ °C and $K_b = 2.79$ K kg mol^{-1}, so

$$\Delta T_b = (2.79 \text{ K kg mol}^{-1})(0.050 \text{ mol kg}^{-1}) = 0.14 \text{ K} \equiv 0.14 \text{ °C}$$

and the calculated boiling point is

$$T_b = T_b^0 + \Delta T_b = 80.9 + 0.14 = 81.0 \text{ °C}$$

note: ΔT_b may be in kelvins or degrees Celsius, since it is simply an interval, $T - T_b^0$, on either temperature scale.

EXAMPLE 9-11: Determine the molar mass of an unknown compound if the solution formed when 3.28 g of the compound dissolves in 100.0 g of benzene boils at 80.9 °C (see Table 9-1).

Solution: We can determine the molality of the solution using Eq. (9-13), then calculate the molar mass from the weights given. First, we rearrange Eq. (9-13) and solve for *m*:

$$m = \frac{\Delta T_b}{K_b} = \frac{T - T_b^0}{K_b} = \frac{(80.9 - 80.2) \text{ K}}{2.53 \text{ K kg mol}^{-1}} = 0.3 \text{ mol (solute) kg}^{-1} \text{ (solvent)}$$

TABLE 9-1: Freezing-Point–Depression and Boiling-Point–Elevation Constants

Solvent	T_f (°C)	K_f (K kg mol^{-1})	T_b (°C)	K_b (K kg mol^{-1})
Benzene	5.5	5.12	80.2	2.53
Camphor	178.4	37.7	208.3	5.95
Cyclohexane	6.5	20.0	80.9	2.79
Naphthalene	80.2	6.9	218.0	5.65
Water	0	1.86	100.0	0.512

Notice the loss in significant figures when we subtract two numbers of similar magnitude. We could proceed in several ways but will demonstrate only two:

(a) We now have the *moles* of solute per kg of benzene, and we can calculate the *weight* of solute per kg of benzene from the data:

$$\frac{3.28 \text{ g}}{100.0 \text{ g}} = 32.8 \text{ g (solute) kg}^{-1} \text{ (solvent)}$$

Thus 0.3 mol and 32.8 g refer to the same mass of solute and 0.3 mol = 32.8 g or

$$1 \text{ mol} = \frac{32.8}{0.3} \text{ g} = 1 \times 10^2 \text{ g} = 0.1 \text{ kg}$$

(b) We can use the definition of molality. Letting W represent weight in kg, we have

$$m = \frac{W_2/M_{\text{m},2}}{W_1} = \frac{W_2}{M_{\text{m},2}W_1}$$

Solving for $M_{\text{m},2}$ gives us

$$M_{\text{m},2} = \frac{W_2}{mW_1} = \frac{0.00328 \text{ kg}}{(0.3 \text{ mol kg}^{-1})(0.100 \text{ kg})} = 0.1 \text{ kg mol}^{-1}$$

(For the weight in grams, $m = 1000W_2/M_{\text{m},2}W_1$, and $M_{\text{m},2} = 1 \times 10^2 \text{ g mol}^{-1}$.)

---

warning: When solving problems involving molalities, be careful not to cancel kg (solute) with kg (solvent). A good practice is to identify both quantities, as in Example 9-11, solution (a). The unit of m really is mol (solute) kg^{-1} (solvent).

C. Freezing-point depression

Explanation of the freezing-point–depression effect is analogous to that of the boiling-point–elevation effect. From the same two viewpoints:

1. Since μ (solution) < μ (liquid solvent) but μ (solid) has remained the same, the temperature must be lowered to the point that μ (solid) equals μ (solution). When they are equal, solvent from the solution can change to solid in a new freezing point equilibrium.
2. The vapor-pressure curve of the solution intersects the solid–vapor curve at a lower temperature, and the solid–liquid curve of the solution will lie to the left of that for the pure solvent. Thus, at any pressure, the temperature for equilibrium between solid and liquid phases is lower for the solution than for the pure solvent.

Empirically, we express the change of freezing point with concentration by

$$\Delta T_{\text{f}} = K_{\text{f}}m \tag{9-15}$$

By a complex derivation, analogous to that for the boiling-point–elevation curve, we can obtain an expression for the **molal freezing-point–depression constant** (or the *cryoscopic constant*):

MOLAL FREEZING-POINT–DEPRESSION CONSTANT

$$K_{\text{f}} = \frac{M_{\text{m}}R(T_{\text{f}}^0)^2}{\Delta H_{\text{f}}} \tag{9-16}$$

Again the assumption of a dilute solution is made more than once, and we should expect calculated values of K_{f} to agree with experimental values at low concentrations only.

EXAMPLE 9-12: Calculate the theoretical value of K_{f} for camphor ($C_{10}H_{16}O$) and compare the result to the experimental value ($\Delta H_{\text{f}} = 6839 \text{ J mol}^{-1}$; also see Table 9-1).

Solution: We merely need to determine the necessary constants and substitute in Eq. (9-16). From Table 9-1, $T_{\text{f}}^0 = 178.4\ °\text{C} = 451.6 \text{ K}$; $M_{\text{m}} = 0.152 \text{ kg mol}^{-1}$. Thus

$$K_{\mathrm{f}} = \frac{(0.152\ \mathrm{kg\ mol^{-1}})(8.314\ \mathrm{J\ K^{-1} mol^{-1}})(451.6)^2}{6839\ \mathrm{J\ mol^{-1}}} = 37.7\ \mathrm{K\ kg\ mol^{-1}}$$

This result agrees perfectly with the experimental value from Table 9-1, 37.7 K kg mol^{-1}.

EXAMPLE 9-13: Calculate the freezing point of the solution described in Example 9-10.

Solution: The concentration is 0.050 molal, and the solvent is cyclohexane. The change in freezing point is given by Eq. (9-15). From Table 9-1, $T_{\mathrm{f}}^{0} = 6.5$ °C, and $K_{\mathrm{f}} = 20.0$ K kg mol^{-1}. Again, the identity of the solute isn't needed. Substituting and evaluating, we get

$$\Delta T_{\mathrm{f}} = 20.0\ \mathrm{K\ kg\ mol^{-1}}(0.050\ \mathrm{mol\ kg^{-1}}) = 1.0\ °\mathrm{C}$$

$$T_{\mathrm{f}} = T_{\mathrm{f}}^{0} - \Delta T_{\mathrm{f}} = 5.5\ °\mathrm{C}$$

Note that the change here is 1.0 °C, whereas in Example 9-10 it was only 0.14 °C. The larger change means that the percent error is less, a good reason to use freezing-point depression for determinations of molar mass.

EXAMPLE 9-14: The calculations in a determination of molar mass could be carried out as in Example 9-11, but they can be simplified by obtaining an expression for $M_{\mathrm{m},2}$. Derive such an expression containing only experimental values and K_{f}.

Solution: We use Eq. (9-15) and replace m with an expression from its definition, i.e.,

$$m = \frac{n\ (\text{solute})}{\text{kg (solvent)}} = \frac{W_2/M_{\mathrm{m},2}}{W_1} = \frac{W_2}{M_{\mathrm{m},2} W_1}$$

Substituting in Eq. (9-15), we have

$$\Delta T_{\mathrm{f}} = K_{\mathrm{f}} m = K_{\mathrm{f}} \frac{W_2}{M_{\mathrm{m},2} W_1}$$

so

$$M_{\mathrm{m},2} = \frac{K_{\mathrm{f}} W_2}{\Delta T_{\mathrm{f}} W_1} \qquad \textbf{(9-17)}$$

EXAMPLE 9-15: Determine the molar mass of an unknown compound if a solution made up of 1.82 g of the compound and 20.0 g naphthalene melts at 73.6 °C. Use Eq. (9-17) and compare the calculations to those in Example 9-11.

Solution: From Table 9-1, $T_{\mathrm{f}}^{0} = 80.2$ °C, and $K_{\mathrm{f}} = 6.9$ K kg mol^{-1}. Substituting in Eq. (9-17), we get

$$M_{\mathrm{m},2} = \frac{6.9\ \mathrm{K\ kg\ mol^{-1}}(0.001\ 82\ \mathrm{kg})}{[(80.2 - 73.6)\ \mathrm{K}]\ (0.020\ \mathrm{kg})} = 0.095\ \mathrm{kg\ mol^{-1}}$$

Comparing this method to that used in Example 9-11, we see that the calculation here is more convenient. The disadvantage is that we must memorize another isolated expression, whereas in Example 9-11 we worked from basic relationships and definitions.

- As a general rule, you should not memorize equations like Eq. (9-17). You should learn only the fundamental equations, such as Eq. (9-15), and rearrange them or substitute in them to obtain subsidiary expressions when needed.

D. Osmotic pressure

At first glance, osmotic pressure seems quite different from the changes in freezing and boiling points. An obvious difference is that osmotic pressure involves different pressures at constant T, whereas freezing and boiling points involve different temperatures at constant P. The major difference is in the phase of the pure state: At the freezing point, the pure solvent is solid; at the boiling point, it is vapor; and, for osmotic pressure, it is liquid. However, the equilibria established in a system are similar. In each case, the equilibrium is between solvent in solution and solvent in the pure state, and, to have equilibrium, μ (solution) must equal μ (pure) for the *solvent*.

The explanation of the phenomenon is analogous to the first of those of freezing-point depression and boiling-point elevation. The second explanation for them doesn't apply to osmotic pressure. Thus

1. Since μ (solution) is less than μ (pure), μ (solution) can be raised to equal μ (pure) by increasing the pressure on the solution only. The system accomplishes this by the process of **osmosis**: Solvent molecules flow from the pure phase into the solution phase through a semipermeable membrane. When the μ's become equal, the process will stop at the new dynamic equilibrium.
2. An explanation based on the phase diagram isn't possible because it would require a curve for liquid–liquid equilibria.

Osmotic pressure, Π, is defined as the pressure that must be applied to the solution to make the chemical potentials of the solvent the same in solution and in pure liquid. A simple empirical expression relating Π to concentration, similar to Eqs. (9-13) and (9-15), is

$$\Pi = kM \qquad \textbf{(9-18)}$$

where k is a proportionality constant, and M is the molarity.

The mathematical derivation that leads to the van't Hoff equation and an expression for k is found in most texts. Like the derivations for K_b and K_f, it starts with the equilibrium condition. We bring in Eq. (9-8), where the pressure on the solution is $P + \Pi$. To deal with pressure changes, we use $d\mu = V\,dP$. Further steps involve changing from X_1 to X_2 and then n_2, expanding the ln in series, and assuming a dilute, ideal solution. Another basic assumption is that the solvent is not compressible. The result is the **van't Hoff equation**:

VAN'T HOFF EQUATION

$$\Pi V = n_2RT \qquad \textbf{(9-19)}$$

where V is the volume of the solution and R must be in L atm. Rearranging, we have

$$\Pi = \frac{n_2}{V}RT = MRT$$

Thus the k of Eq. (9-18) is RT, obviously depending on temperature.

EXAMPLE 9-16: If 100 mL of an aqueous solution contains 20.0 g of sucrose (342 g mol^{-1}), what is its osmotic pressure at 25 °C?

Solution: We make a straightforward substitution to illustrate Eq. (9-19) and the large pressures involved:

$$\Pi = \frac{n_2}{V}RT = \left(\frac{20.0\text{ g}/342\text{ g mol}^{-1}}{0.100\text{ L}}\right)(0.082\,05\text{ L atm K}^{-1}\text{mol}^{-1})(298.2\text{ K}) = 14.3\text{ atm}$$

As with the other colligative properties, we can use osmotic pressure to determine molar masses. Since this method is much less accurate than the others, we use it only in special cases where its characteristic of a large effect is useful.

EXAMPLE 9-17: From Eq. (9-19), derive an expression for calculating the molar mass of a solute $M_{m,2}$.

Solution: To introduce $M_{m,2}$, we can replace n_2 by $W^2/M_{m,2}$ and get

$$\Pi V = \frac{W_2}{M_{m,2}}RT$$

Rearranging gives

$$M_{m,2} = \frac{W_2RT}{\Pi V} \qquad \textbf{(9-20)}$$

From Eq. (9-20), we see that, since Π is rather large, it could be measured even if $M_{m,2}$ is very large or W_2 is very small. The average molecular weights of polymers or other macromolecules often are determined by osmotic pressure measurements.

EXAMPLE 9-18: What is the average molecular weight, MW, of a polymer if 50.0 mL of a solution that contains 1.45 g of the polymer exhibits an osmotic pressure of 3.13 mm Hg at 25 °C?

Solution: Substituting in Eq. (9-20), we get

$$\text{MW} = \frac{1.45\text{ g }(0.082\,05\text{ L atm K}^{-1}\text{mol}^{-1})(298.2\text{ K})}{(3.13\text{ mm Hg}/760\text{ mm Hg atm}^{-1})(0.0500\text{ L})} = 1.72 \times 10^5\text{ g mol}^{-1}$$

9-5. Liquid Solutions with Two Volatile Components

A. Pressure–composition diagrams: Ideal case

We now consider the situation where the solute, as well as the solvent, may be volatile. We will restrict ourselves to solutions of two liquid components that are miscible in all proportions. In this case, either liquid may be the solvent (more than 50 percent), so we will not use 1 and 2 for subscripts as before but will designate the components A and B.

1. The liquid phase

For now, we are considering only ideal solutions, which means that both components obey Raoult's law. Thus

$$P_A = X_A P_A^0 \qquad \textbf{(9-11a)}$$

$$P_B = X_B P_B^0 \qquad \textbf{(9-21)}$$

From Eq. (9-5), the free energy of mixing for the production of one mole of ideal solution is

$$\Delta G\,(\text{mixing}) = RT\left(X_A \ln \frac{P_A}{P_A^0} + X_B \ln \frac{P_B}{P_B^0}\right) \qquad \textbf{(9-22)}$$

The total pressure must be

$$P = P_A + P_B = X_A P_A^0 + X_B P_B^0 \qquad \textbf{(9-23)}$$

A significant relationship is revealed if we replace X_B with $1 - X_A$:

$$P = X_A P_A^0 + (1 - X_A)P_B^0 = X_A P_A^0 + P_B^0 - X_A P_B^0$$

or

$$P = (P_A^0 - P_B^0)X_A + P_B^0 \qquad \textbf{(9-24)}$$

This is the equation of a straight line because $P_A^0 - P_B^0$ is constant. A plot of P versus X_A is the solid line in Figure 9-1. The dotted lines are plots of Eqs. (9-11a) and (9-21). The pressure at any point X on the solid line is the sum of the pressures at the same X on the two dotted lines; in other words, the solid line also is a plot of Eq. (9-23). Note that the x-axis represents both mole fractions: X_A, increasing from left to right, and X_B the reverse. Thus, for any X_A, $X_B = 1 - X_A$.

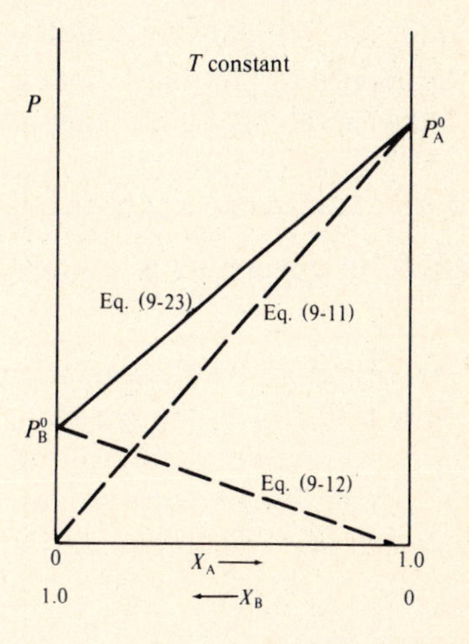

FIGURE 9-1. Ideal Solution: Partial pressures of each component and total pressures.

The solid line represents the equilibrium between the liquid and the vapor phases. The area above the line represents liquid, and the area below represents vapor. If the total pressure is increased to a point above the line, all vapor must disappear.

2. The vapor phase

The composition of the vapor phase, by Dalton's law of partial pressures, is

$$Y_A = \frac{P_A}{P} \tag{9-25}$$

$$Y_B = \frac{P_B}{P} \tag{9-26}$$

where Y represents a mole fraction in the vapor phase.

EXAMPLE 9-19: Calculate the composition of the vapor phase in equilibrium with a solution containing 40.0 g benzene and 60.0 g toluene at 30 °C. The vapor pressures at 30 °C are 118.2 mm Hg for benzene and 36.7 mm Hg for toluene.

Solution: We must obtain the partial pressures in order to get mole fractions in the vapor, but we can calculate them from the mole fractions in the liquid:

$$n_b = \frac{40.0 \text{ g}}{78.1 \text{ g mol}^{-1}} = 0.512 \text{ mol} \qquad n_t = \frac{60.0 \text{ g}}{92.1 \text{ g mol}^{-1}} = 0.651 \text{ mol}$$

$$\text{total } n = 1.163 \text{ mol}$$

Then, from Raoult's law, Eq. (9-11),

$$P_b = X_b P_b^0 = \left(\frac{0.512}{1.163}\right)(118.2) = 52.0 \text{ mm Hg} \qquad P_t = X_t P_t^0 = \left(\frac{0.651}{1.163}\right)(36.7) = 20.5 \text{ mm Hg}$$

and from Dalton's law, Eq. (9-23),

$$P = P_b + P_t = 72.5 \text{ mm Hg}$$

Thus
$$Y_b = \frac{P_b}{P} = \frac{52.0}{72.5} = 0.717 \quad \text{and} \quad Y_t = \frac{P_t}{P} = \frac{20.5}{72.5} = 0.283$$

---

3. The phases together

Note in Example 9-19 that the substance with the larger pure vapor pressure (benzene) has a larger mole fraction in the vapor than in the liquid. In other words, the vapor phase is richer in the high-vapor-pressure component than is the liquid phase. This difference is shown graphically in Figure 9-2, which contains the solid line from Figure 9-1 and an additional line obtained by plotting Y_A for each P. The straight line is the **liquid-composition curve** (X_A), and the curved line is the **vapor-composition curve** (Y_A).

EXAMPLE 9-20: Figure 9-2 is a different sort of phase diagram. What do the areas and lines represent?

Solution: In the area above the liquid-composition curve, the pressure is great enough that there is no vapor, and X_A describes the system. If the pressure is decreased until it equals a point on the liquid-composition curve, vapor will begin to appear, but X_A alone is no longer sufficient to describe the system. Two concentrations are involved: The concentration of component A in the new vapor phase is Y_A; and the concentration of component A in the new liquid phase is X_A but not the original X_A.

In the area below the vapor-composition curve, the system is in the vapor phase. If the pressure is increased until it equals a point on the vapor-composition curve, liquid will begin to form. Again, one concentration isn't sufficient to describe the system. The area between the curves represents the equilibrium between two phases, in contrast to the one-component phase diagram where an equilibrium is represented by a line.

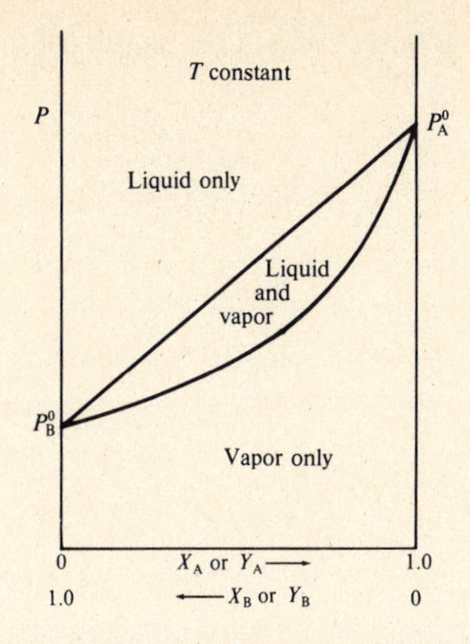

FIGURE 9-2. Ideal Solution: Composition curves for liquid and vapor versus *P*.

The area between the two curves needs further description. A given pressure is represented by a horizontal line. Any horizontal line that intersects both curves is called a **tie line**. The points of intersection, which are "tied" by the line, represent the compositions of the liquid and vapor phases that are in equilibrium with each other.

EXAMPLE 9-21: Consider Figure 9-3, which is that portion of Figure 9-2 between mole fractions 0.30 and 0.60. A solution in which $X_A = 0.42$ was mixed at pressure P_c. (**a**) What happens as the pressure is lowered progressively from P_c to P_h? (**b**) What can be said about the compositions of the vapor and liquid phases at pressures P_e and P_f?

Solution:

(**a**) When the pressure is lowered to P_d, a tiny amount of vapor will appear. As the pressure continues to drop, the amount of vapor increases; the amount of liquid, of course, decreases. At P_g, the last of the liquid will disappear; below P_g, only vapor is present.

(**b**) At P_e, the tie line connects points *m* and *n*. Point *m* corresponds to $X_A = 0.39$; point *n* corresponds to $Y_A = 0.52$. Remember that the total mixture still has a mole fraction of 0.42 for A. At P_f, there is even more vapor and less liquid, and the compositions of each phase are different: At *s*, $X_A = 0.34$; at *t*, $Y_A = 0.47$. At P_g, when the last of the liquid evaporates, the composition of the remaining phase is $Y_A = 0.42$. When all molecules are in the vapor phase, their relative proportion is the same as when all were in the liquid phase.

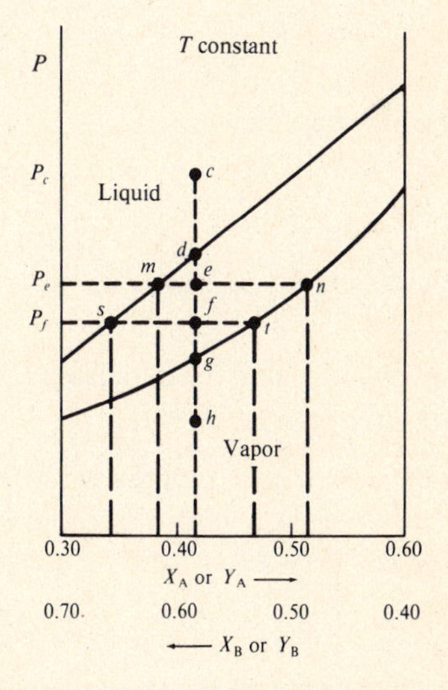

FIGURE 9-3. Section of Figure 9-2 with tie lines.

---

4. The lever rule

In Example 9-21, we saw that the relative amount of the vapor phase increases as the pressure decreases. We can calculate the ratio of the amounts of liquid and vapor from the lengths of the two segments of the tie line. We see in Figure 9-3 that, as the pressure decreases, the segment to the left of $X_A = 0.42$ grows longer. Thus this segment corresponds to the vapor phase. The **lever rule** tells us that the ratio of the *total* number of moles (both A and B) in each phase is equal to the ratio of the lengths of the corresponding segments of the tie line:

LEVER RULE

$$\frac{n_v}{n_l} = \frac{\text{Length } me}{\text{Length } en} \tag{9-27}$$

EXAMPLE 9-22: Demonstrate algebraically that the ratio of the tie-line segments at P_e in Figure 9-3 *does* equal the ratio of the moles in the two phases.

Solution: We represent the length of a segment by the difference between the mole fractions of A at its end points, so

$$en = X_n - X_e = X_A \text{ (vapor)} - X_A \text{ (total mixture)}$$
$$me = X_e - X_m = X_A \text{ (total mixture)} - X_A \text{ (liquid)}$$

We now use the definition of mole fraction to introduce moles, using *a* for moles of A and *n* for total moles, and obtain

$$X_A \text{ (vapor)} = \frac{\text{Moles A}}{\text{Moles A + moles B}} \text{(vapor)} \quad \text{or} \quad \frac{a_v}{n_v}$$

$$X_A \text{ (liquid)} = \frac{\text{Moles A}}{\text{Moles A + moles B}} \text{(liquid)} \quad \text{or} \quad \frac{a_l}{n_l}$$

$$X_A \text{ (total)} = \frac{a_v + a_l}{n_v + n_l}$$

We substitute these expressions for the X_A's into the equations for *en* and *me* and combine the two parts over a common denominator:

$$en = \frac{a_v}{n_v} - \frac{a_v + a_l}{n_v + n_l} = \frac{a_v n_v + a_v n_l - a_v n_v - a_l n_v}{n_v(n_v + n_l)} = \frac{a_v n_l - a_l n_v}{n_v(n_v + n_l)}$$

$$me = \frac{a_v + a_l}{n_v + n_l} - \frac{a_l}{n_l} = \frac{a_v n_l + a_l n_l - a_l n_v - a_l n_l}{n_l(n_v + n_l)} = \frac{a_v n_l - a_l n_v}{n_l(n_v + n_l)}$$

Dividing *me* by *en* gives us

$$\frac{me}{en} = \left[\frac{a_v n_l - a_l n_v}{n_l(n_v + n_l)}\right]\left[\frac{n_v(n_v + n_l)}{a_v n_l - a_l n_v}\right] = \frac{n_v}{n_l}$$

which is Eq. (9-27), the lever rule.

EXAMPLE 9-23: Using the data from Example 9-21, calculate and compare the relative amounts of vapor to liquid at P_e and at P_f.

Solution: As shown in Example 9-22, the lengths of the segments can be represented by the difference between the *x*-coordinates. For P_e at *m*, $X = 0.39$; at *e*, $X = 0.42$; and at *n*, $X = 0.52$. Substituting in Eq. (9-27), we get

$$P_e = \frac{n_v}{n_l} = \frac{0.42 - 0.39}{0.52 - 0.42} = \frac{0.03}{0.10} = 0.3$$

For P_f at *s*, $X = 0.34$; at *f*, $X = 0.42$; and at *t*, $X = 0.47$. We substitute and get

$$P_f = \frac{n_v}{n_l} = \frac{0.42 - 0.34}{0.47 - 0.42} = \frac{0.08}{0.05} = 1.6$$

Comparing these results, we see that there is considerably more vapor at P_f, as already described in Example 9-21.

B. Pressure–composition diagrams: Nonideal cases

Most pairs of liquids don't have diagrams like Figs. 9-1 and 9-2. Only in dilute solutions does the solvent approximate Raoult's-law behavior and the solute exhibit Henry's-law behavior. Three categories of deviations from ideal behavior are based on whether the total vapor pressure of a solution becomes higher or lower than that of the pure components. Most textbooks include figures to illustrate:

1. *Small deviation*—The plot of P versus X [Eq. (9-23)] isn't straight but doesn't rise above the larger P^0 or fall below the smaller P^0. Such solutions are almost ideal. Examples include CCl_4 and $SiCl_4$, benzene and toluene, and water and methanol.
2. *Maximum in curve*—The plot rises above the larger P^0. Examples include acetone and carbon disulfide, benzene and ethanol, and water and ethanol.
3. *Minimum in curve*—The plot falls below the smaller P^0. Examples include acetone and chloroform, and water and several acids, especially hydrochloric acid.

These deviations are explained by the attractive forces between the molecules. In type 1, the attraction between A and B molecules is about the same as that between A and A or B and B. Positive deviations (type 2) occur when the attraction between A and B is smaller than that between A and A or B and B. Negative deviations (type 3) occur because the A–B attraction is greater than the A–A or B–B attractions. This condition can be caused by hydrogen bond formation or by large dipole moments.

C. Temperature–composition diagrams: Ideal case

Now we hold the pressure constant and consider the boiling behavior of a solution. Boiling occurs when the total vapor pressure of the solution equals the confining pressure, but each possible concentration of the solution has a different vapor pressure, which changes with temperature. For each possible solution, the vapor pressure that equals the confining pressure will occur at a different temperature; in other words, each solution will have a different boiling point. At the boiling point, the concentrations of the liquid and vapor phases will be different. Figure 9-4 is a typical **temperature-composition** (or **boiling point**) **diagram.**

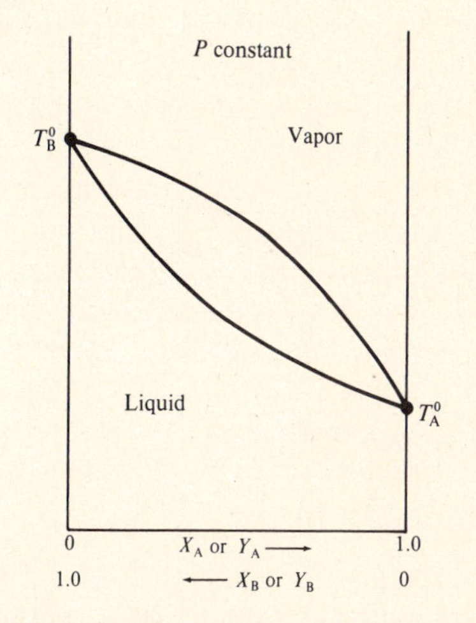

FIGURE 9-4. Ideal Solution: Composition curves for liquid and vapor versus T.

Note the similarities and differences between Figures 9-2 and 9-4. The area between the curves again represents two phases in equilibrium, and a tie line connects the concentrations of the phases. The slope is in the opposite direction because the substance with the lower vapor pressure has the higher normal boiling point. Neither curve is straight, and the single phase regions are interchanged, with the liquid at the bottom. We would expect the system to be liquid at low temperatures. Analysis of what happens as the temperature is raised is useful in understanding fractional distillation.

EXAMPLE 9-24: Suppose that in an ideal two-liquid solution $X_A = 0.40$ and the temperature is lower than T_A^0. **(a)** What would happen as the temperature is raised? **(b)** What can be said about the compositions and the amounts of the phases when there are two in equilibrium?

Solution: The analysis is analogous to that of Example 9-21, in which the pressure was decreased for a constant overall mixture concentration. We haven't identified points on Figure 9-4 as we did on Figure 9-3, because you won't find them on the usual phase diagrams. You have to be able to interpret a diagram without such crutches.

(a) When the rising temperature equals a point on the liquid-composition curve, or slightly above it, a small amount of vapor will appear. More vapor will appear as the temperature continues to rise, and the amount of liquid will decrease. When the temperature equals that of the vapor-composition curve at $X_A = 0.40$, the last of the liquid will evaporate. At higher temperatures, only vapor will be present.

(b) The concentrations of each phase at a certain temperature are given by the ends of the tie line drawn for that temperature. The first bit of vapor will have a high concentration of A, but, as the temperature rises, the concentration in the vapor will decrease until Y_A (vapor) equals 0.40 just as the last liquid disappears. We use the lever rule to obtain the relative amounts of the two phases at any temperature. At first, the amount of vapor is quite small; at the very last, the amount of liquid is quite small.

D. Temperature–composition diagrams: Nonideal case

In part B of this section, we saw that some systems deviate from ideality so much that their vapor-pressure curves exhibit a maximum or a minimum. These mixtures have boiling-point curves that, essentially, are the vapor-pressure curves turned upside down. Type 2 (maximum in vapor-pressure curve) will have a minimum in the boiling-point curve and may be called a **minimum-boiling mixture**. A type 3 system, conversely, is a **maximum-boiling mixture**. Examples of type 2 and type 3 mixtures were given in part B.

A solution having exactly the concentration of the maximum or minimum point is called an **azeotrope** or **azeotropic mixture**. At these points, the boiling temperature remains constant, so azeotropes are also called **constant-boiling mixtures**. The liquid and vapor phases in equilibrium have the same composition, which is a characteristic of compounds. The distinction between a compound and an azeotrope is that the composition of the latter changes with total pressure.

Textbooks usually contain figures to illustrate these systems, so none are included here. If we divide a temperature–composition diagram at the maxima or minima, each half would be like Figure 9-4, except that one end is not pure component; it is the azeotropic mixture having the composition at the maximum or minimum boiling point. The analysis for raising the temperature is analogous to that of Example 9-24.

E. Fractional distillation

1. Ideal case

The analysis in Example 9-24 was based on an assumed closed system. Now suppose that the system in Example 9-24 is in a state between the curves and that the vapor phase is removed and condensed. The new liquid will be richer in A than the original solution (X_A = 0.40). If this new liquid is warmed to a temperature between the curves and again the vapor phase is removed and condensed, the third liquid is even richer in A. If the process is continued, eventually some nearly pure A is obtained. Going the other way, the remaining liquid is saved and reheated, and again the vapor is removed. If the process is continued, eventually some nearly pure B is obtained.

In practice, this separation process is made continuous in a distillation column. **Fractional distillation** is carried out in a vertical column containing many platforms or *plates*. The temperature is higher at the bottom and decreases to the top. Liquid collects on each plate as evaporation and condensation proceed. The liquids on the lower plates become richer in the high-boiling component, and vapor at the top is almost pure low-boiling component. See your textbook for a more thorough description.

2. Nonideal case

If a type 2 or type 3 mixture is subjected to fractional distillation, we will find nearly pure component at one end of the column but, at the other end, a mixture having the composition of the maximum or minimum. For a maximum-boiling system, the azeotrope will collect at the bottom of the column; for a minimum-boiling system, the azeotrope will be the vapor, i.e., the distillate removed at the top. Which component is obtained in pure form is determined by whether its initial concentration lies to the right or left of the maximum or minimum on the temperature–composition diagram. Again, see your textbook for a more thorough description.

9-6. Immiscible Volatile Liquids

In the Section 9-5, we dealt with liquid pairs that were completely miscible in all proportions. If deviations from ideal behavior become large enough, we have to deal at the opposite extreme with pairs that are completely immiscible. Intermediate cases of partial miscibility will be considered in Chapter 10. For now, we will look at two aspects of systems of immiscible and volatile liquids, which have practical applications.

A. Steam distillation

The vapor phase over two immiscible and volatile liquids will contain both compounds. Since neither liquid affects the other, both achieve their "pure" vapor pressure and

$$P(\text{total}) = P_A^0 + P_B^0 \qquad \textbf{(9-28)}$$

When P (total) equals the confining pressure, the mixture will boil; boiling must occur at a temperature lower than the boiling point of either A or B. The boiling temperature will remain constant so long as both liquids are present. These conditions are the basis for the process of **steam distillation**, which is often used in organic chemistry to recover compounds that decompose near their boiling points or have inconveniently high boiling points. Steam distillation also serves as a means of determining molar masses.

EXAMPLE 9-25: Derive an expression that involves partial pressures and molar masses for the ratio of the weights of the liquids in the distillate from a steam distillation.

Solution: We can bring in weights and molar masses by using moles, which suggests starting with mole fractions. The mole fractions in the vapor will be

$$Y_A = \frac{P_A^0}{P} \qquad \text{and} \qquad Y_B = \frac{P_B^0}{P}$$

The ratio is

$$\frac{Y_A}{Y_B} = \frac{P_A^0/P}{P_B^0/P} = \frac{P_A^0}{P_B^0}$$

which we can also express as

$$\frac{Y_A}{Y_B} = \frac{n_A/n}{n_B/n} = \frac{n_A}{n_B} = \frac{W_A/M_{m,A}}{W_B/M_{m,B}}$$

Thus

$$\frac{P_A^0}{P_B^0} = \frac{W_A M_{m,B}}{W_B M_{m,A}}$$

Rearranging, we obtain

$$\frac{W_A}{W_B} = \frac{M_{m,A}}{M_{m,B}} \frac{P_A^0}{P_B^0} \qquad \textbf{(9-29)}$$

EXAMPLE 9-26: When an unknown organic compound was steam distilled at an atmospheric pressure of 753.6 mm Hg, the boiling temperature was 84.5 °C. The layers in the distillate were separated and weighed; the water weighed 8.61 g and the organic compound, 38.45 g. Calculate the molar mass of the unknown compound.

Solution: We will use Eq. (9-29), but one piece of data is missing. We need the vapor pressures of the pure liquids at 84.5 °C but have only the total pressure. The value for water can be found in tables and is 425.2 mm Hg. If we let B represent the unknown,

$$P_B^0 = 753.6 - 425.2 = 328.4 \text{ mm Hg}$$

Rearranging Eq. (9-29), we get

$$M_{m,B} = M_{m,A} \frac{P_A^0}{P_B^0} \frac{W_B}{W_A}$$

Substitution yields

$$M_{m,B} = 18.0 \text{ g mol}^{-1} \left(\frac{425.2 \text{ mm Hg}}{328.4 \text{ mm Hg}}\right)\left(\frac{38.45 \text{ g}}{8.61 \text{ g}}\right) = 104 \text{ g mol}^{-1}$$

B. Distribution of solute between two solvents

If two immiscible liquids are in contact with a solute that is soluble in both of them, the solute will dissolve in both. Qualitatively, more solute will be in the solvent in which its solubility is greater. Quantitatively, the relative amounts in each solvent are determined by the equilibrium condition

$$\mu \text{ (solute in solvent 1)} = \mu \text{ (solute in solvent 2)}$$

We can shorten this expression by designating the two solvents α and β, so that

$$\mu_\alpha = \mu_\beta \qquad \textbf{(9-30)}$$

EXAMPLE 9-27: From Eq. (9-30), derive an expression for the ratio of the mole fractions of the solute in the two solvents, assuming that the solutions formed are ideal.

Solution: In Section 9-1C we obtained Eq. (9-8):

$$\mu = \mu^0 + RT \ln X$$

Substituting for μ_α and μ_β in Eq. (9-30), we have

$$\mu_\alpha^0 + RT \ln X_\alpha = \mu_\beta^0 + RT \ln X_\beta$$

Rearranging gives us

$$RT \ln X_\alpha - RT \ln X_\beta = \mu_\beta^0 - \mu_\alpha^0 \qquad \textbf{(9-31)}$$

Since the μ^0's are constant,

$$RT \ln \frac{X_\alpha}{X_\beta} = \text{Constant}$$

At a fixed temperature,

$$\frac{X_\alpha}{X_\beta} = \text{Constant} \qquad \textbf{(9-32)}$$

The constant is a true equilibrium constant, and Eq. (9-32) is one form of the **Nernst distribution law**. To put it in its usual form, we let the equilibrium be

$$\text{Solute in } \alpha \rightleftharpoons \text{Solute in } \beta$$

In the constant, X_β should be in the numerator, so we rewrite Eq. (9-31) as

$$RT \ln X_\beta - RT \ln X_\alpha = -(\mu_\beta^0 - \mu_\alpha^0)$$

and simplify to obtain the

NERNST DISTRIBUTION LAW

$$\frac{X_\beta}{X_\alpha} = K \qquad \textbf{(9-33)}$$

As for any equilibrium situation [see Eq. (7-22)], we have

$$RT \ln K = -\Delta G^\circ \qquad \textbf{(9-34)}$$

When the solutions are very dilute, the mole fractions may be replaced by molarities or molalities. In general, activities should be used.

The constant K is known as the **distribution coefficient** or, sometimes, as the **partition coefficient**. Like other equilibrium constants, it is dependent on temperature.

EXAMPLE 9-28: A common application of the distribution phenomenon is the test for iodide ion, in which the ion is oxidized to iodine. Adding CCl_4 and shaking will remove most of the iodine from water, as can be seen by the characteristic red-violet color of I_2 in CCl_4. If we measure the concentrations of I_2 in both solvents and find I_2 in CCl_4 is 0.0560 M and I_2 in H_2O is 0.000 655 M, what is the value of the distribution coefficient?

Solution: Since we are going from water to CCl_4 in this case, we can write the equilibrium as

$$I_2 \text{ in } H_2O \rightleftharpoons I_2 \text{ in } CCl_4$$

and thus will write K, using molarities, as

$$K = \frac{\text{Concentration in } CCl_4}{\text{Concentration in } H_2O} = \frac{0.0560}{0.000\,655} = 85.5$$

The test in Example 9-28 illustrates an important process known as **extraction**. Often a desired substance may be extracted from one solvent into another, as in the iodine test, or impurities may be removed. Extraction is usually carried out in several steps. It's possible to calculate the amount of solute remaining in the original solute after an extraction and to show that a series of extractions with small volumes of the second solvent is more efficient than one extraction using the same total volume. Most of the current textbooks omit this topic, so we'll stop here (but see Problem 9-26).

SUMMARY

The major equations of this chapter and connections between them are shown in Table S-9. You should determine the conditions that are necessary for each equation to be valid or the conditions that are imposed to make the connections. Other important things you should know are

1. Deviations from ideal behavior in solutions result primarily from the domination of attractive forces between molecules of solute and solvent.
2. The solubility of a gas increases with pressure (Henry's law: $X_2 = kP_2$) and usually decreases with temperature.
3. Solutions of liquids usually are accompanied by a vapor phase. At equilibrium, the partial pressure of a compound is given by Raoult's law, $P_i = X_i P_i^0$.
4. A solution differs from a pure solvent because it has the colligative properties of

   (a) vapor-pressure lowering
   (b) boiling-point elevation
   (c) freezing-point depression
   (d) osmotic pressure

5. The colligative properties result from the fact that the chemical potential of a solvent is less in solution than in its pure state. Their magnitude depends on the number of particles of solute.
6. Boiling-point elevations and freezing-point depressions are proportional to molality in dilute solutions, or $T = Km$. The K's may be evaluated empirically or calculated from known constants.
7. Determination of molar mass is an important application of colligative properties. In general, freezing-point depression is the most convenient and accurate method; the osmotic pressure method is useful for very large molecules.
8. For a solution to be ideal, each volatile component must obey Raoult's law.
9. The composition of the vapor phase over a solution is different from that of the liquid phase and, in ideal solutions, can be calculated from the mole fractions of the liquid phase and the vapor pressures of the pure components.

TABLE S-9: Summary of Important Equations

$G(\text{mixture}) = \Sigma\mu_i n_i$ (6-81)

$\quad\rightarrow \Delta G(\text{mixing}) = RT\,\Sigma n_i \ln\frac{P_i}{P}$ (9-4)

$X_2 = kP_2$ (9-9; Henry's Law)

$\updownarrow$

$P_2 = k'X_2$ (9-10; Henry's Law)

$P_1 = X_1 P_1^0$ (9-11; Raoult's Law)

$\quad\rightarrow \Delta P = X_2 P_1^0$ (9-12)

$\Delta T_b = K_b m$ (9-13)

$\quad\rightarrow K_b = \frac{MR(T_b^0)^2}{\Delta H_v}$ (9-14)

$|\Delta T_f| = K_f m$ (9-15)

$\quad\rightarrow K_f = \frac{MR(T_f^0)^2}{\Delta H_f}$ (9-16)

$\Pi = kM$ (9-18)

$\quad\rightarrow \Pi V = n_2 RT$ (9-19)

$\frac{W_A}{W_B} = \frac{M_{m,A}}{M_{m,B}}\frac{P_A^0}{P_B^0}$ (9-29; steam distillation)

$K = \frac{X_\beta}{X_\alpha}$ (9-33; Nernst distribution law)

10. Phase diagrams for the vapor–liquid portion of two-component solutions may be plotted as pressure versus concentration at constant temperature or as temperature versus concentration at constant pressure (a boiling-point diagram). The latter is approximately the invert of the former.

11. In a phase diagram, a tie line connects the compositions of the two phases that are in equilibrium; the lever rule gives the relative amounts of the two phases.

12. Because of differences in attractive forces between the molecules of the components, nonideal liquid mixtures often exhibit a maximum or a minimum in the pressure–composition diagram and a corresponding minimum or maximum in the temperature–composition diagram.

13. An azeotropic mixture (or constant-boiling mixture) has the composition of one of the maxima or minima.

14. Fractional distillation is a continuous process of evaporating and recondensing the volatile liquid components of a solution and is carried out in a vertical column. This process separates the components of an ideal solution or of one component and the constant-boiling mixture of an azeotropic system.

15. Steam distillation may be used to recover high-boiling or unstable compounds and to determine molar mass.

16. The Nernst distribution law states that the ratio of the activities (or concentrations) of a common solute in two immiscible solvents is a constant. The constant is called the distribution coefficient but is truly an equilibrium constant.

SOLVED PROBLEMS

Mixtures and Solutions

PROBLEM 9-1 Give an illustration of a solution of a solid in a gas.

Solution: The sublimation of a solid produces vapor, which mixes with air. You could call the mixture a solution of gases, but, in order to show the usual state of the pure solute, a better description is a solution of a solid in a gas. An illustration is paradichlorobenzene, used in moth balls. (You might think of a fine dust in air as a possible illustration, but this mixture is heterogeneous and is called a *suspension.*)

PROBLEM 9-2 Supply the steps in the derivation of Eq. (9-3) for ΔG (mixing).

Solution: In Section 9-1B, we stated that you should substitute Eqs. (6-81) and (9-1) into Eq. (9-2), or

$$\Delta G\text{ (mixing)} = G\text{ (mixture)} - G^0\text{ (unmixed components)}$$

Remember that the μ's will be different before and after mixing, whereas the n's are the same. Before mixing, both gases are at pressure P^0. After mixing, they are at partial pressures, P_A and P_B, but the total pressure is still P^0 $(= P = P_A + P_B)$. So

$$\begin{aligned}\Delta G\text{ (mixing)} &= n_A\mu_A\text{ (mixture)} + n_B\mu_B\text{ (mixture)} - n_A\mu_A^0 - n_B\mu_B^0 \\ &= n_A[\mu_A\text{ (mixture)} - \mu_A^0] + n_B[\mu_B\text{ (mixture)} - \mu_B^0]\end{aligned}$$

Substituting Eq. (9-1) for the μ (mixture) terms,

$$\Delta G\text{ (mixing)} = n_A\left[\mu_A^0 + RT\ln\frac{P_A}{P^0} - \mu_A^0\right] + n_B\left[\mu_B^0 + RT\ln\frac{P_B}{P^0} - \mu_B^0\right]$$

All of the pure state terms cancel, which leaves

$$\Delta G\text{ (mixing)} = RT\left(n_A\ln\frac{P_A}{P^0} + n_B\ln\frac{P_B}{P^0}\right)$$

PROBLEM 9-3 Show that (**a**) ΔV (mixing) $= 0$ and (**b**) ΔH (mixing) $= 0$ for ideal systems. Use thermodynamic relationships previously introduced.

Solution:

(**a**) You may write expressions like Eq. (6-40), $\left(\frac{\partial G}{\partial P}\right)_T = V$, for changes, i.e., $\left(\frac{\partial \Delta G}{\partial P}\right)_T = \Delta V$. For mixing,

$$\left(\frac{\partial \Delta G\text{ (mixing)}}{\partial P}\right)_T = \Delta V\text{ (mixing)}$$

But ΔG(mixing) is independent of P; thus the derivative $= 0$.

(**b**) Two approaches are possible for obtaining ΔH (mixing).

1. Using $\Delta G = \Delta H - T\,\Delta S$, or

$$\Delta H\text{ (mixing)} = \Delta G\text{ (mixing)} - T\,\Delta S\text{ (mixing)}$$

you substitute Eqs. (9-5) and (9-6) to get ΔH(mixing) $= 0$.

2. You could start with the Gibbs–Helmholtz equation, Eq. (6-56):

$$\left(\frac{\partial \Delta G/T}{\partial T}\right)_P = -\frac{\Delta H}{T^2}$$

From Eq. (9-5),

$$\frac{\Delta G\text{ (mixing)}}{T} = nR\,\Sigma\,X_i\ln X_i$$

This expression shows that $\Delta G/T$ is independent of T; thus the derivative $= 0$, and ΔH (mixing) $= 0$.

Henry's Law

PROBLEM 9-4 The solubility of krypton in benzene at 25 °C and 1 atm is 2.73×10^{-5} (mole fraction). Calculate the mole fraction of krypton in benzene if the partial pressure is 0.450 atm.

Solution: You should recognize that this is a Henry's-law problem. From the data given, you can determine k and use it to calculate the new concentration. From Eq. (9-9),

$$k = \frac{X_2}{P_2} = \frac{2.73 \times 10^{-5}}{1\text{ atm}} = 2.73 \times 10^{-5}\text{ atm}^{-1}$$

For $P = 0.45$,

$$X_2 = kP_2 = 2.73 \times 10^{-5}(0.45) = 1.23 \times 10^{-5}$$

PROBLEM 9-5 From the following data for three experimental runs, would you say that this system follows Henry's law? Make calculations to support your answer.

	1	2	3
molality	0.078	0.18	0.28
pressure of HCl, atm	0.15	0.34	0.55

Solution: You should calculate k for each set of data and see whether the values are constant. From Eq. (9-9), $k = m/P$; thus

Run	m	P	k
1	0.078	0.15	0.52
2	0.18	0.34	0.53
3	0.28	0.55	0.51

The values of k are close, so you should conclude that Henry's law does apply.

Raoult's Law

PROBLEM 9-6 As mentioned previously, Raoult's law can be considered a special case of Henry's law, and real solutions obey these laws only in dilute solutions. Show that the solvent must obey Raoult's law over the range of concentrations for which the solute obeys Henry's law. Assume a two-component solution and recall some equations from Chapter 6.

Solution: The question is how to manipulate $X_2 = kP_2$ [Eq. (9-9)] to obtain $P_1 = X_1 P_1^0$ [Eq. (9-11)]. The pressures and mole fractions of two components of a mixture are involved. In Chapter 6, we obtained a relationship between moles and chemical potentials of two components, the Gibbs–Duhem equation, Eq. (6-82), and pointed out that it is useful when dealing with solutions. This brings in chemical potential, but (also in Chapter 6) we obtained a relationship between chemical potential and pressures of an ideal gas, Eq. (6-74). If you eliminate chemical potential between the two, you may be able to relate solute to solvent.

In very dilute solutions, $X_i \simeq n_i$, so you may rewrite Eq. (6-82) as

$$X_1\, d\mu_1 = -X_2\, d\mu_2$$

and differentiate with respect to X_1 to get

$$X_1 \frac{\partial \mu_1}{\partial X_1} = -X_2 \frac{\partial \mu_2}{\partial X_1}$$

Since $dX_1 = -dX_2$, this becomes

$$X_1 \frac{\partial \mu_1}{\partial X_1} = X_2 \frac{\partial \mu_2}{\partial X_2}$$

Now eliminate the μ terms. Assume that the vapors are ideal, and use Eq. (6-74), rewritten as Eq. (9-1) to obtain

$$\mu = \mu^0 + RT \ln \frac{P}{P^0}$$

Differentiating with respect to X and remembering that μ^0 and P^0 are constants, you have

$$\frac{\partial \mu}{\partial X} = RT \frac{\partial \ln P}{\partial X}$$

The modified Gibbs–Duhem equation now becomes

$$X_1 RT \frac{\partial \ln P_1}{\partial X_1} = X_2 RT \frac{\partial \ln P_2}{\partial X_2}$$

Canceling RT and substituting $\dfrac{1}{\partial \ln X}$ for $\dfrac{X}{\partial X}$, you have

$$\frac{\partial \ln P_1}{\partial \ln X_1} = \frac{\partial \ln P_2}{\partial \ln X_2}$$

If Henry's law is obeyed up to a certain X_2, then $P_2 = kX_2$, or

$$\ln P_2 = \ln k + \ln X_2 \qquad \text{and} \qquad \frac{\partial \ln P_2}{\partial \ln X_2} = 0 + 1$$

Thus

$$\frac{\partial \ln P_1}{\partial \ln X_1} = 1$$

The solution to such an equation is

$$\ln P_1 = \ln X_1 + \ln(\text{constant}) \qquad \text{or} \qquad P_1 = \text{constant}(X_1)$$

To evaluate the constant, set $X_1 = 1$, which requires that $P = P^0$. Thus the constant $= P^0$, or

$$P_1 = X_1 P_1^0$$

which is Raoult's law.

PROBLEM 9-7 Calculate the vapor pressure at 30 °C over a solution of 40.0 g of benzyl alcohol ($C_6H_5CH_2OH$) in 50.0 g of benzene. The vapor pressure of pure benzene at 30 °C is 118.2 mm Hg. Assume that the solute is nonvolatile at this temperature and that the solution behaves ideally.

Solution: This is a Raoult's-law problem. You could use either Eq. (9-11) or Eq. (9-12). From the data given and the answer asked for, you should use Eq. (9-11). It gives P_1 directly, and one mole fraction is as easy to calculate as the other.

$$X_1 = \frac{n_b}{n_b + n_{alc}} = \frac{50.0/78.0}{50.0/78.0 + 40.0/108.0} = 0.634$$

$$P_1 = X_1 P_1^0 = 0.634(118.2) = 74.9 \text{ mm Hg}$$

Note that this solution would probably deviate from ideality because of H-bonding in the solute.

PROBLEM 9-8 Calculate (**a**) the free energy of mixing and (**b**) the entropy of mixing for the solution in Problem 9-7.

Solution:

(**a**) The solution is of the solid-in-liquid type, so you must use the expression that applies to all ideal solutions, Eq. (9-5):

$$\Delta G\,(\text{mixing}) = nRT\,\Sigma\, X_i \ln X_i$$

You found the mole fraction of benzene in Problem 9-7: $X_b = 0.634$. Thus $X_{alc} = 0.366$. You also need n, which you can obtain from $X_i = n_i/n$. In this case, you can use $n = n_1 + n_2$, so

$$n = \frac{50.0}{78.0} + \frac{40.0}{108.0} = 0.641 + 0.370 = 1.011$$

and

$$\Delta G\,(\text{mixing}) = 1.011(8.314)(303.2)[0.634(\ln 0.634) + 0.366(\ln 0.366)] = -1.67 \text{ kJ}$$

(**b**) Since $\Delta G = -T\,\Delta S$,

$$\Delta S = -\frac{\Delta G}{T} = -\frac{-1.67 \times 10^3 \text{ J}}{303.2 \text{ K}} = 5.51 \text{ e.u.}$$

PROBLEM 9-9 In order to determine the molar mass of an unknown nonvolatile compound, 3.50 g of the compound was dissolved in 10.0 g of carbon tetrachloride. At 25 °C, the vapor pressure of the solution was found to be 80.8 mm Hg. Calculate the molar mass. (Vapor pressure of CCl_4 at 25 °C = 114.5 mm Hg.)

Solution: Again, you must assume ideal behavior in order to apply Raoult's law. Since the solute is the substance of concern, Eq. (9-12) is the form you should use. Solve for X_2, n_2, and M_2, in that order. Rearranging Eq. (9-12) gives you

$$X_2 = \frac{P_1^0 - P_1}{P_1^0} = \frac{114.5 - 80.8}{114.5} = 0.294$$

Now,

$$X_2 = \frac{n_2}{n_1 + n_2} = 0.294$$

$$n_2 = 0.294 n_1 + 0.294 n_2$$

$$0.706 n_2 = 0.294\left(\frac{10.0}{154.0}\right) = 0.0191$$

so
$$n_2 = \frac{0.0191}{0.706} = 0.0271 = \frac{W_2}{M_{m,2}}$$

thus
$$M_{m,2} = \frac{W_2}{0.0271} = \frac{3.50}{0.0271} = 129.2 \text{ g mol}^{-1}$$

PROBLEM 9-10 The calculations in Problem 9-9 were taken in steps but could have been combined by using one large expression. Derive an expression for ΔP involving masses and molar masses.

Solution: You should start with Eq. (9-12) and replace X_2 with its equivalent, where the numbers of moles are expressed by mass divided by molecular mass. Thus

$$\Delta P = X_2 P_1^0$$

$$X_2 = \frac{n_2}{n_1 + n_2} = \frac{W_2/M_{m,2}}{W_1/M_{m,1} + W_2/M_{m,2}} = \frac{W_2(M_{m,1}M_{m,2})}{M_{m,2}(M_{m,2}W_1 + M_{m,1}W_2)}$$

$$P = P_1^0\left(\frac{M_{m,1}W_2}{M_{m,2}W_1 + M_{m,1}W_2}\right)$$

This expression would be useful if you want ΔP but is rather awkward to solve if you want M_2. If the solution were very dilute, you could neglect W_2 in the denominator and have

$$\Delta P = P_1^0\left(\frac{M_{m,1}}{M_{m,2}}\right)\left(\frac{W_2}{W_1}\right)$$

Colligative Properties

PROBLEM 9-11 In an experiment to determine K_b of ethanol, 0.978 g of glucose ($C_6H_{12}O_6$) was dissolved in 10.0 g of ethanol. This solution boiled at 79.07 °C. Calculate the experimental K_b. (T_b^0 of ethanol = 78.4 °C.)

Solution: You should use Eq. (9-13) after determining the molality of the solution:

$$m = \frac{0.978 \text{ g}/180 \text{ g mol}^{-1}}{0.0100 \text{ kg}} = 0.543 \text{ mol kg}^{-1}$$

Rearranging Eq. (9-13) gives

$$K_b = \frac{\Delta T_b}{m} = \frac{T_b - T_b^0}{m} = \frac{79.07 - 78.4}{0.543} = 1.2 \text{ K kg mol}^{-1}$$

PROBLEM 9-12 Derive an expression for calculating the molecular weight of a solute directly from experimental data and K_b.

Solution: You may start with Eq. (9-13) and substitute for m its equivalent in weights and molar mass:

$$m = \frac{n\text{ (solute)}}{\text{kg (solvent)}} = \frac{W_2/M_{m,2}}{W_1} = \frac{W_2}{M_{m,2}W_1}$$

Substituting in Eq. (9-13), you have

$$\Delta T_b = K_b m = K_b \frac{W_2}{M_{m,2}W_1}$$

so
$$M_{m,2} = \frac{K_b W_2}{\Delta T_b W_1}$$

PROBLEM 9-13 The empirical (simplest) formula of oxalic acid is CHO_2. After 3.70 g of the compound was dissolved in 10.0 g of water, the solution boiled at 102.16 °C. What is the true molecular formula of oxalic acid? Assume negligible dissociation, and use the expression from Problem 9-11. (See Table 9-1.)

Solution: The statement of the problem indicates that the true molecular formula isn't the simplest formula. It would have to be some multiple, however, which you can determine by finding the molar mass and comparing it to the empirical formula weight. From Problem 9-12,

$$M_{m,2} = \frac{K_b W_2}{\Delta T_b W_1} = \frac{(0.512 \text{ K kg mol}^{-1})(0.00370 \text{ kg})}{(102.16 - 100.00) \text{ K } (0.0100 \text{ kg})} = 0.0877 \text{ kg mol}^{-1}$$

The empirical formula weight (CHO_2) = 0.045 kg mol^{-1}. Clearly, the true molecular weight is twice this, so the true molecular formula is $C_2H_2O_4$.

PROBLEM 9-14 Estimate the heat of vaporization of cyclohexane from its K_b. (See Table 9-1.)

Solution: The heat of vaporization is related to K_b by Eq. (9-14), which you should rearrange to get

$$\Delta H_v = \frac{M_m R (T_b^0)^2}{K_b}$$

From Table 9-1, $T_b^0 = 354.1$ K and $K_b = 2.79$. Thus

$$\Delta H_v = \frac{0.0841(8.314)(354.1)^2}{2.79} = 31.4 \text{ kJ mol}^{-1}$$

PROBLEM 9-15 How many grams of sucrose ($C_{12}H_{22}O_{11}$) must be added to 50.0 g of water to create a solution with a freezing point of -1.00 °C?

Solution: The most convenient procedure is to rearrange Eq. (9-17) to obtain

$$W_2 = \frac{M_{m,2}|\Delta T_f| W_1}{K_f}$$

From Table 9-1, $T_f^0 = 0.00$ °C, and $K_f = 1.86$ K kg mol^{-1}. Remember that W_1 must be expressed in kg. Substituting gives you

$$W_2 = \frac{342(|-1.00|)(0.0500)}{1.86} = 9.19 \text{ g}$$

PROBLEM 9-16 A solution of 4.80 g urea [$CO(NH_2)_2$] in 50.0 g 1,4-dioxane freezes at 2.7 °C. Calculate K_f for the solvent. ($T_f^0 = 10.5$ °C.)

Solution: You can get an expression for K_f by rearranging Eq. (9-15) to

$$K_f = \frac{|\Delta T_f|}{m}$$

Since the M_m of urea = 60.0 g mol^{-1},

$$m = \frac{(480/60.0) \text{ mol}}{0.0500 \text{ kg}} = 1.60 \text{ mol kg}^{-1}$$

$$K_f = \frac{10.5 - 2.7}{1.6} = 4.9 \text{ K kg mol}^{-1}$$

PROBLEM 9-17 What is the freezing point of an aqueous solution which by weight is 10 percent methanol and 20 percent ethanol?

Solution: Recall that molality is based on the total moles of solute and that the nature of the solute is unimportant. You must calculate the moles of each solute, then m, and finally ΔT, using Eq. (9-15). First, however, you decide how many grams of each substance are present. You can simplify the calculations if you choose a total weight of 100 g of solution; then you have 10 g CH_3OH, 20 g C_2H_5OH, and 70 g H_2O. Any other choice, however, will also give the correct molality.

$$\text{Moles } CH_3OH = \frac{10.0}{32.0} = 0.313 \qquad \text{Moles } C_2H_5OH = \frac{20.0}{46.0} = 0.435$$

$$m = \frac{0.313 + 0.435}{0.070} = 10.7$$

Substituting for m in Eq. (9-15), you get

$$\Delta T_f = 1.86(10.7) = 19.9$$

$$T_f = T_f^0 - 19.9 = 0 - 19.9 = -19.9 \text{ °C}$$

PROBLEM 9-18 A student was asked to find the molar mass and formula of an organic compound using freezing-point depression. She found that the compound contained 85.7 percent carbon and 14.3 percent hydrogen. Then she dissolved 1.12 g of the compound in 20.0 g benzene and measured the freezing point of this solution to be 1.6 °C. Calculate (**a**) the experimental molecular weight and (**b**) the true molecular formula and the true molecular weight.

Solution: The calculation of the experimental molecular weight is straightforward if you use Eq. (9-17). For the rest of the problem, however, you need to recall how to calculate an empirical formula from the percent composition. One way is to divide each percentage by the corresponding atomic weight and then divide each result by the smallest, which gives the subscripts in the empirical formula.

	Step 1	Step 2	Simplest formula
For C:	$\frac{85.7}{12.0} = 7.14$	$\div 7.14 = 1$	C_1H_2
For H:	$\frac{14.3}{1.0} = 14.3$	$\div 7.14 = 2$	

(**a**) From Table 9-1, $T_f^0 = 5.5$ °C, and $K_f = 5.12$ K kg mol^{-1}. Substituting in Eq. (9-17), you get

$$M_{m,2} = \frac{K_f W_2}{\Delta T_f W_1} = \frac{5.12(1.12)}{(5.5 - 1.6)(20.0)} = 0.074 \text{ kg mol}^{-1}$$

(**b**) The empirical formula weight is 0.0140 kg mol^{-1}, so the ratio is

$$\frac{0.074}{0.0140} = 5.3$$

The true molecular formula must be an integer multiple of the simplest formula; you can assume that there was some experimental error and take 5 as the multiple. Thus the true molecular formula is C_5H_{10}, and the true molecular weight is 0.0700 kg mol^{-1}.

√ **PROBLEM 9-19** If a solution has an osmotic pressure of 10.0 atm at 300 K, what is its concentration?

Solution: Osmotic pressure is given by Eq. (9-19). If you rearrange to introduce molarity, you get

$$\Pi = \frac{n_2}{V} RT = MRT \qquad \text{and} \qquad M = \frac{\Pi}{RT}$$

Substitution yields

$$M = \frac{10.0 \text{ atm}}{0.08205 \text{ L atm K}^{-1}\text{mol}^{-1}(300 \text{ K})} = 0.406 \text{ mol L}^{-1}$$

Note that in an ideal solution, Π does not depend on the nature of either the solute or the solvent.

√ **PROBLEM 9-20** What weight of naphthalene is present in 80.0 mL of a benzene solution if the osmotic pressure is 14.6 atm at 20 °C?

Solution: You can get an expression for weight by rearranging Eq. (9-20):

$$M_{m,2} = \frac{W_2 RT}{\Pi V} \qquad \text{and} \qquad W_2 = \frac{M_{m,2}\Pi V}{RT}$$

Substituting, you get

$$W_2 = \frac{128(14.6)(0.0800)}{0.08205(293.2)} = 6.21 \text{ g}$$

PROBLEM 9-21 A certain dilute ideal aqueous solution has an osmotic pressure of 12.2 atm at 20 °C. How much does the chemical potential of the water in the solution differ from that of pure water at 20 °C? Assume that the density of the solution is 1.000 g/mL and that the volume of solution is the same as the volume of the pure water.

Solution: The mention of chemical potential should remind you of the expression for all ideal solutions, Eq. (9-8), or

$$\mu = \mu^0 + RT \ln X$$

The problem now becomes one of calculating the mole fraction of water. Osmotic pressure is given, so you should think of the van't Hoff equation, Eq. (9-19), from which you can obtain the mole fraction of the solute:

$$\Pi V = n_2 RT$$

When mole fraction is used for concentrations, the volume doesn't matter, so you can choose any volume. Your first thought might be to choose 1 L, but the calculations will be simpler if you choose 1 mol H_2O, i.e., 18.0 mL when $d = 1.000$ g/mL. Now,

$$n_2 = \frac{\Pi V}{RT} = \frac{12.2 \text{ atm } (0.0180 \text{ L})}{0.0821 \text{ L atm K}^{-1}\text{mol}^{-1}(293.2 \text{ K})} = 9.12 \times 10^{-3}$$

Since you chose $n_1 = 1$,

$$X_1 = \frac{n_1}{n_1 + n_2} = \frac{1}{1 + 0.0091} = 0.991$$

Rearranging Eq. (9-8) and substituting, you get

$$\mu_1 - \mu_1^0 = RT \ln X_1 = 8.314(293.2) \ln 0.991 = -22.0 \text{ J}$$

Liquid Solutions with Two Volatile Components

PROBLEM 9-22 What are the vapor pressures of the pure components of an ideal two-liquid system if the mole fractions of one of them are 0.600 in the liquid phase and 0.677 in the vapor phase at a total pressure of 155 mm Hg?

Solution: This problem is a variation of the usual type illustrated by Example 9-19. You can calculate the partial pressures in the vapor using Eqs. (9-25) and (9-26) and then calculate the pure vapor pressures using Eqs. (9-11) and (9-21). Rearranging Eqs. (9-25) and (9-26) so that $P_A = Y_A P$ and $P_B = Y_B P$ and substituting gives you

$$P_A = 0.677(155) = 105 \qquad P_B = (1 - 0.677)(155) = 50.1$$

Rearranging Eqs. (9-11) and (9-21) so that $P_A^0 = P_A/X_A$ and $P_B^0 = P_B/X_B$ and substituting, you get

$$P_A^0 = \frac{105}{0.600} = 175 \text{ mm Hg} \qquad P_B^0 = \frac{50.1}{0.400} = 125 \text{ mm Hg}$$

PROBLEM 9-23 The lever rule is applicable if concentrations are plotted as weight fractions (or weight percents) instead of mole fractions. A solution was prepared by mixing 15.0 g $SiCl_4$ in 22.0 g CCl_4. After liquid–vapor equilibrium was established, analysis showed that the liquid was 72.2 percent CCl_4 and that the vapor was 53.8 percent CCl_4. What are relative weights of the two phases?

Solution: In order to use the lever rule, you need the lengths of the two segments of the tie line. You can calculate the lengths from the data given. You also need the initial weight fraction, which is

$$W_{CCl_4} = \frac{22.0}{22.0 + 15.0} = 0.595$$

Although you don't have the pressure–composition diagram, you can guess from the data the relative positions of the liquid and vapor curves. Since the vapor phase is richer in $SiCl_4$ than the liquid phase, $SiCl_4$ must have the higher P^0. If 100% CCl_4 is at the right-hand end of the *x*-axis, the plots rise to the left and the vapor curve is to the left of the liquid curve. The weight of vapor is proportional to the right-hand segment of the tie line, or 0.722 − 0.595. The weight of liquid then is proportional to the left-hand segment, or 0.595 − 0.538. Writing an expression like Eq. (9-27) for weights, you have

$$\frac{W_{vap}}{W_{liq}} = \frac{0.722 - 0.595}{0.595 - 0.538} = \frac{0.127}{0.057} = 2.2{:}1.00$$

Immiscible Volatile Liquids

PROBLEM 9-24 On a day when the barometric pressure was 756.1 mm Hg, *n*-octane was steam distilled, boiling at 89.0 °C. What weight of *n*-octane per 10.0 g of water would you expect in the distillate? At 89.0 °C, the vapor pressure of H_2O = 506.1 mm Hg.

Solution: The weight ratio is given by Eq. (9-29). Using the subscripts "o" for octane and "w" for water and substituting, you obtain

$$\frac{W_o}{W_w} = \frac{M_{m,o}\,P_o^0}{M_{m,w}\,P_w^0}$$

$$= \left(\frac{114.2}{18.0}\right)\left(\frac{756.1 - 506.1}{506.1}\right) = 3.13{:}1.00$$

For 10.0 g water, the weight of octane would be 31.3 g.

PROBLEM 9-25 A mixture of equal moles of water and cyclohexane was prepared at 60 °C. Assume that the two are completely immiscible. After equilibrium was established, the vapor pressure was 325.7 mm Hg. The vapor pressures of pure cyclohexane and water at 60 °C are 385.0 and 149.4 mm Hg, respectively. **(a)** What are the phases in the system and what are their components? **(b)** What is the composition of the vapor?

Solution:

(a) You'd expect the pressure over the mixture of two immiscible liquids to equal the sum of the individual vapor pressures. Since it doesn't, the liquid with the higher vapor pressure must have evaporated completely; if it were still present as liquid, it would be "producing" its pressure of 385.0. Thus you have one liquid phase (water) and one vapor phase (both compounds).

(b) The water is still maintaining its pressure of 149.4; thus its mole fraction must be

$$Y_w = \frac{P_w^0}{P} = \frac{149.4}{325.7} = 0.4587$$

so

$$Y_c = 1 - 0.4587 = 0.5413$$

PROBLEM 9-26 Using the distribution coefficient from Example 9-28 for H_2O and CCl_4, calculate the fraction by weight of iodine that would be left in the water, if 100 mL of a 0.0100 *M* solution of iodine in water is extracted with 10.0 mL of CCl_4.

Solution: Let's derive a general expression for problems of this sort. If you let w be the initial weight in solvent α and w_1 be the weight remaining after equilibrium is attained, the weight in solvent β is $w - w_1$.

$$C_\alpha = \frac{w_1}{V_\alpha} \quad \text{and} \quad C_\beta = \frac{w - w_1}{V_\beta}$$

In Example 9-28, the extracting solvent is in the numerator, so

$$K = \frac{C_\beta}{C_\alpha} = \frac{(w - w_1)/V_\beta}{w_1/V_\alpha} = \frac{(w - w_1)V_\alpha}{w_1 V_\beta}$$

$$KV_\beta w_1 = wV_\alpha - w_1 V_\alpha$$

$$(KV_\beta + V_\alpha)w_1 = wV_\alpha$$

$$\frac{w_1}{w} = \frac{V_\alpha}{KV_\beta + V_\alpha}$$

For this problem, α is H_2O and β is CCl_4. Substituting the respective volumes and $K = 85.5$ from Example 9-28, you have

$$\frac{w_1}{w} = \frac{100}{85.5(10.0) + 100} = \frac{100}{955} = 0.105$$

Note that a second extraction with another 10 mL would have the same ratio; thus

$$\frac{w_2}{w} = 0.105(0.105) = 0.0110$$

For n steps,

$$\frac{w_n}{w} = \left(\frac{V_\alpha}{KV_\beta + V_\alpha}\right)^n$$

Supplementary Exercises

PROBLEM 9-27 An example of a solution of a liquid in a gas is ________. ΔH is (positive) (negative) (zero) when an ideal solution is formed. The k of Henry's law for a given system (increases) (decreases) (stays the same) when (**a**) the pressure increases; (**b**) the temperature increases. If the pressure of a gas over an ideal gas-in-liquid solution is tripled, the concentration of gas would ________. The boiling point of a 0.10 molal aqueous solution is ________. If the osmotic pressure of a solution is 6.0 atm at 300 K, the pressure would be ________ if the concentration were doubled. If the pressure and overall composition of an ideal two-liquid solution fall in the area between the two composition curves, the amount of vapor will (increase) (decrease) when the pressure is increased slightly. As the pressure is decreased along line *ch* in Figure 9-3, X_A = ________ just before the last bit of liquid disappears. If the vapor over a boiling solution of a certain azeotrope ($X_A = 0.42$) is removed and condensed, X_B = ________ in the condensate. The physical property of ________ is essential for a compound if it is to be recovered by steam distillation.

PROBLEM 9-28 For a solution of H_2S in water under a pressure of 1.5 atm, the molality was found to be 0.155 mol kg^{-1}. Calculate the pressure to be expected over a solution that is 0.50 molal.

PROBLEM 9-29 If two ideal gases are mixed so that the final pressure is the sum of their initial pressures, what is ΔG (mixing)?

PROBLEM 9-30 Look back at Problem 9-2 and its answer. If the temperature were given, is it possible to calculate ΔG (mixing)? If not, what further information is needed?

PROBLEM 9-31 How much is the vapor pressure of water lowered at 20 °C if enough nonvolatile solute is added to give a solution in which the mole fraction of water is 0.73? The vapor pressure of pure water at 20 °C is 17.54 mm Hg.

PROBLEM 9-32 Repeat Problem 9-31 for a solution that is 1.0 molal.

PROBLEM 9-33 Calculate the theoretical value of K_b for ethanol. (See Table 8-1.) Compare the result to that of Problem 9-11.

PROBLEM 9-34 A solution of 0.15 g naphthalene ($C_{10}H_8$) and 6.5 g carbon tetrachloride boiled at 77.7 °C. Calculate K_b of CCl_4. (T_b^0 of CCl_4 = 76.8 °C.)

PROBLEM 9-35 Using the answer to Problem 9-34, calculate the heat of vaporization of CCl_4.

PROBLEM 9-36 When 25.0 g of an unknown compound was dissolved in 200 g benzene, the resulting solution boiled at 82.2 °C. What is the molar mass of the compound?

PROBLEM 9-37 Derive Eq. (9-16).

PROBLEM 9-38 What is the molality of an aqueous solution that freezes at −2.00 °C?

PROBLEM 9-39 Calculate the heat of fusion of naphthalene from freezing-point data.

PROBLEM 9-40 Ethylene glycol ($C_2H_6O_2$) is often used as an antifreeze. What weight would be needed for a 10 °C decrease of the freezing point per kilogram of water?

PROBLEM 9-41 When 2.34 g of urea, $CO(NH_2)_2$, is dissolved in 100 g of nitrobenzene, the solution freezes at 2.54 °C. When 2.12 g of an unknown substance is dissolved in 50.0 g of nitrobenzene, the solution freezes at 3.14 °C. Calculate the molar mass of the unknown substance. (T_f^0 of nitrobenzene = 5.7 °C.)

PROBLEM 9-42 Calculate the osmotic pressure at 300 K of 100 mL of an aqueous solution that contains 10.0 g of ethanol.

PROBLEM 9-43 What is the molecular weight of an unknown substance if a solution of 2.13 g of the substance in 45.2 mL of solution was found to have an osmotic pressure of 2.43 mm Hg at 300 K?

PROBLEM 9-44 In a certain dilute ideal aqueous solution at 30 °C, the chemical potential of the water is 30 J less than that of pure water. Calculate the osmotic pressure of the solution, assuming that the volume of the solution is the same as the original volume of the water and that the density is 1.000 g/mL.

PROBLEM 9-45 Methyl alcohol and water form a nearly ideal solution. What is the mole fraction of water in the vapor phase at 30 °C if 0.50 mol of alcohol is dissolved in 3.0 mol of water? Vapor pressures at 30 °C are 31.8 for water and 160 for methyl alcohol.

PROBLEM 9-46 Assume that Figure 9-3 applies to the benzene–toluene system, with benzene on the right-hand side. Estimate the mole fraction of benzene in the vapor phase when the system is in equilibrium at a pressure halfway between points f and g.

PROBLEM 9-47 Continuing Problem 9-46, estimate the ratio of moles of vapor to moles of liquid at P_d if 4 moles of each substance were mixed originally.

PROBLEM 9-48 In a steam distillation of an unknown organic compound, the layer of unknown compound from the distillate weighed 27.45 g and the water layer weighed 10.84 g. The conditions were 745.4 mm Hg and 90.2 °C, at which temperature the vapor pressure of H_2O = 529.8 mm Hg. What is the molar mass of the unknown compound?

PROBLEM 9-49 Under a pressure of 753.2 mm Hg, iodobenzene steam distills at 98.0 °C. How much steam will be needed to collect 1.0 kg of C_6H_5I in the distillate? (Vapor pressure of H_2O at 98.0 °C = 707.3 mm Hg.)

PROBLEM 9-50 Redo Problem 9-26 for one extraction using 20.0 mL of CCl_4. Compare the fraction remaining to that for two steps of 10.0 mL each. Which procedure gives greater extraction?

Answers to Supplementary Exercises

9-27 mercury in air; zero; (**a**) stays the same, (**b**) decrease; triple; 100.052 °C; 12.0 atm; decrease; 0.31; 0.58; limited solubility in water **9-28** 4.84 **9-29** zero
9-30 no; moles of components **9-31** 4.7 mm Hg **9-32** 0.31 mm Hg
9-33 1.20 K kg mol^{-1} **9-34** 5.0 K kg mol^{-1} **9-35** 31 kJ mol^{-1} **9-36** 0.158 kg mol^{-1}
9-37 see derivation of Eq. (9-14) **9-38** 1.08 m **9-39** 19 kJ mol^{-1} **9-40** 333 g
9-41 134 g mol^{-1} **9-42** 53.5 atm **9-43** 3.63 × 10^5 g mol^{-1} **9-44** 16.5 atm **9-45** 0.54
9-46 0.43 **9-47** 1.6 **9-48** 112 g mol^{-1} **9-49** 1.36 kg **9-50** 0.055; two steps

FINAL EXAM (Chapters 1–9)

Time: 100 minutes

Compound	T_f (K)	ΔH_f (kJ mol^{-1})	T_b (K)	ΔH_v (kJ mol^{-1})	K_f (K kg mol^{-1})	K_b (K kg mol^{-1})
Water	273.2	9.82	373.2	40.63	1.86	0.51
Benzene	278.7		353.4	30.77	5.12	2.53

Total Points: 200

Compound	ΔH_f° (kJ mol^{-1})	ΔG_f° (kJ mol^{-1})
CH_4 (g)	−74.8	−50.7
C_2H_4 (g)	52.3	68.2
C_2H_6 (g)	−84.7	−32.8
C_2H_5OH (l)	−277.7	−174.8
H_2O (g)	−241.8	−228.6
H_2O (l)	−285.8	−237.1
H_2O_2 (l)	−187.8	−120.4

A. Short Answers (2 points per answer = 26)

1. Give the value of the gas constant in SI energy units to four significant figures. ____________

2. If bottles of NH_3 and SO_2, placed the same distance from you, are opened simultaneously, which will you smell first? ____________

3. What important property of gases is directly related to the mean square velocity of the molecules? ____________

4. Give the SI equivalent of 1 atmosphere to three significant figures. ____________

5. If a chemical reaction produces a gas, as when an acid reacts with a metal, what is the sign of the work? ____________

6. What property of a system remains constant in an adiabatic reversible process? ____________

7. Write the partial derivative known as the coefficient of thermal expansion. ____________

8. What property of a system equals the coefficient of change of G with P at constant T? ____________

9. Write the partial derivative known as the Gibbs chemical potential. ____________

10. Using chemical potential, write the expression that states the criterion for chemical equilibrium in a closed system. ____________

11. What is the name for two or more solid forms of the same element? ____________

12. What is the approximate value of the entropy of vaporization of most substances? ____________

13. What is the name usually used for constant-boiling mixtures? ____________

B. Long Answers (Total points = 69)

1. State two of the three properties of exact differentials. (2)
2. What is meant by the statement that two gases are in "corresponding states"? (3)
3. If the radius of atom A is one-half the radius of atom B, which has the larger mean free path? Describe how you arrived at your answer. (3)
4. Define *specific heat*. How is *heat capacity* different? (3)
5. State the first, second, and third laws of thermodynamics as principles of impotence, i.e., as the *impossibility* of doing something. (6)
6. Derive the expressions for calculating the work done by an ideal gas (**a**) in a constant-pressure expansion from V_1 to V_2 and (**b**) in a reversible expansion from V_1 to V_2 (both are isothermal). (8)
7. Why is ΔS not a satisfactory criterion of chemical spontaneity? What properties are satisfactory and under what conditions? (3)
8. Describe the relationships between the two free energies and work. (4)
9. From the Gibbs equation for dA, derive the corresponding Maxwell relation. (4)
10. Derive a general expression for K_c in terms of degree of dissociation and initial concentration for a reaction of the type $AB_2 \rightleftharpoons A + 2B$ (6)
11. Explain why it is incorrect to try to calculate the heat of sublimation of water by adding the tabulated values ΔH_v and ΔH_f. (3)
12. Sketch an approximate phase diagram for carbon dioxide and use it to explain why "dry ice" is possible. Label important points in the diagram for reference in your discussion. (6)
13. Derive the Clausius–Clapeyron equation from the Clapeyron equation, stating clearly all assumptions you make. (8)
14. Describe the thermodynamic basis of the colligative properties of solutions, using boiling-point elevation as an illustration. (5)
15. For two substances A and B, sketch a pressure–composition diagram that illustrates positive deviation from Raoult's law. Label the major areas and describe briefly the cause of positive deviations. (5)

C. Problems (Total Points = 105) To save time, do not do the arithmetic unless the problem is labeled CALCULATE; otherwise just *set up* the final calculation. Solve all equations for the unknown, and substitute numerical values for all symbols. Put the proper units on the final answer. Data you might need are tabulated above.

1. Using the van der Waals equation, find the pressure exerted by 100 g of NO in a 10-L vessel at 1000 °C ($a = 1.36 \times 10^5$ Pa L^2 mol^{-2} and $b = 2.79 \times 10^{-3}$ L mol^{-1}). (6)
2. At constant pressure, how much heat is needed to raise the temperature of 4.0 mol of methane (CH_4) from 300 K to 1000 K? Use $C_P = a + bT$, where $a = 14.15$ and $b = 75.5 \times 10^{-3}$. (6)
3. When 1.24 g of liquid benzene is burned completely at 25 °C in oxygen in a typical bomb calorimeter, 12.40 kcal of energy is released. Find the enthalpy of combustion. (8)
4. When 2 mol of an ideal gas ($C_V = \frac{5}{2}R$), initially at STP, was heated to 400 K, the volume increased to 100 L. What was the change in entropy? (6)
5. The pressure on one mole of an ideal gas, initially at 400 K and 4 atm, is reduced reversibly and isothermally to 2 atm. Find the values of q, w, ΔE, ΔH, ΔS, ΔA, and ΔG. (10)
6. If 100 g of liquid benzene at 1 atm and 353.4 K is converted to vapor at the same temperature and pressure, what are ΔH, ΔS, and ΔG for the process? (8)

7. CALCULATE. If you bubble ethylene gas through water at 298.2 K, is it possible that ethyl alcohol will be formed by the reaction

$$C_2H_4(g) + H_2O(l) \longrightarrow C_2H_5OH(l)?$$

Show calculations to support your answer. (8)

8. For the gaseous reaction $2\,NO \rightleftharpoons N_2 + O_2$, $K_c = 2.25 \times 10^2$. Find the concentration of N_2 at equilibrium if 3.0 mol of NO is placed in a 100-L vessel. (8)

9. Find K_p at 298.2 K for the reaction

$$2H_2O_2(l) \rightleftharpoons 2H_2O(l) + O_2(g) \qquad (8)$$

10. At 298.2 K, $K_p = 5.0 \times 10^{17}$ for the reaction $C_2H_4\,(g) + H_2\,(g) \rightleftharpoons C_2H_6\,(g)$. Estimate an approximate K_p at 400 K. (8)

11. Remembering that the vapor pressure of water is 760 mm at its boiling point, find the vapor pressure at 85 °C, assuming that ΔH_v remains constant over this temperature range. (7)

12. If the Henry's-law constant for oxygen in water at 25 °C is $2.3 \times 10^{-5}\ \text{atm}^{-1}$, what is the concentration of oxygen in a solution when the partial pressure of oxygen is 0.60 atm? (4)

13. When 1.35 g of an unknown solid is dissolved in 100 g of benzene, the resulting solution freezes at 278.18 K. What is the molecular weight of the unknown? (8)

14. CALCULATE. Determine the mole fraction of toluene ($CH_3C_6H_5$) in the vapor phase in equilibrium with a solution containing 30.0 g of benzene and 70.0 g of toluene at 30 °C. The vapor pressures at 30 °C are benzene, 118.2 mm Hg, and toluene, 36.7 mm Hg. (10)

Solutions to Final Exam

A. Short Answers

1. $8.314\ J\ K^{-1}\ mol^{-1}$ ($Pa\ dm^3\ K^{-1}\ mol^{-1}$) [Table 2-1]
2. NH_3 [by Graham's law, Eq. (2-11)]
3. temperature [Section 2-7, see especially Eq. (2-40)]
4. 1.01×10^5 Pa [Section 1-2C]
5. Negative, because work is done by the system. [Section 3-3D]
6. entropy [Section 5-2C]
7. $\alpha = \left(\frac{1}{V}\right)\left(\frac{\partial V}{\partial T}\right)_P$ [Section 6-3C, Eq. (6-36)]
8. Volume [Eq. (6-40)]
9. $\mu_i = \left(\frac{\partial G}{\partial n_i}\right)_{T,P,n_j}$ [Section 6-8, Eq. (6-69)]
10. $\sum_i \nu_i \mu_i = 0$ [Section 7-1, Eq. (7-5)]
11. allotropes [Section 8-1A]
12. $88\ J\ K^{-1}\ mol^{-1}$ or $21\ cal\ K^{-1}\ mol^{-1}$ [Section 8-1D; if you forgot, you could calculate it by using Trouton's rule, Eq. (8-5), and the tabulated data.]
13. azeotrope or azeotropic mixture [Section 9-5D]

B. Long Answers

1. The three properties are (**a**) the line integral is independent of path, (**b**) the cyclic integral is zero, and (**c**) a function exists which can be differentiated to give the differential [Section 1-4E].

2. Corresponding states are the conditions of pressure, volume, and temperature under which the reduced variables for each of the gases are the same. In different words, the gases obey the same equation of state if their pressure, volume, and temperature are expressed as reduced variables [Section 2-5B].

3. The smaller atom, A, has the longer mean free path. In the equation for mean free path, Eq. (2-45), the collision cross-section, σ, appears in the denominator. Since σ is a function of radius, a smaller radius makes the denominator smaller and the fraction larger [Section 2-6C].

4. The specific heat of a substance is the amount of heat needed to raise the temperature of 1 gram of the substance by 1 °C. Heat capacity is the amount of heat needed to raise the temperature of 1 mole of the substance by 1 °C [Sections 3-3 and 3-5B].

5. First: It is impossible to create energy from nothing [Section 3-2B].
   Second: It is impossible to convert heat into work in a cyclic process without creating some other effect. Or: It is impossible to transfer heat from a cold to a warm region without doing work on a system [Section 5-1].
   Third: It is impossible to reach the absolute zero of temperature [Section 5-5A].

6. For both cases, start with Eq. (3-11), $w = -\int P_{ext}\, dV$ [Sections 3-3D and E].
   (**a**) At constant pressure, $P_{ext} = P_{op} =$ constant, so
   $$w = -P(V_2 - V_1)$$
   (**b**) For a reversible expansion, $P_{ext} = P_{int} = nRT/V$, so
   $$w = -nRT \int \frac{dV}{V} = -nRT \ln \frac{V_2}{V_1}$$

7. To indicate spontaneity, ΔS is satisfactory in an isolated system, but an isolated system implies that E is constant, which is not usually a condition of chemical change. At constant T and V, ΔA always is equal to or less than zero in a spontaneous change, and the same is true of ΔG at constant T and P [Sections 6-1 and 6-2].

8. In a reversible process, ΔA equals the maximum work, or we could say that maximum work is a thermodynamic property which we designate ΔA. The maximum work often includes work of expansion, $\Delta(PV)$, which cannot be harnessed to run an engine. The remaining work (non-PV) also is a thermodynamic property and is given the symbol ΔG: $\Delta G =$ non-PV work. In a reversible process, ΔG is the maximum non-PV work. [Section 6-2]

9. Apply the Euler reciprocity theorem [Section 1-4F] to the Gibbs equation, $dA = -S\, dT - P\, dV$ [Eq. (6-34)]. The theorem tells you that the partial derivative of the coefficient of the first variable with respect to the second variable equals the partial derivative of the second coefficient with respect to the first variable. In this case
   $$\left(\frac{\partial(-S)}{\partial V}\right)_T = \left(\frac{\partial(-P)}{\partial T}\right)_V$$
   or
   $$\left(\frac{\partial S}{\partial V}\right)_T = \left(\frac{\partial P}{\partial T}\right)_V$$
   which is the corresponding Maxwell relation, Eq. (6-45) [Section 6-3].

10. The degree of dissociation is $\alpha = x/C_0$ so $x = \alpha C_0$. You can use the usual procedure of setting up a table to get expressions for equilibrium concentrations [Section 7-3D].

	AB_2	$\rightleftharpoons$	A	+	2B
Start	C_0		0		0
Change	$-\alpha C_0$		$+\alpha C_0$		$+2\alpha C_0$
Equilibrium	$C_0(1-\alpha)$		αC_0		$2\alpha C_0$

$$K_c = \frac{[A][B]^2}{[AB_2]} = \frac{\alpha C_0(2\alpha C_0)^2}{C_0(1-\alpha)} = \frac{4\alpha^3}{1-\alpha} C_0^2$$

11. The values in the table are of ΔH_f and ΔH_v at the normal melting and boiling points, which are different temperatures. Since ΔH depends on T, ΔH_s cannot be calculated by adding the tabulated values [Section 8-1C].

12. The phase diagram of CO_2 has the triple point above a pressure of 1 atm, and the melting-point curve should slope to the right as it rises. Your sketch should look like Figure 8-4. At a constant pressure of 1 atm, as the temperature increases, a system consisting of solid CO_2 reaches the sublimation-point curve, hence the solid sublimes. The system becomes vapor without liquid appearing [Section 8-2C2].

13. The Clapeyron equation, Eq. (8-10), is

$$\frac{dP}{dT} = \frac{\Delta H_{\text{trans}}}{T_{\text{trans}}(\Delta V)}$$

When it is applied to the liquid–gas or solid–gas transitions, we can assume
(1) The volume of liquid or solid is negligible compared to the volume of gas, so $\Delta V = V_g - V_l = V_g$ and $\Delta V = V_g - V_s = V_s$.
(2) The gas is an ideal gas, so $V_g = RT/P$ if $n = 1$ mol [Section 8-4A].
Now

$$\frac{dP}{dT} = \frac{\Delta H}{T}\left(\frac{P}{RT}\right)$$

Since $\frac{dP}{P} = d \ln P$, we can rearrange to

$$\frac{d \ln P}{dT} = \frac{\Delta H}{RT^2}$$

which is the Clausius–Clapeyron equation, Eq. (8-13).

14. All of the colligative effects are due to the fact that the chemical potential of a solution is less than that of the pure solvent. Boiling occurs when μ (solution) = μ (vapor). At the normal boiling point, μ (solvent) = μ (vapor). Since μ (solution) < μ (solvent), the temperature must be raised to increase μ (solution) until it equals μ (vapor), which has remained constant [Section 9-4B].

15. Your pressure–composition diagram should resemble Figure E-1. The essential feature is that the liquid-composition curve and the vapor-composition curve start at the vapor pressures of pure A and B and meet at a maximum that lies above the vapor pressure of either pure substance. The cause of a maximum, i.e., a positive deviation, is that the attractive force between molecules of A and B is less than that between molecules of A and A or between B and B [Section 9-5B].

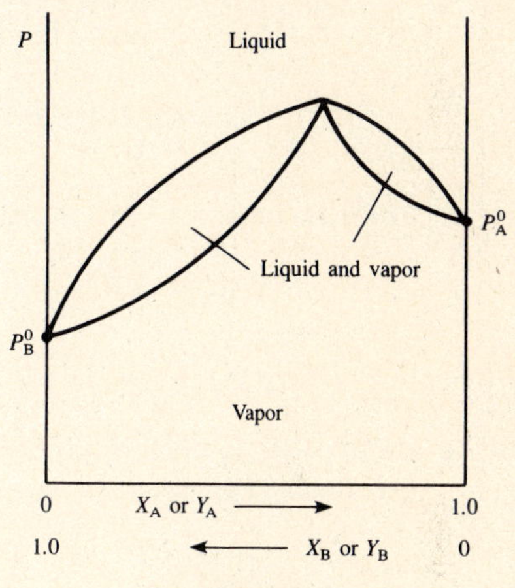

FIGURE E-1

C. Problems

1. Solve the van der Waals equation, Eq. (2-13), for P [Section 2-5A].

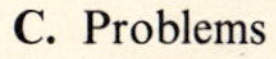

$$\left(P + \frac{an^2}{V^2}\right)(V - nb) = nRT$$

$$P = \frac{nRT}{V - nb} - \frac{an^2}{V^2}$$

Convert T to kelvins and n to moles.

$$1000\ ^\circ\text{C} = 1273\ \text{K}$$

$$n = \frac{100\ \text{g}}{30\ \text{g mol}^{-1}} = \frac{100}{30}\ \text{mol}$$

$$P = \left[\frac{\frac{100}{30}(8314)(1273)}{10 - \frac{100}{30}(2.79 \times 10^{-3})} - \frac{1.36 \times 10^5\left(\frac{100}{30}\right)^2}{(10)^2}\right] \text{Pa}$$

2. To find the amount of heat q_P, integrate the relationship between heat capacity, change in temperature, and change in heat, $dH = dq_P = C_P\, dT$ [Eq. (3-30)]. Substitute the expression for C_P as a function of T and include it in the integration [see Section 3-7A].

$$q_P = \int_{T_1}^{T_2} C_P\, dT = \int_{T_1}^{T_2} (a + bT)\, dT = a\int_{T_1}^{T_2} dT + b\int_{T_1}^{T_2} T\, dT = a(T_2 - T_1) + \frac{b}{2}(T_2^2 - T_1^2)$$

$$= \left[14.15(1000 - 300) + \frac{75.5 \times 10^{-3}}{2}(1000^2 - 300^2)\right] \text{J}$$

3. The measured energy per mole is ΔE, and you want the change in enthalpy ΔH, which you calculate from $\Delta H = \Delta E + RT\, \Delta n$ [Eq. (3-24)]. Convert the measured value of ΔE to kilocalories per mole:

$$\Delta E = 12.4\ \text{kcal}\left(\frac{78\ \text{g mol}^{-1}}{1.24\ \text{g}}\right) = 780\ \text{kcal mol}^{-1}$$

You need the balanced equation to determine Δn [see Section 4-1B].

$$C_6H_6(l) + \tfrac{15}{2}O_2(g) \longrightarrow 6CO_2(g) + 3H_2O(g)$$

Remember to count only gaseous substances when you calculate Δn.

$$\Delta n_g = n_g\,(\text{products}) - n_g\,(\text{reactants}) = 9 - 7.5 = 1.5$$

Now you can substitute values in the equation for ΔH.

$$\Delta H = [780 \times 10^3 + (1.987)(298)(1.5)]\ \text{cal}$$

4. You could start with Eq. (5-12) and integrate it to derive Eq. (5-14),

$$\Delta S = nC_V \ln \frac{T_2}{T_1} + nR \ln \frac{V_2}{V_1}$$

Once you arrive at this equation, substitute into it. The problem gives values for all variables except V_1, but you should remember that the volume of one mole of an ideal gas at STP is 22.4 L. For two moles, $V_1 = 44.8$ L.

$$\Delta S = \left[2(\tfrac{5}{2}R)\ln\frac{400}{273} + 2R\ln\frac{100}{44.8}\right]\text{J K}^{-1}$$

5. This is a problem in which you don't want to calculate the quantities in the order they appear in the question. Instead, think of what the conditions tell you. Since this is an isothermal ideal-gas process, ΔE and $\Delta H = 0$ [Section 3-6A]. Therefore $q = -w$. For a reversible isothermal process with an ideal gas,

$$w = -nRT\ln\frac{V_2}{V_1} = nRT\ln\frac{P_2}{P_1} = (1)(8.314)(400)\ln(\tfrac{2}{4})\ \text{J} \qquad \text{[Eq. (3-40)]}$$

$$q = -(8.314)(400)\ln(\tfrac{1}{2})\ \text{J}$$

Also

$$w = \Delta A = \Delta G = (8.314)(400)\ln(\tfrac{1}{2})\ \text{J}$$

The reason that $\Delta A = \Delta G$ is that $\Delta(PV) = 0$ [Eq. (6-30)] or you may have memorized Eq. (6-49):

$$\Delta G = nRT\ln\frac{P_2}{P_1} \text{ at constant } T.$$

$$\Delta S = \frac{q}{T} = \frac{-w}{T} = -\frac{nRT}{T}\ln\frac{P_2}{P_1} = (-8.314)\ln(\tfrac{1}{2})\ \text{J K}^{-1} \qquad \text{[Eq. (5-4)]}$$

6. From the data provided, ΔH_v of benzene = 30.77 kJ mol^{-1} and the normal boiling point is 353.4 K. One hundred grams of benzene is equivalent to

$$\frac{100\ \text{g}}{78\ \text{g mol}^{-1}} = \frac{100}{78}\ \text{mol}$$

$$q = \Delta H = n\,\Delta H_v = \frac{100}{78}(30.77)\ \text{kJ}$$

$$\Delta S = \frac{\Delta H}{T} = \frac{100}{78}(30.77)\left(\frac{1}{353.4}\right)\text{J K}^{-1} \qquad \text{[Eq. (5-16)]}$$

$$\Delta G = \Delta H - T\,\Delta S = \Delta H - T\frac{\Delta H}{T} = 0 \qquad \text{[Eq. (6-29)]}$$

Probably you wrote $\Delta G = 0$ immediately, because evaporation at the normal boiling point is an equilibrium process.

7. You should calculate ΔG° for the reaction to learn what its sign is [Section 6-5].

$$\Delta G = \Sigma n\,\Delta G_f^\circ(\text{products}) - \Sigma n\,\Delta G_f^\circ(\text{reactants}) \qquad \text{[Eq. (6-52)]}$$

Now substitute values from the table. All $n = 1$, so

$$\Delta G = \Delta G_f^\circ(C_2H_5OH, l) - \Delta G_f^\circ(C_2H_4, g) - \Delta G_f^\circ(H_2O, l) = -174.8 - 68.2 - (-237.1)$$
$$= -243.0 + 237.1 = -5.9\ \text{kJ}$$

Since ΔG is negative, the reaction is spontaneous. There must be other factors involved, since this reaction is not in wide use for producing alcohol.

8. Write the balanced equation and construct a table of terms to represent the equilibrium concentrations [Section 7-3B]. As in Example 7-20 and Problem 7-9, the sum of the coefficients on each side of the equation is equal, so all volume terms in the expression for K_c cancel, and you can use moles in the table instead of concentrations. Then, as the final step, divide by the volume to get the concentration of N_2.

	$2\,NO$	$\rightleftharpoons$ N_2	+ O_2
at start	3	0	0
change	$-2x$	$+x$	$+x$
at equilibrium	$3-2x$	x	x

$$K_c = 2.25 \times 10^2 = \frac{[N_2][O_2]}{[NO]^2} = \frac{(x)(x)}{(3-2x)^2}$$

Since K_c is large, x cannot be neglected, so

$$x^2 = 225(9 - 12x + 4x^2) = 900x^2 - 2700x + 2025 \qquad 899x^2 - 2700x + 2025 = 0$$

Solve for x by using the quadratic formula:

$$x = \frac{+2700 \pm \sqrt{(-2700)^2 - 4(899)(2025)}}{2(899)}\ \text{mol} \qquad [N_2] = \frac{x\ \text{mol}}{100\ \text{L}} = x \times 10^{-2}\ M$$

9. Determine ΔG, then solve Eq. (7-22), $\Delta G^\circ = -RT \ln K_P$, for K_P [Section 7-4B]. Since ΔG_f° for elements = 0, ΔG_f° for O_2 drops out, leaving

$$\Delta G = 2\,\Delta G_f^\circ(H_2O, l) - 2\,\Delta G_f^\circ(H_2O_2, l) = 2(-237.1) - 2(-120.4) = -116.7\ \text{kJ}$$

$$\ln K_P = -\frac{\Delta G^\circ}{RT} = -\frac{-116.7 \times 10^3}{(8.314)(298.2)}$$

$$K_P = \text{antiln}\, \frac{116.7 \times 10^3}{(8.314)(298.2)}$$

10. The dependence of K_P on T involves ΔH [Section 7-4D]. Since the temperature range is small, you can assume that ΔH is constant and use Eq. (7-31),

$$\ln \frac{K_{P_2}}{K_{P_1}} = \frac{\Delta H^\circ}{R}\left(\frac{T_2 - T_1}{T_2 T_1}\right)$$

You must determine $\Delta H^\circ(298)$ from the data provided

$$\Delta H = \Sigma n\, \Delta H_f^\circ(\text{products}) - \Sigma n\, \Delta H_f^\circ(\text{reactants}) \qquad \text{[Eq. (4-5)]}$$

All $n = 1$ and ΔH_f° of $H_2 = 0$, so

$$\Delta H = \Delta H_f^\circ(C_2H_6, g) - \Delta H_f^\circ(C_2H_4, g) = -84.7 - 52.3 = -142.0\ \text{kJ}$$

$$\ln K_{P_2} - \ln(5.0 \times 10^{17}) = \frac{-142 \times 10^3}{8.314}\left(\frac{400 - 298}{400(298)}\right)$$

$$K_{P_2} = \text{antiln}\left[-\frac{142 \times 10^3}{8.314}\left(\frac{400 - 298}{400(298)}\right) + \ln(5.0 \times 10^{17})\right]$$

11. You need to use the integrated form of the Clausius–Clapeyron equation, Eq. (8-15) or (8-16). (Perhaps you noticed the similarity to the previous problem!)

$$\ln \frac{P_2}{P_1} = \frac{\Delta H}{R}\left(\frac{T_2 - T_1}{T_2 T_1}\right) \qquad \text{[Eq. (8-16)]}$$

Let the lower P and T be P_1 and T_1. Substituting, you get

$$\ln 760 - \ln P_1 = \frac{40.63 \times 10^3}{8.314}\left(\frac{373.2 - 358.2}{373.2(358.2)}\right)$$

$$P_1 = \text{antiln}\left[\ln 760 - \frac{40.63 \times 10^3}{8.314}\left(\frac{15}{(373.2)(358.2)}\right)\right]\ \text{mm Hg}$$

12. From the units of k, you should recognize that the concentration is in mole fractions. Then solve Henry's law, Eq. (9-9), for the concentration of oxygen.

$$X_2 = kP_2 = (2.3 \times 10^{-5}\ \text{atm}^{-1})(0.60\ \text{atm}) = 1.4 \times 10^{-5}$$

13. If you remember Eq. (9-17), $M_{m,2} = K_f W_2/\Delta T_f W_1$, substitute into it directly. From the table, for benzene, $K_f = 5.12$ K kg mol^{-1} and $T_f^\circ = 278.7$ K. Then

$$M_{m,2} = \frac{(5.12)(1.35)}{(278.7 - 278.18)(100)}\ \text{kg mol}^{-1}$$

An alternative method is to solve for m in $\Delta T_f = K_f m$ [Eq. (9-15)], then solve for $M_{m,2}$ from $m = \dfrac{n}{W_1} = \dfrac{W_2/M_{m,2}}{W_1}$, as in Example 9-14.

14. First find the number of moles of both benzene and toluene. Then, for each, solve Raoult's law, Eq. (9-11), for the respective vapor pressures. The mole fraction of toluene in the vapor is equal to the fraction of the vapor pressure due to toluene [Section 9-5A2].

$$n\,(\text{benzene}) = \frac{30\text{ g}}{78\text{ g mol}^{-1}} = 0.385\text{ mol}$$

$$n\,(\text{toluene}) = \frac{70\text{ g}}{92\text{ g mol}^{-1}} = 0.761\text{ mol}$$

$$\text{total } n = 1.146\text{ mol}$$

$$P\,(\text{benzene}) = X_b P_b^0 = \frac{0.385}{1.146}(118.2) = 103.1\text{ mm Hg}$$

$$P\,(\text{toluene}) = X_t P_t^0 = \frac{0.761}{1.146}(36.7) = 24.4\text{ mm Hg}$$

$$\text{total } P = 127.5\text{ mm Hg}$$

$$X_t \text{ in vapor} = \frac{P_t}{P_{\text{total}}} = \frac{24.4}{127.5} = 0.19$$

10 PHYSICAL EQUILIBRIUM: LIQUID AND SOLID MIXTURES

THIS CHAPTER IS ABOUT

- ☑ The Phase Rule
- ☑ Partially Miscible Liquids
- ☑ Solubility of Solids in Liquids
- ☑ Solid–Liquid Mixtures: Simple Case
- ☑ Solid–Liquid Mixtures: Formation of a Compound
- ☑ Solid–Solid Solutions

10-1. The Phase Rule

A. Terminology

The Gibbs phase rule expresses the relationships among three characteristics of a system: the number of phases, the number of components, and the number of degrees of freedom. Although *phase* has been introduced already, let's define our terms:

- A **phase** is a physically distinct, mechanically separable, homogeneous portion of a system.
- The **number of components** is the smallest number of independently variable chemical species needed to describe the chemical composition of each phase.
- The **number of degrees of freedom** is the smallest number of intensive variables that must be specified in order to fix the values of all the intensive variables, i.e., the number of independent intensive variables.

EXAMPLE 10-1: In the system Na_2SO_4–H_2O, many different phases are possible, e.g., liquid water/water vapor, a variety of solutions, or several solid phases ($Na_2SO_4 \cdot 7H_2O$/$Na_2SO_4 \cdot 10H_2O$/Na_2SO_4). What is the number of components?

Solution: We can describe some of these phases by one substance, e.g., water; but others require two, e.g., for a solution, we need the amounts of water and Na_2SO_4. None, however, requires more than water and Na_2SO_4, so the number of components is two.

EXAMPLE 10-2: The usual independent variables are concentration, pressure, and temperature. Refer to Figure 10-1, which is the phase diagram for the one-component system, water (presented earlier as Figure 8-3). How many degrees of freedom does the system have when its state is (**a**) in an area of the diagram, (**b**) on a line, and (**c**) at the triple point?

Solution: Concentration isn't a variable when there's only one component, so we have only pressure and temperature.

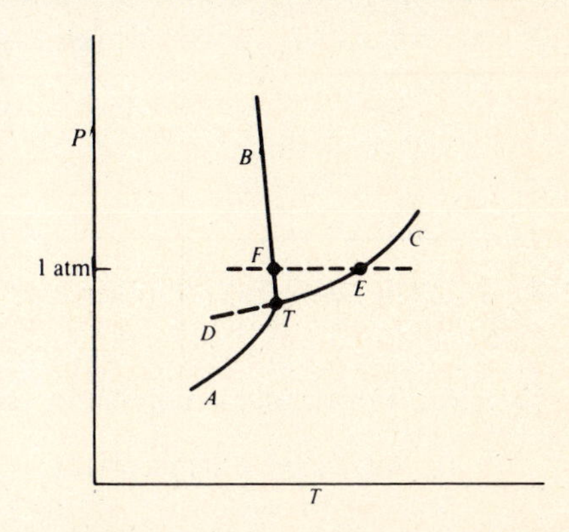

FIGURE 10-1. Phase diagram of water.

(**a**) There's only one phase in each area of the diagram. In order to plot a point that describes the system completely, we have to know both P and T. Hence there are two degrees of freedom, or $f = 2$.

(b) On a line, two phases are in equilibrium, so we can describe the system by specifying either P or T. Therefore $f = 1$.

(c) At the triple point, three phases are present. This condition means that we don't have to specify either P or T, because they can have only one possible set of values. In this case $f = 0$.

The term *variance* is often used to express degrees of freedom. In Example 10-2 we can describe the system in **(a)** as *bivariant* ($f = 2$); that in **(b)** as *monovariant* or *univariant* ($f = 1$); and that in **(c)** as *invariant* ($f = 0$).

B. Derivation of the phase rule

Derivation of the phase rule is based on a system of c components in p phases, with each component appearing in every phase. The total possible number of variables doesn't need to be specified; instead, by considering restrictive relationships between variables (i.e., some are dependent on others), we can obtain an expression for the smallest number of independent variables that must be specified. The total number t is

$$t = pc + 2 \tag{10-1}$$

where the 2 represents pressure and temperature.

Restriction 1: For a given phase, if we know the mole fractions of all but one of the components, we can obtain that one by subtraction from 1. Thus, if one variable in each phase is dependent, or if p variables for all phases are dependent, we can subtract p variables from the total to get the number of independent variables t' for the phase:

$$t' = pc + 2 - p \tag{10-2}$$

Restriction 2: For equilibrium, the chemical potentials of a component in every phase must be equal, as previously expressed by Eq. (6-75). For a given component, there are $p - 1$ equalities. For c components, there are $c(p - 1)$ equalities, which means that $c(p - 1)$ variables are dependent and may be subtracted from the phase total to get the smallest number of variables t'':

$$t'' = pc + 2 - p - c(p - 1) = pc + 2 - p - pc + c \tag{10-3}$$

Since t'' represents the smallest number of variables, it equals f, and we have the **Gibbs phase rule**, i.e.,

GIBBS PHASE RULE

$$f = c - p + 2 \tag{10-4}$$

C. Application of the phase rule

The phase rule tells us the number of independent variables that we have to specify in order to describe a system. This rule can help us to analyze or construct phase diagrams for complex systems or to decide how to plot data. Usually, we know the number of components present, so we can classify systems as one-, two-, or three-component systems. We can then restate the phase rule:

For $c = 1$, $f = 3 - p$; for $c = 2$, $f = 4 - p$; for $c = 3$, $f = 5 - p$

EXAMPLE 10-3: Take another look at Example 10-2. Do the conclusions there agree with the phase rule?

Solution: For the one-component system, $f = 3 - p$.

(a) In an area, there is one phase, so $f = 3 - 1 = 2$. This tells us that both P and T are needed to describe the system.

(b) On a line, there are two phases, so $f = 3 - 2 = 1$. Either P or T describes the system; the other variable is now dependent.
(c) At a point, there are three phases, so $f = 3 - 3 = 0$. The system is invariant; P and T can have only one set of values.

Thus the answer in all three cases is *yes*.

EXAMPLE 10-4: What does the phase rule tell us about plotting a phase diagram for a two-component system?

Solution: Because $c = 2$, Eq. (10-4) becomes $f = 4 - p$. There must be at least one phase, so $f = 3$ when the system is entirely in one phase. Plotting a phase diagram for this system requires a three-dimensional graph, which is inconvenient. We can by-pass this requirement by holding one variable constant, usually either P or T. Then $f = 2$, and we can plot the two variables on a two-dimensional graph.

The method of plotting suggested in Example 10-4 was used to obtain the diagrams in Chapter 9. Recall the pressure–composition diagrams at constant T (Figure 9-2) and temperature–composition diagrams at constant P (Figure 9-4). Those diagrams actually are cross-sectional slices through three-dimensional graphs, which amounts to holding one degree of freedom constant. In many textbooks you'll see one or more figures showing a three-dimensional phase diagram and indicating where these slices are made.

EXAMPLE 10-5: **(a)** What difficulty arises if we try to plot a phase diagram for a three-component system? **(b)** What can we do about it?

Solution:

(a) Since $c = 3$, Eq. (10-4) becomes $f = 5 - p$, which would require a four-dimensional graph for one phase.
(b) If we hold P and T constant, then $f = 2$, permitting use of a two-dimensional graph. The remaining variables, however, are the three concentrations. We can plot all three by using a graph in the shape of an equilateral triangle, letting each side of the triangle represent 0 to 100% of one component. Further details are given in most textbooks. We use only two-component systems in this Outline.

10-2. Partially Miscible Liquids

A. Liquid phase only

If the pressure on a system of two liquids is high enough, no vapor will form, and we can consider only the liquid phase. Two liquids A and B may mix completely when only a small amount of one is present; but when their mole fractions X_A and X_B are closer to equal, the liquids form two solutions, which appear as two layers in a container. Such a system is shown in Figure 10-2. To the left of the curve, where X_B is small, and to the right, where X_A is small, there is one liquid phase. Within the curve, where X_A is closer to X_B, there are two liquid phases, having compositions corresponding to the ends of a tie line such as the one shown at T_a. These phases at a and a' are called **conjugate solutions**.

At higher temperatures, miscibility increases until only one phase remains. The maximum of the curve (point C) is called the **upper consolute temperature**, or the *critical solution temperature*, which occurs when the components of the mixture approach their critical states at similar temperature and pressure. Some systems will have curves that are the inverse of that in Figure 10-2, although they are less common. (Such systems usually result from weak interactions between unlike particles.) The minimum point in their curves is called the **lower consolute temperature**. For other systems, the curve forms a loop, which means that they have both upper and lower consolute temperatures. Many cases of partially miscible systems

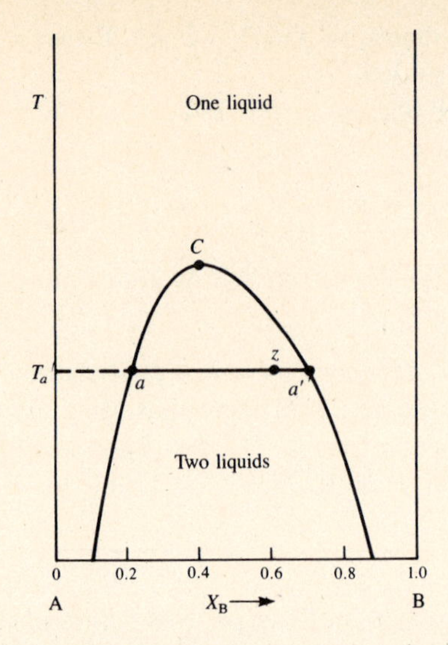

FIGURE 10-2. Two partially miscible liquids at constant high pressure.

involve water and an organic compound. Examples of the three types of systems just mentioned are

(1) Upper consolute temperature: water–phenol, water–aniline
(2) Lower consolute temperature: water–triethylamine, water–diethylamine
(3) Loop: water–β-picoline, water–nicotine

EXAMPLE 10-6: Refer to Figure 10-2. **(a)** What would be the compositions of the conjugate solutions at T_a if we mix 6.0 mol of B with 4.0 mol of A? **(b)** What are their relative amounts?

Solution:

(a) The only thing we need to know about the actual overall composition is that it falls between a and a'. Why? Because, since P is constant, we can simplify the phase rule to $f = c - p + 1$. Then, since $c = 2$ and $p = 2$, $f = 1$. When the temperature is given, nothing else can vary. At T_a, the layers will have compositions a and a' for all overall compositions that fall between these two points. We can estimate from the graph that the mole fractions are about 0.22 at a and 0.71 at a'.

(b) Now we *do* need the overall composition in order to determine the segments of the tie line and to use the lever rule. Clearly, $X_B = 0.60$, which is designated z on the tie line. From Eq. (9-27),

$$\frac{\text{amount of } a'}{\text{amount of } a} = \frac{za}{za'} = \frac{0.60 - 0.22}{0.71 - 0.60} = \frac{0.38}{0.11}$$

$$= 3.5$$

B. Liquid and vapor phases

If the pressure on two liquids is reduced to the point at which vapor appears, we can represent the situation by the diagram in Figure 10-3. In Figure 10-3, the top of Figure 10-2 has been reproduced, and a boiling-point diagram has been added. This form is typical for a minimum-boiling azeotrope, which we described but didn't illustrate in Section 9-5A. If the pressure is reduced further, the boiling points decrease. At some pressure, the curves just meet; and, at even lower pressures, the system can be represented by a phase diagram like that in Figure 10-4. Below the horizontal line *mnp*, we interpret the diagram as we do Figure 10-2; above the line, each loop is like that shown in Figure 9-4. The horizontal line *mnp* is the interesting new feature.

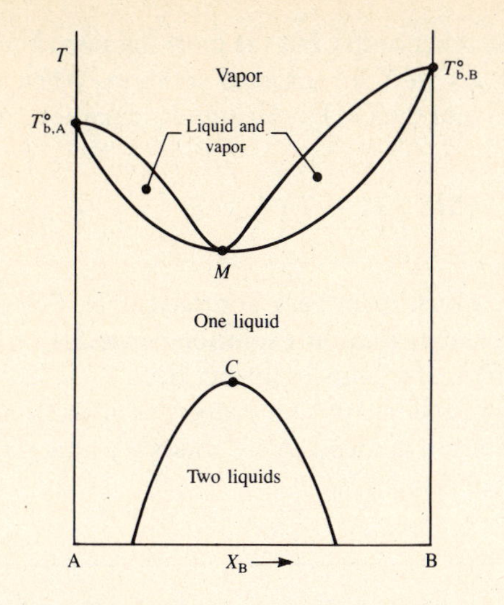

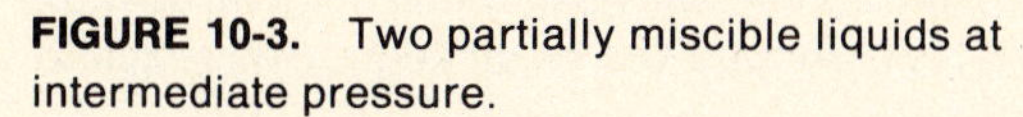
FIGURE 10-3. Two partially miscible liquids at intermediate pressure.

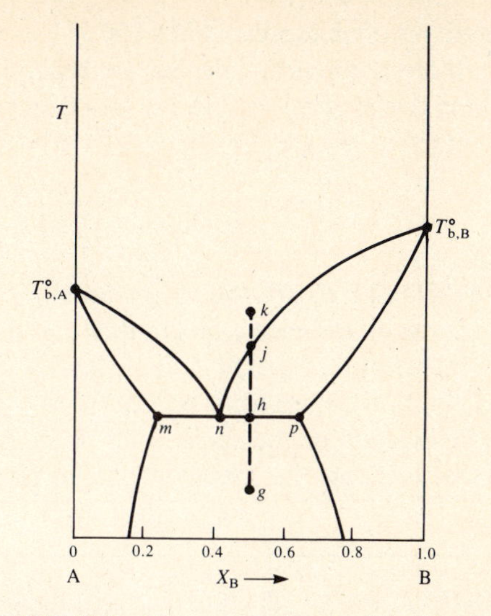

FIGURE 10-4. Two partially miscible liquids at low pressure.

EXAMPLE 10-7: Describe what phases are present and what will happen if a system of two liquids, represented by Figure 10-4 and having an overall composition of $X_B = 0.50$, is slowly warmed from temperature g to temperature k at constant pressure.

Solution: At T_g, there are two liquid phases (two solutions); one is rich in B, the other, rich in A. The situation is like that in Example 10-6. As T rises, the composition of each liquid solution changes slightly, becoming m and p at T_h. Line mp, however, contains the third point n, which represents vapor. At T_h, a vapor phase appears and it has composition n; the system has begun to boil. Now three phases are present, P is constant, and $c = 2$, so Eq. (10-4) tells us that $f = 0$; i.e., the system is at an invariant point, analogous to the triple point of a one-component system. The difference is that the three phases in equilibrium are represented by three points instead of a single point.

In order for the temperature to rise, one of the phases must disappear. But which one? Since the tie line from h goes to n and p, then m must disappear, i.e., the solution rich in A with composition m. At temperatures between T_h and T_j, solution rich in B will be in equilibrium with a vapor phase that is richer in A than is the solution. At T_j the last of the solution will disappear, and above T_j there will be only vapor, having composition $X_B = 0.5$.

10-3. Solubility of Solids in Liquids

note: The topic of solubility of solids in liquids is treated with colligative properties in many textbooks. We include it here as an introduction to solid–liquid mixtures, although we'll limit our discussion to nonelectrolytes.

Previously, we dealt with solutions of nonvolatile substances in liquids from the point of view of the solvent, i.e., the liquid. In describing colligative properties, we said that solvent in solution is in equilibrium with solvent in a pure state; for freezing-point depression, it is the solid state. Now we look at the solute and, analogously, consider the situation where solute in solution is in equilibrium with solute in the solid state.

The term **solubility** refers to a characteristic of a solute in a saturated solution. The solubility of a substance at a particular temperature is the concentration of a saturated solution at that temperature. To ensure that the solution is saturated, it should be in contact with solid solute long enough for equilibrium to be established, or

$$\text{solute (in solution)} \rightleftharpoons \text{solute (solid)}$$

Then

$$\mu(\text{in solution}) = \mu(\text{solid}) \tag{10-5}$$

Solubility is often expressed in g 100 g^{-1} solvent (and in other units), but the mole fraction should be used for theoretical calculations because $\mu = \mu^\circ + RT \ln X$ for an ideal solution. When we combine this expression with Eq. (10-5), bringing in ΔG and the Gibbs–Helmholtz equation, and then integrate, we get

$$\ln X_2 = -\frac{\Delta H_f}{R}\left(\frac{1}{T} - \frac{1}{T_f^\circ}\right) = -\frac{\Delta H_f}{R}\left(\frac{T_f^\circ - T}{T_f^\circ T}\right) \quad \textbf{(10-6)}$$

where ΔH_f and T_f° are the values for the pure solute, which must be a nonelectrolyte. For the integration, we assumed a constant ΔH_f, but we haven't assumed a dilute solution. Since ΔH varies with temperature, the expression is less valid as the temperature moves further from T_f°.

Qualitatively, we see from Eq. (10-6) that solubility depends only on properties of the solute and thus should be the same in all solvents (if the solution is ideal). Also, solubility at a given temperature should be greater for solutes having lower T_f° and ΔH_f.

EXAMPLE 10-8: Calculate the theoretical solubility of biphenyl $(C_6H_5)_2$ at 25 °C. The melting point is 70 °C, and the heat of fusion is 18.56 kJ mol^{-1}. Compare the result to an experimental value of 0.39 for a saturated solution in benzene.

Solution: We simply substitute into Eq. (10-6):

$$\ln X_2 = -\frac{\Delta H_f}{R}\left(\frac{1}{T} - \frac{1}{T_f^\circ}\right) = -\frac{18.56 \text{ kJ mol}^{-1}}{8.314 \text{ J K}^{-1}\text{mol}^{-1}}\left(\frac{343.2 \text{ K} - 298.2 \text{ K}}{(343.2 \text{ K})(298.2 \text{ K})}\right) = -0.982$$

$$X_2 = 0.375$$

$$\text{Deviation from experimental} = 0.39 - 0.375 = 0.015$$

$$\text{Percent of deviation} = \frac{0.015}{0.39}(100) = 3.8\%$$

The solubility of most solids increases with temperature, which is predictable from the fact that the process is usually endothermic (Le Chatelier's principle). Energy is required to overcome the crystal lattice energy. Pressure has a negligible effect on the solubility of solids.

10-4. Solid–Liquid Mixtures: Simple Case

A. Cooling curves

We usually obtain the phase diagrams for solid–liquid systems from **thermal analysis**: Mixtures of two solids having a range of compositions are heated above the higher melting point and are allowed to become completely intermingled. As each mixture cools slowly, the temperature is observed at small intervals of time. Typical plots of temperature versus time, called **cooling curves**, are shown in Figure 10-5; curves 1 and 6 are for the pure components, and curves 2–5 are for mixtures. In a real case, the sloping lines would be curved. Note that the horizontal axis is not continuous but begins at zero for each curve.

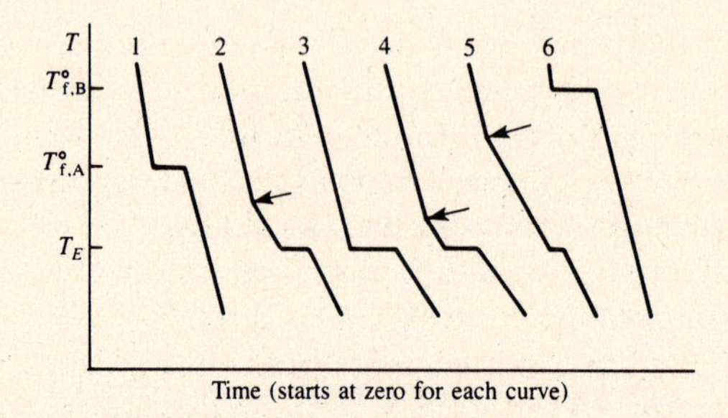

FIGURE 10-5. Cooling curves for a system with a simple eutectic point.

Features on the cooling curves correspond to features on the phase diagram and permit construction of the complete diagram. A horizontal portion, or *plateau* (slope = 0), indicates an invariant point. The temperature remains constant for a length of time (measured by the length of the plateau) until one phase has disappeared. A change in slope without a plateau, i.e., a *break* in the curve, indicates that a new phase has appeared. The break occurs because the system now has a different specific heat, and the rate of cooling changes. The breaks are shown by arrows in Figure 10-5.

B. Eutectic diagrams

The simplest phase diagram is one in which the solids are completely immiscible and the liquids are completely miscible. The diagram of an ideal system, constructed from the set of cooling curves in Figure 10-5, is shown in Figure 10-6. The interpretation of Figure 10-6 is analogous to that of Figure 10-4 (Example 10-7) but is simpler because the solid phases are pure substances. Actually, complete immiscibility of solids is unusual; we describe solid solutions in Section 10-6.

Let's consider how Figure 10-6 was derived from Figure 10-5. The number of each curve in Figure 10-5 is placed at the proper mole fraction in Figure 10-6. Clearly, the plateaus of curves 1 and 6 identify the freezing points of pure A and pure B. The plateaus of curves 2–5 represent another invariant point. Note that the time at this point is longest when the original solution has exactly the eutectic composition (curve 3). The breaks in curves 2, 4, and 5 give us the temperatures to be plotted at the corresponding mole fractions. We can then sketch in the curves representing the liquid–solid equilibria. The more cooling curves we have, the greater will be the accuracy of the curves in the eutectic diagram.

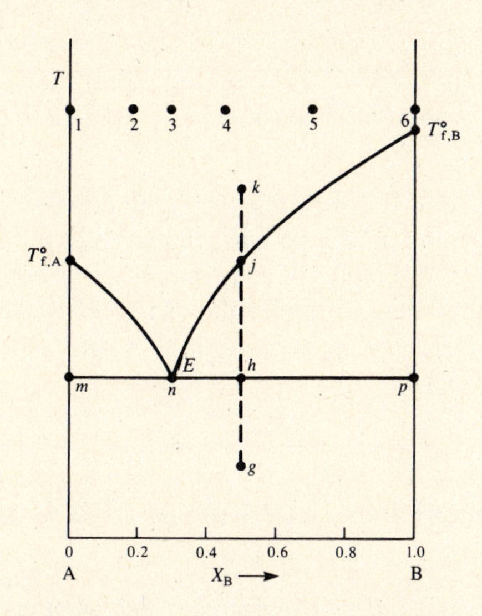

FIGURE 10-6. Simple eutectic diagram.

EXAMPLE 10-9: Describe what phases will be present and what will happen if a system of two solids, consisting of one mole of each and represented by Figure 10-6, is slowly warmed from temperature T_g to temperature T_k at constant pressure.

Solution: For one mole of each component, the mole fraction of each is 0.50, which is the concentration at which line *gk* is drawn. At temperature T_g, there are two solid phases (not solutions, but two pure substances), as shown by a tie line connecting the sides of the graph. When the temperature reaches T_h, a liquid phase appears, consisting of both components and having the composition of point *n* ($X_B = 0.30$). Since three phases are present, *P* is constant, and $c = 2$, Eq. (10-4) tells us that $f = 0$, and so we've identified an invariant point.

In order for the temperature to rise further, one phase must disappear. Since the tie line from *h* ends at *n* and *p*, the phase *m* must be consumed, i.e., solid A disappears. When the temperature is between T_h and T_j, liquid solution is in equilibrium with solid B. As the temperature rises, B slowly disappears, and the amount of liquid increases. At T_j the last bit of solid melts, and above T_j there is only liquid. This liquid phase is often called the *melt*. (The process as a whole is the reverse of a cooling curve starting at point *k*.)

The principal feature of Figure 10-6 is point *n*, which is the **eutectic point**, so it is also labeled point *E*. The temperature at point *n* is the **eutectic temperature** T_E, which is the lowest temperature at which the liquid can exist. The composition at *E* (0.30 in Figures 10-5 and 10-6) is the **eutectic composition**, and the time at T_E in a cooling curve is the **eutectic halt**. The curved lines in Figure 10-6 may be called (1) the solubility curves or (2) the freezing-point curves, depending on whether a component is considered to be the solute or the solvent. For instance, in Figure 10-6, consider component B to be the *solute*. Point *j* represents the solubility of B in A at T_j, i.e., the concentration of a saturated solution in equilibrium with undissolved solute (solid B). Any point on the curve from *n* to $T^\circ_{f,B}$ gives the solubility of B in A at the corresponding temperature. This solubility (concentration) may be calculated using Eq. (10-6). On the other hand, consider B in Figure 10-6 to be the *solvent*. If a solution at T_k ($X_B = 0.5$) is cooled, B will *freeze out* at T_j. The slope of the cooling curve for a mixture of $X_B = 0.5$ would change at this temperature. The curve from *n* to $T^\circ_{f,B}$ shows the depression of the freezing point of solvent B as solute A is added.

Solid substances whose mixtures may form simple eutectic systems include ionic and covalent compounds and elements. Some examples of such systems are

1. Elements, e.g., aluminum–tin, lead–antimony.
2. Ionic compounds, e.g., potassium chloride–silver chloride.
3. Covalent compounds, e.g., benzene–methyl chloride

EXAMPLE 10-10: A system of two metals is cooled from point *k* to point *g* in Figure 10-6. What does the system look like at T_g?

Solution: Between T_j and T_E, solid B is forming. At *E*, the remaining liquid, having the eutectic composition, solidifies. Usually, the solid is so fine-grained that it looks like a compound. Below T_E, solid A, as well as B, appears. The pieces of solid may be large and clearly distinguishable, and the observer probably thinks that there are three solid phases. However, there are only two phases, because the eutectic solid is merely a mixture of solid A and solid B. The different states of subdivision don't change the fact that the only substances present are A and B, in equimolar amounts.

10-5. Solid–Liquid Mixtures: Formation of a Compound

A. Congruent melting point

Sometimes two substances will form a compound which itself creates a eutectic mixture with each of the initial substances. The compounds formed in these systems do not often fit our usual view of a compound, but look like an addition complex. Examples are calcium chloride–potassium chloride ($CaCl_2 \cdot KCl$) and urea–phenol. More conventional formulas come from such systems as gold–tellurium ($AuTe_2$) and mercury–thallium ($TlHg_2$).

A typical phase diagram for a system forming a 1:1 compound is shown in Figure 10-7. The maximum at $X_B = 0.5$ represents the formation of a compound having the formula $A \cdot B$ and melting point T_C. This melting point is called *congruent*, because the composition of the liquid formed when the compound melts is the same as that of the compound; in other words, the compound melts *congruently*. We can easily interpret phase diagrams such as Figure 10-7 by recognizing each half as a simple eutectic system. Each component dissolves in compound to lower the compound's freezing point (curves *CE* and *CF*), and compound dissolves in each component to lower the component's freezing point (curves $ET^\circ_{f,A}$ and $FT^\circ_{f,B}$).

EXAMPLE 10-11: Cover the right half of Figure 10-7 and note the similarity to Figure 10-6.

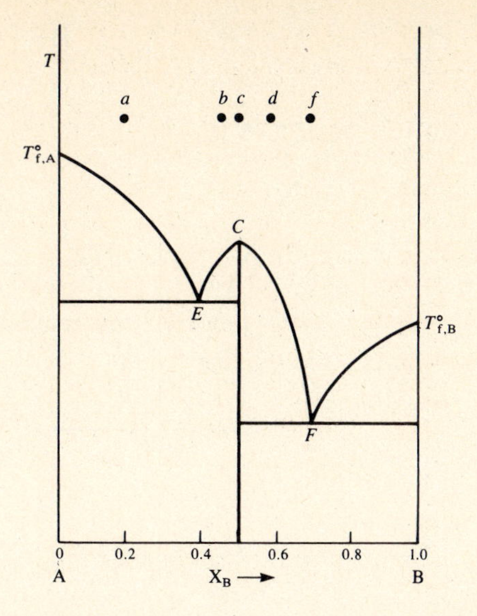

FIGURE 10-7. Formation of compound with congruent melting point.

Describe what would happen if a melt having $X_B = 0.45$ (point *b*) and temperature T_b were cooled slowly.

Solution: The situation is quite similar to that in Example 10-10. The difference is that solid B is the pure metal in Example 10-10, whereas now there is a solid C, which is the compound A·B. At the point on curve *CE* corresponding to $X_B = 0.45$, solid C will appear. As the temperature continues to fall, more C appears, and liquid disappears. The composition of the liquid also changes: As C is removed, the liquid necessarily becomes richer in A. At T_E, solid A first appears. It is in equilibrium with solid C and the melt of composition *E*. As heat is removed, the temperature remains constant until all melt has solidified. Then the temperature can drop, and the system is in the two-phase region below T_E and to the left of $X_B = 0.5$. The solid phases are pure A and pure A·B.

Some systems can form several compounds, and their phase diagrams look like several simple eutectic diagrams placed side by side. For example, the Na–Sb system forms two congruent-melting compounds at 65% and 85% Sb, and the ferric chloride–water system forms four hydrates with congruent melting points.

B. Incongruent melting point

In many systems, the components form a compound which, upon melting, gives a liquid of a different composition and forms another solid. This compound is said to melt *incongruently*; i.e., the compound is not stable up to the temperature that would be its congruent melting point but decomposes at lower temperatures:

$$C_1 \rightleftharpoons C_2 + \text{solution} \qquad \textbf{(10-7)}$$

Incongruent melting occurs in systems of metal alloys (Na–K), salts (CaF_2–$CaCl_2$), silicates (Al_2O_3–SiO_2), and hydrates ($MgSO_4$–H_2O). A typical phase diagram for a system forming a 1:1 compound with an incongruent melting point is shown in Figure 10-8. The equilibrium reaction of Eq. (10-7) becomes

PERITECTIC REACTION

$$C \rightleftharpoons A + \text{solution}_{X_B = P} \qquad \textbf{(10-8)}$$

which is represented by the three points *R*, *C*, and *P*. Point *P* is called the **peritectic point,** T_P is the **peritectic temperature**, and Eq. (10-7) is the **peritectic reaction**. The broken lines and point *M* indicate where the melting point would be if the compound did not decompose before reaching that point; the curves *ME* and *BE* are like those of an ordinary eutectic diagram. If a liquid of composition *f* (or any composition to the right of *P*) were cooled, its behavior would be like that obtained by cooling from point *d* in Figure 10-7. For compositions to the left of *P*, however, there is a difference.

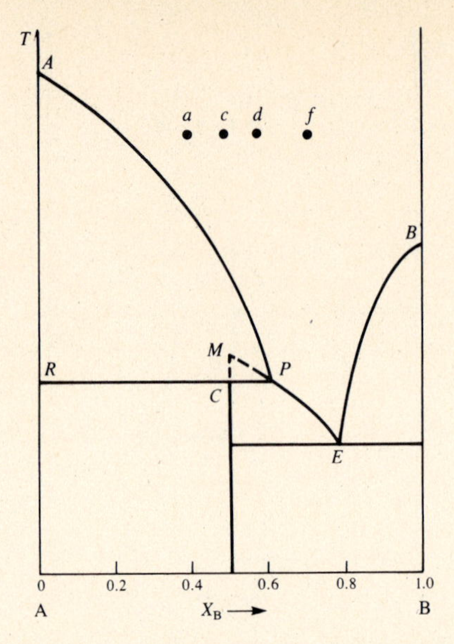

FIGURE 10-8. Formation of compound with incongruent melting point.

EXAMPLE 10-12: In the system represented by Figure 10-8, a solution of composition d ($X_B = 0.59$) is cooled slowly to a temperature below T_E. Describe what happens.

Solution: When the temperature reaches a point on the curve AP, solid A begins to freeze out. As the temperature continues to decrease, more A appears, and the amount of solution decreases while the solution becomes richer in B. When the temperature reaches T_P, a third phase appears: compound C. By the phase rule, $f = 0$, and the system is invariant. The equilibrium expressed by Eq. (10-8) is established; thus the temperature must remain constant, even though heat is being removed, until one of the phases disappears. Since d lies between C and P, solid A must disappear, which it does by reacting with component B from the solution to form compound C.

When all of solid A is gone, the temperature can decrease. Between T_P and T_E, solid C is in equilibrium with the solution. At T_E, solid B appears. The temperature remains constant until all the liquid is gone, then decreases again. Now there are two solid phases: C and B.

C. Complex cases

Some systems exhibit both types of compound formation; their phase diagrams contain both eutectic and peritectic points. For instance, aluminum and calcium form Al_2Ca with a congruent melting point and Al_3Ca with an incongruent melting point.

10-6. Solid–Solid Solutions

A. General features

In Sections 10-4 and 10-5, we assumed that the solids are completely immiscible. Frequently, however, one solid dissolves in another to form a true solid–solid solution. As in liquid solutions, in solid–solid solutions there must be homogeneity, a complete dispersion of molecules or atoms throughout. There are two ways of achieving this homogeneity:

(1) *Substitutional*, in which atoms or molecules of one species (the solute) occupy sites in the crystal lattice of the other (the solvent). This requires that the fundamental particles of each component be similar—they must be about the same in size and bonding strengths and have comparable lattice interactions.

(2) *Interstitial*, in which atoms or molecules of one component occupy the open spaces, or interstices, between the atoms or molecules of the other component arranged in its usual crystal lattice. This requires that the solute particles be appreciably smaller than those of the solvent.

In a solid solution, there is only one phase, i.e., only one type of crystal.

B. Completely miscible solids

As with two completely miscible liquids, two solids may form three types of solution. The phase diagrams of these solutions and their interpretation are like those of equivalent liquid solutions described in Chapter 9.

(1) *Intermediate:* The melting points at all compositions lie between the melting points of the pure components. Examples are Cu–Ni, NH_4CNS–KCNS, and naphthalene–β-naphthol.
(2) *Minimum melting:* The melting-point curve passes through a minimum that is lower than the melting point of either component. Examples are Na_2CO_3–K_2CO_3, Ag–Sb, and Cu–Au.
(3) *Maximum melting:* The melting-point curve exhibits a maximum that is higher than the melting point of either component. One example of this rare situation occurs with TaC and HfC. Mixtures in various proportions of these very refractory compounds are even more refractory than either individual compound.

C. Partially miscible solids

Introduction of partial miscibility of solids complicates phase diagrams. In addition to the usual eutectic points and formation of compounds, there is a new type of equilibrium, as well as the possibility of the formation several solid solutions (e.g., six in the Cu–Zn system). Figure 10-4, which we presented for partially miscible liquids, shows the situation. Frequently, however, points *m* and *p* are very close to the vertical axes, and the diagram is then simplified to the form of Figure 10-6. The interpretation for solids is similar to that for liquids. Examples are Cd–Zn, NaCl–CuCl, and naphthalene–chloracetic acid.

EXAMPLE 10-13: If A represents Ag and B represents Cu in Figure 10-4, what phases and species are present at: **(a)** point *k*; **(b)** a point between *j* and *h*; **(c)** line *mnp*; **(d)** the area to the left of point *m*; **(e)** point *g*?

Solution:

(a) One phase: a liquid solution of Ag and Cu (melt).
(b) Two phases: melt and a solid solution of Ag in Cu.
(c) Three phases: melt, solid solution of Ag in Cu, and solid solution of Cu in Ag.
(d) One phase: solid solution of Cu in Ag.
(e) Two phases: solid solution of Cu in Ag and solid solution of Ag in Cu.

The new type of equilibrium, illustrated in Figure 10-9, is similar to that described by Eq. (10-7), i.e.,

$$\text{solid solution } \alpha \rightleftharpoons \text{solid solution } \beta + \text{melt} \qquad \textbf{(10-9)}$$

Note that solution α is rich in A and solution β is rich in B.

The temperature of line *mnp* may be called a *transition point*, although it is sometimes called a *peritectic point*. It differs from our definition of a peritectic point because the temperature of the equilibrium is higher than the melting point of one of the components and there is no new compound. Examples are Hg–Cd, MnO–FeO, and *p*-iodochlorbenzene–*p*-diiodobenzene.

EXAMPLE 10-14: Describe what happens and what phases are present as a liquid at point k ($X_B = 0.30$) in Figure 10-9 is cooled slowly to point *e*.

Solution: The analysis is similar to those in previous examples, but some new features have to be added. When the temperature reaches T_j, solid solution β appears. The amount of β in equilibrium with melt increases as T decreases to T_h, but then solution α, having composition *n*, appears. Three phases are in equilibrium [Eq. (10-9)], P is constant, and $c = 2$; therefore the system is invariant,

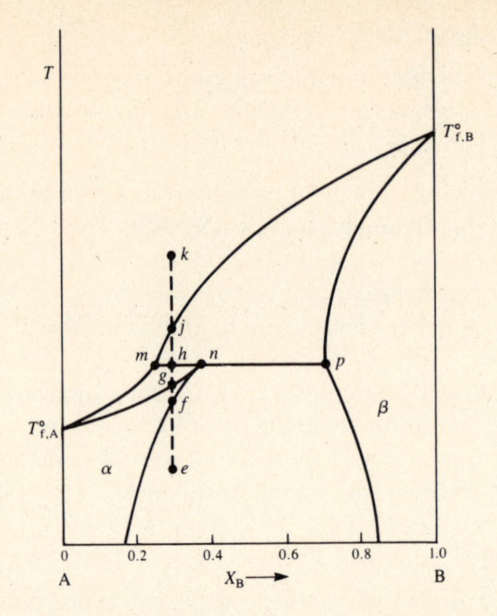

FIGURE 10-9. Transition point in system of partially miscible solids.

and T must remain constant. One phase must disappear before T can decrease. Since h lies between m and n, the phase represented by p must go. The β that formed between T_j and T_h is now consumed by reacting with the melt to produce α, which is the reverse reaction of Eq. (10-9). When all of β is gone, T will again decrease. Between h and g, there is a two-phase region: α in equilibrium with melt. At T_g, the melt disappears, and between g and f, there is one phase: solution α. At T_f, solution β appears again. At point e, two phases are present: α and β.

SUMMARY

You should examine many phase diagrams and learn how to identify the common features, how to interpret a diagram by using the phase rule, and how to describe what happens in a system of specified overall composition as the temperature is changed.

1. The phase rule gives the number of independent variables (degrees of freedom) that must be specified in order to describe the state of a system; it relates degrees of freedom to number of components, number of phases, pressure, and temperature:

$$f = c - p + 2 \qquad \text{[Eq. (10-4)]}$$

2. Systems of partially miscible liquids, at intermediate compositions, exist as two conjugate solutions, one rich in one liquid, the other rich in the other liquid. Above or below the appropriate consolute temperature, partially miscible liquids become completely miscible.
3. The general phase diagram for a system of two partially miscible liquids can be considered an overlap of the solubility curve and the boiling-point diagram of the liquids.
4. The solubility of a nonvolatile, solid nonelectrolyte that forms an ideal solution is independent of the solvent and can be calculated, for a given temperature, from its heat of fusion and normal melting temperature:

$$\ln X_2 = -\frac{\Delta H_f}{R}\left(\frac{1}{T} - \frac{1}{T_f^\circ}\right) \qquad \text{[Eq. (10-6)]}$$

5. Phase diagrams are plotted from the breaks and plateaus of cooling curves; the breaks identify invariant points and points where new phases appear.
6. In a simple system of two solids, the eutectic point is the lowest temperature at which the liquid can exist.
7. The curves in the solid–liquid phase diagram may be regarded either as solubility curves or as freezing-point curves.
8. If a compound is formed in a solid–liquid system, two situations are possible: congruent

melting and incongruent melting (also called peritectic):

$$\text{compound 1} \rightleftharpoons \text{compound 2} + \text{solution} \qquad \text{[Eq. (10-7)]}$$

9. The phase diagrams of systems of partially miscible solids show the same features as those of other systems and introduce a new type of transition equilibrium:

$$\text{solid solution } \alpha \rightleftharpoons \text{solid solution } \beta + \text{melt} \qquad \text{[Eq. (10-9)]}$$

SOLVED PROBLEMS

PROBLEM 10-1 What is the number of components in the following system?

$$CaCO_3(s) \rightleftharpoons CaO(s) + CO_2(g)$$

Solution: This is an equilibrium, so only two species are needed—the amount of the third can be calculated. If, for example, you choose $CaCO_3$ and CO_2, you can obtain CaO by $CaCO_3 - CO_2$.

PROBLEM 10-2 Consider two systems that may seem to be the same: **(a)** $NH_3(g) + HCl(g)$; **(b)** $NH_4Cl(s)$. What is the number of components for each of these systems?

Solution:

(a) The two gases will react to produce NH_4Cl, but an equilibrium state will be reached with all three species present. If the amounts of any two are known, the third can be determined using K_{eq}, so $c = 2$.

(b) In this case, NH_4Cl will decompose to NH_3 and HCl. Again, an equilibrium will be established with all three species present. The difference from part (a) is that the number of moles of NH_3 and HCl will always be equal. Thus, with K_{eq}, their concentrations can be calculated from the amount of NH_4Cl only, so $c = 1$.

PROBLEM 10-3 Refer to Figure 10-2 and identify those portions of the diagram representing the **(a)** bivariant, **(b)** monovariant, and **(c)** invariant conditions of the system.

Solution:

(a) Bivariant: the one-liquid area, anywhere above or to the left of the curve. In order to plot a point, you need to know both temperature and mole fraction.

(b) Monovariant: the two-liquid area, inside the curve. Designation of the temperature or of the mole fraction of one of the liquids is sufficient to describe the system completely.

(c) Invariant: none.

PROBLEM 10-4 Refer to Figure 10-4 and describe what would happen if a system having $X_B = 0.72$ and a temperature corresponding to the x axis is warmed slowly to a temperature above the boiling point of B.

Solution: To see the significant points, place a straight-edge perpendicular to the x axis at $X_B = 0.72$. It should be clear that the two liquid phases present at the start blend into one phase when the rising temperature (vertical line from 0.72) crosses the solubility curve and enters the right-hand area. Vapor appears when the temperature crosses the liquid curve ($pT^\circ_{b,B}$). As the temperature rises, both liquid and vapor are present, but the relative amounts change continuously as liquid becomes vapor. When the vertical line (temperature) crosses the vapor curve ($nT^\circ_{b,B}$), the last of the liquid disappears, and the system is entirely vapor.

PROBLEM 10-5 A saturated solution of naphthalene in benzene at 25 °C contains 57.7 g of naphthalene per 100 g of benzene. Estimate the heat of fusion of naphthalene at its normal melting point of 80.2 °C.

Solution: The relationship between solubility and heat of fusion is expressed by Eq.(10-6). Rearranging to obtain ΔH_f, you get

$$\Delta H_f = -R(\ln X_2)\left(\frac{TT^\circ_f}{T^\circ_f - T}\right)$$

You must calculate X_2 from the weights of the substances, so you need the numbers of moles:

$$\text{naphthalene:}\quad n_2 = \frac{57.7\ \text{g}}{128.2\ \text{g mol}^{-1}} = 0.450\ \text{mol} \qquad \text{benzene:}\quad n_1 = \frac{100\ \text{g}}{78.1\ \text{g mol}^{-1}} = 1.28\ \text{mol}$$

Thus
$$X_2 = \frac{0.450}{0.45 + 1.28} = 0.260$$

Now,
$$\Delta H_f = -8.314(\ln 0.260)\left(\frac{298.2(353.4)}{353.4 - 298.2}\right) = 21.4\ \text{kJ mol}^{-1}$$

PROBLEM 10-6 Gold and thallium form a simple eutectic system with $T_E = 131\ ^\circ\text{C}$ at $X_{\text{Tl}} = 0.66$. If you start with a melt containing 15.0 mol Au and 5.0 mol Tl, what is the maximum number of moles of pure gold that you could obtain by crystallization?

Solution: You could visualize the situation without a diagram, but a look at Figure 10-6 may help. The Au–Tl diagram would be similar to that in Figure 10-6, except that the eutectic point would be at 0.66. The maximum removal of solid gold is achieved by cooling *almost* to 131 °C. If the temperature reaches 131 °C, however, solid Tl will appear, contaminating the gold. If you calculate the amount at 131 °C, you will have the upper limit, which cannot quite be reached. The trick is to remember that, just before the temperature falls to T_E, all the Tl is still in the melt. Thus you know one of the terms in the eutectic mole fraction (moles of Tl), and you can calculate moles of Au:

$$X_E = X_{\text{Tl}} = 0.66 = \frac{n_{\text{Tl}}}{n_{\text{Au}} + n_{\text{Tl}}}$$

Then $n_{\text{Tl}} = 0.66 n_{\text{Au}} + 0.66 n_{\text{Tl}}$, $0.34 n_{\text{Tl}} = 0.66 n_{\text{Au}}$, and

$$n_{\text{Au}} = \left(\frac{0.34}{0.66}\right) n_{\text{Tl}} = \left(\frac{0.34}{0.66}\right) 5.0 = 2.6\ \text{mol}$$

The amount of solid gold is the original amount less the amount still in solution, or

$$n_{\text{Au(s)}} = 15.0 - 2.6 = 12.4\ \text{mol}$$

note: You could have calculated the number of moles using the lever rule: $(20 - x)(0.41) = 0.25x$. Note also that the expression for the final calculation can be stated generally as

$$\frac{n_\text{A}}{n_\text{B}} = \frac{X_\text{A}}{X_\text{B}}$$

PROBLEM 10-7 Sometimes phase diagrams are plotted using weight percent rather than mole percent (mole fraction). In the simple eutectic diagram of the Si–Au system, the eutectic composition is 94% gold by weight. What are the mole fraction and the mole percent of gold in the eutectic mixture?

Solution: This is a straightforward conversion of grams to moles. If you assume that 100 g is the total weight, you have 6 g of Si and 94 g of Au. The amounts, in moles, and the mole fraction are

$$n_{\text{Si}} = \frac{6}{28} = 0.21 \qquad n_{\text{Au}} = \frac{94}{197} = 0.48 \qquad X_{\text{Au}} = \frac{0.48}{0.48 + 0.21} = 0.70$$

Mole percent is simply mole fraction times 100:

$$\text{mol}\ \%\ \text{Au} = 0.70(100) = 70\%$$

PROBLEM 10-8 The system Na–Sb forms a compound with a congruent melting point and a composition of 65.0% Sb by weight. What is the formula of the compound?

Solution: You need to know the relative number of moles of each element in the compound, so you should use the usual procedure for determining the empirical formula from percent by weight. Again, assume 100 g and calculate the weight of each element. Divide each weight by the atomic weight (g mol^{-1}), and then divide each number of moles by the smaller:

$$\text{Sb:}\quad \frac{65.0}{121.8} = .534\ \text{mol} \qquad \text{Na:}\quad \frac{35.0}{23.0} = 1.52\ \text{mol}$$

Dividing by 0.534 gives 1.00 for Sb and 2.85 for Na. Thus the empirical formula is Na_3Sb.

PROBLEM 10-9 You have a liquid of composition *d* in Figure 10-7. **(a)** Can you obtain pure A by cooling? If not, why not? **(b)** If you have a finite source of pure A, what might you do to obtain more pure A from the liquid?

Solution:

(a) No, you can't obtain pure A simply by cooling, because pure C will freeze out when the temperature falls below curve CF. The tie line would not reach the left-hand side of the diagram.

(b) You could add pure A to the liquid, while keeping the temperature up. If you increase the concentration of A until it falls to the left of E, i.e., $X_A > 0.60$, and then cool, pure A will freeze out when T falls below curve $ET^\circ_{f,A}$.

PROBLEM 10-10 Use Figure 10-7 again. You have a solution at point b ($X_B = 0.45$) that contains 1.50 mol of A. You cool until 0.20 mol of compound C freezes out. What is the new concentration of B?

Solution: To calculate the new mole fraction, you need to determine the amount of each substance after C is removed. Thus you need to find the starting amount for substance B. You can use $n_A/n_B = X_A/X_B$ (from Problem 10-6), so that

$$n_B = \left(\frac{X_B}{X_A}\right)n_A = \left(\frac{0.45}{0.55}\right)(1.50) = 1.23$$

Since components of the compound have a 1:1 ratio, removal of 0.20 mol of compound removes 0.20 mol A and 0.20 mol B. The new amounts and new X_B are

$$n_A = 1.50 - 0.20 = 1.30 \text{ mol} \qquad n_B = 1.23 - 0.20 = 1.03 \text{ mol} \qquad X_B = \frac{1.03}{1.30 + 1.03} = 0.44$$

Stated differently, the solution becomes richer in A as the compound is formed.

PROBLEM 10-11 Refer to Figure 10-8. How does cooling a liquid of composition a ($X_B = 0.4$) differ from cooling one of composition d ($X_B = 0.59$)? [See Example 10-12.]

Solution: In both cases, solid A appears when T falls to curve AP. When T reaches RP, more solid A is present in the system at point a than at point d. Before the temperature can decrease below the invariant point, one of the phases in the peritectic reaction must disappear. (*1*) Since a lies between R and C, the phases represented by R and C must be present at lower temperatures, so the solution must disappear. Below T_P, solids A and C are mixed. (*2*) Since d lies between C and P, solid A must disappear. Between T_P and T_E, the system consists of solid compound C and solution. At T_E, solid B appears and solution disappears. Below T_E, solids B and C are mixed.

PROBLEM 10-12 Compare what happens in each system if a liquid at point c in Figure 10-7 and a liquid at point c in Figure 10-8 are cooled slowly. Both points c are at $X_B = 0.5$.

Solution: In Figure 10-7, when the temperature drops to T_C, solid C appears. The temperature remains constant until all liquid has disappeared; the composition of the solid is the same as that of the liquid. Below T_C, only one solid phase is present: compound A·B. Neither A nor B ever appears as the pure solid.

In Figure 10-8, when the temperature decreases to a point on curve AP, solid A appears, then increases in amount as T continues to decrease. At T_P, solid C appears, and the three phases remain at equilibrium at constant T until both the liquid and solid A disappear simultaneously. Below T_P, there is only one solid phase: compound A·B.

PROBLEM 10-13 When you use mole fractions to plot phase diagrams, the compounds that are formed have the same mole fraction in every system. For example, a 1:1 compound must have $X_B = 0.50$. Calculate the mole fraction of B for compounds having the general formulas A_2B, AB_3, and A_2B_3.

Solution: In each case, $X_B = n_B/(n_A + n_B)$. Thus

$$\text{For } A_2B,\ X_B = \tfrac{1}{3} = 0.33; \qquad \text{for } AB_3,\ X_B = \tfrac{3}{4} = 0.75; \qquad \text{for } A_2B_3,\ X_B = \tfrac{3}{5} = 0.60$$

The calculations are trivial; the point is that you can get the formula of a compound from a phase diagram quickly. For example, if a congruent melting point occurs at $X_B = 0.75$, you know immediately that the compound must be the AB_3 type.

Supplementary Exercises

PROBLEM 10-14 In a two-component system, the maximum number of degrees of freedom is ________. The number of components in the system $NH_3(g)$ is ________. The maximum point on the solubility curve of partially miscible liquids is called the ________ ________ ________. For the system whose phase diagram is shown in Figure 10-4, fractional distillation, starting at point g, (will)(will not) produce pure A at any stage of the distillation. Neglecting second-order effects, doubling the pressure would (increase)(decrease)(not affect) the solubility of naphthalene in benzene at 25 °C. Raising the temperature from 25 to 35 °C would ________ the solubility of naphthalene in benzene. Adding more thallium to the original Au–Tl melt described in Problem 10-6 would ________ the amount of solid gold. If you want to obtain pure solid compound from the system whose phase diagram is shown in Figure 10-7, the concentration of the starting melt must be ________. The cooling curve you would obtain for a solution at point d in Figure 10-8 would have ________ plateaus and ________ breaks. In Figure 10-9, ________ (how many?) areas represent two phases in equilibrium.

PROBLEM 10-15 At a temperature near the bottom of the diagram in Figure 10-4, 7 mol of A is mixed with 3 mol of B. When the temperature is raised slowly, what is the ratio of the amount of solution m to solution p just before vapor appears? (Use Figure 10-4 to estimate values.)

PROBLEM 10-16 Calculate the solubility of phenol in any ideal solution at 30 °C. At the normal melting point of 42 °C, the heat of fusion is 11.42 kJ mol^{-1}.

PROBLEM 10-17 Plot the following data and determine the eutectic temperature and eutectic composition for this system of two immiscible solids. The temperatures are the first break in the cooling curve of each solution.

Mole fraction of B	T (K)	Mole fraction of B	T (K)	Mole fraction of B	T (K)
0	520	0.30	459	0.60	518
0.10	505	0.40	474	0.80	549
0.20	485	0.50	498	1.00	570

PROBLEM 10-18 The system Cd–Bi has a simple eutectic point at $T = 413$ K and $X_{Bi} = 0.45$. How much pure Bi (in grams) is present just before Cd appears, when 100.0 g of a melt having $X_{Bi} = 0.85$ has been cooled to 413 K?

PROBLEM 10-19 The immiscible solids Be and Si exhibit a simple eutectic point at 1363 K and $X_{Si} = 0.32$. The melting points of the pure elements are Be = 1555 K and Si = 1685 K. Estimate the temperature at which a solution of $X_{Si} = 0.80$ will be in equilibrium with pure Si. [*Hint:* Sketch an approximate phase diagram.]

PROBLEM 10-20 The system Mg–Bi forms a compound that melts congruently and contains 14.9% Mg by weight. What is the formula of the compound?

PROBLEM 10-21 In Figure 10-8, a solution of composition c is cooled slowly. Estimate the ratio of melt to solid A just before solid C appears. At P, $X_B = 0.62$.

PROBLEM 10-22 Use the following cooling-curve data to answer the questions in parts **(a)**–**(f)**. The system has two solid components, which form no solid solutions. The temperatures are those at which the indicated features occur in each cooling curve. [*Hint:* Plot the phase diagram: You'll have to approximate some of its features, but you should make reasonable assumptions, e.g., for the composition of a compound.]

X_B	First break	First plateau	Second plateau
0		780	
0.20	680	480	
0.40	535	480	330
0.60	425	330	
0.70	345	330	300
0.80	312	300	
0.90	335	300	
1.00		410	

(a) What are the temperatures and compositions of any eutectic points?
(b) What are the temperatures and compositions of any peritectic points?
(c) How many two-phase areas are there in the phase diagram?
(d) What are the probable formulas of any compounds formed?
(e) What phases are present at $X_B = 0.15$ and $T = 400$ K?
(f) What phases are present at $X_B = 0.75$ and $T = 310$ K?

Answers to Supplementary Exercises

10-14 3; 1; upper consolute temperature; will not; not affect; increase; decrease; $> X_{B,E}$, $< X_{B,F}$; 2, 1; 3 **10-15** 4.0 **10-16** 0.84 **10-17** 0.32, 451 K **10-18** 78.2 g
10-19 not lower than 1588 K **10-20** Mg_3Bi_2 **10-21** 4.2 **10-22** **(a)** 300 K, 0.84; **(b)** 480 K, 330 K; **(c)** 7; **(d)** A_2B, AB_2; **(e)** solid A and solid A_2B; **(f)** solid AB_2 and melt

11 SOLUTIONS OF ELECTROLYTES

THIS CHAPTER IS ABOUT

- ☑ Solutions That Conduct Electricity
- ☑ Electrolytic Deposition
- ☑ Nonideal Colligative Properties
- ☑ The Process of Conduction
- ☑ Activity and Activity Coefficients
- ☑ Debye–Hückel Theory

11-1. Solutions That Conduct Electricity

Substances whose solutions conduct electric current are known as **electrolytes**, and the solutions are called **electrolytic solutions**. Such solutions are fundamental to most biological processes, to chemical operations in laboratories and industry, and to such commonplace yet essential devices as the lead storage battery. In the nineteenth century, chemists began to search for an explanation of the difference between the properties of electrolytic and nonelectrolytic solutions. Extensive investigations led to a shaky acceptance of Arrhenius's theory that dissociation or nondissociation into ions was the answer—electrolytes dissociate and nonelectrolytes don't. Arrhenius's theory was the first step toward the complex theory of Debye and Hückel.

Three properties of solutions are significant: colligative properties, conductivity, and activity coefficients. In some texts, these properties are discussed in two or three places, with parts of the theory presented in each place. Following an introductory section on the laws of electrolysis, we outline the features of all three properties and then summarize the theory.

note: Table 11-1 contains two lists: one of electrical terms and related information, and one of ancillary terms and symbols. We'll use these symbols here, but textbooks do differ—especially in the use of subscripts. For example, you may see Λ_0, Λ°, Λ_m°, or Λ_m^∞ used for molar conductivity at infinite dilution.

11-2. Electrolytic Deposition

A. Laws of electrolysis

Soon after Volta discovered that chemical reaction could produce electricity, others showed that electric current could cause chemical reaction. Then Faraday studied the quantitative relationships between amount of current and amount of material released from solution. His results are stated in two laws of electrolysis:

1. The amount (mass m) of a substance produced is directly proportional to the quantity of electricity Q passed through the electrolyte, or $\Delta m \propto Q$.
2. The mass of a substance produced is directly proportional to the equivalent weight of the substance, or $\Delta m \propto M_m$/electrons gained or lost.

We can express **Faraday's first law of electrolysis** as

FARADAY'S FIRST LAW

$$\frac{\Delta m_1}{Q_1} = \frac{\Delta m_2}{Q_2} \tag{11-1}$$

TABLE 11-1: Electrical Terms

SI term	SI symbol	Definition	SI unit	Other units and symbols	Other terms
Electric current	I		ampere (A)		
Quantity of electricity	Q	$\int I\,dt$	coulomb (C)	A s	
Electric potential	V	wQ^{-1}	volt (V)		
Electric resistance	R	ΔVI^{-1}	ohm (Ω)		
Resistivity	ρ	$RA\ell^{-1}$			Specific resistance, specific resistivity
Conductance	G	R^{-1}	siemens (S)	mho (Ω^{-1})	
Electrolytic conductivity	κ	$G\ell A^{-1}$	$S\,m^{-1}$	$\Omega^{-1}m^{-1}$, $\Omega^{-1}cm^{-1}$	Specific conductance, specific conductivity
Molar conductivity	Λ	κC^{-1}	$S\,m^2\,mol^{-1}$	$\Omega^{-1}cm^2\,mol^{-1}$	Molar conductance, equivalent conductance, equivalent conductivity
Molar conductivity of ion	λ	$\mathscr{F}u$	$S\,m^2\,mol^{-1}$		Molar ionic conductance
Electric mobility of ion	u		$m^2\,V^{-1}s^{-1}$		
Transport number of ion	t				Transference number

Other Symbols and Terms

Symbol	Term	Symbol	Term
w	Work	α	Degree of dissociation
ℓ	Length	ν	Number of ions produced by the complete dissociation of the formula unit of a compound; $\nu = p + q$, where p is the number of positive ions and q is the number of negative ions
C	Molar concentration, mol dm^{-3} (with Λ: mol m^{-3})		
m	Molal concentration		
X	Mole fraction		
$\mathscr{F}$	Faraday constant		
n	Amount of matter	K_{cell}	Cell constant
M_m	Molar mass	Λ_0	Molar conductivity at infinite dilution
L_A	Avogadro's constant	μ	Chemical potential
e	Charge on one electron	a	Activity
z	Charge on an ion	γ	Activity coefficient
i	Van't Hoff factor	I	Ionic strength

EXAMPLE 11-1: In an experiment, one coulomb of electricity deposited 0.001 117 97 g of silver. How many grams of silver will be obtained if a current of exactly two amperes flows for exactly one hour through a solution containing silver ions?

Solution: We first calculate the quantity of current that passed through the solution and then apply Faraday's first law. By definition, the quantity of electricity is the electric current I multiplied by the elapsed time, or $Q = It$. In units, one coulomb is one ampere flowing for one second, assuming constant current. So, substituting $I = 2$ A and $t = 3600$ s, we obtain

$$Q = (2\text{ A})(3600\text{ s}) = 7200\text{ A s} = 7200\text{ C}$$

Then rearranging Eq. (11-1) and substituting gives

$$\Delta m_2 = Q_2\frac{\Delta m_1}{Q_1} = (7200\text{ C})\left(\frac{0.001\,117\,97\text{ g}}{1\text{ C}}\right) = 8.049\,38\text{ g}$$

EXAMPLE 11-2: What quantity of electricity is required to deposit one gram equivalent weight of Ag from a solution containing Ag^+ ions? Use data from Example 11-1.

Solution: The equivalent weight of Ag is equal to its atomic weight, or 107.868. From Example 11-1, we have the conversion factor $\frac{1\ \text{C}}{1.11797 \times 10^{-3}\ \text{g}}$, so, for 107.868 g,

$$Q = (107.868\ \text{g})\left(\frac{1\ \text{C}}{1.11797 \times 10^{-3}\ \text{g}}\right) = 9.6486 \times 10^{4}\ \text{C}$$

The accepted numerical value of the quantity of electricity needed to deposit one gram equivalent weight of any element is 9.6485×10^4 C (or, where only three significant digits are required, 96 500). The SI system recommends eliminating "equivalent" and "equivalent weight" and using only "mole," so the units of this quantity, which is called the **Faraday constant** $\mathscr{F}$, are C mol^{-1}.

EXAMPLE 11-3: How should we now describe "one gram equivalent weight of H_2SO_4" or "one equivalent of H_2SO_4"?

Solution: To follow the IUPAC recommendations, we should write "one mole of $\frac{1}{2}H_2SO_4$." Unfortunately, this expression implies that there exists an entity that somehow is one-half of a sulfuric acid molecule. Some authors have followed this IUPAC suggestion, but others reject it and write "$\frac{1}{2}$ mol H_2SO_4" to replace "1 equivalent of H_2SO_4." We use the latter method.

We can express **Faraday's second law of electrolysis** as

FARADAY'S SECOND LAW

$$\Delta m = \left(\frac{Q}{\mathscr{F}}\right)\left(\frac{M_{\text{m}}}{|z|}\right) \qquad \textbf{(11-2)}$$

where the Faraday constant is the proportionality constant and $|z|$ is the absolute value of the number of charges on the ion reacting, either z_+ or z_-. In electrolysis, the ion usually goes to the oxidation state of zero, so the number of electrons gained or lost equals $|z|$. We could consider the denominator of Eq. (11-2), $|z|\mathscr{F}$, to be simply the Faraday constant in moles, i.e., multiplying $\mathscr{F}$ in C eq^{-1} by the number of equivalents per mole gives $\mathscr{F}$ in C mol^{-1}. Thus whenever $\mathscr{F}$ is used, it must be accompanied by the appropriate factor for converting equivalents to moles.

EXAMPLE 11-4: How many grams of copper would be deposited by electrolysis of a solution of $Cu(NO_3)_2$ when 10 000 C is passed through it?

Solution: We apply Faraday's second law, Eq. (11-2). Since the reaction is $Cu^{2+} \rightarrow Cu^0$, two electrons are involved. We could also note that $z_+ = +2$, or $|z_+| = 2$. Substituting in Eq. (11-2), we have

$$\Delta m = \left(\frac{Q}{\mathscr{F}}\right)\left(\frac{M_{\text{m}}}{|z|}\right) = \left(\frac{10\,000\ \text{C}}{96\,486\ \text{C eq}^{-1}}\right)\left(\frac{63.54\ \text{g mol}^{-1}}{2\ \text{eq mol}^{-1}}\right) = 3.293\ \text{g}$$

Another interpretation of the Faraday constant is that it's the quantity of electricity carried by one mole of electrons.

EXAMPLE 11-5: Given the charge on one electron, $e = 1.6022 \times 10^{-19}$ C, calculate the value of the Faraday constant and compare it to the accepted value given previously.

Solution: If the Faraday constant is the charge on one mole of electrons, we can multiply the charge on one electron by Avogadro's constant, so

$$\mathscr{F} = L_{\text{A}}e = (6.0221 \times 10^{23}\ \text{electrons mol}^{-1})(1.6022 \times 10^{-19}\ \text{C electron}^{-1}) = 9.6486 \times 10^4\ \text{C mol}^{-1}$$

The accepted value is 9.6485×10^4, which is very close. (The discrepancy probably is due to different sources of values that are averaged in establishing values for basic quantities.)

B. Measurement of quantity of electricity

The quantity of electricity can be measured quite accurately by a *coulometer*, which is an instrument based on the application of Faraday's laws. When a coulometer is placed in a circuit, the electricity deposits or releases the element on which the instrument is based. The element's mass is measured, and the quantity of electricity is calculated using Faraday's second law, Eq. (11-2).

EXAMPLE 11-6: How much electricity passes through a silver coulometer for every gram of silver deposited?

Solution: First, solve Eq. (11-2) for Q:

$$\Delta m = \left(\frac{Q}{\mathscr{F}}\right)\left(\frac{M_{\mathrm{m}}}{|z|}\right) \qquad \text{and} \qquad Q = \frac{|z|\mathscr{F}\,\Delta m}{M_{\mathrm{m}}}$$

The atomic weight of Ag is 107.9 g mol^{-1}, so, for the deposition of 1 g of Ag,

$$Q = 1(96\,500\ \mathrm{C\,mol^{-1}})\left(\frac{1\ \mathrm{g}}{107.9\ \mathrm{g\,mol^{-1}}}\right) = 894.3\ \mathrm{C}$$

11-3. Nonideal Colligative Properties

A. Experimental evidence

We showed in Chapter 9 that the various colligative properties of dilute solutions of nonelectrolytes were proportional to the molality; for example,

$$\Delta T_{\mathrm{f}} = K_{\mathrm{f}} m \qquad \text{[Eq. (9-15)]}$$

For the solvent water, $K_{\mathrm{f}} = 1.86$ K m^{-1}; thus any 0.010 m solution should have a freezing point of -0.0186 K. But this relationship doesn't hold for electrolytic solutions.

EXAMPLE 11-7: Consider the following K_{f} data for several solutes at two concentrations; the K_{f} values were calculated using Eq. (9-15) and experimental ΔT_{f} values for the molalities shown.

m	NaCl	$CuSO_4$	K_2SO_4
0.010	3.60	2.70	5.01
0.10	3.48	2.08	4.32

How do these values correlate with **(a)** the K_{f} for nonelectrolytic solutions? **(b)** concentration of the solution?

Solution:

(a) Recall that K_{f} for a nonelectrolyte would be 1.86 and that the value of a colligative property depends solely on the number of moles of solute. For NaCl, the value of K_{f} is almost twice the expected 1.86 of nonelectrolytic solutions, suggesting that there are twice as many particles. For K_2SO_4, the value of K_{f} is almost three times as great. The factors 2 and the 3 represent the number of ions obtained from complete dissociation of one formula unit of each compound.

(b) The K_{f} values decrease as m increases and, clearly, aren't constant.

B. Empirical treatment

Van't Hoff, in studying osmotic pressures, found the general effect illustrated in Example 11-7. To allow for such nonideal behavior, he introduced a correction term, now known as the **van't**

Hoff *i* factor, which is the ratio of the actual K_f value to that of a nonelectrolyte. Thus

VAN'T HOFF *i* FACTOR

$$i = \frac{\Delta T_f}{\Delta T_{f,\,non}} = \frac{mK_f}{mK_{f,\,non}} = \frac{K_f}{K_{f,\,non}} \tag{11-3}$$

EXAMPLE 11-8: Calculate the *i* factors for NaCl and K_2SO_4 from the data in Example 11-7.

Solution: We use Eq. (11-3):

For 0.010 m NaCl: $i = \frac{3.60}{1.86} = 1.94$; for 0.10 m NaCl: $i = \frac{3.48}{1.86} = 1.87$

For 0.010 m K_2SO_4: $i = \frac{5.01}{1.86} = 2.69$; for 0.10 m K_2SO_4: $i = \frac{4.32}{1.86} = 2.32$

From Example 11-8, we see that *i* is approximately the number of ions but that it decreases as concentration increases. This *seems* to be another situation in which we get better agreement for dilute solutions. But the situation is not so simple: At higher concentrations than those used here, *i* actually increases. Finally, *i* applies to all colligative properties, i.e.,

$$i = \frac{\Delta P}{\Delta P_{non}} = \frac{\Delta T_f}{\Delta T_{f,\,non}} = \frac{\Pi}{\Pi_{non}} = \cdots \tag{11-4}$$

EXAMPLE 11-9: What boiling point could we expect for a 0.10 m K_2SO_4 solution?

Solution: First, we rewrite Eq. (11-4) for boiling-point elevation as

$$T_b = i\Delta T_{b,\,non} = iK_b m$$

From Example 11-8, $i = 2.32$, and we know that K_b for water is 0.52 K kg mol^{-1}, or 0.52 K m^{-1}; thus

$$\Delta T_b = 2.32(0.52 \text{ K } m^{-1})(0.10\ m) = 0.12 \text{ K}$$

so

$$T_b = 100.00 + \Delta T_b = 100.12\ ^\circ\text{C}$$

C. Theoretical treatment

Arrhenius, having theorized that electrolytes don't dissociate completely at higher concentrations, introduced the concept of degree of dissociation, α. [See Section 7-3D and Eq. (7-16).] Thus, if we let the new *i* factor equal the ratio of concentration after dissociation to that of the undissociated compound, we see that *i* is related to α by

DEGREE OF DISSOCIATION

$$\alpha = \frac{i - 1}{\nu - 1} \tag{11-5}$$

where ν is the number of ions obtained if dissociation is complete. At infinite dilution, $i = \nu$ and $\alpha = 1$. Good agreement between values from Eq. (11-5) and α values from conductance strongly support the Arrhenius theory. The Arrhenius theory works quite well for weak electrolytes, but we now know that strong electrolytes always exist as ions—even in the solid state—so we need a different theory to explain their deviations from expected behavior.

EXAMPLE 11-10: Use the Arrhenius theory to calculate the degree of dissociation of $CuSO_4$ in a 0.010 m solution.

Solution: We know that for $CuSO_4$ $\nu = 2$, and we know that $K_f = 2.70$ from Example 11-7. We use Eq. (11-3) to calculate *i*:

$$i = \frac{K_f}{K_{f,\,non}} = \frac{2.70}{1.86} = 1.45$$

Now, using Eq. (11-5), we obtain

$$\alpha = \frac{i-1}{\nu-1} = \frac{1.45-1}{2-1} = \frac{0.45}{1} = 0.45$$

11-4. The Process of Conduction

A. Terminology

The basic electrical quantities are current (I), potential (V), and resistance (R), with units of ampere (A), volt (V), and ohm (Ω), respectively. (See Table 11-1.) These quantities are related by **Ohm's law**:

OHM'S LAW

$$R = \frac{\Delta V}{I} \tag{11-6}$$

The reciprocal of the resistance is the *conductance* (G), which is directly proportional to the cross-sectional area (A) and inversely proportional to the length (ℓ) of the medium, i.e.,

$$G = \frac{1}{R} \propto \frac{A}{\ell} = \kappa \frac{A}{\ell} \tag{11-7}$$

The unit of conductivity is the reciprocal of the resistance unit, or Ω^{-1} (once called *mho*, but now officially called *siemens*, S). The proportionality constant (κ) is the *conductivity* (formerly called *specific conductivity* or *specific conductance*). For our purposes in connection with electrolytic solutions, we designate it the *electrolytic conductivity*, whose SI unit is $\mathrm{S\ m^{-1}}$ ($\Omega^{-1}\mathrm{m}^{-1}$ is also acceptable). We can also use $\mathrm{cm^{-1}}$ rather than $\mathrm{m^{-1}}$ in this unit. Values of κ increase with temperature and concentration.

note: To avoid confusion, be careful when using terms ending in *-ance* and *-ivity*. Usage in textbooks differs, and you may even find inconsistencies in the same book.

Since the SI recommends dropping *equivalent* and using only *mole*, the term *molar conductivity* replaces such terms as *equivalent conductance* and *equivalent conductivity* for the quantity defined by

$$\Lambda = \frac{\kappa}{C} \quad \left(\frac{\Omega^{-1}\mathrm{m}^{-1}}{\mathrm{mol\,m^{-3}}}\right) \tag{11-8}$$

The SI unit of Λ is $\mathrm{S\,m^2\,mol^{-1}}$ (or $\Omega^{-1}\mathrm{m^2\,mol^{-1}}$). [Note that C is in $\mathrm{mol\,m^{-3}}$. Usually, however, concentrations are in molarity with units of $\mathrm{mol\ L^{-1}}$ (or $\mathrm{mol\ dm^{-3}}$). The unit used in older tables is usually $\Omega^{-1}\ \mathrm{cm^2\ mol^{-1}}$; to convert to SI units, you must remember that $1\ \mathrm{m^2} = 10^4\ \mathrm{cm^2}$ and convert accordingly.]

EXAMPLE 11-11: From tables, we find that the molar conductivity of 0.0050 M KCl is $144\ \Omega^{-1}\mathrm{cm^2\,mol^{-1}}$. Calculate its electrolytic conductivity in SI units.

Solution: Rearranging Eq. (11-8) gives $\kappa = \Lambda C$. Next, converting to SI units, we have

$$\Lambda = (144\ \Omega^{-1}\mathrm{cm^2\,mol^{-1}})\left(\frac{1\ \mathrm{m^2}}{10^4\ \mathrm{cm^2}}\right) = 1.44 \times 10^{-2}\ \Omega^{-1}\mathrm{m^2\,mol^{-1}}$$

and

$$C = (5.0 \times 10^{-3}\ \mathrm{mol\,L^{-1}})\left(\frac{1\ \mathrm{L}}{10^3\ \mathrm{cm^3}}\right)\left(\frac{10^6\ \mathrm{cm^3}}{1\ \mathrm{m^3}}\right) = 5.0\ \mathrm{mol\,m^{-3}}$$

So

$$\kappa = (1.44 \times 10^{-2}\ \Omega^{-1}\mathrm{m^2\,mol^{-1}})(5.0\ \mathrm{mol\,m^{-3}}) = 7.2 \times 10^{-2}\ \mathrm{S\,m^{-1}}$$

Molar conductivity is a difficult concept to visualize. It may help to picture a rectangular cell with two of its parallel sides serving as the electrodes of the cell. These large plates are one unit of distance apart and have an area great enough to contain the volume of the solution that would have 1 mole of the electrolyte in it. The reciprocal of the resistance measured between the plates is the molar conductivity, Λ. For a given electrolyte, Λ decreases with an increase in concentration: slightly for strong electrolytes and considerably for weak electrolytes.

B. Measurement

The conductivity of a solution is measured in a cell that is one arm of a Wheatstone bridge. Although such cells have been called *conductance cells* and *conductivity cells*, they actually measure resistance. Alternating current is used, so there is no net chemical reaction. The electrolytic conductivity is calculated from Eq. (11-7), rearranged to

$$\kappa = \frac{1}{R}\left(\frac{\ell}{A}\right) \tag{11-9}$$

Since ℓ and A are difficult to measure, the usual procedure is to treat ℓ/A as a cell constant and to calibrate the cell by using a solution, usually of KCl, having a known molar conductivity. Then

$$\kappa = \frac{1}{R} K_{\text{cell}} \tag{11-10}$$

EXAMPLE 11-12: In a certain conductivity cell, the resistance of a 0.01 000 M KCl solution is 150.0 Ω. The known molar conductivity of the solution is 141.27 $\Omega^{-1}\text{cm}^2\,\text{mol}^{-1}$. Calculate the cell constant.

Solution: There are two parts to the problem. First, we need to determine κ from Eq. (11-8):

$$\kappa = C\Lambda = (0.01\,000 \text{ mol L}^{-1})(141.27\ \Omega^{-1}\text{cm}^2\,\text{mol}^{-1}) = 1.413\ \Omega^{-1}\text{cm}^2\,\text{dm}^{-3}$$

But 1 $\text{dm}^3 = 10^3\ \text{cm}^3$, so $\text{cm}^2\,\text{dm}^{-3} = 10^{-3}\ \text{cm}^{-1}$; thus

$$\kappa = 1.413 \times 10^{-3}\ \Omega^{-1}\text{cm}^{-1}$$

Second, we can substitute values after rearranging Eq. (11-10):

$$K_{\text{cell}} = R\kappa = (150.0\ \Omega)(1.413 \times 10^{-3}\ \Omega^{-1}\text{cm}^{-1}) = 0.2120\ \text{cm}^{-1}$$

EXAMPLE 11-13: Using the cell described in Example 11-12, a student measured the resistance of a 0.100 M NaCl solution to be 19.9 Ω. Calculate the experimental value of the molar conductivity of this solution.

Solution: Essentially, this is the reverse of the previous example. Knowing the cell constant, we can use Eq. (11.10) to calculate the electrolytic conductivity, so

$$\kappa = \frac{1}{R} K_{\text{cell}} = \frac{0.2120\ \text{cm}^{-1}}{19.9\ \Omega} = 1.07 \times 10^{-2}\ \Omega^{-1}\text{cm}^{-1}$$

We now use Eq. (11-8) to calculate the molar conductivity:

$$\Lambda = \frac{\kappa}{C} = \left(\frac{1.07 \times 10^{-2}\ \Omega^{-1}\text{cm}^{-1}}{0.100\ \text{mol dm}^{-3}}\right)\left(\frac{10^3\ \text{cm}^3}{1\ \text{dm}^3}\right) = 1.07 \times 10^2\ \Omega^{-1}\text{cm}^2\,\text{mol}^{-1}$$

C. Molar conductivity at infinite dilution

We know that molar conductivity increases as concentration decreases. In fact, the plot of Λ versus $\sqrt{C}$ is practically a straight line for strong electrolytes; but, for weak electrolytes, the curve rises sharply as it approaches $C = 0$. Thus, for strong electrolytes in dilute solution, the data are represented fairly well by

$$\Lambda = S\sqrt{C} + \Lambda_0 \tag{11-11}$$

where the slope, S, is negative and the y intercept, Λ_0, is the **molar conductivity at infinite dilution** (also, **limiting molar conductivity**), which is obtained by extrapolation of the line to $\sqrt{C} = 0$. For weak electrolytes, extrapolation is unreliable because of the almost asymptotic approach of the curve to the y axis.

Consider a strong electrolyte that yields ions A and B in solution, according to

$$A_pB_q \longrightarrow pA^{z+} + qB^{z-}$$

(some authors use ν_+ for p and ν_- for q). Kohlrausch assumed that in such a system the molar conductivity at infinite dilution is simply the sum of the independent contributions of the ions. **Kohlrausch's law of independent migration of ions** is

KOHLRAUSCH'S LAW
$$\Lambda_0 = p\lambda_{+,0} + q\lambda_{-,0} \tag{11-12}$$

where the lowercase lambda represents a **molar ionic conductivity at infinite dilution.** Comparison of Λ_0 values of electrolytes containing common ions shows that one ion does make about the same contribution in every case.

EXAMPLE 11-14: The Λ_0 value is 126.5 for NaCl, 149.9 for KCl, 128.5 for NaBr, and 151.9 for KBr. What does a comparison of these values tell us about λ_0 for Na^+ and K^+?

Solution: In order to compare values, pair the solutions that contain the same anion and subtract:

$$\begin{aligned} \Lambda_0(\text{KCl}) &= 149.9 = \lambda_0(K^+) + \lambda_0(Cl^-) \\ \Lambda_0(\text{NaCl}) &= 126.5 = \lambda_0(Na^+) + \lambda_0(Cl^-) \\ \text{Difference} &= 23.4 = \lambda_0(K^+) - \lambda_0(Na^+) \end{aligned} \qquad \begin{aligned} \Lambda_0(\text{KBr}) &= 151.9 = \lambda_0(K^+) + \lambda_0(Br^-) \\ \Lambda_0(\text{NaBr}) &= 128.5 = \lambda_0(Na^+) + \lambda_0(Br^-) \\ \text{Difference} &= 23.4 = \lambda_0(K^+) - \lambda_0(Na^+) \end{aligned}$$

The constant difference implies that the K^+ and Na^+ ions contribute the same amount to Λ_0 in all cases; i.e., their contributions are independent of the associated anion.

We'll present a method for obtaining individual λ values later, but we can still obtain Λ_0 values for electrolytes—even weak ones—by applying Eq. (11-12).

EXAMPLE 11-15: Calculate Λ_0 for the weak electrolyte NH_4OH from the Λ_0 values for these strong electrolytes: NH_4Cl, 149.7; NaCl, 126.5; and NaOH, 248.1

Solution: As expressed by Eq. (11-12), each Λ_0 has two ionic parts. So, by adding and subtracting appropriately, we can cancel the parts we don't want and keep only the $NH_4{}^+$ and OH^- parts, i.e.,

$$\begin{aligned} \Lambda_0(NH_4OH) = \lambda_0(NH_4{}^+) + \lambda_0(OH^-) &= \lambda_0(NH_4{}^+) + \lambda_0(Cl^-) + \lambda_0(Na^+) \\ &\quad + \lambda_0(OH^-) - \lambda_0(Na^+) - \lambda_0(Cl^-) \\ &= \Lambda_0(NH_4Cl) + \Lambda_0(NaOH) - \Lambda_0(NaCl) \end{aligned}$$

When we substitute numbers, we get

$$\Lambda_0(NH_4OH) = 149.7 + 248.1 - 126.5 = 271.3\ \Omega^{-1}\text{cm}^2\,\text{mol}^{-1}$$

D. Degree of dissociation

The Arrhenius theory (Sections 7-3 and 11-3) of the dissociation of electrolytes into ions offers an explanation of the decrease in conductivity as concentration increases. Arrhenius assumed incomplete dissociation at higher concentrations, but complete dissociation at infinite dilution. He therefore considered the lower Λ at higher concentration to be a good measure of the number of ions, or **degree of dissociation**, which he defined as

DEGREE OF DISSOCIATION
$$\alpha = \frac{\Lambda}{\Lambda_0} \tag{11-13}$$

[Recall that α was defined by Eq. (7-16) and that Eq. (11-5) provides a method of calculating α, using the colligative property behavior of electrolytes.] Calculations of α using Eqs. (11-5) and (11-13) usually agree well enough to support the idea of partial dissociation. However, we now know that salts are completely ionic and that the conductance mechanism of molecular strong electrolytes is complex.

EXAMPLE 11-16: (**a**) Use information from Examples 11-13 and 11-15 to calculate the apparent degree of dissociation of a 0.100 M NaCl solution. (**b**) Use information from Example 11-7 to calculate the same quantity by the van't Hoff method, and compare the result to that obtained in part (**a**).

Solution:

(**a**) In order to use Eq. (11-13), we need Λ and Λ_0. From Example 11-13, Λ for 0.100 M NaCl is 107 $\Omega^{-1}\text{cm}^2\,\text{mol}^{-1}$; and from Example 11-15, Λ_0 for NaCl is 126.5. Substituting into Eq. (11-13), we obtain

$$\alpha = \frac{\Lambda}{\Lambda_0} = \frac{107}{126.5} = 0.846$$

(**b**) From the tabulation in Example 11-7, we find that the calculated K_f for a 0.10 M NaCl solution is 3.48. Substituting into Eq. (11-3), we get

$$i = \frac{K_f}{K_{f,\,non}} = \frac{3.48}{1.86} = 1.87$$

Substitution into Eq. (11-5) yields

$$\alpha = \frac{i-1}{\nu-1} = \frac{1.87-1}{2-1} = 0.87$$

The agreement is good enough to support the assumptions upon which the two methods are based.

E. Molar conductivity of ions at infinite dilution

1. Velocity

The individual and independent conductivities of ions, postulated by Kohlrausch [Eq. (11-12)], are determined by the velocities of those ions. The speed of an ion as it moves under the influence of an electric field is called its *mobility*, u; i.e., $u = vE^{-1}$, where E is the potential gradient.

EXAMPLE 11-17: What are the SI units of mobility?

Solution: Velocity is measured in meters per second and potential gradient in volts per meter. Thus

$$u = \frac{v}{E} = \frac{\text{m s}^{-1}}{\text{V m}^{-1}} = \text{m}^2\,\text{V}^{-1}\text{s}^{-1}$$

warning: The unit $\text{cm}^2\,\text{V}^{-1}\text{s}^{-1}$ is used in most tables.

Again, we consider the strong electrolyte A_pB_q yielding its ions in solution. If the concentration of the solution is C, the concentration of cations is pC and the current I_+ carried by cations through a unit area A in unit time is $z_+u_+pC\mathscr{F}EA$. Combining Eq. (11-6) with Eq. (11-9) and substituting for I gives

$$\kappa_+ = \frac{I_+}{\Delta V}\left(\frac{\ell}{A}\right) = \frac{I_+}{A}\left(\frac{1}{E}\right) = \frac{z_+u_+pC\mathscr{F}EA}{AE} = z_+u_+pC\mathscr{F} \qquad \textbf{(11-14)}$$

The molar ionic conductivity, from Eq. (11-8), is

$$\lambda_+ = \frac{\kappa_+}{C_+} = \frac{z_+u_+pC\mathscr{F}}{pC}$$

Letting i represent both cations and anions, we have

$$\lambda_i = |z_i|u_i\mathscr{F} \qquad \textbf{(11-15)}$$

EXAMPLE 11-18: If the mobility of a certain univalent cation is 3.78×10^{-4} cm² V⁻¹s⁻¹, what is its molar ionic conductivity?

Solution: We simply substitute into Eq. (11-15); any challenge is in juggling the units. Thus

$$\lambda_+ = 1(96\,500 \text{ C mol}^{-1})(3.78 \times 10^{-4} \text{ cm}^2\text{ V}^{-1}\text{s}^{-1}) = 36.5 \text{ C mol}^{-1}\text{cm}^2\text{ V}^{-1}\text{s}^{-1}$$

We then use C = A s and $\text{A V}^{-1} = \Omega^{-1}$ to reduce the units, so that

$$\lambda_+ = 36.5\ \Omega^{-1}\text{cm}^2\text{ s}^{-1}$$

Unfortunately, we can't obtain λ values in this way because we can't measure mobilities. In fact, Eq. (11-15) is used to calculate u from λ, once λ has been determined.

2. Transport number

Since the cation and anion usually have different mobilities, they usually carry different fractions of the current. The fraction of the current carried by an ion is called the **transport number**, t, of that ion (formerly called the *transference number*), i.e.,

$$t_+ = \frac{I_+}{I_+ + I_-} \qquad t_- = \frac{I_-}{I_+ + I_-} \tag{11-16}$$

Clearly,

$$t_+ + t_- = 1 \tag{11-17}$$

The transport number is simply a *current fraction*, which is analogous to a mole fraction. If we substitute $I_+ = z_+u_+pC\mathscr{F}EA$ for I_+ and the equivalent term for I_- into Eqs. (11-16), and cancel terms, we obtain

$$t_+ = \frac{z_+pu_+}{z_+pu_+ + z_-qu_-} \quad \text{and} \quad t_- = \frac{z_-qu_-}{z_+pu_+ + z_-qu_-} \tag{11-18}$$

In the special case where $z_+ = z_-$ and, necessarily, $p = q$, Eq. (11-18) reduces to

$$t_+ = \frac{u_+}{u_+ + u_-} \quad \text{and} \quad t_- = \frac{u_-}{u_+ + u_-} \tag{11-19}$$

EXAMPLE 11-19: The mobilities of K^+ and acetate ion are 7.61×10^{-4} and 4.23×10^{-4} cm² V⁻¹s⁻¹, respectively, at 25 °C. What are their transport numbers in a potassium acetate solution?

Solution: Since $p = q$, we can use Eq. (11-19). Substitution gives

$$t_+ = \frac{7.61 \times 10^{-4}}{(7.61 + 4.23) \times 10^{-4}} = \frac{7.61}{11.84} = 0.643 \quad \text{and} \quad t_- = 1 - t_+ = 1 - 0.643 = 0.357$$

Recall that we can't measure mobilities, so we don't obtain transport numbers from them. However, we can determine transport numbers by several methods. Most textbooks describe the **Hittorf method** or the **moving boundary method** or both. These methods determine t_+ and t_- at several concentrations; then extrapolation to infinite dilution gives $t_{+,0}$ and $t_{-,0}$.

The importance of transport numbers is that they give a simple way to obtain molar ionic conductivities. Equations (11-15) and (11-18) show that t and λ are related through u. It may be an oversimplification, but we get a valid result if we solve Eq. (11-15) for u and substitute into Eq. (11-18). Thus

$$t_+ = \frac{p\lambda_+}{p\lambda_+ + q\lambda_-} \quad \text{and} \quad t_- = \frac{q\lambda_-}{p\lambda_+ + q\lambda_-} \tag{11-20}$$

For infinite dilution, we use Eq. (11-12) to get

$$t_{+,0} = \frac{p\lambda_{+,0}}{\Lambda_0} \quad \text{and} \quad t_{-,0} = \frac{q\lambda_{-,0}}{\Lambda_0} \qquad \textbf{(11-21)}$$

Using these expressions we can calculate ionic conductivities from experimental values of t and Λ.

EXAMPLE 11-20: In Example 11-19, we obtained the transport numbers of K^+ and Ac^- in a KAc solution. If an experiment yields 114.3 $\Omega^{-1}cm^2\,mol^{-1}$ as the value of Λ_0 of KAc, what are the experimental molar ionic conductivities at infinite dilution?

Solution: Rearranging Eq. (11-21) (note that $p = q = 1$), we have

$$\lambda_{+,0} = t_{+,0}\Lambda_0 \quad \text{and} \quad \lambda_{-,0} = t_{-,0}\Lambda_0$$

Substituting, we obtain

$$\lambda_{+,0} = 0.643(114.3) = 73.5\ \Omega^{-1}cm^2\,mol^{-1} \quad \text{and} \quad \lambda_{-,0} = 0.357(114.3) = 40.8\ \Omega^{-1}cm^2\,mol^{-1}$$

---

3. Alternative method of calculation

Another way to calculate the second λ($\lambda_{-,0}$ or $\lambda_{+,0}$, as the case may be) is subtraction from Λ_0. This method is an application of Eq. (11-12).

EXAMPLE 11-21: In Example 11-20, we found that $\lambda_0(K^+)$ is 73.5 $\Omega^{-1}cm^2\,mol^{-1}$. If Λ_0 for KI is 150.3 and that for KNO_3 is 144.9, what are the values of $\lambda_{-,0}$ for I^- and $NO_3{}^-$?

Solution: Rearranging Eq. (11-12), we have $q\lambda_{-,0} = \Lambda_0 - p\lambda_{+,0}$. Thus

for I^-: $1(\lambda_{-,0}) = 150.3 - 1(73.5)$
$\lambda_{-,0} = 76.8\ \Omega^{-1}cm^2\,mol^{-1}$

for NO_3^-: $1(\lambda_{-,0}) = 144.9 - 1(73.5)$
$\lambda_{-,0} = 71.4\ \Omega^{-1}cm^2\,mol^{-1}$

---

4. Tables of values

From a few transport numbers and many molar conductivities, molar ionic conductivities at infinite dilution have been calculated for most ions, and are available in tables. Table 11-2 contains only a few such values, including one set to illustrate the dependence of

TABLE 11-2: Molar Ionic Conductivities at Infinite Dilution

At 298 K			
Cation	$\lambda_{+,0}$ ($\Omega^{-1}m^2\,mol^{-1}$)	Anion	$\lambda_{-,0}$ ($\Omega^{-1}m^2\,mol^{-1}$)
$NH_4{}^+$	0.007 34	$C_2H_3O_2{}^-$	0.004 08
Ba^{2+}	0.0127	$C_2O_4{}^{2-}$	0.0146
H^+	0.035 00	Cl^-	0.007 63
K^+	0.007 35	$NO_3{}^-$	0.007 14
La^{3+}	0.0209	OH^-	0.0199
Na^+	0.005 01	$SO_4{}^{2-}$	0.0160
Ag^+	0.006 19		
Variation with Temperature			
	273 K	323 K	373 K
K^+	0.004 04	0.0115	0.0206

conductivity on temperature. (You must multiply the listed values by 10^4 when units of cm^2 are needed.)

F. Applications

1. Calculation of Λ_0 and mobility

With molar ionic conductivities available, we can calculate Λ_0 for any compound by using Eq. (11-12), and the mobility of any ion by using Eq. (11-15).

EXAMPLE 11-22: Calculate (**a**) the molar conductivity at infinite dilution of $La(OH)_3$ and (**b**) the mobility of each ion at infinite dilution.

Solution:

(**a**) Recall that we are dealing with 1 mol of $La(OH)_3$ when using molar conductivity. By Eq. (11-12),

$$\begin{aligned}\Lambda_0 &= p\lambda_{+,0} + q\lambda_{-,0} \\ &= 1(\lambda_0)(La^{3+}) + 3\lambda_0(OH^-) = 0.0209 + 3(0.0199) \\ &= 0.0806\ \Omega^{-1}m^2\,mol^{-1}\end{aligned}$$

(**b**) Rearranging Eq. (11-15) for La^{3+},

$$\begin{aligned}u_{+,0} &= \frac{\lambda_{+,0}}{z_+\mathscr{F}} = \left(\frac{0.0209\ \Omega^{-1}m^2\,mol^{-1}}{3(96\,500\ C\,mol^{-1})}\right)\left(\frac{10^4\ cm^2}{1\ m^2}\right) \\ &= (7.22 \times 10^{-4}\ \Omega^{-1}cm^2\,C^{-1})\left(\frac{C}{A\ s}\right)\left(\frac{A\ V^{-1}}{\Omega^{-1}}\right) \\ &= 7.22 \times 10^{-4}\ cm^2\,V^{-1}s^{-1}\end{aligned}$$

and for OH,

$$u_{-,0} = \frac{\lambda_{-,0}}{z_-\mathscr{F}} = \frac{0.0199}{1(96\,500)} = 20.6 \times 10^{-4}\ cm^2\,V^{-1}s^{-1}$$

---

2. Calculation of equilibrium constants

We can calculate two types of equilibrium constants from conductivity data: ionization constants of weak acids and bases and solubility product constants of slightly soluble salts.

EXAMPLE 11-23: Calculate the ionization constant K_a of acetic acid if the molar conductivity of a 0.005 00 M solution is 22.9 $\Omega^{-1}cm^2\,mol^{-1}$.

Solution: We first have to calculate the degree of dissociation because it is the only connection between conductivity and ionization. This step involves solving Eq. (11-13) for α. We substitute Eq. (11-12) into Eq. (11-13), obtain values of λ from Table 11-2, and convert to cm^2:

$$\alpha = \frac{\Lambda}{\Lambda_0} = \frac{\Lambda}{p\lambda_{+,0} + q\lambda_{-,0}}$$

The equation for the ionization of acetic acid is $HC_2H_3O_2 \rightleftharpoons H^+ + C_2H_3O_2^-$, so $p = q = 1$, and

$$\alpha = \frac{22.9}{1(350) + (40.8)} = \frac{22.9}{390.8} = 0.0586$$

We now look back to Chapter 7 and specifically to Example 7-21, where we calculated equilibrium constants using α. In Example 7-21a, we derived the expression

$$K_C = \frac{\alpha^2}{1-\alpha}C_0$$

for the general equilibrium system $AB \rightleftharpoons A^+ + B^-$. Substituting our calculated value for α, we have

$$K_a = \left[\frac{(0.0586)^2}{1 - 0.0586}\right](0.005\,00) = 1.82 \times 10^{-5}$$

which rounded to 1.8×10^{-5} is the value we all remember for this common weak acid.

---

3. Solubility

The application of molar conductivity to solubility requires some description. The conductivity κ of solutions of slightly soluble salts can be measured (see Section 11-4B). But the solutions are so dilute that the contribution of the water is significant and must be subtracted:

$$\kappa(\text{salt}) = \kappa(\text{solution}) - \kappa(H_2O) \qquad \textbf{(11-22)}$$

Because the solutions are so dilute, their molar conductivities are approximately the same as the value at infinite dilution, i.e., $\Lambda \approx \Lambda_0$. Then the concentration can be calculated from the rearranged Eq. (11-8):

$$C = \frac{\kappa(\text{salt})}{\Lambda_0(\text{salt})} \qquad \textbf{(11-23)}$$

EXAMPLE 11-24: In a conductivity cell, a saturated solution of $BaSO_4$ has a conductivity of $2.904 \times 10^{-4}\ \Omega^{-1}\text{m}^{-1}$ at 298 K. The water used has a conductivity of $1.20 \times 10^{-6}\ \Omega^{-1}\text{m}^{-1}$. What is the value of K_{sp} for $BaSO_4$?

Solution: We use Eq. (11-22), to calculate the conductivity of the salt, obtaining

$$\kappa(BaSO_4) = 2.904 \times 10^{-4} - 1.20 \times 10^{-6} = 2.892 \times 10^{-4}\ \Omega^{-1}\text{m}^{-1}$$

We substitute the ionic conductivities from Table 11-2 into Eq. (11-12) and have

$$\Lambda_0(BaSO_4) = p\lambda_0(Ba^{2+}) + q\lambda_0(SO_4^{\ 2-}) = 1(0.0127) + 1(0.0160) = 0.0287\ \Omega^{-1}\text{m}^2\,\text{mol}^{-1}$$

Next, we substitute into Eq. (11-23), converting C from m^3 to dm^3, to get

$$C = \left(\frac{2.892 \times 10^{-4}\ \Omega^{-1}\text{m}^{-1}}{0.0287\ \Omega^{-1}\text{m}^2\,\text{mol}^{-1}}\right)\left(\frac{\text{m}^3}{10^3\ \text{dm}^3}\right) = 1.01 \times 10^{-5}\ \text{mol dm}^{-3}$$

Then, since K_{sp} is the product of the molar concentrations of the ions in a saturated solution, we can write

$$K_{sp}(BaSO_4) = [Ba^{2+}][SO_4^{\ 2-}] = [Ba^{2+}]^2 = (1.01 \times 10^{-5})^2 = 1.02 \times 10^{-10}$$

11-5. Activity and Activity Coefficients

A. Modifications for ions

The concept of *activity* was introduced in Section 6-7 as a ratio of fugacities [Eq. (6-64)]. For solutions, activity may be considered an *effective* concentration, which is used to adjust "ideal" thermodynamic expressions for the deviations from ideal behavior exhibited by real systems. For either solute or solvent, the expression for the chemical potential, Eq. (6-74), becomes

$$\mu = \mu^\circ + RT \ln \frac{P}{P^\circ} \qquad \textbf{(11-24)}$$

where P is vapor pressure and the superscript, as usual, indicates the pure substance. In an ideal solution, Raoult's law applies to both components, so $P/P^\circ = X$, and

$$\mu = \mu^\circ + RT \ln X \qquad \textbf{(11-25)}$$

For real solutions, we replace X with the effective concentration, a, so

$$\mu = \mu^\circ + RT \ln a \qquad \textbf{(11-26)}$$

The relationship between X and a is the **activity coefficient**, γ. Analogous to Eq. (6-66), $\gamma = a/X$, or

$$a = \gamma X \tag{11-27}$$

Now,

$$\mu = \mu^\circ + RT \ln \gamma X \tag{11-28}$$

Equation (11-28) applies to either solvent or solute, but a different standard state must be chosen for each component. For the solvent, the standard state is the pure solvent. For the solute, it is a hypothetical state: the pure solute is imagined to behave as though it were in an infinitely dilute solution; i.e., Henry's law is obeyed.

If other units of concentration are used, the activity coefficient has a different value. For molality, m, typically used with colligative properties, Eq. (11-27) becomes

$$a = \gamma m \tag{11-29}$$

For molarity, typically used with equilibrium constants, Eq. (11-27) becomes

$$a = \gamma C \tag{11-30}$$

note: Some authors assign different symbols for the activity coefficients in each of the three equations (11-27), (11-29), and (11-30); others leave it up to you to keep in mind which unit of concentration you're using.

The usual unit of concentration for electrolytic solutions is molality, so we'll use Eq. (11-29) at first.

On the basis of Kohlrausch's law, Eq. (11-12), we define (for ionic compounds) separate **ionic activities**: a_+ for the cation and a_- for the anion. We can now write Eq. (11-26) for both ions in a 1:1 electrolyte, i.e.,

$$\mu_+ = \mu_+^\circ + RT \ln a_+ \qquad \mu_- = \mu_-^\circ + RT \ln a_-$$

If we let $\mu = \mu_+ + \mu_-$ and $\mu^\circ = \mu_+^\circ + \mu_-^\circ$ and add the ion expressions, we get

$$\mu = \mu^\circ + RT \ln a_+ a_- \tag{11-31}$$

Comparing Eq. (11-31) to Eq. (11-26), we see that $a_+ a_-$ is the activity of the electrolyte, a_2 (where 2 represents solute, as usual). For a solute of the general formula A_pB_q, the general definition of a_2 is

$$a_2 = a_+^p a_-^q \tag{11-32}$$

If A_pB_q is in a solution of concentration m, then, from Eq. (11-29),

$$a_+ = \gamma_+ m_+ \qquad a_- = \gamma_- m_-$$

where $m_+ = pm$ and $m_- = qm$. Now,

$$a_2 = (\gamma_+ pm)^p (\gamma_- qm)^q \tag{11-33}$$

Since we can't measure either the activity or the activity coefficient for single ions, we can't apportion responsibility for nonideal behavior between the ions. Thus we have to assume that each ion is equally responsible, and we average their activities as a geometric mean:

$$a_\pm = (a_+^p a_-^q)^{1/v} \tag{11-34}$$

Also

$$\gamma_\pm = (\gamma_+^p \gamma_-^q)^{1/v} \tag{11-35}$$

where $v = p + q$, as in Eq. (11-5). We use these quantities, called the **mean ionic activity** and the **mean ionic activity coefficient**, for either ion. Rearranging Eq. (11-34) and comparing it to Eq. (11-32), we see that

$$a_2 = a_\pm^v \tag{11-36}$$

Another average, the **mean ionic molality**, also is useful:

$$m_\pm = [(pm)^p (qm)^q]^{1/v} \tag{11-37}$$

or

$$m_\pm = m(p^p q^q)^{1/v} \tag{11-38}$$

Combining Eqs. (11-33)–(11-37), we obtain a new form of Eq. (11-29):

$$a_\pm = \gamma_\pm m_\pm \tag{11-39}$$

EXAMPLE 11-25: Obtain expressions for (**a**) the mean activity coefficients of electrolytes of the common types AB and A_2B in solutions of concentration m (note the difference); and (**b**) the activity of each type as a function of the mean activity coefficient.

Solution: We want to replace $m_\pm$ in Eq. (11-39) with its value from Eq. (11-38).

(**a**) For the 1:1 case, $m_\pm = m$, so we have

$$\gamma_\pm = \frac{a_\pm}{m}$$

and for the 2:1 case,

$$m_\pm = m[(2)^2(1)^1]^{1/3} = 4^{1/3}m$$

Thus

$$\gamma_\pm = \frac{a_\pm}{4^{1/3}m}$$

(**b**) The activity of an electrolyte solute is given by Eq. (11-36). We substitute the $a_\pm$ expressions obtained in part (a) into that equation. For the 1:1 case,

$$a_2 = a_\pm^{(p+q)} = a_\pm^{(1+1)} = (\gamma_\pm m)^2 = \mathrm{m}^2\gamma_\pm^2$$

and for the 2:1 case,

$$a_2 = a_\pm^{(2+1)} = (4^{1/3}m\gamma_\pm)^3 = 4m^3\gamma_\pm^3$$

B. Ionic strength

The mean activity coefficient is closely related to the number of charged particles rather than to the chemical properties of different ions. A new quantity that represents a solution by including both concentration and charge is the **ionic strength**, I (μ is often used instead):

$$I = \frac{1}{2}\sum_i m_i z_i^2 \tag{11-40}$$

or

$$I = \frac{1}{2}\sum_i C_i z_i^2 \tag{11-41}$$

Here m and C are the molal or molar concentrations of ion i, and z is its charge. *All* ions present—not just those of an electrolyte of concern—must be included. This concept was proposed by Lewis and Randall, who also made the generalization that a strong electrolyte has the same activity in all solutions having equal ionic strengths (for dilute aqueous cases). More specifically, they found that

$$\log \gamma_\pm = k\sqrt{I} \tag{11-42}$$

EXAMPLE 11-26: Calculate the ionic strength of a 0.010 M solution of Na_2SO_4.

Solution: The compound will yield 2 mol Na^+ and 1 mol SO_4^{2-} per mol of compound. Using Eq. (11-41), we have $C(Na^+) = 0.020\ M$ and $C(SO_4^{2-}) = 0.010\ M$. Thus

$$I = \tfrac{1}{2}[C(\mathrm{Na}^+) \times z(\mathrm{Na}^+)^2 + C(\mathrm{SO}_4^{2-}) \times z(\mathrm{SO}_4^{2-})^2]$$
$$= \tfrac{1}{2}[0.020(1)^2 + 0.010(-2)^2] = 0.030$$

C. Determination of activity coefficients

We can measure activity coefficients by using the properties involved in studies of electrolytic solutions: freezing-point depression, osmotic pressure, dissociation constants, and solubility of slightly soluble salts (the last two involve conductivity). Perhaps the best method is to measure the emf of electrochemical cells, which we describe in the next chapter.

To develop the method involving solubility of slightly soluble salts, we begin by writing the true K_{sp}, using activities, and assuming the special case of a 1:1 electrolyte (AB). Now we'll use molarity, so Eq. (11-39) becomes $a_\pm = C_\pm \gamma_\pm$. Then

$$K_{sp} = a_+ a_- = a_\pm^2 = C_\pm^2 \gamma_\pm^2$$

For the 1:1 case, $C_\pm = C$, which is the solubility of AB, i.e., the concentration of a saturated solution, often determined from conductivity measurements as in Example 11-24. Next we take the square root and then the logarithm of both sides:

$$\log \sqrt{K_{sp}} = \log C + \log \gamma_\pm$$

$$\log C = \log \sqrt{K_{sp}} - \log \gamma_\pm \qquad \textbf{(11-43)}$$

From Eq. (11-42), we replace the activity coefficient with the ionic strength:

$$\log C = \log \sqrt{K_{sp}} - k\sqrt{I} \qquad \textbf{(11-44)}$$

which is the equation of a straight line. If we measure C at various I and plot the data as log C versus $\sqrt{I}$, we can extrapolate to obtain the intercept, $\log \sqrt{K_{sp}}$. We could calculate K_{sp} but we don't need it; we can use the intercept value as a constant in Eq. (11-43) in order to calculate $\gamma_\pm$.

The technique for varying the ionic strength is to add a soluble salt that doesn't share a common ion with the slightly soluble salt.

EXAMPLE 11-27: Calculate the ionic strengths of solutions of CuCl and $NaNO_3$ when **(a)** $[NaNO_3] = 0.0100\ M$ and $[CuCl] = 0.001\,31\ M$; **(b)** $[NaNO_3] = 0.0500\ M$ and $[CuCl] = 0.001\,75\ M$.

Solution: The ionic strength for molarity is given by

$$I = \frac{1}{2}\sum_i C_i z_i^2$$

(a) $I = \frac{1}{2}[0.00131(1)^2 + 0.00131(1)^2 + 0.0100(1)^2 + 0.0100(1)^2] = 0.0113$

(b) $I = \frac{1}{2}[0.00175(1)^2 + 0.00175(1)^2 + 0.0500(1)^2 + 0.0500(1)^2] = 0.0518$

EXAMPLE 11-28. Let's expand Example 11-27 into a laboratory experiment to determine the mean activity coefficient of CuCl. We would prepare five or six more solutions and determine their concentrations. **(a)** Convert the data of Example 11-27 into a form suitable for plotting. **(b)** Assume that the straight line drawn through all the points plotted crosses the *y* axis at -2.972 and calculate K_{sp}. **(c)** Calculate $\gamma_\pm$ for the two concentrations.

Solution:

(a) The method of conversion is based on Eq. (11-44): $\log C = -k\sqrt{I} + \log \sqrt{K_{sp}}$, which is in slope-intercept form. Thus we want to plot log C against $\sqrt{I}$. Using the data in Example 11-27, we have

C	I	$\log C$	$\sqrt{I}$
1.31×10^{-3}	1.13×10^{-2}	-2.883	1.063×10^{-1}
1.75×10^{-3}	5.18×10^{-2}	-2.757	2.276×10^{-1}

(b) The intercept is $\log \sqrt{K_{sp}} = -2.972$. Thus $K_{sp} = 1.14 \times 10^{-6}$.

(c) Rearranging Eq. (11-43), we get

$$\log \gamma_\pm = \log \sqrt{K_{sp}} - \log C = -2.972 - \log C$$

For $C = 1.31 \times 10^{-3}$,

$$\log \gamma_\pm = -2.972 - (-2.883) = -0.089$$

$$\gamma_\pm = 0.815$$

For $C = 1.75 \times 10^{-3}$,

$$\log \gamma_\pm = -2.972 - (-2.757) = -0.215$$
$$\gamma_\pm = 0.610$$

11-6. Debye–Hückel Theory

The Arrhenius theory of partial dissociation of electrolytes in solution applies only to weak electrolytes. The theory is supported by good agreement of dissociation constants (K_a and K_b) calculated from conductivities and from colligative properties at all concentrations. For strong electrolytes, however, such agreement is not found; furthermore, two significant properties—the *i* factor and the activity coefficient—decrease at first as concentration increases but then increase. The clinching argument against applying the Arrhenius theory to strong electrolytes is our knowledge that salts always exist as separate ions, even in the solid state. Thus, during the dissolving process, the solute compound can't really be dissociating, but rather the solvent overcomes the electrostatic forces of attraction between its ions. This effect depends primarily on the dielectric constant, because the force of attraction between two charges is inversely proportional to the dielectric constant of the medium.

The Debye–Hückel theory accounts for many of the phenomena observed in dilute solutions of strong electrolytes. It is, however, very complicated mathematically, so we'll outline only its basic assumptions and those results that explain the experimental phenomena described in this chapter. Mathematical simplifications in the development of the theory limit its validity to very dilute solutions, i.e., concentrations below about 0.01 *M*. Modifications by Onsager and others have extended the range to somewhat higher concentrations.

There are three basic assumptions in the Debye–Hückel theory:

- Strong electrolytes are completely dissociated into ions.
- Deviations from ideal behavior result from electrostatic attractions between the charges of the ions.
- A given ion will have more ions of the opposite sign close to it than ions of the same sign; this cluster of ions is called the **ionic atmosphere**.

A. Conductivity behavior

When an electrical potential is applied across a solution, two effects produced by the ionic atmosphere prevent the ion from moving at the expected speed and hence from conducting the expected amount of current:

1. The **relaxation effect** (also called the *asymmetry effect*) occurs because the central ion tries to move out of its atmosphere. The symmetry present before application of the potential is distorted in such a way that an unbalanced force acts on the central ion, tending to hold it back.
2. The **electrophoretic effect** occurs because the atmosphere and the central ion are pulled in opposite directions. The ions are usually solvated, so solvent molecules are pulled along. Again, the central ion is held back by the flow of solvent against which it is trying to move.

The mathematical development of the Debye–Hückel theory represents an attempt to express quantitatively the electrical properties of the ionic atmosphere. One significant quantity is the *thickness of the ionic atmosphere*, usually written as $1/\kappa$ (this κ is not the conductivity!):

$$\frac{1}{\kappa} = \left(\frac{\varepsilon_0 \varepsilon_r kT}{L_A e^2 \sum_i C_i z_i^2}\right)^{1/2} \qquad \textbf{(11-45)}$$

where ε_0 is the permittivity of a vacuum, ε_r is the dielectric constant (relative permittivity) of the medium, and e is the charge on an electron. The term $\Sigma_i C_i z_i^2$ shows that thickness of the atmosphere is related to ionic strength [Eq. (11-41)]. Again, we see that the significant factors are the dielectric constant, concentration, and electron charge at a given temperature. The term $(kT)^{1/2}$ is proportional to the average speed of the ions, so this property is also a factor.

EXAMPLE 11-29: Calculate the thickness of the ionic atmospheres about the ions in a 0.001 00 *M* solution of KCl at 298 K.

Solution: We first need values for the constants in Eq. (11-45):

$$\varepsilon_0 = 8.8541 \times 10^{-12}\ \mathrm{C^2\,J^{-1}m^{-1}} \qquad \varepsilon_r = 78.5 \quad (\text{for } H_2O \text{ at } 298\ \mathrm{K})$$
$$k = 1.3807 \times 10^{-23}\ \mathrm{J\,K^{-1}} \qquad e = 1.6022 \times 10^{-19}\ \mathrm{C}$$

Next, we evaluate the summation. Since both $z = 1$,

$$\sum_i C_i z_i^2 = \sum C = 2(0.001\,00\ \mathrm{mol\,dm^{-3}})(10^3\ \mathrm{dm^3\,m^{-3}}) = 2.00\ \mathrm{mol\,m^{-3}}$$

Finally, we substitute into Eq. (11-45):

$$\frac{1}{\kappa} = \left[\frac{(8.854 \times 10^{-12}\ \mathrm{C^2\,J^{-1}m^{-1}})(78.5)(1.381 \times 10^{-23}\ \mathrm{J\,K^{-1}})(298\ \mathrm{K})}{(6.022 \times 10^{23}\ \mathrm{mol^{-1}})(1.602 \times 10^{-19}\ \mathrm{C})^2(2.00\ \mathrm{mol\,m^{-3}})}\right]^{1/2}$$
$$= (92.539 \times 10^{-18}\ \mathrm{m^2})^{1/2}$$
$$= 9.62 \times 10^{-9}\ \mathrm{m}$$

Note that the value is the same for every 1:1 electrolyte at this concentration.

The final result of the Debye–Hückel–Onsager theory for conductivity is the Debye–Hückel–Onsager equation:

DEBYE–HÜCKEL–ONSAGER EQUATION

$$\Lambda = \Lambda_0 - (M + N\Lambda_0)\sqrt{C} \qquad \textbf{(11-46)}$$

where M and N are complex expressions that include the quantities from Eq. (11-45) (except concentration), the viscosity of the solvent, the Faraday constant, and a factor for the type of electrolyte. Notice that Eq. (11-46) has the same form as the experimental expression, Eq. (11-11), where the slope S now is $-(M + N\Lambda_0)$. For a 1:1 electrolyte in water at 298 K, the calculated values of the complex expressions are

$$M = \frac{60.20\ \Omega^{-1}\mathrm{cm^2\,mol^{-1}}}{(\mathrm{mol\,dm^{-3}})^{1/2}} \qquad \text{and} \qquad N = 0.229\left(\frac{1}{\mathrm{mol\,dm^{-3}}}\right)^{1/2}$$

This odd way of writing the units is useful in calculations, given Λ_0 in $\Omega^{-1}\mathrm{cm^2\,mol^{-1}}$ and C in units of molarity. Predicted and experimental slopes for various electrolytes in water agree well, so long as concentrations are low. Deviations begin to appear at concentrations around 0.01 M, or somewhat lower when larger ionic charges are involved.

EXAMPLE 11-30: Use the Debye–Hückel–Onsager equation to calculate the Λ of a 0.005 000 M $AgNO_3$ aqueous solution and compare the result to the experimental value of 127.20 $\Omega^{-1}\mathrm{cm^2\,mol^{-1}}$.

Solution: We calculate Λ_0 using Eq. (11-12) and λ_0 values from Table 11-2:

$$\Lambda_0 = p\lambda_0(Ag^+) + q\lambda_0(NO_3^-) = 1(61.9) + 1(71.4) = 133.3\ \Omega^{-1}\mathrm{cm^2\,mol^{-1}}$$

If we solve the slope term of Eq. (11-46) first, we can avoid a long mathematical expression. Using the values for M and N given in the preceding text paragraph, we have

$$M + N\lambda_0 = 60.20\,\frac{\Omega^{-1}\mathrm{cm^2\,mol^{-1}}}{(\mathrm{mol\,dm^{-3}})^{1/2}} + 0.229\left(\frac{1}{\mathrm{mol\,dm^{-3}}}\right)^{1/2}(133.3\ \Omega^{-1}\mathrm{cm^2\,mol^{-1}})$$
$$= 90.73\,\frac{\Omega^{-1}\mathrm{cm^2\,mol^{-1}}}{(\mathrm{mol\,dm^{-3}})^{1/2}}$$

Substituting in Eq. (11-46) yields

$$\Lambda = 133.3\ \Omega^{-1}\mathrm{cm^2\,mol^{-1}} - 90.73\,\frac{\Omega^{-1}\mathrm{cm^2\,mol^{-1}}}{(\mathrm{mol\,dm^{-3}})^{1/2}}(0.005\,000\ \mathrm{mol\,dm^{-3}})^{1/2}$$
$$= 126.9\ \Omega^{-1}\mathrm{cm^2\,mol^{-1}}$$

The agreement with 127.2 is quite satisfactory.

B. Activity coefficients

The thermodynamic properties of a solution are affected by interionic attractions because the ionic atmosphere causes an electric potential at the surface of the ion. This electric potential is believed to be the cause of all deviations from ideal behavior of a solution; it is related to the mean ionic activity coefficient, the factor we use to account for such deviations. The additional **electrical free energy** arising from the ionic atmosphere of one ion is expressed as

$$G_e = kT \ln \gamma_i \qquad \textbf{(11-47)}$$

This free energy represents the energy needed, or work done, to charge the ion, if the uncharged species is in its final position in solution. Debye and Hückel derived an expression for G_e and hence $\ln \gamma_i$, which contains the thickness of the ionic atmosphere $1/\kappa$ [Eq. (11-45)] and a repetition of some of the quantities that appear in Eq. (11-45), i.e., the dielectric constant ε_r, permittivity of vacuum ε_0, and ionic charge z_i. When the restriction of very dilute solution is imposed, the ionic strength is introduced and ln is changed to log, so the expression becomes

$$\log \gamma_i = -Az_i^2\sqrt{I} \qquad \textbf{(11-48)}$$

where A contains many constants. For water at 298 K, $A = 0.5091$.

EXAMPLE 11-31: Calculate the values of the activity coefficient of the potassium ion in a 0.001 00 M aqueous solution of K_2SO_4 at 298 K, as predicted by Eq. (11-48).

Solution: First, we need a value for I. The concentrations for use in Eq. (11-41) are $[K^+]=$ 0.001 00 M aqueous solution of K_2SO_4 at 298 K as predicted by Eq. (11-48).

$$I = \tfrac{1}{2}[0.002\,00(1)^2 + 0.001\,00(-2)^2]$$
$$= \tfrac{1}{2}(0.006\,00) = 0.003\,00$$

Substituting in Eq. (11-48), we have

$$\log \gamma_i = -0.5091(1)^2\sqrt{0.003\,00} = -0.027\,885$$
$$\gamma_i = 0.94$$

Since ionic activity coefficients can't be measured, we normally need to know the mean activity coefficient. So we write Eq. (11-48) for the positive and negative ions and substitute in Eq. (11-35) to get

$$\log \gamma_\pm = -0.5091 z_+|z_-|\sqrt{I} \qquad \textbf{(11-49)}$$

Note the absolute value of the z_-. Sometimes the leading minus sign on the right-hand side is omitted; in that case, z_- must be used as a negative number. (You should attempt the derivation of Eq. (11-49) before proceeding to the Solved Problems.) Because of the assumption of very dilute solution, Eq. (11-49) applies only at low concentrations and must be considered a limiting law. Usually, it is called the **Debye–Hückel limiting law**.

EXAMPLE 11-32: Calculate the mean activity coefficient for the solution in Example 11-31.

Solution: The value of I is the same, so we substitute it into Eq. (11-49):

$$\log \gamma_\pm = -0.5091(1)(2)\sqrt{0.003\,00} = -0.055\,769$$
$$\gamma_\pm = 0.88$$

Equation (11-49) represents a straight line, so $\log \gamma_\pm$ should decrease as $\sqrt{I}$ increases. In fact, for most strong electrolytes, the relationship is a curve, which slopes up at higher concentrations. One contributing factor is *association*, i.e., the formation of ion pairs. Various attempts to obtain mathematical expressions that are valid for higher concentrations increase the complexity of the calculations and don't extend the range much. The Debye–Hückel equations are what we use for most applications, recognizing the limitation to very dilute solutions.

SUMMARY

In Table S-11 we show the major equations used in this chapter and connections between them. You should determine the conditions that make each equation valid and the conditions imposed to make the connections. Other important things you should know are

1. The development of the current theory of electrolytic solutions involved studies of colligative properties, conductivity, and activity coefficients.
2. Faraday's laws of electrolysis state that the mass produced is proportional both to the amount of electricity and to the equivalent weight of the substance: $\Delta m \propto Q$, and $\Delta m \propto M_m$/electrons gained or lost.
3. The Faraday constant is the quantity of electricity that will deposit the equivalent weight of any substance and also is the charge of one mole of electrons: $\mathcal{F} = 9.6485 \times 10^4 \text{ C eq}^{-1}$.
4. The van't Hoff i factor is a method of expressing the colligative properties of electrolytic solutions relative to solutions of nonelectrolytes and provides a means of calculating the degree of dissociation [Eqs. (11-3), (11-4), (11-5)].
5. Conductivity is the proportionality constant between conductance, area, and reciprocal length [Eq. (11-7)]. It increases with both temperature and concentration.
6. Molar conductivity is the conductivity of one mole of the electrolyte between plates that are one unit distance apart. It increases with temperature but decreases with concentration [Eq. (11-8)].
7. Molar conductivity at infinite dilution is obtained by extrapolation for strong electrolytes but must be calculated for weak electrolytes [Eqs. (11-11), (11-12)].
8. Kohlrausch's law of independent migration of ions states that, at infinite dilution, each ion of an electrolyte moves independently of the oppositely charged ion and makes its own constant contribution to the molar conductivity for every compound [Eq. (11-12)].
9. The molar conductivity at infinite dilution of a weak electrolyte can be calculated from molar conductivities of strong electrolytes [Eq. (11-12)].
10. The degree of dissociation may be calculated as the ratio of the molar conductivity to the molar conductivity at infinite dilution [Eq. (11-13)].
11. Molar ionic conductivities are obtained from molar conductivities of compounds by using transport numbers, which are expressions of the relative mobilities of the ions of the compound [Eq. (11-20)].
12. With ionic conductivities available, conductivities of compounds may be calculated as needed. From conductivities, it is possible to calculate ionic mobilities, degree of dissociation, ionization (dissociation) constant, and concentrations of very dilute solutions (and then K_{sp}).
13. Activities and activity coefficients for electrolytes are expressed using geometric means of the ion values [Eqs. (11-34), (11-35)].
14. Ionic strength incorporates the interaction of all ions in a solution, including both concentration and charge, and is proportional to the mean ionic activity coefficient [Eqs. (11-41), (11-42)].
15. Activity coefficients may be determined from measurements involving freezing-point depression, osmotic pressure, dissociation constants, and solubilities of slightly soluble salts.
16. The basic assumptions of the Debye–Hückel theory specify (**a**) complete dissociation of strong electrolytes and (**b**) electrostatic interionic attractive forces that are responsible for deviations from ideal behavior, primarily by creating an ionic atmosphere about an ion.
17. The conductivity behavior of strong electrolytes is explained by a relaxation effect and an electrophoretic effect. The thickness of the ionic atmosphere is an important factor.
18. The result of the Debye–Hückel theory of conductance is expressed succinctly in the Debye–Hückel–Onsager equation [Eq. (11-46)].
19. The behavior of activity coefficients in solution is explained by the electrical properties of the ionic atmosphere. The results are expressed in the Debye–Hückel limiting law, which is reliable at concentrations below 0.01 M [Eq. (11-48)].

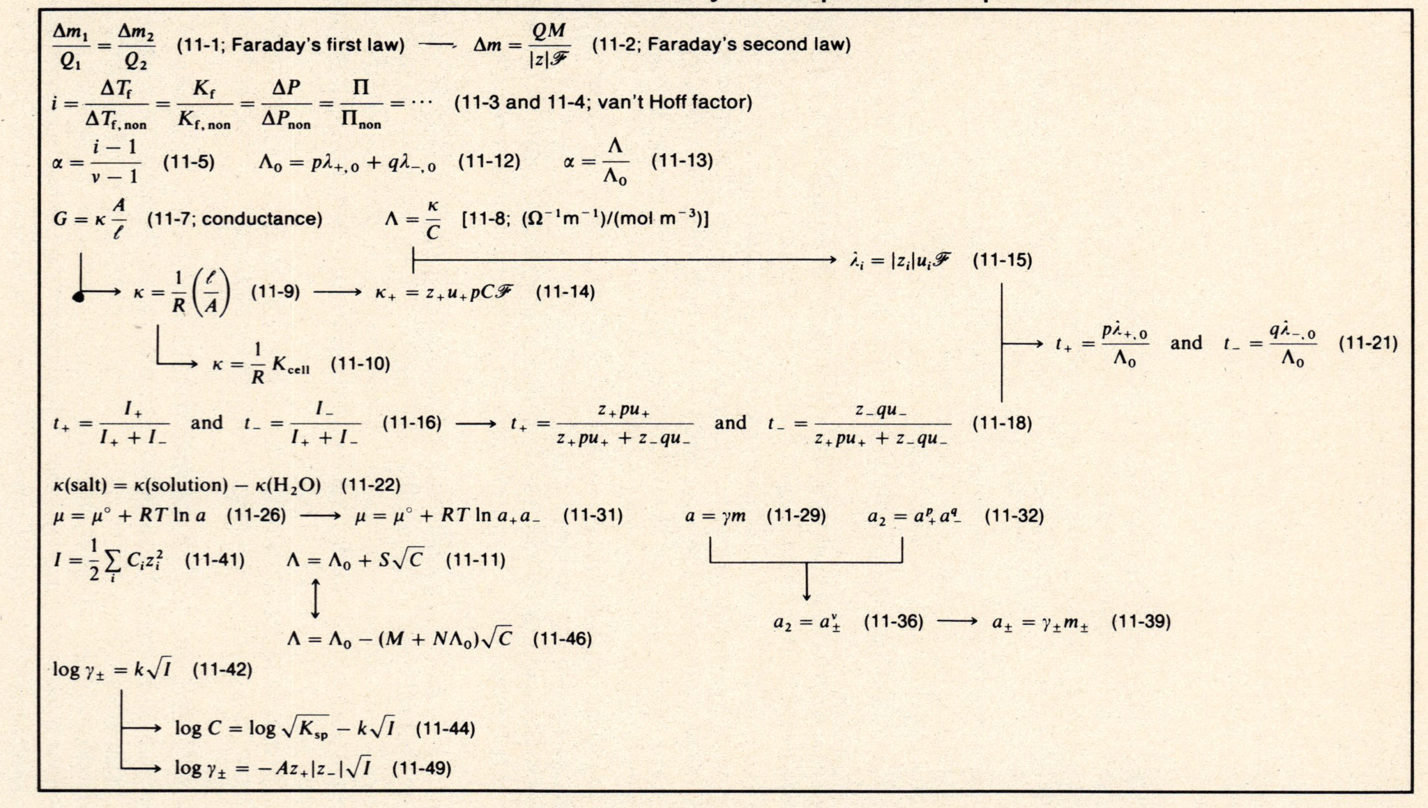

TABLE S-11: Summary of Important Equations

$\frac{\Delta m_1}{Q_1} = \frac{\Delta m_2}{Q_2}$ (11-1; Faraday's first law) $\longrightarrow$ $\Delta m = \frac{QM}{|z|\mathscr{F}}$ (11-2; Faraday's second law)

$i = \frac{\Delta T_f}{\Delta T_{f,\text{non}}} = \frac{K_f}{K_{f,\text{non}}} = \frac{\Delta P}{\Delta P_{\text{non}}} = \frac{\Pi}{\Pi_{\text{non}}} = \cdots$ (11-3 and 11-4; van't Hoff factor)

$\alpha = \frac{i-1}{\nu-1}$ (11-5) $\quad \Lambda_0 = p\lambda_{+,0} + q\lambda_{-,0}$ (11-12) $\quad \alpha = \frac{\Lambda}{\Lambda_0}$ (11-13)

$G = \kappa \frac{A}{\ell}$ (11-7; conductance) $\quad \Lambda = \frac{\kappa}{C}$ [11-8; $(\Omega^{-1}\text{m}^{-1})/(\text{mol m}^{-3})$]

$\longrightarrow \lambda_i = |z_i| u_i \mathscr{F}$ (11-15)

$\longrightarrow \kappa = \frac{1}{R}\left(\frac{\ell}{A}\right)$ (11-9) $\longrightarrow \kappa_+ = z_+ u_+ pC\mathscr{F}$ (11-14)

$\longrightarrow t_+ = \frac{p\lambda_{+,0}}{\Lambda_0}$ and $t_- = \frac{q\lambda_{-,0}}{\Lambda_0}$ (11-21)

$\longrightarrow \kappa = \frac{1}{R} K_{\text{cell}}$ (11-10)

$t_+ = \frac{I_+}{I_+ + I_-}$ and $t_- = \frac{I_-}{I_+ + I_-}$ (11-16) $\longrightarrow$ $t_+ = \frac{z_+ p u_+}{z_+ p u_+ + z_- q u_-}$ and $t_- = \frac{z_- q u_-}{z_+ p u_+ + z_- q u_-}$ (11-18)

$\kappa(\text{salt}) = \kappa(\text{solution}) - \kappa(H_2O)$ (11-22)

$\mu = \mu^\circ + RT \ln a$ (11-26) $\longrightarrow$ $\mu = \mu^\circ + RT \ln a_+ a_-$ (11-31) $\quad a = \gamma m$ (11-29) $\quad a_2 = a_+^p a_-^q$ (11-32)

$I = \frac{1}{2}\sum_i C_i z_i^2$ (11-41) $\quad \Lambda = \Lambda_0 + S\sqrt{C}$ (11-11)

$\updownarrow$

$\Lambda = \Lambda_0 - (M + N\Lambda_0)\sqrt{C}$ (11-46)

$\longrightarrow a_2 = a_\pm^\nu$ (11-36) $\longrightarrow$ $a_\pm = \gamma_\pm m_\pm$ (11-39)

$\log \gamma_\pm = k\sqrt{I}$ (11-42)

$\longrightarrow \log C = \log \sqrt{K_{sp}} - k\sqrt{I}$ (11-44)

$\longrightarrow \log \gamma_\pm = -Az_+|z_-|\sqrt{I}$ (11-49)

SOLVED PROBLEMS

Electrolytic Deposition

PROBLEM 11-1 How long must a current of 3.0 A flow in order to deposit 0.10 g of Cu from a solution of $CuSO_4$?

Solution: You should probably solve this problem in two steps: (*1*) Find the number of coulombs; and (*2*) find the time.

(1) To find the number of coulombs, apply Faraday's second law, Eq. (11-2), and calculate coulombs from moles. Solving Eq. (11-2) for Q, you get

$$Q = \frac{|z|\mathcal{F}\,\Delta m}{M_{\mathrm{m}}}$$

Since $Cu^{2+} \rightarrow Cu^0$, $|z| = 2$, and

$$Q = \frac{2(96\,500\text{ C mol}^{-1})(0.10\text{ g})}{63.54\text{ g mol}^{-1}} = 303.7\text{ C}$$

(2) To find the time, use $Q = It$:

$$t = \frac{Q}{I} = \frac{303.7\text{ A s}}{3.0\text{ A}} = 101.2\text{ s}$$

PROBLEM 11-2 If 10^{21} electrons flowed through a circuit in 30 s, what is the magnitude of the current?

Solution: You know that the Faraday constant can be interpreted as the quantity of electricity carried by one mole of electrons. Thus one mole of electrons is equivalent to 96 500 coulombs, and the quantity of electricity is

$$Q = (10^{21}\text{ electrons})\left(\frac{96\,500\text{ C mol}^{-1}}{6.022 \times 10^{23}\text{ electrons mol}^{-1}}\right) = 160.2\text{ C}$$

Again, $Q = It$, so

$$I = \frac{Q}{t} = \frac{160.2\text{ A s}}{30\text{ s}} = 5.34\text{ A}$$

PROBLEM 11-3 When a solution of a tin salt is subjected to a current of 4.0 A for 10 minutes, 0.731 g of tin is deposited. What is the charge of the tin ion in this compound?

Solution: You can determine the charge by dividing equivalent weight ($M_{\mathrm{m}}/|z|$) into molar mass, if you know the equivalent weight. To obtain the equivalent weight, first calculate Q:

$$Q = It = (4.0\text{ A})(600\text{ s}) = 2400\text{ A s} = 2400\text{ C}$$

Solving Eq. (11-2) for $M_{\mathrm{m}}/|z|$ yields

$$\frac{M_{\mathrm{m}}}{|z|} = \frac{\Delta m\mathcal{F}}{Q} = \frac{(0.731\text{ g})(96\,500\text{ C eq}^{-1})}{2400\text{ C}} = 29.39\text{ g mol}^{-1}$$

Now solve for $|z|$, using the molar mass of tin, 118.7 g mol^{-1}:

$$|z| = \frac{M_{\mathrm{m}}}{29.39} = \frac{118.7\text{ g mol}^{-1}}{29.39\text{ g eq}^{-1}} = 4.04\text{ eq mol}^{-1}$$

Thus the charge of the ion must be +4.

PROBLEM 11-4 You have 100 mL of a 0.0500 *M* solution of CrO_4^{2-}, which you want to reduce completely to Cr^{3+}. If you use electrolysis with a constant current of 2.50 A, how long will it take?

Solution: The amount of CrO_4^{2-} you have is (0.100 L)(0.0500 mol L^{-1}) = 0.005 00 mol. The reduction reaction is $Cr^{6+} + 3e^- \rightarrow Cr^{3+}$. For this reaction $|z|$ must be interpreted as the number of electrons gained. From Eq. (11-2)

$$Q = \frac{|z|\mathscr{F}\,\Delta m}{M_m}$$

Substituting this expression into $t = Q/I$, with $\Delta m/M_m = 0.005\,00$, you get

$$t = \frac{|z|\mathscr{F}\,\Delta m}{M_m I} = \frac{3(96\,000 \text{ A s mol}^{-1})(0.005\,00 \text{ mol})}{2.50 \text{ A}} = 579 \text{ s}$$

PROBLEM 11-5 One type of coulometer is based on the reduction of iodide ion to iodine. In one experiment, 20.3 g of I_2 was released by a constant current that flowed for 1.50 h. What was the current?

Solution: Combining $Q = It$ and Eq. (11-2) as in Problem 11-4 gives

$$I = \frac{|z|\mathscr{F}\,\Delta m}{M_m t}$$

The molar mass of I_2 is 253.8 g, and since $2I^- \rightarrow I_2$, 2 electrons are lost and $|z| = 2$:

$$I = \frac{2(96\,500)(20.3)}{253.8(1.50)(3600)} = 2.86 \text{ A}$$

Nonideal Colligative Properties

PROBLEM 11-6 A 0.010 m solution of $Pb(NO_3)_2$ was found experimentally to freeze at -0.049 °C. Calculate the van't Hoff i factor and the degree of dissociation α.

Solution: The expected freezing-point depression [from Eq. (9-15)] is

$$\Delta T_f = K_f m = 0.0186$$

Using Eq. (11-4), you get

$$i = \frac{\Delta T_f}{\Delta T_{f,\text{non}}} = \frac{0.049}{0.0186} = 2.6$$

and from Eq. (11-5),

$$\alpha = \frac{i-1}{\nu-1} = \frac{2.6-1}{3-1} = \frac{1.6}{2} = 0.80$$

PROBLEM 11-7 Derive Eq. (11-5).

Solution: In Section 11-3C, the clue was given that you should let i equal the concentration after dissociation divided by the concentration before dissociation. The procedure for getting the required algebraic expressions is like that used in Chapter 7, where we introduced degree of dissociation. Dissociation can be represented in general as

$$A_pB_q \rightleftharpoons pA + qB$$

Begin by setting up the usual table, as in Example 7-21:

Start	C_0	0	0
Change	$-\alpha C_0$	$p\alpha C_0$	$q\alpha C_0$
Equilibrium	$C_0 - \alpha C_0$	$p\alpha C_0$	$q\alpha C_0$

The total concentration of all particles is the sum of the equilibrium concentrations, or

$$C = C_0 - \alpha C_0 + p\alpha C_0 + q\alpha C_0$$

Factoring out C_0 and recognizing that $p + q = \nu$, you get

$$C = C_0(1 - \alpha + \alpha\nu)$$

Now let i be the ratio of C to C_0:

$$i = \frac{C}{C_0} = 1 - \alpha + \alpha\nu$$

and rearrange: $$i - 1 = \alpha\nu - \alpha = \alpha(\nu - 1)$$

Thus $$\alpha = \frac{i - 1}{\nu - 1}$$

The Process of Conduction

PROBLEM 11-8 In order to determine the molar conductivity of a 0.0500 M solution of $AgNO_3$, you measured the solution's resistance in a conductivity cell and found that $R = 75.8\ \Omega$. Then, in the same cell, a 0.0200 M KCl solution had a resistance of 157.9 Ω. Given that the accepted molar conductance of the KCl solution is 0.013 834 $\Omega^{-1}m^2\,mol^{-1}$, calculate the molar conductance of the $AgNO_3$ solution.

Solution: You need to calculate the cell constant from the data for the KCl solution, but first you need the value of κ. From Eq. (11-8),

$$\kappa = C\Lambda = (0.0200\text{ mol dm}^{-3})(0.013\,834\ \Omega^{-1}\text{m}^2\text{ mol}^{-1}) = (2.77 \times 10^{-4}\ \Omega^{-1}\text{m}^2\text{ dm}^{-3})\left(\frac{10^3\text{ dm}^3}{\text{m}^3}\right)$$
$$= 0.277\ \Omega^{-1}\text{m}^{-1}$$

Now obtain the cell constant from Eq. (11-10):

$$K_{\text{cell}} = R\kappa = (157.9\ \Omega)(0.277\ \Omega^{-1}\text{m}^{-1}) = 43.7\text{ m}^{-1}$$

Then, for the unknown solution,

$$\kappa = \frac{1}{R}K_{\text{cell}} = \frac{43.7\text{ m}^{-1}}{75.8\ \Omega} = 0.577\ \Omega^{-1}\text{m}^{-1}$$

$$\Lambda = \frac{\kappa}{C} = \frac{0.577\ \Omega^{-1}\text{m}^{-1}}{0.0500\text{ mol dm}^{-3}}\left(\frac{\text{m}^3}{10^3\text{ dm}^3}\right) = 1.15 \times 10^{-2}\ \Omega^{-1}\text{m}^2\text{ mol}^{-1}$$

PROBLEM 11-9 For a certain strong electrolyte, Λ_0 is known to be 0.008 93 $\Omega^{-1}m^2\,mol^{-1}$. For a 0.0200 M solution, Λ is found to be 0.007 86. What Λ would you expect for a 0.005 00 M solution?

Solution: For a problem involving Λ at different concentrations, you must assume that the plot of Λ versus $\sqrt{C}$ is a straight line and use Eq. (11-11): $\Lambda = S\sqrt{C} + \Lambda_0$. First calculate the slope:

$$S = \frac{\Lambda - \Lambda_0}{\sqrt{C}} = \frac{0.007\,86 - 0.008\,93}{(0.0200)^{1/2}} = \frac{-0.001\,07}{0.1414} = -0.007\,57$$

Now use Eq. (11-11) to determine Λ at the new concentration:

$$\Lambda = (-0.007\,57)(0.005\,00)^{1/2} + 0.008\,93 = -0.000\,535 + 0.008\,93 = 0.008\,39\ \Omega^{-1}\text{m}^2\text{ mol}^{-1}$$

PROBLEM 11-10 Calculate the molar conductivity at infinite dilution of H_2CO_3 using the first step of ionization and values for appropriate strong electrolytes.

Solution: You could choose any three strong electrolytes that meet the following requirements: one contains H^+, one contains HCO_3^-, the third contains the two "spectator" ions combined with H^+ and with HCO_3^-, and the Λ_0 values for all are available. You could look up some compounds in a table, but here are three that meet the requirements stated: HBr, 428.1; $KHCO_3$, 118.0; and KBr, 151.8. You want to keep H^+ and HCO_3^- and eliminate K^+ and Br^-; thus you should add the first two and subtract the third, so

$$\Lambda_0(H_2CO_3) = \Lambda_0(HBr) + \Lambda_0(KHCO_3) - \Lambda_0(KBr) = 428.1 + 118.0 - 151.8 = 394.3\ \Omega^{-1}\text{cm}^2\text{ mol}^{-1}$$

PROBLEM 11-11 The bicarbonate ion is not listed in Table 11-2, so, using information from that table and Problem 11-10, calculate the value for HCO_3^-.

Solution: You need to calculate $\lambda_{-,0}$ for HCO_3^- using Eq. (11-12), rearranged to

$$q\lambda_{-,0} = \Lambda_0 - p\lambda_{+,0}$$

In this case,

$$1[\lambda_0(HCO_3^-)] = \Lambda_0(H_2CO_3) - 1[\lambda_0(H^+)]$$

From Problem 11-10, $\Lambda_0(H_2CO_3) = 394.3\ \Omega^{-1}\text{cm}^2\text{ mol}^{-1}$, and from Table 11-2, $\lambda_0(H^+) =$

$0.0350\ \Omega^{-1}m^2\ mol^{-1}$. Changing cm^2 to m^2 and substituting, you have

$$\lambda_0(HCO_3^-) = 0.039\,43 - 0.0350 = 0.0044\ \Omega^{-1}m^2\ mol^{-1}$$

PROBLEM 11-12 The molar conductivity of a 0.0200 M solution of $BaCl_2$ is found to be $238\ \Omega^{-1}cm^2\ mol^{-1}$. Calculate the apparent degree of dissociation.

Solution: By Eq. (11-13), $\alpha = \Lambda/\Lambda_0$. Thus you need Λ_0 for $BaCl_2$, which you can calculate by adding ionic molar conductivities from Table 11-2 and then using Eq. (11-12). Thus

$$\Lambda_0(BaCl_2) = p\lambda_0(Ba^{2+}) + q\lambda_0(Cl^-) = 1(0.0127) + 2(0.007\,63) = 0.0280\ \Omega^{-1}m^2\ mol^{-1}$$

Then
$$\alpha = \frac{\Lambda}{\Lambda_0} = \left(\frac{238\ \Omega^{-1}cm^2\ mol^{-1}}{0.0280\ \Omega^{-1}m^2\ mol^{-1}}\right)\left(\frac{m^2}{10^4\ cm^2}\right) = 0.850$$

PROBLEM 11-13 Calculate (**a**) the mobilities and (**b**) the transport numbers at infinite dilution of both ions of barium chloride, $BaCl_2$.

Solution:

(**a**) You may calculate mobility from Eq. (11-15), rearranged to

$$u_i = \frac{\lambda_i}{z_i\mathscr{F}}$$

You can obtain values of λ from Table 11-2. Then, for Ba^{2+},

$$u_{+,0} = \left(\frac{0.0127\ \Omega^{-1}m^2\ mol^{-1}}{2(96\,500\ C\ mol^{-1})}\right)\left(\frac{C}{A\,s}\right)\left(\frac{A\,V^{-1}}{\Omega^{-1}}\right)\left(\frac{10^4\ cm^2}{m^2}\right) = 6.58 \times 10^{-4}\ cm^2\,V^{-1}s^{-1}$$

and for Cl^-,

$$u_{-,0} = \frac{0.007\,63}{1(96\,500)} = 7.91 \times 10^{-4}\ cm^2\,V^{-1}s^{-1}$$

(**b**) You could calculate transport numbers from mobilities, using Eq. (11-18), or from ionic conductivities, using Eq. (11-20) or Eq. (11-21). Equation (11-21) is probably the better one to remember, because usually you will not have values of mobilities. However, you can try it both ways and see if they agree. First, substitute the mobilities calculated in (a) into Eqs. (11-18):

$$t_+ = \frac{z_+pu_+}{z_+pu_+ + z_-qu_-} \qquad t_- = \frac{z_-qu_-}{z_+pu_+ + z_-qu_-}$$

For Ba^{2+}: $\quad z_+pu_+ = 2(1)(6.59 \times 10^{-4}) = 13.18 \times 10^{-4}$

For Cl^-: $\quad z_-qu_- = 1(2)(7.91 \times 10^{-4}) = 15.82 \times 10^{-4}$

The 10^{-4} will cancel, so you have

$$t_+ = \frac{13.18}{13.18 + 15.82} = 0.4545 \quad \text{and} \quad t_- = \frac{15.82}{13.18 + 15.82} = 0.5455$$

Now, in order to use Eq. (11-20) or Eq. (11-21), you must obtain λ_0 values from Table 11-2. You can simplify the math by changing the units to cm^2, moving the decimal point four places. You should use Eq. (11-21), because the required data are available in Problem 11-12:

$$t_{+,0} = \frac{p\lambda_{+,0}}{\Lambda_0} \qquad t_- = \frac{q\lambda_{-,0}}{\Lambda_0}$$

For Ba^{2+}: $t_{+,0} = \dfrac{1(127)}{280} = 0.454$; $\qquad$ for Cl^-: $t_{-,0} = \dfrac{2(76.3)}{280} = 0.545$

The agreement is excellent.

PROBLEM 11-14 Derive an expression for K_a in terms of Λ.

Solution: In Example 11-23, we used an expression for K in terms of α, which was obtained in Example 7-21a. You can start with it and replace α with Λ/Λ_0 from Eq. (11-13):

$$K_a = \frac{\alpha^2}{1-\alpha}C_0 = \frac{(\Lambda/\Lambda_0)^2}{1-(\Lambda/\Lambda_0)}C_0$$

Incidentally, this expression is called the *Ostwald dilution law*. If you rearrange it to a simpler form, the coefficient of C_0 becomes

$$\frac{(\Lambda/\Lambda_0)^2}{1-(\Lambda/\Lambda_0)} = \frac{\Lambda^2}{\Lambda_0^2}\frac{1}{(\Lambda_0-\Lambda)/\Lambda_0} = \frac{\Lambda^2}{\Lambda_0(\Lambda_0-\Lambda)}$$

and

OSTWALD DILUTION LAW

$$K_a = \frac{\Lambda^2}{\Lambda_0(\Lambda_0-\Lambda)}C_0$$

PROBLEM 11-15 Using the result of Problem 11-14, calculate K_a for NH_4OH if the electrolytic conductivity of a 0.0010 *M* solution is measured to be 0.0034 $\Omega^{-1}m^{-1}$.

Solution: You first need to find λ_0 in Table 11-2 and then calculate Λ_0 by using Eq. (11-12):

$$\Lambda_0 = p\lambda_0(NH_4^-) + q\lambda_0(OH^-)$$

And since $p = q = 1$,

$$\Lambda_0 = 0.007\,34 + 0.0199 = 0.0272\ \Omega^{-1}m^2\,mol^{-1}$$

From Eq. (11-8) and the given data,

$$\Lambda = \frac{\kappa}{C} = \frac{0.0034\ \Omega^{-1}m^{-1}}{0.0010\ M} = \left(3.4\frac{\Omega^{-1}m^{-1}}{mol\ dm^{-3}}\right)\left(\frac{10^{-3}\ m^3}{dm^3}\right) = 3.4\times10^{-3}\ \Omega^{-1}m^2\,mol^{-1}$$

From Problem 11-14,

$$K_a = \frac{\Lambda^2}{\Lambda_0(\Lambda_0-\Lambda)}C_0 = \left(\frac{(3.4\times10^{-3})^2}{(27.2\times10^{-3})(27.2-3.4)(10^{-3})}\right)(1\times10^{-3}) = 1.8\times10^{-5}$$

PROBLEM 11-16 The electrolytic conductivity of a saturated solution of barium oxalate ($BaC_2O_4\cdot2H_2O$) was $9.46\times10^{-3}\ \Omega^{-1}m^{-1}$ greater than that of the water used. Calculate the solubility product constant.

Solution: To calculate K_{sp}, you need the molar concentration of the salt, which you can calculate using Eq. (11-23):

$$C = \frac{\kappa}{\Lambda_0}$$

You obtain λ_0 values from Table 11-2 and calculate Λ_0 by using Eq. (11-12):

$$\Lambda_0 = p\lambda_0(Ba^{2+}) + q\lambda_0(C_2O_4^{2-})$$

Again, $p = q = 1$, so

$$\Lambda_0 = 0.0127 + 0.0146 = 0.0273\ \Omega^{-1}m^2\,mol^{-1}$$

The conductivity of the water has already been subtracted in the information given; i.e., the κ given is the κ(solution). Thus

$$C = \left(\frac{9.46\times10^{-3}\ \Omega^{-1}m^{-1}}{0.0273\ \Omega^{-1}m^2\,mol^{-1}}\right)\left(\frac{m^3}{10^3\ dm^3}\right) = 3.47\times10^{-4}\ mol\ dm^{-3}$$

Now,

$$K_{sp} = [Ba^{2+}][C_2O_4^{2-}] = (3.47\times10^{-4})^2 = 12.0\times10^{-8} = 1.20\times10^{-7}$$

Activity and Activity Coefficients

PROBLEM 11-17 A 0.050 *m* $CaCl_2$ solution has a mean activity coefficient of 0.58. Calculate the mean ionic activity, the mean molality, and the activity of the compound.

Solution: You need the mean molality first, so use Eq. (11-38):

$$m_\pm = m[p^pq^q]^{1/\nu}$$

For $CaCl_2$,

$$m_\pm = m[(1)^1\,(2)^2]^{1/3} = 4^{1/3}\,m = 1.587\,m$$

For $m = 0.050$, $m_{\pm} = 1.587(0.050) = 0.079$

For mean ionic activity, use Eq. (11-39):

$$a_{\pm} = \gamma_{\pm} m_{\pm} = 0.58(0.079) = 0.046$$

For solute activity, use Eq. (11-36):

$$a_2 = (a_{\pm})^{\nu}$$

For $CaCl_2$, $\nu = 2 + 1 = 3$, so

$$a_2 = (0.046)^3 = 9.7 \times 10^{-5}$$

PROBLEM 11-18 Calculate the ionic strength of a solution that is 0.2 m in K_2SO_4, 0.4 m in Na_3PO_4, and 0.3 m in NaCl.

Solution: You should determine the charges and concentrations of all ions and substitute into Eq. (11-40). A handy procedure is to set up a table:

Ion	m	z_i	z_i^2	mz_i^2
K^+	0.4	1	1	0.4
SO_4^{2-}	0.2	2	4	0.8
Na^+ (1) (2)	1.2, 0.3 } = 1.5	1	1	1.5
PO_4^{3-}	0.4	3	9	3.6
Cl^-	0.3	1	1	0.3
				$\Sigma mz_i^2 = 6.6$

Note that the two sources of Na^+ combine because all Na^+ are alike. Now

$$I = \tfrac{1}{2}(\Sigma mz_i^2) = \tfrac{1}{2}(6.6) = 3.3$$

PROBLEM 11-19 Derive an expression for plotting data in order to determine activity coefficients on the basis of dissociation of weak electrolytes.

Solution: The derivation is similar to that for the method based on solubilities. You should write the true equilibrium constant, based on activities. For a weak acid, such as HA,

$$K_a = \frac{a_{H^+} a_{A^-}}{a_{HA}}$$

Then use Eq. (11-30) to write ionic activities: $a_+ = \gamma_+ C_+$ and $a_- = \gamma_- C_-$. Now

$$K_a = \frac{(\gamma_+ C_+)(\gamma_- C_-)}{\gamma_2 C_a} = \frac{C_+ C_-}{C_a}\left(\frac{\gamma_+ \gamma_-}{\gamma_2}\right) = \frac{[H^+][A^-]}{[HA]}\left(\frac{\gamma_+ \gamma_-}{\gamma_2}\right)$$

Since the solution is dilute, you can let $\gamma_2 = 1$; then replace $\gamma_+ \gamma_-$ with $\gamma_{\pm}$ and take the logarithm:

$$\log K_a = \log \frac{[H^+][A^-]}{[HA]} - \log \gamma_{\pm}^2$$

Now you can replace the concentration log term by $C\alpha^2/(1 - \alpha)$, which was obtained in Chapter 7 and used in Example 11-24, or by its equivalent in terms of conductivity, the Ostwald dilution law given in Problem 11-14 as $\Lambda^2 C/\Lambda_0(\Lambda_0 - \Lambda)$. Rearranging gives

$$\log \frac{\Lambda^2 C}{\Lambda_0(\Lambda_0 - \Lambda)} = \log K_a - 2 \log \gamma_{\pm}$$

From Eq. (11-42), $\log \gamma_{\pm} = k\sqrt{I}$. When you plot the left-hand term against $\sqrt{I}$, a straight line with intercept $\log K_a$ should result. Then, when you substitute this value back into the equation, you can calculate $\gamma_{\pm}$ at various concentrations.

PROBLEM 11-20 The equivalent conductance of a 0.0100 M acetic acid solution is 16.3 $\Omega^{-1}\text{cm}^2\ \text{mol}^{-1}$. Assume that K_a from extrapolation is approximately the same as the activity K_a, or 1.75×10^{-5}, and use the expression obtained in Problem 11-19 to calculate the mean activity coefficient of the acid at this concentration.

Solution: You need Λ_0 for HAc; use Eq. (11-12) and obtain λ_0 values from Table 11-2. Then

$$\Lambda_0 = p\lambda_0(\mathrm{H^+}) + q\lambda_0(\mathrm{Ac^-}) = (0.0350 + 0.004\,08)\ \Omega^{-1}\mathrm{m^2\ mol^{-1}} = 391\ \Omega^{-1}\mathrm{cm^2\ mol^{-1}}$$

Rearranging the expression, you get

$$\begin{aligned} 2\log\gamma_\pm &= \log K_a - \log\frac{\Lambda^2 C}{\Lambda_0(\Lambda_0 - \Lambda)} \\ &= \log(1.75\times 10^{-5}) - \log\frac{(16.3)^2(1.00\times 10^{-2})}{391(391-16.3)} = -4.7570 - (-4.7418) = -0.0152 \\ \log\gamma_\pm &= -0.0076 \\ \gamma_\pm &= 0.983 \end{aligned}$$

Debye–Hückel Theory

PROBLEM 11-21 Calculate the thickness of the ionic atmosphere at 298 K for KCl solutions of concentrations 0.0100 M and 0.100 M. Compare your results to those in Example 11-29.

Solution: The question is: "How does changing the concentration by multiples of 10 affect $1/\kappa$?" The calculation is like that of Example 11-29, involving substitution into Eq. (11-45):

$$\frac{1}{\kappa} = \frac{\varepsilon_0\varepsilon_r kT}{(L_A e^2 \sum_i C_i z_i^2)^{1/2}}$$

To simplify the second calculation, you can factor out the summation and evaluate the constant:

$$\begin{aligned} \left(\frac{1}{\kappa}\right)^2 &= \frac{(8.854\times 10^{-12}\ \mathrm{C^2\ J^{-1}m^{-1}})(78.5)(1.381\times 10^{-23}\ \mathrm{J\ K^{-1}})(298\ \mathrm{K})}{(6.022\times 10^{23}\ \mathrm{mol^{-1}})(1.602\times 10^{-19})^2}\left(\frac{1}{\sum_i C_i z_i^2}\right) \\ &= (1.851\times 10^{-16})\left(\frac{1}{\sum_i C_i z_i^2}\right) \end{aligned}$$

You can further simplify this expression because $z = 1$ in both cases, so the summation equals $2C$. Thus, for $C = 0.0100$, $2C = (0.0200\ \mathrm{mol\ dm^{-3}})(10^3\ \mathrm{dm^3\ m^{-3}}) = 20\ \mathrm{mol\ m^{-3}}$, and

$$\frac{1}{\kappa} = \left[(1.851\times 10^{-16})\left(\frac{1}{20}\right)\mathrm{m^2}\right]^{1/2} = (9.25\times 10^{-18}\ \mathrm{m^2})^{1/2} = 3.04\times 10^{-9}\ \mathrm{m}$$

For $C = 0.100$, $2C = 200\ \mathrm{mol\ m^{-3}}$, and

$$\frac{1}{\kappa} = \left[(1.851\times 10^{-16})\left(\frac{1}{200}\right)\mathrm{m^2}\right]^{1/2} = (92.5\times 10^{-20}\ \mathrm{m^2})^{1/2} = 9.62\times 10^{-10}\ \mathrm{m}$$

Changing the notation to 10^{-9} for comparison to Example 11-29, you get

$$\frac{1}{\kappa} = 0.962\times 10^{-9}\ \mathrm{m}$$

Predictably, because the square root is involved, you have found that $1/\kappa$ decreases by 10 when C increases by 100.

PROBLEM 11-22 Supposedly, the Debye–Hückel–Onsager equation for conductivity becomes unreliable for concentrations greater than about 0.010 M. Check that prediction against experimental results for a concentration of 0.020 M of two types of electrolyte, AB and AB_2. At $C = 0.0200\ M$, $\Lambda(\mathrm{KCl}) = 0.013\,83$, and $\Lambda(\mathrm{BaCl_2}) = 0.023\,82\ \Omega^{-1}\mathrm{m^2\,mol^{-1}}$. Use the values of constants for 1:1 electrolytes given in this chapter.

Solution: Use the Debye–Hückel–Onsager equation, Eq. (11-46):

$$\Lambda = \Lambda_0 - (M + N\Lambda_0)\sqrt{C}$$

Begin by calculating Λ_0 in the usual way [Eq. (11-12) and Table 11-2]:

$$\Lambda_0 = p\lambda_{+,0} + q\lambda_{-,0}$$

For KCl: $\Lambda_0 = 1(0.007\,35) + 1(0.007\,63) = 0.014\,98\ \Omega^{-1}\text{m}^2\ \text{mol}^{-1}$

For $BaCl_2$: $\Lambda_0 = 1(0.0127) + 2(0.007\,63) = 0.0280\ \Omega^{-1}\text{m}^2\ \text{mol}^{-1}$

Now calculate a value for $M + N\Lambda_0$ for KCl:

$$M + N\Lambda_0 = 60.20\,\frac{\Omega^{-1}\text{cm}^2\ \text{mol}^{-1}}{(\text{mol dm}^{-3})^{1/2}} + 0.229\left(\frac{1}{\text{mol dm}^{-3}}\right)^{1/2}(149.8\ \Omega^{-1}\text{cm}^2\ \text{mol}^{-1}) = 94.50\,\frac{\Omega^{-1}\text{cm}^2\ \text{mol}^{-1}}{(\text{mol dm}^{-3})^{1/2}}$$

Substituting into Eq. (11-46), you get

$$\Lambda = 149.8 - 94.50(0.0200)^{1/2} = 149.8 - 13.4 = 136.4\ \Omega^{-1}\text{cm}^2\ \text{mol}^{-1}$$

Doing the same calculations for $BaCl_2$, you have

$$M + N\Lambda_0 = 60.20 + 0.229(280) = 124.32\,\frac{\Omega^{-1}\text{cm}^2\ \text{mol}^{-1}}{(\text{mol dm}^{-3})^{1/2}}$$

and

$$\Lambda = 280 - 124.32(0.0200)^{1/2} = 280 - 17.6 = 262\ \Omega^{-1}\text{cm}^2\ \text{mol}^{-1}$$

The predicted value for KCl is about 2 percent lower than the experimental value and that for $BaCl_2$ is about 10 percent higher. You can see that appreciable deviations *do* appear as concentrations exceed 0.01 M. (The value for $BaCl_2$ is questionable since we used M and N for a 1:1 electrolyte, rather than a 1:2 electrolyte.)

PROBLEM 11-23 Derive Eq. (11-49) from Eq. (11-48).

Solution: Begin with Eq. (11-35):

$$\gamma_\pm = (\gamma_+^p\,\gamma_-^q)^{1/v}$$

In logarithmic form, the expression becomes

$$v \log \gamma_\pm = p \log \gamma_+ + q \log \gamma_-$$

Now write Eq. (11-48) for each ion:

$$\log \gamma_+ = -Az_+^2\sqrt{I} \qquad \log \gamma_- = -Az_-^2\sqrt{I}$$

and substitute:

$$v \log \gamma_\pm = p(-Az_+^2\sqrt{I}) + q(-Az_-^2\sqrt{I}) = -(pz_+^2 + qz_-^2)A\sqrt{I} \qquad \textbf{(a)}$$

Since the solution must be neutral overall,

$$pz_+ = qz_- \qquad \textbf{(b)}$$

Multiplying eq. (b) by z_+,

$$pz_+^2 = qz_+|z_-| \qquad \textbf{(c)}$$

Multiplying eq. (b) by $|z_-|$,

$$pz_+|z_-| = q|z_-|^2 \qquad \textbf{(d)}$$

Substituting eqs. (c) and (d) into (a),

$$v \log \gamma_\pm = -(qz_+|z_-| + pz_+|z_-|)A\sqrt{I}$$

Factoring and dividing by $v = p + q$,

$$\log \gamma_\pm = -\left(\frac{z_+|z_-|(q+p)}{p+q}\right)A\sqrt{I} = -Az_+|z_-|\sqrt{I} \qquad \textbf{[Eq. (11-49)]}$$

PROBLEM 11-24 Use the Debye–Hückel limiting law to calculate the mean ionic activity coefficient for Sr^{2+} and $C_2O_4^{2-}$ in a saturated solution of SrC_2O_4 in a 0.050 M solution of $Na_2C_2O_4$; $K_{sp}(SrC_2O_4) = 5.6 \times 10^{-8}$.

Solution: In order to use Eq. (11-49), you need the ionic strength and thus the concentrations of all ions. A complication arises in determining the concentrations of Sr^{2+} and $C_2O_4^{2-}$. The Na salt contributes $C_2O_4^{2-}$ in such large concentration that the contribution of the same ion from the Sr salt is negligible. Thus $[C_2O_4^{2-}]$ is from the $Na_2C_2O_4$. Rearranging the K_{sp} expression and substituting $[C_2O_4^{2-}] = 0.050$, you have

$$K_{sp} = [Sr^{2+}][C_2O_4^{2-}]$$

$$[Sr^{2+}] = \frac{K_{sp}}{[C_2O_4^{2-}]} = \frac{5.6 \times 10^{-8}}{0.050} = 1.1 \times 10^{-6}$$

The sodium ion concentration is 0.10. Substituting in Eq. (11-41) gives

$$2I = \Sigma C z^2 = (1.1 \times 10^{-6})(2)^2 + (0.050)(2)^2 + (0.10)(1)^2 \cong 0.20 + 0.10 = 0.30$$
$$I = \tfrac{1}{2}(0.30) = 0.15$$

Now, using the Debye–Hückel limiting law, Eq. (11-49), you obtain

$$\log \gamma_\pm = -0.5091 z_+ |z_-| \sqrt{I} = -0.5091(2)(2)\sqrt{0.15} = -0.7887$$
$$\gamma_\pm = 0.16$$

PROBLEM 11-25 If the Debye–Hückel limiting law applies, what is the solubility of $Pb(IO_3)_2$ at 298 K in a 0.0100 M KNO_3 solution? $K_{sp}[Pb(IO_3)_2] = 2.6 \times 10^{-13}$ at 298 K.

Solution: You need an equation for a 1:2 electrolyte like Eq. (11-43), which was derived for the 1:1 case. In general terms, for a 1:2 electrolyte, $K_{sp} = a_+ a_-^2$. Combining Eq. (11-32) with Eq. (11-36) gives $a_+^p a_-^q = a_\pm^\nu$. For the 1:2 case, $a_+ a_-^2 = a_\pm^3$. Converting to molarity, $a_\pm = C_\pm \gamma_\pm$. Thus

$$K_{sp} = C_\pm^3 \gamma_\pm^3$$

Take the logarithm:

$$\log K_{sp} = 3 \log C_\pm + 3 \log \gamma_\pm$$

and rearrange to get

$$3 \log C_\pm = \log K_{sp} - 3 \log \gamma_\pm$$

or

$$\log C_\pm = \tfrac{1}{3} \log K_{sp} - \log \gamma_\pm \qquad \textbf{(a)}$$

You can calculate $\log \gamma_\pm$ from Eq. (11-49) and assume that the contributions of Pb^{2+} and IO_3^- to the ionic strength are negligible (very small concentrations). For a 1:1 electrolyte (KNO_3), $I = C = 0.0100$ [from Eq. (11-41)]. Substitute into Eq. (11-49):

$$\log \gamma_\pm = -0.5091(2)(1)\sqrt{0.0100} = -0.1018$$

and into eq. (a):

$$\log C_\pm = \tfrac{1}{3} \log(2.6 \times 10^{-13}) + 0.1018 = -4.195 + 0.1018 = -4.093$$
$$C_\pm = 8.07 \times 10^{-5}$$

Write Eq. (11-38) for C:

$$C_\pm = C(p^p q^q)^{1/\nu}$$

For a 1:2 electrolyte, $p = 1$ and $q = 2$, so

$$C_\pm = C[(1)^1(2)^2]^{1/3} = 4^{1/3} C$$
$$C = \frac{C_\pm}{1.587} = \frac{8.07 \times 10^{-5}}{1.587} = 5.09 \times 10^{-5}\ M$$

PROBLEM 11-26 The Debye–Hückel theory is applicable to weak as well as strong electrolytes. **(a)** Derive an expression for the dissociation constant of a weak acid, HA, in terms of the mean activity coefficient; then modify it to include degree of dissociation. **(b)** Make simplifying assumptions to obtain a convenient expression for calculating an approximate value of the degree of dissociation. **(c)** What additional assumptions would be necessary in order to use the Debye–Hückel limiting law to calculate the correct degree of dissociation? Derive an expression relating α to α_0.

Solution:

(a) The equilibrium constant in activities is

$$K_a = \frac{a_{H^+} a_{A^-}}{a_{HA}}$$

You can write Eq. (11-30) for the individual ions:

$$a_+ = \gamma_+ C_+ \qquad a_- = \gamma_- C_-$$

and you can write Eq. (11-30) for the molecule:

$$a_{HA} = \gamma_{HA} C_{HA}$$

Substituting into the expression for K and collecting related terms, you get

$$K_a = \left(\frac{\gamma_+\gamma_-}{\gamma_{HA}}\right)\left(\frac{C_+C_-}{C_{HA}}\right)$$

As in Problem 11-25, $a_+^p a_-^q = a_\pm^\nu$. For $p = q = 1$, $\gamma_+\gamma_- = \gamma_\pm^2$. Then, to introduce the degree of dissociation, use the expression obtained in Chapter 7 and used in Example 11-23:

$$K_C = \frac{C_+C_-}{C_{HA}} = \left(\frac{\alpha^2}{1-\alpha}\right)C_0$$

Now

$$K_a = \left(\frac{\gamma_\pm^2}{\gamma_{HA}}\right)\left(\frac{\alpha^2}{1-\alpha}\right)C_0$$

(b) Two useful simplifications can be made: (1) Assume a dilute solution, so $\gamma_{HA} = 1$. (2) Assume that K is small. A small dissociation constant means that there are fewer ions, i.e., α is small, so $1 - \alpha = 1$. Then

$$\gamma_\pm^2\alpha^2 = \frac{K}{C_0}$$

If either $\gamma_\pm$ or α is known, the other can be calculated from this expression by using either

$$\alpha = \left(\frac{K}{C_0}\right)^{1/2}\left(\frac{1}{\gamma_\pm}\right) \quad \text{or} \quad \gamma_\pm = \left(\frac{K}{C_0}\right)^{1/2}\left(\frac{1}{\alpha}\right)$$

If you had a situation in which there were no interionic attractions, there would be no deviations from ideality, and $\gamma_\pm$ would equal 1. Making this assumption, you have

$$\alpha = \left(\frac{K}{C_0}\right)^{1/2}$$

This expression gives an approximate value for α, which can be labeled α_0. Then

$$\gamma_\pm = \frac{\alpha_0}{\alpha} \quad \text{or} \quad \alpha = \frac{\alpha_0}{\gamma_\pm}$$

Since $\gamma_\pm$ must be less than 1, the correct α must be greater than α_0.

(c) If the solution is very dilute, you can calculate $\gamma_\pm$ from the Debye–Hückel limiting law, Eq. (11-49). In order to obtain I, however, you first need C. The concentrations of the ions will be αC_0, but in the very dilute solution $\alpha \approx \alpha_0$, so you can use $\alpha_0 C_0$; i.e.,

$$I = \tfrac{1}{2}[C_+(1)^1 + C_-(1)^1] = \tfrac{1}{2}(\alpha_0C_0 + \alpha_0C_0) = \alpha_0C_0$$

From Eq. (11-49), $\log \gamma_\pm = -0.5091\sqrt{\alpha_0C_0}$; from part (b), $\alpha = \alpha_0/\gamma_\pm$, so

$$\log \alpha = \log \alpha_0 - \log \gamma_\pm = \log \alpha_0 + 0.5091\sqrt{\alpha_0C_0}$$

PROBLEM 11-27 Use the results of Problem 11-26 to calculate the correct degree of dissociation in a 0.001 00 M solution of acrylic acid, $K_a = 5.50 \times 10^{-5}$. Compare this result to the value obtained if interionic attractions are ignored.

Solution: You need to use the expression for α_0 and the last expression in Problem 11-26.

$$\alpha_0 = \left(\frac{K}{C_0}\right)^{1/2} = \left(\frac{5.50 \times 10^{-5}}{10^{-3}}\right)^{1/2} = 0.235$$

$$\begin{aligned}\log \alpha &= \log \alpha_0 + 0.5091\sqrt{\alpha_0C_0}\\ &= \log(0.235) + 0.5091(0.235 \times 10^{-3})^{1/2}\\ &= -0.6289 + 0.007\,80 = -0.6211\\ \alpha &= 0.239\end{aligned}$$

The increase in $\alpha = 0.004$, or only 1.7%. You can see that the usual practice of ignoring activities introduces little error.

Supplementary Exercises

PROBLEM 11-28 The length of time required for a current of 2 A to yield 1000 C is ________. The instrument that measures quantity of electricity is called the ________. If you know the boiling point of a 0.010 *M* salt solution and if you double the amount of salt in the solution, the boiling-point elevation (will) (will not) be doubled. The SI unit for conductance is the ________, its symbol is ________, and the commonly used alternative symbol is ________. The Hittorf method measures ________.The values of λ_0 in Table 11-2 show that (potassium) (barium) has the greater mobility. For ________ ________ aqueous solutions, a strong electrolyte has the same activity in all solutions that have the same ionic strength. If the mean activity coefficient of the ions of $Al(NO_3)_3$ in a solution is 0.20, the activity coefficient of the salt is ________. The ionic strength of a 0.010 *m* KI solution is ________.

PROBLEM 11-29 If you wish to deposit silver from a solution of silver nitrate at the rate of 0.500 g min^{-1}, what current must you use?

PROBLEM 11-30 A certain metal forms divalent ions. When a solution of the metal is electrolyzed by a current of 2.00 A for exactly 30 minutes, 2.10 g is deposited. What is the metal?

PROBLEM 11-31 In the electrolysis of a NaCl solution, assume that the only reaction of the chloride ion is to form chlorine gas. What volume of gas, at 298 K and 1 atm, would be produced if a current of 3.5 A were to flow for 5.0 hours?

PROBLEM 11-32 In the electrolysis of a solution of $Cu(CN)_2^-$, a silver coulometer in the circuit released 2.47 g Ag. How much Cu metal was deposited?

PROBLEM 11-33 A 0.050 *m* solution of $CoCl_2$ freezes at -0.25 °C. Calculate the boiling point of the solution, assuming that *i* is independent of temperature.

PROBLEM 11-34 For the solution in Problem 11-33, calculate the degree of dissociation (according to the Arrhenius theory).

PROBLEM 11-35 In a certain conductivity cell, the plates of the electrodes are exactly 4.00 cm apart. What is the cross-sectional area if a 0.001 00 *M* NaCl solution ($\Lambda = 123.7\ \Omega^{-1}cm^2\,mol^{-1}$) has a resistance of 4200 Ω?

PROBLEM 11-36 In the cell used in Problem 11-35, what resistance would be measured for a 0.0200 *M* solution of $BaCl_2$? The Λ of this solution is 238 $\Omega^{-1}cm^2\,mol^{-1}$.

PROBLEM 11-37 A common standard solution for conductivity is 0.0100 *M* KCl, which has a molar conductivity of 0.014 127 $\Omega^{-1}m^2\,mol^{-1}$. When a conductivity cell is filled with this solution, the resistance is 184.4 Ω. When a 0.100 *M* NaAc solution is placed in the same cell, the resistance is 35.7 Ω. Calculate the molar conductivity of the second solution.

PROBLEM 11-38 Calculate the apparent degree of dissociation of $BaCl_2$ in a 0.0200 *M* solution, if Λ_0 is 280.0 $\Omega^{-1}cm^2\,mol^{-1}$. (See Problem 11-36 for additional information.)

PROBLEM 11-39 In an investigation of solutions of a strong electrolyte, the molar conductivities at concentrations of 0.0320 *M* and 0.000 540 *M* were 35.6 and 165 $\Omega^{-1}cm^2\,mol^{-1}$, respectively. Assume that these points lie on a straight line obtained by plotting all of the data in the usual manner. Calculate the molar conductivity at infinite dilution.

PROBLEM 11-40 Assume that the electrolyte in Problem 11-39 is a 2:1 compound. Calculate its van't Hoff *i* factor at each concentration.

PROBLEM 11-41 Calculate the mobility at infinite dilution of OH^-.

PROBLEM 11-42 Calculate the transport numbers at infinite dilution of the ammonium ion in ammonium acetate and in ammonium chloride.

PROBLEM 11-43 The molar conductivity of a 0.0100 M solution of chloroacetic acid was found by a student to be 0.0125 $\Omega^{-1}m^2\,mol^{-1}$. If Λ_0 is known to be 0.0390, what is the experimental value of the ionization constant?

PROBLEM 11-44 Water having an electrolytic conductivity of $7.8 \times 10^{-6}\,\Omega^{-1}m^{-1}$ is used to prepare a saturated solution of silver chloride at 298 K. If the conductivity of the solution is $1.869 \times 10^{-4}\,\Omega^{-1}m^{-1}$, what is the experimental value for K_{sp} of AgCl?

PROBLEM 11-45 Obtain an expression for the activity of $Ca_3(PO_4)_2$ in a solution of concentration m in terms of m and $\gamma_\pm$.

PROBLEM 11-46 Calculate the activity of sulfuric acid in a 0.20 M solution if the mean activity coefficient is 0.21.

PROBLEM 11-47 What is the ionic strength of a solution that contains 50.0 g of $MgCl_2$ and 50.0 g of $Al(NO_3)_3$ per liter of solution? (Retain three significant figures.)

PROBLEM 11-48 Find the thickness of the ionic atmosphere of the ions at 298 K in (**a**) 0.001 00 M and (**b**) 0.100 M aqueous solutions of a 2:1 electrolyte.

PROBLEM 11-49 Use the Debye–Hückel–Onsager equation to calculate the molar conductivity of a 0.005 00 M aqueous solution of $NH_4C_2H_3O_2$ at 298 K.

PROBLEM 11-50 Use the Debye–Hückel limiting law to calculate the solubility of $BaSO_4$ at 298 K in a 0.005 00 M $NaNO_3$ solution; $K_{sp}(BaSO_4)$ at 298 K $= 1.08 \times 10^{-10}$.

Answers to Supplementary Exercises

If your answers differ from those given here, check the units.

11-28 500 s; coulometer; will not; siemens, S, Ω^{-1}; t_+; potassium; very dilute; 1.6×10^{-3}; 0.010
11-29 7.45 A **11-30** Cd **11-31** 8.0 L **11-32** 1.45 g **11-33** 100.070 °C **11-34** 0.84
11-35 7.70 cm^2 **11-36** 109 Ω **11-37** 0.007 30 $\Omega^{-1}m^2\,mol^{-1}$ **11-38** 0.851
11-39 184 $\Omega^{-1}cm^2\,mol^{-1}$ **11-40** 1.386; 2.790 **11-41** $2.06 \times 10^{-3}\,cm^2\,V^{-1}s^{-1}$
11-42 0.643; 0.490 **11-43** 1.51×10^{-3} **11-44** 1.68×10^{-10} **11-45** $108\,m^5\gamma_\pm^5$
11-46 3.0×10^{-4} **11-47** 2.98 **11-48** (**a**) 5.55×10^{-9}; (**b**) 5.55×10^{-10}
11-49 0.0108 $\Omega^{-1}m^2\,mol^{-1}$ **11-50** 1.45×10^{-5} M

12 ELECTROCHEMICAL CELLS

THIS CHAPTER IS ABOUT

- ☑ Description of Cells
- ☑ Potentials of Cells and Electrodes
- ☑ Thermodynamics of Cells
- ☑ Types of Electrodes and Galvanic Cells
- ☑ Applications of Galvanic Cell Potentials

12-1. Description of Cells

A. Chemistry and electricity

Electricity becomes a concern in chemistry in two situations: In the first, electrical energy can bring about chemical change, as you saw in Chapter 11, where we described electrolysis and Faraday's laws. A device for carrying out such chemical changes is called an **electrolytic cell**. Practical examples of electrolysis include decomposition of water into its elements, recharging a lead storage battery, and electroplating iron or steel objects with silver.

In the second situation, chemical change can produce electricity. A device for obtaining electrical energy from chemical energy is called a **galvanic cell** (also, *voltaic cell*). (We use the term *electrochemical cell* for both electrolytic and galvanic cells, although some authors restrict this term to galvanic cells.) Practical examples of galvanic cells include the Daniell cell (now used primarily for laboratory and demonstration purposes), the dry cell (flashlights), and the lead storage battery (automobiles). A *battery* is merely a combination of several cells. The work done by a galvanic cell yields ΔG, from which we can evaluate equilibrium constants and other thermodynamic properties. We can determine activity coefficients, solubility product constants, and pH as well as the end point of a titration.

Although it's convenient to make a distinction between these two types of electrochemical cells, you should bear in mind that they are fundamentally the same. If a cell of either type is operating at low current, a small change in applied voltage may reverse the direction of flow of current and convert the cell into the other type.

B. Components of a cell

If two reactive chemicals are mixed in a beaker, a chemical change occurs but no electricity is produced, and no useful work is done.

EXAMPLE 12-1: (**a**) Describe what you would see if mossy Zn is placed in a solution of $CuSO_4$. (**b**) Explain what happens, using chemical equations.

Solution:

(**a**) Surely, you've done this exercise in an early lab course. At first, a black coating appears on the zinc; after some time, pieces of copper metal lie on the bottom of the beaker. The blue color of the solution diminishes or disappears.

(**b**) We explain this reaction as the transfer of two electrons from an atom of Zn to a Cu^{2+} ion, causing the ion to become an atom of Cu while the Zn atom changes to a Zn^{2+} ion. We can represent the gain and loss of electrons by two partial equations, which can be added to give the equation for the net reaction:

$$\begin{array}{rcll} Zn(s) & \longrightarrow & Zn^{2+}(aq) + 2e^- & \text{(oxidation)} \\ Cu^{2+}(aq) + 2e^- & \longrightarrow & Cu(s) & \text{(reduction)} \\ \hline Cu^{2+}(aq) + Zn(s) & & Cu(s) + Zn^{2+}(aq) & \end{array}$$

If the reactants are separated and suitably connected, the transfer of electrons will occur through a wire. This flow of electrons, of course, is an electric current, which can light a bulb or do other work. To force the electrons through the wire, the reactants are separated by a medium through which electrons can't flow (usually a solution of ions). The ions complete the circuit by moving through the solution and gaining or losing electrons at two electrodes. An **electrode** often is a metal; one end rests in the solution, and the other end is attached to a second electrode by wires through the device that is to be operated. Ideally, the external system from electrode to electrode would be made of the same material, and the entire system would be at the same temperature. Otherwise, thermocouple voltages will occur. In general, there will be two electrolytic solutions, each containing one of the ions corresponding to one of the electrodes. These usually are separated by a porous material or a salt bridge. Thus a typical galvanic cell can be represented by

terminal–electrode–solution–salt bridge–solution–electrode–terminal **(12-1)**

In some special types of cells, one or more of these parts is missing. Usually this means that one component is serving two functions.

EXAMPLE 12-2: Describe a galvanic cell for the reaction used in Example 12-1 and represent it by an expression like Eq. (12-1).

Solution: Let's use an electrode of Zn with a solution of $ZnSO_4$ and an electrode of Cu with a solution of $CuSO_4$. A typical salt bridge contains a solution of KCl, and the terminals may be of copper. This type of cell was one of the first to be used widely; a Zn–Cu cell is still called a *Daniell cell*, after its developer. Now, following Eq. (12-1), we have

$$Cu - Zn - ZnSO_4(aq) - KCl(aq) - CuSO_4(aq) - Cu - Cu \qquad \textbf{(12-2)}$$

We could omit the terminal on the right, since it's made of the same metal as the electrode.

C. Conventional representation of a cell

An expression like Eq. (12-2) is written merely for the purpose of illustration. Cells normally are represented by similar expressions but more information is included: usually, concentrations of the ions, physical states, and symbols such as commas and vertical lines to represent physical features of the cell. A properly written expression contains all the information that would appear on a sketch of the cell. The written expression is even called a *cell diagram.* Note from Eq. (12-2) that soluble materials are in the middle, insoluble substances and gases are on each side, moving out from the center, and the metal terminals are at the extremes of the expression. Symbols commonly used include

Symbol	Meaning
|	An interface between two phases.
||	An interface between two liquid phases, where the junction potential has been removed, as by a salt bridge.
,	Different species in same solution.

EXAMPLE 12-3: Write the cell diagram for the Daniell cell of Example 12-2, assuming that the ionic concentrations are 1 *m*.

Solution: We include the Cu terminals, although they often are omitted. The salt bridge is indicated by the double vertical line, and only the participating ions are included, i.e.,

$$Cu(s)\,|\,Zn(s)\,|\,Zn^{2+}(m = 1)\,||\,Cu^{2+}(m = 1)\,|\,Cu(s)\,|\,Cu(s) \qquad \textbf{(12-3)}$$

We can shorten the diagram if there's no chance of confusion. Usually, the solid metals are obvious, and the terminals are omitted; thus we can also describe the Daniell cell as

$$Zn\,|\,Zn^{2+}\ (1\ m)\,||\,Cu^{2+}\ (1\ m)\,|\,Cu \qquad \textbf{(12-4)}$$

D. Names and signs of the electrodes

The electrode at which oxidation occurs is called the **anode** and that at which reduction occurs is the **cathode**. By convention, the anode is placed at the left in diagrams or expressions like Eq. (12-1), which means that electrons flow from left to right in the external circuit. These rules apply to *both* electrolytic and galvanic cells. The signs of the electrodes, however, are different in the two types of cell, i.e.,

electrolytic cell: anode = +; cathode = −
galvanic cell: anode = −; cathode = +

EXAMPLE 12-4: Explain why the signs of the electrodes are different in the two types of cell.

Solution: Instead of looking at the same electrode in both cells, let's look at the negative electrode in each cell and restate the question: "What causes one electrode in a cell to have a more negative potential than the other electrode?"

(a) In the galvanic cell the process is spontaneous. When the terminals are connected, a path is available, but electrons are needed. Consequently, oxidation begins in the solution, placing electrons on the anode. This makes the anode more negative than the cathode.

(b) In the electrolytic cell the process isn't spontaneous; it must be forced to occur. An external source of electrons (e.g., a battery) is placed in the circuit. It forces electrons onto the cathode, making it negative, and reduction begins in the solution. The positive cations flow through the solution toward the cathode, replacing those that have been reduced, while the negative anions move toward the now-positive anode. There the anions are oxidized, giving up some of their electrons to replenish those taken from the anode by the battery. The cathode remains more negative, because the battery continues to pump electrons to it.

12-2. Potentials of Cells and Electrodes

A. Electromotive force

The vital feature of electrochemical cells is that the **electric potential** (V) of each electrode is different. When the electrodes are connected, a current will flow spontaneously from the electrode of higher potential (more positive) to that of lower potential. (Electrons flow in the opposite direction, i.e., from more negative to less negative.)

If the potential difference of the cell is opposed by an external potential that just stops the flow of current, the cell is operating reversibly. The potential difference under reversible conditions is the **electromotive force (emf)** of the cell, often called the *electric potential* of the cell or *cell potential*. The symbol E (or $\mathcal{E}$) is used, and the unit is volts.

B. Measurement of emf

Voltmeters draw current from the cell, so the measured voltage is not the emf of the cell. A *potentiometer* opposes the cell emf with a known emf, so that, at the balance point (zero deflection on a galvanometer), no current flows; thus the true value of the unknown emf is obtained.

The opposing, known emf is applied across the full length of a uniform slide wire in the potentiometer. The unknown emf, E_x, which must be less than the known emf, is applied across a portion of the slide wire (see Figure 12-1). Each voltage is proportional to the length of the

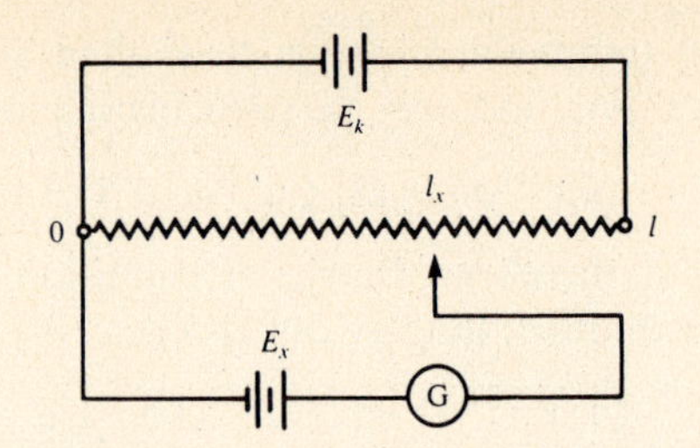

FIGURE 12-1. A Potentiometer Circuit

wire (the resistance), and at balance,

$$\frac{E_x}{E_k} = \frac{\ell_x}{\ell} \tag{12-5}$$

EXAMPLE 12-5: A certain cell was balanced against a known emf of 1.018 V in a potentiometer having a slide wire 9.50 cm long. At the balance point, the sliding contact was at 6.23 cm. What is the emf of the cell?

Solution: We rearrange Eq. (12-5) to solve for E_x:

$$E_x = \frac{\ell_x}{\ell} E_k = \frac{6.23 \text{ cm}}{9.50 \text{ cm}} (1.018 \text{ V}) = 0.668 \text{ V}$$

C. Cell potential and electrode potential

Combinations of various electrodes yield different cell potentials; the cell potential varies with concentration (activity) of the ions in the electrolytic solution. Because we can't possibly measure and tabulate all the combinations, we need a procedure for calculating all emfs from a controllable number of values. As in other cases, we break the whole into contributions by the parts and assume that each part contributes equally. We define an **electrode potential** as the contribution an electrode makes to the emf of any cell. This seems quite reasonable, because chemically the net cell reaction is the sum of two *half-reactions*, each occurring at one of the electrodes. Thus the electrode potential (or *half-cell potential*) represents the potential owing to a half-reaction; e.g., E_{Zn} refers to the reaction $Zn^{2+}(aq) + 2e^- \rightarrow Zn(s)$. If the cell diagram has been written according to convention, with the oxidation reaction on the left, the cell potential is

$$E_{cell} = E_r - E_l \tag{12-6}$$

D. Standard electrode potentials

It isn't possible to measure the potential of a single electrode, so we resort to another familiar procedure: choosing a reference point against which electrode potentials can be measured. The **standard hydrogen electrode** (SHE) is the official reference point, being assigned a potential of zero at all temperatures. Hydrogen gas at 1 atm is bubbled around one end of a sheet of platinum in a solution of $H^+(a_\pm = 1)$ at 298 K. It is assumed that oxidation occurs; thus the SHE is always placed at the left in cell diagrams. If a cell is composed of the SHE and some other electrode, X, the measured voltage of the cell is the electrode potential of the second electrode, because Eq. (12-6) becomes

$$E_{cell} = E_X - 0$$

or

$$E_X = E_{cell} \tag{12-7}$$

If all components of electrode X are at unit activity, then E_X is called the **standard electrode potential**, $E°$. These values have been determined and tabulated for many half-reactions.

EXAMPLE 12-6: A galvanic cell made up of a standard hydrogen electrode and a $Zn|Zn^{2+}$ (1 *m*) electrode has a measured cell voltage of -0.763 V. What is the standard electrode potential of the Zn electrode?

Solution: The E° of the cell is the algebraic sum of the individual electrode potentials, Eq. (12-6). In Eq. (12-7), however, we have already substituted 0.0 for the hydrogen electrode, so

$$E^\circ_{Zn} = E^\circ_{cell} = -0.763 \text{ V}$$

Note that, in calculating E°, the number of moles doesn't matter. The E° for $Zn^{2+} | Zn$ is the same for

$$Zn^{2+} + 2e^- \longrightarrow Zn \qquad \text{and} \qquad \frac{1}{2}Zn^{2+} + 1e^- \longrightarrow \frac{1}{2}Zn$$

Because we can't conveniently set up and control a hydrogen electrode, we often use secondary electrodes. A common one is calomel (Hg_2Cl_2) in a paste with Hg and a solution of KCl saturated with Hg_2Cl_2. One type of *calomel electrode* uses 1 *M* KCl and has a potential of 0.2802 V with respect to the hydrogen electrode; another type uses a saturated KCl solution and has a potential of 0.2415 V. The half-reaction for the latter is

$$Hg_2Cl_2(s) + 2e^- \longrightarrow 2Hg(l) + 2Cl^- \text{ (saturated)}$$

This type of electrode involves the use of an insoluble salt in intimate contact with the metal of the salt. The solution must contain the anion.

E. Oxidation potential versus reduction potential

In order to obtain Eq. (12-7), we placed the standard hydrogen electrode on the left; i.e., it was the anode, or the electrode where oxidation occurred. Thus reduction occurred at the other electrode; consequently, E° is often referred to as the *standard reduction potential* (see Table 12-1). From the result of Example 12-6, the standard reduction potential of $Zn^{2+} | Zn$ is -0.763 V, referring to the half-reaction Zn^{2+} $(a = 1) + 2e^- \rightarrow Zn(s)$. This is the approved international convention.

warning: In the past, a different convention was used widely in the United States and is found in older textbooks and tables, although most recent textbooks have dropped it. This convention places the hydrogen electrode on the right in Eq. (12-6), changing the sign of the single electrode potentials and making the half-reactions oxidations. Thus E° of $Zn | Zn^{2+}$ is $+0.763$ V for the half-reaction $Zn(s) \rightarrow Zn^{2+}$ $(a = 1) + 2e^-$. Be sure to check the table you're using. Most tables include the equation for the half-reaction, which will help you if the title of the table is unclear.

EXAMPLE 12-7: Using Table 12-1, calculate E° for the Daniell cell described in Examples 12-1 and 12-2, and write the equation of the net reaction.

Solution: We already know that Zn and Cu electrodes are involved. In Table 12-1, we find the entries

$$Cu^{2+} + 2e^- \longrightarrow Cu \qquad E^\circ = 0.337 \text{ V}$$
$$Zn^{2+} + 2e^- \longrightarrow Zn \qquad E^\circ = -0.763 \text{ V}$$

Method 1: We also know from Example 12-1, and observation, that Zn changes to Zn^{2+} (the opposite of the reaction shown in Table 12-1). If we reverse the half-reaction for Zn, we can add the two, as we already did in Example 12-1:

$$\begin{array}{rcl} Zn & \longrightarrow & Zn^{2+} + 2e^- \\ Cu^{2+} + 2e^- & \longrightarrow & Cu \\ \hline Cu^{2+} + Zn & \longrightarrow & Zn^{2+} + Cu \end{array}$$

We use Eq. (12-6) to calculate E°_{cell}, but which electrode is on the right? Since Zn is oxidized, it must be the anode and therefore on the left. Then

$$E^\circ_{cell} = E^\circ_{Cu} - E^\circ_{Zn} = 0.337 - (-0.763)\ V = 1.100\ V$$

Method 2: Another way to set up this calculation is to treat it like a Hess's law problem: The net reaction is the sum of several steps. To add the two reactions, we must reverse one. Let's choose the Cu reaction, changing the sign of E°. Thus

Cu	$\longrightarrow$	$Cu^{2+} + 2e^-$	$E^\circ = -0.337\ V$
$Zn^{2+} + 2e^-$	$\longrightarrow$	Zn	$E^\circ = -0.763\ V$
$Cu + Zn^{2+}$	$\longrightarrow$	$Cu^{2+} + Zn$	$E^\circ = -1.100\ V$

We know that Zn is oxidized in this cell, so the answer is backwards. We simply chose the wrong half-reaction to reverse, so we merely reverse the result and change the sign to obtain

$$Cu^{2+} + Zn \longrightarrow Cu + Zn^{2+} \qquad 1.100\ V$$

If we hadn't known that Zn should be oxidized, we still would have known that the answer was backwards, because E°_{cell} was negative. For a spontaneous reaction, E°_{cell} must be positive. We discuss this further in Section 12.3.

F. Guideline for using a table of standard electrode potentials

A simple rule will help you to get the proper sign for E°_{cell} and the equation of the net reaction for a cell composed of any two electrodes: *The half-reaction with the more negative E° is reversed.* This rule applies whether the table lists reduction potentials or oxidation potentials and makes method 2 in Example 12-7 virtually foolproof.

EXAMPLE 12-8: Determine the net cell reaction and the cell potential if the two electrodes are $Fe^{2+}|Fe$ and $Ni^{2+}|Ni$.

Solution: In Table 12-1, we see that the Fe reaction is more negative, so we reverse it and change the sign of E°; we copy the Ni reaction as is and have

Fe	$\longrightarrow$	$Fe^{2+} + 2e^-$	$E^\circ = +0.440\ V$
$Ni^{2+} + 2e^-$	$\longrightarrow$	Ni	$E^\circ = -0.250\ V$
$Ni^{2+} + Fe$	$\longrightarrow$	$Ni + Fe^{2+}$	$E^\circ = +0.190\ V$

12-3. Thermodynamics of Cells

A. Work and free energy

Electrical work equals the quantity of electricity multiplied by the electric potential, or emf:

$$w = QE = ItE \tag{12-8}$$

As in Chapter 11, Q is the quantity (coulombs), I is the current (amperes), and t is the time (seconds). In an electric cell, Q equals the number of moles of electrons transferred multiplied by the Faraday constant, so

$$w = n\mathscr{F}E \tag{12-9}$$

If all components of the cell are at unit activity, $E = E^\circ$, and if the cell operates reversibly, $w = w_{max}$. Since electrical work is non-PV work (useful work), $w_{elec,\,max} = -\Delta G$, and Eq. (12-9) becomes

$$\Delta G^\circ = -n\mathscr{F}E^\circ \tag{12-10}$$

Equation (12-10) permits us to obtain any thermodynamic property that can be calculated from ΔG for any reaction for which E° can be measured or can be calculated from single-electrode potentials.

EXAMPLE 12-9: Calculate ΔG° for the reaction in Example 12-8 and express the result in joules.

Solution: We simply substitute into Eq. (12-10); the purpose here is to practice manipulating units. Note that $n = 2$ mol. Thus

$$\Delta G^\circ = -n\mathscr{F}e$$
$$= (-2\ \text{mol})(9.65 \times 10^4\ \text{C mol}^{-1})(0.190\ \text{V}) = -3.67 \times 10^4\ \text{C V}$$

For the SI units,

$$\text{C} = \text{A s} \qquad \text{V} = \text{kg m}^2\,\text{s}^{-3}\,\text{A}^{-1} \qquad \text{J} = \text{kg m}^2\,\text{s}^{-2}$$

Thus

$$\text{C V} = \text{A s}\,(\text{kg m}^2\,\text{s}^{-3}\,\text{A}^{-1}) = \text{kg m}^2\,\text{s}^{-2} = \text{J}$$

and

$$\Delta G^\circ = -3.67 \times 10^4\ \text{J}$$

- For future problems, keep in mind that $\text{C} = \text{JV}^{-1}$.

B. Calculation of single-electrode potentials

The method we have been using—reverse one equation, multiply one or both by a factor to balance electrons, if necessary, and add—is based on Eq. (12-10). We can also use Eq. (12-10) to calculate a new single-electrode potential from two listed in Table 12-1 when the electrons

TABLE 12-1: Standard Electrode Potentials at 298 K

Half-reaction	E° (V)
$Cl_2 + 2e^- \rightarrow 2Cl^-$	1.360
$Tl^{3+} + 2e^- \rightarrow Tl^{1+}$	1.211
$Br_2 + 2e^- \rightarrow 2Br^-$	1.065
$Hg^{2+} + 2e^- \rightarrow Hg$	0.854
$Ag^+ + e^- \rightarrow Ag$	0.799
$Hg_2^{2+} + 2e^- \rightarrow 2Hg$	0.79
$Fe^{3+} + e^- \rightarrow Fe^{2+}$	0.771
$Hg_2SO_4 + 2e^- \rightarrow 2Hg + SO_4^{2-}$	0.615
$I_2 + 2e^- \rightarrow 2I^-$	0.536
$Cu^+ + e^- \rightarrow Cu$	0.521
$Cu^{2+} + 2e^- \rightarrow Cu$	0.337
$HgCl_2 + 2e^- \rightarrow Hg + 2Cl^-$	0.268
$AgCl + e^- \rightarrow Ag + Cl^-$	0.222
$Sn^{4+} + 2e^- \rightarrow Sn^{2+}$	0.15
$AgBr + e^- \rightarrow Ag + Br^-$	0.071
$2H^+ + 2e^- \rightarrow H_2$	0.00
$Pb^{2+} + 2e^- \rightarrow Pb$	−0.126
$Sn^{2+} + 2e^- \rightarrow Sn$	−0.136
$AgI + e^- \rightarrow Ag + I^-$	−0.152
$CuI + e^- \rightarrow Cu + I^-$	−0.185
$Ni^{2+} + 2e^- \rightarrow Ni$	−0.250
$Tl^+ + e^- \rightarrow Tl$	−0.34
$PbSO_4 + 2e^- \rightarrow Pb + SO_4^{2-}$	−0.355
$Fe^{2+} + 2e^- \rightarrow Fe$	−0.440
$Zn^{2+} + 2e^- \rightarrow Zn$	−0.763
$Sn(OH)_6^{2-} + 2e^- \rightarrow HSnO_2 + 3OH^- + H_2O$	−0.90
$Al^{3+} + 3e^- \rightarrow Al$	−1.66
Calomel Electrodes	
$Hg_2Cl_2 + 2e^- \rightarrow 2Hg + 2Cl^-$ (1 *M*)	0.2802
$Hg_2Cl_2 + 2e^- \rightarrow 2Hg + 2Cl^-$ (saturated)	0.2415

don't cancel. In this case we calculate $\Delta G°$ for each half-reaction, subtract left from right [as in Eq. (12-6)], and then calculate $E°_{cell}$ from $\Delta G°_{cell}$.

EXAMPLE 12-10: Determine $E°$ for the half-reaction, $Fe^{3+} + 3e^- \rightarrow Fe$.

Solution: In Table 12-1, we find two half-reactions that include some of the desired species and then calculate $\Delta G°$ using Eq. (12-10), $\Delta G° = -n\mathcal{F}E°$:

Half-reaction	$E°$	$\Delta G°$
$Fe^{3+} + e^- \longrightarrow Fe^{2+}$	0.771 V	$-1\mathcal{F}(0.771) = -0.771\mathcal{F}$
$Fe^{2+} + 2e^- \longrightarrow Fe$	−0.440 V	$-2\mathcal{F}(-0.440) = 0.880\mathcal{F}$
$Fe^{3+} + 3e^- \longrightarrow Fe$		$0.109\mathcal{F}$

Note that we can simplify by not multiplying by the Faraday constant because we know it will cancel in the next step. Next, we rearrange Eq. (12-10):

$$E° = \frac{\Delta G°}{-n\mathcal{F}}$$

For the new half-reaction, $n = 3$,

$$E°(Fe^{3+}\,|\,Fe) = -\frac{0.109\mathcal{F}}{3\mathcal{F}} = -0.0363 \text{ V}$$

C. Sign of cell potential

We know that ΔG must be negative for a spontaneous process. From Eq. (12-10), it follows that $E°$ of a cell must be positive for a spontaneous process. The signs for the $E°$ values in Table 12-1 were determined on this basis, as was the guideline presented in Section 12-2F.

D. Equilibrium constants

In Chapter 7, we obtained the so-called reaction isotherm:

$$\Delta G = \Delta G° + RT \ln Q \qquad \textbf{[Eq. (7-19)]}$$

and saw that, at equilibrium, it reduces to

$$\Delta G° = -RT \ln K \qquad \textbf{[Eq. (7-22)]}$$

Combining Eqs. (12-10) and (7-22), we have

$$E° = \frac{RT}{n\mathcal{F}} \ln K \qquad \textbf{(12-11)}$$

At $T = 298.15\ K$,

$$E° = \frac{0.025\,69 \text{ V mol}}{n} \ln K \qquad \text{or} \qquad E° = \frac{0.059\,16 \text{ V mol}}{n} \log K$$

EXAMPLE 12-11: Calculate the equilibrium constant for the reaction $Zn + Sn^{2+} \rightarrow Zn^{2+} + Sn$. Assume unit activities for the ions.

Solution: In Table 12-1 we see that the Zn electrode is more negative, so we would reverse its equation and change the sign of its potential to obtain $E°_{cell}$, if we follow the method we have been using. This time, however, we don't need the equation for the cell; we can merely change the sign of $E°_{Zn}$ and add:

$$E°_{cell} = E°_{Zn} + E°_{Sn} = +0.763 + (-0.136) = 0.627 \text{ V}$$

alternative: We could use Eq. (12-6):

$$E^\circ_{\text{cell}} = E^\circ_{\text{r}} - E^\circ_{\text{l}}$$

Now we don't change any signs, but we must recognize that the Zn electrode is more negative and is the anode, i.e., on the left. Then

$$E^\circ_{\text{cell}} = E^\circ_{\text{Sn}} - E^\circ_{\text{Zn}} = -0.136 - (-0.763) = 0.627 \text{ V}$$

We rearrange Eq. (12-11) to get an expression for K:

$$\log K = \frac{n\mathcal{F}E^\circ}{RT} = \frac{n}{0.059\,16}E^\circ = \left(\frac{2 \text{ mol}}{0.059\,16 \text{ V mol}}\right)(0.627 \text{ V}) = 21.1968$$

$$K = 1.57 \times 10^{21}$$

---

note: For E° the ions are at unit activity, whereas for K the ions are at equilibrium activity. Nevertheless the two constants are related by Eq. (12-11).

E. The Nernst equation

In Section 12-3D we dealt only with standard conditions. In the more general case, where activities are not unity, Eq. (7-19) applies. Here Q represents the usual quotient of activities (concentrations) raised to the powers of the coefficients in the balanced equation. If we substitute Eq. (12-10) for ΔG and ΔG° into Eq. (7-19) and rearrange, we obtain the **Nernst equation** for the general reaction, $a\text{A} + b\text{B} \rightarrow c\text{C} + d\text{D}$:

$$-n\mathcal{F}E = -n\mathcal{F}E^\circ + RT \ln Q$$

NERNST EQUATION

$$E = E^\circ - \frac{RT}{n\mathcal{F}} \ln \frac{a_{\text{C}}^c a_{\text{D}}^d}{a_{\text{A}}^a a_{\text{B}}^b} \qquad \textbf{(12-12)}$$

When we calculate the constants for 298 K, the Nernst equation becomes

$$E = E^\circ - \frac{0.025\,69}{n} \ln Q \qquad \text{or} \qquad E = E^\circ - \frac{0.059\,16}{n} \log Q$$

EXAMPLE 12-12: Calculate the voltage of the cell composed of $Ni^{2+}|Ni$ and $Cu^{2+}|Cu$ electrodes at 298 K, if $a_{Ni^{2+}} = 0.85$ and $a_{Cu^{2+}} = 0.22$.

Solution: There are two possible approaches, both using the Nernst equation. We could calculate E for each electrode and then the E of the cell, or we could calculate E° of the cell and then E. The latter involves somewhat less calculation, so we'll use it this time. The E° for Ni is more negative, so we reverse its equation and sign and then add the Cu equation:

$$\begin{array}{lll} \text{Ni} \longrightarrow \text{Ni}^{2+} + 2e^- & & E^\circ = 0.250 \text{ V} \\ \text{Cu}^{2+} + 2e^- \longrightarrow \text{Cu} & & E^\circ = 0.337 \text{ V} \\ \hline \text{Ni} + \text{Cu}^{2+} \longrightarrow \text{Ni}^{2+} + \text{Cu} & & E^\circ = 0.587 \text{ V} \end{array}$$

Substituting into Eq. (12-12) and recalling that the activities of solids are 1, we get

$$E = 0.587 \text{ V} - \frac{0.059\,16 \text{ V mol}}{2 \text{ mol}} \log \frac{a_{Ni^{2+}}}{a_{Cu^{2+}}} = 0.587 - 0.029\,58 \log \frac{0.85}{0.22} = 0.587 - 0.017 = 0.570 \text{ V}$$

The Nernst equation indicates that we can adjust activities (concentrations) to make the second term in it numerically larger than—and opposite in sign to—E°. The cell reaction would then be reversed.

EXAMPLE 12-13: Using a cell of $Pb^{2+}|Pb$ and $Sn^{2+}|Sn$, calculate E at 298 K, if the activity of the Sn^{2+} is 0.20 and that of Pb^{2+} is 0.010.

Solution: We can find E°_{cell} quickly now, using Eq. (12-6):

$$E^\circ_{\text{cell}} = E^\circ_{\text{r}} - E^\circ_{\text{l}}$$

The more negative Pb goes on the left; therefore

$$E^\circ_{\text{cell}} = E^\circ_{\text{Pb}} - E^\circ_{\text{Sn}} = -0.126 - (-0.136) = 0.010 \text{ V}$$

The cell reaction is $Sn + Pb^{2+} \rightarrow Sn^{2+} + Pb$. We calculate E from Eq. (12-12):

$$E = E^\circ - \frac{0.025\,69 \text{ V mol}}{2 \text{ mol}} \ln \frac{a_{\text{Sn}^{2+}}}{a_{\text{Pb}^{2+}}} = 0.010 - 0.012\,85 \ln \frac{0.20}{0.010} = 0.010 - 0.038 = -0.028 \text{ V}$$

The negative E tells us that the spontaneous reaction is in the direction opposite to the reaction that would occur if the activities were unity. The equation for this reaction is

$$Sn^{2+}\ (a = 0.20) + Pb(s) \longrightarrow Sn(s) + Pb^{2+}\ (a = 0.010)$$

The result in Example 12-13 implies that E must be zero at some intermediate point between a positive E and a negative E. An $E = 0$ value means no net reaction, i.e., a state of equilibrium. The corresponding activities are those of equilibrium, so $Q = K$, and Eq. (12-12) becomes

$$0 = E^\circ - \frac{RT}{n\mathcal{F}} \ln K$$

which rearranges to Eq. (12-11).

F. Temperature dependence of E

The expression for the temperature dependence of ΔG obtained in Chapter 6 is

$$\left(\frac{\partial \Delta G}{\partial T}\right)_P = -\Delta S \qquad \textbf{[Eq. (6-54)]}$$

Substituting Eq. (12-10) for ΔG gives

$$n\mathcal{F}\left(\frac{\partial E}{\partial T}\right)_P = \Delta S$$

Thus the change in cell potential with temperature may be calculated from

$$\left(\frac{\partial E}{\partial T}\right)_P = \frac{\Delta S}{n\mathcal{F}} \tag{12-13}$$

A familiar definition of ΔG is $\Delta G = \Delta H - T\Delta S$. We solve for ΔH, then substitute Eq. (12-10) for ΔG and Eq. (12-13) for ΔS:

$$\Delta H = \Delta G + T\Delta S = -n\mathcal{F}E + T\left[n\mathcal{F}\left(\frac{\partial E}{\partial T}\right)_P\right]$$

and

$$\Delta H = n\mathcal{F}\left[T\left(\frac{\partial E}{\partial T}\right)_P - E\right] \tag{12-14}$$

Now we can calculate the temperature coefficient of the cell potential if we know either the entropy change or the heat of reaction. Typical values of the coefficient for systems not involving gases are about 10^{-5} V K^{-1}, i.e., essentially negligible until ΔT is 100 K or more. Integration of Eq. (12-13), assuming a constant ΔS, gives

$$E_{T_2} = E_{T_1} + \frac{\Delta S}{n\mathcal{F}}(T_2 - T_1) \tag{12-15}$$

At $T_1 = 298$ K,

$$E_T = E_{298} + \frac{\Delta S}{n\mathcal{F}}(T - 298)$$

We sometimes use this expression and the values of E_{298} and $\Delta S/n\mathcal{F}$ to tabulate electrode potentials over a range of temperatures.

EXAMPLE 12-14: If $\Delta S°$ at 298 K for the reaction of the Daniell cell is $-30.9\ \mathrm{J\,K^{-1}}$, (**a**) what is the temperature coefficient of the cell? (**b**) What is $E°_{\text{cell}}$ at 50 °C? (Refer to Example 12-7.)

Solution: The concepts appearing in this problem are ΔS, the temperature coefficient of E, and $E°$ at a temperature other than 298 K. These should remind you of the equations just derived: Eq. (12-13) and Eq. (12-15). Using Eq. (12-13), we can calculate the temperature coefficient from ΔS. We could use ΔS again in Eq. (12-15) to calculate E, but the math is simpler if we replace $\Delta S/n\mathscr{F}$ with the temperature coefficient:

$$\frac{\Delta S}{n\mathscr{F}} = \left(\frac{\partial E}{\partial T}\right)_P \qquad \text{and} \qquad E_{T_2} = E_{T_1} + \frac{\Delta S}{n\mathscr{F}}(T_2 - T_1)$$

Substituting yields

$$E_{T_2} = E_{T_1} + \left(\frac{\partial E}{\partial T}\right)_P (T_2 - T_1)$$

(**a**) When we wrote the equation for the cell reaction, using the half-reactions in Table 12-1, we found $n = 2$ (Example 12-7). Thus from Eq. (12-13) we get

$$\left(\frac{\partial E}{\partial T}\right)_P = \frac{-30.9\ \mathrm{J\,K^{-1}}}{(2\ \mathrm{mol})(9.65 \times 10^4\ \mathrm{C\,mol^{-1}})} = -1.60 \times 10^{-4}\ \mathrm{V\,K^{-1}}$$

(**b**) The $E°$ at 25 °C is 1.100 V (Example 12-7). Now, we can calculate $E°$ at 50 °C from the combined expression:

$$E°_{50} = 1.100\ \mathrm{V} - 1.60 \times 10^{-4}\ \mathrm{V\,K^{-1}}(50\ \mathrm{K} - 25\ \mathrm{K}) = 1.100 - 4.0 \times 10^{-3} = 1.096\ \mathrm{V}$$

Since potentials are relatively constant with temperature, calculations like those in Example 12-14 are made only for precise work. Equations (12-13) and (12-14) are used more often in calculations of ΔS and ΔH for cell reactions or of heats of formation and entropies of ions.

EXAMPLE 12-15: For the Daniell cell, calculate ΔH at 298 K. (Refer to Example 12-14.)

Solution: Use Eq. (12-14) and the value of $(\partial E/\partial T)_P$ from Example 12-14:

$$\begin{aligned}\Delta H &= (2\ \mathrm{mol})(9.65 \times 10^4\ \mathrm{C\,mol^{-1}})[298\ \mathrm{K}(-1.60 \times 10^{-4}\ \mathrm{V\,K^{-1}}) - 1.100\ \mathrm{V}] \\ &= 19.30 \times 10^4\ \mathrm{J\,V^{-1}}(-0.048\ \mathrm{V} - 1.100\ \mathrm{V}) = -2.216 \times 10^5\ \mathrm{J} = -221.6\ \mathrm{kJ}\end{aligned}$$

12-4. Types of Electrodes and Galvanic Cells

So far, we have discussed several types of electrodes and only one type of cell. In this section, we summarize the variety of types of each that are used.

A. Types of electrodes

1. Metal | metal ion electrode

A metal is in equilibrium with a solution of ions of the metal (the electrode is reversible to these ions).

Half-reaction: $\quad M^{x+}\,(a_{M^{x+}}) + ne^- \longrightarrow M(s)$

Nernst equation: $\quad E = E° - \dfrac{RT}{n\mathscr{F}} \ln \dfrac{a_M}{a_{M^{x+}}}$

Since $a_M = 1$,

$$E = E° - \frac{RT}{n\mathscr{F}} \ln (a_{M^{x+}})^{-1} = E° + \frac{RT}{n\mathscr{F}} \ln a_{M^{x+}}$$

For the copper electrode in previous examples, we have

$$Cu^{2+}\,(a_{Cu^{2+}}) + 2e^- \longrightarrow Cu(s) \qquad \text{and} \qquad E = 0.337 + \frac{RT}{2\mathscr{F}} \ln (a_{Cu^{2+}})$$

2. **Gas electrode**

A gas bubbles around an inert metal inserted into a solution of ions to which the gas is reversible.

Half-reaction: $2G^{x+}\ (a_{G^{x+}}) + ne^- \longrightarrow G_2(g, P_{G_2})$

Nernst equation: $E = E^\circ - \dfrac{RT}{n\mathscr{F}} \ln \dfrac{P_{G_2}}{a^2_{G^{x+}}}$

For the standard hydrogen electrode described in Section 12-2, we have

$$2H^+\ (a_{H^+}) + 2e^- \longrightarrow H_2(g, 1\ \text{atm}) \quad \text{and} \quad E = 0 - \frac{RT}{2\mathscr{F}} \ln\left(\frac{1}{a^2_{H^+}}\right) = \frac{RT}{\mathscr{F}} \ln a_{H^+}$$

(For the situation of a gas yielding a negative ion, see Problem 12-11.)

3. **Amalgam electrode**

An amalgam of the metal is in equilibrium with a solution of ions of the metal to which the electrode is reversible.

Half-reaction: $M^{x+}\ (a_{M^{x+}}) + ne^- \longrightarrow M(Hg)$

Nernst equation: $E = E^\circ - \dfrac{RT}{n\mathscr{F}} \ln \dfrac{a_M}{a_{M^{x+}}}$

The activity of the metal is lowered by dissolving it in Hg; it doesn't normally equal 1. In the Weston standard cell ($E = 1.018\,07$ V), one electrode is a Cd amalgam with a saturated solution of $CdSO_4$, and we have

$$Cd^{2+}\ (a_{Cd^{2+}}) + 2e^- \longrightarrow Cd(Hg) \quad \text{and} \quad E = 1.018\,07 - \frac{298.2\ R}{2\mathscr{F}} \ln \frac{a_{Cd(Hg)}}{a_{Cd^{2+}}}$$

4. **Metal/insoluble salt electrode**

A metal and a slightly soluble salt of the metal are in equilibrium with a solution of the anion of the salt to which the electrode is reversible.

Half-reaction: $M_pA_q(s) + ne^- \longrightarrow pM(s) + qA^{r-}\ (a_{A^{r-}})$

Nernst equation: $E = E^\circ - \dfrac{RT}{n\mathscr{F}} \ln \dfrac{a^p_M a^q_{A^{r-}}}{a_{MA}}$

Since $a = 1$ for solids,

$$E = E^\circ - \frac{RT}{n\mathscr{F}} \ln (a_{A^{r-}})^q$$

For mercury and mercury (I) sulfate with a solution of potassium sulfate (see Table 12-1), we have

$$Hg_2SO_4(s) + 2e^- \longrightarrow 2Hg(l) + SO_4^{2-}\ (a_{SO_4^{2-}}) \quad \text{and} \quad E = 0.615 - \frac{RT}{2\mathscr{F}} \ln a_{SO_4^{2-}}$$

5. **Oxidation–reduction electrode**

Ions of the same element are in different oxidation states. Here x must be larger than y.

Half-reaction: $M^{x+}\ (a_{M^{x+}}) + ne^- \longrightarrow M^{y+}\ (a_{M^{y+}})$

Nernst equation: $E = E^\circ - \dfrac{RT}{n\mathscr{F}} \ln \dfrac{a_{M^{y+}}}{a_{M^{x+}}}$

For the tin ions Sn^{4+} and Sn^{2+}, we have

$$Sn^{4+}\ (a_{Sn^{4+}}) + 2e^- \longrightarrow Sn^{2+}\ (a_{Sn^{2+}}) \quad \text{and} \quad E = 0.15 - \frac{RT}{2\mathscr{F}} \ln \frac{a_{Sn^{2+}}}{a_{Sn^{4+}}}$$

B. General form of Nernst equation for an electrode

We can write a general half-reaction for any electrode:

$$\text{oxidized form} + ne^- \longrightarrow \text{reduced form}$$

Then the Nernst equation (Eq. 12-12) becomes

$$E = E^\circ - \frac{RT}{n\mathscr{F}} \ln \frac{a(\text{reduced form})}{a(\text{oxidized form})} \tag{12-16}$$

This expression applies to every electrode written in the conventional direction, i.e., as a reduction process.

C. Types of galvanic cells

In the typical galvanic cell, the cell potential is a direct consequence of the net chemical reaction. This type is called a *chemical cell.* Cells can be created, however, in which the chemical reaction is secondary, and the cell potential is a consequence of a difference in concentration. This type is called a *concentration cell.* Either of these types may involve an interface between two solutions, called a *liquid junction,* that contributes to the total difference in potential. A cell with such an interface is called a *cell with transference.* A *cell without transference* has no interface between solutions or has had the effect of such an interface removed by some device, perhaps a salt bridge. Consequently, there are four major categories of galvanic cell.

1. Chemical cell without transference

This is the type of cell we've been illustrating. The simplest is two metal | metal ion electrodes (e.g., Sn and Cu) with a salt bridge. The cell diagram is

$$Sn\,|\,Sn^{2+}\,||\,Cu^{2+}\,|\,Cu$$

The salt bridge isn't needed if there's no liquid junction. One way to avoid a liquid junction is to use a solution containing the cation for one electrode and the anion for the other.

EXAMPLE 12-16: A chemical cell without transference can be prepared from an Ag | AgBr electrode and the standard hydrogen electrode, but something else is needed. Determine what else we need and then write the net equation, the cell diagram, and the Nernst equation for this cell.

Solution: In addition to the electrodes, we need a solution of HBr to provide H^+ and Br^- ions. The half-reactions are

$$2H^+ + 2e^- \longrightarrow H_2 \qquad E^\circ = 0.0\ \text{V}$$

$$AgBr + e^- \longrightarrow Ag + Br^- \qquad E^\circ = 0.071\ \text{V}$$

Since the hydrogen electrode is more negative, it involves oxidation. Reversing its equation, taking 2AgBr, and adding, we obtain the cell reaction:

$$2AgBr(s) + H_2(g, P_{H_2}) \longrightarrow 2Ag(s) + 2H^+\ (a_{H^+}) + 2Br^-\ (a_{Br^-})$$

The cell diagram is

$$Ag\,|\,AgBr\,|\,HBr\ (a_{HBr})\,|\,H_2,\ Pt$$

The Nernst equation for this cell is

$$E = 0.071 - \frac{RT}{2\mathscr{F}} \ln \frac{a_{H^+}^2 a_{Br^-}^2}{P_{H_2}}$$

---

2. Chemical cell with transference

Since ions move at different speeds, diffusion through a liquid junction leads to an imbalance of charge on each side of the junction, developing a junction potential, E_j, that will be part of the measured cell potential. The cell diagram for the tin/copper cell without the salt bridge is

$$Sn\,|\,Sn^{2+}\,|\,Cu^{2+}\,|\,Cu$$

3. Concentration cell without transference

The *electrode type* depends on differences in concentrations at the electrodes, in contrast to the *electrolyte type*, which depends on differences in concentrations of the solutions into which the electrodes dip.

(a) Electrode type

(1) gas electrodes: Two gas electrodes at different pressures dip into the same solution of ions. A typical cell diagram is

$$\text{Pt, Cl}_2(P_A)\,|\,\text{Cl}^-\,|\,\text{Cl}_2(P_B)\text{, Pt}$$

(2) amalgam electrodes: The amounts of metal mixed with mercury are different. A typical cell diagram is

$$\text{Na(Hg)}(a_{\text{Na, A}})\,|\,\text{Na}^+\,|\,\text{Na(Hg)}(a_{\text{Na, B}})$$

EXAMPLE 12-17: Calculate the potential of a cell composed of two chlorine electrodes, if the gas pressure in the left electrode is 0.20 atm and that in the right electrode is 0.55 atm.

Solution: To calculate E_{cell} we use the Nernst equation, but we must write the cell reaction first. To make clear what is happening, we show the complete process of obtaining the overall equation. Since oxidation occurs at the anode (to the left), we have to reverse the equation from Table 12-1:

anode:	2Cl^-	$\longrightarrow$	$\text{Cl}_2(P_A) + 2e^-$	$E° = -1.360$
cathode:	$\text{Cl}_2(P_B) + 2e^-$	$\longrightarrow$	2Cl^-	$E° = 1.360$
overall:	$\text{Cl}_2(P_B)$	$\longrightarrow$	$\text{Cl}_2(P_A)$	$E° = 0.0$

We see that two moles of electrons are exchanged for each mole of chlorine (i.e., $n = 2$) and that $E°_{\text{cell}}$ must be zero for all concentration cells. Now we can write the Nernst equation as

$$E_{\text{cell}} = -\frac{RT}{n\mathscr{F}} \ln \frac{P_A}{P_B}$$

For the reaction Cl_2 (0.55 atm) $\rightarrow$ Cl_2 (0.20 atm), we have

$$E_{\text{cell}} = \frac{0.059\,16}{2} \log \frac{0.20}{0.55} = 0.013 \text{ V}$$

(b) Electrolyte type (combination of chemical cells): Two chemical cells without transference, identical except for the concentrations of their solutions, are connected in opposition. A typical cell diagram is

$$\text{Pt, H}_2\,|\,\text{HBr}\,(a_{\text{HBr},1})\,|\,\text{AgBr}\,|\,\text{Ag} - \text{Ag}\,|\,\text{AgBr}\,|\,\text{HBr}\,(a_{\text{HBr},2})\,|\,\text{H}_2\text{, Pt}$$

EXAMPLE 12-18: Calculate the cell potential of the chemical cell just described (two cells composed of an AgBr | Ag electrode and a standard hydrogen electrode) if the concentration of HBr is 1.00×10^{-3} *M* in the left cell and 1.00×10^{-4} *M* in the right cell. Assume that the activity coefficients are unity.

Solution: We're dealing with two cells, *not* two electrodes in a single cell, so we have the same chemical reaction in both cells. In Example 12-16 we obtained the equation and the value of E_{cell} but will repeat it here for the cell on the left, or cell 1:

anode:	$\text{H}_2\text{(1 atm)}$	$\longrightarrow$	$2\text{H}^+\,(a_{\text{H}^+,1}) + 2e^-$	$E° = 0.0$ V
cathode:	$\text{AgBr}(s) + e^-$		$\text{Ag}(s) + \text{Br}^-\,(a_{\text{Br}^-,1})$	$E° = 0.071$ V
overall:	$\text{H}_2\text{(1 atm)} + 2\text{AgBr}(s)$	$\longrightarrow$	$2\text{H}^+\,(a_{\text{H}^+,1}) + 2\text{Br}^-\,(a_{\text{Br}^-,1}) + 2\text{Ag}(s)$	$E° = 0.071$ V

For the cell on the right, or cell 2, the reverse occurs, except that the concentration of HBr is different:

$$2H^+(a_{H^+,2}) + 2Br^-(a_{Br^-,2}) + 2Ag(s) \longrightarrow H_2(1\ \text{atm}) + 2AgBr(s) \qquad E^\circ = -0.071$$

Adding the two cell reactions gives the equation for the chemical cell:

$$2HBr(a_{HBr,2}) \longrightarrow 2HBr(a_{HBr,1}) \qquad E^\circ = 0$$

As it should, E° for this concentration cell comes out zero. We divide the equation by 2 to let $n = 1$ and write the Nernst equation as

$$E_{cell} = -\frac{RT}{\mathscr{F}} \ln \frac{a_{HBr,1}}{a_{HBr,2}}$$

Since the activity coefficients are unity, $a = C$, and we have

$$E_{cell} = -\frac{0.059\,16}{1} \log \frac{10^{-4}}{10^{-3}}$$

$$= -0.059\,16(-1) = 0.059\,16\ \text{V}$$

---

4. Concentration cell with transference

In this case the two electrodes are identical, except for the concentrations of the solutions. A typical cell diagram is

$$\text{Pt, } Cl_2(1\ \text{atm})\,|\,HCl(a_{HCl,1})\,|\,HCl(a_{HCl,2})\,|\,Cl_2(1\ \text{atm}), \text{Pt}$$

The cell might be arranged vertically with the less-dense solution floating on top of the more-dense solution (greater a). The cell reaction is

$$\begin{array}{lrcl} \text{anode:} & Cl^-(a_{Cl^-,1}) & \longrightarrow & Cl(1\ \text{atm}) + e^- \\ \text{cathode:} & Cl(1\ \text{atm}) + e^- & \longrightarrow & Cl^-(a_{Cl^-,2}) \\ \hline & Cl^-(a_{Cl^-,1}) & \longrightarrow & Cl^-(a_{Cl^-,2}) \end{array} \tag{1}$$

Ions are moving across the liquid junction (H^+ from left to right and Cl^- from right to left), and the transport numbers become factors. When one faraday passes through the cell, t_+ moles of H^+ move across the junction, i.e.,

$$t_+H^+(a_{H^+,1}) \qquad t_+H^+(a_{H^+,2}) \tag{2}$$

Similarly, for Cl^-, replacing t_- with $1 - t_+$,

$$(1 - t_+)Cl^-(a_{Cl^-,2}) \longrightarrow (1 - t_+)Cl^-(a_{Cl^-,1}) \tag{3}$$

We add eqs. (1), (2), and (3) to obtain the net amount of material transported. Clearing the parentheses, we have

$$t_+H^+(a_{H^+,1}) + Cl^-(a_{Cl^-,2}) - t_+Cl^-(a_{Cl^-,2}) + Cl^-(a_{Cl^-,1}) \longrightarrow$$
$$t_+H^+(a_{H^+,2}) + Cl^-(a_{Cl^-,1}) - t_+Cl^-(a_{Cl^-,1}) + Cl^-(a_{Cl^-,2})$$

Canceling like terms and collecting a_1 on one side and a_2 on the other,

$$t_+H^+(a_{H^+,1}) + t_+Cl^-(a_{Cl^-,1}) \longrightarrow t_+H^+(a_{H^+,2}) + t_+Cl^-(a_{Cl^-,2})$$
$$t_+HCl(a_{HCl,1}) \longrightarrow t_+HCl(a_{HCl,2})$$

The Nernst equation, Eq. (12-12), becomes

$$E_{cell} = -\frac{RT}{n\mathscr{F}} \ln \frac{a_2^{t_+}}{a_1^{t_+}}$$

Using 1 and 2 for the two solutions gives us the general expression

$$E_{cell} = -\frac{t_+RT}{n\mathscr{F}} \ln \frac{a_2}{a_1} \tag{12-17}$$

Recall from Eqs. (11-36) and (11-39) that the activity of an ionic solute is $a_{\pm}^{\nu} = (m_{\pm}\gamma_{\pm})^{\nu}$. Substituting for a gives

$$E_{\text{cell}} = -\frac{\nu t_{+}RT}{n\mathscr{F}} \ln \frac{m_2\gamma_2}{m_1\gamma_1}$$

These expressions apply to cases when the anion reacts at the electrodes. For the opposite cases, when the cation reacts at the electrodes, a similar derivation yields the general expression

$$E_{\text{cell}} = -\frac{t_{-}RT}{n\mathscr{F}} \ln \frac{a_1}{a_2} \tag{12-18}$$

and

$$E_{\text{cell}} = -\frac{\nu t_{-}RT}{n\mathscr{F}} \ln \frac{m_1\gamma_1}{m_2\gamma_2}$$

EXAMPLE 12-19: Write the cell diagram and calculate the cell potential for a concentration cell with transference using hydrogen electrodes and HCl solutions of $a = 0.30$ and $a = 0.50$. Use 0.20 for the transport number of Cl^-.

Solution: Before writing the cell diagram, we need to think about which solution to put on the left. Since H^+ is the ion that reacts at the electrode, we will use Eq. (12-18). To make E positive, the ln term must be negative, meaning that the denominator must be larger than the numerator. Consequently, we let $a_{\text{HCl},2} = 0.50$. The cell diagram is

$$\text{Pt, } H_2\,(1 \text{ atm})\,|\,\text{HCl}\,(a = 0.30)\,|\,\text{HCl}\,(a = 0.50)\,|\,H_2\,(1 \text{ atm}), \text{ Pt}$$

Using Eq. (12-18), we get

$$E_{\text{cell}} = -\frac{t_{\text{Cl}^-}RT}{n\mathscr{F}} \ln \frac{a_{\text{HCl},1}}{a_{\text{HCl},2}} = -(0.20)(0.059\,16) \log \frac{0.30}{0.50} = 0.0026 \text{ V}$$

12-5. Applications of Galvanic Cell Potentials

A. Activity coefficients

Concentration cells are useful for determining activity coefficients. For the electrolyte type, i.e., two chemical cells in opposition, we can change the expression obtained for the potential in Example 12-18 by replacing the activities with $(m\gamma)^{\nu}$, as we just did above for Eqs. (12-17) and (12-18):

$$E = -\frac{RT}{n\mathscr{F}} \ln \frac{(m_1\gamma_1)^{\nu}}{(m_2\gamma_2)^{\nu}}$$

and

$$E = -\frac{\nu RT}{n\mathscr{F}} \ln \frac{m_1}{m_2} - \frac{\nu RT}{n\mathscr{F}} \ln \frac{\gamma_1}{\gamma_2} \tag{12-19}$$

EXAMPLE 12-20: An electrolyte concentration cell without transference is prepared from two chemical cells composed of hydrogen gas electrodes and Ag | AgCl electrodes. **(a)** Write the cell diagram. **(b)** Calculate the activity coefficient of HCl in a 0.100-m solution if the activity coefficient in a 0.0100-m solution is 0.905. When the electrolyte concentrations in the two chemical cells were 0.0100 m and 0.100 m, the measured emf at 25 °C was 0.0559 V.

Solution:

(a) We must write two chemical cells in opposition, putting the more concentrated solution on the right so that E_{cell} will be positive when Eq. (12-19) is used:

$$H_2 \mid HCl\,(a_{HCl,\,1}) \mid AgCl, Ag - Ag \mid AgCl \mid HCl\,(a_{HCl,\,2}) \mid H_2$$

(b) We obtained the equation for E for this cell in Example 12-18 and changed it to Eq.(12-19). Now we rearrange Eq. (12-19) to the form:

$$\ln \frac{\gamma_1}{\gamma_2} = -\frac{n\mathscr{F}}{\nu RT} E - \ln \frac{m_1}{m_2}$$

In this case, $n = 1$ and $\nu = p + q = 1 + 1 = 2$, so

$$\log \frac{0.905}{\gamma_2} = -\frac{1}{2(0.059\,16)}(0.0559) - \log \frac{0.0100}{0.100} = -0.4724 + 1$$

$$\frac{0.905}{\gamma_2} = \text{antilog}\,(0.5276) = 3.370$$

$$\gamma_2 = \frac{0.905}{3.370} = 0.269$$

We simplified this example to illustrate the calculation. In practice, we probably wouldn't know the value of any activity coefficient and would have to plot measurements of E at several concentrations, using an appropriate version of the Nernst equation.

B. Equilibrium constants and solubility product constants

We illustrated the calculation of equilibrium constants in Example 12-11. Since the solubility product constant is a special equilibrium constant, the calculation is similar, and Eq. (12-11) becomes

$$E^\circ = \frac{RT}{n\mathscr{F}} \ln K_{sp} \qquad \textbf{(12-20)}$$

If we can find two electrodes for which the half-reaction equations add to give the equation for dissolving the desired slightly soluble salt, we can calculate K_{sp}.

EXAMPLE 12-21: Calculate the solubility product constant of AgI.

Solution: The desired reaction is

$$AgI(s) \rightleftharpoons Ag^+ + I^-$$

In Table 12-1, we find

$$Ag^+ + e^- \longrightarrow Ag \qquad E^\circ = 0.799\text{ V}$$
$$AgI + e^- \longrightarrow Ag + I^- \qquad E^\circ = -0.152\text{ V}$$

In this problem, we are not interested in identifying the oxidation electrode but rather in getting the desired equation. Consequently, we reverse the first equation:

$$Ag \longrightarrow Ag^+ + e^- \qquad E^\circ = -0.799\text{ V}$$

Adding it to the second gives

$$AgI \longrightarrow Ag^+ + I^- \qquad E^\circ = -0.951\text{ V}$$

Substituting into Eq. (12-20), after rearranging it, yields

$$\log K_{sp} = \frac{1}{0.059\,16}(-0.951) = -16.075$$

$$K_{sp} = 8.41 \times 10^{-17}$$

C. pH

An electrode that is reversible to hydrogen ions can be combined with one of the reference electrodes to determine the hydrogen ion concentration (or pH). We use the difference between the measured emf and the reference emf to calculate a_{H^+} or pH directly. The hydrogen electrode is the obvious choice, but it is inconvenient to use.

EXAMPLE 12-22: When a hydrogen electrode at 1 atm and a saturated calomel electrode were placed in a solution containing hydrogen ions and connected through a potentiometer (with the hydrogen electrode on the left), the observed emf was 0.0955 V. What is the pH of the solution? For the saturated calomel electrode, $E = 0.2415$ V from Table 12-1.

Solution: We find the potential of the hydrogen electrode and then apply the Nernst equation. Rearranging Eq. (12-6) gives

$$E_l = E_r - E_{cell}$$

$$E_{H_2} = 0.2415 - (0.0955) = 0.1460 \text{ V}$$

If we write the half-reactions for the exchange of one electron, the Nernst equation for the H_2 electrode is

$$E = E^\circ - 0.059\,16 \log \frac{a_{H^+}}{p_{H^2}^{1/2}}$$

Since $P = 1$ atm and $E^\circ = 0$, $E = -0.059\,16 \log a_{H^+}$. Recall that pH $= -\log a_{H^+}$. Thus

$$\text{pH} = \frac{E}{0.059\,16} \qquad \textbf{(12-21)}$$

In this case,

$$\text{pH} = \frac{0.1460}{0.059\,16} = 2.47$$

Normally, the *glass electrode* is used instead of a hydrogen electrode. Hydrogen ions pass through a thin glass bulb to reach an electrode inside. If the inside electrode is Ag | AgCl, the entire electrode is

$$\text{Ag} \,|\, \text{AgCl} \,|\, \text{HCl}\ (a) \,|\, \text{glass} \,|$$

The second electrode usually is a calomel electrode, and the calculations are similar to those of Example 12-22. Normally a pH-meter, which gives pH directly, is used.

SUMMARY

In Table S-12, we show the major equations presented in this chapter and the connections between. You should determine the conditions that are necessary for the validity of each equation and the conditions that are imposed to make the connections. Other important things you should know are

1. Electrical energy causes chemical change in an electrolytic cell; chemical change produces electrical energy in a galvanic cell.
2. The most general galvanic cell would contain two terminals, two electrodes, two solutions, and a salt bridge between the solutions.
3. A cell is represented by a cell diagram, in which accepted conventions are used to describe a cell on one line. A cell diagram

   (a) defines a specific chemical reaction which goes to the right;
   (b) presumes that oxidation occurs in the left half-cell and reduction in the right half-cell; and
   (c) assumes that electrons flow from left to right in an external circuit.

TABLE S-12: Summary of Important Equations

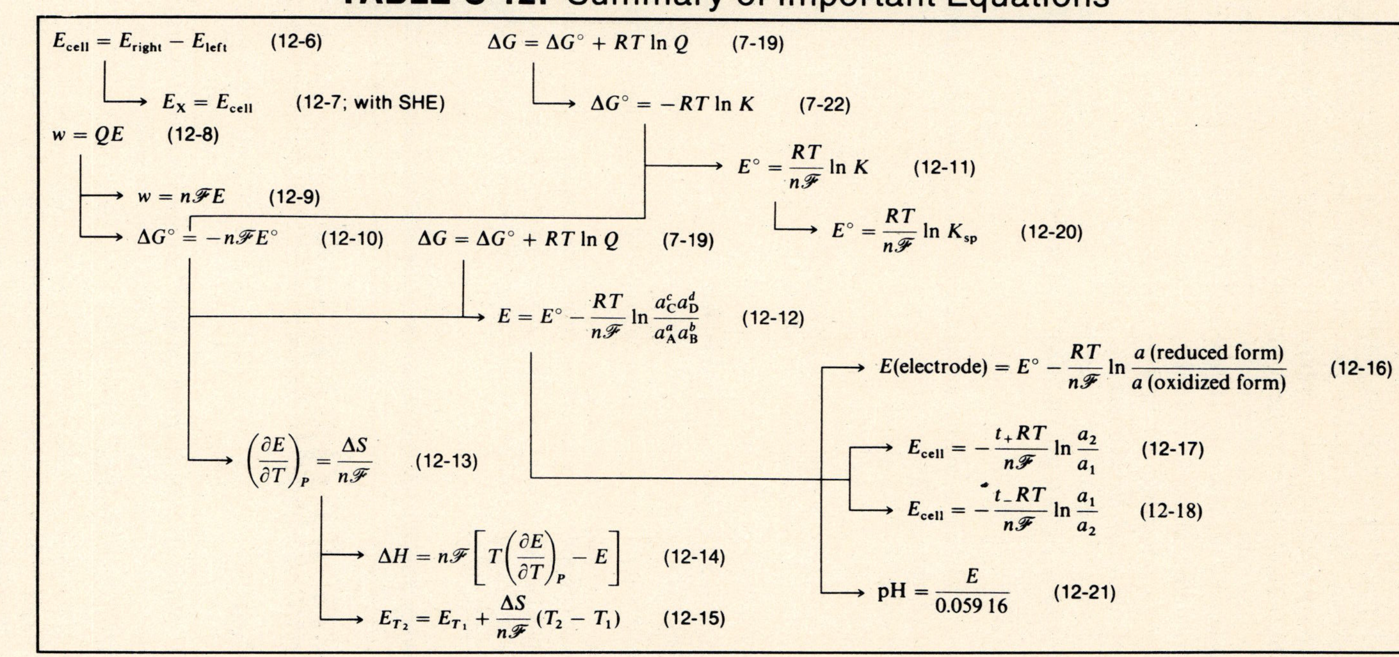

4. The electrode where oxidation occurs always is the anode; the one where reduction occurs is always the cathode. In a galvanic cell, the anode is negative; in an electrolytic cell, it is positive.
5. The electromotive force, or electric potential, of a cell is the difference between the potentials of the two electrodes.
   (a) Positive E means that the reaction defined by the cell diagram is spontaneous; negative E means nonspontaneous
   (b) Positive E means that the cell is the galvanic type; negative E means the electrolytic type
6. Single-electrode potentials can be calculated from cell potentials only if an arbitrary reference point is established. The potential of the standard hydrogen electrode is defined to be 0 V.
7. Standard electrode potential is the potential of the electrode when all components have an activity of 1 and usually refers to the reduction reaction.
8. From tables of standard electrode potentials, the cell potential of any cell may be calculated.
9. The electrode having the more negative reduction potential is forced, in the cell, to oxidize.
10. Electrical work under reversible conditions equals $-\Delta G°$, or $\Delta G° = -n\mathscr{F}E°$.
11. In calculating cell potentials, $E°$ may be used instead of $\Delta G°$, but, for single electrode potentials (electrons don't cancel out), $\Delta G°$ must be used.
12. For a spontaneous cell reaction, $E°$ must be positive so that $\Delta G°$ will be negative.
13. The Nernst equation permits calculation of cell and electrode potentials when activities are not unity: $E = E° - \dfrac{RT}{n\mathscr{F}} \ln Q$.
14. The temperature coefficient of the cell potential is useful for determining thermodynamic properties noncalorimetrically.
15. The usual types of electrodes are

metal | metal ion	gas
metal | insoluble salt	amalgam
oxidation–reduction	

16. Four major categories of galvanic cell are
   (a) chemical cell with and without transference; and
   (b) concentration cell with and without transference
17. A chemical cell without transference can be prepared by using only one electrolyte solution or approximated by using a salt bridge.
18. A concentration cell without transference can be obtained by using gas or amalgam electrodes or by combining two chemical cells without transference in opposition.
19. Cell potentials may be used to determine activity coefficients, equilibrium constants, solubility product constants, and pH.

SOLVED PROBLEMS

Description of Cells

PROBLEM 12-1 Write the cell diagram for a cell composed of one electrode of Pb and a 1-m solution of Pb^{2+} ions and a second electrode of Ag and a 1-m solution of Ag^{+} ions. Let the Pb be the anode and include a salt bridge.

Solution: You should write an expression like Eq. (12-3). Terminals are not mentioned in the problem, but they are needed. Platinum is commonly used, rather than the copper shown in Eq. (12-3). Use a single vertical line for interfaces and a double vertical line for the salt bridge. The anode goes on the left. The cell diagram is

$$\text{Pt}(s)\,|\,\text{Pb}(s)\,|\,\text{Pb}^{2+}\,(1\ m)\,\|\,\text{Ag}^{+}\,(1\ m)\,|\,\text{Ag}(s)\,|\,\text{Pt}(s)$$

Potentials of Cells and Electrodes

PROBLEM 12-2 An electrode consisting of Cd and a 1-m solution of Cd^{2+} was combined into a cell with a standard hydrogen electrode as the anode. The emf measured by a potentiometer was

-0.403 V. Calculate the standard reduction potential of the $Cd^{2+}|Cd$ electrode and write the equations for the half-reactions and net cell reaction.

Solution: Since the hydrogen electrode is on the left (the anode), Eq. (12-7) applies:

$$E_X = E_{cell}$$
$$E(Cd^{2+}|Cd) = -0.403 \text{ V}$$

The half-reaction for the reduction of Cd^{2+} is

$$Cd^{2+}(a_{Cd^{2+}} = 1) + 2e^- \longrightarrow Cd(s) \qquad \textbf{(a)}$$

For the hydrogen electrode, as the anode, oxidation occurs, so you must write

$$H_2(g) \longrightarrow 2H^+(a_{H^+} = 1) + 2e^- \qquad \textbf{(b)}$$

Adding eqs. (a) and (b), you get

$$Cd^{2+}(a_{Cd^{2+}} = 1) + H_2(g) \longrightarrow Cd(s) + 2H^+(a_{H^+} = 1)$$

PROBLEM 12-3 An electrode sometimes used as a standard is the silver|silver chloride electrode, another of the metal|insoluble salt type. The AgCl is coated on an Ag wire and dips into a Cl^- solution (not AgCl). Write the cell diagram and the equation for the net cell reaction when this electrode is made into a cell with an $Fe^{2+}|Fe$ electrode and all activities are 1.

Solution: You write the half-cell diagram for the standard electrode as $Ag|AgCl|Cl^-(a_{Cl^-} = 1)$. If you put the Fe electrode on the left, i.e.,

$$Pt|Fe|Fe^{2+}(a_{Fe^{2+}} = 1)||Cl^-(a_{Cl^-} = 1)|AgCl(s)|Ag|Pt$$

you have assumed that the anion with Fe^{2+} is not Cl^- and that the cation with Cl^- is not Fe^{2+}. What if you used an $FeCl_2$ solution? Then one solution serves both electrodes, so there is no liquid junction and the salt bridge isn't needed. Omitting the terminals, the cell diagram is

$$Fe|FeCl_2(a_{FeCl_2} = 1)|AgCl(s)|Ag$$

The net cell reaction from the half reactions is

$$\begin{array}{lrl} \text{anode:} & Fe & \longrightarrow Fe^{2+} + 2e^- \\ \text{cathode:} & AgCl + e^- & \longrightarrow Ag + Cl^- \\ \text{net:} & Fe + 2AgCl & \longrightarrow Fe^{2+} + 2Ag + 2Cl^- \end{array}$$

PROBLEM 12-4 What are the net cell reaction and the potential of a cell prepared, according to convention, from the electrodes $I^-|I_2$ and $Br^-|Br_2$ when the solutions are at unit activity?

Solution: In Table 12-1, both $E°$ values are positive. The lower, i.e., $I^-|I_2$, is the more negative, however, so you reverse its equation and change its sign.

$$\begin{array}{rll} 2I^- & \longrightarrow I_2 + 2e^- & E° = -0.536 \text{ V} \\ Br_2 + 2e^- & \longrightarrow 2Br^- & E° = 1.065 \text{ V} \\ \hline Br_2 + 2I^- & \longrightarrow 2Br^- + I_2 & E° = 0.529 \text{ V} \end{array}$$

PROBLEM 12-5 One use of $E°$ tables is to predict, without calculation, which of two possible reactions will occur. In which situation would the metal be oxidized: Sn in a 1-*m* solution of Fe^{2+} ions or Fe in a 1-*m* solution of Sn^{2+} ions?

Solution: In Table 12-1, you can see that the $Fe^{2+}|Fe$ electrode has a more negative potential than does $Sn^{2+}|Sn$. If a cell were made of these two electrodes, the Fe half-reaction would be reversed, and the reaction would be

$$Fe + Sn^{2+} \longrightarrow Fe^{2+} + Sn$$

Tin would not be oxidized by a 1-*m* solution of Fe^{2+}.

Thermodynamics of Cells

PROBLEM 12-6 Determine $E°$ for the electrode for which the reaction is $2Hg^{2+} + 2e^- \longrightarrow Hg_2^{2+}$.

Solution: You need to find two electrodes with half-reactions that can be combined to give this one, calculate ΔG° for the new reaction, and then calculate E°. There are two appropriate electrodes in Table 12-1:

$$Hg^{2+} + 2e^- \longrightarrow Hg \qquad E^\circ = 0.854 \text{ V}$$
$$Hg_2^{2+} + 2e^- \longrightarrow 2Hg \qquad E^\circ = 0.79 \text{ V}$$

Before setting up a table (as in Example 12-10), you should recognize that the first equation must be multiplied by 2 and the second equation must be reversed in order to get the desired equation by adding. (This has nothing to do with anode and cathode; you merely want to get the specified equation.)

Half-reaction	E°	ΔG°
$2Hg^{2+} + 4e^- \longrightarrow 2Hg$	0.854 V	$4(0.854)\mathscr{F} = 3.416\mathscr{F}$
$2Hg \longrightarrow Hg_2^{2+} + 2e^-$	−0.79 V	$2(-0.79)\mathscr{F} = -1.58\mathscr{F}$
$2Hg^{2+} + 2e^- \longrightarrow Hg_2^{2+}$		$1.836\mathscr{F}$

Rearranging Eq. (12-10), you have

$$E^\circ = -\frac{\Delta G^\circ}{n\mathscr{F}} = \frac{1.836\mathscr{F}}{2\mathscr{F}} = 0.918 \text{ V}$$

PROBLEM 12-7 Calculate the equilibrium constant at 298 K for the net cell reaction of the cell composed of the $Ni^{2+}|Ni$ and $HSnO_2^-|Sn(OH)_6^{2-}$ electrodes.

Solution: In order to calculate K, you need the value of E° to substitute in Eq. (12-11). You first have to reverse the half-reaction for $Sn(OH)_6^{2-}$ and then add the two half-reactions to obtain the net cell reaction:

$$HSnO_2^- + 3OH^- + H_2O \longrightarrow Sn(OH)_6^{2-} + 2e^- \qquad E^\circ = +0.90 \text{ V}$$
$$Ni^{2+} + 2e^- \longrightarrow Ni \qquad E^\circ = -0.250 \text{ V}$$
$$HSnO_2^- + 3OH^- + H_2O + Ni^{2+} \longrightarrow Ni + Sn(OH)_6^{2-} \qquad E^\circ = 0.65 \text{ V}$$

Now, solve Eq. (12-11) for $\log K$:

$$\log K = \frac{n}{0.059\,16}E^\circ = \frac{2}{0.059\,16}(0.65) = 21.974$$
$$K = 9.43 \times 10^{21}$$

PROBLEM 12-8 For the cell in Problem 12-7, calculate E if the activity of each ion is 0.1.

Solution: For E, you need to obtain the value of Q and then to substitute it into Eq. (12-12). From the cell reaction in Problem 12-7,

$$Q = \frac{a_{Sn(OH)_6^{2-}}}{(a_{HSnO_2^-})(a_{OH^-})^3(a_{Ni^{2+}})} = \frac{0.1}{0.1(0.1)^3(0.1)} = \frac{1}{(10^{-1})^4} = 10^4$$

Thus

$$E = E^\circ - \frac{RT}{n\mathscr{F}}\ln Q$$
$$= 0.65 - \frac{0.059\,16}{2}\log(10^4) = 0.65 - 0.118 = 0.53 \text{ V}$$

PROBLEM 12-9 Calculate the equilibrium constant for the net cell reaction at 298 K when a cell has been assembled of an aluminum | aluminum ion electrode and a lead | lead ion electrode.

Solution: Proceeding in the usual way, you should determine E°_{cell} and then substitute into Eq. (12-11). Since $Al^{3+}|Al$ has the more negative reduction potential, it involves oxidation and is on the left. Consequently, from Eq. (12-6),

$$E^\circ_{cell} = E^\circ_r - E^\circ_l = -0.126 - (-1.66) = +1.53 \text{ V}$$

Rearranging Eq. (12-11) and substituting, you have

$$\log K = \frac{n}{0.059\,16}(1.53)$$

Now you have hit a snag: What is the value of n? The answer depends on how you write the balanced equation for the net reaction. Going back to the half-reactions,

$$Pb^{2+} + 2e^- \longrightarrow Pb$$
$$Al \longrightarrow Al^{3+} + 3e^- \qquad \text{(reversed)}$$

To balance when you add, you could multiply the first half-reaction by 3 and the second by 2. Then

$$2Al + 3Pb^{2+} \longrightarrow 2Al^{3+} + 3Pb$$

and $n = 6$. Alternatively, you could multiply the first by $\frac{1}{2}$ and the second by $\frac{1}{3}$, i.e., letting one electron be exchanged. Then, you would have

$$\frac{1}{3}Al + \frac{1}{2}Pb^{2+} \longrightarrow \frac{1}{3}Al^{3+} + \frac{1}{2}Pb$$

and $n = 1$ (see Section 7-2). Thus, when $n = 6$,

$$\log K_6 = \frac{6}{0.059\,16}(1.53) = 155.172$$
$$K_6 = 1.49 \times 10^{155}$$

and when $n = 1$,

$$\log K_1 = \frac{1}{0.059\,16}(1.53) = 25.862$$
$$K_1 = 7.28 \times 10^{25}$$

Observe that K_1 is the sixth root of K_6.

PROBLEM 12-10 The temperature coefficient of the standard Weston cell is -5.00×10^{-5} V K^{-1}, and its emf at 298 K is 1.018 07 V. Calculate ΔH, ΔS, and ΔG for the cell reaction.

Solution: The electrodes of a Weston cell are $Hg|Hg_2SO_4$ and $Cd|CdSO_4$. From Table 12-1, the Hg electrode has an $E°$ of 0.615 V. Since the cell $E°$ is 1.018 V, you can deduce that the Cd half-reaction must be lower in the table; therefore it's more negative, and you have to reverse it when writing the cell reaction:

$$Cd(s) + Hg_2SO_4(s) + \frac{8}{3}H_2O(l) \longrightarrow CdSO_4 \cdot \frac{8}{3}H_2O(s) + 2Hg(l)$$

What you need from this equation is the number of electrons, which is 2 since $Cd^0 \rightarrow Cd^{2+}$. You calculate ΔG from Eq. (12-10):

$$\Delta G° = -n\mathcal{F}E°$$
$$= (-2\text{ mol})(9.65 \times 10^4\text{ J V}^{-1}\text{mol}^{-1})(1.018\,07\text{ V}) = -19.65 \times 10^4\text{ J} = -196.5\text{ kJ}$$

From Eq. (12-13),

$$\Delta S° = n\mathcal{F}\left(\frac{\partial E°}{\partial T}\right)_P$$
$$= (2\text{ mol})(9.65 \times 10^4\text{ J V}^{-1}\text{mol}^{-1})(-5.00 \times 10^{-5}\text{ V K}^{-1}) = -9.65\text{ J K}^{-1}$$

And from $\Delta H = \Delta G + T\Delta S$,

$$\Delta H° = -196.5\text{ kJ} + 298(-9.65)\text{ J} = (-196.5 - 2.9)\text{ kJ} = -199.4\text{ kJ}$$

alternative: You could also get $\Delta H°$ from Eq. (12-14):

$$\Delta H° = n\mathcal{F}\left[T\left(\frac{\partial E°}{\partial T}\right)_P - E°\right]$$

Types of Electrodes and Galvanic Cells

PROBLEM 12-11 In the chlorine gas electrode, a negative ion is formed by reduction of the gas. (**a**) Write general equations like those in Section 12-4 for this type of electrode. (**b**) Write the specific equations for chlorine at 1 atm.

Solution:

(**a**) The general half-reaction would be

$$G_2(g, P_{G_2}) + ne^- \longrightarrow 2G^{r-}(a_{G^{r-}})$$

The general Nernst equation would be

$$E = E^\circ - \frac{RT}{n\mathscr{F}} \ln \frac{a^2_{\mathrm{Gr}^-}}{P_{\mathrm{G}_2}}$$

(b) For the chlorine electrode, you have

$$\mathrm{Cl_2}(g, P_{\mathrm{Cl_2}}) + 2e^- \longrightarrow 2\mathrm{Cl}^-\ (a_{\mathrm{Cl}^-})$$

Assume that $a = P = 1$ atm and use Table 12-1 to solve for E:

$$E = 1.360 - \frac{RT}{2\mathscr{F}} \ln a^2_{\mathrm{Cl}^-} = 1.360 - \frac{RT}{\mathscr{F}} \ln a_{\mathrm{Cl}^-}$$

PROBLEM 12-12 One metal | insoluble salt electrode is $Pb|PbSO_4$. Calculate the electrode potential if this electrode is immersed in a saturated solution of $PbSO_4$, $K_{sp} = 1.06 \times 10^{-8}$. Assume that activities are equal to concentrations.

Solution: The key word in this problem is *saturated*. Recall that E° values are for 1 m solutions (or $a = 1$). You should determine the concentration and apply the Nernst equation. From Table 12-1, the half-cell reaction is

$$\mathrm{PbSO_4}(s) + 2e^- \longrightarrow \mathrm{Pb}(s) + \mathrm{SO_4^{2-}}\ (a = 1) \qquad E^\circ = -0.355\ \mathrm{V}$$

The only concentration of concern is that of the SO_4^{2-}. You calculate it from K_{sp}:

$$K_{sp} = [\mathrm{Pb^{2+}}][\mathrm{SO_4^{2-}}] = 1.06 \times 10^{-8}$$

For this salt, the two concentrations are equal, so

$$[\mathrm{SO_4^{2-}}]^2 = 1.06 \times 10^{-8} \qquad \text{and} \qquad [\mathrm{SO_4^{2-}}] = 1.03 \times 10^{-4}$$

Then

$$E = E^\circ - \frac{RT}{2\mathscr{F}} \ln \frac{a_{\mathrm{SO_4^{2-}}}}{1}$$

$$= -0.355 - \frac{0.059\,16}{2} \log(1.03 \times 10^{-4}) = -0.355 + 0.118 = -0.237\ \mathrm{V}$$

PROBLEM 12-13 Using Table 12-1, design a chemical cell without transference that has at least one metal | insoluble salt electrode and doesn't use a salt bridge. Write the cell reaction and calculate the standard cell potential.

Solution: To eliminate transference, you should use only one electrolyte solution, which would have to contain ions of both electrodes. The metal | insoluble salt electrode will involve an anion; Br^-, Cl^-, and SO_4^{2-} are the only ones in Table 12-1. The cation would come from a metal | metal ion electrode or from the hydrogen electrode. If you were to choose $Hg|Hg_2SO_4$ and $Zn|Zn^{2+}$, the Zn is more negative; you reverse its half-reaction and obtain

$$\begin{array}{lll} \mathrm{Zn} \longrightarrow \mathrm{Zn^{2+}} + 2e^- & & E^\circ = +0.763\ \mathrm{V} \\ \mathrm{Hg_2SO_4} + 2e^- \longrightarrow 2\mathrm{Hg} + \mathrm{SO_4^{2-}} & & E^\circ = +0.615\ \mathrm{V} \\ \hline \mathrm{Hg_2SO_4} + \mathrm{Zn} \longrightarrow \mathrm{Zn^{2+}} + 2\mathrm{Hg} + \mathrm{SO_4^{2-}} & & E^\circ = 1.378\ \mathrm{V} \end{array}$$

Applications of Galvanic Cell Potentials

PROBLEM 12-14 Calculate the cell potential at 25 °C of the cell

$$\mathrm{Pb}|\mathrm{PbSO_4}(s)|\mathrm{H_2SO_4}(m = 0.100, \gamma = 0.265)|\mathrm{H_2}(1\ \mathrm{atm})$$

Solution: You should ask yourself which type cell this is, although you may not need to know in order to solve the problem. Do you recognize it as a chemical cell without transference? (See Example 12-16.) While you're learning, it's best to write out the half-reactions and the cell reactions, so that you get the correct reactants for Q and the correct value of n. In Table 12-1, you can see that the Pb electrode is below the H_2 and must be reversed:

$$\begin{array}{lll} \mathrm{Pb}(s) + \mathrm{SO_4^{2-}} \longrightarrow \mathrm{PbSO_4}(s) + 2e^- & & E^\circ = +0.355\ \mathrm{V} \\ 2\mathrm{H^+} + 2e^- \longrightarrow \mathrm{H_2} & & E^\circ = 0.00\ \mathrm{V} \\ \hline \mathrm{Pb}(s) + \mathrm{SO_4^{2-}} + 2\mathrm{H^+} \longrightarrow \mathrm{PbSO_4}(s) + \mathrm{H_2}(g) & & E^\circ = +0.355\ \mathrm{V} \end{array}$$

Now you can use the Nernst equation:

$$E = 0.355 - \frac{0.059\,16}{2}\log\frac{P_{H_2}}{a_{H^+}^2 a_{SO_4^{2-}}}$$

From Chapter 11, you'll recall two equations involving activity:

$$a = \gamma m \quad \textbf{[Eq. (11-29)]} \quad \text{and} \quad a_2 = a_+^p a_-^q \quad \textbf{[Eq. (11-32)]}$$

For this problem, the denominator of the log term becomes

$$a_+^p a_-^q = a_{H^+}^2 a_{SO_4^{2-}} = a_2 = a_{H_2SO_4}$$

and, from Eq. 11-29,

$$a_{H_2SO_4} = \gamma m = 0.265(0.100) = 0.0265$$

So

$$E = 0.355 - \frac{0.059\,16}{2}\log\frac{1}{0.0265} = 0.355 - 0.0466 = 0.308 \text{ V}$$

PROBLEM 12-15 Concentration cells with transference in which the cation reacts at the electrodes are illustrated by a cell composed of two hydrogen electrodes dipping in solutions having different concentrations of hydrogen ions. Also the solutions must be in contact:

$$\text{Pt, } H_2 \text{ (1 atm)} | \text{HCl } (a_1) | \text{HCl } (a_2) | H_2 \text{ (1 atm), Pt}$$

For this cell, derive Eq. (12-18), which expresses E_{cell} as a function of transport number and activity.

Solution: As usual, you should determine the cell reaction.

anode:	H_2 (1 atm)	$\longrightarrow$	$2H^+ (a_1) + 2e^-$
cathode:	$2H^+ (a_2) + 2e^-$	$\longrightarrow$	H_2 (1 atm)
overall:	$2H^+ (a_2)$	$\longrightarrow$	$2H^+ (a_1)$
or	$H^+ (a_2)$	$\longrightarrow$	$H^+ (a_1)$ **(i)**

In the solutions, H^+ will be moving from left to right across the liquid junction and Cl^- will be moving in the opposite direction. The amount of Cl^- crossing the junction when one Faraday passes through the cell is the transport number, t_-. You can show this situation by the reaction

$$t_- \, Cl^- (a_2) \longrightarrow t_- \, Cl^- (a_1) \quad \textbf{(ii)}$$

The amount of H^+ is given by t_+ or $1 - t_-$:

$$(1 - t_-) H^+ (a_1) \longrightarrow (1 - t_-) H^+ (a_2) \quad \textbf{(iii)}$$

The net amount of material transferred is expressed by the sum of eqs. (i), (ii), and (iii), so

$$H^+ (a_2) + t_- \, Cl^- (a_2) + H^+ (a_1) - t_- \, H^+ (a_1) \longrightarrow H^+ (a_1) + t_- \, Cl^- (a_1) + H^+ (a_2) - t_- \, H^+ (a_2)$$

Canceling and rearranging to place a_2 on the left [because it is on the left in eq. (i)] gives

$$t_- \, Cl^- (a_2) + t_- \, H^+ (a_2) \longrightarrow t_- \, Cl^- (a_1) + t_- \, H^+ (a_1)$$

or

$$t_- \, \text{HCl} (a_2) \longrightarrow t_- \, \text{HCl} (a_1)$$

Again, you use the Nernst equation to obtain E_{cell}. Since this is a concentration cell, $E^\circ_{cell} = 0$, so

$$E_{cell} = -\frac{RT}{n\mathscr{F}}\ln\frac{a_1^{t_-}}{a_2^{t_-}}$$

or

$$E_{cell} = -\frac{t_- RT}{n\mathscr{F}}\ln\frac{a_1}{a_2} \quad \textbf{[Eq. (12-18)]}$$

PROBLEM 12-16 Find the solubility product constant of Hg_2SO_4.

Solution: Since K_{sp} is an equilibrium constant, you can use the relationship between E° and K. We used two expressions for ΔG° to obtain Eq. (12-11):

$$\Delta G^\circ = n\mathscr{F}E^\circ \quad \textbf{[Eq. (12-10)]} \quad \text{and} \quad \Delta G = \Delta G^\circ + RT\ln Q \quad \textbf{[Eq. (7-19)]}$$

At equilibrium, $\Delta G = 0$ and $Q = K$, so

$$\Delta G^\circ = -RT \ln K \quad \textbf{[Eq. (7-22)]} \qquad \text{and} \qquad E^\circ = \frac{RT}{n\mathcal{F}} \ln K \quad \textbf{[Eq. (12-11)]}$$

For K_{sp}, Eq. (12-11) becomes

$$E^\circ = \frac{RT}{n\mathcal{F}} \ln K_{sp} \quad \textbf{[Eq. (12-20)]}$$

As usual, you should write the equation of the process:

$$Hg_2SO_4 \longrightarrow Hg_2^{2+} + SO_4^{2-}$$

Now look in Table 12-1 for two electrodes that will yield this expression as the net cell reaction. You may find

$$Hg_2^{2+} + 2e^- \longrightarrow 2Hg \qquad E^\circ = 0.79 \text{ V}$$
$$Hg_2SO_4 + 2e^- \longrightarrow 2Hg + SO_4^{2-} \qquad E^\circ = 0.615 \text{ V}$$

To obtain the desired reaction, reverse the first half-reaction and add:

$$Hg_2SO_4 \longrightarrow Hg_2^{2+} + SO_4^{2-} \qquad E^\circ = -0.17 \text{ V}$$

Rearranging Eq. (12-20) and then substituting, you obtain

$$\ln K_{sp} = \frac{n\mathcal{F}}{RT} E^\circ$$

$$\log K_{sp} = \frac{2}{0.059\,16}(-0.17) = -5.747$$

$$K_{sp} = 1.79 \times 10^{-6}$$

Supplementary Exercises

PROBLEM 12-17 A flashlight battery is a ________ dry cell. The substance usually used in salt bridges is ________. The cathode in a galvanic cell carries a (positive) (negative) sign. If some silver is dropped into a cupric sulfate solution, copper (will) (will not) be released. If the activity of the ion in the electrode reaction $M^+ + e^- \rightarrow M$ is less than unity, the cell potential will be (more) (less) negative than E°. In general, you can vary the ________ ________ ________ in order to change a small negative cell potential to a positive one. Name a standard electrode that is the metal | insoluble salt type. In Table 12-1, an example of an oxidation–reduction electrode (other than the one used in Example 12-20) is ________. The $Ni|Ni^{2+}|Ag^+|Ag$ cell is a ________ ________ ________ cell.

PROBLEM 12-18 The voltage of the cell in Problem 12-3 was measured to be 0.662 V. The E° for the standard Ag | AgCl electrode is 0.222 V. Calculate E° for the $Fe^{2+}|Fe$ electrode.

PROBLEM 12-19 Calculate the electrode potential of the AgBr electrode if an emf of +0.147 V is obtained when this electrode is measured against a saturated calomel electrode, which is on the right in the conventional cell diagram.

PROBLEM 12-20 Calculate the cell potential at unit activities for a galvanic cell composed of $Tl^+|Tl$ and $Sn^{+4}|Sn^{+2}$ electrodes, with the more negative electrode on the left.

PROBLEM 12-21 Calculate the equilibrium constant for the net reaction in the cell of Problem 12-20 if the equation is $2Tl + Sn^{4+} \rightarrow 2Tl^+ + Sn^{2+}$ and the temperature is 298 K.

PROBLEM 12-22 Calculate the electrode potential at unit activities for the half-reaction $Tl^{3+} + 3e^- \rightarrow Tl$.

PROBLEM 12-23 Calculate the electrode potential of the $Fe^{2+} | Fe$ electrode if the mean ionic activity of the solution is 0.20.

PROBLEM 12-24 Calculate the potential of a cell with $Br_2 | Br^-$ and $Zn^{2+} | Zn$ electrodes if the concentration of Br^{-1} is 0.40 *m* and that of Zn^{2+} is 0.20 *m*. Assume that $a = 1$ for Br_2 and $\gamma_\pm = 0.72$ for 0.20 *m* $ZnBr_2$. Use the half-reactions as they appear in Table 12-1.

PROBLEM 12-25 A cell composed of the standard electrodes $Al^{3+} | Al$ and $Ag^+ | Ag$ is found to have a temperature coefficient of the cell potential of -1.37×10^{-5} V K^{-1}. Calculate ΔS and ΔH for the reaction.

PROBLEM 12-26 Calculate the electrode potential at 25 °C of the $Tl^{3+} | Tl^{1+}$ electrode if the activity of Tl^{1+} is 0.35 and that of Tl^{3+} is 0.48.

PROBLEM 12-27 A potential of 1.521 V was measured at 25 °C for the cell

$$Zn | ZnSO_4\ (m = 0.0100) | Hg_2SO_4 | Hg$$

Calculate the mean ionic activity coefficient of the ions.

PROBLEM 12-28 Calculate the solubility product constant for CuI.

PROBLEM 12-29 A concentration cell with transference was prepared using two H_2 electrodes dipping into solutions of HCl. When the concentrations of the solutions were 0.0100 *m* ($\gamma_\pm = 0.904$) and 0.100 ($\gamma_\pm = 0.796$), the cell voltage was 0.0190 V. (**a**) Calculate the transport number of Cl^- in this system. (**b**) Calculate ΔG° for the cell.

PROBLEM 12-30 A cell composed of a $Zn^{2+} | Zn$ electrode dipping in a 5.00×10^{-3} *m* $ZnSO_4$ solution and of an $Hg_2SO_4 | Hg$ electrode was found to have a voltage of 1.533 V. (**a**) What type cell is this? (**b**) Calculate the mean ionic activity coefficient of $ZnSO_4$ at this concentration.

PROBLEM 12-31 In a cell reaction involving the exchange of one electron, E° is found to be 1.233 V and the dependence of E on T at constant P is -3.26×10^{-5} V K^{-1}. Calculate ΔG°, ΔH°, and ΔS°.

Answers to Supplementary Exercises

12-17 galvanic; KCl; positive; will not; more; concentrations of ions; calomel, $Hg | Hg_2SO_4$, or $Ag | AgCl$; $Fe^{3+} | Fe^{2+}$; chemical with transference **12-18** -0.44 V **12-19** $+0.095$ V **12-20** $+0.49$ V
12-21 3.67×10^{16} **12-22** $+0.694$ V **12-23** -0.461 V **12-24** $+1.894$ V
12-25 -3.97 J K^{-1}; -713 kJ mol^{-1} **12-26** 1.215 V **12-27** 0.383 **12-28** 1.16×10^{-12}
12-29 (**a**) 0.17; (**b**) 0 **12-30** (**a**) chemical without transference; (**b**) 0.480
12-31 -119 kJ mol^{-1}; -120 kJ mol^{-1}; -3.15 J mol^{-1}

13 CHEMICAL KINETICS

THIS CHAPTER IS ABOUT

- ☑ Terminology
- ☑ Rate of Reaction
- ☑ Rate Equations
- ☑ Integrated Forms of Rate Equations
- ☑ Half-life
- ☑ Determination of Order
- ☑ Complicating Processes
- ☑ Temperature Dependence of Rate
- ☑ Theories of Reaction Rate
- ☑ Mechanisms
- ☑ Catalysis

13-1. Terminology

A. Thermodynamics versus kinetics

Thermodynamics deals with systems in equilibrium and the energy relationships between reactants and products, whereas **chemical kinetics**, or **reaction kinetics**, is concerned with systems before equilibrium is reached. The two subdivisions of chemical kinetics are best described by the questions: (1) How fast is a reaction going? and (2) What chemical process is occurring in the system to produce the net result? The answers to (2) are based primarily on the results of measurements made in order to answer (1).

B. Homogeneous versus heterogeneous

As with our previous use of terms, a reaction is kinetically **homogeneous** when all participants are in the same phase. The presence of two phases makes the reaction **heterogeneous**.

C. Mechanism and elementary reactions

Many chemical reactions don't proceed in a straightforward single step as suggested by their balanced equations. In the reaction $2C_2H_6 + 7O_2 \rightarrow 4CO_2 + 6H_2O$, it's virtually impossible for 9 molecules to come together simultaneously, break up into 30 atoms, and then form new molecules. Actually, the process is a complex series of steps that add up to the overall change. This process is called the **mechanism** of the reaction and answers question (2). Several mechanisms may provide reasonable explanations of a particular reaction.

Each step in a mechanism is an **elementary reaction** (also *simple reaction*), meaning that it cannot be reduced further. For example, an elementary reaction identified from a sequence of reactions is $CCl_3^- \rightarrow CCl_2 + Cl^-$. The reaction occurs as described by its balanced equation and usually involves only one or two species, but these species may be short-lived free radicals.

The **molecularity** of a reaction is the number of reacting species in an elementary reaction. When the number of reacting species is one, we say that reaction is *unimolecular*; for two species, *bimolecular*; and for three species, *trimolecular*. Note that molecularities must be integers.

EXAMPLE 13-1: The decomposition of HI, $2HI \rightleftharpoons H_2 + I_2$, is an elementary reaction. What is its molecularity?

Solution: Since two HI molecules are reacting, the molecularity is two. Or we say that the reaction is bimolecular.

D. Factors affecting the rate of reaction

The time required for a particular process to occur is affected by several variables:

1. Nature of the reactants—The reaction of NO with O_2 is rapid, whereas the reaction of N_2 with O_2 is negligible at room temperature.
2. Degree of subdivision of reactants—You can't get a log to burn with a match, but you can ignite thin shavings and twigs. A cloud of sawdust can explode from a spark. An *explosion* is merely a very rapid reaction.
3. Presence of a catalyst—The usual illustration is a mixture of hydrogen and oxygen gases. Even after several years, no water could be detected. If the right kind of catalyst is sprinkled in, however, the mixture will explode.
4. Temperature—A reaction usually goes faster as the temperature rises. A familiar rule of thumb is that a reaction goes twice as fast for a rise of 10 °C in temperature.
5. Concentration—An increase in the concentration of one of the reactants usually makes the reaction go faster, although in some complicated systems, the relationship is inverse.

13-2. Rate of Reaction

A. Definition

When we choose a system to be studied, we've established the first three factors listed in Section 13-1D. If the temperature also is held constant, concentration is the only variable. Then the rate of reaction at that temperature expresses the change in concentration of a reactant or product per unit time. The general definition of rate is expressed in terms of the **extent of reaction,** (ξ) (also called the *advancement* of the reaction). In Section 7-1D, we defined ξ as

$$\xi = \frac{n_i - n_{i,0}}{\nu_i} \qquad \textbf{[Eq. (7-3)]}$$

Also, we had

$$\sum_i \nu_i A_i = 0 \qquad \textbf{[Eq. (7-2)]}$$

where the coefficient ν is negative for reactants.

The **rate of reaction** is defined as the rate of increase of ξ:

RATE OF REACTION

$$\text{rate} = \frac{d\xi}{dt} \qquad \textbf{(13-1)}$$

i.e., the speed at which the reaction is advancing. From Eq. (7-3), since $n_{i,0}$ is constant,

$$\text{rate} = \frac{d\xi}{dt} = \frac{1}{\nu_i}\frac{dn_i}{dt} \qquad \textbf{(13-2)}$$

EXAMPLE 13-2: Write the expressions for the rate for the three participating species in the reaction

$$aA \longrightarrow cC + dD$$

Solution: Remember that ν_A is negative, i.e., $\nu_A = -a$, so Eq. (13-2) becomes

$$\text{rate} = -\frac{1}{a}\frac{dn_A}{dt} = \frac{1}{c}\frac{dn_C}{dt} = \frac{1}{d}\frac{dn_D}{dt}$$

Because A is being used up, dn_A is negative; all three expressions give a positive rate, as they must if ξ is to increase.

EXAMPLE 13-3: Derive an expression relating the rate of change of concentration to the rate of reaction, assuming that the volume is constant.

Solution: The concentration usually used is molarity (represented by c). When V is constant, $c_i = n_i/V$, or $n_i = c_iV$. Then $dn_i = Vdc_i$ and Eq. (13-2) becomes

$$\frac{d\xi}{dt} = \frac{V}{\nu_i}\frac{dc_i}{dt} \qquad \textbf{(13-3)}$$

We rearrange, placing the rate of change of concentration on the left, so that

$$\frac{dc_i}{dt} = \frac{\nu_i}{V}\frac{d\xi}{dt} = v_i \qquad \textbf{(13-4)}$$

Some authors use the symbol v_i for rate of change of concentration; e.g., v_A represents the rate of disappearance of A in the reaction in Example 13-2. Molar concentration frequently is shown in brackets, so we could write

$$v_A = \frac{d[A]}{dt} = \frac{-a}{V}\frac{d\xi}{dt}$$

for the disappearance of A. The customary procedure is to assume that the volume of the system remains constant and to study rates of change of concentration. In this case the rate of reaction is often used as the rate per unit volume, $\frac{1}{V}\frac{d\xi}{dt}$, which is equal to $\frac{v_i}{\nu_i}$, from Eq. (13-4). We often write Eq. (13-4) as

$$\frac{\text{rate}}{V} = \frac{1}{\nu_i}\frac{dc_i}{dt} = \frac{1}{\nu_i}v_i$$

EXAMPLE 13-4: We can summarize the concepts presented in this section by using the reaction $2N_2O_5 \rightarrow 4NO_2 + O_2$, because the coefficients are not all 1. Write the expression for the rate of reaction at constant volume in terms of the concentration changes for each species in the reaction.

Solution: We obtain the rate at constant volume by substituting into Eq. (13-4):

$$\frac{\text{rate}}{V} = \frac{1}{\nu_i}\frac{dc_i}{dt} = -\frac{1}{2}\frac{dc_{N_2O_5}}{dt} = \frac{1}{4}\frac{dc_{NO_2}}{dt} = \frac{1}{1}\frac{dc_{O_2}}{dt}$$

Thus we see that the rate of reaction equals the rate of formation of O_2 but only one-half the rate of disappearance of N_2O_5 and one-fourth the rate of formation of NO_2. Note that we could have balanced this equation as follows:

$$N_2O_5 \longrightarrow 2NO_2 + \tfrac{1}{2}O_2$$

Then the rate of reaction would equal the rate of disappearance of N_2O_5.

note: It is essential that a balanced equation be given with a rate of reaction.

B. Measurement of rates

Chemical methods of determining concentrations in a reacting mixture generally are unsatisfactory. If samples are removed, the system may be disturbed, causing different behavior than if it were left alone. If the analysis takes much time, the concentration measured will not be that at the time of removal because the concentration continues to change.

Physical methods, which can be adapted to continuous and instantaneous measurement, are preferred. Each system must be analyzed to determine which property is best suited for measurement. The first kinetic study (Wilhelmy, 1850) explored the change in the rotation of polarized light by a sugar solution as reaction proceeded. Other physical properties that have been measured are conductivity, pressure, and absorption of light. Because the rate changes as concentration changes, we must specify *when* a stated rate was measured and *how* it was calculated.

EXAMPLE 13-5: The following data were obtained for the reaction A → products. (**a**) Calculate the instantaneous rate of the reaction at 17.5 min if the volume remains constant. (**b**) Calculate an approximate rate.

Measurement	1	2	3	4	5	6
Time (min)	0	5	10	15	20	25
[A]	0.5	0.36	0.26	0.19	0.14	0.10

Solution:

(**a**) To obtain the rate of reaction from concentration, we must use Eq. (13-3). Since the volume is constant and $v_A = -1$, this becomes

$$\frac{\text{rate}}{V} = -\frac{dc_A}{dt}$$

which is the slope of the plot of c_A versus t, as shown in Figure 13-1. The slope is the tangent to the curve at a particular point and changes continuously. A tangent is drawn at 17.5 min, with end points at 0.25 M and 8.4 min and 0.05 M and 29.4 min. Its slope is

$$m = \frac{y_2 - y_1}{x_2 - x_1} = \frac{(0.25 - 0.05)M}{(8.4 - 29.4)\text{min}} = -0.0095\ M\ \text{min}^{-1} = \frac{dc_A}{dt}$$

and

$$\frac{\text{rate}}{V} = -\frac{dc_A}{dt} = 0.0095\ \text{min}^{-1}$$

This rate is the *instantaneous* rate at 17.5 min.

(**b**) We can calculate an approximate rate from two of the data points. In doing so, we are assuming that the curve is linear between those points. For measurements 2 and 3,

$$\frac{dc_A}{dt} = \frac{0.36 - 0.26}{5 - 10} = -0.020\ M\ \text{min}^{-1}$$

$$\frac{\text{rate}}{V} = 0.020\ \text{min}^{-1}$$

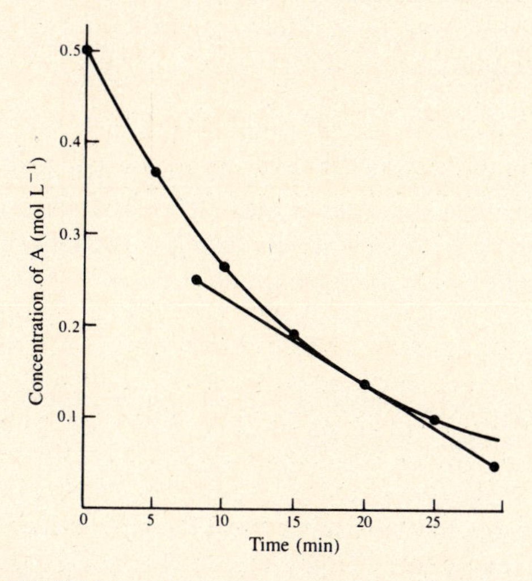

FIGURE 13-1. Plot of concentration versus time.

For measurements 5 and 6,

$$\frac{dc_A}{dt} = \frac{0.14 - 0.10}{20 - 25} = -0.0080\ M\ \text{min}^{-1}$$

$$\frac{\text{rate}}{V} = 0.0080\ \text{min}^{-1}$$

These values are the *average rate* over the particular time interval.

Because the rate changes with the elapsed time of reaction, studies frequently are made by comparing the *initial rate* under different conditions, e.g., different temperatures or different initial concentrations. The initial rate is the slope of the curve after a short time, perhaps at 1 min in Example 13-5.

EXAMPLE 13-6: Calculate the initial rate $v_{A,0}$ of the reaction in Example 13-5.

Solution: In the first minute, the curve is very close to a straight line, so we can calculate the slope from the coordinates at 0 and 1 min. From the curve, the concentration at 1 min is 0.47, so

$$v_{A,0} = m = \frac{y_2 - y_1}{x_2 - x_1} = \frac{0.5 - 0.47}{0 - 1} = -0.030\ M\ \text{min}^{-1}$$

This is an average rate, but it is very close to the instantaneous initial rate. If the reaction is complex, this approach may not yield the correct rate law.

13-3. Rate Equations

A. Rate and concentration

The rate of a reaction is usually proportional to the concentrations of the reactants raised to small powers that are frequently (but not necessarily) integers. For the general reaction A + B → products,

$$\text{rate} \propto c_A^p c_B^r$$

When we insert a proportionality constant,

$$v_A = \frac{dc_A}{dt} = k c_A^p c_B^r \qquad \textbf{(13-5)}$$

The constant k is called the **rate constant**, or *specific rate constant*, and equations of this type are called *rate equations* or *rate laws*. The rate constant varies with temperature but is independent of concentration.

B. Order of a reaction

The exponents of the concentrations in Eq. (13-5) are known as the *order of the reaction with respect to each reactant*, i.e., p is the order of the reaction with respect to A, and r is the order of the reaction with respect to B. Somewhat confusingly, the sum of the exponents of the rate equation is called the *order of the reaction*, i.e.,

$$\text{order} = p + r \qquad \textbf{(13-6)}$$

Sometimes authors try to avoid confusion by calling this expression the *overall order*. Three important points to remember about order are that it

1. may be a fraction;
2. isn't necessarily related to the coefficients in the balanced equation (a, b, etc.); and
3. must be determined experimentally.

EXAMPLE 13-7: Specify the orders with respect to each reactant and the overall order for the reaction having a rate equation of

$$\text{rate} = kc_A^{1/2}c_B$$

Solution: The individual orders are the exponents, so the order with respect to A is $\frac{1}{2}$, and the order with respect to B is 1. The overall order is the sum of the exponents [Eq. (13-6)], so the order of the reaction is $\frac{3}{2}$. A shorter way to state this information is to say the reaction is half-order in A, first-order in B, and three-halves-order overall.

C. Integral orders

Many reactions, especially elementary ones, have integral orders, and their rate equations take on special significance because they can be integrated conveniently. Thus we have

First-order reactions: $$\frac{1}{\nu_A}v_A = k_1c_A \tag{13-7}$$

Second-order reactions: $$\frac{1}{\nu_A}v_A = k_2c_A^2 \tag{13-8}$$

or $$\frac{1}{\nu_A}v_A = k_2c_Ac_B \tag{13-9}$$

Third-order reactions: $$\frac{1}{\nu_A}v_A = k_3c_A^3 \tag{13-10}$$

or $$\frac{1}{\nu_A}v_A = k_3c_A^2c_B \tag{13-11}$$

or $$\frac{1}{\nu_A}v_A = k_3c_Ac_Bc_C \tag{13-12}$$

Third-order reactions are rare, so attention usually is limited to first- and second-order cases. Note that unimolecular and bimolecular reactions must be first-order and second-order, respectively, but the reverse isn't true.

13-4. Integrated Forms of Rate Equations

agreement: In this section, we assume that the volume of the system remains approximately constant and use c rather than brackets for concentration.

A. First-order reactions

The rate equation for a first-order reaction, $A \rightarrow B + C$, for which the rate doesn't depend on the products, is obtained from Eqs. (13-4) and (13-7):

$$\frac{\text{rate}}{V} = \frac{1}{\nu_A}\frac{dc_A}{dt} = k_1c_A$$

$$-\frac{dc_A}{c_A} = k_1\,dt \qquad (\text{since } \nu_A = -1)$$

Integrating between the limits t at c_A and $t = 0$ at $c_{A,0}$, we get

$$-\ln\frac{c_A}{c_{A,0}} = k_1t \tag{13-13}$$

or

$$\ln\frac{c_{A,0}}{c_A} = k_1t$$

Another form of Eq. (13-13) is

$$c_A = c_{A,0}e^{-k_1t}$$

which brings out the exponential relationship. Without limits, integration gives

$$-\ln c_A = k_1 t + C \quad \textbf{(13-14)}$$

which is the equation of a straight line. A plot of $\ln c_A$ versus t for a first-order reaction should give a straight line with slope of $-k_1$. If it's more convenient to use base-10 logarithms, the slope is $-k_1/2.303$.

EXAMPLE 13-8: Plot the data of Example 13-5 as $\log c$ against t and evaluate the rate constant. What is the order of the reaction?

Solution: The first step is to determine the logarithm of each concentration. Rearranging and adding to the table in Example 13-5, we have

Time (min)	0	5	10	15	20	25
c_A	0.50	0.36	0.26	0.19	0.14	0.10
$\log c_A$	-0.301	-0.444	-0.585	-0.721	-0.854	-1.0

We plotted these data in Figure 13-2, and, since the plot produced a straight line, we know that the order of the reaction is 1. In order to calculate k_1, we need the slope of the line as represented by two points on the line. The curve of best fit may not touch any of the experimental points, so as a general rule, they shouldn't be used for obtaining the slope. However, in this case the 5-min and 15-min points are on the line, so we'll use them.

$$m = \frac{-0.444 - (-0.721)}{5 - 15 \text{ min}} = \frac{0.277}{-10 \text{ min}} = -0.0277 \text{ min}^{-1} = -\frac{k_1}{2.303}$$

and

$$k_1 = 0.064 \text{ min}^{-1}$$

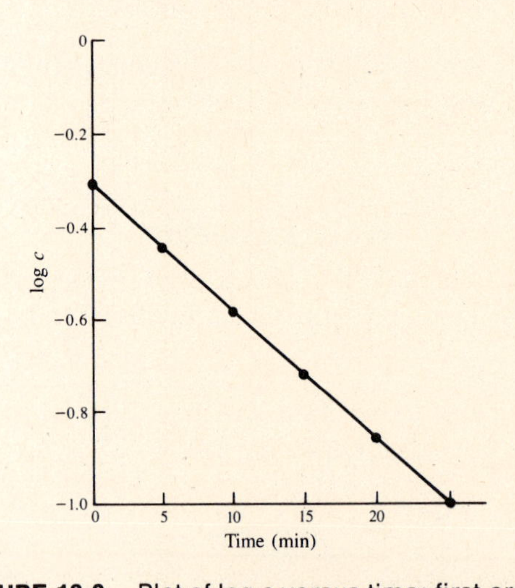

FIGURE 13-2. Plot of log c versus time: first-order.

EXAMPLE 13-9: Using the result obtained in Example 13-8, calculate the concentration of A after the reaction has run for 30 min. Does this value agree with the data plotted in Figure 13-1?

Solution: From Example 13-8, we have the rate constant $k_1 = 0.0638 \text{ min}^{-1}$. For calculations involving concentrations at different times, we'll want an integrated form of the first-order rate equation:

$$-\ln \frac{c_A}{c_{A,0}} = k_1 t$$

Let's rearrange this equation before substitution:

$$-\ln c_A + \ln c_{A,0} = k_1 t \quad \text{or} \quad \ln c_A = \ln c_{A,0} - k_1 t$$

Now substituting the given values, we get

$$\ln c_A = \ln 0.50 - (0.0638\ \text{min}^{-1})(30\ \text{min}) = -0.693 - 1.914 = -2.607$$
$$c_A = 0.074\ M$$

If we extend the curve in Figure 13-1 smoothly, it would give a value of about 0.07 at 30 min.

---

note: Many textbook authors use a different approach: They let x represent the concentration of reactant that has reacted after time t; then $c = c_0 - x = a - x$, where a represents the initial concentration of reactant A. We can obtain an equation analogous to Eq. (13-13) by a similar derivation:

$$\ln \frac{a}{a - x} = k_1 t \qquad \textbf{(13-15)}$$

B. Second-order reactions

Second-order reactions present two possibilities for single-step reactions:

Type I	$A \longrightarrow$	products	$v_A = k_2 c_A^2$; Eq. (13-8)
Type II	$A + B \longrightarrow$	products	$v_A = k_2 c_A c_B$; Eq. (13-9)

For the Type I case, we integrate Eq. (13-8) without limits:

$$-\frac{1}{v_A}\frac{dc_A}{dt} = k_2 c_A^2 \qquad \text{or} \qquad -\frac{dc_A}{c_A^2} = k_2\, dt$$

and

$$\frac{1}{c_A} = k_2 t + C \qquad \textbf{(13-16)}$$

which is the equation of a straight line. A plot of $1/c_A$ versus t for a second-order reaction of Type I should give a straight line with slope of k_2. Integrating with the limits (c_A, t) and $(c_{A,0}, 0)$ yields

$$\frac{1}{c_A} - \frac{1}{c_{A,0}} = k_2 t \qquad \textbf{(13-17)}$$

We see that the integration constant of Eq. (13-16) equals $1/c_{A,0}$.

EXAMPLE 13-10: A certain reaction, $A \rightarrow B + C$, is second-order. From the following data, calculate the rate constant.

Measurement	1	2	3	4	5
Time (hr)	0	1	2	4	6
Concentration (M)	0.60	0.45	0.36	0.26	0.20

Solution: Because the reaction is Type I ($A \rightarrow$ products), we use Eq. (13-16), plot the data obtained, and determine the slope of the line. In order to do this, we need the reciprocals of the concentrations:

Time (hr)	0	1	2	4	6
$1/c_A$ (M^{-1})	1.67	2.22	2.78	3.85	5.00

These data are plotted in Figure 13-3. To calculate the slope, let's choose 1 hr and 6 hrs:

$$m = \frac{(5.00 - 2.22)\ M^{-1}}{(6 - 1)\ \text{hr}} = k_2 = 0.556\ M^{-1}\,\text{hr}^{-1}$$

EXAMPLE 13-11: Use the result obtained in Example 13-10 to calculate the concentration after 3.0 hr. Does the result agree with the line in Figure 13-3?

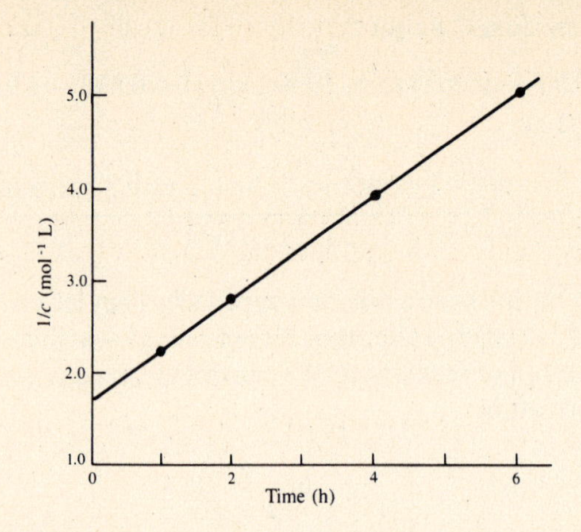

FIGURE 13-3. Plot of $1/c$ versus time: second-order.

Solution: For a single calculation, the integrated form [Eq. (13-17)] is more convenient:

$$\frac{1}{c_A} = k_2 t + \frac{1}{c_{A,0}}$$

$$= (0.556\ M^{-1}\text{hr}^{-1})(3.0\ \text{hr}) + \frac{1}{0.60} M^{-1} = 1.67 + 1.67 = 3.34\ M^{-1}$$

$$c_A = 0.30\ M$$

Reading $1/c_A$ at 3.0 hr from Figure 13-3 gives approximately 3.35 for $1/c_A$, so the agreement is fine.

We can handle a Type II second-order reaction ($A + B \rightarrow C + D$) conveniently by letting x be the concentration that has reacted in time t. Then $(a - x)$ and $(b - x)$ are the concentrations of A and B at t. The rate equation for this case is Eq. (13-9):

$$\frac{1}{\nu_A}\frac{dc_A}{dt} = k_2 c_A c_B$$

Substituting $c_A = a - x$, $c_B = b - x$, and $\nu_A = -1$, we obtain

$$-\frac{d(a-x)}{dt} = k_2(a-x)(b-x)$$

We separate the differential term, recalling that a is a constant:

$$-\frac{da}{dt} + \frac{dx}{dt} = 0 + \frac{dx}{dt}$$

Thus

$$\frac{dx}{dt} = k_2(a-x)(b-x) \qquad \textbf{(13-18)}$$

After rearrangement, we can integrate, using the method of partial fractions. Integration without limits yields

$$\frac{1}{a-b}[\ln(a-x) - \ln(b-x)] = k_2 t + C$$

or

$$\frac{1}{a-b}\ln\frac{a-x}{b-x} = k_2 t + C \qquad \textbf{(13-19)}$$

Integration with the limits (x, t) and $(0, 0)$ yields

$$\frac{1}{a-b}\ln\frac{a-x}{b-x} - \frac{1}{a-b}\ln\frac{a}{b} = k_2 t$$

We see that the integration constant of Eq. (13-19) is $1/(a-b)\ln(a/b)$, which is the intercept of the straight line obtained by plotting $1/(a-b)\ln[(a-x)/(b-x)]$ versus t. The plot would look like that in Figure 13-3. Combining the ln terms gives the equation in the form usually found in textbooks:

$$\frac{1}{a-b}\ln\frac{b(a-x)}{a(b-x)} = k_2 t \tag{13-20}$$

If we plot the left-hand side of Eq. (13-20) against t, we again obtain a straight line. The plot would look like that in Figure 13-3, but it would pass through the origin.

EXAMPLE 13-12: A second-order reaction of Type II was carried out in a solution that initially was 0.40 M in A and 0.70 M in B. After 1 hr, the concentration of A was found to be 0.15 M. Calculate the rate constant.

Solution: Second-order reactions of Type II are represented by Eq. (13-20). In order to use it, we need the value of x; we are given $(a - x)$ but must calculate $(b - x)$:

$$a - x = 0.15\ M$$

$$x = a - 0.15 = 0.40 - 0.15 = 0.25\ M$$

Thus

$$b - x = 0.70 - 0.25 = 0.45\ M$$

Substituting into Eq. (13-20), we get

$$\frac{1}{(0.40 - 0.70)M}\ln\frac{0.70(0.15)}{0.40(0.45)} = k_2\,(1\ \text{hr})$$

$$k_2 = \frac{1}{1\ \text{hr}}\left(-\frac{1}{0.30\ M}\right)\ln\frac{0.70(0.15)}{0.40(0.45)} = 1.8\ M^{-1}\,\text{hr}^{-1}$$

C. Third-order reactions

Elementary third-order reactions are rare, so we include a brief discussion of this category only for completeness. Three types of reaction are possible; the rate equations are Eqs. (13-10), (13-11), and (13-12). We present the integrated expression for the simplest type only. Its rate equation is Eq. (13-10):

$$\frac{1}{\nu_A}\frac{dc_A}{dt} = k_3 c_A^3$$

Since $\nu_A = -1$,

$$-\frac{dc_A}{dt} = k_3 c_A^3 \qquad \text{or} \qquad -\frac{dc_A}{c_A^3} = k_3\,dt$$

Integrating with the usual limits gives

$$\frac{1}{2}\left(\frac{1}{c_A^2} - \frac{1}{c_{A,0}^2}\right) = k_3 t \tag{13-21}$$

The other types of reactions give the same result, if the initial concentrations of the reactants are in their stoichiometric ratios. Curiously, nitric oxide is involved in many of the known third-order reactions in the homogeneous gas phase, e.g.,

$$2NO + H_2 \longrightarrow N_2O + H_2O$$

Another case, $O + O \rightarrow O_2$, involves a third species, M, which acts as a catalyst. The rate depends on c_O^2 and c_M.

D. Zero-order reaction

For some A $\rightarrow$ product reactions, the rate is independent of concentration; i.e., it is governed by some other factor. The rate law is simply

$$\frac{1}{\nu_A}v_A = -\frac{dc_A}{dt} = k_0 \tag{13-22}$$

Rearranging and integrating with the usual limits gives

$$c_{A,0} - c_A = k_0 t \quad \textbf{(13-23)}$$

Integrating without limits gives

$$c_A = -k_0 t + C \quad \textbf{(13-24)}$$

which shows us that a plot of c_A versus t should give a straight line of slope $-k_0$. Common examples of zero-order reactions are photochemical reactions, in which the amount of light is the controlling factor, and certain catalyzed reactions, in which the amount of catalyst determines the rate.

EXAMPLE 13-13: The catalytic decomposition of a certain gas is found to proceed at a constant rate of 0.050 M min^{-1}. If the initial concentration was 0.50 M, determine the concentrations at 2, 4, and 6 min and plot the data as c against t.

Solution: For a constant rate, the order is zero and the rate law is Eq. (13-22). Integration with limits is more useful here, so, from Eq. (13-23),

$$c_A = c_{A,0} - k_0 t$$

From Eq. (13-22), k_0 = rate = 0.050 M min^{-1}. Then

$$c_A = 0.50\ M - (0.050\ M\ \text{min}^{-1})(t)$$

We tabulate the data as follows:

t (min)	0	2	4	6
c (M)	0.50	0.40	0.30	0.20

The graph of these data (Figure 13-4) clearly shows that all the gas decomposed in 10 minutes.

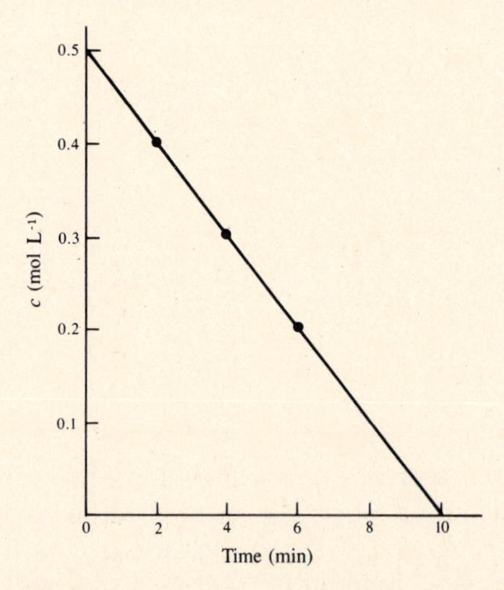

FIGURE 13-4. Plot of c versus time: zero-order.

13-5. Half-life

A useful means of expressing the rate of reactions is the **half-life**, i.e., the time required for one-half of some initial amount to react. Half-life is a well-known identifying characteristic of radioactive substances and is used to indicate their persistence. The expressions relating half-life to rate constant are different for each order.

A. First-order half-life

EXAMPLE 13-14: Derive an expression for the length of time required for the concentration of a reactant to become half the initial concentration in a first-order reaction.

Solution: The first-order rate equation is Eq. (13-7), which we integrated to obtain Eq. (13-13):

$$-\ln\frac{c_A}{c_{A,0}} = k_1 t \qquad \text{or} \qquad \ln\frac{c_{A,0}}{c_A} = k_1 t$$

We want the time at which $c_A = 0.5\, c_{A,0}$, identified as $t_{1/2}$, so

$$\ln\frac{c_{A,0}}{0.5c_{A,0}} = k_1 t_{1/2}$$

$$t_{1/2} = \frac{1}{k_1}\ln 2 = \frac{0.693}{k_1} \qquad \textbf{(13-25)}$$

(A useful rearrangement is $k_1 = 0.693/t_{1/2}$.)

EXAMPLE 13-15: In Example 13-8, the rate constant of the first-order reaction was found to be 0.0638 min^{-1}. Determine the half-life of this reaction.

Solution: For a first-order reaction, the half-life is given by Eq. (13-25), so

$$t_{1/2} = \frac{0.693}{k_1} = \frac{0.693}{0.0638\ \text{min}^{-1}} = 10.9\ \text{min}$$

EXAMPLE 13-16: How long will it take for the concentration of A to be reduced to $\frac{1}{16}$ of the initial concentration in the reaction in Example 13-8?

Solution: In general, a "how long" question should be answered with an integrated rate equation; but, when we recognize that the fraction of the initial concentration is a power of $\frac{1}{2}$, we can shorten the calculation by using the half-life. Since $1/16 = (1/2)^4$, the time required is four half-lives, or

$$t = 4t_{1/2} = 4(10.9\ \text{min}) = 43.6\ \text{min}$$

B. Second-order half-life

EXAMPLE 13-17: Derive an expression for calculating the half-life of a second-order reaction of Type I.

Solution: The integrated form of the second-order rate equation is Eq. (13-17). Again, we substitute $c_A = 0.5\, c_{A,0}$:

$$\frac{1}{0.5c_{A,0}} - \frac{1}{c_{A,0}} = k_2 t_{1/2}$$

$$t_{1/2} = \frac{1}{k_2}\left(\frac{2}{c_{A,0}} - \frac{1}{c_{A,0}}\right)$$

or

$$t_{1/2} = \frac{1}{k_2 c_{A,0}} \qquad \textbf{(13-26)}$$

C. Other orders

Note in Eq. (13-26) that the power of $c_{A,0}$ is 1 less than the order. In general,

$$t_{1/2} \propto \frac{1}{c_{A,0}^{p-1}} \qquad \textbf{(13-27)}$$

For zero-order, $t_{1/2} = \dfrac{c_{A,0}}{2k_0}$ for third-order, $t_{1/2} = \dfrac{3}{2k_3 c_{A,0}^2}$

Note that $c_{A,0}$ will appear in every half-life expression, except for the first-order case ($p = 1$). The calculated values of these half-lives depend on when the concentration is measured in the course of the reaction, because $c_{A,0}$ must represent the concentration at the start of the time interval over which one-half of $c_{A,0}$ disappears. For example, if we used Eq. (13-26) to calculate two successive half-lives, the second would be different from the first. Thus the concept of half life is actually useful only for first-order reactions.

13-6. Determination of Order

A. Graphic method

Experimental rate data may be plotted from calculations involving the various integrated forms of the rate equations. If one plot is a straight line, the order is established (see Example 13-8). Figures 13-2, 13-3, and 13-4 are typical plots for first-, second-, and zero-order reactions. This method will not work if the order is not integral.

B. Initial rate method

The reaction is carried out with different concentrations of one reactant, while the concentrations of other reactants are held constant. The order with respect to the reactant that is varied is determined by comparison of the rate equations, using the initial rate (see Section 13-2B). The underlying assumption is that, for a short time, all concentrations are essentially constant. This method won't work if an "induction period" occurs before a steady state is established.

EXAMPLE 13-18: From the following data, determine the order with respect to each reactant for the reaction $A + B \rightarrow$ products:

Trial	1	2	3
c_A	1.0	0.5	0.5
c_B	0.8	0.8	1.6
initial rate	0.08	0.02	0.04

Solution: The rate equation for the reaction, from Eq. (13-5), is

$$v_A = -kc_A^p c_B^r$$

For reactant A, we compare trials 1 and 2, for which c_B is constant. Taking the ratio of the two rate equations,

$$\frac{\text{rate 2}}{\text{rate 1}} = \frac{kc_{A,2}^p c_{B,2}^r}{kc_{A,1}^p c_{B,1}^r} = \frac{c_{A,2}^p}{c_{A,1}^p}$$

$$\frac{0.02}{0.08} = \frac{(0.5)^p}{(1.0)^p}$$

$$p = 2$$

note: Usually the power will not be so obvious, and the general procedure for calculating p is to take the logarithm of both sides of the equation:

$$\log(0.25) = p \log(0.5)$$

$$p = \frac{\log(0.25)}{\log(0.5)} = \frac{-0.6021}{-0.3010} = 2$$

For reactant B, we use trials 2 and 3, for which c_A is constant. As before, we have

$$\frac{\text{rate 3}}{\text{rate 2}} = \frac{kc_{A,3}^p c_{B,3}^r}{kc_{A,2}^p c_{B,2}^r} = \frac{c_{B,3}^r}{c_{B,2}^r}$$

$$\frac{0.04}{0.02} = \frac{(1.6)^r}{(0.8)^r}$$

$$r = 1$$

The experimental rate equation is rate $= kc_A^2 c_B$.

C. Half-life method

For two runs with different initial concentrations, both half-lives are calculated. If the reaction is first-order, they should be equal. If not, the order is calculated by taking the ratio of the two.

EXAMPLE 13-19: Derive a general expression for calculating order from half-lives determined for different initial concentrations.

Solution: The general expression for half-life is Eq. (13-27). Writing it for two runs and taking the ratio gives

$$\frac{t_{1/2,2}}{t_{1/2,1}} = \frac{\left(\dfrac{1}{c_{A,0}^{p-1}}\right)_2}{\left(\dfrac{1}{c_{A,0}^{p-1}}\right)_1} = \left(\frac{c_{A,0,1}}{c_{A,0,2}}\right)^{p-1}$$

For convenience, let's use T for the half-life ratio and C for the concentration ratio. Then

$$\ln T = (p-1)\ln C = p \ln C - \ln C$$

$$p \ln C = \ln T + \ln C$$

$$p = \frac{\ln T + \ln C}{\ln C} = \frac{\ln T}{\ln C} + 1$$

Putting back the ratios,

$$p = \frac{\ln (t_{1/2,2}/t_{1/2,1})}{\ln (c_{A,0,1}/c_{A,0,2})} + 1 \tag{13-28}$$

D. Excess reactant

The concentration of one reactant may be kept essentially constant throughout the reaction by starting with a large excess amount of it. The excess usually lowers the reaction order by eliminating the order with respect to that reagent, e.g., a second-order reaction is made to appear first-order. The result is called a **pseudo-first-order** reaction.

13-7. Complicating Processes

Some of the complexities that can lead to nonintegral orders fall into convenient categories. We list four common categories and describe them briefly.

1. In an *opposing reaction*, also known as a *reversible reaction*, product molecules react to form reactant molecules. For the forward reaction $A + B \xrightarrow{k} C + D$, after some C and D have been produced, $C + D \xrightarrow{k'} A + B$. Eventually a state of equilibrium is reached, with $K = k/k'$.
2. In a *consecutive reaction* one step follows another, i.e., $A + B \xrightarrow{k} C \xrightarrow{k'} D + E$. There is a rate equation for each step. If one k is much larger than the other, its reaction will determine the overall rate, but if both k are about the same magnitude, a complex rate equation results.
3. In a *parallel reaction*, also known as a *simultaneous reaction*, a set of reactants may be able to yield more than one set of products simultaneously, i.e., $A + B \xrightarrow{k} C + D$ and $A + B \xrightarrow{k'} E + F$. The relative rates of the two reactions determine the relative amounts of products in the resulting mixture. Parallel reactions are common with organic compounds.
4. In a *chain reaction* the product of one step in a complex series of reactions is itself a reactant in a previous step. Thus the previous reaction is repeated, creating a cycle, or closed sequence. These propagation reactions usually involve very active species, such as free atoms and free radicals, which are products in some steps and reactants in others. Some steps destroy some of the intermediate reactants; i.e., they are *termination* reactions, which serve to limit the amount of reaction.

13-8. Temperature Dependence of Rate

As we mentioned in Section 13-1D, reaction rates increase rapidly with temperature. Quantitatively, if the rate constant of a reaction has been determined at different temperatures, a straight line is obtained when ln k is plotted against $1/T$, i.e.,

$$\ln k = \mathrm{A}(1/T) + \mathrm{B}$$

This exponential relationship was expressed by van't Hoff and later by Arrhenius as

ARRHENIUS EQUATION $$k = Ae^{-E_a/RT} \tag{13-29}$$

where E_a is the **activation energy** and A is the **preexponential factor** (formerly called the FREQUENCY FACTOR). Taking the logarithm of each side, we get

$$\ln k = \ln A - \frac{E_a}{RT}$$

Rearranging to linear form, we have

$$\ln k = -\frac{E_a}{R}(1/T) + \ln A$$

The slope of the plot of ln k versus $1/T$ is $-E_a/R$, and the intercept is ln A. We assume that E_a doesn't depend on temperature. Differentiating, we get the rate of change of k with T:

$$\frac{d \ln k}{dT} = \frac{E_a}{RT^2} \tag{13-30}$$

Integration without limits would regenerate a linear equation. Integration with limits gives

$$\ln \frac{k_{T_2}}{k_{T_1}} = \frac{E_a}{R}\left(-\frac{1}{T_2} + \frac{1}{T_1}\right) = \frac{E_a}{R}\left(\frac{1}{T_1} - \frac{1}{T_2}\right) = \frac{E_a}{R}\left(\frac{T_2 - T_1}{T_1 T_2}\right) \tag{13-31}$$

Note that this equation has the same form as the van't Hoff equation for the temperature dependence of the equilibrium constant, Eq. (7-31).

EXAMPLE 13-20: Using the rule of thumb that at about room temperature a temperature increase of 10 °C will double the reaction rate, determine E_a.

Solution: For two temperatures, we want the integrated form of the Arrhenius equation [Eq. (13-31)]. If the rate doubles, the ratio of the rate constants is 2. We could use any temperatures near 298 K and choose 290 K and 300 K. Solving for E_a and substituting, we have

$$E_a = R \ln \left(\frac{k_{T_2}}{k_{T_1}}\right)\left(\frac{T_2 T_1}{T_2 - T_1}\right) = 8.314\ \mathrm{J\,K^{-1}mol^{-1}}\ (\ln 2)\left[\frac{300\ \mathrm{K}(290\ \mathrm{K})}{300\ \mathrm{K} - 290\ \mathrm{K}}\right] = 50.1\ \mathrm{kJ\,mol^{-1}}$$

13-9. Theories of Reaction Rate

Two major theories offer explanations of reaction rates and their dependence on temperature. We summarize the main features and, for simplicity, consider only bimolecular gaseous reactions. Both theories include an activation energy, although neither yields accurate E_a values. The activation energy is interpreted as a barrier between reactants and products; i.e., when two reactant species come together, they must have energy greater than this barrier and overcome it if they are to become products.

A. Collision theory

The main features of the **collision theory** for simple molecules in a bimolecular reaction are

1. the reacting species must come in contact, or collide;
2. every collision doesn't result in reaction;
3. the colliding species must have energy greater than some minimum energy (the activation energy); and
4. the species must collide in the proper orientation; i.e., the sites at which reaction will occur must be close to each other. [This feature has been called the steric factor and the probability factor; a better term is *orientation probability factor*.]

The math involved uses concepts from mechanics. The frequency of collision, Eq. (2-43), is modified by an exponential term that represents the fraction of the collisions that have energy greater than E_a and by the orientation factor, P. The rate constant is

$$k_2 = Pze^{-E_a/RT} \tag{13-32}$$

which has the same form as the Arrhenius equation, Eq. (13-29).

For unimolecular reactions, the picture of two species reacting while in contact isn't appropriate. One suggestion is that a molecule is activated by a collision but doesn't react immediately. Some time later, it may dissociate into products or may lose its activation by collision. An important feature in development of the theory is the *steady-state approximation*, i.e., the assumption that the concentration of the activated intermediate reactant becomes essentially constant within a short time and remains constant for the duration of the reaction.

B. Activated-complex theory

The **activated-complex theory**, also referred to as the theory of *absolute reaction rates* and the *transition-state theory*, uses concepts from thermodynamics and statistical mechanics. The major features of the activated-complex theory for bimolecular reactions are that the

1. reacting species must come in contact;
2. potential energy of the system along a path connecting the reactant state to the product state contains a maximum (in three dimensions, a saddle-like shape); and
3. configuration of atoms of the two species in contact at this maximum (the low point of the saddle) is called an *activated complex*. The state of the system at this point is called the *transition state*.

We assume that reactants are in equilibrium with the activated complex:

$$\mathrm{A} + \mathrm{B} \rightleftharpoons (\mathrm{AB})^{\ddagger} \qquad K^{\ddagger} = \frac{[(\mathrm{AB})^{\ddagger}]}{[\mathrm{A}][\mathrm{B}]}$$

This K is not a true equilibrium constant but can be considered a proportionality constant. The activated complex may break up into products:

$$(\mathrm{AB})^{\ddagger} \longrightarrow \text{products} \qquad (\text{rate constant} = k_1^{\ddagger})$$

Another way to show the complete process is

$$\mathrm{AB} + \mathrm{C} \overset{K^{\ddagger}}{\rightleftharpoons} (\mathrm{ABC})^{\ddagger} \overset{k_1^{\ddagger}}{\longrightarrow} \mathrm{A} + \mathrm{BC}$$

The overall rate constant is

$$k_2 = k_1^{\ddagger} K^{\ddagger} \tag{13-33}$$

The math involves partition functions and translational and vibrational degrees of freedom. In order to use these factors, we have to make assumptions about the nature of the activated complex. Equation (13-33) becomes

$$k_2 = \frac{kT}{h}\frac{K_c^{\ddagger}}{c^{\circ}} \tag{13-34}$$

where k is Boltzmann's constant and c° is the standard-state concentration, usually 1. Concentration units are mol dm^{-3}. Equation (13-34) is a general expression; $K^{\ddagger}$ is determined by the nature of the activated complex and the activation energy in each case. Calculations are complex, simplifications are needed, and the results must be considered approximations. Yet, agreement with collision theory is good.

A different interpretation of the factors in the Arrhenius equation results from replacing $K^{\ddagger}$ with $\Delta G^{\circ\ddagger}$, the **free energy of activation**, and $\Delta G^{\circ\ddagger}$ with $\Delta S^{\circ\ddagger}$ and $\Delta H^{\circ\ddagger}$:

$$\Delta G^{\circ\ddagger} = -RT \ln K_c^{\ddagger} = \Delta H^{\circ\ddagger} - T\,\Delta S^{\circ\ddagger}$$

$$\ln K_c^{\ddagger} = -\frac{\Delta H^{\circ\ddagger}}{RT} + \frac{\Delta S^{\circ\ddagger}}{R}$$

Equation (13-34) becomes

$$k_2 = \frac{kT}{h}\left(\frac{1}{c^\circ}\right)e^{\Delta S^{\circ\ddagger}/R}e^{-\Delta H^{\circ\ddagger}/RT} \tag{13-35}$$

We have seen that ΔH approximates ΔE in most cases. If $\Delta H^{\circ\ddagger}$ is equivalent to E_a, the ΔS term becomes part of the orientation probability factor of collision theory. This result is expected, since entropy is related to probability.

13-10. Mechanisms

In Section 13-1A, we stated that the deduction of the mechanism (or path) of a chemical reaction is based primarily on the experimental rate equation, or rate law. However, we also have to apply our general knowledge of chemistry, stoichiometric relationships, molecular geometry, and ingenuity in arriving at a plausible (or the most plausible) mechanism. In Section 13-1C, we went on to state that the mechanism usually involves a complex series of steps, each of which is an elementary reaction. However, one step is slower than the others and is called the **rate-determining step**. The proposed mechanism must yields a rate equation that agrees with the experimental rate equation.

EXAMPLE 13-21: List at least 10 chemical species, other than reactants and products, that might appear in intermediate reactions in a mechanism for the reaction

$$2Br^- + 2H^+ + H_2O_2 \longrightarrow Br_2 + 2H_2O$$

Which are reasonable possibilities?

Solution: Possible ordinary species are BrO^-, BrO_2^-, BrO_3^-, BrO_4^-, and their corresponding molecular acids HBrO, $HBrO_2$, $HBrO_3$, $HBrO_4$; also H_2, O_2, O_2^{2-}, OH^-, and HO_2^-. More exotic species might include Br, O, H, OH, $H_3O_2^-$. We can eliminate many of these possibilities as being highly unlikely, based on our knowledge of chemistry. Since the solution is acidic, any OH^- would be converted to H_2O. The higher oxidation states of bromines are unlikely, because the Br^- would reduce them. There is no reason for the hydrogen and oxygen atoms to be converted to H, H_2, O_2, or O. Thus we are left with the possibilities of low oxidation states of bromine and species involving gain or loss of H^+, such as BrO^-, HBrO, HO_2^-, and $H_3O_2^-$.

EXAMPLE 13-22: Propose a mechanism for the reaction $2NO_2 + F_2 \rightarrow 2NO_2F$. The reaction is found experimentally to be second-order; its rate equation is

$$-\frac{1}{2}\frac{d[NO_2]}{dt} = k_2[NO_2][F_2]$$

Solution: Two compounds (each to the first power) appear in the rate equation, suggesting that the rate-determining step might involve them. If they give a product molecule directly, the reaction would start with

$$NO_2 + F_2 \xrightarrow{k_1} NO_2F + F \qquad \text{(step 1)}$$

We need to eliminate the free F atom, and we need a second NO_2F. Both needs are satisfied by a possible next step:

$$NO_2 + F \xrightarrow{k_2} NO_2F \qquad \text{(step 2)}$$

Adding steps 1 and 2, we have

$$2NO_2 + F_2 \longrightarrow 2NO_2F$$

which is the net equation. Thus these two steps represent a plausible mechanism but are not necessarily the only one. Since step 1 gives the experimental rate equation, we conclude that it's the rate-determining step. Further support for that conclusion is the fact that a free F atom would be very reactive, suggesting that step 2 would be fast.

A. Rate constant from mechanism

One of the tests of a mechanism is that the derived rate constant should equal the experimental one. Deriving the rate equation from the mechanism usually is a complicated procedure, leading to expressions that are combinations of rate constants and equilibrium constants from the individual steps. We examine only three of the less complex cases in the following examples.

EXAMPLE 13-23: Consider the reaction that has the mechanism

$$A + B \underset{k_{-1}}{\overset{k_2}{\rightleftharpoons}} C \qquad \text{(step 1)}$$

$$C \xrightarrow{k_1} D \qquad \text{(step 2)}$$

Note that the subscripts of k show the order and that the minus sign indicates a reverse reaction. We know the order because these are elementary reactions, for which order equals molecularity. Either step might be the rate-determining one. For each possibility, obtain a rate equation for the production of D. Express the equations in terms of concentrations of A and B.

Solution: Let's begin with a general analysis of the system. The rate of formation of D is expressed by

$$\frac{dc_D}{dt} = k_1 c_C \tag{i}$$

The concentration of D is determined by all three reactions, so

$$\frac{dc_C}{dt} = k_2 c_A c_B - k_{-1} c_C - k_1 c_C$$

If we make the steady-state approximation, c_C is constant and

$$\frac{dc_C}{dt} = 0 = k_2 c_A c_B - k_{-1} c_C - k_1 c_C \tag{ii}$$

Now we solve for c_C:

$$c_C = \frac{k_2}{k_{-1} + k_1} c_A c_B$$

Substituting c_C in eq. (i), we have an expression for the overall situation, i.e.,

$$\frac{dc_D}{dt} = k_1 \frac{k_2}{k_{-1} + k_1} c_A c_B \tag{iii}$$

If step 2 is slow, we can neglect $k_1 c_C$ in eq. (ii) and k_1 in eq. (iii), which gives

$$\frac{dc_D}{dt} = k_1 \left(\frac{k_2}{k_{-1}}\right) c_A c_B$$

The ratio k_2/k_{-1} is the equilibrium constant for the reaction in step 1, so we have

$$\frac{dc_D}{dt} = k_1 K c_A c_B$$

If step 1 is slow, we can neglect $k_{-1} c_C$. In other words, as soon as A + B yields any C, C immediately changes to D. Thus eq. (iii) becomes

$$\frac{dc_D}{dt} = k_2 c_A c_B$$

Experimental measurement of rates over the time of the reaction should reveal the relative magnitudes of k_1 and k_{-1}, thus permitting a decision as to the rate-determining step.

EXAMPLE 13-24: Suggest two possible two-step mechanisms and predict the expected rate equation for each for the reaction

$$2NOCl \longrightarrow 2NO + Cl_2$$

Solution: An obvious starting reaction is

$$NOCl \longrightarrow NO + Cl \qquad \text{(step 1)}$$

To eliminate the Cl and add the needed molecules, we can use

$$Cl + NOCl \longrightarrow NO + Cl_2 \qquad \text{(step 2)}$$

The sum of steps 1 and 2 is the net reaction. Which step is rate-determining? Since Cl is very active, step 2 should be fast and step 1 slow, the overall order would be 1, and the rate = $k_1[NOCl]$. Another possibility for getting rid of Cl is

$$Cl + Cl \longrightarrow Cl_2 \qquad \text{(step 2')}$$

Then step 1 would have to occur twice for each occurrence of step 2′, and the sum of 2(step 1) and step 2′ is the net reaction. Step 2′ would be fast, at least at high pressure, so again step 1 is rate-determining and predicts first-order and the same rate equation.

Another possible mechanism is

$$2NOCl \longrightarrow NOCl_2 + NO \qquad \text{(step 1)}$$
$$NOCl_2 \longrightarrow NO + Cl_2 \qquad \text{(step 2)}$$

This mechanism may be preferable, because it avoids the high-energy species Cl. If step 1 were slow, the reaction would be second-order; if step 2 were slow, first-order.

EXAMPLE 13-25: Experimental studies show that the rate law for the reaction in Example 13-24 is

$$\frac{d[NO]}{dt} = k[NOCl]^2$$

which is not as predicted in Example 13-24. Propose a mechanism that explains this result.

Solution: Since the reaction is second-order, the guesses in Example 13-24 are eliminated, and we must try others. One approach is to try to find a plausible reaction that fits the rate equation and then look for plausible additional steps that add up to the net equation. The formation of a dimer of NOCl is a reasonable possibility, and it would give a rate equation similar to the experimental one:

$$2NOCl \longrightarrow N_2O_2Cl_2 \qquad \text{(step 1; slow)}$$

To eliminate this intermediate complex, we could have

$$N_2O_2Cl_2 \longrightarrow N_2O_2 + Cl_2 \qquad \text{(step 2; fast)}$$
$$N_2O_2 \rightleftharpoons 2NO \qquad \text{(step 3; fast equilibrium)}$$

The sum of steps 1–3 *is* the net reaction, $2NOCl \rightarrow 2NO + Cl_2$.

The requirement that a mechanism yield a rate equation that agrees with the experimental equation serves as a check on free-wheeling imaginations. We begin the development of the rate equation by writing an expression with the differential term that appears in the experimental rate equation:

$$\frac{1}{\nu_{NO}}\frac{d[NO]}{dt} = k_3[N_2O_2] \qquad \text{(from step 3)}$$

Note that, in this example, we let the subscript on k designate the step. Since $\nu_{NO} = +2$,

$$\frac{d[NO]}{dt} = 2k_3[N_2O_2] \qquad \textbf{(i)}$$

We need an expression for $[N_2O_2]$. This intermediate reactant is produced in step 2 and consumed in Step 3; thus its rate of change of concentration is

$$\frac{d[N_2O_2]}{dt} = k_2[N_2O_2Cl_2] - k_3[N_2O_2] \qquad \textbf{(ii)}$$

Now we apply the steady-state approximation (see Section 13-9B); i.e., we can assume that the concentration of an intermediate reactant is constant over most of the reaction time. So the rate of change = 0, and eq. (ii) becomes

$$0 = k_2[N_2O_2Cl_2] - k_3[N_2O_2]$$

or

$$[N_2O_2] = \frac{k_2}{k_3}[N_2O_2Cl_2] \qquad \textbf{(iii)}$$

Next, we need an expression for $[N_2O_2Cl_2]$. This intermediate reactant is produced in step 1 and consumed in step 2, so

$$\frac{d[N_2O_2Cl_2]}{dt} = k_1[NOCl]^2 - k_2[N_2O_2Cl_2] \qquad \textbf{(iv)}$$

When we apply the steady-state approximation, eq. (iv) becomes

$$0 = k_1[NOCl]^2 - k_2[N_2O_2Cl_2]$$

or

$$[N_2O_2Cl_2] = \frac{k_1}{k_2}[NOCl]^2 \qquad \textbf{(v)}$$

We substitute eq. (v) in eq. (iii),

$$[N_2O_2] = \frac{k_2}{k_3}\left(\frac{k_1}{k_2}[NOCl]^2\right)$$

and, finally, we put the result in eq. (i) and simplify to get

$$\frac{d[NO]}{dt} = 2k_3\left(\frac{k_1}{k_3}[NOCl]^2\right) = 2k_1[NOCl]^2$$

This result is the experimental rate equation if $k_{ex} = 2k_1$, so our proposed mechanism has passed this test. Of course, others may also.

---

warning: Mechanisms are speculations. As we saw in Examples 13-24 and 13-25, several mechanisms may explain the same reaction. Intuition may make some unlikely; studies of intermediate reactions may eliminate some; an incorrect rate equation will eliminate others. Often, however, we remain uncertain about the actual mechanism.

13-11. Catalysis

Textbooks differ on the emphasis placed on catalysis. We summarize the concept only briefly.

A **catalyst** is a substance that changes the rate of a reaction and is recovered unchanged at the end. A *positive catalyst* increases the rate; a *negative catalyst*, or *inhibitor*, reduces the rate. In *homogeneous catalysis*, reactants and catalyst are in the same phase, often gaseous. In *heterogeneous catalysis*, the reaction occurs at the interface of two phases; the catalyst is often a solid and adsorbs a liquid or a gas. The **substrate** is the reacting material.

Catalysts and catalytic actions have the following characteristics:

1. A catalyst doesn't change energy factors. Specifically, the overall ΔG is the same with and without a catalyst.
2. A catalyst doesn't affect the position of equilibrium. The value of K_{eq} stays the same, and the rates of the forward and reverse reactions are affected identically.
3. A positive catalyst provides an alternative path (mechanism) between reactants and products and usually has a lower E_a than the uncatalyzed mechanism. Sometimes the preexponential term is increased.
4. A negative catalyst (inhibitor) functions differently: It doesn't affect E_a, but it interferes with

the mechanism (perhaps by destroying intermediate free radicals) or poisons a natural catalyst.

5. The concentration of the catalyst may enter the rate equation. In general, rate = k[cat.] + rate (no cat.). If the rate without the catalyst is negligible, rate = k[cat.].
6. In homogeneous catalysis, the catalyst usually enters the reaction process by forming a complex that breaks down at a later step.
7. A small amount of catalyst has a large effect because it's regenerated and can function again with more of the reactant.

SUMMARY

In Table S-13, we show the major equations presented in this chapter and the connections between them. You should determine the conditions that are necessary for the validity of each equation and the conditions that are imposed to make the connections. Other important things you should know are

1. Chemical kinetics deals with the rates and the mechanisms of reactions. The rate of a reaction is the change in the extent or the advancement of a reaction in unit time. A mechanism for a reaction is a series of elementary reactions that add up to the net reaction.
2. Elementary reactions occur in one step, with one or two species (rarely more than two) reacting as indicated by the balanced equation. The number of species is the molecularity.
3. The factors that affect the rate of reaction are the nature of the reactants, degree of subdivision, catalyst, temperature, and concentration.
4. The rate of reaction is defined in terms of the extent of reaction. Practically, it is stated in terms of concentration: the change in concentration in unit time, or $\frac{dc}{dt}$. Reaction rates change with time, so the initial rate often is used for comparisons.
5. The rate equation (or rate law) expresses the experimentally determined dependence of rate on concentration. The order of a reaction is the sum of the exponents of the concentration terms in the rate equation.
6. The proportionality constant in the rate equation is called the rate constant. A rate constant may be evaluated by plotting data calculated from the integrated forms of the rate equations, using coordinates that yield a straight line.
7. The concept of half-life is a useful indication of rate for first-order reactions but not for other orders.
8. Orders may be determined by several methods: graphic, initial rate, half-life, and excess reagent.
9. Several complicating processes lead to complex rate equations and nonintegral reaction orders: opposing, consecutive, parallel, and chain reactions.
10. Reaction rates increase exponentially with temperature, as expressed by the Arrhenius equation [Eq. (13-29)].
11. The collision theory of bimolecular reactions is based on the assumption that all collisions don't result in reaction, because many don't involve sufficient energy or because the molecules aren't properly oriented when they collide.
12. The activated-complex theory of bimolecular reactions is based on the assumption that a relatively stable complex having energy equal to the top of the energy barrier is in equilibrium with reactants and that some of the complex changes into products.
13. A plausible reaction mechanism is deduced from the experimental rate equation with the aid of standard types, chemical information, and intuition. Rate equations and rate constants may be derived from mechanisms.
14. A complete kinetic study of a reaction includes

    (a) experimental determination of rate law;
    (b) deduction of a plausible mechanism;
    (c) demonstration that the mechanism is consistent (leads to the experimental rate constant); and
    (d) determination of dependence on temperature.

TABLE S-13: Summary of Important Equations

$$\xi = \frac{n_i - n_i^0}{\nu_i} \quad (7\text{-}3)$$

$$\rightarrow \frac{d\xi}{dt} = \frac{1}{\nu_i}\frac{dn_i}{dt} \quad (13\text{-}2)$$

$$\text{rate} = \frac{d\xi}{dt} \quad (13\text{-}1)$$

$$\rightarrow \frac{dc_i}{dt} = \frac{\nu_i}{V}\frac{d\xi}{dt} = v_i \quad (13\text{-}4)$$

$$v_A = \frac{dc_A}{dt} = kc_A^p c_B^r \quad (13\text{-}5)$$

$$\rightarrow \frac{\text{rate}}{V} = \frac{1}{\nu_i}\frac{dc_i}{dt}$$

$$\rightarrow \text{reaction order} = p + r \quad (13\text{-}6)$$

$$\frac{1}{\nu_A} v_A = k_1 c_A \quad (13\text{-}7)$$

$$\rightarrow -\ln\frac{c_A}{c_{A,0}} = k_1 t \quad (13\text{-}13)$$

$$\rightarrow -\ln c_A = k_1 t + C \quad (13\text{-}14)$$

$$\rightarrow \ln\frac{a}{a - x} = k_1 t \quad (13\text{-}15)$$

$$\rightarrow t_{1/2} = 0.693/k_1 \quad (13\text{-}25)$$

$$\frac{1}{\nu_A} v_A = k_2 c_A^2 \quad (13\text{-}8)$$

$$\rightarrow \frac{1}{c_A} = k_2 t + C \quad (13\text{-}16)$$

$$\rightarrow \frac{1}{c_A} - \frac{1}{c_{A,0}} = k_2 t \quad (13\text{-}17)$$

$$\rightarrow t_{1/2} = 1/k_2 c_{A,0} \quad (13\text{-}26)$$

$$\rightarrow t_{1/2} \propto \frac{1}{c_{A,0}^{p-1}} \quad (13\text{-}27)$$

$$\frac{1}{\nu_A} v_A = k_2 c_A c_B \quad (13\text{-}9)$$

$$\rightarrow \frac{1}{a - b}\ln\frac{a - x}{b - x} = k_2 t + C \quad (13\text{-}19)$$

$$\rightarrow \frac{1}{a - b}\ln\frac{b(a - x)}{a(b - x)} = k_2 t \quad (13\text{-}20)$$

$$\frac{1}{\nu_A} v_A = k_3 c_A^3 \quad (13\text{-}10)$$

$$\rightarrow \frac{1}{2}\left(\frac{1}{c_A^2} - \frac{1}{c_{A,0}^2}\right) = k_3 t \quad (13\text{-}21)$$

$$\frac{1}{\nu_A} v_A = k_0 \quad (13\text{-}22)$$

$$\rightarrow c_{A,0} - c_A = k_0 t \quad (13\text{-}23)$$

$$\rightarrow c_A = -k_0 t + C \quad (13\text{-}24)$$

$$k = Ae^{-E_a/RT} \quad (13\text{-}29)$$

$$\rightarrow \frac{d\ln k}{dT} = \frac{E_a}{RT^2} \quad (13\text{-}30)$$

$$\rightarrow \ln\frac{k_{T_2}}{k_{T_1}} = \frac{E_a}{R}\left(\frac{T_2 - T_1}{T_1 T_2}\right) \quad (13\text{-}31)$$

$$k_2 = pze^{-E_a/RT} \quad (13\text{-}32)$$

$$k_2 = k_1^{\ddagger} K^{\ddagger} \quad (13\text{-}33)$$

SOLVED PROBLEMS

Rate Equations

PROBLEM 13-1 (**a**) Determine the order of the reaction having the experimental rate equation, rate $= -k[A]^2[B]^{1/2}[C]$. (**b**) What can you say about the molecularity of this reaction?

Solution:

(a) The order is the sum of the exponents in the rate equation, Eq. (13-6), which, in this case, is 3.5.

(b) Since the rate depends on three reactants and has a nonintegral order, it is impossible that the reaction is elementary. Consequently, it cannot have a molecularity.

PROBLEM 13-2 For a certain first-order reaction, $A \rightarrow B + C$, the initial concentration of A was 0.35 *M*. After 30 seconds, the concentration is 0.31 *M*. Calculate the rate constant.

Solution: The integrated form of the first-order rate equation is Eq. (13-13):

$$-\ln \frac{c_A}{c_{A,0}} = k_1 t$$

$$k_1 = \frac{1}{t} \ln \frac{c_{A,0}}{c_A} = \frac{1}{30 \text{ s}} \ln \frac{0.35}{0.31} = 4.0 \times 10^{-3} \text{ s}^{-1}$$

PROBLEM 13-3 Calculate the concentration of A in the reaction in Problem 13-3 after 3 minutes.

Solution: This problem illustrates another application of rate equations, i.e., when the rate constant is known. In Example 13-9, we solved Eq. (13-13) for ln *c* before substituting. This time, you should handle the math differently by finding the ln term, then *c*:

$$-\ln \frac{c_A}{c_{A,0}} = k_1 t$$

$$\ln \frac{c_{A,0}}{c_A} = k_1 t = (4.0 \times 10^{-3} \text{ s}^{-1})(3 \text{ min})(60 \text{ s min}^{-1}) = 0.72$$

$$\frac{c_{A,0}}{c_A} = 2.05$$

$$c_A = \frac{c_{A,0}}{2.05} = \frac{0.35}{2.05} = 0.17 \, M$$

PROBLEM 13-4 To illustrate a third type of calculation involving rates, calculate how long it would take for 90 percent of A to be consumed in the system in Problem 13-2.

Solution: Start with the integrated form of the rate equation, rearrange, and substitute:

$$-\ln \frac{c_A}{c_{A,0}} = k_1 t$$

$$t = \frac{1}{k_1} \ln \frac{c_{A,0}}{c_A}$$

If 90 percent of A is used up, $c_A = c_{A,0} - 0.90c_{A,0} = 0.10c_{A,0}$. Then

$$t = \frac{1}{4.0 \times 10^{-3} \text{ s}^{-1}} \ln \frac{c_{A,0}}{0.10c_{A,0}} = \frac{1}{4.0 \times 10^{-3} \text{ s}^{-1}} \ln \frac{1}{0.10} = 5.8 \times 10^2 \text{ s}$$

PROBLEM 13-5 For the Type II second-order reaction, $A + B \rightarrow$ products, integration of the rate equation gave Eq. (13-19):

$$\frac{1}{a - b} \ln \frac{a - x}{b - x} = k_2 t + I$$

If $a = b$, the left-hand term becomes 0/0, which is indeterminate. This also occurs if A = B, so that the reaction becomes 2A $\rightarrow$ products. Integrate the rate equation for this situation.

Solution: When the coefficients in the balanced equation are not 1, you have to return to Eq. (13-3), the general expression for the rate of reaction:

$$\frac{d\xi}{dt} = \frac{V}{\nu_i} \frac{dc_i}{dt}$$

For constant volume,

$$\frac{\text{rate}}{V} = \frac{1}{\nu_i}\frac{dc_i}{dt}$$

For 2A → products,

$$\frac{\text{rate}}{V} = \frac{1}{\nu_A}\frac{dc_A}{dt} = k_2 c_A^2$$

Now rearrange and integrate:

$$\int_0^c \frac{dc_A}{c_A^2} = \nu_A k_2 \int_0^t dt$$

$$-\frac{1}{c_A} + \frac{1}{c_{A,0}} = \nu_A k_2(t - 0) \quad \text{or} \quad \frac{1}{c_A} - \frac{1}{c_{A,0}} = -\nu_A k_2 t$$

Since A is a reactant, its ν is negative, so you have

$$\frac{1}{c_A} - \frac{1}{c_{A,0}} = 2k_2 t$$

PROBLEM 13-6 A certain second-order reaction of the type 2A → products has a rate constant of 12 $M^{-1}\text{min}^{-1}$. If you start with 0.40 mol A in a 5.0-L flask, how much will remain after 2.0 h?

Solution: From Problem 13-5,

$$\frac{1}{c_A} - \frac{1}{c_{A,0}} = 2k_2 t$$

The initial concentration is 0.40/5.0 = 0.080 mol L^{-1}, so

$$\begin{aligned}\frac{1}{c_A} &= \frac{1}{0.080\ M} + 2(12\ M^{-1}\text{min}^{-1})(2\ \text{hr})\left(\frac{60\ \text{min}}{1\ \text{hr}}\right)\\ &= 12.5 + 2880 = 2892.5\ M^{-1}\\ c &= 3.5 \times 10^{-4}\ M\end{aligned}$$

PROBLEM 13-7 For a Type II second-order reaction, A + B → products, the rate law is $\frac{1}{\nu_A}\frac{dc_A}{dt} = k_2 c_A c_B$, and the rate constant is 27.8 $\text{mol}^{-1}\text{dm}^3\ \text{min}^{-1}$. If the initial concentrations in a particular experiment are 0.050 mol dm^{-3} A and 0.020 mol dm^{-3} B, how long will it take for 95 percent of B to be consumed?

Solution: The rate law given is similar to Eq. (13-9), which we integrated using the $c_A = (a - x)$ method. Substitution for c_A and c_B gave Eq. (13-18):

$$\frac{dx}{dt} = k_2(a - x)(b - x)$$

You need to integrate with limits in a situation involving two concentrations. You should be able to derive Eq. (13-20):

$$\frac{1}{a - b}\ln\frac{b(a - x)}{a(b - x)} = k_2 t$$

You can use x for both A and B because the concentration of A that reacts equals the concentration of B that reacts ($\nu_A = \nu_B$). You must calculate x first:

$$x = 0.020(0.95) = 0.019$$

Then

$$a - x = 0.050 - 0.019 = 0.031\ \text{mol dm}^{-3} \quad \text{and} \quad b - x = 0.020 - 0.019 = 0.001\ \text{mol dm}^{-3}$$

Solving Eq. (13-20) for t and substituting, you get

$$t = \frac{1}{27.8}\left(\frac{1}{0.050 - 0.020}\right)\ln\frac{0.020(0.031)}{0.050(0.001)} = 3.0\ \text{min}$$

Half-life

PROBLEM 13-8 Determine the half-life of the reaction in Problem 13-2.

Solution: The reaction is first-order, so the half-life is given by Eq. (13-25):

$$t_{1/2} = \frac{0.693}{k_1}$$

In Problem 13-2, you found that $k_1 = 4.0 \times 10^{-3}\ \text{s}^{-1}$; thus

$$t_{1/2} = \frac{0.693}{4.0 \times 10^{-3}\ \text{s}^{-1}} = 1.7 \times 10^2\ \text{s}$$

PROBLEM 13-9 Consider a first-order reaction and a second-order reaction of Type I, both of which have a half-life of 10 min, when the initial concentration of A is 0.20 M. How long will it take, in each case, for the concentration to become 0.050 M?

Solution: For half-life problems you should always divide the new concentration by the original concentration to see whether the ratio is a power of $\frac{1}{2}$. In this case, the new concentration is one-fourth of the original, so the time required is two half-lives, i.e., $1/4 = (1/2)^2$. The half-life is constant for the first-order case, so you can see that two half-lives equal 20 min; the second-order case is more complex. Recall that $t_{1/2}$ depends on the concentration at the start of the run, as shown by Eq. (13-27):

$$t_{1/2} \propto \frac{1}{c_{\text{A},0}^{p-1}}$$

When $p = 2$,

$$t_{1/2} = \frac{1}{k_2 c_{\text{A},0}}$$

In the first 10 minutes, the concentration will drop to 0.10 M, the same as in the first-order case. Now, however, $t_{1/2}$ changes. To calculate the new value, you need k_2. Rearrange and substitute:

$$k_2 = \frac{1}{t_{1/2} c_{\text{A},0}} = \frac{1}{(10\ \text{min})(0.20\ M)} = 0.50\ M^{-1}\text{min}^{-1}$$

When $c_{\text{A},0} = 0.10\ M$,

$$t_{1/2} = \frac{1}{(0.50\ M^{-1}\text{min}^{-1})(0.10\ M)} = 20\ \text{min}$$

Consequently, it will take 20 min for the concentration to go from 0.10 M to 0.050 M and 30 min to go from 0.20 M to 0.050 M.

PROBLEM 13-10 In Problem 10-9, you found that the half-life of the second-order reaction doubled in going from one-half to one-fourth of the original concentration. Show that this doubling is a general rule for Type I second-order reactions; i.e., in successive reductions of concentration by one-half, the half-life doubles for each step.

Solution: Use Eq. (13-27) and let primed quantities represent the second step. Because $c'_{\text{A},0} = \frac{1}{2}c_{\text{A},0}$

$$t'_{1/2} = \frac{1}{k_2 c'_{\text{A},0}} = \frac{1}{k_2(c_{\text{A},0}/2)} = \frac{2}{k_2 c_{\text{A},0}} = 2t_{1/2}$$

Each succeeding half-life will be twice as long as the one before.

PROBLEM 13-11 Calculate the order of the reaction for which the following data were obtained:

Run	$c_{\text{A},0}$ (mol dm^{-3})	$t_{1/2}$ (min)
1	0.50	7.3
2	0.30	12.1

Solution: We derived Eq. (13-28) for the half-life method of determining order in Example 13-19:

$$p = \frac{\ln(t_{1/2,2}/t_{1/2,1})}{\ln(c_{A,0,1}/c_{A,0,2})} + 1 = \frac{\ln(12.1/7.3)}{\ln(0.50/0.30)} + 1 = \frac{0.5053}{0.5108} + 1 = 0.989 + 1 \cong 2$$

PROBLEM 13-12 The element actinium, $^{227}_{89}Ac$, decays by emitting a β-particle and has a half-life of 13.5 yr. If you have a pure 1-g sample of the element, how long will it take until only 0.900 g of the original isotope remains?

Solution: To calculate a time interval that isn't a multiple of the half-life, you need the rate constant. All radioactive decay is first-order, so you should use Eq. (13-25):

$$k_1 = \frac{0.693}{t_{1/2}} = \frac{0.693}{13.5\ \text{yr}} = 0.0513\ \text{yr}^{-1}$$

Now turn to Eq. (13-13), the integrated form of the first-order rate equation:

$$\ln\frac{c_{A,0}}{c_A} = k_1 t$$

and

$$t = \frac{1}{k_1}\ln\frac{c_{A,0}}{c_A} = \frac{1}{0.0513\ \text{yr}^{-1}}\ln\frac{1}{0.900} = 2.05\ \text{yr}$$

Temperature Dependence of Rate

PROBLEM 13-13 The following data were obtained for a reaction carried out at two temperatures, but with identical initial concentrations:

Run	T (K)	Rate (mol dm^{-3}s^{-1})
1	273	1.5×10^{-3}
2	293	7.0×10^{-3}

Calculate E_a for the reaction.

Solution: The rate, not the rate constant, is given at two temperatures. Did you think of the Arrhenius equation? You must assume that all concentrations are the same in both runs, so that the ratio of rate constants in the integrated form of Eq. (13-29) equals the ratio of rates:

$$E_a = R\left(\ln\frac{k_{T_2}}{k_{T_1}}\right)\left(\frac{T_2 T_1}{T_2 - T_1}\right) \quad \text{and} \quad \frac{k_{T_2}}{k_{T_1}} = \frac{\text{rate}(T_2)}{\text{rate}(T_1)}$$

$$E_a = 8.314\left(\ln\frac{7.0 \times 10^{-3}}{1.5 \times 10^{-3}}\right)\left[\frac{293(273)}{293 - 273}\right] = 51\ \text{kJ mol}^{-1}$$

PROBLEM 13-14 Determine the rate at 50 °C for the reaction in Problem 13-13.

Solution: Here the ratio of the rates equals the ratio of the rate constants, which you can obtain from the calculated E_a. Use trial 1 for T_1 and 323 K for T_2 and use Eq. (13-31):

$$\ln\frac{k_{323}}{k_{273}} = \left(\frac{51.2 \times 10^3}{8.314}\right)\left(\frac{1}{273} - \frac{1}{323}\right) = 3.492$$

$$\frac{k_{323}}{k_{273}} = 32.85 = \frac{\text{rate}_{323}}{\text{rate}_{273}}$$

$$\text{rate}_{323} = 32.85(1.5 \times 10^{-3}) = 4.9 \times 10^{-4}\ \text{mol dm}^{-3}\text{s}^{-1}$$

PROBLEM 13-15 For a certain reaction, the rate constants are 0.24 dm^3 mol^{-1}s^{-1} at 700 K and 5.2 dm^3 mol^{-1}s^{-1} at 800 K. Calculate the preexponential factor.

Solution: The preexponential factor is A in the Arrhenius equation [Eq. (13-29)]:

$$k = Ae^{-E_a/RT}$$

In order to calculate A, however, you need to find E_a, using the same procedure as in Problem 13-13:

$$E_a = R\left(\ln\frac{k_{T_2}}{k_{T_1}}\right)\left(\frac{T_2T_1}{T_2 - T_1}\right) = 8.314\left(\ln\frac{5.2}{0.24}\right)\left[\frac{800(700)}{800 - 700}\right] = 1.4 \times 10^2 \text{ kJ mol}^{-1}$$

In logarithmic form, Eq. (13-29) becomes

$$\ln A = \ln k + \frac{E_a}{RT}$$

Now use the data at 700 K and the value of E_a:

$$\ln A = \ln(0.24) + \frac{143.2 \times 10^3}{8.314(700)} = -1.427\,12 + 24.605\,66 = 23.178\,54$$

$$A = 1.2 \times 10^{10} \text{ dm}^3\text{ mol}^{-1}\text{s}^{-1}$$

Mechanisms

PROBLEM 13-16 Propose a mechanism for the oxidation of bromide by hydrogen peroxide, introduced in Example 13-21:

$$2Br^- + 2H^+ + H_2O_2 \longrightarrow Br_2 + 2H_2O$$

The experimental rate equation for the formation of Br_2 is

$$\frac{d[Br_2]}{dt} = k[H_2O_2][H^+][Br^-]$$

Solution: Some possible intermediate species are listed in Example 13-21. Several, such as gases, high-energy free atoms, and the higher oxidation states of Br, aren't very likely candidates for the mechanism. As in Example 13-22, the three reactants (each to the first power) in the rate equation suggest that the rate-determining step is a reaction involving all three with coefficients of 1: a termolecular reaction, normally avoided, may satisfy all requirements. You should start by finding reasonable products of the reaction involving $Br^- + H^+ + H_2O_2$. With 3 H atoms, 2 O, and 1 Br, about the only choices are H_2O and HBrO:

$$Br^- + H^+ + H_2O_2 \longrightarrow H_2O + HBrO \qquad \text{(step 1)}$$

You want H_2O as one product but need a coefficient of 2. You now have HBrO to cancel out and still need Br_2. These requirements can be satisfied by a single reaction:

$$HBrO + H^+ + Br^- \longrightarrow Br_2 + H_2O \qquad \text{(step 2)}$$

Adding the two steps gives the net reaction. Step 1 must be slow, since it appears to provide the rate equation. An important check of a proposed mechanism is whether it predicts the rate equation, so you must derive the equation from the above two steps. From step 2,

$$\frac{d[Br_2]}{dt} = k_2[HBrO][H^+][Br^-] \qquad \textbf{(i)}$$

and from steps 1 and 2,

$$\frac{d[HBrO]}{dt} = k_1[Br^-][H^+][H_2O_2] - k_2[H^+][Br^-][HBrO]$$

If you apply the steady-state approximation to the intermediate reaction, i.e., assume that [HBrO] is constant, then the rate = 0 and

$$k_1[Br^-][H^+][H_2O_2] = k_2[H^+][Br^-][HBrO] \qquad \text{or} \qquad [HBrO] = \frac{k_1}{k_2}[H_2O_2]$$

Substituting this result in eq. (i), you get

$$\frac{d[Br_2]}{dt} = k_2[H^+][Br^-]\frac{k_1}{k_2}[H_2O_2] = k_1[H^+][Br^-][H_2O_2]$$

which is the expression for the experimental result. Thus the termolecular reaction seems to be valid, although such reactions are not common.

PROBLEM 13-17 In Problem 13-16, another possibility for the products of step 1 is

$$Br^- + H^+ + H_2O_2 \longrightarrow H_2 + HBrO_2$$

Why isn't this a good choice?

Solution: There are several reasons:

(a) It requires greater oxidation, i.e., Br^{-1} to Br^{+3}.
(b) It includes a reduction in the presence of an oxidizing agent.
(c) The usual residue from hydrogen peroxide when it serves as an oxidizing agent is water.

As you can see, general chemical knowledge often eliminates many possible intermediate reactions from further, time-consuming consideration.

PROBLEM 13-18 For the reaction

$$A_2 + B_2 \longrightarrow 2AB$$

predict the rate equation and order for the possible mechanisms **(a)** and **(b)**.

(a)
$$A_2 \longrightarrow 2A \quad \text{(slow)}$$
$$A + B_2 \longrightarrow AB + B \quad \text{(fast)}$$
$$B + A \longrightarrow AB \quad \text{(fast)}$$

(b)
$$A_2 + B_2 \longrightarrow A_2B_2 \quad \text{(slow)}$$
$$A_2B_2 \longrightarrow 2AB \quad \text{(fast)}$$

Solution: The slow reaction should be the rate-determining step in both cases.

(a) Rate $= k_1 c_{A_2}$; first order.
(b) Rate $= k_2 c_{A_2} c_{B_2}$; second order.

PROBLEM 13-19 For a process that can be described by two reversible steps, **(a)** obtain an expression for the overall equilibrium constant in terms of the rate constants; and **(b)** state where you have seen the same result previously. The process is

$$A \underset{k_{-1}}{\overset{k_1}{\rightleftharpoons}} B \quad \text{(step 1)}$$

$$B \underset{k_{-2}}{\overset{k_2}{\rightleftharpoons}} C \quad \text{(step 2)}$$

Solution:

(a) When complete equilibrium has been reached, for step 1,

$$K_1 = \frac{c_B}{c_A} = \frac{k_1}{k_{-1}} \qquad \textbf{(a)}$$

and for step 2,

$$K_2 = \frac{c_C}{c_B} = \frac{k_2}{k_{-2}} \qquad \textbf{(b)}$$

You can solve eq. (a) for c_B and substitute in eq. (b):

$$c_B = c_A \frac{k_1}{k_{-1}} \qquad \text{and} \qquad \frac{c_C}{c_A} = \frac{k_1 k_2}{k_{-1} k_{-2}} = K$$

Comparing this result to eqs. (a) and (b), you can see that

$$K = K_1 K_2$$

(b) The expression obtained in part (a) is Eq. (7-13), which states the general rule: The overall equilibrium constant for a series of equilibrium reactions is the product of the constants for the individual reactions.

PROBLEM 13-20 **(a)** Derive the rate equation for the following proposed mechanism for the reaction in Problem 13-16. **(b)** State what the result tells you.

$$H^+ + H_2O_2 \rightleftharpoons H_3O_2^+ \quad \text{(step 1)}$$
$$Br^- + H_3O_2^+ \longrightarrow H_2O + HBrO \quad \text{(step 2)}$$
$$Br^- + HBrO \longrightarrow Br_2 + OH^- \quad \text{(step 3)}$$
$$H^+ + OH^- \rightleftharpoons H_2O \quad \text{(step 4)}$$
$$\overline{2Br^- + 2H^+ + H_2O_2 \longrightarrow Br_2 + 2H_2O}$$

Solution:

(a) The approach you should take is to write rate equations for some of the steps and to find relationships between concentrations of species that can be substituted to try to arrive at the experimental rate equation given in Problem 13-16:

$$\frac{d[\mathrm{Br_2}]}{dt} = k[\mathrm{H_2O_2}][\mathrm{H^+}][\mathrm{Br^-}]$$

You can get the desired differential term by starting with step 3:

$$\frac{d[\mathrm{Br_2}]}{dt} = k_3[\mathrm{Br^-}][\mathrm{HBrO}] \tag{i}$$

From steps 2 and 3, you get an expression for [HBrO]:

$$\frac{d[\mathrm{HBrO}]}{dt} = k_2[\mathrm{Br^-}][\mathrm{H_3O_2^+}] - k_3[\mathrm{Br^-}][\mathrm{HBrO}]$$

If you apply the steady-state approximation to this intermediate reaction, the rate = 0 and

$$k_2[\mathrm{Br^-}][\mathrm{H_3O_2^+}] = k_3[\mathrm{Br^-}][\mathrm{HBrO}]$$

$$[\mathrm{HBrO}] = \frac{k_2}{k_3}[\mathrm{H_3O_2^+}] \tag{ii}$$

From steps 1 and 2, you get an expression for $[\mathrm{H_3O_2^+}]$:

$$\frac{d[\mathrm{H_3O_2^+}]}{dt} = k_1[\mathrm{H^+}][\mathrm{H_2O_2}] - k_2[\mathrm{Br^-}][\mathrm{H_3O_2^+}]$$

Applying the steady-state approximation to this intermediate reaction gives

$$k_1[\mathrm{H^+}][\mathrm{H_2O_2}] = k_2[\mathrm{Br^-}][\mathrm{H_3O_2^+}]$$

$$[\mathrm{H_3O_2^+}] = \frac{k_1}{k_2}\frac{[\mathrm{H^+}][\mathrm{H_2O_2}]}{[\mathrm{Br^-}]} \tag{iii}$$

Now, you substitute eq. (iii) in eq. (ii) and the result in eq. (i). Equation (ii) becomes

$$[\mathrm{HBrO}] = \frac{k_2}{k_3}\left(\frac{k_1}{k_2}\frac{[\mathrm{H^+}][\mathrm{H_2O_2}]}{[\mathrm{Br^-}]}\right)$$

Then eq. (i) becomes

$$\frac{d[\mathrm{Br_2}]}{dt} = k_3[\mathrm{Br^-}]\left(\frac{k_1}{k_3}\frac{[\mathrm{H^+}][\mathrm{H_2O_2}]}{[\mathrm{Br^-}]}\right) = k_1[\mathrm{H^+}][\mathrm{H_2O_2}]$$

(b) The result isn't the experimental rate equation. This disagreement tells you that the proposed mechanism can't be correct.

Supplementary Exercises

PROBLEM 13-21 The molecularity for the elementary reaction $H_2 + I_2 \rightarrow 2HI$ is __________; the order of the reaction is __________. When the plot of concentration versus time is a straight line, the order of the reaction is __________. Write the expression for the half-life of a first-order reaction. If a certain radioactive isotope has a half-life of 3.6 days, 75 percent of a given amount will decay in __________ days. The order of the reaction $2A \rightarrow$ products, based on the following data, is __________.

Run	$c_{A,0}$	Initial rate
1	0.74	0.0080
2	0.37	0.0040

The order of the reaction when the concentration of A decreased 0.004 M in 20 s and 0.004 M more in the next 20 s is ________. The two contributing effects in the preexponential factor of collision theory for large molecules are ________ ________ and ________ ________. For any reaction in a properly written mechanism, the order and the molecularity must be (equal) (unequal). For a reaction that takes place in more than one step, the rate (is) (isn't) essentially equal to the rate of the slowest step, and the overall rate constant (is) (isn't) always equal to the rate constant of the rate-determining step.

PROBLEM 13-22 Determine the rate constant of the first-order reaction for which the following data were measured. Use the graphic method.

Time (s)	0	20	40	60	100
Concentration ($mol\ dm^{-3}$)	0.200	0.165	0.142	0.121	0.089

PROBLEM 13-23 Continuing Problem 13-22, calculate the length of time required for the concentration to drop to 0.010 $mol\ dm^{-3}$.

PROBLEM 13-24 A second-order reaction of Type I ($A \rightarrow$ products) has a rate constant of 18.4 $mol^{-1} dm^3 min^{-1}$. If you start with 0.75 mol A in a 20-L flask, determine the concentration after 10 min.

PROBLEM 13-25 Continuing Problem 13-24, calculate the length of time for 90 percent of A to react.

PROBLEM 13-26 Acetaldehyde decomposes on heating, i.e.,

$$CH_3CHO \xrightarrow{\Delta} CH_4 + CO$$

From the following experimental data, determine the rate law, including the value of the rate constant.

Run	$[CH_3CHO]$ (M)	Initial rate ($M\ s^{-1}$)
1	1.2×10^{-2}	2.1×10^{-3}
2	3.1×10^{-2}	8.6×10^{-3}

PROBLEM 13-27 If the initial concentration of reactant A was 0.015 $mol\ dm^{-3}$ and the concentration after 20 s is 0.011 $mol\ dm^{-3}$, calculate the average rate of disappearance of A.

PROBLEM 13-28 What is the half-life of the reaction in Problem 13-22?

PROBLEM 13-29 How long will it take for 0.80 mol of the reactant in Problem 13-22 to become 0.10 mol? (See Problem 13-28.)

PROBLEM 13-30 How long will it take for the concentration of A in Problem 13-24 to fall to $\frac{1}{8}$ of the starting value?

PROBLEM 13-31 One radioactive isotope of lead has a half-life of 36.0 min. Determine the length of time required for 60 percent of a given quantity to decay.

PROBLEM 13-32 A first-order reaction has a half-life of 30 min. Calculate the fraction of the original concentration that will remain after 15 min.

PROBLEM 13-33 For a certain reaction, the rate increased by a factor of 50 when the temperature was raised from 20 °C to 70 °C. Calculate the activation energy for the reaction.

PROBLEM 13-34 Continuing Problem 13-33, at what temperature would the rate be 75 times the rate at 20 °C, all other factors being equal?

PROBLEM 13-35 A reaction has an activation energy of 48 kJ mol^{-1}, and the rate at 10 °C is 1.4 mol $dm^{-3}min^{-1}$. Determine the rate at 35 °C, if all other factors are the same.

PROBLEM 13-36 A reaction is found to have an activation energy of 110 kJ mol^{-1} and, at 500 K, a rate constant of 4.3×10^{-2} $dm^3\ mol^{-1}s^{-1}$. Calculate the preexponential factor.

PROBLEM 13-37 For the reaction in Problem 13-36, calculate the rate constant at 600 K.

PROBLEM 13-38 The reaction $2A + B_2 \rightarrow 2AB$ is found by experiment to be second-order in A and first-order in B_2. (**a**) Which of the following mechanisms are plausible for this reaction? (**b**) Which mechanism is the most plausible?

(1) $2A + B_2 \longrightarrow 2AB$

(2)
$B_2 \rightleftharpoons 2B$ (fast)
$A + B \longrightarrow AB$ (slow)

(3)
$2A \rightleftharpoons A_2$ (fast)
$A_2 + B_2 \longrightarrow 2AB$ (slow)

(4)
$B_2 \rightleftharpoons 2B$ (fast)
$2A + B \longrightarrow A_2B$ (slow)
$A_2B + B \longrightarrow 2AB$ (fast)

Answers to Supplementary Exercises

13-21 bimolecular, second; zero; $t_{1/2} = 0.693/k_1$; 7.2 days; first; zero; collision frequency, orientation factor; equal; is, isn't **13-22** 7.7×10^{-3} s^{-1} **13-23** 389 s **13-24** 4.7×10^{-3} *M* **13-25** 13 min **13-26** rate = $1.6[CH_3CHO]^{1.5}$ **13-27** 2×10^{-4} *M* s^{-1} **13-28** 90 s **13-29** 270 s **13-30** 10.15 min **13-31** 47.6 min **13-32** 0.707 **13-33** 65.4 kJ **13-34** 349 K **13-35** 7.3 mol $dm^{-3}min^{-1}$ **13-36** 1.34×10^{10} $dm^3\ mol^{-1}s^{-1}$ **13-37** 3.5 $dm^3\ mol^{-1}s^{-1}$ **13-38** (**a**); 1, 3 (**b**) 3

MID-TERM EXAM (Chapters 10–13)

Time: 50 minutes

Total Points: 100

A. Short Answers (2 points per answer = 22)

1. In a three-component system, what is the maximum number of degrees of freedom?

2. What is the number of components, after equilibrium has been attained, in the system prepared by placing pure NH_3 in a flask?

3. What do we call the temperature above which partially miscible liquids become completely miscible?

4–5. Name the two types of solid solution.

6. Name the SI unit of conductance.

7–8. Name the two effects that are results of the ionic atmosphere in the Debye–Hückel theory.

9. What is the sign of the cathode in a galvanic (also voltaic) cell?

10. Refer to the table in Part C following Problem 5. Will I_2 reduce Sn^{4+} to Sn^{2+} when all species are at unit activity?

11. What is the order of a reaction for which the plot of concentration vs. time is a straight line?

B. Long Answers (Total points = 26)

1. Figure E-2 is a phase diagram for two solids A and B.

(a) What is point b called?
(b) What is point c called?
(c) If the system initially is at point n ($X_B = 0.6$) and is cooled slowly,
(1) what will the first change be?
(2) describe what happens when the temperature reaches T_c.
(3) what is the system like at temperatures below T_c? (8)

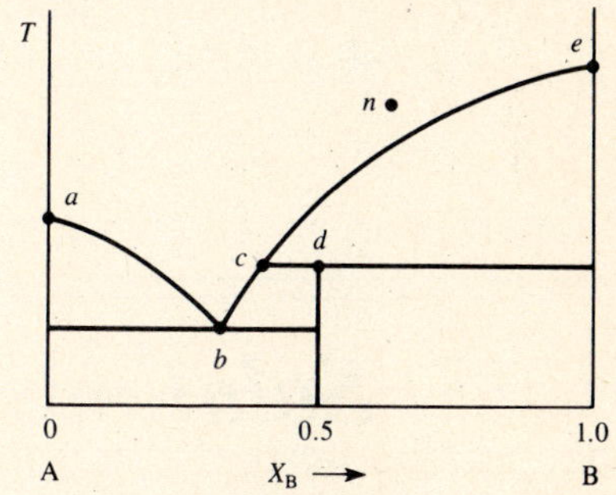

FIGURE E-2

2. What is meant by "transference" in a chemical cell "with transference"? What causes it, and why is it undesirable? What is usually done to remove or minimize it? (5)

3. Describe the major features of an "elementary reaction" as used in chemical kinetics. (3)

4. **(a)** Write the Arrhenius equation for the dependence of rate constants on temperature.
(b) Derive from it an expression for calculating the rate constant at a second temperature if the rate constant is known at one temperature. (6)

5. Describe briefly how the activated-complex theory of reaction rates explains the fact that all reactions do not go instantly to completion. (4)

C. Problems (Total points = 52). To save time, do not do the arithmetic unless the problem is labeled CALCULATE; otherwise just *set up* the final calculation. Solve all equations for the unknown, and substitute numerical values for all symbols. Put the proper units on the final answer.

1. The heat of fusion of dibromophenol is 3.53 kcal mol^{-1}, and its melting point is 40 °C. Find its theoretical solubility in alcohol at 25 °C. Assume that the solution is ideal. (4)

2. What mass of barium will be released if a current of 5.0 A passes through a $Ba(NO_3)_2$ solution for 20 minutes? (6)

3. In a certain conductivity cell, the resistance of a 0.020 *M* AgAc solution is 79.9 Ω. In the same cell, the resistance of a 0.010 *M* KCl solution is 127.0 Ω. The known molar conductivity of this KCl solution is 128.8 $\Omega^{-1}cm^2\,mol^{-1}$. Find the molar conductivity of AgAc. (8)

4. The mean activity coefficient of a 0.010 *M* solution of $CaCl_2$ is 0.732. Calculate the activity of the compound. (6)

5. Find the ionic strength of a solution that is 0.020 *M* in $Mg(NO_3)_2$ and 0.040 *M* in Na_2SO_4. (4)

Some Standard Reduction Potentials $E°$ (V) at 298 K

Reaction	$E°$ (V)
$I_2 + 2e^- \longrightarrow 2I^-$	0.536
$Cu^{2+} + 2e^- \longrightarrow Cu$	0.337
$Sn^{4+} + 2e^- \longrightarrow Sn^{2+}$	0.15
$Sn^{2+} + 2e^- \longrightarrow Sn$	−0.136
$Cr^{3+} + 1e^- \longrightarrow Cr^{2+}$	−0.400
$Cr^{2+} + 2e^- \longrightarrow Cr$	−0.557

6. CALCULATE. Find $E°$ at 298 K for the half-cell $Cr^{3+} + 3e^- \rightarrow Cr$. (8)

7. Find the cell potential at 298 K of a cell composed of a Sn^{2+}–Sn electrode with the activity of $Sn^{2+} = 0.15$ and a Cu^{2+}–Cu electrode with the activity of $Cu^{2+} = 0.70$. (6)

8. In a certain first-order reaction, the concentration changed from 0.200 *M* to 0.078 *M* in 75 minutes. What is the half-life of the reaction? (6)

9. For a second-order reaction of the type 2A → products, the initial concentration of A was 0.40 *M*. After 2 minutes, the concentration was 0.38 *M*. Find the rate constant. (4)

Solutions to Mid-Term Exam

A. Short Answers

1. Four; $f = 3 - p + 2$, and p must be at least 1. [Section 10-1C]
2. One; the concentrations of N_2 and H_2 can be calculated from K_{eq} and the concentration of NH_3. [Problem 10-2b]
3. upper consolute temperature [Section 10-2A]

4–5. substitutional and interstitial [Section 10-6A]

6. siemens [Section 11-4A]

7–8. relaxation (or asymmetry) effect, electrophoretic effect [Section 11-6C]

9. positive [Section 12-1]

10. no [Section 12-2F]

11. zero [Section 13-4]

B. Long Answers

1. (a) eutectic point [Section 10-4B]
   (b) peritectic point (also, incongruent melting point) [Section 10-5B]

(c) (1) Solid B will appear. (2) Solid compound AB will appear. The temperature will remain constant while the reaction B + solution → AB proceeds. All solution will disappear, and the temperature again will decrease. (3) There will be two solid phases: B and compound AB. [Section 10-5B]

2. Transference is the movement of ions across a liquid junction (the boundary between two electrolyte solutions) at unequal speeds so that an inequality of charge develops. This causes a potential—the liquid junction potential—which is included in the measured emf. Usually a salt bridge, containing ions which move at about the same speed, is used to connect the two solutions. [Section 12-4]

3. An elementary reaction occurs in a single step in accord with its balanced equation. The molecularity and the order are equal and are obtained from the balanced equation. [Section 13-1]

4. **(a)** The Arrhenius equation is Eq. (13-29):

$$k = Ae^{-E_a/RT}$$

(b) Taking the logarithm of both sides gives

$$\ln k = \ln A - E_a/RT$$

Differentiating gives

$$d \ln k = 0 + \frac{E_a}{R}\left(\frac{1}{T^2}\right) dT$$

Integrating with limits (k_2, T_2) and (k_1, T_1), you get

$$\ln \frac{k_2}{k_1} = \frac{E_a}{R}\left(-\frac{1}{T_2} + \frac{1}{T_1}\right) = \frac{E_a}{R}\left(\frac{T_2 - T_1}{T_1 T_2}\right) \qquad \text{[Section 13-8]}$$

5. The activated-complex theory postulates that the two reacting species form an activated complex that is in equilibrium with the separate species: $A + B \rightleftharpoons (AB)^{\ddagger}$. A particular $(AB)^{\ddagger}$ unit may simply return to A and B, but others will break up differently, forming products, e.g., C and D. [Section 13-9B]

C. Problems

1. Solubility is given by Eq. (10-6):

$$\ln X_2 = -\frac{\Delta H_f}{R}\left(\frac{T_f^\circ - T}{T_f^\circ T}\right) = -\frac{3.53 \times 10^3 \text{ cal mol}^{-1}}{1.987 \text{ cal K}^{-1}\text{mol}^{-1}}\left[\frac{(313 - 298)\text{ K}}{(313\text{ K})(298\text{ K})}\right]$$

$$X_2 = \text{antiln}\left[-\frac{3.53 \times 10^3(15)}{1.987(313)(298)}\right]$$

2. The number of moles deposited is obtained from the number of coulombs and the Faraday constant [Section 11-2]:

$$n = Q/\mathcal{F} = \frac{(5.0\text{ A})(20\text{ min})(60\text{ s min}^{-1})(\text{C A}^{-1}\text{s}^{-1})}{96\,500\text{ C mol}^{-1}}$$

The mass is n times the mass of $\frac{1}{2}Ba^{2+}$:

$$m = \frac{(5.0)(20)(60)}{96\,500}\text{ mol}\left(\frac{137.3}{2}\text{ g mol}^{-1}\right) = \frac{(5.0)(20)(60)(137.3)}{2(96\,500)}\text{ g}$$

3. Obtain the cell constant from the data for the KCl solution, then use it with data for the unknown solution. [Section 11-4]

For KCl:

$$K_{cell} = \kappa R = \Lambda C R \qquad \text{[Eqs. (11-8) and (11-10)]}$$
$$= (128.8\ \Omega^{-1}\text{cm}^2\text{ mol}^{-1})(0.010\text{ mol L}^{-1})(10^{-3}\text{ L cm}^{-3})(127\ \Omega)$$

For AgAc:

$$\Lambda = \frac{K_{cell}}{CR} = \frac{(128.8)(0.010)(10^{-3})(127)\text{ cm}^{-1}}{(0.020\text{ mol L}^{-1})(10^{-3}\text{ L cm}^{-3})(79.7\ \Omega)}$$
$$= \frac{(128.8)(0.010)(127)}{(0.020)(79.7)}\ \Omega^{-1}\text{cm}^2\text{ mol}^{-1}$$

4. The activity of a compound that yields three ions is the third power of the mean ionic activity: $a = a_\pm^3$ [Section 11-5]. Since $a_\pm = \gamma_\pm m_\pm$ [Eq. (11-39)], you must determine the mean ionic molality, $m_\pm$. For an AB_2 compound,

$$m_\pm = [(1m)^1(2m)^2]^{1/3} = 4^{1/3}m = 4^{1/3}(0.010)$$
$$a_\pm = \gamma_\pm m_\pm = (0.732)(4)^{1/3}(0.010)$$
$$a = a_\pm^3 = 4(0.732)^3(0.010)^3$$

5. In this case, you have

$$Mg(NO_3)_2 \longrightarrow Mg^{2+} + 2NO_3^- \quad \text{and} \quad Na_2(SO_4) \longrightarrow 2Na^+ + SO_4^{2-}$$

Ionic strength includes the concentrations and the squares of the charges of all ions [Section 11-5]. Using Eq. (11-41), $I = \frac{1}{2}\Sigma_i C_i z_i^2$,

$$I = \tfrac{1}{2}[(0.02)(2)^2 + (0.02)(1)^2 + (0.04)(1)^2 + 0.04(2)^2] \text{ mol kg}^{-1}$$

6. You must use the two half-cell potentials given in the table but you cannot add them directly [Section 12-3]. You must calculate and add $\Delta G°$: $\Delta G° = -n\mathscr{F}E°$ [Eq. (12-10)].

$$Cr^{3+} + 1e^- \longrightarrow Cr^{2+} \qquad \Delta G° = -1\mathscr{F}(-0.400) = +0.400\,\mathscr{F}$$

$$Cr^{2+} + 2e^- \longrightarrow Cr \qquad \Delta G° = -2\mathscr{F}(-0.557) = 1.114\mathscr{F}$$

$$Cr^{3+} + 3e^- \longrightarrow Cr \qquad \Delta G° = 0.400\mathscr{F} + 1.114\mathscr{F} = 1.514\mathscr{F}$$

$$E° = -\frac{\Delta G°}{n\mathscr{F}} = -\frac{1.514\mathscr{F}}{3\mathscr{F}} = -0.505 \text{ V}$$

7. Since the activities are not unity, use the Nernst equation, Eq. (12-12), but first write the balanced equation for the cell reaction [Section 12-3E]. Since the Sn half-reaction is lower in the table, it is reversed, and you have

$$Cu^{2+} + 2e^- \longrightarrow Cu \qquad E° = 0.337 \text{ V}$$

$$Sn \longrightarrow Sn^{2+} + 2e^- \qquad E° = 0.136 \text{ V}$$

$$Cu^{2+} + Sn \longrightarrow Cu + Sn^{2+} \qquad 0.473 \text{ V}$$

Then

$$E = E° - \frac{RT}{n\mathscr{F}}\ln\frac{a_{Sn^{2+}}}{a_{Cu^{2+}}} = 0.473 \text{ V} - \frac{0.059\,16 \text{ V mol}}{2 \text{ mol}}\log\frac{0.15}{0.70} = \left(0.473 - \frac{0.059\,16}{2}\log\frac{0.15}{0.70}\right)\text{V}$$

8. First, use Eq. (13-13) to determine the rate constant and then obtain $t_{1/2}$ by using Eq. (13-25). Rearranging Eq. (13-13):

$$k_1 = \frac{1}{t}\left(\ln\frac{c_0}{c}\right) = \frac{1}{75}\left(\ln\frac{0.20}{0.078}\right)$$

Substituting in Eq. (13-25):

$$t_{1/2} = \frac{0.693}{k_1} = \frac{(0.693)(75)}{\ln(0.20/0.078)} \text{ min}$$

9. For a second-order reaction when the two reactants are identical, the integrated rate expression is Eq. (13-17),

$$\frac{1}{c} - \frac{1}{c_0} = k_2 t \quad \text{or} \quad k_2 = \frac{1}{t}\left(\frac{1}{c} - \frac{1}{c_0}\right)$$

$$k_2 = \frac{1}{2 \text{ min}}\left(\frac{1}{0.38 \text{ M}} - \frac{1}{0.40 \text{ M}}\right) = \frac{1}{2}\left(\frac{1}{0.38} - \frac{1}{0.40}\right) M^{-1} \text{ min}^{-1}$$

14 INTRODUCTION TO QUANTUM THEORY

THIS CHAPTER IS ABOUT

- ☑ **The Upheaval in Physics**
- ☑ **The Old Quantum Theory**
- ☑ **Wave–Particle Duality**
- ☑ **The New Quantum Theory**
- ☑ **Applications to Simple Systems**

The discoveries and related theories concerning the nature of matter that appeared between 1895 and 1930—and their significance for chemistry—are usually presented in general chemistry texts. Physical chemistry textbooks tend to omit those that have been superseded or have become "givens," e.g., the nuclear atom. Most such textbooks move quickly to activity after 1926: the application to atomic and molecular structure of the new theory of quantum mechanics based on waves. We briefly discuss the earlier events and their significance, giving greater attention to those that lead into wave mechanics.

In Section 14-1, we mention the discoveries that led to recognition of the major features of the atom and supported the quantum hypothesis. The most significant with respect to the quantum theory is the experimental evidence from spectroscopy, for which the theory must account. In Section 14-2, we describe in some detail the Bohr theory of the electronic structure of the hydrogen atom, because it was a major contribution at the time. It served as a stepping stone to the new, wave-mechanical quantum theory, and the picture of electron configuration it provides is still useful for many applications in chemistry. In Section 14-3, we present the concept of waves associated with matter, a topic also significant in development of the new quantum theory.

14-1. The Upheaval in Physics

Scientists couldn't explain several discoveries made near the end of the nineteenth century. Those discoveries and subsequent developments based on them upset the picture then held by scientists concerning the nature of matter and required major changes in theoretical and applied physics. Before describing these developments, however, we present the results from spectroscopy, dating from about 1870–1890, which eventually led to the idea of energy levels in an atom.

A. Spectroscopy

Experimental evidence from spectroscopy provides significant information about the extranuclear structure of atoms. Before 1895, many spectroscopic measurements had led to empirical formulas from which the wave lengths of the spectral "lines" could be calculated. Lyman, Balmer, and others had identified the several series of lines for hydrogen. The general equation for the wavelengths of the hydrogen lines is

RITZ–RYDBERG EQUATION

$$\tilde{\nu} = \frac{1}{\lambda} = R_{\mathrm{H}}\left(\frac{1}{n_1^2} - \frac{1}{n_2^2}\right) \quad \textbf{(14-1)}$$

where $\tilde{\nu}$ is the **wave number**, λ is the **wavelength**, R_{H} is the **Rydberg constant** for hydrogen (109 678 cm^{-1}, or 1.097×10^7 m^{-1}), $n_1 = 1, 2, 3, \ldots$, and $n_2 = n_1 + 1, n_1 + 2, n_1 + 3, \ldots$. The Lyman series results when $n_1 = 1$, the Balmer series when $n_1 = 2$, etc. The frequency of radiation is $\nu = c/\lambda = c\tilde{\nu}$, where c is the velocity of light.

EXAMPLE 14-1: **(a)** Calculate the wavelength of the first line of the Lyman series in the hydrogen spectrum. **(b)** In what region of electromagnetic radiation does it fall?

Solution:

(a) For the Lyman series, n_1 in Eq. (14-1) always is 1. The first line appears when $n_2 = 2$. Substituting in Eq. (14-1), we have

$$\frac{1}{\lambda} = (1.097 \times 10^7 \text{ m}^{-1})\left(\frac{1}{1^2} - \frac{1}{2^2}\right) = (1.097 \times 10^7 \text{ m}^{-1})\left(1 - \frac{1}{4}\right) = 0.822\,75 \times 10^7 \text{ m}^{-1}$$

$$= 1.215 \times 10^{-7} \text{ m} = 121.5 \text{ nm}$$

The proper SI unit is the meter or the subunit nanometer (10^{-9} m). Before adoption of the SI, the unit Angstrom was commonly used; 1 Å = 10^{-8} cm = 10^{-10} m. In Angstroms

$$\lambda = 1.215 \times 10^3 \text{ Å} = 1215 \text{ Å}$$

(b) This magnitude places the series in the ultraviolet region of radiation.

EXAMPLE 14-2: **(a)** Derive a general expression for the wavelengths in the Balmer series. **(b)** Calculate the wavelength of the first line and state the region of radiation in which it falls.

Solution:

(a) For the Balmer series, n_1 always equals 2. Using n for all n_2, we obtain from Eq. (14-1)

$$\frac{1}{\lambda} = R_{\text{H}}\left(\frac{1}{2^2} - \frac{1}{n^2}\right)$$

Changing the fractions to a common denominator,

$$\frac{1}{\lambda} = R_{\text{H}}\left(\frac{n^2 - 4}{4n^2}\right)$$

Taking the reciprocal of both sides, we obtain

$$\lambda = \frac{1}{R_{\text{H}}}\left(\frac{4n^2}{n^2 - 4}\right) = \left(\frac{4}{1.097 \times 10^7 \text{ m}^{-1}}\right)\left(\frac{n^2}{n^2 - 4}\right)$$

or the

BALMER EQUATION $$\lambda = (3.646 \times 10^{-7})\left(\frac{n^2}{n^2 - 4}\right) \text{m} \qquad \textbf{(14-2)}$$

(b) The first line is represented by $n = 3$, i.e., $n_1 + 1 = 2 + 1$. Then

$$\lambda = (3.646 \times 10^{-7})\left(\frac{9}{9 - 4}\right) \text{m} = 6.563 \times 10^{-7} \text{ m} = 656.3 \text{ nm} = 6563 \text{ Å}$$

This wavelength falls in the visible region of radiation (red).

B. X-rays

In 1895, Roentgen discovered **x-rays**. They played no role in the development of the new atomic theory, although it was eventually used to explain their origin.

C. Radioactivity

In 1896, Becquerel discovered **radioactivity**. As did x-rays, radioactivity indicated some internal structure of atoms but had no direct theoretical impact. Indirectly, however, it provided an important tool for exploring the structure of atoms—a beam of alpha-particles revealed that atoms have nuclei (Rutherford's α-scattering experiments, 1908).

D. The electron

In 1897, Thomson discovered the **electron**. What he actually measured was the ratio of charge to mass (e/m), but the existence of such a ratio indicates that there is a particle. Measurements in cathode-ray tubes showed that the charge is negative. It seemed clear that an atom is not indivisible nor is it the smallest particle. When Millikan measured the charge (in 1909), the mass could at last be calculated.

E. Black-body radiation

A **black body** is an idealized substance that absorbs and emits all frequencies of radiation equally. A close approximation is a particle of soot; in practice, a container with a tiny hole is used. Radiation escaping through the hole when the container is heated is equivalent to black-body radiation. In 1900, Rayleigh and Jeans derived an equation based on the then-current physical view of equipartition of energy, showing that the amount of energy of a particular frequency radiated by a black body should increase as frequency increases. This is impossible, since an infinite amount of energy would be approached at high frequencies. In addition, experimental measurements showed that the amount of energy emitted passes through a maximum, dropping back to zero at high frequencies (low wavelengths), and that the maximum shifts to lower wavelengths at higher temperatures. The **Wien displacement law** expresses the inverse proportionality between temperature and the wavelength at the maximum in the plot of intensity (energy density per unit wavelength) versus wavelength:

WIEN DISPLACEMENT LAW $$\lambda \text{ (max emission)} = kT^{-1} \qquad \textbf{(14-3)}$$

where the proportionality constant $k = 2.898 \times 10^{-3}$ m K.

EXAMPLE 14-3: Determine the temperature required for the peak of the emission curve of a black body to be in the visible region of the spectrum.

Solution: Let's choose 5500 Å as about the midpoint of the visible region. Rearranging the Wien law, we have

$$T = \frac{k}{\lambda} = \left(\frac{2.898 \times 10^{-3}\text{ m K}}{5.5 \times 10^{3}\text{ Å}}\right)\left(\frac{1\text{ Å}}{10^{-10}\text{ m}}\right) = 0.5269 \times 10^{-6}(10^{10})\text{ K} = 5269\text{ K}$$

Stefan's law states that the total energy emitted over all wavelengths is proportional to T^4, where T is the temperature of the container:

STEFAN'S LAW $$U = aT^4 \qquad \textbf{(14-4)}$$

The proportionality constant $a = 7.566 \times 10^{-16}$ J m^{-3}K^{-4}. The rate of emission, or power emitted per unit area, is called the **excitance**, M_e. It is proportional to U, or

$$M_e = \sigma T^4 \qquad \textbf{(14-5)}$$

where σ, the **Stefan–Boltzmann constant**, is 5.670×10^{-8} J m^{-2}s^{-1}K^{-4}.

EXAMPLE 14-4: How many watts would be emitted by a surface area of 1 m^2 at 500 °C if the surface behaved like a black body?

Solution: Since one watt is one joule per second, the units of the Stefan–Boltzmann constant also are W m^{-2}K^{-4}. Substituting in Eq. (14-5), we have

$$M_e = (5.670 \times 10^{-8}\text{ W m}^{-2}\text{K}^{-4})(773\text{ K})^4 = (5.670 \times 10^{-8})(3.570 \times 10^{11}) = 2.024 \times 10^{4}\text{ W m}^{-2}$$

F. Quantum hypothesis

The disturbing failure of the Rayleigh–Jeans equation to fit experimental results at high frequencies (short wavelengths) became known as the *ultraviolet catastrophe.* Its positive result was the concept of the quantum of energy. In 1900, Planck proposed an equation that *did* fit the data and which he explained with a revolutionary new idea. He assumed that a black body doesn't emit radiation continuously but in small packets of energy, which he called *quanta*. This concept is the foundation of **quantum mechanics**. The amount of energy in one quantum is proportional to the frequency, i.e.,

$$\varepsilon = h\nu \qquad \textbf{(14-6)}$$

where h is the proportionality constant, now known as the **Planck constant** (6.6262×10^{-34} J s). The total amount of energy at a given frequency depends on how many small bodies within the solid are oscillating at that frequency.

G. Explanation of the photoelectric effect

In 1905, Einstein explained the **photoelectric effect** (the ejection of electrons from a surface by light), extending Planck's quantum concept from the process of emission by oscillators to all electromagnetic radiation. One of the perplexing features of the photoelectric effect had been the *threshold frequency*; i.e., electrons aren't emitted until the frequency of the incident light exceeds some value (characteristic of each metal), usually in the ultraviolet region.

EXAMPLE 14-5: Why should ultraviolet light cause emission of an electron when red light of the same intensity won't?

Solution: The wave theory of light predicts that there should be no difference. However, Eq. (14-6) tells us that a quantum having a frequency in the range of ultraviolet light has more energy than a quantum of red light (lower frequency). Consequently, ultraviolet light imparts more energy to the metal and thus may increase the kinetic energy of an electron to the point that it can escape from the metal.

With the extension of the quantum concept to all radiant energy, the packets of energy, or "atoms of radiation," came to be called **photons**. The total energy of the incident photon is divided, some overcoming the forces that tended to hold the electron in the metal and the remainder providing kinetic energy to the departing electron:

EINSTEIN PHOTOELECTRIC EQUATION

$$\varepsilon = h\nu = W + \frac{1}{2}mv^2 \qquad \textbf{(14-7)}$$

where W is called the *work function.*

EXAMPLE 14-6: Calculate the energy of a photon of light having a wavelength of 4500 Å.

Solution: According to Einstein's application of the quantum concept, the energy of a photon, Eq. (14-6), is

$$\varepsilon = h\nu = h\left(\frac{c}{\lambda}\right)$$

Substituting values and converting Å to m, we get

$$\varepsilon = (6.6262 \times 10^{-34}\ \text{J s})\left(\frac{2.998 \times 10^{8}\ \text{m s}^{-1}}{4500 \times 10^{-10}\ \text{m}}\right) = 4.415 \times 10^{-19}\ \text{J}$$

EXAMPLE 14-7: The work function for Ni is 8.0×10^{-19} J. **(a)** Would the photon in Example 14-6 release an electron? **(b)** What is the threshold frequency of Ni? **(c)** Calculate the wavelength corresponding to this frequency.

Solution:

(a) No, since the energy calculated for 4500 Å is less than the work function.
(b) Since the threshold frequency would just release an electron, the electron would have no

kinetic energy. Then Eq. (14-7) becomes

$$h\nu_0 = W$$

$$\nu_0 = \frac{W}{h} = \frac{8.0 \times 10^{-19}\ \text{J}}{6.6262 \times 10^{-34}\ \text{J s}} = 1.207 \times 10^{-15}\ \text{s}^{-1}$$

(c) The threshold wavelength $\lambda_0 = c/\nu_0$, so

$$\lambda_0 = \left(\frac{2.998 \times 10^8\ \text{m s}^{-1}}{1.207 \times 10^{15}\ \text{s}^{-1}}\right)(10^{10}\ \text{Å m}^{-1}) = 2484\ \text{Å}$$

H. Charge and mass of electron

In 1909, Millikan measured the **electron charge** (or **proton charge**) e by using drops of water. Using drops of oil in 1912, he obtained a value close to the currently accepted value: $e = 1.6022 \times 10^{-19}$ C. (From accurate measurements of e/m, the **mass of the electron** m_e is 9.1095×10^{-31} kg.)

I. Nucleus of the atom

In 1911, Rutherford showed that an atom contains a **nucleus** and is mostly empty space. Calculations based on the angles and relative intensities of the scattering of alpha particles by a gold foil led to several postulates about the interior of atoms:

1. An atom contains a nucleus with diameter about 1/10 000 of the diameter of the atom.
2. The nucleus contains most of the mass of the atom.
3. The nucleus contains the entire positive charge of the atom.
4. The negative charge is distributed spherically around the nucleus.

The shortcoming of the theory was that such an atom would collapse because of electrostatic attraction. Even if an electron were assumed to be moving with sufficient velocity that it could overcome the nuclear attraction, other theories predicted that it would emit energy and spiral into the nucleus.

14-2. The Old Quantum Theory

In 1913, Bohr provided an answer to the unstable Rutherford atom and to the unexplained atomic spectra by employing Planck's and Einstein's quantum concept to develop a model of the hydrogen atom.

A. Bohr's assumptions

1. **The electron moves around the nucleus in certain circular orbits, but not all orbits are permitted.**

   Simplifying Bohr's argument, we can say that an orbit is permitted if the angular momentum of the electron in that orbit equals an integer times $h/2\pi$, where h is Planck's constant, i.e.,

$$m_e vr = n\frac{h}{2\pi} \qquad n = 1, 2, 3, \ldots \qquad \textbf{(14-8)}$$

2. **When the electron is in the lowest available permitted orbit, the atom is stable and does not radiate energy.**

   For each permitted orbit, the atom has a specific energy E. In other words, the energy of an atom is "quantized."

3. **The electron may jump from one permitted orbit to another.**

   When an electron jumps, the atom gains or loses energy equal to the difference between the energies of the two orbits, or $\Delta E = E_f - E_i$. This energy comes or goes as one photon with energy $= |\Delta E|$. By Eq. (14-6),

$$|\Delta E| = \varepsilon = h\nu = |E_f - E_i|$$

and $$\nu = \frac{1}{h}|(E_f - E_i)| \tag{14-9}$$

This frequency should correspond to one line in the spectrum of hydrogen.

B. Bohr's results

Bohr derived two expressions that provided checks on his proposal.

1. The radius of the atom

is expressed as

$$r = \frac{(4\pi\varepsilon_0)n^2h^2}{4\pi^2 m_e e^2} = \frac{(4\pi\varepsilon_0)n^2\hbar^2}{m_e e^2} \tag{14-10}$$

where $\hbar$ is a symbol often used for $h/2\pi$.

aside: You won't find the quantity $4\pi\varepsilon_0$ in the equations in older textbooks; it's a proportionality constant needed in SI. The permittivity of vacuum ε_0 appeared previously in a situation of electrical attraction: the expression for the thickness of the ionic atmosphere, Eq. (11-45). Two useful values are

$$4\pi\varepsilon_0 = 1.1127 \times 10^{-10}\ \text{C}^2\,\text{J}^{-1}\text{m}^{-1}\ (\text{or C}^2\,\text{N}^{-1}\text{m}^{-2})$$

$$\frac{1}{4\pi\varepsilon_0} = 8.9871 \times 10^{9}\ \text{J m C}^{-2}\ (\text{or N m}^2\,\text{C}^{-2})$$

To be strictly accurate, we should correct for the finite mass of the nucleus (a proton) by substituting for m the *reduced mass* μ, i.e.,

$$\mu = \frac{m_e m_p}{m_e + m_p} \tag{14-11}$$

EXAMPLE 14-8: Calculate the reduced mass of the hydrogen atom.

Solution: For m_e to make a contribution to the denominator, we need to use more than the usual four significant figures, or

$$m_e = 9.109\,534 \times 10^{-31}\ \text{kg} \qquad \text{and} \qquad m_p = 1.672\,649 \times 10^{-27}\ \text{kg}$$

Substituting into Eq. (14-11), we obtain

$$\mu = \frac{(9.109\,534 \times 10^{-31}\ \text{kg})(1.672\,649 \times 10^{-27}\ \text{kg})}{9.109\,534 \times 10^{-31}\ \text{kg} + 16\,726.49 \times 10^{-31}\ \text{kg}} = 9.104\,90 \times 10^{-31}\ \text{kg}$$

In Eq. (14-10), when $n = 1$, we write r as a_0, which is called the **Bohr radius** and is sometimes used as an atomic unit. Substituting in Eq. (14-10), we get

BOHR RADIUS
$$a_0 = \frac{(4\pi\varepsilon_0)h^2}{4\pi^2 \mu e^2} \tag{14-12}$$

EXAMPLE 14-9: Calculate the value of a_0 (**a**) using m_e and (**b**) using μ.

Solution:

(**a**) Substituting in Eq. (14-12), but with m_e instead of μ, we get

$$a_0 = \frac{(1.1127 \times 10^{-10}\ \text{C}^2\,\text{J}^{-1}\text{m}^{-1})(6.6262 \times 10^{-34}\ \text{J s})^2}{4(3.1416)^2(9.1095 \times 10^{-31}\ \text{kg})(1.6022 \times 10^{-19}\ \text{C})^2} = 5.2920 \times 10^{-11}\ \text{J s}^2\,\text{kg}^{-1}\text{m}^{-1}$$

Since $1\ \text{J} = 1\ \text{kg m}^2\,\text{s}^{-2}$, $a_0 = 5.2920 \times 10^{-11}\ \text{m} = 52.920\ \text{pm}$.

(b) Using μ from Example 14-8 in Eq. 14-12, we have

$$a_0 = \frac{(1.1127 \times 10^{-10})(6.6262 \times 10^{-34})^2}{4(3.1416)^2(9.1049 \times 10^{-31})(1.6022 \times 10^{-19})^2} = 5.2947 \times 10^{-11}\ \text{m}$$

The number that you should remember is 0.529 Å. The Angstrom was defined as 10^{-8} cm to approximate the diameter of the smallest atom.

---

2. The energy of the atom
is expressed as

$$E = \frac{1}{4\pi\varepsilon_0}\left(-\frac{e^2}{2r}\right) \tag{14-13}$$

Substituting Eq. (14-10) gives

$$E_n = -\frac{e^2(4\pi^2 m_e e^2)}{2(4\pi\varepsilon_0)^2 n^2 h^2} = -\frac{4\pi^2 m_e e^4}{2(16\pi^2\varepsilon_0^2)h^2}\left(\frac{1}{n^2}\right) = -\frac{m_e e^4}{8\varepsilon_0^2 h^2}\left(\frac{1}{n^2}\right)$$

Sometimes, we write this expression in terms of a_0, or

$$E_n = -\frac{e^2}{2(4\pi\varepsilon_0)a_0}\left(\frac{1}{n^2}\right) \tag{14-14}$$

EXAMPLE 14-10: Calculate the energy of the photon emitted when an electron changes from an orbit where $n = 4$ to an orbit where $n = 2$.

Solution: The energy is $\varepsilon = |\Delta E_n|$. For emission, the final level has the lower n, so $\Delta E = E_2 - E_4$. Instead of calculating each E, let's substitute Eq. (14-14) for each E_n and factor out the constants:

$$\varepsilon = |\Delta E_n| = |E_2 - E_4|$$

$$E_2 - E_4 = -\frac{e^2}{(4\pi\varepsilon_0)2a_0}\left(\frac{1}{2^2} - \frac{1}{4^2}\right)$$

$$\Delta E = -\frac{(1.602 \times 10^{-19}\ \text{C})^2}{(1.113 \times 10^{-10}\ \text{C}^2\,\text{J}^{-1}\text{m}^{-1})(2)(0.5292 \times 10^{10}\ \text{m})}\left(\frac{1}{4} - \frac{1}{16}\right) = -4.085 \times 10^{-19}\ \text{J}$$

and
$$\varepsilon = |\Delta E| = 4.085 \times 10^{-19}\ \text{J}$$

If we use the reduced mass, we get 4.083×10^{-19} J. For most purposes, the use of m_e is acceptable.

C. Confirmation of Bohr's results

The calculated radius of the hydrogen atom was about what was expected from other considerations, but the convincing argument for accepting Bohr's assumptions was that he calculated the value of the Rydberg constant. You may have noticed that the equation obtained in Example 14-10 looks a lot like Eq. (14-1). If we make the equation general by using n_1 and n_2, and if we substitute it into Eq. (14-9), we get

$$\nu = \frac{1}{h}\left[\frac{e^2}{(4\pi\varepsilon_0)2a_0}\left(\frac{1}{n_1^2} - \frac{1}{n_2^2}\right)\right]$$

To change to wave number, we substitute $\nu = c/\lambda = c\tilde{\nu}$, obtaining

$$\tilde{\nu} = \frac{1}{hc}\left[\frac{e^2}{(4\pi\varepsilon_0)2a_0}\left(\frac{1}{n_1^2} - \frac{1}{n_2^2}\right)\right]$$

Comparing this result to Eq. (14-1), we see that the Rydberg constant is

$$R_{\text{H}} = \frac{1}{hc}\frac{e^2}{(4\pi\varepsilon_0)2a_0}$$

The calculated value agrees nicely with the experimental value.

D. Shortcomings of Bohr's theory

In spite of its impressive results, the new theory had faults:

1. It couldn't explain the spectra of larger atoms, even when modified by including elliptical orbits and additional quantum numbers.
2. It didn't attempt to predict intensities or polarizations of spectral lines.
3. The assumption of quantization was arbitrary; there wasn't any reason for it, except that it worked.
4. The assumption of orbital angular momentum of the electron was shown to be invalid by the Stern–Gerlach experiment.

E. Stern–Gerlach experiment

In an experiment to study space quantization, Stern and Gerlach verified the prediction of quantum theory that the orientation in space of a rotating body isn't arbitrary but that only certain orientations are allowed. For the hydrogen-like silver atom, their results showed that the electron had a spin angular momentum, not an orbital angular momentum. In other words, the electron wasn't moving in an orbit.

14-3. Wave–Particle Duality

By 1920, scientists had generally accepted the notion that radiant energy behaved like waves in some experiments and like particles (i.e., matter) in other experiments (see Table 14-1). The manifestation of particlelike behavior was the quantum restriction that the only solutions of equations allowed are those that involve integers. Another area of physics in which mathematical solutions involve integers is wave theory, especially the equations for standing waves.

A. Wave nature of matter proposed

In 1924, de Broglie suggested that matter also had a dual nature and that the electron in the Bohr atom could be treated like a standing wave circling the nucleus. He proposed that the wavelength of a particle is inversely proportional to its momentum, i.e., the particle's mass multiplied by its velocity:

DE BROGLIE EQUATION
$$\lambda = \frac{h}{mv} \qquad \textbf{(14-15)}$$

For a standing wave to form a circle, there must be an integral number of wavelengths in the circumference, i.e.,

$$n\lambda = 2\pi r \qquad \textbf{(14-16)}$$

TABLE 14-1: Waves and Particles

	Light	Matter
Evidence for		
Wave behavior	Diffraction Interference Polarization	Diffraction of electrons Success of quantum mechanics
Particle behavior	Black-body radiation Photoelectric effect Compton effect	Scintillations from α and β rays Conformance to laws of classical mechanics
Distinguishing characteristics		
Wave behavior	Speed of light (c)	Speed less than c
Particle behavior	Zero rest mass	Finite rest mass

EXAMPLE 14-11: (**a**) Combine Eqs. (14-15) and (14-16). (**b**) What does the resulting expression represent and why is it significant?

Solution:

(a) Substituting λ from Eq. (14-15) into Eq. (14-16), and rearranging, we obtain

$$\frac{nh}{mv} = 2\pi r \qquad \text{or} \qquad mvr = \frac{nh}{2\pi}$$

(b) This expression is identical to Eq. (14-8), or Bohr's postulate concerning the restriction of electrons to specific orbits. This result helps to justify the arbitrary assumption made by Bohr—if an electron *does* have wave properties. But recall that the Stern–Gerlach experiment showed that the electron doesn't have orbital angular momentum.

EXAMPLE 14-12: Calculate the velocity of the electron in a hydrogen atom in the ground state.

Solution: The ground state means that $n = 1$ and $r = a_0$. Rearranging Eq. (14-8) gives

$$v = \frac{nh}{2\pi m_e r} = \frac{1(6.6262 \times 10^{-34}\ \text{J s})}{2(3.1416)(9.110 \times 10^{-31}\ \text{kg})(0.529 \times 10^{-10}\ \text{m})} = 2.19 \times 10^{6}\ \text{m s}^{-1}$$

EXAMPLE 14-13: Calculate the wavelength associated with the electron in Example 14-12.

Solution: Using de Broglie's equation, Eq. (14-15), we have

$$\lambda = \frac{h}{m_e v} = \frac{6.6262 \times 10^{-34}}{(9.110 \times 10^{-31})(2.19 \times 10^{6})} = 3.32 \times 10^{-10}\ \text{m} = 3.32\ \text{Å}$$

We could have used Eq. (14-16) to make the calculation:

$$\lambda = \frac{2\pi r}{n} = 2(3.1416)(0.529 \times 10^{-10}\ \text{m}) = 3.32 \times 10^{-10}\ \text{m}$$

B. Experimental evidence

In 1927, Davisson and Germer discovered the diffraction of electrons from the surface of a nickel crystal, providing the first experimental evidence that electrons exhibit wavelike behavior. In addition, their results confirmed Eq. (14-15).

EXAMPLE 14-14: Calculate the wavelength associated with an electron that has been accelerated by a potential of 100 V.

Solution: The wavelength is given by the de Broglie expression, Eq. (14-15), $\lambda = h/mv$, but we also need an expression for the velocity. The kinetic energy is

$$\frac{1}{2} m_e v^2 = Ve \qquad \text{and} \qquad v^2 = \frac{2Ve}{m_e}$$

so
$$\lambda = \frac{h}{m_e\left(\frac{2Ve}{m_e}\right)^{1/2}} = \frac{h}{(2m_e Ve)^{1/2}}$$

$$= \frac{6.6262 \times 10^{-34}\ \text{J s}}{[2(9.110 \times 10^{-31}\ \text{kg})(100\ \text{V})(1.602 \times 10^{-19}\ \text{C})]^{1/2}} = 1.23 \times 10^{-10}\ \text{m} = 1.23\ \text{Å}$$

C. Uncertainty principle

In 1927, Heisenberg proposed that the accuracy with which we can specify a microscopic particle's position is limited. Until then, it was assumed that measurement of the direction, momentum, and current position of a moving object allowed the exact calculation of its future position. This method works with cars, stars, and projectiles, but with electrons and other very small particles, the act of measuring disturbs the object and changes its motion. Specifically, if the position is determined very accurately, the momentum is changed and can't be determined accurately. This INDETERMINACY, now called the **uncertainty principle**, was stated by Heisenberg as

UNCERTAINTY PRINCIPLE $$\Delta x \, \Delta p_x \geq h \tag{14-17}$$

where Δx is the uncertainty in the position's x coordinate, and Δp_x is the uncertainty in the momentum's x component at the same instant. Similar expressions apply for the y and z directions and for energy and time, i.e., $\Delta E \, \Delta t \geq h$. Some authors use $h/2$, $h/2\pi$ ($\hbar$), or $h/4\pi$ ($\hbar/2$). These terms are of the same order of magnitude, and the significant point is that the accuracy with which we can describe the state of a system is limited; the more accurately we measure x, the less accurately we can measure p_x.

EXAMPLE 14-15: A beam of electrons is diffracted on passing through a slit, which imparts an angular momentum perpendicular to the beam. If the slit is narrowed, what can we say about the uncertainty in the momentum?

Solution: If we let the direction of the beam be the y direction, the slit is in the x direction (perpendicular to the beam). The slit width determines the uncertainty in x. The narrower the slit, the more accurately we know x, i.e., Δx becomes smaller. According to Eq. (14-17), $\Delta p_x \cong h/\Delta x$; thus, if Δx becomes smaller, Δp_x must become larger.

EXAMPLE 14-16: Suppose that we want to determine the position of the electron in Example 14-14 with an error no greater than 1 Å. Calculate the uncertainty in the momentum and compare it to the momentum.

Solution: If we assume minimum uncertainties, Eq. (14-17) becomes $\Delta x \, \Delta p_x = h$ and

$$\Delta p_x = \frac{h}{\Delta x} = \frac{6.6262 \times 10^{-34}\ \text{J s}}{1 \times 10^{-10}\ \text{m}} = 6.63 \times 10^{-24}\ \text{kg m s}^{-1}$$

We can calculate the momentum from Eq. (14-15), using λ from Example 14-14, so

$$p = mv = \frac{h}{\lambda} = \frac{6.6262 \times 10^{-34}\ \text{J s}}{1.23 \times 10^{-10}\ \text{m}} = 5.39 \times 10^{-24}\ \text{kg m s}^{-1}$$

We see that the uncertainty is slightly larger than the momentum itself.

14-4. The New Quantum Theory

A. Quantum mechanics

In the area of physics called **mechanics**, we try to describe the state of a system, usually by specifying particle position and momentum. Momentum involves mass and velocity, and we can use these two properties to distinguish four types of mechanics:

Type	Mass	Velocity
Newtonian (classical)	Large	Ordinary
Relativistic Newtonian	Large	Close to speed of light
Quantum (wave)	Small	Ordinary
Relativistic quantum	Small	Close to speed of light

The particles of concern in chemistry fall into the third type, and quantum mechanics provides the theoretical basis for all of chemistry.

In 1925, Heisenberg, Born, and Jordan developed a quantum mechanics based on matrix algebra, and, in 1926, Schrödinger developed **wave mechanics** based on equations similar to the wave equations of classical mechanics. Chemists generally prefer the latter approach (perhaps because the mathematics is familiar), and it is used almost universally in chemistry textbooks. The word *wave* is usually dropped, in favor of *quantum* mechanics.

B. Operator algebra and eigenvalues

An **operator** changes a function by a mathematical process into another function. Some authors designate operators by boldface type, some use script type, and some use a caret over the operator symbol. We'll use boldface type to specify quantum operators.

EXAMPLE 14-17: List some examples of operators.

Solution: The simplest case is multiplication: in ab, b is to be multiplied by the operator a. Others are $\sqrt{\ }$ (take the square root) and $\frac{d}{dx}$ (take the first derivative). An operator we'll find convenient is the **Laplacian** or **del-squared operator**:

$$\nabla^2 \equiv \frac{\partial^2}{\partial x^2} + \frac{\partial^2}{\partial y^2} + \frac{\partial^2}{\partial z^2}$$

which tells us to take the second derivatives with respect to each variable and add them.

In quantum mechanics, we encounter a special type of equation known as an **eigenvalue equation**:

$$\mathbf{O}f = af \tag{14-18}$$

where $\mathbf{O}$ is an operator, a is a constant called the **eigenvalue**, and f is the **eigenfunction**. The essential characteristic of this expression is that, when the operation is performed, we get back the function, but times a constant. Usually, there is more than one solution (eigenfunction) and more than one eigenvalue. There may be several eigenfunctions for the same eigenvalue (an **eigenstate**); such an eigenstate is called *degenerate*.

EXAMPLE 14-18: Take the first derivative with respect to x of e^{2x}. Identify the result, the eigenfunction, and the eigenvalue.

Solution: First, we write the operation and answer as an equation:

$$\frac{d}{dx}(e^{2x}) = e^{2x}\frac{d}{dx}(2x) = 2e^{2x}$$

This is an eigenequation, since the operation gives the function times a constant. The eigenfunction is e^{2x}; the eigenvalue is 2.

Since physical properties are real (not imaginary), eigenvalues must be real, which means that the operators in quantum mechanics must be of the type called *Hermitian*. An operator $\mathbf{O}$ is Hermitian if the following relationship holds:

$$\int u_m^* \mathbf{O} u_n \, d\tau = \int u_n \mathbf{O}^* u_m^* \, d\tau \tag{14-19}$$

where u_m and u_n are two eigenfunctions, the asterisk designates the *complex conjugate*, obtained by replacing each i ($\sqrt{-1}$) in a complex quantity by $-i$, and $d\tau$ is the volume element ($dx\,dy\,dz$). The equation also holds if $m = n$.

EXAMPLE 14-19: Verify that the operator $-i\hbar\frac{d}{dx}$ is Hermitian for the case $m = n$, using the eigenvalue $u_m = Ae^{ikx/\hbar}$. This is the solution to an actual eigenequation for a one-dimensional case. (A is an arbitrary constant.)

Solution: Let's begin by writing the complex conjugates we need. Replacing i by $-i$ gives

$$\mathbf{O}^* = \left(-i\hbar \frac{d}{dx}\right)^* = i\hbar \frac{d}{dx} \qquad \text{and} \qquad u_m^* = Ae^{-ikx/\hbar}$$

Substituting in Eq. (14-19), we get

$$\int Ae^{-ikx/\hbar}(-i\hbar)\frac{d}{dx}Ae^{ikx/\hbar}\,dx = \int Ae^{ikx/\hbar}(i\hbar)\frac{d}{dx}Ae^{-ikx/\hbar}\,dx$$

In general, $\frac{d}{dx}(ke^u) = ke^u \frac{du}{dx}$. Then

$$\frac{d}{dx}Ae^{ikx/\hbar} = Ae^{ikx/\hbar}\left(\frac{ik}{\hbar}\right) \qquad \text{and} \qquad \frac{d}{dx}Ae^{-ikx/\hbar} = Ae^{-ikx/\hbar}\left(\frac{-ik}{\hbar}\right)$$

Now, the main equation becomes

$$\int Ae^{-ikx/\hbar}(-i\hbar)(ik/\hbar)Ae^{ikx/\hbar}\,dx = \int Ae^{ikx/\hbar}(i\hbar)(-ik/\hbar)Ae^{-ikx/\hbar}\,dx$$

After canceling $\hbar$, we see that the two integrals are identical. Thus the operator is Hermitian.

C. The general Schrödinger wave equation

Schrödinger combined de Broglie's matter waves, the classical second-order differential equation for wave motion, and other classical concepts into an equation for the total energy of a system. This equation includes the potential energy and a function designated Ψ, which is related to the amplitude of the wave. The total energy of a system is the sum of the kinetic energy and the potential energy. In classical mechanics these energies are

$$E = E_k + E_p = \frac{1}{2}mv^2 + V(x, y, z, t) \qquad \textbf{(14-20)}$$

It is more convenient to use momentum ($p = mv$) than velocity, so we substitute $v^2 = p^2/m^2$ into Eq. (14-20) and get

$$E = \frac{1}{2m}p^2 + V \qquad \textbf{(14-21)}$$

Equation (14-21) is the classical expression for total energy.

To obtain a quantum equation from a classical equation, we replace the variables with operators. For Eq. (14-21), we use

$$\text{for } E, \qquad -\frac{h}{2\pi i}\frac{\partial}{\partial t} = \mathbf{E}$$

$$\text{for } p_x, \qquad \frac{h}{2\pi i}\frac{\partial}{\partial x} = \mathbf{p}_x \qquad (p_y \text{ and } p_z \text{ are similar})$$

Now, Eq. (14-21) becomes

$$-\frac{h}{2\pi i}\frac{\partial}{\partial t} = \frac{1}{2m}\left[\left(\frac{h}{2\pi i}\right)^2\frac{\partial^2}{\partial x^2} + \left(\frac{h}{2\pi i}\right)^2\frac{\partial^2}{\partial y^2} + \left(\frac{h}{2\pi i}\right)^2\frac{\partial^2}{\partial z^2}\right] + V$$

Simplifying, we obtain

$$-\frac{h}{2\pi i}\frac{\partial}{\partial t} = -\frac{h^2}{8\pi^2 m}\left(\frac{\partial^2}{\partial x^2} + \frac{\partial^2}{\partial y^2} + \frac{\partial^2}{\partial z^2}\right) + V \qquad \textbf{(14-22)}$$

Equation (14-22) is the quantum expression for total energy. The right-hand side of the equation is called the **Hamiltonian operator**. We can simplify it by recognizing the quantity in parentheses as the Laplacian operator (∇^2) and substituting to obtain

$$\mathbf{H} = -\frac{h^2}{8\pi^2 m}\nabla^2 + V$$

Introducing Ψ as the function to be operated on (the *operand*), we have

SCHRÖDINGER WAVE EQUATION

$$\mathbf{H}\Psi = -\frac{h}{2\pi i}\frac{\partial}{\partial t}\Psi \tag{14-23}$$

which is time-dependent. The function Ψ is called the *wave function.*

D. The postulates of quantum mechanics

The Schrödinger equation can't be derived in the usual sense. The customary approach to the wave version of quantum mechanics is to use Eq. (14-23) as a starting point, stating a few postulates and building on them. Although the approach is arbitrary, the results are invariably consistent with experimental evidence. Some texts list five or six postulates, some less fundamental than the others; other texts list three. Let's choose the following three:

Postulate I: For a given assembly of particles, a function, $\Psi(x, y, z, t)$, called the wave function, defines the state of the system. (Thus it sometimes is called a state function.)
Postulate II: The wave function is determined by Eq. (14-23):

$$\mathbf{H}\Psi = -\frac{h}{2\pi i}\frac{\partial}{\partial t}\Psi$$

Postulate III: The *average value*, $\bar{A}$, of any measurable property A of the system is given by

$$\bar{A} = \frac{\int \Psi^*\mathbf{A}\Psi \, d\tau}{\int \Psi^*\Psi \, d\tau} \tag{14-24}$$

where Ψ^* is the complex conjugate of Ψ. Usually, $\Psi^*\Psi$ is normalized (which we describe later), so the denominator equals 1, and Eq. (14-24) becomes

$$\langle A \rangle = \int \Psi^*\mathbf{A}\Psi \, d\tau$$

where the symbol $\langle \ \rangle$ is used to designate the average value, or **expectation value**, which is the value we expect to obtain when we measure the property.

E. Properties of wave functions (Postulate I)

1. A satisfactory wave function must be *well-behaved* mathematically.

A satisfactory wave function must be single-valued, finite, and continuous; its first and second derivatives must also be continuous. Its square must be integrable, meaning that an integral such as $\int \Psi^2 \, d\tau$ is finite.

2. In general, a wave function is *complex.*

A wave function has a *complex conjugate* Ψ^*. The product $\Psi^*\Psi$ has no imaginary parts and is equivalent to the square of the absolute value of Ψ, i.e., $\Psi^*\Psi = |\Psi|^2$. This product also is a **probability density**. Multiplied by $dx\,dy\,dz$ (a volume element that equals $d\tau$), the product gives the probability that the system will be in $d\tau$ at (x, y, z) at time t. Because of this property, Ψ has been called a **probability amplitude** (analogous to the amplitude of a wave).

3. The sum of the probabilities must equal 1.

Over the entire coordinate space, the sum of the probabilities is represented by

$$\int \Psi^*\Psi \, d\tau = 1 \tag{14-25}$$

Equation (14-25) is the expression for the **normalization condition.**

EXAMPLE 14-20: For a certain wave function u, $\int u^*u \, d\tau = A$. How can we accommodate the requirement of Eq. (14-25)?

Solution: We have to multiply the wave function by a normalizing factor that will yield a value of 1 for the integral. We begin by dividing the equation by A,

$$\int \frac{u^*u}{A} d\tau = 1$$

If we write A as $\sqrt{A}\sqrt{A}$, then

$$\int \left(\frac{u}{\sqrt{A}}\right)^* \left(\frac{u}{\sqrt{A}}\right) d\tau = 1$$

The new $u/\sqrt{A}$ is a *normalized function*, and $1/\sqrt{A}$ is the *normalizing factor*.

---

4. **Wave functions can be *orthogonal*.**

   If two wave functions, Ψ_m and Ψ_n, are eigenfunctions of the same operator and have different eigenvalues, they are said to be **orthogonal** to each other. This condition is expressed by

$$\int \Psi_m^* \Psi_n \, d\tau = 0 \qquad \textbf{(14-26)}$$

   We can combine Eq. (14-26) and the normalization condition, Eq. (14-25), as

$$\int \Psi_m^* \Psi_n \, d\tau = \delta_{mn} \qquad \textbf{(14-27)}$$

   where δ_{mn} is called the *Kronecker delta*. When $m = n$, $\delta_{mn} = 1$; when $m \neq n$, $\delta_{mn} = 0$. Wave functions that satisfy Eq. (14-27) are called **orthonormal**.

5. **A wave function may be divided.**

   A wave function may be divided into two functions:

$$\Psi(x, y, z, t) = \psi(x, y, z)e^{-2\pi iEt/h} \qquad \textbf{(14-28)}$$

   Then the normalization condition, Eq. (14-25), becomes

$$\int \psi^*\psi \, d\tau = 1 \qquad \textbf{(14-29)}$$

EXAMPLE 14-21: Show why the exponential term disappears between Eqs. (14-25) and (14-29).

Solution: When Ψ from Eq. (14-28) is substituted into Eq. (14-25), we need its complex conjugate also, so we replace i with $-i$ in the exponential term. Then the product of the two exponential terms is

$$e^{+2\pi iEt/h}(e^{-2\pi iEt/h}) = e^0 = 1$$

We are left with ψ and ψ^* only. This new time-independent wave function, $\psi(x, y, z)$, has all the properties described in this section for Ψ.

F. The time-independent Schrödinger equation (Postulate II)

Most properties of interest in chemistry don't depend on time; such situations are called **stationary states**. As a result, we can substitute Eq. (14-28) into Eq. (14-23) to obtain

$$\mathbf{H}\psi = E\psi \qquad \textbf{(14-30)}$$

EXAMPLE 14-22: Derive Eq. (14-30) using Eqs. (14-23) and (14-28).

Solution: Let's begin by copying Eq. (14-23):

$$\mathbf{H}\Psi = -\frac{h}{2\pi i}\frac{\partial}{\partial t}\Psi$$

We'll carry out the differentiation on the right-hand side, substituting Eq. (14-28) for Ψ:

$$\frac{\partial}{\partial t}\Psi = \frac{\partial}{\partial t}(\psi e^{-2\pi iEt/h})$$

Using $d(uv) = u\,dv + v\,du$, we get

$$\frac{\partial}{\partial t}\Psi = \psi\frac{\partial}{\partial t}e^{-2\pi iEt/h} + e^{-2\pi iEt/h}\frac{\partial}{\partial t}\psi$$

Since ψ is independent of t, the second term equals 0, and we have

$$\frac{\partial}{\partial t}\Psi = \psi e^{-2\pi iEt/h}\left(-\frac{2\pi iE}{h}\right)$$

Returning to Eq. (14-23), we substitute this result on the right and Eq. (14-28) on the left:

$$\mathbf{H}(\psi e^{-2\pi iEt/h}) = -\frac{h}{2\pi i}(\psi e^{-2\pi iEt/h})\left(-\frac{2\pi iE}{h}\right)$$

Canceling like terms, we have

$$\mathbf{H}\psi = E\psi$$

which is Eq. (14-30), the time-independent Schrödinger equation. Although Eqs. (14-23) and (14-30) are similar, in the latter we have replaced the imaginary aspect of the terms and the differential with respect to time by the eigenvalue. Also, ψ does not depend on time.

EXAMPLE 14-23: Show that the function $\psi = Ae^{i(8\pi^2 mE)^{1/2}x/h}$ is a solution of the time-independent Schrödinger equation if the potential energy is zero for a free particle moving in one dimension.

Solution: In order to use Eq. (14-30), we substitute for $\mathbf{H}$ its definition but have to reduce ∇^2 to $\frac{\partial^2}{\partial x^2}$ and drop V because it's given as zero. Then we have

$$-\frac{h^2}{8\pi^2 m}\frac{\partial^2}{\partial x^2}\psi = E\psi$$

We can simplify the equations by representing the constants in the exponent of ψ by b, i.e., $b = i(8\pi^2 mE)^{1/2}/h$. Rearranging the equation and inserting ψ, we get

$$\frac{\partial^2}{\partial x^2}(Ae^{bx}) = -\frac{8\pi^2 m}{h^2}EAe^{bx}$$

Taking the first derivative,

$$\frac{\partial}{\partial x}(Ae^{bx}) = Ae^{bx}\frac{d}{dx}(bx) = Abe^{bx}$$

then the second derivative,

$$\frac{\partial}{\partial x}(Abe^{bx}) = Abe^{bx}\frac{d}{dx}(bx) = Ab^2e^{bx}$$

gives us the left-hand side of our equation. Substituting, we get

$$Ab^2e^{bx} = -\frac{8\pi^2 m}{h^2}EAe^{bx}$$

Canceling Ae^{bx} on each side and inserting the square of b gives

$$-\frac{8\pi^2 m}{h^2}\mathrm{E} = -\frac{8\pi^2 m}{h^2}\mathrm{E}$$

Thus the wave function is a solution.

G. Real properties (Postulate III)

Postulate III is particularly significant. It ties the mathematical structure built on the other postulates to actual observation and experiment, permitting calculation of the expected value of an observable property. At the same time, it sets a limit: the average value of the property is all that we can expect to predict. The value of an observable property of a system, obtained as an average, or expectation, value from Eq. (14-24) or as an eigenvalue from Eq. (14-30), must be real, i.e., must not have imaginary components. This condition is guaranteed by requiring that all quantum-mechanical operators be Hermitian, as defined by Eq. (14-19).

EXAMPLE 14-24: Show that the expectation value (eigenvalue) obtained for a wave function (eigenfunction) ψ, using Eq. (14-24) is real if the operator **A** is Hermitian.

Solution: The Hermitian condition is Eq. (14-19). In this case, $m = n$, so we have

$$\int \psi^* \mathbf{A} \psi \, d\tau = \int \psi \mathbf{A}^* \psi^* \, d\tau \qquad \textbf{(a)}$$

Now we write the normalized form of Eq. (14-24) for both the expectation value and its complex conjugate:

$$\langle A \rangle = \int \psi^* \mathbf{A} \psi \, d\tau \qquad \textbf{(b)}$$

$$\langle A \rangle^* = \int \psi \mathbf{A}^* \psi^* \, d\tau \qquad \textbf{(c)}$$

From eq. (**a**), we can see that the integrals of eqs. (**b**) and (**c**) are equal; therefore $\langle A \rangle = \langle A \rangle^*$, and for this to be true, $\langle A \rangle$ must be real.

If ψ is an eigenfunction of A, we know from Eq. (14-18) that $\mathbf{A}\psi = A\psi$. Then we can substitute $A\psi$ for $\mathbf{A}\Psi$ in Eq. (14-24) to obtain

$$\bar{A} = \langle A \rangle = \frac{\int \psi^* A \psi \, d\tau}{\int \psi^* \psi \, d\tau}$$

or, when normalized,

$$\bar{A} = \langle A \rangle = \int \psi^* A \psi \, d\tau \qquad \textbf{(14-31)}$$

EXAMPLE 14-25: The normalized wave function for the lowest energy state of a harmonic oscillator is $(2b/\pi)^{1/4} e^{-bx^2}$. Obtain an expression for the expectation value of x^2.

Solution: Since the wave function is normalized, we can use Eq. (14-31), with $A = x^2$. Since there is no i, $\psi^* = \psi$, and

$$\overline{x^2} = \langle x^2 \rangle = \int_{-\infty}^{\infty} \left(\frac{2b}{\pi}\right)^{1/4} (e^{-bx^2})(x^2)\left(\frac{2b}{\pi}\right)^{1/4} (e^{-bx^2}) \, dx$$

$$= \left(\frac{2b}{\pi}\right)^{1/2} \int_{-\infty}^{\infty} x^2 e^{-2bx^2} \, dx$$

Using a table of definite integrals, we find that

$$\int_0^{\infty} x^{2n} e^{-ax^2} \, dx = \frac{1(3)(5)\cdots(2n-1)}{2^{n+1} a^n} \sqrt{\frac{\pi}{a}}$$

which is one-half of our integral, when $n = 1$ and $a = 2b$. Then

$$\int_0^\infty x^2 e^{-2bx^2}\,dx = \frac{1}{2^2(2b)}\left(\frac{\pi}{2b}\right)^{1/2}$$

and

$$\overline{x^2} = 2\left[\left(\frac{2b}{\pi}\right)^{1/2}\left(\frac{1}{8b}\right)\left(\frac{\pi}{2b}\right)^{1/2}\right] = \frac{1}{4b}$$

14-5. Applications to Simple Systems

A. Free particle

We used one eigenfunction of the free particle moving in one dimension in Example 14-23, i.e.,

$$\psi_1 = Ae^{i(8\pi^2 mE)^{1/2}x/h}$$

Another wave function that's a solution is

$$\psi_2 = Be^{-i(8\pi^2 mE)^{1/2}x/h}$$

For a free particle, the potential energy V is zero, and E may have any value because it isn't quantized. The particle has the same probability of being at any point on its path.

EXAMPLE 14-26: Prove that the probability of finding a free particle at any point on its path is constant.

Solution: The probability density is $\psi^*\psi = |\psi|^2$. We use the expression for ψ_1 and represent the constant terms in the exponential by b:

$$\psi_1^* = A^*e^{-bx}$$

Then

$$\psi_1^*\psi_1 = A^*e^{-bx}Ae^{bx} = A^*A = |A|^2$$

We obtain the same result by using the expression for ψ_2, i.e., $|B|^2$. Since A and B are constants (independent of x), the probability is equal at every x. However, we can say nothing about the position of the particle.

B. Particle in a one-dimensional box

We now restrict the free particle by requiring that it move between $x = 0$ and $x = a$. The potential energy is still zero within these limits but immediately becomes infinite beyond them. In other words, we are stipulating a box with infinitely high walls. Furthermore, any wave function must equal zero at $x = 0$, at $x = a$, and anywhere outside the box. Since $V = 0$, ψ_1 and ψ_2 are in the correct form, but neither of them can meet the boundary conditions. In order for $\psi = 0$ at $x = a$, A and B would have to be 0; this would mean that $\psi = 0$ at all x.

aside: Two eigenfunctions that are solutions of the same eigenvalue equation can be combined to give a new eigenfunction, which will also be a solution to the equation. Each eigenfunction is "weighted" by a constant in what is called a *linear combination*:

$$\psi_3 = c_1\psi_1 + c_2\psi_2 \qquad \text{and} \qquad \psi_4 = c_1\psi_1 - c_2\psi_2$$

This method is frequently used to obtain new descriptions of a system.

A linear combination of ψ_1 and ψ_2 avoids the problem just described. We can omit the constants, because we can change A and B to accommodate them, and write

$$\psi_3 = Ae^{i(8\pi^2 mE)^{1/2}x/h} + Be^{-i(8\pi^2 mE)^{1/2}x/h}$$

From the restrictions imposed by the boundary conditions, it can be shown that

$$E_n = \frac{n^2h^2}{8ma^2} \qquad \textbf{(14-32)}$$

In the process, the wave function is reduced to a sine function which, when normalized, is

$$\psi_n = \left(\frac{2}{a}\right)^{1/2} \sin\frac{n\pi x}{a} \tag{14-33}$$

The quantum number n appeared because the sine function equals 0 only for integral multiples of π. Note that n can't equal zero (the particle would vanish!) but can be any integer: $n = 1, 2, 3, \ldots \infty$. The energy may have only certain values, depending on n^2, because it is quantized. The permitted energies usually are called **energy levels**. These values of E inserted in ψ_3 give acceptable solutions to the Schrödinger equation.

EXAMPLE 14-27: The diameter of a hydrogen atom is about 1 Å. (**a**) Calculate the energy of an electron in a one-dimensional box of that width for $n = 1$ and $n = 2$. (**b**) If the electron jumps from the second level to the first, emitting one photon, determine the wavelength of the emitted radiation.

Solution:

(**a**) The energies for a particle in a one-dimensional box are given in Eq. (14-32). When $a = 10^{-10}$ m (or 1 Å),

$$E_n = \frac{n^2(6.6262 \times 10^{-34}\ \text{J s})^2}{8(9.1095 \times 10^{-31}\ \text{kg})(10^{-10}\ \text{m})^2} = (6.025 \times 10^{-18}\ \text{J})(n^2)$$

For $n = 1$, $\quad E_1 = 6.025 \times 10^{-18}$ J

For $n = 2$, $\quad E_2 = (6.025 \times 10^{-18}\ \text{J})(2)^2 = 24.10 \times 10^{-18}$ J

(**b**) This calculation is just like those we did for the Bohr atom in order to predict spectral wavelengths (see Example 14-10). For the frequency, from Eq. (14-9),

$$\nu = \frac{1}{h}|E_f - E_i| = \frac{c}{\lambda}$$

Then $\lambda = hc\,|E_f - E_i|$

$$E_f - E_i = E_1 - E_2 = (6.03 - 24.10) \times 10^{-18}\ \text{J} = -18.07 \times 10^{-18}\ \text{J}$$
$$|E_f - E_i| = 1.8 \times 10^{-17}\ \text{J}$$

and

$$\lambda = \frac{(6.6262 \times 10^{-34}\ \text{J s})(2.9979 \times 10^{8}\ \text{m s}^{-1})}{1.8 \times 10^{-17}\ \text{J}} = 11.07 \times 10^{-9}\ \text{m} = 10.99\ \text{nm} = 1.099\ \text{Å}$$

Note that this wavelength is in the x-ray region of radiation; it's too short to match the shortest line of the hydrogen spectrum. The particle-in-the-box model isn't a good one for hydrogen, but it seems to be a step in the right direction.

C. Particle in a three-dimensional box

The box has sides a, b, and c. Again, the potential energy is zero inside the box and infinite outside. The Schrödinger equation must reflect the three dimensions, so we write Eq. (14-30) in the form

$$-\frac{h^2}{8\pi^2 m}\nabla^2\psi = E\psi$$

or

$$\frac{\partial^2\psi}{\partial x^2} + \frac{\partial^2\psi}{\partial y^2} + \frac{\partial^2\psi}{\partial z^2} = -\frac{8\pi^2 m}{h^2}E\psi \tag{14-34}$$

We can solve this type of linear differential equation by letting ψ be the product of three wave functions, each dependent on one variable, i.e.,

$$\psi(x, y, z) = \psi_x(x)\psi_y(y)\psi_z(z)$$

We find that $E = E_x + E_y + E_z$. The solution breaks up into three parts, one for each coordinate, and the solution of each proceeds similarly to that for the particle in a one-dimensional box. The three wave functions have the same form as Eq. (14-33); for x, it's

$$\psi_x(x) = \left(\frac{2}{a}\right)^{1/2} \sin \frac{n_x \pi x}{a} \qquad \textbf{(14-35)}$$

There are three quantum numbers, n_x, n_y, n_z, and the three values of E have the same form as Eq. (14-32); for E_x, it's

$$E_x = \frac{n_x^2 h^2}{8ma^2}$$

The total E is

$$E = \frac{h^2}{8m}\left(\frac{n_x^2}{a^2} + \frac{n_y^2}{b^2} + \frac{n_z^2}{c^2}\right)$$

D. Harmonic oscillator

Some textbooks introduce the analysis of a harmonic oscillator early, as an illustration of quantization, and others present it later, as an extension of the particle-in-a-box problem. The importance of the harmonic oscillator is that it serves as a model for molecular vibrations. It's a very good approximation for diatomic molecules in their lower energy states. The potential energy doesn't jump instantly to infinity at a point (as with the particle in the box) but approaches infinity gradually, following a parabola. For a one-dimensional oscillator, the simplest potential energy function is $V = \frac{1}{2}kx^2$, where k is the force constant from Hooke's law, Eq. (3-10). We now use x instead of the r in Eq. (3-10) to represent the displacement from the origin or equilibrium point. The classical total energy is expressed by Eq. (14-20) and is constant:

$$E = E_k + E_p = \frac{p^2}{2m} + \frac{kx^2}{2} \qquad \textbf{(14-36)}$$

For the quantum-mechanical treatment, we replace p with the momentum operator **p** to get the Hamiltonian operator **H**, which in turn gives the Schrödinger equation. In one dimension, Eq. (14-36) becomes

$$-\frac{h^2}{8\pi^2 m} + \frac{\partial^2\psi}{\partial x^2} + \frac{kx^2}{2} = E\psi \qquad \textbf{(14-37)}$$

When dealing with two objects, such as the two atoms of a diatomic molecule, we must replace the mass with the reduced mass, or

$$\mu = \frac{m_1 m_2}{m_1 + m_2}$$

We can obtain proper solutions to Eq. (14-37) only when

$$E = (v + \tfrac{1}{2})hv \qquad \textbf{(14-38)}$$

where v is the **vibrational quantum number**.

For the harmonic oscillator, we have a different situation than for the particle in the box in two respects: (1) the energy levels are equally spaced; and (2) the lowest energy level isn't zero. When $v = 0$, $E_0 = \frac{1}{2}hv_0$, which is called the *zero-point energy*. This means that some vibration is occurring even at $T = 0$ K; i.e., $v = 0$ doesn't imply the absence of vibration.

EXAMPLE 14-28: (**a**) What principle tells us that we shouldn't expect a state of no vibration in a harmonic oscillator? (**b**) Develop the reasoning.

Solution:

(**a**) The uncertainty principle.

(**b**) If the vibrating oscillator or molecule were to come to a complete stop, the position and the momentum would become known precisely, which is forbidden by the principle.

The maximum extension of a *classical* oscillator is set by clear limits: x must fall between $\pm L$. The wave functions, however, approach zero exponentially at $x = \pm\infty$, so there's a definite probability that the oscillator may be at a large x (larger than L). We say that the system has *tunneled* through a barrier and call the phenomenon the **tunnel effect**. The effect is more likely to occur as the mass becomes smaller, so it is particularly significant for subatomic particles. For example, the energy of an α particle in the nucleus is less than the potential barrier, yet α-particle emission does occur.

E. Rigid rotor

This model is usually applied to diatomic (dumbbell-like) molecules, but may be extended to any linear molecule such as acetylene. There can be no movement along the interparticle axis. The free rotation means that the potential energy is zero; in other words, all the energy is kinetic:

$$E = E_k = \tfrac{1}{2}I\omega^2 \qquad \textbf{(14-39)}$$

where I is the moment of inertia and ω is the angular velocity. The moment of inertia is the reduced mass times the square of the interparticle distance, or $I = \mu r_0^2$. The angular momentum is the moment of inertia times the angular velocity, or $L = I\omega$. Using these definitions with Eq. (14-39), we get

$$E = \frac{L^2}{2I} \qquad \textbf{(14-40)}$$

The math involves a set of internal coordinates and the coordinates of the center of mass. We ignore the translational motion of the center of mass and the associated energy and transform the Cartesian coordinates into spherical coordinates (r, Θ, Φ). Since r is constant, only the two angles are variables in the Hamiltonian operator. There are two coordinates, so two quantum numbers are required. One appears in the expression for E:

$$E_J = J(J+1)\left(\frac{h^2}{8\pi^2 I}\right) \qquad J = 0, 1, 2, \ldots \qquad \textbf{(14-41)}$$

We see that the energy of the rigid rotor is quantized and may equal 0. The zero value is possible because neither coordinate is specified. The other quantum number appears as a constant in solutions to the Schrödinger equation. It must be an integer at zero and can have all values up to J:

$$M = 0, \pm 1, \pm 2, \ldots, \pm J \qquad \textbf{(14-42)}$$

EXAMPLE 14-29: **(a)** What are the values of M for a rigid rotor when $J = 1$? **(b)** What is this situation called?

Solution:

(a) According to Eq. (14-42), M can equal 0, +1, and −1 when $J = 1$.

(b) The three values of M mean that there are three wave functions for the energy E_1. This is the *degeneracy* situation referred to in Section 14-4B. Stated differently, the energy level for $J = 1$ is three-fold degenerate. In general, the degeneracy is $2J + 1$.

SUMMARY

In Table S-14, we show the major equations of this chapter and the connections between them. You should determine the conditions that are necessary for the validity of each equation and the conditions imposed to make the connections. Other important things you should know are

1. Failure of the accepted physics of his time to explain the frequency distribution of black-body radiation led to Planck's proposal of quantization of oscillators [Eq. (14-6)].
2. Einstein assumed that radiant energy is quantized and explained the photoelectric effect [Eq. (14-7)].

TABLE S-14: Summary of Important Equations

$\tilde{\nu} = \frac{1}{\lambda} = R_H\left(\frac{1}{n_1^2} - \frac{1}{n_2^2}\right)$ (14-1)

$\lambda = \frac{1}{R_H}\left(\frac{n^2}{n^2 - 4}\right)$ (14-2)

$\lambda\,(\text{max emission}) = kT^{-1}$ (14-3)

$U = aT^4$ (14-4)

$M_e = \sigma T^4$ (14-5)

$\varepsilon = h\nu$ (14-6)

→ $\varepsilon = W + \frac{1}{2}mv^2$ (14-7)

→ $\nu = \frac{1}{h}|E_f - E_i|$ (14-9)

$mvr = n\frac{h}{2\pi}$ (14-8)

→ $r = \frac{4\pi\varepsilon_0 n^2 \hbar^2}{me^2}$ (14-10)

→ $a_0 = \frac{4\pi\varepsilon_0 \hbar^2}{\mu e^2}$ (14-12)

$E = \frac{1}{4\pi\varepsilon_0}\left(-\frac{e^2}{2r}\right)$ (14-13)

→ $E_n = -\frac{e^2}{4\pi\varepsilon_0(2a_0)}\left(\frac{1}{n^2}\right)$ (14-14)

$\lambda = h/mv$ (14-15)

→ $n\lambda = 2\pi r$ (14-16)

$\Delta x \Delta p_x \geq h$ (14-17; uncertainty)

$\mathbf{O}f = af$ (14-18)

$\int u_m^* \mathbf{O} u_n \, d\tau = \int u_n \mathbf{O}^* u_m^* \, d\tau$ (14-19; def. of Hermitian operator)

$E = E_k + E_p = \frac{1}{2}mv^2 + V(x, y, z, t)$ (14-20)

→ $E = \frac{p^2}{2m} + V$ (14-21)

→ $-\frac{h}{2\pi i}\frac{\partial}{\partial t} = -\frac{h^2}{8\pi^2 m}\left(\frac{\partial^2}{\partial x^2} + \frac{\partial^2}{\partial y^2} + \frac{\partial^2}{\partial z^2}\right) + V$ (14-22)

→ $\mathbf{H}\Psi = -\frac{h}{2\pi i}\frac{\partial}{\partial t}\Psi$ (14-23)

→ $\mathbf{H}\psi = E\psi$ (14-30)

→ $\bar{A} = \psi^* A \psi \, d\tau$ (14-31)

$\bar{A} = \frac{\int \Psi^* \mathbf{A} \Psi \, d\tau}{\int \Psi^* \Psi \, d\tau}$ (14-24)

$\int \Psi_m^* \Psi_n \, d\tau = 0$ (14-26; orthogonality)

→ $\int \Psi_m^* \Psi_n \, d\tau = \delta_{mn}$ (14-27; orthonormality)

$\int \Psi^* \Psi \, d\tau = 1$ (14-25; normalization)

→ $\int \psi^* \psi \, d\tau = 1$ (14-29; normalization)

$\Psi(x, y, z, t) = \psi(x, y, z)e^{-2\pi i E t/h}$ (14-28)

Applications

One-dimensional box

$E_n = \frac{n^2 h^2}{8ma^2}$ (14-32)

$\psi_n = \left(\frac{2}{a}\right)^{1/2} \sin\frac{n\pi x}{a}$ (14-33)

Three-dimensional box

$\frac{\partial^2\psi}{\partial x^2} + \frac{\partial^2\psi}{\partial y^2} + \frac{\partial^2\psi}{\partial z^2} = -\frac{8\pi^2 m}{h^2}E\psi$ (14-34)

$\psi_x(x) = \left(\frac{2}{a}\right)^{1/2} \sin\frac{n_x \pi x}{a}$ (14-35)

Harmonic oscillator

$E = \frac{p^2}{2m} + \frac{kx^2}{2}$ (14-36; from 14-20)

$E = (v + \frac{1}{2})h\nu$ (14-38)

→ $-\frac{h^2}{8\pi^2 m}\frac{\partial^2\psi}{\partial x^2} + \frac{kx^2}{2} = E\psi$ (14-37)

Rigid rotor

$E_{\text{total}} = E_k = \frac{1}{2}I\omega^2$ (14-39)

→ $E = \frac{L^2}{2I}$ (14-40)

$E_J = J(J + 1)\frac{h^2}{8\pi^2 I}; \; J = 0, 1, 2, \ldots$ (14-41)

3. Bohr explained atomic spectra and the stability of the Rutherford atom by using the quantum concept [Eq. (8-14)].
4. De Broglie suggested that particles could behave like waves [Eq. (14-15)].
5. The uncertainty principle proposed by Heisenberg places a limit on the accuracy with which the state of a particle can be specified [Eq. (14-17)].
6. The quantum mechanics usually used by chemists is based on the Schrödinger wave equation [Eq. (14-22)].
7. In an eigenequation, the operation gives back the function times a constant [Eq. (14-18)].
8. All quantum-mechanical operators must be Hermitian [Eq. (14-19)].
9. The basic postulates of quantum mechanics are:

   - A wave function defines the state of a system [Eqs. (14-33), (14-35), and (14-37)].
   - The Schrödinger equation determines the wave function.
   - The average value of a property can be calculated from the wave function and the operator for that property [Eq. (14-31)].

10. Some significant properties of the wave function are:

    - It must be single-valued, finite, and well-behaved mathematically.
    - The absolute value squared is the probability density.
    - It must be normalized [Eq. (14-29)].
    - It may be separated into a function of time and a function independent of time [Eq. (11-22)].

11. Wave functions of a system must be orthogonal. [Eq. (14-26)].
12. For the usual physical systems, the time-independent Schrödinger equation is used [Eq. (14-30)].
13. Hamiltonian operators are obtained by adding a potential energy function to the momentum operator [Eq. (14-20)].
14. The energy of a free particle is not quantized.
15. The energy of a particle confined within boundaries is quantized. For a box with infinite walls, the energy levels depend on the square of the integer n [Eq. (14-32)].
16. The treatment of a particle in a three-dimensional box is analogous to that in one dimension but with three quantum numbers. [Eqs. (14-33) and (14-35)].
17. The simple harmonic oscillator is like a particle in a parabolic well. Its energy levels are quantized and, for a given frequency, depend on the first power of an integer, v, plus one-half [Eqs. (14-36), (14-37), and (14-38)].
18. The linear rigid rotor involves two quantum numbers: $J = 0, 1, 2, \ldots,$ and $M = 0, \pm 1, \pm 2, \ldots, \pm J$ [Eqs. (14-39) and (14-40)].

SOLVED PROBLEMS

Spectra, Wavelength, and Black-Body Radiation

PROBLEM 14-1 For the Paschen series of lines in the hydrogen spectrum, $n_1 = 3$. **(a)** Calculate the wavelength of the line for which $n_2 = 5$. **(b)** In which region of the electromagnetic spectrum does it fall?

Solution:

(a) The Ritz–Rydberg expression, Eq. (14-1), relates the wavelength to the integers, n, so

$$\frac{1}{\lambda} = R_H\left(\frac{1}{n_1^2} - \frac{1}{n_2^2}\right) = (1.097 \times 10^7\ \text{m}^{-1})\left(\frac{1}{3^2} - \frac{1}{5^2}\right) = (1.097 \times 10^7)\left(\frac{1}{9} - \frac{1}{25}\right)\text{m}^{-1}$$
$$= 7.801 \times 10^5\ \text{m}^{-1}$$
$$\lambda = 1.282 \times 10^{-6}\ \text{m} = 12\,820\ \text{Å}$$

(b) This large a wavelength is in the infrared region.

PROBLEM 14-2 In each spectral series, the lines begin to crowd together, approaching a limit. Calculate the wavelength of that limit for the Balmer series.

Solution: The limit is the point at which $n = \infty$. Then the Ritz–Rydberg equation becomes

$$\frac{1}{\lambda} = R_H\left(\frac{1}{2^2} - 0\right) = (1.097 \times 10^7\ \mathrm{m^{-1}})\left(\frac{1}{4}\right) = 0.2742 \times 10^7\ \mathrm{m^{-1}}$$

$$\lambda = 3.647 \times 10^{-7}\ \mathrm{m} = 364.7\ \mathrm{nm} = 3647\ \text{Å}$$

PROBLEM 14-3 A student repeated Thomson's experiment and obtained $1.69 \times 10^8\ \mathrm{C\,g^{-1}}$ for e/m of an electron. Then she repeated the Millikan experiment and found e to be 1.55×10^{-19} C. Calculate **(a)** the experimental mass of the electron and **(b)** the percent error from the accepted value.

Solution:

(a) The calculation is straightforward; if you have e/m and want m, you must take the reciprocal and multiply by e:

$$\frac{1}{e/m}(e) = \frac{1}{1.69 \times 10^8\ \mathrm{C\,g^{-1}}}(1.55 \times 10^{-19}\ \mathrm{C}) = 9.17 \times 10^{-28}\ \mathrm{g} = 9.17 \times 10^{-31}\ \mathrm{kg}$$

(b) The accepted value is 9.1095×10^{-31} kg, so

$$\%\ \text{error} = \frac{9.17 - 9.11}{9.11}(100) = \frac{0.06}{9.11}(100) = 0.66\%$$

PROBLEM 14-4 Rewrite the Wien displacement law in terms of frequency and calculate the new constant.

Solution: Since frequency is c/λ, $\lambda = c/\nu$, and Eq. (14-3) becomes

$$\frac{c}{\nu_{max}} = k\left(\frac{1}{T}\right) \qquad \text{or} \qquad \nu_{max} = \frac{cT}{k} = k'T$$

The value of k' is

$$k' = \frac{c}{k} = \frac{2.998 \times 10^8\ \mathrm{m\,s^{-1}}}{2.898 \times 10^{-3}\ \mathrm{m\,K}} = 1.305 \times 10^{11}\ \mathrm{s^{-1}K^{-1}}$$

PROBLEM 14-5 Calculate the wavelength at which the radiation from a black body at 1000 K has the maximum intensity.

Solution: The relationship between temperature and the maximum in the wavelength-distribution curve is the Wien displacement law, Eq. (14-3), so

$$\lambda_{max} = k\left(\frac{1}{T}\right) = (2.898 \times 10^{-3}\ \mathrm{m\,K})\left(\frac{1}{1000}\right) = 2.898 \times 10^{-6}\ \mathrm{m} = 28\,980\ \text{Å}$$

PROBLEM 14-6 Calculate the excitance at 1000 K of a black body having a surface of 10 cm^2.

Solution: The excitance is given by Eq. (14-5):

$$M_e = \sigma T^4$$

When $T = 1000$ K,

$$M_e = (5.670 \times 10^{-8}\ \mathrm{J\,m^{-2}s^{-1}K^{-4}})(10^3\ \mathrm{K})^4 = 5.670 \times 10^4\ \mathrm{J\,m^{-2}s^{-1}}$$

For an area of 10 cm^2,

$$M_e = (10\ \mathrm{cm^2})(10^{-4}\ \mathrm{m^2\,cm^{-2}})(5.670 \times 10^4)\ \mathrm{J\,m^{-2}s^{-1}} = 56.7\ \mathrm{W}$$

PROBLEM 14-7 Calculate the velocity of an electron emitted by cesium due to a photon of wavelength 4500 Å. The work function of Cs is 2.98×10^{-19} J.

Solution: The Einstein equation for the photoelectric effect is Eq. (14-7):

$$h\nu = W + \tfrac{1}{2}mv^2$$

Replacing ν by c/λ and solving for v^2, you obtain

$$v^2 = \frac{2}{m}\left(\frac{hc}{\lambda} - W\right)$$

The mass of the electron is 9.1095×10^{-31} kg, so

$$v^2 = \left(\frac{2}{9.11 \times 10^{-31}\ \text{kg}}\right)\left[\frac{(6.6262 \times 10^{-34}\ \text{J s})(2.998 \times 10^{8}\ \text{m s}^{-1})}{(4.50 \times 10^{3}\ \text{A})(10^{-10}\ \text{m A}^{-1})} - 2.98 \times 10^{-19}\ \text{J}\right]$$

$$= \left(\frac{2}{9.11\ \text{kg}} \times 10^{31}\right)(4.41 - 2.98) \times 10^{-19}\ \text{kg m}^2\,\text{s}^{-2} = 31.4 \times 10^{10}\ \text{m}^2\,\text{s}^{-2}$$

$$v = 5.60 \times 10^{5}\ \text{m s}^{-1}$$

The Bohr Atom

PROBLEM 14-8 Calculate the energy of a "Bohr-type" hydrogen atom in its ground state ($n = 1$). Assume that the radius has already been calculated.

Solution: When you can use the calculated Bohr radius a_0, the general expression for E_n becomes Eq. (14-14):

$$E_n = -\frac{e^2}{(4\pi\varepsilon_0)2a_0}\left(\frac{1}{n^2}\right)$$

$$E_1 = -\left[\frac{(1.602 \times 10^{-19}\ \text{C})^2}{(1.113 \times 10^{-10}\ \text{C}^2\,\text{J}^{-1}\text{m}^{-1})(2)(0.5292 \times 10^{-10}\ \text{m})}\right]\left(\frac{1}{1^2}\right) = -2.179 \times 10^{-18}\ \text{J}$$

PROBLEM 14-9 The hydrogen spectrum has a visible line at 4101.7 Å. Determine the n of the orbit from which the electron transition that produces this line originates.

Solution: This may seem to involve Bohr's equations but, except for the Bohr terminology and concepts, it is a spectroscopy problem, which you can use Eq. (14-1) to solve:

$$\frac{1}{\lambda} = R_{\text{H}}\left(\frac{1}{n_1^2} - \frac{1}{n_2^2}\right)$$

Recall that n_1 refers to the final level of the transition. You need n_2, but what do you use for n_1? The clue is that the line is in the visible region. This means that it's part of the Balmer series and that $n_1 = 2$. Let's first rearrange Eq. (14-1) to isolate n_2:

$$\frac{1}{\lambda R_{\text{H}}} = \frac{1}{n_1^2} - \frac{1}{n_2^2}$$

$$\frac{1}{n_2^2} = \frac{1}{n_1^2} - \frac{1}{\lambda R_{\text{H}}} = \frac{1}{2^2} - \frac{1}{(4101.7 \times 10^{-10}\ \text{m})(1.097 \times 10^{7}\ \text{m}^{-1})} = 0.2500 - 0.2222 = 0.0278$$

$$n_2^2 = \frac{1}{0.0278} = 35.97$$

$$n_2 = 6$$

PROBLEM 14-10 (**a**) Calculate the ionization energy of hydrogen, i.e., the energy required to just remove an electron from the influence of the nucleus. (**b**) Compare your result to that of Problem 14-8.

Solution:

(**a**) In the Ritz–Rydberg expression, ionization is comparable to moving an electron from $n = 1$ to $n = \infty$. Using the Bohr approach, you'd write $\Delta E = E_\infty - E_1$. The next step is to substitute Eq. (14-14) for the E's, but you should recognize that, for $n = \infty$, $1/\infty = 0$, making $E_\infty = 0$. In other words, ionization occurs at the point where the electron is free from the influence of the nucleus; then $\Delta E = -E_1$. This energy requirement is the *ionization energy*:

$$E_{\text{i}} = \Delta E = -\left[-\frac{e^2}{2(4\pi\varepsilon_0)a_0}\left(\frac{1}{1^2}\right)\right] = \frac{(1.602 \times 10^{-19}\ \text{C})^2}{2(1.113 \times 10^{-10}\ \text{C}^2\,\text{J}^{-1}\text{m}^{-1})(0.5292 \times 10^{-10}\ \text{m})}$$

$$= 2.179 \times 10^{-18}\ \text{J}$$

(b) In Problem 14-8, the energy is the potential energy of the electron, i.e., the energy required to bring it from infinity to its distance from the nucleus. In this problem, you're taking the electron from its position back to infinity. Thus the energies are equal but opposite in sign.

PROBLEM 14-11 Convert your result in Problem 14-10 to **(a)** $kJ\,mol^{-1}$ and **(b)** electron volts.

Solution:

(a) The result in Problem 14-10 is for one atom. For 1 mol, you must multiply by Avogadro's number:

$E_i = (2.179 \times 10^{-18}\ J\,atom^{-1})(6.022 \times 10^{23}\ atom\,mol^{-1}) = 13.122 \times 10^5\ J\,mol^{-1} = 1.312 \times 10^3\ kJ\,mol^{-1}$

(b) Since $1\ eV = 1.602 \times 10^{-19}\ J$,

$$E_i = \frac{2.179 \times 10^{-18}\ J}{1.602 \times 10^{-19}\ J\,eV^{-1}} = 13.60\ eV$$

Wave–Particle Duality

PROBLEM 14-12 The de Broglie hypothesis should apply to all particles. Calculate the wavelength of a hydrogen molecule, which is the lightest particle that exists independently at room temperature.

Solution: The de Broglie hypothesis is $\lambda = h/mv$, Eq. (14-15), so you need the velocity. You can calculate the rms velocity from Eq. (2-39):

$$\overline{u^2} = \frac{3RT}{L_A m}$$

where $L_A m$ is the molar weight or 2.016×10^{-3} kg for H_2. Taking 298.2 K as room temperature, you have

$$\overline{u^2} = \frac{3(8.314\ J\,K^{-1}\,mol^{-1})(298.2\ K)}{2.016 \times 10^{-3}\ kg\,mol^{-1}}$$

Recall that $1\ J = 1\ kg\,m^2\,s^{-2}$. Then

$$\overline{u^2} = 3.689 \times 10^6\ m^2\,s^{-2} \qquad \text{and} \qquad (\overline{u^2})^{1/2} = 1.921 \times 10^3\ m\,s^{-1}$$

Now substitute $v = (\overline{u^2})^{1/2}$ and $m = 2$ (mass of proton) into Eq. (14-15):

$$\lambda = \frac{(6.6262 \times 10^{-34}\ J\,s)(kg\,m^2\,s^{-2}\,J^{-1})}{2(1.6726 \times 10^{-27}\ kg)(1.921 \times 10^3\ m\,s^{-1})} = 1.031 \times 10^{-10}\ m = 1.031\ \text{Å}$$

This wavelength is about the diameter of one H atom.

PROBLEM 14-13 Refer to Examples 14-14 and 14-16. If the voltage is increased to 10 000 V, which is a possible voltage for an electron microscope, calculate the values of the wavelength, momentum, and uncertainty in momentum.

Solution: If you happen to notice that the voltage has been increased 100 times and that λ is inversely proportional to $V^{1/2}$ (Example 14-14), you'll see immediately that the new λ is $\frac{1}{10}$ of the previous value, or 1.23×10^{-11} m. The momentum is inversely proportional to the wavelength ($p = h/\lambda$), so the new p must be 10 times the old (Example 14-16), or $5.39 \times 10^{-23}\ kg\,m\,s^{-1}$. The uncertainty in p remains the same: $\Delta p_x = h/\Delta x = 6.63 \times 10^{-24}\ kg\,m\,s^{-1}$ (Example 14-16), but now the uncertainty is only $\frac{1}{10}$ the magnitude of the momentum.

Quantum Theory

PROBLEM 14-14 In Chapter 3, we introduced Hooke's law, $F = -kr$ [Eq. (3-10)]. Using $F = ma$, show that this law is an eigenvalue equation and identify the eigenvalue and the eigenfunction.

Solution: You have two expressions for the force, so you can set them equal: $-kr = ma$. Rewriting the acceleration as the second derivative of displacement with respect to time, and rearranging, you get

$$\frac{d^2r}{dt^2} = -\frac{kr}{m}$$

which is of the form of Eq. (14-18). The eigenfunction is r, and the eigenvalue is $-k/m$.

PROBLEM 14-15 Find an acceptable solution, i.e., an expression for r, to the equation in Problem 14-14.

Solution: One approach to eigenvalue problems is simple trial and error. In this case, you need a function that you can differentiate twice and get back the same function. You should recall that this reversal occurs with the sine and cosine functions:

$$\frac{d}{dx}\sin u = \cos u \frac{du}{dx} \qquad \frac{d}{dx}\cos u = -\sin u \frac{du}{dx}$$

Let the angle be a function of t, at, and $r = \sin(at)$; then perform the operation required by the Hooke's law expression in Problem 14-14. Doing it in steps, you get

$$\frac{d}{dt}\sin(at) = \cos(at)\frac{d}{dt}(at) = a\cos(at)$$

and

$$\frac{d^2}{dt^2}\sin(at) = \frac{d}{dt}[a\cos(at)] = a[-\sin(at)]\frac{d}{dt}(at) = -a^2\sin(at)$$

Since $r = \sin(at)$, this result agrees with the Hooke's law expression, if $a^2 = k/m$. Now, the eigenfunction becomes $\sin[\sqrt{k/m}(t)]$.

PROBLEM 14-16 Give a reason for each of the conditions a wave function must meet to be an acceptable solution to the Schrödinger equation.

Solution: Mathematically well-behaved functions are single-valued, finite, and continuous. Also, you must be able to integrate $\psi^*\psi$. Some of these conditions are required by the nature of physical properties. If the wave function had more than one value, you'd have two different probabilities for the same segment of space and two expectation values; if it were infinite, there wouldn't be a wave (infinite amplitude); and if it weren't continuous, its value might change suddenly with a change in position. In fact, observable properties change smoothly through all values. If the integral of $\psi^*\psi$ weren't finite, the wave function couldn't be normalized to make the total probability of locating the electron equal to 1.

PROBLEM 14-17 Find the normalization constant for the wave function e^{-bx^2} (a one-dimensional function).

Solution: A wave function is normalized when the integral of its probability density over all space equals 1, as expressed by Eq. (14-25):

$$\int_{-\infty}^{\infty} \psi^*\psi \, d\tau = 1$$

In this case, you want the factor to multiply e^{-bx^2} by, in order to satisfy Eq. (14-25). If N is the normalizing factor, the wave function is

$$\psi = Ne^{-bx^2} \tag{a}$$

There is no i, so $\psi^* = \psi$, or $\psi^*\psi = \psi^2$. Equation (14-25) now becomes

$$N^2 \int_{-\infty}^{\infty} e^{-2bx^2}\, dx = 1 \tag{b}$$

In a table of definite integrals, you'll find

$$\int_0^{\infty} e^{-a^2x^2}\, dx = \frac{1}{2a}(\sqrt{\pi}) \tag{c}$$

If you let $a^2 = 2b$ (or $a = \sqrt{2b}$), then the integral in eq. (b) is twice the integral in eq. (c):

$$2[\text{eq. (c)}] = 2\left[N^2\left(\frac{1}{2\sqrt{2b}}\right)(\sqrt{\pi})\right] = N^2\left(\frac{\pi}{2b}\right)^{1/2} = 1$$

Then $$N^2 = \left(\frac{2b}{\pi}\right)^{1/2} \quad \text{and} \quad N = \left(\frac{2b}{\pi}\right)^{1/4}$$

The normalized wave function is $(2b/\pi)^{1/4}e^{-bx^2}$. We identified it in Example 14-25 as the lowest energy state of a harmonic oscillator.

Applications

PROBLEM 14-18 In Example 14-26, we found that the probability of finding a free particle was the same for every x; i.e., the position was completely undetermined. (**a**) What does this suggest with regard to the uncertainty principle? (**b**) Find an expression for the momentum. Does it confirm the suggestion in (**a**)?

Solution:

(**a**) The uncertainty principle, Eq. (14-17), requires that when the position is completely unknown, the momentum must be precisely known.

(**b**) The momentum operator is

$$\mathbf{p}_x = \frac{h}{2\pi i}\frac{\partial}{\partial x}$$

One eigenfunction for the free particle is

$$\psi_1 = Ae^{i(8\pi^2 mE)^{1/2}x/h}$$

The eigenequation for momentum is

$$\mathbf{p}_x\psi = p_x\psi \qquad \text{[Eq. (14-18)]}$$

Substituting for $\mathbf{p}_x$ and ψ, you get

$$\frac{h}{2\pi i}\frac{\partial}{\partial x}Ae^{i(8\pi^2 mE)^{1/2}x/h} = p_x Ae^{i(8\pi^2 mE)^{1/2}x/h}$$

$$\frac{h}{2\pi i}Ae^{i(8\pi^2 mE)^{1/2}x/h}\frac{d}{dx}\left[\frac{i(8\pi^2 mE)^{1/2}}{h}\right]x = p_x Ae^{i(8\pi^2 mE)^{1/2}x/h}$$

Canceling the A's and the exponentials, and completing the differentiation, you have

$$p_x = \frac{h}{2\pi i}\left[\frac{i(8\pi^2 mE)^{1/2}}{h}\right] = (2mE)^{1/2}$$

This tells you that the momentum has a definite value for a particular ψ, as predicted in part (a).

PROBLEM 14-19 (**a**) Calculate the first and second energy levels of a 10-g ball in a one-dimensional box, 1 m long. (**b**) If the ball jumps from the second level to the first, what is the wavelength of the emitted radiation? (**c**) Compare your answer to Example 14-27.

Solution:

(**a**) According to Eq. (14-32), the energies for a particle in a one-dimensional box are given by

$$E_n = \frac{n^2h^2}{8ma^2} = \frac{n^2(6.626 \times 10^{-34}\text{ J s})^2}{8(10 \times 10^{-3}\text{ kg})(1\text{ m})^2} = (5.49 \times 10^{-66}\text{ J})(n^2)$$

$$\text{For } n = 1, \qquad E_1 = 5.49 \times 10^{-66}\text{ J}$$

$$\text{For } n = 2, \qquad E_2 = 4E_1 = 22.0 \times 10^{-66}\text{ J}$$

(**b**) The change in energy is

$$\Delta E = |E_1 - E_2| = 3E_1 = 16.5 \times 10^{-66}\text{ J}$$

Using Eq. (14-9), you get

$$\lambda = \frac{hc}{\Delta E} = \frac{(6.626 \times 10^{-34}\text{ J s})(2.998 \times 10^8\text{ m s}^{-1})}{16.5 \times 10^{-66}\text{ J}} = 1.20 \times 10^{40}\text{ m}$$

(**c**) The value of ΔE is so small and that of λ so large that they're impossible to measure, whereas in Example 14-27 the values for an electron could be measured. Thus quantization is not observed for the ball, and its behavior may be calculated using classical (Newtonian) mechanics.

PROBLEM 14-20 For a particle in the lowest energy level of a one-dimensional box 1 Å wide, calculate the probability of finding the particle between 0.40 Å and 0.60 Å.

Solution: The probability is always given by $\psi^*\psi = |\psi|^2$ and the normalized wave function for the one-dimensional box is given by Eq. (14-33). Since ψ is not imaginary

$$\int \psi^*\psi\, dx = \int \psi^2\, dx = \int \left[\left(\frac{2}{a}\right)^{1/2} \sin\frac{n\pi x}{a}\right]^2 dx$$

Squaring and substituting $n = 1$ and $a = 1$ gives

$$\int \psi^2\, dx = \frac{2}{1}\int \sin^2(\pi x)\, dx \qquad \textbf{(a)}$$

From a table of integrals, you obtain

$$\int \sin^2 u\, du = \frac{u}{2} - \frac{1}{4}\sin 2u + C \qquad \textbf{(b)}$$

When you let $u = \pi x$, $du = \pi\, dx$, and $dx = \frac{du}{\pi}$. The integral in eq. (a) becomes

$$\int \frac{1}{\pi}\sin^2 u\, du + C \qquad \textbf{(c)}$$

Substituting eq. (c) into eq. (a) and inserting the limits to remove C, you get

$$\int_{0.40}^{0.60} \psi^2\, dx = 2\left(\frac{1}{\pi}\right)\int_{0.40}^{0.60} \sin^2 u\, du$$

Applying eq. (b) and substituting $u = \pi x$, you have

$$\int_{0.40}^{0.60} \psi^2\, dx = \frac{2}{\pi}\left[\frac{\pi x}{2} - \frac{1}{4}\sin 2\pi x\right]_{0.40}^{0.60} = \left[x - \frac{1}{2\pi}\sin 2\pi x\right]_{0.40}^{0.60} = 0.20 - \frac{1}{2\pi}\sin 0.40\pi = 0.20$$

PROBLEM 14-21 In Example 14-25, we gave the normalized wave function for the lowest energy state of an harmonic oscillator as $(2b/\pi)^{1/4}e^{-bx^2}$. By substitution in the Schrödinger equation, determine the value of b.

Solution: The Schrödinger equation for the one-dimensional harmonic oscillator is Eq. (14-37). You can simplify the math by using letters for terms that remain constant until they are needed. Let

$$-\frac{h^2}{8\pi^2 m} = M \qquad \text{and} \qquad \left(\frac{2b}{\pi}\right)^{1/4} = N$$

Now substitute these terms and ψ into Eq. (14-37):

$$M\frac{\partial^2}{\partial x^2}(Ne^{-bx^2}) + \frac{1}{2}kx^2(Ne^{-bx^2}) = ENe^{-bx^2} \qquad \textbf{(a)}$$

Later you should replace E with $(v + \frac{1}{2})hv$ from Eq. (14-38); $v = 0$ in this problem. But first, let's do the differentiation separately. Recall that $de^u = e^u\, du$ and take the first derivative:

$$\frac{\partial}{\partial x}(Ne^{-bx^2}) = Ne^{-bx^2}d(-bx^2) = Ne^{-bx^2}(-2bx)$$

The second derivative is of the form $d(uv) = v\, du + u\, dv$, so

$$\frac{\partial}{\partial x}(-2bNxe^{-bx^2}) = -2bN[e^{-bx^2}\, dx + xde^{-bx^2}]$$

and

$$\frac{\partial^2}{\partial x^2}(Ne^{-bx^2}) = -2bN[e^{-bx^2} - 2bx^2e^{-bx^2}]$$

Substitute the latter expression in eq. (a) and change E to $\frac{1}{2}hv = E_0$:

$$M(-2bNe^{-bx^2}) + M(4b^2x^2Ne^{-bx^2}) + \frac{1}{2}kx^2Ne^{-bx^2} = \frac{1}{2}hvNe^{-bx^2}$$

Now, Ne^{-bx^2} cancels, and you have

$$-2bM + 4b^2Mx^2 + \frac{1}{2}kx^2 - \frac{1}{2}hv = 0$$

This equation can be valid if both the coefficients of x^2 and the constant terms cancel; i.e., if

$$-4b^2M = \frac{1}{2}k \qquad \text{and} \qquad 2bM = -\frac{1}{2}h\nu$$

Using the second equality,

$$b = -\frac{h\nu}{4M} = -\frac{h\nu}{4}\left(-\frac{8\pi^2 m}{h^2}\right) = \frac{2\pi^2 \nu m}{h}$$

You can obtain the same result from the first equality. The complete wave function is

$$\psi_0 = \left(\frac{4\pi\nu m}{h}\right)^{1/4} e^{-2\pi^2 \nu m x^2/h}$$

PROBLEM 14-22 The bond length of an O_2 molecule is 1.21 Å. Treating the molecule as a rigid rotor with this distance as its fixed length, calculate the energies of the first two rotational energy levels. Assume that both atoms are ^{16}O.

Solution: The energy levels are given by Eq. (14-41). You must determine I and then solve for E_0 and E_1. Actually, you need to calculate only E_1 since E_0 clearly equals 0. The moment of inertia is $I = \mu r_0^2$, where μ is the reduced mass. Replacing μ by its definition you get

$$I = \frac{m_1 m_2}{m_1 + m_2} r_0^2$$

but $m_1 = m_2 = m$, so

$$I = \frac{m^2}{2m} r_0^2 = \frac{m}{2} r_0^2 = \left[\frac{16.0 \times 10^{-3}\ \text{kg mol}^{-1}}{2(6.02 \times 10^{23}\ \text{mol}^{-1})}\right](1.21 \times 10^{-10}\ \text{m})^2 = 1.95 \times 10^{-46}\ \text{kg m}^2$$

You can now substitute in Eq. (14-41):

$$E_J = J(J+1)\left(\frac{h^2}{8\pi^2 I}\right)$$

When $J = 1$,

$$E_1 = 1(2)\left[\frac{(6.626 \times 10^{-34}\ \text{J s})^2}{8\pi^2(1.95 \times 10^{-46}\ \text{kg m}^2)}\right] = 5.70 \times 10^{-23}\ \text{J}$$

Supplementary Exercises

PROBLEM 14-23 The value of the Rydberg constant for hydrogen to four significant figures is __________. It is also the wave number of the __________ series limit for the hydrogen spectrum. Name the two completely unexpected atomic phenomena discovered just before 1900. The Wien displacement law says that the frequency of greatest intensity (is) (isn't) directly proportional to the absolute temperature. The value of the Bohr radius to four significant figures is __________. The __________ __________ phenomenon illustrates the wavelike behavior of electrons. An increase in temperature (increases) (decreases) the de Broglie wavelength of a small molecule. When the accelerating voltage of a beam of electrons is increased, the associated wavelength of the electrons (increases) (decreases). __________ __________ mechanics applies to a particle having a velocity near the speed of light. Write the symbol and the expression for *del squared*. The term cos x (is) (isn't) an eigenfunction of the operator $\frac{d}{dx}$. When $J = 3$, M can have as many as __________ values.

PROBLEM 14-24 Calculate the wavelength of the limit of the Paschen series ($n_1 = 3$) in the hydrogen spectrum.

PROBLEM 14-25 Calculate the frequency having the greatest intensity for radiation from a black body at 1500 K.

PROBLEM 14-26 Calculate the surface area required to obtain 100 W of power from a surface radiating at 800 K. Assume black-body behavior.

PROBLEM 14-27 Calculate the energy of a photon of wavelength 4800 Å.

PROBLEM 14-28 The work function of sodium is 3.65×10^{-19} J. Determine the kinetic energy of the electron released by a photon having a wavelength of 300 nm.

PROBLEM 14-29 Determine the wavelength of the photon emitted when the electron in a hydrogen atom drops from an orbit where $n = 5$ to an orbit where $n = 3$.

PROBLEM 14-30 Calculate the voltage required to give an electron the velocity needed for its de Broglie wavelength to be 200 pm.

PROBLEM 14-31 If the voltage in Problem 14-30 is accurate only to ± 0.3 V (i.e., 37.6 ± 0.3 V), find the minimum uncertainty in the position of the electron. $\left[\textit{Hint: } \Delta(E_k) = \frac{p}{m}\Delta p.\right]$

PROBLEM 14-32 Refer to Problem 14-15. Show that the function $a \cos(bt + \phi)$ is an acceptable eigenfunction. Determine the eigenvalue and the value of b.

PROBLEM 14-33 Assume that the F_2 molecule is a rigid rotor with a fixed length of 1.42 Å. Calculate the energy of the third rotational energy level.

Answers to Supplementary Exercises

14-23 1.097×10^7 m^{-1}; Lyman; x-rays, radioactivity; is; 0.5292×10^{-10} m; electron diffraction; decreases; decreases; Relativistic quantum; $\nabla^2 = \frac{\partial^2}{\partial x^2} + \frac{\partial^2}{\partial y^2} + \frac{\partial^2}{\partial z^2}$; isn't; 7 **14-24** 820.4 nm
14-25 1.552×10^{14} s^{-1} **14-26** 43.1 cm^2 **14-27** 4.14×10^{-19} J **14-28** 2.97×10^{-19} J
14-29 1.28 μm **14-30** 37.6 V **14-31** 50.1 nm **14-32** $-k/m$; $(k/m)^{1/2}$
14-33 2.10×10^{-22} J

15 QUANTUM THEORY OF ATOMS

THIS CHAPTER IS ABOUT

- ☑ **Hydrogen-Like Atoms**
- ☑ **Angular Momentum and Magnetic Moment**
- ☑ **Approximation Methods**
- ☑ **Helium-Like Atoms**
- ☑ **Multi-Electron Atoms**
- ☑ **Experimental Verification**

15-1. Hydrogen-Like Atoms

A. The Schrödinger equation

For any atom or ion having one electron, i.e., a hydrogen-like species, we can solve the Schrödinger equation exactly. The Hamiltonian operator, Eq. (14-22), requires the potential energy caused by the attraction between the electron and the nucleus: q_1q_2/r. In SI units,

$$E_{\mathrm{p}} = V = -\frac{Ze^2}{4\pi\varepsilon_0 r} \tag{15-1}$$

where e is the charge of the electron, Ze is the charge of the nucleus, ε_0 is the permittivity of a vacuum, and r is the distance of the electron from the nucleus. When we use the reduced mass μ (although it's essentially the same as the mass of the electron) with the substitution of Eq. (15-1) for V and with the expanded expression for the Hamiltonian operator, the Schrödinger equation, Eq. (14-30), becomes

$$\left(-\frac{h^2}{8\pi^2\mu}\nabla^2 - \frac{Ze^2}{4\pi\varepsilon_0 r}\right)\psi = E\psi \tag{15-2}$$

The solution is easier and the interpretation as a model is more useful if we transform the equation to spherical coordinates, i.e., r, θ, ϕ. (We mentioned this technique in Section 14-5E for treating the rigid rotor.)

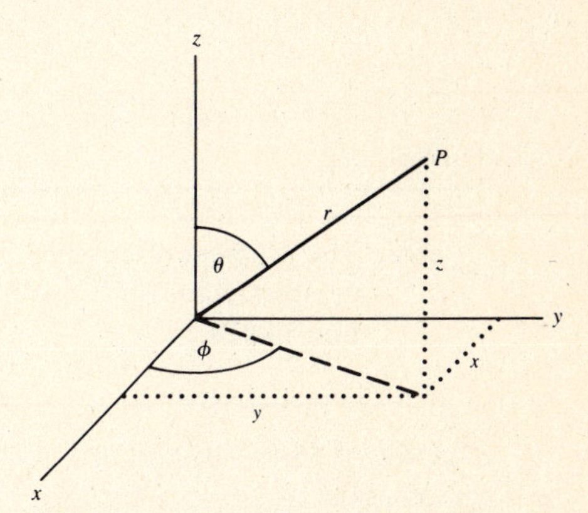

FIGURE 15-1. Comparison of Cartesian and Spherical Coordinates

In Figure 15-1, the relationship between spherical and Cartesian coordinates is shown for a point P. The dashed line represents the projection of P on the xy plane; the dotted lines represent the x, y, z coordinates. The line from the origin to P is r, and the arcs show where the angles θ and ϕ are measured. You can see that ranges of the spherical coordinates are for r, 0 to ∞; θ, 0 to π; and ϕ, 0 to 2π. (You should be aware, however, that in some books the axes and the angles may be labeled differently.)

From Figure 15-1, we can work out the trigonometric relationships for transforming one system to the other, i.e.,

$$x = r \sin\theta \cos\phi \qquad y = r \sin\theta \sin\phi \qquad z = r\cos\theta \qquad x^2 + y^2 + z^2 = r^2 \qquad \textbf{(15-3)}$$

The infinitesimal element of volume, $dv = dx\,dy\,dz$, becomes

$$d\tau = r^2 \sin\theta \, dr\, d\theta\, d\phi$$

in spherical coordinates. When we convert Eq. (15-2) to sphericai coordinates and rearrange the constants, we get

$$\frac{1}{r^2}\frac{\partial}{\partial r}\left(r^2\frac{\partial\psi}{\partial r}\right) + \frac{1}{r^2 \sin\theta}\frac{\partial}{\partial\theta}\left(\sin\theta\frac{\partial\psi}{\partial\theta}\right) + \frac{1}{r^2\sin^2\theta}\frac{\partial^2\psi}{\partial\phi^2} + \frac{8\pi^2\mu}{h^2}\left(E + \frac{Ze^2}{4\pi\varepsilon_0 r}\right)\psi = 0 \qquad \textbf{(15-4)}$$

We can separate this equation into three parts, analogous to the solution for the particle in the three-dimensional box (Section 14-5C):

$$\psi(r, \theta, \phi) = [R(r)][\Theta(\theta)][\Phi(\phi)] \qquad \textbf{(15-5)}$$

When we substitute Eq. (15-5) into Eq. (15-4), we can divide the result into three separate equations, one for each coordinate. Each equation produces one quantum number, and the energy eigenvalues appear in the equation for r, the radial equation. Usually, such a separation isn't possible for atoms having more than one electron, because V usually depends on the angles as well as on the radial distance.

B. The Φ equation

The solution of the Φ part of the spherical Schrödinger equation is

$$\Phi = Ne^{im_\ell\phi} \qquad \textbf{(15-6)}$$

where m_ℓ is an undetermined integer known as the **magnetic quantum number**, because it plays a role when the atom is in a magnetic field. Its values are $m_\ell = 0, \pm 1, \pm 2, \ldots$. The normalization constant N is $1/\sqrt{2\pi}$.

EXAMPLE 15-1: Write the wave functions Φ for the first three values of m. Do you see any difficulty with them?

Solution: We merely write Eq. (15-6), inserting the value of N and $m_\ell = 0, +1$, and -1, and obtain

$$\Phi_0 = \frac{1}{\sqrt{2\pi}}e^0 = \frac{1}{\sqrt{2\pi}} \qquad \textbf{(15-7)}$$

$$\Phi_{+1} = \frac{1}{\sqrt{2\pi}}e^{i\phi} \qquad \textbf{(15-8)}$$

$$\Phi_{-1} = \frac{1}{\sqrt{2\pi}}e^{-i\phi} \qquad \textbf{(15-9)}$$

Since the last two expressions are complex, they are "unreal" and have no direct physical significance.

---

note: The problem of complex wave functions is solved by making linear combinations of complex wave functions in such a way that real wave functions are created (see Section 14-5B).

EXAMPLE 15-2: Obtain two real wave functions from Eqs. (15-8) and (15-9) by linear combination.

Solution: We add and subtract the two functions of Eqs. (15-8) and (15-9), as described in Section 14-5B. For the constants, we use $1/\sqrt{2}$ to normalize the new functions:

$$\Phi_x = \frac{1}{\sqrt{2}}(\Phi_{+1} + \Phi_{-1}) \qquad \text{and} \qquad \Phi_y = \frac{1}{\sqrt{2}}(\Phi_{+1} - \Phi_{-1})$$

Substituting from Eqs. (15-8) and (15-9), we have

$$\Phi_x = \frac{1}{\sqrt{2}}\left(\frac{e^{i\phi}}{\sqrt{2\pi}} + \frac{e^{-i\phi}}{\sqrt{2\pi}}\right) = \frac{1}{2\sqrt{\pi}}(e^{i\phi} + e^{-i\phi}) = \frac{\cos\phi}{\sqrt{\pi}}$$

and

$$\Phi_y = \frac{1}{\sqrt{2}}\left(\frac{e^{i\phi}}{\sqrt{2\pi}} - \frac{e^{-i\phi}}{\sqrt{2\pi}}\right) = \frac{1}{2\sqrt{\pi}}(e^{i\phi} - e^{-i\phi}) = \frac{i\sin\phi}{\sqrt{\pi}}$$

The new wave functions are designated x and y, because $\cos\phi$ is a maximum along the x axis and $\sin\phi$ is a maximum along the y axis.

C. The Θ equation

The solution of the Θ part of the spherical Schrödinger equation involves **associated Legendre equations**, with solutions that are **associated Legendre polynomials**. These expressions include both the quantum number m_ℓ and a new parameter ℓ, called the **azimuthal quantum number**. Its values are restricted to the integers and zero, or $\ell = 0, 1, 2, 3, \ldots$. These values are usually represented, in order, by the letters *s, p, d, f, g, h*,.... A restriction is now placed on m_ℓ: $|m_\ell|$ can't be larger than ℓ. Thus $m = 0, \pm 1, \pm 2, \ldots, \pm \ell$. Some normalized values of $\Theta(\theta)$ are

ℓ	$\lvert m_\ell \rvert$	$\Theta_{\ell,m} = \Theta(\theta)$		Orbital	Other symbols
0	0	$\Theta_{0,0} = \frac{1}{2}\sqrt{2}$	**(15-10)**	*s*	Θ_{00} or Θ_{s0}
1	0	$\Theta_{1,0} = \frac{1}{2}\sqrt{6}\cos\theta$	**(15-11)**	*p*	Θ_{10} or Θ_{p0}
	1	$\Theta_{1,1} = \frac{1}{2}\sqrt{3}\sin\theta$	**(15-12)**	*p*	$\Theta_{1\pm1}$ or $\Theta_{p\pm1}$
2	0	$\Theta_{2,0} = \frac{1}{4}\sqrt{10}(3\cos^2\theta - 1)$		*d*	Θ_{20} or Θ_{d0}
	1	$\Theta_{2,1} = \frac{1}{2}\sqrt{15}\sin\theta\cos\theta$		*d*	$\Theta_{2\pm1}$ or $\Theta_{d\pm1}$
	2	$\Theta_{2,2} = \frac{1}{4}\sqrt{15}\sin^2\theta$		*d*	$\Theta_{2\pm2}$ or $\Theta_{d\pm2}$

D. The *R* equation

Solution of the *R* part of the spherical Schrödinger equation requires considerable manipulation to put it in the form known as the **associated Laguerre equation**, with solutions that are **associated Laguerre polynomials**. These expressions include both the quantum number ℓ and a new parameter n, which is known as the **principal quantum number** and is restricted to integral values excluding zero, or $n = 1, 2, 3, \ldots, \infty$. A restriction is now placed on ℓ: it can't be larger than $(n - 1)$. Thus $\ell = 0, 1, 2, 3, \ldots, (n - 1)$. Some normalized values of $R(r)$ are

n	ℓ	$R(r)$	Orbital	
1	0	$R_{1,0} = 2\left(\frac{Z}{a_0}\right)^{3/2} e^{-Zr/a_0}$	1*s*	**(15-13)**
2	0	$R_{2,0} = \frac{1}{2\sqrt{2}}\left(\frac{Z}{a_0}\right)^{3/2}\left(2 - \frac{Zr}{a_0}\right)e^{-Zr/2a_0}$	2*s*	**(15-14)**
2	1	$R_{2,1} = \frac{1}{2\sqrt{6}}\left(\frac{Z}{a_0}\right)^{3/2}\frac{Zr}{a_0}e^{-Zr/2a_0}$	2*p*	**(15-15)**

E. Allowed energy levels

Another result of the solution of the *R* equation is an expression for the allowed energy levels, which are determined by the quantum number n:

$$E_n = -\frac{2\pi^2\mu Z^2 e^4}{(4\pi\varepsilon_0)^2 h^2 n^2} = -\frac{e^2}{(4\pi\varepsilon_0)2a_0}\left(\frac{Z^2}{n^2}\right) = -\frac{Z^2}{n^2}E_{\mathrm{H}} \qquad \textbf{(15-16)}$$

This result is identical to that obtained by Bohr for hydrogen [$Z = 1$; see Eq. (14-14)]. For a typical calculation, see Example 14-10. Since n is involved only in the *R* equation, E_n must be independent of angle; it's dependent only on the distance of the electron from the nucleus. Note that this is not true for atoms with more than one electron. The constant term in Eq. (15-16) is represented by the symbol E_{H}, because its negative is the ground state energy of the hydrogen atom (when $Z = 1$ and $n = 1$). Two useful values for E_{H} are 2.1798×10^{-18} J and 13.605 eV.

F. Total wave function

The total, or complete, wave function of a hydrogen-like atom is the product of the three parts shown in Eq. (15-5) and is called an **orbital**. In order to show clearly the dependence on quantum numbers, this equation often is written as

$$\psi(r, \theta, \phi) = [R_{n,\ell}(r)][\Theta_{\ell, m_\ell}(\theta)][\Phi_{m_\ell}(\phi)] \qquad \textbf{(15-17)}$$

EXAMPLE 15-3: Write the total wave function for a hydrogen atom in the ground state.

Solution: For the ground state, the quantum numbers must be the smallest possible: $n = 1, \ell = 0, m_\ell = 0$. Then Eq. (15-17) becomes

$$\psi_{n,\ell,m_\ell} = \psi_{1,0,0} = [R_{1,0}(r)][\Theta_{0,0}(\theta)][\Phi_0(\phi)]$$

Combining Eqs. (15-7), (15-10), and (15-13) gives

$$\psi_{1,0,0} = 2\left(\frac{Z}{a_0}\right)^{3/2} e^{-Zr/a_0}\left(\frac{1}{\sqrt{2}}\right)\left(\frac{1}{\sqrt{2\pi}}\right)$$

For Z = 1,

$$\psi_{1,0,0} = \psi_{1s} = \left(\frac{1}{\sqrt{\pi}}\right)\left(\frac{1}{a_0}\right)^{3/2} e^{-r/a_0} \qquad \textbf{(15-18)}$$

G. Angular probability distribution

For visualizing the probability of finding an electron in different locations around a nucleus, it's convenient to treat Θ and Φ, the angular parts, separately from R, the radial part. Thus we could write an **angular wave function**

$$\psi_a = \Theta\Phi \qquad \textbf{(15-19)}$$

and an **angular probability density** $|\psi_a|^2 = |\Theta\Phi|^2$.

EXAMPLE 15-4: **(a)** Write the equations for the angular quantities $\psi_{a,1s}$ and $|\psi_{a,1s}|^2$. **(b)** What can we say about the symmetry of the plot of these expressions?

Solution:

(a) As in previous Examples,

$$\psi_{a,1s} = [\Theta_{0,0}(\theta)][\Phi_0(\phi)] = \left(\frac{1}{\sqrt{2}}\right)\left(\frac{1}{\sqrt{2\pi}}\right) = \frac{1}{2\sqrt{\pi}}$$

Then the angular probability density is

$$|\psi_{a,1s}|^2 = \frac{1}{4\pi}$$

(b) Neither quantity is dependent on the angle, so all angles are equally probable, and the plot is spherically symmetrical. Also, there is no dependence on n, so all s orbitals have the same symmetry.

Plots of ψ_a and ψ_a^2 values give the *shapes of the orbitals*. For the s-orbital case, the two plots look the same. For the p-orbital case, however, $\Theta\Phi$ is dependent on $\cos\theta$, $|\Theta\Phi|^2$ is dependent on $\cos^2\theta$, and both are independent of ϕ. Thus the plot of ψ_a gives two spheres, but the plot of ψ_a^2 gives two elongated shapes, sometimes described as *tear drops*. Sketches of two-dimensional plots for the p_z orbital are shown in Figures 15-2 and 15-3. In three dimensions the shape is that formed by rotating the two-dimensional cross-section through 360° about the z axis. These shapes—the identical but perpendicular ones for p_x and p_y and the more complex ones for the five d orbitals—should be familiar to you from previous chemistry courses. The same shapes can be used for every p_x orbital (or any orbital), regardless of energy level, because only the R portion of the wave function is dependent on the energy level (i.e., only R is dependent on n), and R is omitted from these plots. In effect, we hold the coordinate r constant,

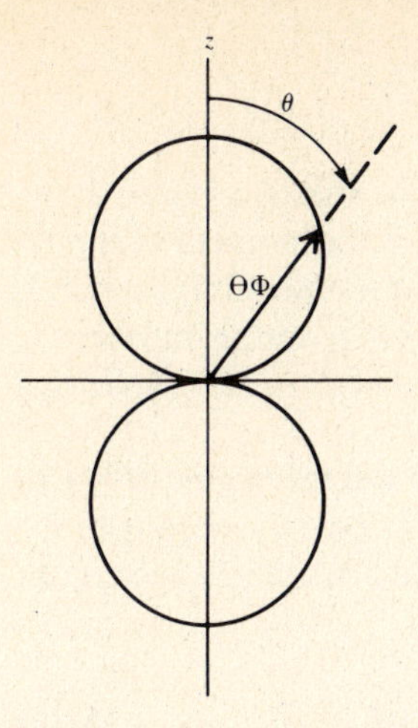

FIGURE 15-2. Sketch of ΘΦ for p_z Orbital

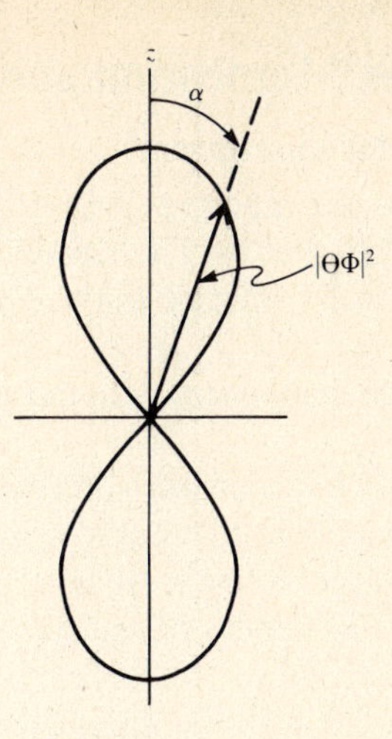

FIGURE 15-3. Sketch of $|\Theta\Phi|^2$ for p_z Orbital

so we can compare the relative magnitudes of $|\Theta\Phi|^2$ at different angles. The probability of finding the electron at a particular angle θ from the z axis is proportional to the length of the line drawn at that angle from the origin to the curve. In Figure 15-3, the length of the arrow represents the probability that the electron is at angle α from the z axis. The maximum probability is along the axis where $\theta = \alpha = 0$.

H. Radial probability distribution

The probability that the electron is a certain distance from the nucleus is usually expressed as the probability that it's located in a small spherical shell at distance r; the **radial probability distribution** is defined as the probability per unit radius, or $r^2R^2(r)$. When we plot it against r (or against r/a_0), we obtain curves like those in Figure 15-4 for the 1s and 2s orbitals. The number of maxima equals n, so 3s orbitals have three peaks, etc. The smaller peaks near the origin show that there's a significant probability that the electron will be near the nucleus, which is called the *penetration effect*. The maximum in the 1s curve falls at 0.529Å, which is exactly the radius (a_0) predicted by Bohr. For p orbitals, 2p has one maximum, 3p has two, etc.

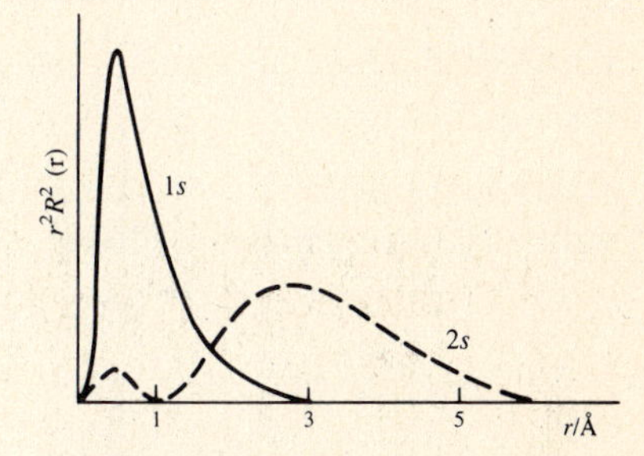

FIGURE 15-4. Sketch of $R(r)$ for 1s and 2s Orbitals

I. Orbital probability density

To represent the complete probability density of an orbital, we must plot $|\psi(r,\theta,\phi)|^2$. Essentially, we combine plots like Figures 15-3 and 15-4 into one plot. Then all p_z orbitals won't be the same. For example, 3p_z will be larger than 2p_z and will have a different inner structure, because it has two maxima in the r^2R^2 portion of ψ^2 instead of one. One way to represent the situation is by a *contour diagram*, i.e., a planar cross-section containing lines that connect points having the same value of ψ^2. In Figure 15-5, one contour is shown, but if we rotate it about the z axis, we create a three-dimensional shape; its surface is called the **boundary surface**. The procedure commonly used is to choose the dimensions of the plot so that, for a particular orbital, the probability is 0.9 of finding the electron inside the surface. Then the *90 percent boundary surface* represents the size as well as the shape of the orbital.

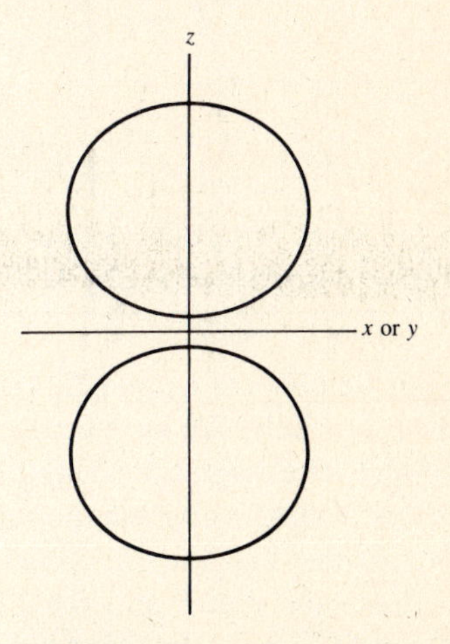

FIGURE 15-5. Contour Diagram of Probability Density (90%) of 2p_z Orbital

15-2. Angular Momentum and Magnetic Moment

A. Orbital angular momentum

We introduced the important property of angular momentum L in the brief discussion of the rigid rotor in Chapter 14. The classical definition is $L = I\omega$, where I is the moment of inertia and ω is the angular velocity. In three dimensions, the classical angular momentum is a vector, $\mathbf{L}$, with components L_x, L_y, and L_z. Using the fundamental postulates of quantum mechanics, we replace each component with the appropriate operator, e.g., $\mathbf{L}_x$, transformed to spherical coordinates.

However, we can measure precisely only one of the components and the magnitude of $\mathbf{L}$. We choose component L_z because it has a less complex operator, involving only one angle. We apply the operators to the wave functions for hydrogen-like atoms and calculate expectation values for the eigenvalues L^2 and L_z:

$$L^2 = \ell(\ell + 1)\left(\frac{h^2}{4\pi^2}\right) \quad \text{or} \quad L = \sqrt{\ell(\ell + 1)}\hbar \tag{15-20}$$

and

$$L_z = m_\ell \frac{h}{2\pi} = m_\ell \hbar \tag{15-21}$$

We see that these quantities are quantized, and the quantum numbers are those we had before, ℓ and m_ℓ. Note that Eq. (15-20) is the same as the expression for the angular momentum of a rigid rotor, a result of the combination of Eqs. (14-40) and (14-41), although we used J instead of ℓ.

EXAMPLE 15-5: Show that the normalized wave function for the Φ coordinate leads to Eq. (15-21).

Solution: Equation (15-6) is the solution of the Φ part of the Schrödinger equation. Inserting the normalization constant, we have

$$\Phi = \frac{1}{\sqrt{2\pi}} e^{im_\ell\phi}$$

The quantum-mechanical operator for $\mathbf{L}_z$ is $\frac{h}{2\pi i}\frac{\partial}{\partial\phi}$. Thus we have

$$\begin{aligned}\mathbf{L}_z\Phi &= \frac{h}{2\pi i}\frac{\partial}{\partial\phi}\left(\frac{1}{\sqrt{2\pi}} e^{im_\ell\phi}\right) = \frac{h}{2\pi i}\frac{1}{\sqrt{2\pi}} e^{im_\ell\phi}\frac{\partial}{\partial\phi}(im_\ell\phi)\\ &= \frac{h}{2\pi i}(\Phi)(im_\ell) = m_\ell\frac{h}{2\pi}\Phi\end{aligned}$$

If we compare this result to the general eigenequation, Eq. (14-18), $\mathbf{O}f = af$, we see that the eigenvalue a is represented by $L_z = m_\ell\hbar$, which is Eq. (15-21).

The components L_x and L_y are not quantized and, as previously stated, we can't determine them precisely. We have to think of the vector $\mathbf{L}$ as precessing about the z axis in a random manner.The magnitude (length) of $\mathbf{L}$ is limited to the values predicted by Eq. (15-20). When a magnetic field is applied, the randomness is removed, and the position of the vector is limited. We assume that the direction of the field is along the z axis. Then we determine the angles between the vector and the z axis by calculating the values of L_z, using Eq. (15-21). The precessing vector describes cones around the z axis. This vector model of the atom is used to explain atomic and molecular spectra.

EXAMPLE 15-6: (**a**) How many cones will be generated by the different positions of the angular momentum vector for a d-orbital electron? (**b**) What is L_z for each? (**c**) What is L for each?

Solution:

(a) For $\ell = 2$, there are five $(2\ell + 1)$ values of m, but one of these is zero. When $m_\ell = 0$, $L_z = 0$, by Eq. (15-21), the vector is perpendicular to the z axis and describes a circle as it rotates. Thus there are only 4 cones.

(b) The values of L_z are $\pm 2\hbar$ and $\pm\hbar$:

$$\frac{h}{2\pi} = \frac{6.6262 \times 10^{-34}}{2\pi} = 1.0546 \times 10^{-34}\ \text{J s}$$

For $m = \pm 2$, $\quad L_z = \pm 2(1.055 \times 10^{-34}\ \text{J s}) = 2.1092 \times 10^{-34}\ \text{J}$

For $m = \pm 1$, $\quad L_z = \pm 1.0546 \times 10^{-34}\ \text{J}$

(c) The length is constant, since it isn't dependent on m. By Eq. (15-20),

$$L = \sqrt{\ell(\ell + 1)}\hbar = \sqrt{1(1 + 1)}\hbar = \sqrt{2}\hbar = 1.4142(1.0546 \times 10^{-34}\ \text{J s}) = 1.4914 \times 10^{-34}\ \text{J s}$$

B. Orbital magnetic moment

A moving charged particle generates a magnetic field. In the case of an electron, the direction of the magnetic field is opposite that of the angular momentum. The associated property is the **magnetic dipole moment** $\boldsymbol{\mu}_\ell$, a vector that points in the direction opposite that of and is proportional to the vector **L**. Thus

$$\boldsymbol{\mu}_\ell = -\frac{e}{2m_e}\mathbf{L} \tag{15-22}$$

The proportionality constant is called the **magnetogyric ratio** (or *gyromagnetic ratio*). Substituting the magnitude of **L** from Eq. (15-20), we get the magnitude of $\boldsymbol{\mu}_\ell$:

$$\mu_\ell = -\frac{e\hbar}{2m_e}\sqrt{\ell(\ell + 1)} = -\mu_B\sqrt{\ell(\ell + 1)} \tag{15-23}$$

The constant term μ_B is called the *Bohr magneton.* It is used as the unit of magnetic dipole moment.

EXAMPLE 15-7: Determine the magnitude and units of the Bohr magneton.

Solution: Substitute values for the constants in the definition:

$$\mu_B = \frac{eh}{4\pi m_e} = \frac{(1.6022 \times 10^{-19}\ \text{C})(6.6262 \times 10^{-34}\ \text{J s})}{4\pi(9.1095 \times 10^{-31}\ \text{kg})} = 9.2742 \times 10^{-24}\ \text{C J s kg}^{-1}$$

To convert the units to the usual ones, we use C = A s and J = kg m^2 s^{-2}. Then

$$\text{C J s kg}^{-1} = (\text{A s})(\text{kg m}^2\,\text{s}^{-2})(\text{s kg}^{-1}) = \text{A m}^2$$

If we want to retain joules, C J s kg^{-1} = J(A s^2 kg^{-1}). The inverse of the quantity in parentheses is the SI unit of **magnetic flux density**, the **tesla**: 1T = 1 kg s^{-2}A^{-1}. Thus $\mu_B = 9.2742 \times 10^{-24}\ \text{J T}^{-1}$.

In Section 15-2B, we saw that the orientation of the angular momentum vector in a magnetic field is determined by its component on the z axis, L_z. Alternatively, we could say that the orientation is determined by the z component of the magnetic moment vector. Since the two vectors are related by Eq. (15-22), we can also write the components as

$$\mu_{\ell z} = -\frac{e}{2m_e}L_z$$

From Eq. (15-21), $L_z = m_\ell\hbar$, so

$$\mu_{\ell z} = -m_\ell\hbar\frac{e}{2m_e} = -m_\ell\mu_B$$

If the magnetic field strength is B, the energy change due to the magnetic field is given by

$$\Delta E_{\text{orb}} = \mu_{\ell z} B = -m_\ell \mu_B B \qquad \textbf{(15-24)}$$

Without the magnetic field, the energy level is degenerate. In the magnetic field, the level is split into several new levels, which are not degenerate. This behavior explains the phenomenon known as the *Zeeman effect*: the splitting of a line in a spectrum into several closely-spaced lines when a strong magnetic field is applied.

EXAMPLE 15-8: Obtain expressions for the energies of the new levels arising from the $n = 2$, $\ell = 1$ level of a hydrogen atom when the atom is placed in a strong magnetic field.

Solution: With no magnetic field, the energy of the atom is given by Eq. (15-16), so, for $z = 1$ and $n = 2$, $E_2 = -\frac{1}{4}E_H$. The change in energy is given by Eq. (15-24), $\Delta E_{\text{orb}} = E_2 - m_\ell \mu_B B$, so for $\ell = 1$, $m_\ell = 0, +1, -1$. In the magnetic field B there will be three levels having the following energies:

$$E_{2,1,0} = E_2 \qquad E_{2,1,1} = E_2 - \mu_B B \qquad E_{2,1,-1} = E_2 + \mu_B B$$

C. Spin angular momentum

Certain early experiments suggested that elementary particles have an angular momentum. Dirac's relativistic development of quantum mechanics showed that the operator for L_z (see Example 15-5) had to include a second term, i.e.,

$$\mathbf{M}_z = \mathbf{L}_z \pm \frac{1}{2}\left(\frac{h}{2\pi}\right)$$

Thus there are two energy levels, one slightly above and one slightly below the level given by L_z alone. This effect would be caused by an electron spinning on its axis, although this interpretation probably isn't correct. Nevertheless, we use the terms **electron spin, spin angular momentum** vector S, and **spin magnetic moment** vector $\boldsymbol{\mu}_s$. Continuing the analogy to orbital angular momentum, we express the magnitude of **S** by an equation similar to Eq. (15-20):

$$S = \sqrt{s(s+1)}\hbar \qquad \textbf{(15-25)}$$

Unlike ℓ, the quantum number s may have only one value for a particular elementary particle. Similarly, we express the component of **S** in the z direction in the form of Eq. (15-21):

$$S_z = m_s \hbar \qquad \textbf{(15-26)}$$

Note that we have introduced two quantum numbers:

1. s is the **spin quantum number**, or *spin*, and has only one value;
2. m_s is the quantum number for the z component and may have values $m_s = -s, -s+1, \ldots, s$. These numbers are analogous to ℓ and m_ℓ for orbital angular momentum.

EXAMPLE 15-9: Determine the values of the new quantum numbers for an electron.

Solution: Experimental results can be explained if we take $s = \frac{1}{2}$ for an electron (also for protons and neutrons). Since the range of m_s is $-s$ to $+s$, its values for an electron are restricted to $-\frac{1}{2}$ and $+\frac{1}{2}$.

As with L, S_x and S_y can't be measured, so the vector **S** must precess about the z axis. In an external magnetic field, the position of the vector is determined by the component on the z axis. Only two values of S_z may be calculated from Eq. (15-26) because there are only two values of m_s. Thus two orientations are possible, and as the two vectors rotate, they describe two cones. By Eq. (15-25), the magnitude (length) of this vector is $\sqrt{3}\,\hbar/2$.

D. Spin magnetic moment

The equations for the various aspects of spin magnetic moment are analogous to those for orbital magnetic moment, except that the magnetogyric ratio is twice as large as for orbital motion. Thus the **spin magnetic moment vector** is

$$\boldsymbol{\mu}_s = -\frac{e}{m_e}\mathbf{S} \tag{15-27}$$

Substituting the magnitude of S from Eq. (15-25), we get the magnitude of $\boldsymbol{\mu}_s$

$$\mu_s = -\frac{e}{m_e}\frac{h}{2\pi}\sqrt{s(s+1)} = -2\mu_B\sqrt{s(s+1)} \tag{15-28}$$

The component in the z direction is

$$\mu_{sz} = -\frac{e}{m_e}S_z = -2m_s\mu_B$$

For the orbital magnetic moment, the change of energy due to electron spin in a magnetic field is

$$\Delta E_{\text{spin}} = \mu_{sz}B = -2m_s\mu_B B \tag{15-29}$$

Since it's impossible to apply a magnetic field to either the orbital magnetic moment or the spin magnetic moment but not the other, the two energy effects always combine. The energies obtained from Eq. (15-24) in Example 15-8 are only approximations; the calculation must also include Eq. (15-29) to obtain the correct energy value. Combining the two equations gives

$$\Delta E_{\text{mag}} = -m_\ell\mu_B B - 2m_s\mu_B B$$

or

$$\Delta E_{\text{mag}} = -\mu_B B(m_\ell + 2m_s) \tag{15-30}$$

E. Spin wave function

We can maintain the Schrödinger equation with three degrees of freedom and the corresponding three quantum numbers and conveniently account for the fourth degree of freedom by introducing a **spin wave function** as a fourth part of the equation. The symbols α and β are used for positive m_s and negative m_s, respectively. Thus Eq. (15-17) becomes two equations:

$$\psi_{n\ell m_\ell m_s} = R_{n\ell}(r)\Theta_{\ell m_\ell}(\theta)\Phi_{m_\ell}(\phi)\alpha \quad \textbf{spin up} \tag{15-31}$$

and

$$\psi_{n\ell m_\ell m_s} = R_{n\ell}(r)\Theta_{\ell m_\ell}(\theta)\Phi_{m_\ell}(\phi)\beta \quad \textbf{spin down} \tag{15-32}$$

EXAMPLE 15-10: Write symbols for the two possible wave functions of a hydrogen atom in the ground state when spin is included.

Solution: The two wave functions could be simply $1s\alpha$ and $1s\beta$. Electrons are usually numbered for identification (even though they are indistinguishable), and the symbols become $1s(1)\alpha(1)$ and $1s(1)\beta(1)$. Sometimes the symbols used are $\psi_{1s}(1)\alpha(1)$ and $\psi_{1s}(1)\beta(1)$.

15-3. Approximation Methods

As soon as an atom includes two or more electrons, we can no longer solve the Schrödinger equation exactly for it. The difficulty lies in the potential energy, which now must include terms for the repulsion between electrons. Two methods are commonly used to obtain approximate solutions: variation and perturbation.

A. Variation method

The average, or expectation, value for a normalized wave function is expressed by Eq. (14-31). When we use the Hamiltonian operator and a normalized eigenfunction, Eq. (14-31) gives the ground-state energy as

$$E_g = \int \psi^* \mathbf{H} \psi \, d\tau \qquad \textbf{(15-33)}$$

The **variation theorem** says that, if any normalized, well-behaved function Y, which is not an eigenfunction, is used in Eq. 15-33, the energy obtained will be greater than or equal to E_g, i.e.,

VARIATION THEOREM
$$E_Y = \int_{-\infty}^{\infty} Y^* \mathbf{H} Y \, d\tau \geq E_g \qquad \textbf{(15-34)}$$

The *variation method* then consists of choosing various reasonable functions and calculating E_Y, trying to get the smallest value of E_Y that still is greater than E_g. Usually, the chosen functions contain parameters that are adjusted to minimize the calculated E_Y. The disadvantage of this method is that we have to *know* the true E_g in order to tell whether a minimum E and its wave function are close to being correct.

EXAMPLE 15-11: Assume a reasonable wave function for the particle-in-a-one-dimensional-box problem and calculate E. Compare the result to E_g.

Solution: The energy is given by Eq. (14-32). For the ground state, $n = 1$, and

$$E_g = \frac{h^2}{8ma^2} = 0.1250 \frac{h^2}{ma^2} \qquad \textbf{[Eq. (14-32)]}$$

For one dimension, the Hamiltonian operator is the x term of Eq. (14-22):

$$\mathbf{H}_x = -\frac{h^2}{8\pi^2 m} \frac{d^2}{dx^2}$$

Recall that an acceptable wave function must be zero at the walls of the box, i.e., at $x = 0$ and at $x = a$. A function that meets this requirement is $Y = x(a - x)$, or $Y = ax - x^2$. This function isn't normalized, so we must use the more general version of Eq. (15-3), obtained by division by $\int Y^* Y \, d\tau$.

$$E_Y = \frac{\int_{-\infty}^{\infty} Y^* \mathbf{H} Y \, d\tau}{\int_{-\infty}^{\infty} Y^* Y \, d\tau}$$

In this case, the limits are 0 and a, and we substitute for Y:

$$E_Y = \frac{\int_0^a (ax - x^2)\left(-\frac{h^2}{8\pi^2 m} \frac{d^2}{dx^2}\right)(ax - x^2) \, dx}{\int_0^a (ax - x^2)(ax - x^2) \, dx} = \frac{\frac{h^2 a^3}{24\pi^2 m}}{\frac{a^5}{30}} = \frac{5h^2}{4\pi^2 ma^2} = 0.1267 \left(\frac{h^2}{ma^2}\right)$$

We see that E_Y is greater than E_g, although not by much. Our first trial function was a good guess. (This illustration appears in several textbooks, probably because it's relatively simple and straightforward.)

B. Perturbation method

The *perturbation method* is appropriate when we can put the Hamiltonian operator in the form $\mathbf{H} = \mathbf{H}^\circ +$ a small term. The $\mathbf{H}^\circ$ part must be the operator of a system for which we can solve the Schrödinger equation. The extra, small term is a **perturbation** on the known Hamiltonian operator, producing the correct Hamiltonian operator for the system. The term is itself a Hamiltonian operator, $\mathbf{H}'$, so we can write

$$\mathbf{H} = \mathbf{H}^\circ + \mathbf{H}' \qquad \textbf{(15-35)}$$

The energy is given by

$$E = E^\circ + E' \qquad \textbf{(15-36)}$$

where E' is given by the average-value expression, Eq. (14-31), with $\mathbf{H}'$ as the operator and ψ° as the wave function. We illustrate this method in Section 15-4 by application to the helium atom.

15-4. Helium-Like Atoms

The potential energy of the helium atom contains three terms, two for the attraction between each electron and the nucleus and one for the repulsion between the two electrons. Since the nuclear charge is $2e$, $Q_1Q_2 = 2e(e)$, and V is

$$V = -\frac{2e^2}{r_1} - \frac{2e^2}{r_2} + \frac{e^2}{r_{1,2}}$$

Here r_1 and r_2 represent the distance of each electron from the nucleus and $r_{1,2}$ is the distance between the electrons. The general Hamiltonian operator is

$$\mathbf{H} = -\frac{h^2}{8\pi^2 m}(\nabla_1^2 + \nabla_2^2) + \frac{1}{4\pi\varepsilon_0}\left(-\frac{Ze^2}{r_1} - \frac{Ze^2}{r_2} + \frac{e^2}{r_{1,2}}\right) \qquad \textbf{(15-37)}$$

where ∇_1 and ∇_2 are restricted to the first and second electron, respectively.

A. Zero perturbation

If we ignore the repulsion term, i.e., omit $\mathbf{H}'$ and E' from Eqs. (15-35) and (15-36), Eq. (15-37) becomes the sum of two hydrogen-like Hamiltonian operators, one for each electron: $\mathbf{H} = \mathbf{H}(1) + \mathbf{H}(2)$. If we take the wave function as $\psi(1)\psi(2)$ and separate the Schrödinger equation into two equations, one for each ψ, then $E^\circ = E(1) + E(2)$. The expression for each E is the same as for the hydrogen-like atom, Eq. (15-16), so $E_n = -(Z^2/n^2)E_\text{H}$. The two E's are identical; thus

$$E(1) = E(2) = -\frac{Z^2}{n^2}E_\text{H}$$

and

$$E^\circ = E(1) + E(2) = 2\left(-\frac{Z^2}{n^2}E_\text{H}\right) \qquad \textbf{(15-38)}$$

EXAMPLE 15-12: The ground-state energy of helium has been found experimentally to be -79.0 eV. Compare this value to the zero-perturbation value from Eq. (15-38). What does the comparison tell us about the assumption we made?

Solution: With $Z = 2$ for helium and $n = 1$ for the ground state, Eq. (15-38) becomes

$$E^\circ_\text{He} = -2(2)^2E_\text{H} = -8E_\text{H} = -8(13.605\text{ eV}) = -108.84\text{ eV}$$

Comparing this result to -79.0 eV, we see that the discrepancy is large and that we can't ignore the interelectron repulsion.

B. First-order perturbation

We can put back the repulsion term as a small perturbation on $\mathbf{H}^\circ$, as defined in Section 15-4A, and Eq. (15-35) becomes

$$\mathbf{H} = \mathbf{H}^\circ + \frac{e^2}{r_{1,2}}$$

The energy is given by Eq. (15-36), where the perturbation energy E' is the expectation value for the operator $e^2/r_{1,2}$:

$$E' = \frac{\displaystyle\int_{-\infty}^{\infty} \psi^{\circ *}\left(\frac{e^2}{4\pi\varepsilon_0 r_{1,2}}\right)\psi^\circ\, d\tau}{\displaystyle\int_{-\infty}^{\infty} \psi^{\circ *}\psi^\circ\, d\tau} \qquad \text{(i)}$$

We obtain an approximate wave function by multiplying two $1s$ orbitals; i.e., $\psi^\circ \approx 1s(1)[1s(2)]$. When we substitute this function in eq. (i), we obtain

$$E' = \frac{5}{4} Z E_H \qquad \text{(ii)}$$

Substituting Eq. (15-38) and eq. (ii) in Eq. (15-36), we get

$$E^\circ = -2\frac{Z^2}{n^2} E_H + \frac{5}{4} Z E_H = \left(\frac{5}{4} Z - \frac{2Z^2}{n^2}\right) E_H \qquad \textbf{(15-39)}$$

This expression is valid for any helium-like atom or ion.

EXAMPLE 15-13: Calculate E for a helium atom, including the first-order perturbation, and compare the result to the experimental value, -79.0 eV, given in Example 15-12.

Solution: Instead of using Eq. (15-39) directly, let's calculate the perturbation energy from eq. (ii). For $Z = 2$,

$$E' = \frac{5}{4}(2)E_H = 2.5(13.605 \text{ eV}) = 34.01 \text{ eV}$$

Then, using Eq. (15-36), we get

$$E = E^\circ + E' = -108.84 + 34.01 = -74.83 \text{ eV}$$

This result is fairly close to the experimental value.

C. Complete wave function

1. Indistinguishability of electrons

We've pointed out that a spin function must be added to the original three-dimensional wave function, giving two possibilities for the hydrogen atom [Eqs. (15-31) and (15-32)]. With more than one electron, however, the situation becomes complicated. For two-electron spin functions, four combinations are possible:

$$\alpha(1)\alpha(2) \qquad \beta(1)\beta(2) \qquad \alpha(1)\beta(2) \qquad \alpha(2)\beta(1) \qquad \textbf{(15-40)}$$

The difficulty is that the last two forms require that we distinguish between electrons, but this is impossible.

EXAMPLE 15-14: Why is it impossible to say which electron has "spin up" and which has "spin down"?

Solution: The Heisenberg uncertainty principle tells us that we can't follow the path of an electron and thus know where it is, as we can with an object in classical mechanics.

We can obtain spin functions that don't require distinguishing between electrons by combining the last two forms in Eq. (15-40) and inserting a normalization constant of $1/\sqrt{2}$:

$$\frac{1}{\sqrt{2}}[\alpha(1)\beta(2) + \alpha(2)\beta(1)] \qquad \frac{1}{\sqrt{2}}[\alpha(1)\beta(2) - \alpha(2)\beta(1)] \qquad \textbf{(15-41)}$$

2. Symmetric and antisymmetric functions

A function is *symmetric* if interchanging two electrons doesn't change the function. If the function changes, it's *antisymmetric*. The change usually is one of sign.

EXAMPLE 15-15: Examine the first two spin functions in Eq. (15-40) and the two in Eq. (15-41) and determine which are symmetric and which are antisymmetric.

Solution: We interchange the electrons in each function and see whether the function has changed:

$$\alpha(1)\alpha(2) \longrightarrow \alpha(2)\alpha(1) \qquad \text{(no change)}$$

$$\beta(1)\beta(2) \longrightarrow \beta(2)\beta(1) \qquad \text{(no change)}$$

$$\frac{1}{\sqrt{2}}[\alpha(1)\beta(2)+\alpha(2)\beta(1)] \longrightarrow \frac{1}{\sqrt{2}}[\alpha(2)\beta(1)+\alpha(1)\beta(2)] \qquad \text{(no change)}$$

$$\frac{1}{\sqrt{2}}[\alpha(1)\beta(2)-\alpha(2)\beta(1)] \longrightarrow \frac{1}{\sqrt{2}}[\alpha(2)\beta(1)-\alpha(1)\beta(2)] = \frac{-1}{\sqrt{2}}[\alpha(1)\beta(2)-\alpha(2)\beta(1)] \qquad \text{(change)}$$

The first three functions are symmetric. In the last case, the result is the negative of the original, so the original function is antisymmetric.

---

3. Pauli exclusion principle

From relativistic quantum field theory, Pauli derived a rule known as the **Pauli exclusion principle**: *For a system of electrons, the wave function must be antisymmetric with respect to interchange of all coordinates of any two electrons.* This statement applies to particles with half-integral spin quantum numbers (**Fermi particles**, e.g., protons and neutrons). For particles with integral spin quantum numbers (**Bose particles**, e.g., deuterons and alpha particles), the rule is reversed—their wave functions must be symmetric. The practical rule that follows from the Pauli principle and is learned in general chemistry is as follows: *Two electrons in the same atom may not have the same values for all four quantum numbers.*

4. Wave function for ground state

Previously, we suggested a combination of two $1s$ orbitals as a wave function for helium, i.e., $\psi \approx 1s(1)1s(2)$, which is symmetric. The complete function must be antisymmetric, so the only spin function we're permitted to include is the antisymmetric one in Example 15-15. Then the approximate wave function becomes

$$\psi_g \approx 1s(1)1s(2)\frac{1}{\sqrt{2}}[\alpha(1)\beta(2) - \alpha(2)\beta(1)] \qquad \textbf{(15-42)}$$

note: We applied a rule here that may not be obvious at first. The only way to obtain an antisymmetric combined wave function is to combine a symmetric function with an antisymmetric function. If two symmetric or two antisymmetric functions are combined, the result is symmetric.

EXAMPLE 15-16: Obtain approximate wave functions for a helium atom that has one electron in a $1s$ orbital and one in a $2s$ orbital (an *excited state*).

Solution: The general approach (Example 15-10) is to write expressions for ψ's, then add spin functions. We could write

$$\psi = 1s(1)2s(2) \qquad \psi' = 1s(2)2s(1)$$

but these expressions imply distinguishability of electrons, so we must use combinations of them, as in Eq. (15-41):

$$\psi_a = \frac{1}{\sqrt{2}}[1s(1)2s(2) + 1s(2)2s(1)] \quad \text{and} \quad \psi_b = \frac{1}{\sqrt{2}}[1s(1)2s(2) - 1s(2)2s(1)]$$

When we incorporate spin functions with ψ_a and ψ_b, we have to remember the Pauli exclusion principle: The resulting function must be antisymmetric. We see that ψ_b is antisymmetric, so to obtain an antisymmetric result, we must use a symmetric spin function. There are three of these in Example 15-15, so we have three combinations. For convenience, we won't write out ψ_b each time:

$$\psi_1 = \psi_b\alpha(1)\alpha(2) \qquad \psi_2 = \psi_b\beta(1)\beta(2) \qquad \psi_3 = \psi_b\frac{1}{\sqrt{2}}[\alpha(1)\beta(2) + \alpha(2)\beta(1)]$$

These expressions represent the same energy level and are known as a *triplet state*. In a magnetic field, the three-fold degeneracy would be split, and three levels would appear.

What about ψ_a? It is symmetric, which means it requires an antisymmetric spin function. There's only one of these in Example 15-15, so there's only one combined ψ:

$$\psi_4 = \psi_a(2)^{-1/2}[\alpha(1)\beta(2) - \alpha(2)\beta(1)]$$

This expression represents a *singlet state* and is nondegenerate.

15-5. Multi-Electron Atoms

A. Approximation method of calculation

We can treat the lithium atom similarly to helium, but for larger atoms, we need other approximations. Most calculations are based on the *self-consistent-field method*, also known as the *Hartree–Fock method*. Two essential assumptions characterize this approach:

1. The nucleus and all electrons except the one of interest form a spherically symmetric field, which the one electron "sees." (This assumption is related to the concepts of *effective nuclear charge* and *screening effect*.)
2. The complete Schrödinger equation can be divided into one-electron, hydrogen-like equations, one for each electron.

The wave function then is the product of the one-electron, hydrogen-like wave functions, i.e., the orbitals, which are described by the usual four quantum numbers. Solving the equation is a process of successive approximations. We use the potential energy from the calculation for the first electron to improve the wave function for the calculation for the second electron, etc., returning to the first electron, until the orbitals don't change and thus are "self-consistent." The results are approximate, but the calculated energies tend to agree within about 1 percent with experimental values. Part of the discrepancy between the true value and the calculated value is the *correlation energy*, caused by interactions between electrons. The discrepancy is about 1 eV, which is significant in calculations involving chemical reactions.

EXAMPLE 15-17: Compare correlation energy to heats of reaction by converting 1 eV to kcal and to kJ.

Solution: Tables are available, so you could find the values there. Let's do it the hard way: We'll use conversion factors to obtain the values. We begin with

$$1 \text{ eV} = \text{electron charge} \times \text{voltage}$$
$$= (1.6022 \times 10^{-19}\text{ C})(1\text{ V})\left(\frac{\text{A s}}{\text{C}}\right)\left(\frac{\text{m}^2\text{ kg s}^{-3}\text{A}^{-1}}{\text{V}}\right)\left(\frac{\text{J}}{\text{kg m}^2\text{ s}^{-2}}\right) = 1.6022 \times 10^{-19}\text{ J}$$

For one mole of electrons, we multiply by Avogadro's constant,

$$1\text{ eV} = (1.6022 \times 10^{-19}\text{ J electron}^{-1})(6.0221 \times 10^{23}\text{ electron mol}^{-1})$$
$$= 9.648 \times 10^4\text{ J mol}^{-1} = 96.48\text{ kJ mol}^{-1}$$

and convert to calories:

$$1\text{ eV} = (96.48\text{ kJ mol}^{-1})\left(\frac{1}{4.184}\text{ cal J}^{-1}\right) = 23.06\text{ kcal mol}^{-1}$$

Whether you're familiar with heats of reaction expressed in kJ or in kcal, you can easily see that these numbers are a sizable percentage of most heats of reaction. Consequently correlation energies are important factors in chemical energy relationships.

The order of the calculated energy levels of the hydrogen-like orbitals gives us a picture of the arrangement of electrons in atoms. Many general chemistry textbooks include a figure that shows the changes in relative positions of energy levels as the atomic number increases. At low atomic number, the orbitals of the same n lie close together. As Z increases, all energies decrease because the electrons are pulled toward the nucleus. In the middle range of Z, energies of sublevels of different n intermix. At high Z, the energies of sublevels of the same n, in particular those of small n, again collect; the overlapping of principal levels disappears.

EXAMPLE 15-18: Identify and explain the important feature of the Periodic Table that illustrates the overlapping of the sublevel energies of different principal levels.

Solution: You should recall from your general chemistry course that this feature is the appearance of the *transition elements*, i.e., those elements having unfilled inner shells. Thus scandium ($Z = 21$) follows calcium ($Z = 20$) in the table, but Sc doesn't have properties that would place it in Group III under Al. The energy of an atom is less if an electron enters the $4s$ orbital than if it enters a $3d$ orbital, which means that the energy of the $4s$ orbital drops below that of $3d$ at $Z = 6$ or $Z = 7$ and stays there as Z increases. The energy of $3d$ begins to decrease faster than that of $4s$ and crosses it between $Z = 20$ and $Z = 21$. Thus when the nineteenth and twentieth electrons are added, $4s$ is the lowest-energy empty orbital, and they enter it. For the twenty-first electron, the energy of $3d$ is lower than that of $4p$, and the electron enters it.

B. Approximate method of representing electron configurations

The **aufbau principle** says that the hydrogen one-electron orbitals can be used for all atoms. In hydrogen, each n gives only one level (a degenerate one), but with more than one electron, the degeneracy is broken, and there are n energy levels for each n (one for each ℓ). We build an *orbital diagram* (rather than a strict hydrogen energy-level diagram) of the atoms by adding one electron at a time according to several rules. (We assume that the nucleus receives the proper number of protons and neutrons each time.) The rules, with some commentary, are

1. Place the added electron in the lowest available energy level, i.e., the ground state.
2. Place only two electrons in one orbital; i.e., two electrons may not have the same set of quantum numbers (Pauli exclusion principle); the two electrons in the same orbit must have opposite spin.
3. Place the added electron in an empty orbital when permitted by rule 1 (*Hund's Rule*); i.e., don't pair electrons until necessary.
4. Assign the same spin to single electrons in different equivalent orbitals (also Hund's rule).

note: Rules 3 and 4 lead to the configurations that give the lowest energy state for the atom.

Various diagrams are used to present a qualitative orbital diagram in which the energy levels are in proper order and electron placement is shown clearly. The quantitative differences between the actual energies of atoms are ignored; the energy levels decrease as Z increases, and the separations between levels change. We won't reproduce one of these large diagrams but will use a compact presentation, which should be familiar from your general chemistry course. A short-hand method is to write the orbital designations in increasing order of energy and to add superscript numbers to show the number of electrons in each orbital, e.g., for sodium, $1s^2 2s^2 2p^6 3s^1$. If you want to show the distribution of electrons in unfilled degenerate orbitals, you can show each of the equivalent orbitals, e.g., for nitrogen, $1s^2 2s^2 2p_x^1 2p_y^1 2p_z^1$. The complete list of orbitals in order of increasing energy when $Z = 1$, with the maximum number of electrons permitted in each orbital, is

$$1s^2 2s^2 2p^6 3s^2 3p^6 4s^2 3d^{10} 4p^6 5s^2 4d^{10} 5p^6 6s^2 4f^{14} 5d^{10} 6p^6 7s^2 5f^{14}$$

We call such a description of a particular atom its **electron configuration**.

EXAMPLE 15-19: Write the electron configurations for ^{15}P and ^{27}Co. Comment briefly on any special features of the configurations.

Solution: For ^{15}P, $1s^2 2s^2 2p^6 3s^2 3p^3$. We could break up the last term to show the three p orbitals: $3p_x^1 3p_y^1 3p_z^1$. This illustrates Hund's rule that electrons remain unpaired as long as possible.

For ^{27}Co, $1s^2 2s^2 2p^6 3s^2 3p^6 4s^2 3d^7$. Here we illustrate that the $4s$ orbital receives electrons before the $3d$ orbital does. If we wanted to show the d electron arrangement, we could write

$$3d^2_{xy}3d^2_{yz}3d^1_{xz}3d^1_{x^2y^2}3d^1_{z^2}$$

Since there are more than five d electrons, the sixth and seventh are forced to form pairs according to rule 1; i.e., they must go into the lowest energy level available.

Many atoms apparently have greater stability (lower energy) when an inner shell (the d and f shells) is filled or half-filled. This is the superficial explanation for an actual configuration of d^5s^1 or $d^{10}s^1$ when, according to the rules, the configuration should be d^4s^2 or d^9s^2. Similar effects occur with f orbitals. For heavier atoms, you should consult a table because several variations occur. Why are such configurations more stable? Several factors are involved, including electron repulsion, screening, and magnetic moment interactions.

EXAMPLE 15-20: Write the electron configuration of the copper atom, ^{29}Cu. What does it tell us about the expected charge on copper ions?

Solution: We write $1s^2 2s^2 2p^6 3s^2 3p^6 4s^2 3d^9$. However, the configuration beyond $3p^6$ is actually $4s^1 3d^{10}$, which we could have predicted from the general rule about filled and half-filled subshells.

From the actual configuration, we would predict that the atom readily loses one electron to form a Cu^+ ion, which, of course, it does. We also know that Cu^{2+} is common, which tells us that a second electron is lost relatively easily.

---

note: We often shorten the electron configuration by writing only the orbitals beyond the lowest noble gas contained in the sequence. Thus, the Cu atom becomes [Ar]$4s^1 3d^{10}$.

15-6. Experimental Verification

The property of atoms that provides the most clearcut illustration of periodicity also provides energy values that serve as a check against theoretically calculated values and support the electron configurations previously predicted. This property is the **ionization potential**, or the voltage required to create an ion, i.e., to remove an electron completely from an atom. Frequently, it's called the *ionization energy*, E_i; the units are usually eV. For the hydrogen atom in the ground state, we already have seen that $E_i = 13.60$ eV. The value is equal and opposite in sign to E_H, the ground-state energy from Eq. (15-16). When there are several electrons, each may be removed individually as the voltage increases. The voltages at which each electron is pulled off are the first, second, third,... ionization energies. The relative magnitudes of the voltages tell us how tightly the electrons are held, which ties in with the arrangement of electrons.

EXAMPLE 15-21: Consider the ionization energies of Na, Mg, and Al. What conclusions can we draw? Do the ionization energies agree with our energy-level diagram (as shown by the relative energies of the orbitals)?

$1E_i$

	E_i			
Element	1st	2nd	3rd	4th
Na	5.1	47.3	71.7	97.0
Mg	7.6	15.0	80.1	109.3
Al	6.0	18.8	28.4	120.0

Solution: Look at the points where the E_i undergoes a large increase. For Na, the big jump is between the 1st and 2nd E_i. This implies that one electron can be removed easily but that removal of the second is difficult. Thus Na seems to have one electron outside a filled shell. For Mg, the jump is between the 2nd and 3rd E_i, suggesting that Mg has two electrons outside the filled shells. For Al, we conclude that there are three outer electrons. These agree with Mg = [Ne]$2s^2$ and Al = [Ne]$2s^2 2p^1$.

EXAMPLE 15-22: The experimental first ionization energy of helium is 24.6 eV. Use information from previous Examples as needed to determine a theoretical value of $E_{i,\mathrm{He}}$ for comparison.

Solution: The ionization energy is the energy added to the atom to create the ion; i.e.,

$$E_{\mathrm{He}^+} = E_{\mathrm{He}} + E_{i,\mathrm{He}}$$
$$E_{i,\mathrm{He}} = E_{\mathrm{He}^+} - E_{\mathrm{He}}$$

In Example 15-12, assuming zero perturbation, we found $E^{\circ}_{\mathrm{He}} = -108$ eV, using Eq. (15-38). To obtain E_{He^+}, we recognize that the ion is a hydrogen-like species (one electron), which means that we can use Eq. (15-16): $E_n = -(Z^2/n^2)E_{\mathrm{H}}$. For He, $n = 1$, so

$$E_{\mathrm{He}^+} = Z^2 E_{\mathrm{H}} = -(2)^2(13.605 \text{ eV}) = -54.42 \text{ eV}$$

and

$$E_{i,\mathrm{He}} = -54.42 - (-108.84) = 54.42 \text{ eV}$$

This result is much too large, so let's try E_{He} from the first-order perturbation calculation. From Example 15-13, $E_{\mathrm{He}} = -74.83$ eV. Thus

$$E_{i,\mathrm{He}} = -54.42 - (-74.83) = 20.41 \text{ eV}$$

This isn't far from the experimental value of 24.6 eV. We can feel confident that the basis of the calculation is valid. The use of additional perturbation terms should improve the agreement.

SUMMARY

In Table S-15, we show the major equations of this chapter and the connections between them. You should determine the conditions that are necessary for the validity of each equation and the conditions imposed to make the connections. Other important things you should know are

1. The Schrödinger equation for hydrogen-like atoms is converted to spherical coordinates before solving [Eq. (15-4)].
2. The Schrödinger equation may be broken into three parts, one for each coordinate. Each part yields a quantum number and a wave function [Eqs. (15-15) and (15-17)].
3. The two angular wave functions may be combined to give an angular distribution function that provides a "shape" of the orbital [Eq. (15-19)].
4. The radial part yields an expression for the energy levels that is identical to that of Bohr and a most-probable distance of the electron from the nucleus that is identical to Bohr's ground-state radius.
5. The spin quantum number must be introduced arbitrarily, although it does appear when relativistic effects are considered.
6. The defining rules for the four quantum numbers are

$$n = 1, 2, 3, \ldots, \infty$$
$$\ell = 0, 1, 2, \ldots, (n-1)$$
$$m_\ell = 0, \pm 1, \pm 2, \ldots, \pm \ell$$
$$m_s = \pm \tfrac{1}{2}$$

7. The orbital angular momentum and its component in the z direction are quantized by ℓ and m_ℓ [Eqs. (15-20) and (15-21)].
8. The magnetic dipole moment and its component in the z direction are quantized by ℓ and m_ℓ [Eqs. (15-22) and (15-23)].
9. In an applied magnetic field, the degenerate energy levels are separated into several levels of different energies [Eq. (15-24)].
10. The spin angular moment, the spin magnetic moment, and their components in the z direction are quantized by s and m_s [Eqs. (15-25), (15-26), (15-27), and (15-28)].

TABLE S-15: Summary of Important Equations

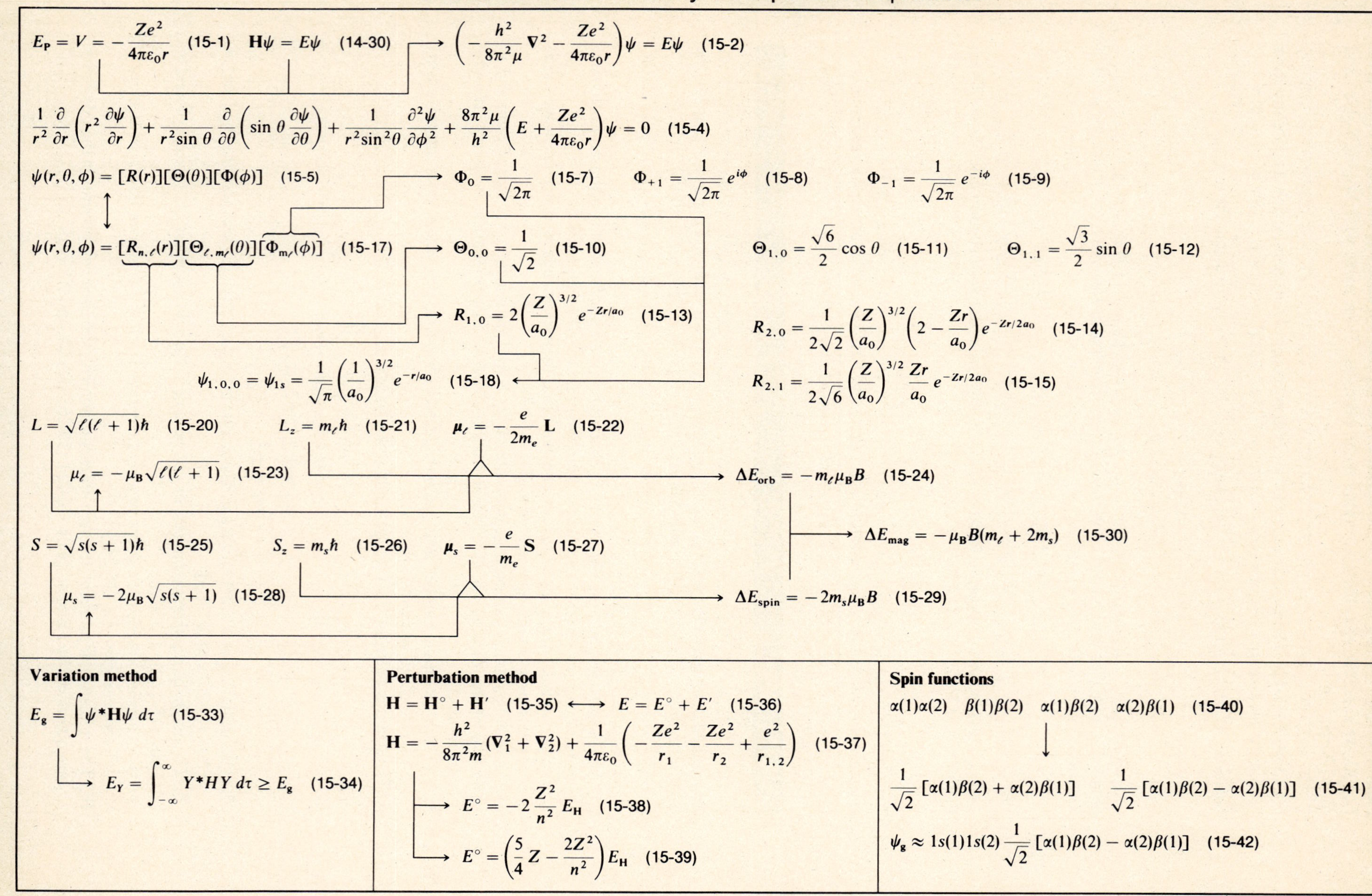

$$E_P = V = -\frac{Ze^2}{4\pi\varepsilon_0 r} \quad (15\text{-}1) \qquad \mathbf{H}\psi = E\psi \quad (14\text{-}30) \longrightarrow \left(-\frac{h^2}{8\pi^2\mu}\nabla^2 - \frac{Ze^2}{4\pi\varepsilon_0 r}\right)\psi = E\psi \quad (15\text{-}2)$$

$$\frac{1}{r^2}\frac{\partial}{\partial r}\left(r^2\frac{\partial\psi}{\partial r}\right) + \frac{1}{r^2\sin\theta}\frac{\partial}{\partial\theta}\left(\sin\theta\frac{\partial\psi}{\partial\theta}\right) + \frac{1}{r^2\sin^2\theta}\frac{\partial^2\psi}{\partial\phi^2} + \frac{8\pi^2\mu}{h^2}\left(E + \frac{Ze^2}{4\pi\varepsilon_0 r}\right)\psi = 0 \quad (15\text{-}4)$$

$$\psi(r,\theta,\phi) = [R(r)][\Theta(\theta)][\Phi(\phi)] \quad (15\text{-}5)$$

$$\Phi_0 = \frac{1}{\sqrt{2\pi}} \quad (15\text{-}7) \qquad \Phi_{+1} = \frac{1}{\sqrt{2\pi}}e^{i\phi} \quad (15\text{-}8) \qquad \Phi_{-1} = \frac{1}{\sqrt{2\pi}}e^{-i\phi} \quad (15\text{-}9)$$

$$\psi(r,\theta,\phi) = [R_{n,\ell}(r)][\Theta_{\ell,m_\ell}(\theta)][\Phi_{m_\ell}(\phi)] \quad (15\text{-}17)$$

$$\Theta_{0,0} = \frac{1}{\sqrt{2}} \quad (15\text{-}10) \qquad \Theta_{1,0} = \frac{\sqrt{6}}{2}\cos\theta \quad (15\text{-}11) \qquad \Theta_{1,1} = \frac{\sqrt{3}}{2}\sin\theta \quad (15\text{-}12)$$

$$R_{1,0} = 2\left(\frac{Z}{a_0}\right)^{3/2} e^{-Zr/a_0} \quad (15\text{-}13) \qquad R_{2,0} = \frac{1}{2\sqrt{2}}\left(\frac{Z}{a_0}\right)^{3/2}\left(2 - \frac{Zr}{a_0}\right)e^{-Zr/2a_0} \quad (15\text{-}14)$$

$$\psi_{1,0,0} = \psi_{1s} = \frac{1}{\sqrt{\pi}}\left(\frac{1}{a_0}\right)^{3/2} e^{-r/a_0} \quad (15\text{-}18) \qquad R_{2,1} = \frac{1}{2\sqrt{6}}\left(\frac{Z}{a_0}\right)^{3/2}\frac{Zr}{a_0}e^{-Zr/2a_0} \quad (15\text{-}15)$$

$$L = \sqrt{\ell(\ell+1)}\hbar \quad (15\text{-}20) \qquad L_z = m_\ell \hbar \quad (15\text{-}21) \qquad \boldsymbol{\mu}_\ell = -\frac{e}{2m_e}\mathbf{L} \quad (15\text{-}22)$$

$$\mu_\ell = -\mu_B\sqrt{\ell(\ell+1)} \quad (15\text{-}23) \qquad \longrightarrow \Delta E_{orb} = -m_\ell \mu_B B \quad (15\text{-}24)$$

$$\longrightarrow \Delta E_{mag} = -\mu_B B(m_\ell + 2m_s) \quad (15\text{-}30)$$

$$S = \sqrt{s(s+1)}\hbar \quad (15\text{-}25) \qquad S_z = m_s\hbar \quad (15\text{-}26) \qquad \boldsymbol{\mu}_s = -\frac{e}{m_e}\mathbf{S} \quad (15\text{-}27)$$

$$\mu_s = -2\mu_B\sqrt{s(s+1)} \quad (15\text{-}28) \qquad \longrightarrow \Delta E_{spin} = -2m_s\mu_B B \quad (15\text{-}29)$$

Variation method

$$E_g = \int \psi^*\mathbf{H}\psi\, d\tau \quad (15\text{-}33)$$

$$\longrightarrow E_Y = \int_{-\infty}^{\infty} Y^*HY\, d\tau \geq E_g \quad (15\text{-}34)$$

Perturbation method

$$\mathbf{H} = \mathbf{H}^\circ + \mathbf{H}' \quad (15\text{-}35) \longleftrightarrow E = E^\circ + E' \quad (15\text{-}36)$$

$$\mathbf{H} = -\frac{h^2}{8\pi^2 m}(\nabla_1^2 + \nabla_2^2) + \frac{1}{4\pi\varepsilon_0}\left(-\frac{Ze^2}{r_1} - \frac{Ze^2}{r_2} + \frac{e^2}{r_{1,2}}\right) \quad (15\text{-}37)$$

$$\longrightarrow E^\circ = -2\frac{Z^2}{n^2}E_H \quad (15\text{-}38)$$

$$\longrightarrow E^\circ = \left(\frac{5}{4}Z - \frac{2Z^2}{n^2}\right)E_H \quad (15\text{-}39)$$

Spin functions

$$\alpha(1)\alpha(2) \quad \beta(1)\beta(2) \quad \alpha(1)\beta(2) \quad \alpha(2)\beta(1) \quad (15\text{-}40)$$

$$\downarrow$$

$$\frac{1}{\sqrt{2}}[\alpha(1)\beta(2) + \alpha(2)\beta(1)] \qquad \frac{1}{\sqrt{2}}[\alpha(1)\beta(2) - \alpha(2)\beta(1)] \quad (15\text{-}41)$$

$$\psi_g \approx 1s(1)1s(2)\frac{1}{\sqrt{2}}[\alpha(1)\beta(2) - \alpha(2)\beta(1)] \quad (15\text{-}42)$$

11. The variation method of finding approximate wave functions and energies involves determining the energies for many trial functions, in order to find the function that gives the minimum energy [Eqs. (15-33) and (15-34)].
12. The perturbation method of finding approximate wave functions and energies involves finding a Hamiltonian operator of a system for which the Schrödinger equation can be solved and adding to it terms to fit the given system [Eqs. (15-35), (15-36), and (15-37)].
13. A spin wave function must be included with the functions for the three coordinates to give the complete wave function. For an electron, there are two spin wave functions, one for each of the opposite directions of spin [Eqs. (15-31) and (15-32)].
14. The Pauli exclusion principle may be stated as follows: (**a**) for a system of electrons, the wave function must be antisymmetric with respect to the interchange of all coordinates of any two electrons; or (**b**) two electrons in an atom can't have the same set of quantum numbers [Eq. (15-42)].
15. In helium-like atoms, the perturbation method gives an expression for the ground-state energy in terms of the atomic number and the ground-state energy of hydrogen [Eqs. (15-38) and (15-39)].
16. When more than two or three electrons are in an atom or ion, approximate calculations are made by the self-consistent field method.
17. Electron configuration diagrams are based on the assumption that the orbitals and energy-level relationships of the hydrogen atom can be extended to all atoms (Aufbau principle).
18. The electron configuration for a particular atom in the ground state is governed by the Pauli exclusion principle and Hund's rules.
19. Experimental ionization energies agree well with those calculated from wave functions by approximation methods and qualitatively verify the picture of the valence-electron arrangement in atoms.

SOLVED PROBLEMS

Hydrogen-Like Atoms

PROBLEM 15-1 Obtain the normalization constant for the Φ wave functions, Eq. (15-6).

Solution: The normalization condition is Eq. (14-25). You may rewrite it for this situation as

$$\int_0^{2\pi} \Phi^*\Phi \, d\phi = 1 \qquad \textbf{(a)}$$

The limits are the range of values of ϕ. From Eq. (15-6), $\Phi = Ne^{im_\ell\phi}$, so $\Phi^* = Ne^{-im_\ell\phi}$. The product is

$$\Phi^*\Phi = Ne^{-im_\ell\phi}Ne^{im_\ell\phi} = N^2$$

and the integral becomes

$$N^2 \int_0^{2\pi} d\phi = 1$$

Integrating yields

$$N^2(2\pi - 0) = 1 = N^2(2\pi)$$

$$N^2 = \frac{1}{2\pi}$$

$$N = \frac{1}{\sqrt{2\pi}}$$

PROBLEM 15-2 Show that $\Phi_x = (\sqrt{\pi})^{-1}\cos\phi$ is normalized. (See Example 15-2.)

Solution: You need to substitute the function and its complex conjugate into the normalization condition,

Eq. (14-25), rewritten as eq. (a) for Problem 15-1:

$$\int_0^{2\pi} \Phi^*\Phi \, d\phi = 1$$

In this case $\Phi^* = \Phi$, so for the product you have

$$\Phi^2 = \left(\frac{\cos\phi}{\sqrt{\pi}}\right)^2 = \frac{1}{\pi}\cos^2\phi$$

The integral becomes

$$\frac{1}{\pi}\int_0^{2\pi} \cos^2\phi \, d\phi = 1 \tag{b}$$

From a table of integrals,

$$\int \cos^2 u \, du = \frac{u}{2} + \frac{1}{4}\sin 2u + C$$

When you substitute, the integral in eq. (b) becomes

$$\frac{\phi}{2} + \frac{1}{4}\sin 2\phi \Big|_0^{2\pi} = \left(\frac{2\pi}{2} + \frac{1}{4}\sin 4\pi\right) - (0 + 0) \tag{c}$$

Since $\sin 4\pi = \sin 0 = 0$, eq. (c) reduces to π, and eq. (b) becomes $(1/\pi)(\pi) = 1$, or $1 = 1$. Thus the wave function is properly normalized.

PROBLEM 15-3 Determine whether the following wave functions are orthogonal (see Example 15-2):

$$\Phi_x = (\sqrt{\pi})^{-1}\cos\phi \qquad \Phi_y = i(\sqrt{\pi})^{-1}\sin\phi$$

Solution: You use Eq. (14-26) to test for orthogonality and can rewrite it for one coordinate, using Φ_x and Φ_y, as

$$\int_0^{2\pi} \Phi_y^*\Phi_x \, d\phi = 0 \tag{a}$$

Since $\Phi_y = (i \sin\phi)/\sqrt{\pi}$, $\Phi_y^* = (\sin\phi)/\sqrt{\pi}$, and the integral in eq. (a) becomes

$$\int_0^{2\pi} \frac{\sin\phi}{\sqrt{\pi}}\frac{\cos\phi}{\sqrt{\pi}} \, d\phi = \frac{1}{\pi}\int_0^{2\pi} \sin\phi\cos\phi \, d\phi = 0 \tag{b}$$

In a table of integrals, you'll find

$$\int \sin mu \cos nu \, du = -\frac{\cos(m+n)u}{2(m+n)} - \frac{\cos(m-n)u}{2(m-n)} + C$$

Here, $m = n = 1$, so

$$\int \sin\phi\cos\phi \, d\phi = -\frac{\cos 2\phi}{2(2)} - \frac{\cos 0}{0} + C$$

The second term is indeterminate, but all terms cancel when you insert the limits. Thus eq. (b) becomes

$$\frac{1}{\pi}\left(-\frac{\cos 2\phi}{4}\right)\Bigg|_0^{2\pi} = 0 - 0 = 0$$

and the wave functions are orthogonal.

PROBLEM 15-4 **(a)** Calculate the difference between the energy levels for principal quantum numbers 4 and 6 in a hydrogen atom. **(b)** If this energy were released as a photon, determine the region of the spectrum where it would lie.

Solution:

(a) You worked similar problems in Chapter 14, using Bohr's formula. This time use Eq. (15-16). As usual, $\Delta E = E_f - E_i$. The release of a photon decreases n, a change from $n = 6$ to $n = 4$, so $\Delta E = E_4 - E_6$.

Substituting for the E's from Eq. (15-16), with $Z = 1$ and $E_H = 2.180 \times 10^{-18}$ J, you get

$$\Delta E = -\frac{1}{4^2}E_H - \left(-\frac{1}{6^2}\right)E_H = \left(\frac{1}{6^2} - \frac{1}{4^2}\right)E_H$$

$$= (2.180 \times 10^{-18}\ \text{J})\left(\frac{1}{36} - \frac{1}{16}\right) = (2.180 \times 10^{-18})(-0.034\,72)\ \text{J} = -7.57 \times 10^{-20}\ \text{J}$$

(b) You could calculate the wavelength using Eq. (14-6), with $\nu = c/\lambda$, but you should recall that spectral series are identified by the lower energy level. For $n_f = 1$, the region is the ultraviolet (Lyman series); for $n_f = 2$, the region is the visible (Balmer series); all others lie in the infrared or far infrared regions. Thus you should predict that the frequencies for $n_f = 4$ fall in the transition region between the infrared and far infrared regions.

PROBLEM 15-5 **(a)** Determine the degree of degeneracy of the $n = 2$ level of the hydrogen atom (omitting spin). **(b)** List the quantum numbers for each of the wave functions involved.

Solution:

(a) For a degenerate energy level, more than one wave function corresponds to the same energy, or eigenvalue. (See Section 14-4B.) If $n = 2$, ℓ can be 0 or 1, but $\ell = 1$ carries with it three $(2\ell + 1)$ values of m_ℓ: $+1, 0, -1$. Omitting spin, four combinations are possible, and the second level is four-fold degenerate. Thus there are four orbitals: one s and three p.

(b)

n	ℓ	m_ℓ
2	0	0
2	1	+1
2	1	0
2	1	−1

(b) Each set gives a different wave function, but the four wave functions have the same energy in the absence of a magnetic field. If spin were included, the two values of m_s would double the number of combinations, giving eight-fold degeneracy.

PROBLEM 15-6 Write the total wave function for the state of a hydrogen atom for which $n = 2$, $\ell = 1$, and $m_\ell = 0$. Ignore the electron spin.

Solution: The total wave function is given by Eq. (15-17), i.e., $\psi_{2,1,0} = (R_{2,1})(\Theta_{1,0})(\Phi_0)$. You need to find the expressions for the three parts and multiply them. In order, these parts are represented by Eqs. (15-15), (15-11), and (15-7). Putting them together and letting $Z = 1$, you have

$$\psi_{2,1,0} = \psi_{2p_z} = \frac{1}{2\sqrt{6}}\left(\frac{1}{a_0}\right)^{3/2}\left(\frac{r}{a_0}\right)e^{-r/2a_0}\left[\frac{\sqrt{6}}{2}\cos\theta\right]\left[\frac{1}{\sqrt{2\pi}}\right]$$

Now, collect the coefficients and combine them:

$$\frac{1}{2\sqrt{6}}\left(\frac{\sqrt{6}}{2}\right)\left(\frac{1}{\sqrt{2\pi}}\right) = \frac{1}{4\sqrt{2\pi}}$$

Substituting, you obtain

$$\psi_{2p_z} = \frac{1}{4\sqrt{2\pi}}\left(\frac{1}{a_0}\right)^{5/2} re^{-r/2a_0}\cos\theta$$

PROBLEM 15-7 **(a)** In Problem 15-6, if you had been asked for the wave function for the state $n = 2$, $\ell = 1$, $m_\ell = 1$, what difficulty would have arisen? **(b)** Show how this complication can be handled by obtaining the angular wave function $\Theta\Phi$. (The radial part, $R_{2,1}$, is still the same.)

Solution:

(a) If you use Eq. (15-8) for Φ, the result has an imaginary part.

(b) You can eliminate this complication by combining the functions for $m_\ell = +1$ and $m_\ell = -1$ into two real functions, as shown in Example 15-2. You can't correlate the new Φ_x and Φ_y with either value of m_ℓ, so you should obtain two answers. Writing Eq. (15-19) for the two possibilities, you get

$$\psi_{a,x} = \Theta_{1,1}\Phi_x \quad \text{and} \quad \psi_{a,y} = \Theta_{1,1}\Phi_y$$

Substituting expressions for Φ_x and Φ_y (Example 15-2), and $\Theta_{1,1}$ [Eq. (15-12)] yields

$$\psi_{a,x} = \frac{\sqrt{3}}{2}\sin\theta\left(\frac{\cos\phi}{\sqrt{\pi}}\right) = \left(\frac{3}{4\pi}\right)^{1/2}\sin\theta\cos\phi$$

and

$$\psi_{a,y} = \frac{\sqrt{3}}{2}\sin\theta\left(\frac{\sin\phi}{\sqrt{\pi}}\right) = \left(\frac{3}{4\pi}\right)^{1/2}\sin\theta\sin\phi$$

You have to multiply these expressions by $R_{2,1}$ to obtain the complete wave functions for the $2p_x$ and $2p_y$ orbitals.

Angular Momentum and Magnetic Moment

PROBLEM 15-8 Determine (**a**) the magnitude of the angular momentum vector and (**b**) the possible values of L_z for a $2p$ electron in a hydrogen-like atom.

Solution:

(**a**) The magnitude of **L** is given by Eq. (15-20). For a p electron, $\ell = 1$, so

$$L = \sqrt{\ell(\ell+1)}\hbar = \sqrt{1(2)}\hbar = \sqrt{2}(1.0546 \times 10^{-34}\text{ J s}) = 1.4914 \times 10^{-34}\text{ J s}$$

(**b**) You calculate the values of the component in the z direction using Eq. (15-21):

$$L_z = m_\ell\hbar = m_\ell(1.0546 \times 10^{-34}\text{ J s})$$

Since $\ell = 1$ (a p electron), there are three values of m: 0, $+1$, and -1. Thus L_z may have the three values $+1.0546 \times 10^{-34}$ J s, 0, and -1.0546×10^{-34} J s.

PROBLEM 15-9 Describe two differences between the Bohr theory and the quantum-mechanical theory of quantization of angular momentum.

Solution: One fundamental difference is that Bohr had to assume (or postulate) that angular momentum is quantized ($L = n\hbar$), whereas in the quantum-mechanical approach, the quantization arises from the mathematics of solving the wave equation ($L = \sqrt{\ell(\ell+1)}\hbar$). Another significant difference is that Bohr used the principal quantum number n. In quantum mechanics, n is associated with the energy, not the angular momentum, and the azimuthal quantum number ℓ quantizes the orbital angular momentum.

Approximation Methods

PROBLEM 15-10 Verify one aspect of the variation theorem by calculating E for a particle in a one-dimensional box, using the correct wave function, and seeing if the derived E agrees with the accepted E_g for $n = 1$ in Chapter 14. [See Eqs. (14-32) and (14-33).]

Solution: The normalized wave function, expressed as a sine function rather than as an exponential one, is given by Eq. (14-33). For $n = 1$, the trial function is

$$Y = \left(\frac{2}{a}\right)^{1/2}\sin\frac{\pi x}{a} = Y^* \qquad \textbf{(a)}$$

Use the Hamiltonian operator from Example 15-11 and, because the wave function is normalized, Eq. (15-34) for E_Y.

$$E_Y = \int_0^a \left(\frac{2}{a}\right)^{1/2}\sin\frac{\pi x}{a}\left(-\frac{h^2}{8\pi^2 m}\frac{d^2}{dx^2}\left(\frac{2}{a}\right)^{1/2}\sin\frac{\pi x}{a}\,dx\right)$$

or

$$E_Y = \frac{2}{a}\left(-\frac{h^2}{8\pi^2 m}\right)\int_0^a \sin\frac{\pi x}{a}\frac{d^2}{dx^2}\sin\frac{\pi x}{a}\,dx \qquad \textbf{(b)}$$

Take the two derivatives separately:

$$\frac{d}{dx}\sin\frac{\pi x}{a} = \cos\frac{\pi x}{a}\frac{d}{dx}\left(\frac{\pi x}{a}\right) = \frac{\pi}{a}\cos\frac{\pi x}{a}$$

and

$$\frac{d}{dx}\left(\frac{\pi}{a}\cos\frac{\pi x}{a}\right) = \frac{\pi}{a}\frac{d}{dx}\left(\cos\frac{\pi x}{a}\right) = -\left(\frac{\pi}{a}\right)^2\sin\frac{\pi x}{a}$$

When you substitute the second derivative into eq. (b), you get

$$E_Y = -\frac{h^2}{4\pi^2 ma}\left(-\frac{\pi}{a}\right)^2 \int_0^a \sin^2\frac{\pi x}{a}\,dx \tag{c}$$

For the integration, let $u = \frac{\pi x}{a}$; then $du = \frac{\pi}{a}\,dx$, and $dx = \frac{a}{\pi}\,du$, and the integral becomes

$$\frac{a}{\pi}\int_0^a \sin^2 u\,du \tag{d}$$

In a table of integrals, you'll find

$$\int \sin^2 u\,du = \frac{u}{2} - \frac{1}{4}\sin 2u + C$$

For $u = (\pi x)/a$, eq. (d) becomes

$$\frac{a}{\pi}\left[\frac{\pi x}{2a} - \frac{1}{4}\sin\frac{2\pi x}{a}\right]_0^a = \frac{a}{\pi}\left(\frac{\pi a}{2a} - \frac{1}{4}\sin\frac{2\pi a}{a} - 0 + \frac{1}{4}\sin 0\right) = \frac{a}{\pi}\left(\frac{\pi}{2} - 0 - 0 + 0\right) = \frac{a}{2}$$

Now
$$E_Y = \left(\frac{h^2}{4ma^3}\right)\left(\frac{a}{2}\right) = \frac{h^2}{8ma^2}$$

From Eq. (14-32), $E_1 = E_g = h^2/8ma^2$.
Thus, using the correct wave function in the variation method gives the correct ground-state energy.

Helium-Like Atoms

PROBLEM 15-11 Obtain the algebraic expression for E', the first-order perturbation energy for the He atom. Use the product of two $1s$ wave functions as the approximate wave function.

Solution: The wave functions of the two hydrogen-like $1s$ orbitals are given by Eq. (15-18), but with $Z = 2$ in the $R_{1,0}$ part. Their product is

$$\psi^\circ = 1s(1)[1s(2)] = \left(\frac{1}{\sqrt{\pi}}\right)\left(\frac{2}{a_0}\right)^{3/2}(e^{-2r_1/a_0})\left(\frac{1}{\sqrt{\pi}}\right)\left(\frac{2}{a_0}\right)^{3/2}(e^{-2r_2/a_0})$$

and
$$\psi^\circ = \psi^{\circ *} = \frac{1}{\pi}\left(\frac{2}{a_0}\right)^3 e^{-2r_1/a_0}e^{-2r_2/a_0} \tag{a}$$

You can calculate the expectation value of E' from Eq. (14-31):

$$\bar{A} = \int \psi^* A\psi\,d\tau$$

and
$$E' = \int_{-\infty}^{\infty} \psi^{\circ *}\left(\frac{e^2}{4\pi\varepsilon_0 r_{1,2}}\right)\psi^\circ\,d\tau \tag{b}$$

When you substitute eq. (**a**) in eq. (**b**) and separate the coordinates for each particle, you get

$$E' = \left(\frac{8}{\pi a_0^3}\right)^2 \int_{-\infty}^{\infty}\int_{-\infty}^{\infty} \frac{e^2}{4\pi\varepsilon_0 r_{1,2}}(e^{-2r_1/a_0}e^{-2r_2/a_0})^2\,d\tau_1\,d\tau_2$$
$$= \frac{5}{16}\left(\frac{e^2}{\pi\varepsilon_0 a_0}\right) = \frac{5}{2}E_H = \frac{5}{2}(13.605\text{ eV}) = 34.013\text{ eV} = \frac{5}{2}(2.1798\times 10^{-18}\text{ J}) = 5.4495\times 10^{-18}\text{ J}$$

PROBLEM 15-12 Use the results of the perturbation treatment of helium to calculate the ground-state energies of the helium-like ions Li^+ and Be^{2+}. Compare your results to the experimental values: -197 eV and -370 eV.

Solution: The expression for E from the perturbation method is Eq. (15-39):

$$E^\circ = \left(\frac{5}{4}Z - 2\frac{Z^2}{n^2}\right)E_H$$

where $E_H = 13.61$ eV and $n = 1$ for both ions. For Li^+, $Z = 3$, so

$$E^\circ = \left[\frac{5}{4}(3) - 2(3)^2\right](13.61) = (3.75 - 18)(13.61) = -193.9\text{ eV}$$

For Be^{2+}, $Z = 4$, so

$$E = \left[\frac{5}{4}(4) - 2(4)^2\right](13.61) = (5 - 32)(13.61) = -367.5 \text{ eV}$$

Both results agree quite well with the experimental values.

PROBLEM 15-13 A helium atom has many possible excited states, one of which is one electron in the $1s$ orbital and the other electron in the $2p_z$ orbital. **(a)** Write two correct wave functions for this state of the atom. **(b)** Which, if either, wave function is asymmetric? If one is, why is it?

Solution:

(a) Your first impulse probably is to write $1s(1)2p_z(2)$ and $1s(2)2p_z(1)$, and you should do so because you need these orbitals. They are not suitable alone, however, because they imply the distinguishability of electrons. You have to combine them into functions that do not require a specific electron to be in a particular orbital. You can usually combine them by adding and subtracting. Therefore you should write, with N for the normalizing factor,

$$\psi_+ = N[1s(1)2p_z(2) + 1s(2)2p_z(1)] \quad \text{and} \quad \psi_- = N[1s(1)2p_z(2) - 1s(2)2p_z(1)]$$

(b) The second expression is asymmetric, because interchanging the electrons gives

$$N[1s(2)2p_z(1) - 1s(1)2p_z(2)]$$

which is the negative of the original.

Multi-Electron Atoms

PROBLEM 15-14 Write the ground-state electron configurations for ^{14}Si, ^{24}Cr, and ^{34}Se, showing the separation of ℓ or d orbitals when they aren't completely full or completely empty.

Solution: Use the electron-configuration method. The number of electrons must equal the atomic number.

For ^{14}Si, $1s^2 2s^2 2p^6 3s^2 3p_x^1 3p_y^1$

For ^{24}Cr, $1s^2 2s^2 2p^6 3s^2 3p^6 4s^1 3d_{xy}^1 3d_{yz}^1 3d_{xz}^1 3d_{x^2y^2}^1 3d_{z^2}^1$

For ^{34}Se, $1s^2 2s^2 2p^6 3s^2 3p^6 4s^2 3d^{10} 4p_x^2 4p_y^1 4p_z^1$

Note that the Cr atom, expected to be $4s^2 3d^4$, changes to the half-filled configuration $4s^1 3d^5$. In Si, the two p electrons remain separate (and have the same spin). In Se, the fourth p electron had to pair with one of the first three (and has an opposite spin).

PROBLEM 15-15 Write the expected electron configurations of ^{46}Pd, ^{57}La, ^{58}Ce, and ^{64}Gd and then look up the actual configurations. Start with the nearest rare gas.

Solution:

	Expected	Actual
^{46}Pd	$[Kr]5s^2 4d^8$	$[Kr]4d^{10}$
^{57}La	$[Xe]6s^2 4f^1$	$[Xe]6s^2 5d^1$
^{58}Ce	$[Xe]6s^2 4f^2$	$[Xe]6s^2 4f^2$
^{64}Gd	$[Xe]6s^2 4f^8$	$[Xe]6s^2 4f^7 5d^1$

In Pd, the stability of the filled subshell is great enough to pull both electrons from $5s$. In La, $5d$ gets an electron before $4f$, but in the next element, Ce, the $5d$ electron switches into $4f$, giving the predicted configuration. In Gd, the half-filled-shell effect pushes an electron into $5d$, so that $4f$ will have only 7.

PROBLEM 15-16 Offer a quantum-mechanical explanation for the fact that, in the Aufbau process, the $4s$ orbital receives its electrons before the $3d$ orbital gets one.

Solution: The explanation involves the radial probability densities for each of the orbitals (see Figure 15-4). For s orbitals, the curve has as many maxima as n. In the $4s$ case, therefore, four maxima mean four regions where the probability of finding the electron is good. These regions lie among the inner energy levels ($n = 1, 2, 3$), so s orbitals are called *penetrating orbitals*. Penetration brings the electron close to the nucleus, reducing the potential energy. However, the 3*d* orbital has one broad maximum, lying below the main maximum of 4*s* but above the three subsidiary maxima. The result is that the atom has a lower energy if the electron enters 4*s* than if it enters 3*d*.

PROBLEM 15-17 Using the first ionization energy of sodium (see Example 15-21), calculate the *effective nuclear charge* (Z_{eff}) of sodium. This is the value of Z that makes the expression derived for the energy of hydrogen give the correct E_i for a hydrogen-like atom.

Solution: The expression for the energy of a hydrogen-like atom is obtained from Eq. (15-16):

$$E_n = -\left(\frac{Z^2}{n^2}\right)E_H = -\left(\frac{Z^2}{n^2}\right)(13.6 \text{ eV})$$

The ionization potential is the negative of E_n. For sodium, the electron is in the third level ($n = 3$), so

$$E_i = -E_n = \frac{Z_{eff}^2}{(3)^2}(13.6 \text{ eV}) = 5.1 \text{ eV}$$

$$Z_{eff}^2 = \frac{5.1(9)}{13.6}$$

$$Z_{eff} = 1.8$$

Supplementary Exercises

PROBLEM 15-18 In spherical coordinates the range of θ is ________ to ________. The letter ________ represents the quantum number $\ell = 4$ in electronic configurations. All the values of m for $\ell = 3$ (are) (aren't) included in 0, ± 1, ± 2, ± 3. The degeneracy of the $n = 3$ level in a hydrogen atom including electron spin is (the same as) (two times) (four times) that of omitting electron spin. A state of the hydrogen atom (can) (can't) have $n = 2, \ell = 2, m = +1, s = -\frac{1}{2}$ as a set of quantum numbers. In the absence of a magnetic field, spin has ________ effect on the energy levels in a hydrogen-like atom. When a system has two or more electrons, the wave function must be (symmetric) (asymmetric). The principal energy level for which $n = 4$ has ________ energy sublevels. The ground-state electron configuration of ^{42}Mo is ________. A large energy-level increase occurs between the ________ and ________ ionization potentials of Ga.

PROBLEM 15-19 The angular wave function for $\ell = 2, m = 0$ is

$$\psi_a = \left(\frac{\sqrt{5}}{4\sqrt{\pi}}\right)(3\cos^2\theta - 1)$$

Is it orthogonal with the angular wave function $\psi_{a,x}$ obtained in Problem 15-7?

PROBLEM 15-20 Write an expression in terms of a_0 for the most probable distance from the nucleus to the electron in a 1*s* orbital of a hydrogen-like atom.

PROBLEM 15-21 (**a**) What is the magnitude of the angular momentum vector of a *d* electron? (**b**) What is the maximum value of L_z?

PROBLEM 15-22 For a hydrogen atom (**a**) calculate the change in energy when the degeneracy of the $n = 2, \ell = 1$ level is split by a magnetic field of 1.00×10^5 gauss. (**b**) Compare this energy change to the energy difference between the first and second principal energy levels by calculating the ratio of the smaller to the larger.

PROBLEM 15-23 Calculate the ground-state energy of the ion B^{3+}, using the result of the perturbation treatment of the helium atom.

PROBLEM 15-24 Write a complete wave function for a helium atom in the excited state ($2s^2$) that satisfies the exclusion principle and the requirement of indistinguishability of electrons.

PROBLEM 15-25 How many orbitals are in the principal energy level for which $n = 4$?

PROBLEM 15-26 Calculate the effective nuclear charge of potassium. ($E_i = 4.3$ eV.)

Answers to Supplementary Exercises

15-18 0, π; g; are; two times (18:9); can't; no; asymmetric; 4; [Kr]$5s4d^5$; third, fourth **15-19** yes **15-20** a_0/Z **15-21** (**a**) 2.58×10^{-34} J s; (**b**) 2.109×10^{-34} J s **15-22** (**a**) 9.274×10^{-23} J; (**b**) 5.67×10^{-5} **15-23** -595.4 eV **15-24** $2s(1)2s(2)\frac{1}{\sqrt{2}}[\alpha(1)\beta(2) - \alpha(2)\beta(1)]$ **15-25** 16 **15-26** 2.25

16 QUANTUM THEORY OF MOLECULES

THIS CHAPTER IS ABOUT

- ☑ The Molecular Problem
- ☑ The Molecular Orbital Method: Hydrogen Molecule Ion, H_2^+
- ☑ Molecular Orbital Theory: Diatomic Molecules
- ☑ Valence Bond Theory
- ☑ Electron Spin
- ☑ Polar Molecules

16-1. The Molecular Problem

A. Potential energy and dissociation energy

When two hydrogen atoms approach each other from infinite separation to form a molecule, the potential energy decreases because of an attraction between the atoms. At close range, a repulsive force dominates, and the potential energy rises rapidly (see Figure 16-1). The minimum E_p occurs at the **equilibrium distance** R_{eq} between the nuclei, and the corresponding energy is the **equilibrium dissociation energy** D_{eq} (also called the *electronic binding energy*.) An experimental value from spectroscopy for the ground state is called the **spectroscopic dissociation energy** D_0. These values differ by the ground-state energy of a harmonic oscillator:

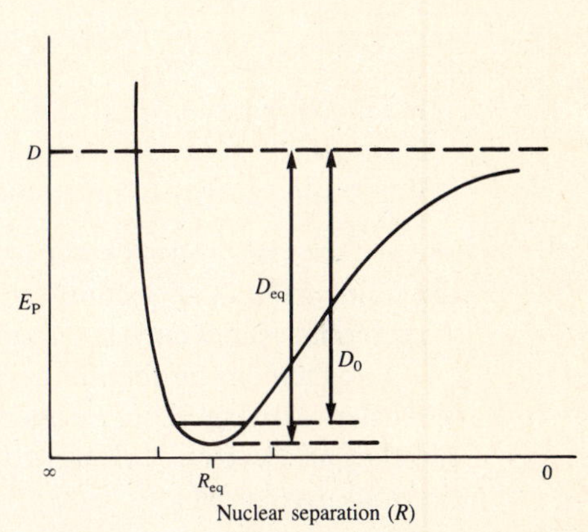

FIGURE 16-1. Potential Energy Curve for H_2

$$D_{eq} = D_0 + E_0 = D_0 + \frac{1}{2}h\nu_0 \qquad \textbf{(16-1)}$$

Here ν_0 represents the **frequency of vibration** of the atoms.

EXAMPLE 16-1: The measured value of D_0 for H_2 is 432 kJ mol^{-1}. (**a**) Calculate D_{eq} if $\nu_0 = 1.31 \times 10^{14}$ s^{-1}. (**b**) Convert both values of D to eV for future reference.

Solution:

(**a**) We merely substitute into Eq. (16-1):

$$D_{eq} = D_0 + \frac{h\nu_0}{2} = 432 \text{ kJ mol}^{-1} + \frac{(6.6262 \times 10^{-34} \text{ J s})(1.31 \times 10^{14} \text{ s}^{-1})(6.02 \times 10^{23} \text{ mol}^{-1})}{2}$$

$$= 432 \text{ kJ mol}^{-1} + 26.1 \times 10^3 \text{ J mol}^{-1}$$

$$= 458 \text{ kJ mol}^{-1}$$

(**b**) The conversion factor for eV (single particle) and J is easy to remember because it's the

Faraday constant, 1 eV = 96 500 J mol^{-1}. Thus for molecular hydrogen, H_2,

$$D_{eq} = \frac{458 \text{ kJ mol}^{-1}}{96.5 \text{ kJ mol}^{-1} \text{eV}^{-1}} = 4.75 \text{ eV} \qquad \text{and} \qquad D_0 = \frac{432}{96.5} = 4.48 \text{ eV}$$

For the hydrogen molecule ion H_2^+, the values are

$$D_{eq} = 269.5 \text{ kJ mol}^{-1} = 2.79 \text{ eV} \qquad \text{and} \qquad D_0 = 255.8 \text{ kJ mol}^{-1} = 2.65 \text{ eV}$$

We'll use these experimental values in the discussions of theory that follow.

B. The Born–Oppenheimer approximation

The complete wave function contains energy terms for the nuclei as well as for the electrons, making it very complex. Because nuclei are so much heavier than electrons and move much more slowly, Born and Oppenheimer proposed that the motions of the two be treated separately. Essentially, the nuclei are considered to be fixed at a distance R. The distances between electrons and nuclei and between electrons and electrons are represented by r. The wave function becomes

$$\psi = [\psi_n(R)][\psi_e(r, R)]$$

Next we find a Hamiltonian operator, $\mathbf{H}_e$, and write the Schrödinger equation for the electronic wave function.

$$\mathbf{H}_e \psi_e(r, R) = E_e \psi_e(r, R) \tag{16-3}$$

In general, $\mathbf{H}_e$ contains three types of energy terms:

1. kinetic energy of electrons;
2. potential energy of interactions between electrons; and
3. potential energy of interactions between electrons and nuclei.

The nuclear separation R enters into the third term. For a specific R, we solve Eq. (16-3) by the variation method (see Section 15-3) to determine the minimum E. Plotting E_p against R yields the **potential energy curve**; its minimum provides R_{eq} and D_{eq}, as shown in Figure 16-1.

Two methods of obtaining trial wave functions for molecules have been developed. In the **molecular orbital** method, we assume that molecular orbitals surround the collection of nuclei and that an electron "belongs to" the molecule, not to the atom from which it came. The electrons may be fed into these orbitals as in the Aufbau procedure with atoms. We obtain the trial wave function by linear combination of atomic orbitals. The **valence bond** method is based on the concept of a shared pair of electrons forming a chemical bond, as proposed by Lewis. We obtain the trial wave function by multiplying the wave functions of the individual atoms. Each approach has its own terminology and pictorial model, but with appropriate refinements, each gives about the same calculated values.

16-2. The Molecular Orbital Method: Hydrogen Molecule Ion, H_2^+

Ionic molecular hydrogen, H_2^+, is the simplest possible molecule, consisting of two protons and one electron. It is commonly used to illustrate the molecular orbital method, even though its Schrödinger equation can be solved exactly.

A. The Schrödinger equation

For the three-body problem posed by H_2^+, there are three potential energy terms; the general Hamiltonian operator for three bodies is

$$\mathbf{H}_e = -\frac{h^2}{8\pi^2 m_e} \nabla^2 + \left[\frac{1}{(4\pi\varepsilon_0)}\right]\left[\frac{e_1(-Z_A e)}{r_A} + \frac{e_1(-Z_B e)}{r_B} + \frac{(-Z_A e)(-Z_B e)}{R}\right] \tag{16-4}$$

where Z_A and Z_B designate the atomic numbers of the two nuclei. All the charges are the same, so we can factor out $-ee$. Then the Schrödinger equation, Eq. (16-3), becomes

$$\left[-\frac{h^2}{8\pi^2 m_e}\nabla^2 - \frac{ee}{(4\pi\varepsilon_0)}\left(\frac{Z_A}{r_A}+\frac{Z_B}{r_B}-\frac{Z_A Z_B}{R}\right)\right]\psi = E\psi \qquad \textbf{(16-5)}$$

We can solve this equation exactly, after transforming coordinates. For H_2^+, the two Z's equal 1, and the potential energy curve has a shape much like that in Figure 16-1, with the minimum at $R_{\text{eq}} = 1.06$ Å and $D_{\text{eq}} = 2.79$ eV.

EXAMPLE 16-2: Explain why the H_2^+ ion is stable, i.e., the reason that a hydrogen atom and a hydrogen ion stick together rather than remain separate.

Solution: The minimum in the $E_p R$ curve shows that there is stability. The values of ψ and ψ^2 are large for the region between the two protons. We may visualize this situation as one where the electron spends a lot of time between the two protons, or as a high density (concentration) of electron charge between the protons. The effect is a decrease in potential energy, but at the same time, the kinetic energy of the electron increases. The net result of these and other energy factors is a decrease in E, producing a minimum at a proton–proton separation of 1.06 Å. The concentration of the electron cloud between the nuclei acts like glue, holding them together.

B. The molecular orbital method

In general, equations such as Eqs. (16-3), (16-4), and (16-5) will contain several terms, owing to more electrons in the molecule, and must be solved by approximation methods. The variation method (see Section 15-3) is usually the choice. We obtain the trial function for H_2^+ by linearly combining the two hydrogen-like, ground-state orbitals that would be associated with two separate protons. The procedure is known as a *linear combination of atomic orbitals* (LCAO) and for H_2^+ gives

$$\psi = c_1\psi_{1s_A} \pm c_2\psi_{1s_B}$$

which we can simplify by representing the right-hand side ψ's by their subscripts. The two wave functions are

$$\psi_g = c_1 1s_A + c_2 1s_B \qquad \textbf{(16-6)}$$

and

$$\psi_u = c_1 1s_A - c_2 1s_B \qquad \textbf{(16-7)}$$

The subscripts g and u refer to inversion of the wave function through a center of symmetry (see Section 17-1). If the sign remains the same, the wave function is said to be symmetrical with respect to inversion (or to have *even* parity). The g comes from the German word for even: *gerade*. The ψ of Eq. (16-6) is symmetrical and has the subscript g, but the ψ of Eq. (16-7) is unsymmetrical (*odd* parity) and is designated u for *ungerade*. We can easily visualize the latter case by using the p_z orbital diagram in Figure 15-2. One lobe should be labeled positive and the other negative. Inversion through the origin of the coordinate system reverses the sign, so the figure (and the wave function) is not symmetrical, thus it is ungerade.

The average value of E is given by Eq. (14-24)

$$E = \frac{\int \psi^* \mathbf{H}_e \psi \, d\tau}{\int \psi^* \psi \, d\tau}$$

When we substitute Eq. (16-6) into Eq. (14-24), the multiplication leads to several types of integrals, which have been given names for easy reference and symbols for the simplification of complex expressions:

ATOMIC INTEGRAL $$\int 1s_A \mathbf{H}_e \, 1s_A \, d\tau = H_{AA} = \int 1s_B \mathbf{H}_e \, 1s_B \, d\tau = H_{BB} \qquad \textbf{(16-8)}$$

RESONANCE INTEGRAL $$\int 1s_A \mathbf{H}_e\, 1s_B\, d\tau = \int 1s_B \mathbf{H}_e\, 1s_A\, d\tau = H_{AB} \tag{16-9}$$

(ORTHOGONALITY) $$\int 1s_A 1s_A\, d\tau = S_{AA} = \int 1s_B 1s_B\, d\tau = S_{BB} = 1 \tag{16-10}$$

OVERLAP INTEGRAL $$\int 1s_A 1s_B\, d\tau = S_{AB} = \int 1s_B 1s_A\, d\tau = S_{BA} = S \tag{16-11}$$

note: The atomic integral is called the Coulomb integral in some textbooks. It represents an atom (here, hydrogen) perturbed slightly by a second atom at a distance R. The resonance integral represents the overlap of electronic charge density and contains contributions from orbitals of both atoms.

EXAMPLE 16-3: **(a)** Make the substitution of Eq. (16-6) in the denominator only of Eq. (14-24) and show how the integrals of Eqs. (16-10) and (16-11) come about. **(b)** Derive a simple expression in terms of S, using the fact that $c_1^2 = c_2^2$ for H_2^+.

Solution:

(a) The denominator is $\int \psi^* \psi\, d\tau$, but $\psi^* = \psi$, so we have

$$\int \psi^2\, d\tau = \int (c_1 1s_A + c_2 1s_B)^2\, d\tau \tag{a}$$

$$= \int [c_1^2 (1s_A)^2 + 2c_1 c_2 1s_A 1s_B + c_2^2 (1s_B)^2]\, d\tau \tag{b}$$

$$= c_1^2 \int (1s_A)^2\, d\tau + 2c_1 c_2 \int 1s_A 1s_B\, d\tau + c_2^2 \int (1s_B)^2\, d\tau \tag{c}$$

In eq. (c), we see that the double integrals of Eqs. (16-10) and (16-11) come from squaring the wave function expression.

(b) We can simplify eq. (c) by using the symbols for the integrals given in Eqs. (16-10) and (16-11): From Eq. (16-10),

$$\int (1s_A)^2\, d\tau = S_{AA} = 1 \qquad \text{and} \qquad \int (1s_B)^2\, d\tau = S_{BB} = 1$$

and from Eq. (16-11),

$$\int 1s_A 1s_B = S_{AB} = S$$

Thus eq. (c) becomes

$$\int \psi^2\, d\tau = c_1^2 + 2c_1 c_2 S + c_2^2 \tag{d}$$

For H_2^+, $c_1^2 = c_2^2$ because the two atoms are identical. If we use only the positive roots, $c_1 = c_2$, and eq. (d) becomes

$$\int \psi^2\, d\tau = c^2(2 + 2S) = c^2 2(1 + S)$$

where c now is included in the normalizing constant N, and we write

$$\int \psi^2\, d\tau = N^2 2(1 + S) \tag{16-12}$$

extra practice: You should use an approach similar to that in part (a) to show the source of the integrals in Eqs. (16-8) and (16-9).

In Example 16-3 we used only $c_1 = c_2$. Recognizing that we must also allow $c_1 = -c_2$, we have to write a second possible expression for the denominator of Eq. (14-24):

$$\int \psi^2 \, d\tau = c^2 2(1 - S) \quad \text{or} \quad N^2 2(1 - S) \tag{16-13}$$

In the two wave functions of Eqs. (16-6) and (16-7), we also include c_1 and c_2 in N:

$$\psi_g = N_g(1s_A + 1s_B) \tag{16-14}$$

and

$$\psi_u = N_u(1s_A - 1s_B) \tag{16-15}$$

Each of these wave functions is a **molecular orbital**. The probability density is ψ^2, so it's the square of Eq. (16-14) and (16-15) that gives the density of electron charge between the protons. In fact, ψ_g^2 predicts an appreciable density, whereas ψ_u^2 predicts a point of zero density; i.e., the plot of ψ_u^2 exhibits a node halfway between the protons.

We obtain expressions for E by substituting into Eq. (14-24). Equations (16-12) and (16-14) give E_g and Eqs. (16-13) and (16-15) give E_u. Using the symbols for the integrals from Eqs. (16-8) and (16-9), we have

$$E_g = \frac{H_{AA} + H_{AB}}{1 + S} \quad \text{and} \quad E_u = \frac{H_{AA} - H_{AB}}{1 - S} \tag{16-16}$$

The $\mathbf{H}_e$ in the integrals is Eq. (16-4). We can solve these equations and plot the results to get a curve similar to that in Figure 16-1, but the values obtained, $R_{eq} = 1.32$ Å and $D_{eq} = 1.77$ eV, don't agree very well with the experimental values of $R_{eq} = 1.06$ Å, $D_{eq} = 2.79$ eV. The wave functions of Eqs. (16-6) and (16-7) are oversimplifications. When we include factors such as the change in size of an orbital for two nuclei and the distortion generated by the potential field of a second nucleus, we get much better agreement.

16-3. Molecular Orbital Theory: Diatomic Molecules

A. Molecular hydrogen H_2

Molecular hydrogen, with two electrons around two protons, has six potential-energy terms in its Hamiltonian operator. The internuclear term is constant (by the Born–Oppenheimer approximation), but the Schrödinger equation is still too complicated to solve exactly. Using the variation method, we can obtain a trial wave function by multiplying two wave functions for H_2^+, one for each electron:

$$\psi = [\psi_M(1)][\psi_M(2)]$$

where ψ_M is given by Eq. 16-6. Dropping the c's for convenience and substituting, we have

$$\psi = [1s_A(1) + 1s_B(1)][1s_A(2) + 1s_B(2)]$$

Multiplying,

$$\psi = 1s_A(1)1s_A(2) + 1s_A(1)1s_B(2) + 1s_B(1)1s_A(2) + 1s_B(1)1s_B(2)$$

and rearranging gives

$$\psi = [1s_A(1)1s_B(2) + 1s_A(2)1s_B(1)] + [1s_A(1)1s_A(2) + 1s_B(1)1s_B(2)] \tag{16-17}$$

The brackets in Eq. (16-17) emphasize the different types of products. Inside the first set of brackets are the atomic orbitals of two separate hydrogen atoms, with the electrons interchanged in the second term. Inside the second set of brackets, each term shows both electrons on one nucleus; i.e., the H_2 molecule would consist of two ions, H^+ and H^-. One of the shortcomings of the simple molecular orbital theory is that it gives equal weight to both sets of terms, i.e., too much emphasis to *ionic* terms. Insertion of a weighting factor improves agreement with experimental results.

B. Bonding and antibonding orbitals

When the charge density between two nuclei is large, a chemical bond forms. An orbital that yields an increase in charge density between nuclei is called a **bonding orbital**, e.g., ψ_g in Eq. (16-14). An orbital such as ψ_u in Eq. (16-15), whose ψ_u^2 predicts a node between the nuclei, is an **antibonding orbital** and is usually designated by an asterisk. The charge density is insufficient to prevent nuclear repulsion.

C. Symbols for molecular orbitals

Just as the hydrogen atom has many atomic orbitals (AOs) that correspond to excited energy states, molecular hydrogen has many molecular orbitals (MOs). These MOs are designated by the Greek letters σ, π, δ, and ϕ (corresponding to the *s*, *p*, *d*, and *f* designations of atomic orbitals). They represent the values of the quantum number λ: 0, 1, 2, ..., where λ quantizes the **molecular angular momentum** components according to λ $(h/2\pi)$. Thus λ is comparable to the m_ℓ quantum number of atoms, but there is nothing comparable to ℓ. If several orbitals have the same λ, they are numbered in order of increasing energy.

EXAMPLE 16-4: Identify the bonding designations of the orbitals in Eqs. (16-14) and (16-15).

Solution: Because the orbitals are symmetrical about the internuclear axis (they have cylindrical symmetry), there is no angular momentum component on the *z* axis, i.e., $\lambda = 0$, and thus they are σ orbitals. Since they represent the lowest energies, they are numbered 1. The normalization factor is often omitted, so we have

$$1\sigma_g = 1s_A + 1s_B \quad \text{and} \quad 1\sigma_u = 1s_A - 1s_B = 1\sigma_u^*$$

Sometimes the AOs from which the MO is formed are included, e.g., $\sigma_g(1s)$ and $\sigma_u^*(1s)$.

---

note: The bonding σ is g and the antibonding σ is u. The opposite is true for π bonds; the bonding π is u, and the antibonding π is g.

D. Molecular orbitals of higher energies

Let's look briefly at the formation of MOs from AOs having energy higher than 1*s*. The resulting **molecular energy-level diagram** developed for H_2^+ is used for any species consisting of two identical atoms, such as H_2, N_2, O_2^+ (called *homonuclear*). We assume that MOs are formed solely from the identical AOs on the two atoms, as in Eq. (16-17), where we used the 1*s* AOs of atoms A and B. Figures 16-2 and 16-3 represent the MOs formed from the 2*p* AOs.

1. 2*s* AOs

The MOs are similar to those from 1*s* but have higher energies: $2\sigma_g(2s)$ and $2\sigma_u^*(2s)$.

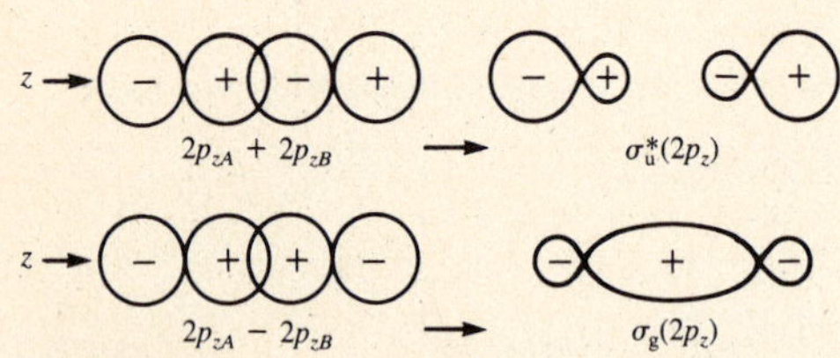

FIGURE 16-2. Molecular Orbitals Formed from Two $2p_z$ Atomic Orbitals

2. $2p_z$ AOs

When two $2p_z$ AOs are brought together (Figure 16-2, with *z* as the internuclear axis), we must recognize that the lobes of a *p* orbital have opposite signs. If we draw them with the positive lobes in the same direction, a positive lobe would overlap a negative lobe, implying repulsion. When the wave functions are added, the MO is antibonding:

$$2p_{zA} + 2p_{zB} = 2p_{zA} + 2p_{zB} = 3\sigma_u^*(2p_z)$$

If we reverse one p_z (change its sign), two positive lobes will overlap, and the MO is bonding:

$$2p_{zA} + (-2p_{zB}) = 2p_{zA} - 2p_{zB} = 3\sigma_g(2p_z)$$

The 3 shows that these are the third set of σ orbitals. This is the only case in which the bonding MO results from subtraction. For all other AOs, addition gives the bonding MO.

EXAMPLE 16-5: Why are the MOs from two p_z AOs labeled σ?

Solution: We chose *z* as the internuclear axis; thus p_z lies along that axis. The two AOs and, consequently, the two MOs are symmetrical about the axis, have no angular momentum component ($\lambda = 0$), and the symbol is σ.

3. $2p_x$ or $2p_y$ AOs

When two p_x or two p_y AOs are brought together with their positive lobes oriented in the same direction and with parallel axes (Figure 16-3), both the two positive lobes and the two negative lobes will overlap. The resulting MO has a positive lobe on one side of the internuclear axis and a negative lobe on the other. However, if one p_y is subtracted from the other, repulsion occurs. Again, we get one bonding MO and one antibonding MO from two AOs. For p_y, these MOs are

$$1\pi_u(2p_y) = 2p_{yA} + 2p_{yB} \quad \text{and} \quad 1\pi_g^*(2p_y) = 2p_{yA} - 2p_{yB}$$

Two analogous orbitals are formed by $2p_{xA}$ and $2p_{xB}$. The MOs, $\pi_u(2p_y)$ and $\pi_u(2p_x)$, have the same energy, so four electrons in the molecule will have this energy. This situation also occurs for $\pi_g^*(2p_y)$ and $\pi_g^*(2p_x)$.

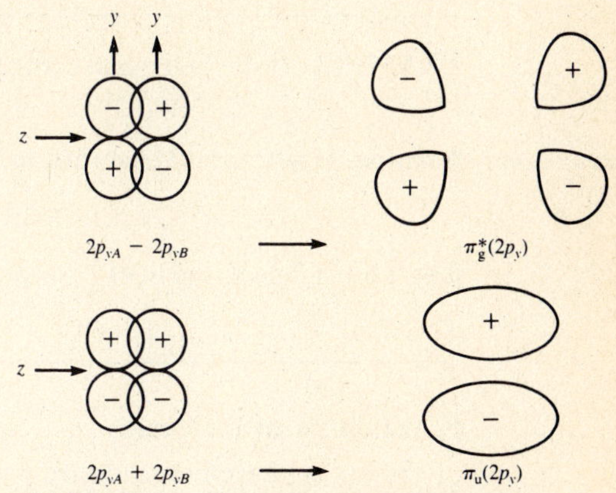

FIGURE 16-3. Molecular Orbitals Formed from Two $2p_y$ Atomic Orbitals

EXAMPLE 16-6: Justify the designations for the MOs formed from two $2p_y$ AOs.

Solution: The symbol π means that the quantum number $\lambda = 1$, analogous to $m_\ell = 1$ for p_y or p_x. The MOs have an angular momentum component because they're not symmetrical with respect to the internuclear axis (see Figure 16-3). The subscript u on the bonding MO identifies it as being antisymmetric with respect to a center of symmetry. In the lower part of Figure 16-3, there is no center of symmetry for π_u. In the upper part of Figure 16-3, we see that subtraction inverted $2p_{yB}$ and that the resulting MO has a center of symmetry. Therefore it is labeled g. When we place a 1 in front of the MOs, we show that these are the π orbitals of lowest energy.

E. Molecular electron configurations

The electron configuration of a molecule can be shown on a horizontal line, as we did for atoms in Chapter 15. The MOs are listed in order of increasing energy, and the number of electrons in each is given as a superscript: $(1\sigma_g)^2\,(1\sigma_u)^2$, etc. The rules for adding electrons are the same as those for atoms (see Section 15-5). The relative energies of $3\sigma_g$ and $1\pi_u$ change between N_2 and O_2, adding a complication. For molecules up to and including N_2, we have

$$N_2: \quad (1\sigma_g)^2(1\sigma_u^*)^2(2\sigma_g)^2(2\sigma_u^*)^2(1\pi_u)^4(3\sigma_g)^2$$

Note that $1\pi_u(2p_x)$ and $1\pi_u(2p_y)$ are written together because they have the same energy (they are degenerate). The term $(1\pi_u)^4$ represents two MOs with two electrons each, *not* one with four.

For O_2 to Ne_2, $3\sigma_g$ has a lower energy than $1\pi_u$, so to complete the second period of the Periodic Table, we have

$$Ne_2: \quad KK(2\sigma_g)^2(2\sigma_u^*)^2(3\sigma_g)^2(1\pi_u)^4(1\pi_g^*)^4(3\sigma_u^*)^2$$

Here we introduce K for the MOs from the inner "closed" shells of the atoms, which remain constant at 4 electrons in the Aufbau process. (For large atoms, the 1 *s* electrons probably don't form MOs but remain close to their original nuclei.) Sometimes the asterisks are omitted, so you must remember which MOs are antibonding.

note: The theoretical explanation for the $3\sigma_g$ MO being pushed above the $1\pi_g$ for molecules of atoms to the left of O in period II is that (for these atoms) the *s* and *p* AOs have nearly the same energy, and mixing occurs. This means that $2\sigma_g$ (from *s* AOs) has some *p* characteristic that lowers its energy, and $3\sigma_g$ (from p_z AOs) has some s characteristic that raises its energy—enough that it's greater than the energy of $1\pi_u$ (from p_x and p_y AOs). For the larger atoms (O and F), the s and p AOs are farther apart in energy. Any mixing is slight, and the energy of $3\sigma_g$ is smaller—below that of $1\pi_u$.

EXAMPLE 16-7: Write and compare the electron configurations for the molecular ions, N_2^+ and O_2^+.

Solution: There are 13 electrons in N_2^+. With 4 in KK, the configuration is

$$N_2^+: \quad KK(2\sigma_g)^2(2\sigma_u^*)^2(1\pi_u)^4(3\sigma_g)^1$$

There are 15 electrons in O_2^+. Since $3\sigma_g$ changes position, we have

$$O_2^+: \quad KK(2\sigma_g)^2(2\sigma_u^*)^2(3\sigma_g)^2(1\pi_u)^4(1\pi_g^*)^1$$

F. Stability of molecules

So far, we haven't mentioned bonds. In MO theory, we speak of **bond order**, which is half the difference between the number of electrons in bonding orbitals and the number in antibonding orbitals, i.e.,

$$\text{bond order} = \frac{1}{2}(\text{electrons in bonding MOs} - \text{electrons in antibonding MOs}) \qquad \textbf{(16-18)}$$

EXAMPLE 16-8: Determine the bond orders of N_2 and N_2^+.

Solution: For N_2, the electron configuration is $(1\sigma_g)^2(1\sigma_u^*)^2(2\sigma_g)^2(2\sigma_u^*)^2(1\pi_u)^4(3\sigma_g)^2$. The sum of the exponents of the antibonding orbitals is 4 and of the bonding orbitals is 10, so the bond order $= (1/2)(10 - 4) = 3$. You probably have noted that some canceling would simplify the process. The 1σ MOs (or KK) cancel, as do the 2σ MOs. There are 6 uncanceled bonding electrons, so the bond order is one-half of 6.

For N_2^+, it's clear that there are only 5 uncanceled bonding electrons, so the bond order is 2.5 (half-orders are permitted).

Because its order is lower, we predict that the N_2^+ ion is less stable than the N_2 molecule. Such statements must be made with caution. In general, when closely related species are compared, a higher bond order does mean greater bond strength. For example, consider three oxygen species:

Species	Bond order	Bond dissociation energy (kJ mol^{-1})
O_2^+	2.5	643
O_2	2	494
O_2^-	1.5	395

However, He_2^+ (bond order -0.5) has been detected in spectra in high energy situations. The species He_2^{2+} has a bond order of 1.0 but hasn't been found. Let's consider the number of electrons in each species: He_2^{2+} has only two compared to three in He_2^+. On this basis, we would predict less shielding of nuclear repulsion in He_2^{2+} and less stability, even though its bond order is greater.

EXAMPLE 16-9: Describe what the MO theory predicts about the existence of a helium molecule, He_2.

Solution: From the Aufbau process, the electron configuration of this species would be

$(1\sigma_g)^2(1\sigma_u^*)^2$. The bond order is zero, and we predict that the species shouldn't exist. It may appear that when the bond order is zero, the bonding and antibonding MOs are counteracting each other exactly. In the full mathematical treatment, however, antibonding orbitals are raised in energy more than bonding orbitals are lowered. The formation of He_2 would be accompanied by an increase in energy, creating an unstable state. From another viewpoint, we could say there is a small repulsive effect between two He atoms.

G. Paramagnetism

A small magnetic effect known as **paramagnetism** occurs when a substance with a permanent magnetic dipole is placed in a magnetic field. Since the effect results from unpaired electrons, it provides a means of studying electron configurations and relative energy levels.

EXAMPLE 16-10: The oxygen molecule is paramagnetic, which contradicts our expectation that there should be two shared pairs of electrons between the atoms (predicted by writing Lewis formulas). (**a**) Show that the paramagnetism of O_2 is consistent with the molecular orbital picture. (**b**) Give the bond order.

Solution:

(**a**) For O_2, the electron configuration is $KK(2\sigma_g)^2(2\sigma_u^*)^2(3\sigma_g)^2(1\pi_u)^4(1\pi_g^*)^2$. Recall that $1\pi_g^*$ is degenerate and consists of two MOs, one from two p_x AOs and one from two p_y AOs. According to Hund's Rule, the last two electrons in O_2 won't pair up; one will go into each π_g^* MO, and these two unpaired electrons cause the paramagnetism.

(**b**) From Eq. (16-18), the bond order is $(1/2)(8 - 4)$, or 2. This result agrees with our expectation of two bonds but disagrees with the Lewis formula. We discuss this problem in Section 16-4.

H. Heteronuclear diatomic molecules

For species involving similar atoms, such as CN and NO, we can use the same energy level arrangement as for homonuclear species, even though the separation between energies will be different.

EXAMPLE 16-11: Write the electron configuration for CN.

Solution: The molecule contains 13 electrons. Placing them in molecular orbitals in the same order as we did for N_2 gives

$$\text{CN:} \quad KK(2\sigma_g)^2(2\sigma_u^*)^2(1\pi_u)^4(3\sigma_g)^1$$

For distinctly different atoms (e.g., H and F), the diagram is different, because there is only one AO from H to form an MO. As atoms get larger, the energy levels decrease. In this case, the $1s$ AO of H has about the same energy as the $2p$ AOs of F. Since p_z has the same symmetry with respect to the internuclear axis as s, we assume that these two AOs (in the molecule HF) form two MOs, σ_g and σ_u^*. The other AOs of F remain essentially unchanged and retain their electrons; they are considered to be MOs and are called *nonbonding*.

16-4. Valence Bond Theory

The **valence bond theory** (also known as the *Heitler–London theory*) preceded molecular orbital theory and has now been replaced by MO theory for calculations. However, VB theory remains useful for the pictorial representation of molecules, especially the large molecules of organic chemistry.

A. The hydrogen molecule

The VB theory starts with the molecule H_2. By multiplying the wave functions of the two atoms [i.e., $1s_A(1)1s_B(2)$], recognizing that the electrons are interchangeable [i.e., $1s_A(2)1s_B(1)$], and finally adding and subtracting, we obtain two trial wave functions; one is symmetric, (ψ_+) and the other antisymmetric (ψ_-):

$$\psi_+ = 1s_A(1)1s_B(2) + 1s_A(2)1s_B(1) \quad \textbf{(16-19)}$$

$$\psi_- = 1s_A(1)1s_B(2) - 1s_A(2)1s_B(1) \quad \textbf{(16-20)}$$

Comparing Eqs. (16-19) and (16-17), we see that the right-hand side of Eq. (16-19) corresponds to the terms in the first set of brackets in Eq. (16-17). The values calculated from Eq. (16-19) are too low and can be improved by adding ionic terms. In this manner, values calculated from the two theories can be brought into close agreement. Using Eqs. (16-19) and (16-20) in Eq. (14-24), we obtain values of E, which, as in Eq. (16-16), we can abbreviate by using the symbols J and K for the types of integrals:

$$E_+ = \frac{J + K}{1 + S^2} \quad \text{and} \quad E_- = \frac{J - K}{1 - S^2} \quad \textbf{(16-21)}$$

Note that the normalizing factor is different. The integrals are also slightly different from those in Eqs. (16-8), (16-9), and (16-11) because we now have two electrons.

COULOMB INTEGRAL $$\int 1s_A(1)1s_B(2)\mathbf{H}1s_A(1)1s_B(2)\,d\tau = J \quad \textbf{(16-22)}$$

EXCHANGE INTEGRAL $$\int 1s_A(1)1s_B(2)\mathbf{H}1s_A(2)1s_B(1)\,d\tau = K \quad \textbf{(16-23)}$$

OVERLAP INTEGRAL $$\int 1s_A(1)1s_B(2)\,d\tau = \int 1s_A(2)1s_B(1)\,d\tau = S \quad \textbf{(16-24)}$$

We can interpret the integrals as follows:

1. The Coulomb integral represents the energy of the separate H atoms, each with its original electron.
2. The exchange integral represents a contribution to the energy that results from the exchange of electrons; it accounts for almost 90 percent of the bonding effect.
3. The overlap integral represents the overlapping of the electron charge cloud around the two nuclei. More specifically, it represents the overlap of two AOs, which may form two MOs only if S is positive.

EXAMPLE 16-12: Consider the overlap of an s AO and a p_y AO, and explain why they won't form a bond.

Solution: Recall that a p_y AO has two lobes, one above and one below the internuclear axis, and that one is positive and one is negative. (See Figure 16-3.) An s AO is spherically symmetrical and positive. When the s overlaps the p_y, part of the positive lobe and part of the negative lobe will be covered. These parts are exactly equal and cancel each other; therefore $S = 0$, and no bonding occurs.

B. Comparison to experimental results

The D_{eq} calculated using Eq. (16-19) is 3.14 eV, which is considerably less than the experimental value of 4.75 eV (see Example 16-1). If we add "ionic terms" to give an equation analogous to Eq. (16-17) for MO theory, D_{eq} becomes 4.00 eV. We can add even more terms and further improve the agreement.

C. Hybridization of atomic orbitals

A significant, and still very useful, device from VB theory is the creation of new *hybrid* AOs by combining two or more standard AOs. The typical, important example is the carbon atom, which forms compounds such as CH_4 and CCl_4 that have four identical bonds arranged in a tetrahedral shape. The same angular arrangement of atoms exists in the *straight chains* of organic compounds. This geometric shape, as well as the planar and linear shapes caused by double and triple bonds between C atoms, can't be explained by the usual s and p AOs. You should be familiar with all of this from general and organic chemistry courses.

EXAMPLE 16-13: (a) Explain why we might expect the compound of C and H to be CH_2. (b) How does the VB method explain the fact that it really is CH_4?

Solution:

(a) The ground-state electron configuration of the C atom is $1s^22s^22p^2$. The two p electrons are unpaired, and we would expect them to be shared with the electrons of two H atoms to form two covalent bonds. The bond angle should be 90°, since the p orbitals are at right angles to each other.

(b) The fact of four bonds to four H atoms is explained by *promotion of an electron*: The assumption is that one of the $2s$ electrons gains enough energy to move to the third p orbital. The source of this promotional energy is the large decrease in energy when five separate atoms come together in a stable molecule. Thus, having four unpaired electrons, C can form four shared pairs with four H atoms.

EXAMPLE 16-14: (a) Explain why we might expect the four bonds of CH_4 not to be identical. (b) How does the VB method explain the fact that they are identical?

Solution:

(a) After promotion, the outer-shell electron configuration of C is $2s^12p^3$. The three bonds between H atoms and p orbitals should be at 90° to each other—along the x, y, z axes. It's hard to visualize just where the fourth bond would be, but the four aren't identical and aren't at the experimentally determined bond angle of 109.5°.

(b) We recognize that the shape of the molecule is tetrahedral and assume that the four original AOs have rearranged somehow into four new AOs that are identical and that point to the corners of a tetrahedron. The new AOs are called **sp^3 hybrids**. We express this **hybridization** mathematically by linearly combining the original AOs. This procedure is merely mathematical juggling in order to get the VB method to predict known molecular shapes.

We can apply the same treatment to NH_3 or H_2O, where one or two of the corners of the tetrahedron are occupied by an unshared pair of electrons rather than an atom. We can explain the discrepancy between the actual bond angles and 109.5° as the result of repulsive forces between the *lone pair(s)* and the shared pairs of electrons.

D. Repulsion between electron pairs

The **valence-shell electron-pair repulsion (VSEPR)** method extends the concept of repulsion between electron pairs to all molecules having a central atom. The assumption is that the pairs of electrons around the central atom take geometric positions that place them as far apart as possible. Then hybrid orbitals can be devised to fit the geometry.

EXAMPLE 16-15: Apply the VSEPR method to predict the shape of a methane molecule, CH_4.

Solution: There are 8 electrons arrayed as 4 electron pairs around the central C atom. One way to visualize how 4 points can be placed equidistant from each other on the surface of a sphere is to use a ball and place the fingers in trial positions on it. When the points are equidistant from each other, they will be at the corners of a tetrahedron. With a H atom at each corner, CH_4 has a tetrahedral shape.

EXAMPLE 16-16: Predict the shape of BeH_2.

Solution: There are two pairs of electrons. To get as far apart as possible, they must be on opposite sides of the Be. Thus the molecule is linear.

EXAMPLE 16-17: What hybridization explains the linear shape of BeH_2?

Solution: The molecule has two bonds at 180° from each other. The ground state of Be is $1s^2 2s^2$. If one of the $2s$ electrons goes to a $2p$ AO, the s and p could form two hybrid orbitals (sp) that point in opposite directions. Mathematically, $\psi(sp)_1 = 2s + 2p_x$ and $\psi(sp)_2 = 2s - 2p_x$.

The five basic molecular shapes, hybridizations, and examples are shown in Table 16-1. (See also Solved Problems 16-8, 16-9, 16-10.) If one or more corners of the basic shape have no atom attached, i.e., are occupied by a lone pair, the arrangement of the actual atoms is described differently.

TABLE 16-1: Hybrid Orbitals and Their Geometry

Basic geometry	Hybridization	Example
Linear	sp	BeH_2
Planar (trigonal)	sp^2	BCl_3
Tetrahedral	sp^3	CCl_4
Trigonal bipyramidal	sp^3d	PCl_5
Octahedral	sp^3d^2	SF_6

EXAMPLE 16-16: Predict the shape of a molecule of H_2O.

Solution: With the VSEPR method, we observe that the O atom has 4 electron pairs around it, so the basic geometry is tetrahedral. Since only two corners are occupied, the three atoms create an angle and the molecule is described as *bent*.

16-5. Electron Spin

For convenience, we have omitted electron spin in all the wave functions in this chapter. The conclusions about energies are not changed by this omission, because spin has no effect on energies. Proper and complete wave functions, however, must contain a spin function. In Section 15-4, we encountered the Pauli exclusion principle for atoms, which requires that the complete wave function of a system be antisymmetric with respect to interchanging the electrons. Symmetric molecular wave functions must be combined with an antisymmetric spin function and vice versa, just as with atomic wave functions.

16-6. Polar Molecules

A. Dipole Moment

Nonpolar molecules exhibit the property of *polarizability*; i.e., in an electric field they may develop a dipole moment. The amount of polarizability is measured by the **dipole moment** μ a vector quantity with a magnitude equal to the product of the distance between two equal charges of opposite sign and the charge, or

DIPOLE MOMENT
$$\mu = rq \tag{16-25}$$

The direction of the vector is from negative to positive. The SI unit of μ is the **debye** D: $1\text{ D} = 3.336 \times 10^{-30}$ C m. (Charge is often measured in esu: $1\text{ D} = 10^{-18}$ esu cm.)

EXAMPLE 16-17: Determine the magnitude of the dipole moment for a proton and an electron separated by the Bohr radius of the hydrogen atom.

Solution: The electron (and proton) charge is 1.6022×10^{-19} C, and the Bohr radius is 0.529 Å, or

5.29×10^{-11} m. By Eq. (16-25),

$$\mu = rq = (5.29 \times 10^{-11}\text{ m})(1.602 \times 10^{-19}\text{ C})\left(\frac{1\text{ D}}{3.336 \times 10^{-30}\text{ C m}}\right) = 2.54\text{ D}$$

Heteronuclear diatomic molecules are *polar*; they have a permanent dipole moment. The explanation is that the buildup of electron charge that constitutes the bond is attracted toward the nucleus of greater charge. This distortion creates effective centers of positive and negative charge that lie on the internuclear axis but not at the same point.

B. Percent ionic character

We can think of a polar bond as an intermediate state between a completely ionic bond and a completely covalent bond. If the magnitude of the dipole moment of the pure ionic case is μ_{ion}, we can define a **percent ionic character** I, using the ratio of the experimentally determined μ to μ_{ion}:

PERCENT IONIC CHARACTER

$$I = \frac{\mu_{\text{exp}}}{\mu_{\text{ion}}}(100)\% \qquad \textbf{(16-26)}$$

EXAMPLE 16-18: What is the percent ionic character in HI(g) if the internuclear distance is 163 pm and the dipole moment is 1.27×10^{-30} C m?

Solution: In order to use Eq. (16-26), we need μ_{ion}, the dipole moment if each atom had the full electron charge. From Eq. (16-25),

$$\mu_{\text{ion}} = (163 \times 10^{-12}\text{ m})(1.6022 \times 10^{-19}\text{ C}) = 261 \times 10^{-31}\text{ C m}$$

From Eq. (16-26),

$$I = \frac{1.27 \times 10^{-30}}{26.1 \times 10^{-30}}(100) = 4.87\%$$

Theoretically, we can include polarity in wave functions by adding ionic and covalent terms [as in Eq. (16-17)] with weighting factors, i.e.,

$$\psi = c_1\psi_{\text{cov}} + c_2\psi_{\text{ion}}$$

If we let $c_1 = 1$ and λ be the new weighting factor,

$$\psi = \psi_{\text{cov}} + \lambda\psi_{\text{ion}} \qquad \textbf{(16-27)}$$

The ionic contribution to the energy is dependent on λ^2, and with $c_1^2 = 1$, we get a theoretical percent ionic character:

$$I_{\text{theor}} = \frac{\lambda^2}{1 + \lambda^2}(100)\% \qquad \textbf{(16-28)}$$

C. Electronegativity

An empirical attempt to deal with polarity is based on the concept of **electronegativity**. Defined as the ability of an atom to attract the electrons in a bond to itself, electronegativity is based on the valence bond method. Electronegativity can't be measured, so several arbitrary scales have been proposed; the one commonly used is that of Pauling, a scheme based on dissociation energies (bond enthalpies; see Section 4-1). Pauling chose to describe the ionic contribution to the energy of a polar bond, A—B, as the difference Δ between the experimental dissociation energy of A—B and the geometric mean of the dissociation energies of A—A and B—B, which may be calculated from Table 4-1. The square root of Δ seemed to be the significant quantity, so it was set equal to the difference between numbers assigned to each element:

$$\Delta^{1/2} = 0.102(\chi_{\text{B}} - \chi_{\text{A}}) \qquad \textbf{(16-29)}$$

TABLE 16-2: Some Pauling Electronegativities in $(\text{eV mol}^{-1})^{1/2}$

H						
2.1						
Li	Be	B	C	N	O	F
1.0	1.6	2.0	2.5	3.0	3.5	4.0
Na	Mg	Al	Si	P	S	Cl
0.9	1.2	1.6	1.9	2.1	2.5	3.0
K	Ca		Ge	As	Se	Br
0.8	1.0		2.0	2.2	2.5	2.8
					Te	I
					2.2	2.5

The number χ is the electronegativity of the element; the constant converts eV to kJ. The units of Pauling's χ are $(\text{eV mol}^{-1})^{1/2}$. Representative values are listed in Table 16-2, but note that values vary from source to source. The values of χ run from 4.0 for F to 0.7 for Cs, so that the difference for any two elements is approximately the dipole moment in debyes, i.e., $\mu = |\chi_B - \chi_A|$. The expression for Δ is often given as the square of Eq. (16-29):

$$\Delta = K(\chi_B - \chi_A)^2 \tag{16-30}$$

Here, $K = 23.1$ kcal eV^{-1} or 96.5 kJ eV^{-1}.

EXAMPLE 16-19: Calculate the ionic contribution to the energy of the N—I bond.

Solution: Let's use Eq. (16-30), since it gives Δ directly. In Table 16-2, we find $\chi_N = 3.0$ and $\chi_I = 2.5$. Substituting, we get

$$\Delta = (125 \text{ kJ mol}^{-1})(3.0 - 2.5)^2 = 125(0.5)^2 = 31 \text{ kJ mol}^{-1}$$

Note that it doesn't matter which element is called A or B; we want the *absolute* value of the difference.

The relationship between electronegativity and ionic character proposed by Pauling is

$$\log(100\% - I) = 2.0 - 0.11(\chi_B - \chi_A)^2 \tag{16-31}$$

where I is the percent ionic character. Another expression that has been used is

$$I = 16(\chi_B - \chi_A) + 3.5(\chi_B - \chi_A)^2 \tag{16-32}$$

EXAMPLE 16-20: Determine the percent ionic character of the P—F bond, using Pauling's relationship.

Solution: Pauling's expression for I is Eq. (16-31). To use it we must find the electronegativities of P and F in Table 16-2 and subtract:

$$\chi_F - \chi_P = 4.0 - 2.1 = 1.9$$

We can rearrange Eq. (16-31) as we substitute:

$$100\% - I = \text{antilog}[2.0 - 0.11(1.9)^2] = \text{antilog}(2 - 0.397)$$

Thus

$$I = 100\% - \text{antilog}(1.603) = 100 - 40.1 = 59.9\%$$

SUMMARY

In Table S-16, we show the major equations of this chapter and the connections between some of them. The situation with this chapter is somewhat different than with previous chapters. Most of the equations are not steps in a derivation with accompanying conditions but are applications of equations from previous chapters. Others are simply definitions of terms. The summary in Table S-16 should help you keep the molecular orbital and valence bond methods straight. Other important things you should know are

1. The equilibrium dissociation energy and the spectroscopic dissociation energy differ by the ground-state vibrational energy.

TABLE S-16: Summary of Important Equations

$D_{eq} = D_0 + \frac{1}{2}h\nu_0$ (16-1) $\Psi = \psi_n(R)\psi_e(r, R)$ (16-2) $\mathbf{H}_e\psi_e(r, R) = E_e\psi_e(r, R)$ (16-3)

$$\left[-\frac{h^2}{8\pi^2 m_e}\nabla^2 - \frac{ee}{4\pi\varepsilon_0}\left(\frac{Z_A}{r_A} + \frac{Z_B}{r_B} - \frac{Z_A Z_B}{R}\right)\right]\psi = E\psi \quad \text{(16-5)}$$

MO method

H_2^+ $\psi_g = c_1 1s_A + c_2 1s_B$ (16-6)

$\longrightarrow E_g = \dfrac{H_{AA} + H_{AB}}{1 + S}$ and $E_u = \dfrac{H_{AA} - H_{AB}}{1 - S}$ (16-16)

$\psi_u = c_1 1s_A - c_2 1s_B$ (16-7)

$$H_{AA} = H_{BB} = \int 1s_A \mathbf{H}_e 1s_A \, d\tau = \int 1s_B \mathbf{H}_e 1s_B \, d\tau \quad \text{(16-8)}$$

$$H_{AB} = \int 1s_A \mathbf{H}_e 1s_B \, d\tau = \int 1s_B \mathbf{H}_e 1s_A \, d\tau \quad \text{(16-9)}$$

$$S_{AA} = S_{BB} = 1 = \int 1s_A 1s_A \, d\tau = \int 1s_B 1s_B \, d\tau \quad \text{(16-10)}$$

$$S_{AB} = S_{BA} = S = \int 1s_A 1s_B \, d\tau = \int 1s_B 1s_A \, d\tau \quad \text{(16-11)}$$

H_2 $\psi = [1s_A(1)1s_B(2) + 1s_A(2)1s_B(1)] + [1s_A(1)1s_A(2) + 1s_B(1)1s_B(2)]$ (16-17)

$$\text{bond order} = \frac{1}{2}\left(\begin{array}{c}\text{electrons in}\\ \text{bonding MO's}\end{array} - \begin{array}{c}\text{electrons in}\\ \text{antibonding MO's}\end{array}\right) \quad \text{(16-18)}$$

VB method

H_2 $\psi_+ = 1s_A(1)1s_B(2) + 1s_A(2)1s_B(1)$ (16-19)

$\longrightarrow E_+ = \dfrac{J + K}{1 + S^2}$ and $E_- = \dfrac{J - K}{1 - S^2}$ (16-21)

$\psi_- = 1s_A(1)1s_B(2) - 1s_A(2)1s_B(1)$ (16-20)

$$J = \int 1s_A(1)1s_B(2)\mathbf{H}1s_A(1)1s_B(2) \, d\tau \quad \text{(16-22)}$$

$$K = \int 1s_A(1)1s_B(2)\mathbf{H}1s_A(2)1s_B(1) \, d\tau \quad \text{(16-23)}$$

$$S = \int 1s_A(1)1s_B(2) \, d\tau = \int 1s_A(2)1s_B(1) \, d\tau \quad \text{(16-24)}$$

Properties

$\mu = rq$ (16-25; dipole moment) $I = \dfrac{\mu_{exp}}{\mu_{ion}}(100)\%$ (16-26) $I_{theor} = \dfrac{\lambda^2}{1 + \lambda^2}(100)\%$ (16-28)

$\Delta^{1/2} = k(\chi_B - \chi_A)$ (16-29)

$\log(100\% - I) = 2.0 - 0.11(\chi_B - \chi_A)^2$ (16-31) $I = 16(\chi_B - \chi_A) + 3.5(\chi_B - \chi_A)^2$ (16-32)

2. The Born–Oppenheimer approximation assumes that nuclear motions have no influence on electronic motions; i.e., the distance between nuclei is constant.
3. The valence bond method assumes that electrons remain with their original atom and that a chemical bond is the sharing of two electrons between two atoms.
4. The molecular orbital method assumes that all electrons belong to the molecule and that a bond results from a large accumulation of electronic charge between the nuclei.
5. Gerade and ungerade describe symmetric and unsymmetric wave functions with respect to inversion through a center of symmetry. The words mean *even* and *odd* parity.
6. Bonding orbitals have a buildup of charge between the nuclei; antibonding orbitals have a node between the nuclei.
7. Bond order is a measure of the amount of bonding, or stability, of the molecule.
8. Paramagnetism, caused by unpaired electrons in a molecule or ion, provides an experimental check on electron configurations predicted by the molecular orbital method.
9. Diatomic molecules composed of similar atoms can be represented by the same electron configuration scheme as homonuclear molecules.
10. The valence bond method starts with the hydrogen molecule and must include "ionic" and other terms to attain agreement with experimental results.
11. Similarities in the VB and MO methods:

    (a) Bonding electrons are associated with both nuclei.
    (b) Bonding results from an increased electron charge between nuclei.
    (c) Bonding electrons have opposite spin.

12. Differences between the VB and MO methods:

    (a) For trial wave functions, VB multiplies two AOs but MO adds two AOs (LCAO).
    (b) For ionic terms, VB ignores but MO overemphasizes them.
    (c) Calculations are difficult for VB but easy for MO with computers.

13. Areas of application for the VB and MO methods:

    (a) VB is used in organic chemistry and for resonance structures and the hybridization concept; its shared-pair picture of a bond is useful.
    (b) MO is used in inorganic chemistry for transition metal complexes and for paramagnetism, three-electron bonds, and spectroscopy (excited states).

14. Valence-shell electron-pair repulsion is a useful method for predicting molecular geometry when there is a central atom.
15. All molecular wave functions must include a factor for electron spin, and the total function must be antisymmetric with respect to an interchange of electrons.
16. Polar bonds may be treated as covalent bonds with some degree of ionic character, expressed as a percent or represented by the dipole moment.
17 Electronegativity is the ability of an atom to attract the electrons in a bond to itself and thus is related to the energy required to break the bond.
18. The Pauling electronegativity scale is a set of numbers, one for each element, chosen so that the difference for any two elements is approximately equal to the dipole moment of a bond between atoms of the elements.

SOLVED PROBLEMS

MO Theory

PROBLEM 16-1 Use MO theory to predict whether the molecule Be_2 should exist.

Solution: Write the molecular electron configuration, i.e.,

$$\text{Be:} \quad KK(2\sigma_g)^2(2\sigma_u^*)^2$$

You can see immediately that the number of filled antibonding orbitals equals the number of filled bonding orbitals. The calculated bond order [Eq. (16-18)] is zero. Either method predicts that Be_2 shouldn't exist.

PROBLEM 16-2 (**a**) Calculate the bond orders of B_2 and B_2^-. (**b**) Determine whether either of these species should exist.

Solution:

(**a**) Bond orders are given by Eq. (16-18), but you should write the electron configurations first.

$$B_2: \quad [He]_2(2\sigma_g)^2(2\sigma_u^*)^2(1\pi_u)^2 \qquad \text{bond order} = \frac{1}{2}(4-2) = 1$$

$$B_2^-: \quad [He]_2(2\sigma_g)^2(2\sigma_u^*)^2(1\pi_u)^3 \qquad \text{bond order} = \frac{1}{2}(5-2) = 1.5$$

(**b**) Yes, both should exist; B_2 has been found to have two unpaired electrons as predicted.

PROBLEM 16-3 If an electron is added to N_2 to form N_2^-, predict whether the ion or the molecule is more stable.

Solution: Does $3\sigma_g$ come before or after $1\pi_u$ in the electron configuration of N_2^-? Since N_2^- has one more electron than N_2, $3\sigma_g$ comes first, i.e.,

$$N_2^-: \quad KK(2\sigma_g)^2(2\sigma_u^*)^2(3\sigma_g)^2(1\pi_u)^4(1\pi_g^*)$$

You can see that the added electron has entered an antibonding orbital, reducing the bond order from 3.0 to 2.5. On this basis you would predict that N_2 is more stable than N_2^-. If you want to be sure, you'd have to measure the dissociation energies of N_2^- and N_2 and compare them.

PROBLEM 16-4 Is the result in Problem 16-3 general, i.e., does adding an electron always decrease the stability? If not, give an example in which adding an electron should increase stability.

Solution: The answer is *no*. If the added electron enters a bonding MO, bond order will increase. We usually interpret this to mean that stability will be greater. In Problem 16-2, you saw that adding an electron to B_2, forming B_2^-, increased the bond order from 1.0 to 1.5. Without evidence to the contrary, you can predict that the species with the higher bond order is more stable.

PROBLEM 16-5 The molecule B_2 is found to be paramagnetic in its ground state. Show that this observation leads to the conclusion that the $1\pi_u$ MOs are lower in energy than the $3\sigma_g$ MO for this species.

Solution: Of the ten electrons in B_2, the first eight will fill the MOs: $KK(2\sigma_g)(2\sigma_u^*)$; the last two will go into either $3\sigma_g$ or $1\pi_u$. If they go into $3\sigma_g$, they will be paired, and there would be no paramagnetism. But there is paramagnetism, which can be explained if the two electrons have entered the two $1\pi_u$ MOs where they can remain unpaired and have parallel spins. For this to happen, $1\pi_u$ must be at a lower energy than $3\sigma_g$.

PROBLEM 16-6 (**a**) Write the electron configuration of NO. (**b**) Is the molecule paramagnetic?

Solution:

(**a**) You can use the MO order for homonuclear molecules. Since these two atoms are similar in size, the MOs from their AOs should be much like those for identical atoms. There are fifteen electrons, so

$$NO: \quad KK(2\sigma_g)^2(2\sigma_u^*)^2(3\sigma_g)^2(1\pi_u)^4(1\pi_g^*)$$

This configuration also holds for N_2^- (Problem 16-3). The species are isoelectronic.

(**b**) Yes, because there is an unpaired electron.

PROBLEM 16-7 State whether each species is paramagnetic: Be_2, C_2, N_2^-, CO.

Solution: You may be able to do these in your head now, but let's write the electron configurations beyond He_2 anyway, i.e.,

$$Be_2: \quad [He_2](2\sigma_g)^2(2\sigma_u)^2 \qquad CO: \quad [He_2](2\sigma_g)^2(2\sigma_u)^2(1\pi_u)^4(3\sigma_g)^2$$

$$C_2: \quad [He_2](2\sigma_g)^2(2\sigma_u)^2(1\pi_u)^4 \qquad N_2^-: \quad [He_2](2\sigma_g)^2(2\sigma_u)^2(1\pi_u)^4(3\sigma_g)^2(1\pi_g^*)$$

Only N_2^- is paramagnetic; all electrons are paired in the others.

VB Theory

PROBLEM 16-8 Substitute Eq. (16-19) into the denominator of Eq. (14-24) and show how the denominator becomes $1 + S^2$ in Eq. (16-21).

Solution: The denominator is $\int \psi^* \psi \, d\tau$, but $\psi^* = \psi$, so you can write $\int \psi_+^2 \, d\tau_1 \, d\tau_2$, introducing a volume element for each atom. We have been omitting the normalizing factor, but now you need it. In Eqs. (16-19) and (16-20), $N = 1/\sqrt{2}$. You substitute ψ_+ from Eq. (16-19) and get

$$\int \psi_+^2 \, d\tau_1 \, d\tau_2 = \frac{1}{2} \int [1s_A(1)1s_B(2) + 1s_A(2)1s_B(1)]^2 \, d\tau_1 \, d\tau_2 \tag{a}$$

The square of the bracketed term has the form $x^2 + 2xy + y^2$. You can simplify by noting that

$$x^2 = y^2 = [1s_A(1)1s_B(2)]^2 \tag{b}$$

Because $1s_A$ and $1s_B$ are normalized AOs, the integral $\int x^2 \, d\tau_1 = \int y^2 \, d\tau_2 = 1$, by Eq. (14-25). Now eq. (a) becomes

$$\int \psi_+^2 \, d\tau_1 \, d\tau_2 = \frac{1}{2}\left[1 + 1 + 2 \int [1s_A(1)1s_B(2)1s_A(2)1s_B(1)] \, d\tau_1 \, d\tau_2\right] = 1 + \int 1s_A(1)1s_B(2) \, d\tau_1 \int 1s_A(2)1s_B(1) \, d\tau_2$$

By Eq. (16-24), each of these integrals equals S, so

$$\int \psi_+^2 \, d\tau_1 \, d\tau_2 = 1 + S^2$$

PROBLEM 16-9 Is it possible for an s AO and a p_z AO to form MOs? If so, why? If they do, what type of MO would it be?

Solution: Yes. An s AO is spherically symmetrical and the p_z AO lies along the internuclear axis; thus both have the same symmetry with respect to that axis. This orientation of p_z means, in addition, that the positive s orbital can overlap the positive lobe of p_z, giving a positive value to the overlap integral. The resulting MOs are of the σ type.

PROBLEM 16-10 Write the combinations of AOs that are the sp^3 hybrids.

Solution: Using t to represent the tetrahedral sp^3 wave function, you have

$$t_1 = N(s + p_x + p_y + p_z) \qquad t_3 = N(s - p_x - p_y + p_z)$$
$$t_2 = N(s + p_x - p_y - p_z) \qquad t_4 = N(s - p_x + p_y - p_z)$$

PROBLEM 16-11 Many carbon compounds, e.g., ethylene and benzene, are planar, as is the elemental form of carbon, graphite. How does the VB method explain this geometry?

Solution: These species are characterized by three atoms around a C atom, so there must be a hybridization that produces three AOs lying in a plane. For them to be equally far apart, the angle between them must be 120°. From the ground state of C ($2s^2 2p^2$), promotion of one s electron is assumed. In the hybridization, one p AO doesn't participate, usually the one designated p_z. Three hybrid AOs are formed from s, p_x, and p_y, designated sp^2. In ethylene, two sp^2 AOs bond with H atoms, and one bonds with another C. The p_z AOs on each C are perpendicular to the plane of the three sp^2 orbitals. The overlap of their lobes produces another bond between the C atoms. Thus you have the usual double bond.

PROBLEM 16-12 Predict the shapes of BCl_3 and NCl_3, using VSEPR.

Solution: In order to use VSEPR you must determine the number of electron pairs around the central atom. In NCl_3, there are 8 valence electrons (4 pairs) that repel each other to positions at the corners of a tetrahedron. With only 3 Cl's, one corner isn't occupied, so it's omitted in describing the geometry, which is pyramidal: The N sits above the plane defined by the three Cl's. In BCl_3, there are only three pairs of electrons. When three points are placed on a sphere with maximum separation, they lie in a plane at 120° from each other. Thus BCl_3 is planar. The hybridization must be sp^2, as for carbon in Problem 16-11.

PROBLEM 16-13 Predict the shape of SF_6. What hybridization explains it?

Solution: The essential information for the VSEPR method is the number of electron pairs around the central atom (in this case, six). Distributing six points equally on the surface of a sphere places them at the ends of the Cartesian axes, i.e., at the corners of an octahedron. The geometry is called octahedral. To get six equal AOs about an atom, you must create hybrids that use *d* AOs. The outer-shell, ground-state configuration of S is $3s^2 3p^4$. If one *s* electron and one *p* electron are promoted to the 3*d* level, you have $s^1p^3d^2$. Hybridization then gives six sp^3d^2 hybrid AOs. Each electron of S shares with an electron from an F, and the pair occupies one of the hybrid AOs.

PROBLEM 16-14 If an F^+ could be removed from SF_6 to yield SF_5^-, what shape would the latter have? (Refer to Problem 16-13.)

Solution: You saw that SF_6 would be octahedral. Removing the F without one of its original electrons (or F^+) leaves six electron pairs around the central atom; thus the basic geometry still is octahedral. Imagine an octahedron oriented with four corners in a horizontal plane, and then remove the corner below the plane. The remaining figure is called a square pyramid.

Electron Spin

PROBLEM 16-15 For the wave functions in Eqs. (16-19) and (16-20), write the complete wave functions, which must include electron spin.

Solution: The Pauli exclusion principle (Section 15-4) tells us that the complete wave function must be antisymmetric. There are four spin functions for two electrons; one is antisymmetric and three are symmetric (see Example 15-15). The function ψ_+ of Eq. (16-19) is symmetric, so you have to combine it with the antisymmetric spin function:

$$\psi_1 = [1s_A(1)1s_B(2) + 1s_A(2)1s_B(1)][\alpha(1)\beta(2) - \alpha(2)\beta(1)]$$

Since there is only one wave function, this state is called a *singlet state*. The ψ_- of Eq. (16-20) is antisymmetric, so you must use the symmetric spin functions. There are three, you get three total wave functions, and the state is called a *triplet state*:

$$\psi_2 = [1s_A(1)1s_B(2) - 1s_A(2)1s_B(1)][\alpha(1)\alpha(2)]$$
$$\psi_3 = [1s_A(1)1s_B(2) - 1s_A(2)1s_B(1)][\beta(1)\beta(2)]$$
$$\psi_4 = [1s_A(1)1s_B(2) - 1s_A(2)1s_B(1)][\alpha(1)\beta(2) + \alpha(2)\beta(1)]$$

Polar Molecules

PROBLEM 16-16 If the percent ionic character of HCl is calculated theoretically to be 17 percent [perhaps from Eq. (16-28)], determine the approximate internuclear distance.

Solution: You should treat the 17% as an experimental value of μ and use Eq. (16-26) to calculate μ_{ion}. Then use Eq. (16-25) to calculate *r*. In order to use Eq. (16-26), however, you need the dipole moment. Although it isn't given, you can estimate it from the values in Table 16-2. Thus

$$\mu = |\chi_{\text{Cl}} - \chi_{\text{H}}| = 3.0 - 2.1 = 0.9\ \text{D}$$

Now, Eq. (16-26) gives

$$17\% = \frac{0.9\ \text{D}}{\mu_{\text{ion}}}(100\%) \qquad \text{or} \qquad \mu_{\text{ion}} = \frac{0.9}{0.17} = 5.3\ \text{D}$$

Rearranging Eq. (16-25) yields

$$r = \frac{\mu}{q} = \left(\frac{5.3\ \text{D}}{1.602 \times 10^{-19}\ \text{C}}\right)\left(\frac{3.336 \times 10^{-30}\ \text{C m}}{\text{D}}\right) = 11.0 \times 10^{-11}\ \text{m} = 110\ \text{pm}$$

The accepted value of *r* is 129 pm.

PROBLEM 16-17 We didn't present the equation for calculating Δ from thermodynamic data. **(a)** Write an equation for Δ from the description in Section 16-6C. **(b)** Use values from Table 4-1 and take the electronegativity of bromine to be 2.8 to calculate the electronegativity of carbon. **(c)** Compare your result to the value in Table 16-2.

Solution:

(a) The energy difference Δ is the difference between the energy of the heteronuclear bond and the geometric average energies of the homonuclear bonds involving the two atoms. You can express this as

$$\Delta = D(A - B) - [D(A - A)D(B - B)]^{1/2}$$

(b) In this problem, C and Br are the only atoms given, so

$$\Delta = D(C - Br) - [D(C - C)D(Br - Br)]^{1/2}$$

When you get a value of D from Table 4-1 in kJ mol^{-1} and substitute,

$$\Delta = 276 - [347(192)]^{1/2} = 276 - 258 = 18 \text{ kJ mol}^{-1}$$

For Δ, you also have Eq. (16-30): $\Delta = K(\chi_B - \chi_A)^2$, where $K = 125$ kJ mol^{-1}, so

$$(18)^{1/2} = (125)^{1/2}(2.8 - \chi_C) = \left(\frac{18}{125}\right)^{1/2} = 2.8 - \chi_C$$

$$\chi_C = 2.8 - \left(\frac{18}{125}\right)^{1/2} = 2.8 - 0.38 = 2.4$$

(c) The agreement with the value from Table 16-3 (2.5) is good.

PROBLEM 16-18 Compare Eqs. (16-31) and (16-32) to each other and to Eq. (16-26) by using each to calculate I for HI.

Solution: From the values in Table 16-2, you find that $\chi_B - \chi_A$ for HI $= 2.5 - 2.1 = 0.4$. Rewriting Eq. (16-31) and substituting, you get

$$100\% - I = \text{antilog}[2.0 - 0.11(0.4)^2]$$

$$I = 100\% - \text{antilog}(1.982) = (100 - 95.9)\% = 4.1\%$$

Substituting 0.4 in Eq. (16-32) gives

$$I = [16(0.4) + 3.5(0.4)^2]\% = (6.4 + 0.56)\% = 7.0\%$$

If you take the 4.87% from Example 16-18 as the reference value, clearly Eq. (16-31) gives better agreement than does Eq. (16-32), for this compound.

Supplementary Exercises

PROBLEM 16-19 The dissociation energy obtained from theoretical calculations is called the ________ ________ ________. Born and Oppenheimer assumed that ________ ________ is negligible in calculations of electron motion. The number of potential energy terms appearing in the Schrödinger equation for H_2^+ is ________. The mathematical procedure for getting trial MOs from AOs is called ________ ________ of ________ ________. Write the expression for a resonance integral. The letter symbol for $\lambda = 2$ (is) (isn't) δ. The σ molecular orbital (is) (isn't) formed from p_x atomic orbitals. All g molecular orbitals (are) (aren't) bonding. The dipole moment of H—F, using electronegativities, is ________.

PROBLEM 16-20 Calculate the ground-state frequency of vibration of the atoms of H_2^+, using the dissociation energies given in Section 16-1A.

PROBLEM 16-21 Write the electron configuration for the C_2^- ion.

PROBLEM 16-22 Which of the following are paramagnetic: O_2^+, O_2^-, NO^+, NO^-?

PROBLEM 16-23 What are the bond orders of the species in Problem 16-4?

PROBLEM 16-24 Which is more stable: O_2 or O_2^-?

PROBLEM 16-25 Predict the shapes of NO_3^- and SO_3.

PROBLEM 16-26 Predict the shapes of NH_3 and IF_5.

PROBLEM 16-27 The internuclear distance in the predominantly covalent compound CO is 124 pm. If the dipole moment is 0.12 D, what is the percent ionic character? Assume that the ionic charges are $+2$ and -2.

PROBLEM 16-28 Using the result from Example 16-18, determine λ for HI in Eq. (16-27).

PROBLEM 16-29 If the percent ionic character of HBr is 12%, find the approximate distance between the H and Br nuclei.

PROBLEM 16-30 From electronegativities, determine the ionic contribution to the energy of a bond between P and Cl.

PROBLEM 16-31 Using Eq. (16-32), calculate the percent ionic character of an S—F bond.

PROBLEM 16-32 In Table 4-1, there is no entry for Cl—Br. Use values from Tables 4-1 and 16-2 to calculate the bond energy of Cl—Br. (See Problem 16-17.)

Answers to Supplementary Exercises

16-19 equilibrium dissociation energy; nuclear motion; three; linear combination, atomic orbitals; $\int 1s_A \mathrm{H} 1s_B \, d\tau$; is; isn't; aren't; 1.9 D **16-20** $6.87 \times 10^{13}\ \mathrm{s}^{-1}$ **16-21** $[\mathrm{He}]_2(2\sigma_g)^2(2\sigma_u^*)^2(1\pi_u)^4(3\sigma_g)$ **16-22** O_2^+, O_2^-, NO^- **16-23** 2.5, 1.5, 3, 2 **16-24** O_2 **16-25** planar; trigonal pyramidal **16-26** trigonal pyramidal; square pyramidal **16-27** 1.0% **16-28** 0.23 **16-29** 121 pm **16-30** 101 kJ mol^{-1} **16-31** 31.9% **16-32** 221 kJ mol^{-1}

17 SYMMETRY AND SPECTROSCOPY

THIS CHAPTER IS ABOUT

☑ **Symmetry**
☑ **Group Theory and Symmetry**
☑ **Molecular Spectroscopy**
☑ **Rotational Spectra**
☑ **Vibrational Spectra**
☑ **Rotational–Vibrational Spectra**
☑ **Other Spectroscopies**

17-1. Symmetry

So far, we've used a couple of types of symmetry with only brief explanation (remember *gerade g* and *ungerade u*?). Now we'll consider the formal aspects of symmetry—a topic that has gained a significant role in chemistry. Theoretically, symmetry helps us determine which AO's will form MO's. Practically, we can relate infrared spectra to structure by symmetry, and we can classify molecules according to their symmetry. Many compounds that are quite different in most properties, such as water and *cis*-dichloroethylene, have some similar properties because they have the same symmetry.

warning: To become facile in recognizing symmetry features, you really must use three-dimensional models: "Hands-on" practice is vital. The standard ball-and-stick models will do for most cases, but sets designed for transition-metal complexes should cover all cases. Use models to duplicate examples given here and in your text and in answering questions until you never miss an improper axis of rotation in a two-dimensional projection.

A. Symmetry elements and symmetry operations

- A **symmetry operation** is an action performed on an object such that the object looks exactly the same to a fixed observer *after* the action as it did *before* the action.
- A **symmetry element** is a geometric feature of an object with respect to which the operation is carried out.

There are only four symmetry elements—a *point*, a *line*, a *plane*, and a *combination of line and plane*—but more than one similar operation may be associated with the elements involving a line (or axis). The elements, their symbols, and a description of the basic operations are listed in Table 17-1.

note: The **unit** or **identity element** E doesn't fit the definition of element, since it isn't strictly a geometric feature. It is, however, a property possessed by *every* object simply because it exists: The most distorted, unsymmetrical object still looks the same if you don't move it. Thus every object has at least *one* symmetry operation. (Mathematically, this operation is needed to fulfill the requirements of group theory.)

TABLE 17-1: Symmetry Elements and Operations

Element	Symbol	Operation
axis of symmetry or **rotation axis**	C_n	Rotation about the axis by $360°/n$ (counterclockwise)
plane of symmetry or **mirror plane**	σ	Reflection through the plane
center of symmetry or **inversion center**	i	Projection through the center to the same distance on the other side
improper axis or **rotation–reflection axis**	S_n	Rotation about the axis by $360°/n$ (counterclockwise) *plus* reflection through a plane
unit element or **identity element**	E	Nonmotion—doing nothing

EXAMPLE 17-1: Suppose that you perform a complete pirouette. **(a)** Is this a symmetry operation? **(b)** If so, which one?

Solution:

(a) If, while you turn, an observer closes his eyes and then opens them after you stop, he cannot tell that you have moved. A pirouette is therefore a symmetry operation.

(b) Since your body rotates about an axis during a pirouette, the action is a C operation. Since it is a 360° rotation, $n = 1$; so it is a C_1 operation.

Note, however, that you can achieve the same result by not moving, which is the E operation. Thus, C_1 is the same as E and is unnecessary or *trivial* as a symmetry operation (ballet notwithstanding). Repeating a C operation for any multiple of 360° has the *net effect* of not moving.

A **rotation axis of symmetry** (or **proper axis**) C_n of n greater than 1 generates n operations, although all those operations may not be distinct from other operations. The number n is the *order* of the rotation. If there are two or more proper axes, the one of largest n is called the *principal axis.*

EXAMPLE 17-2: **(a)** What single operations are generated by a C_4 axis of rotation? **(b)** Which are its distinct operations and which are equivalent to other operations?

Solution:

(a) A C_4 rotation is $360°/4 = 90°$. If the rotation is repeated four times, the object stops at 90°, 180°, 270°, and 360°. Each of these positions could have been reached by a single operation. These are designated

$$C_4^1 = 360°/4 = 90° \quad C_4^2 = 2(360°/4) = 180° \quad C_4^3 = 3(360°/4) = 270° \quad C_4^4 = 4(360°/4) = 360°$$

(b) In Example 17-1 we saw that $E = 360°$; thus $C_4^4 = E$. Similarly, $C_2^1 = 360°/2 = 180°$, which is the same as C_4^2. Thus the C_4 axis contributes two distinct operations: C_4^1 and C_4^3. So the four operations around a C_4 axis would be written C_4^1, C_2^1, C_4^3, E. (Often the superscript 1 is omitted, and we see C_4, C_2, C_4^3, E.)

Another type of equivalency occurs if the object rotates in the clockwise direction. Since 90° clockwise is the same as 270° counterclockwise, C_4^3 could also be represented as $-C_4^1$. Sometimes primes or plus and minus signs are used to show this situation. For example, our four operations might also be written C_4^+, C_2^1, C_4^-, E. (Later, we'll see that C_4^1 and C_4^3 are in the same *class*.)

A **plane of symmetry** σ may be visualized as a *mirror*: Everything on one side must match everything on the other side, as a reflection matches the original. Planes of symmetry fall into three types, depending on which of three specific positions the plane occupies relative to the principal rotation axis:

- **Horizontal plane** σ_h: The plane is perpendicular to the axis.
- **Vertical plane** σ_v: The plane includes the axis and also includes the atoms not located on the axis.
- **Dihedral** (or **diagonal**) **plane** σ_d: The plane includes the axis and bisects the angle between atoms not located on the axis.

EXAMPLE 17-3: All three types of plane are found in the square-planar ion, $PtCl_4^{2-}$. Describe the locations of these planes.

Solution: First, we must identify the principal rotation axis. A square object with identical corners must have a C_4 axis perpendicular to the plane of the square, at the center (the Pt atom). Thus the σ_h plane is the plane of the $PtCl_4^{2-}$ ion. The σ_v plane must include the C_4 axis and atoms not on the axis: Thus one σ_v includes C_4 and two opposite corners; and another σ_v plane, perpendicular to the first, includes C_4 and the other two corners. Two σ_d planes, perpendicular to each other, include C_4 and bisect the sides of the square. (Note that σ_h in a planar molecule bisects the atoms, reflecting the top half into the bottom half.)

A **center of symmetry** or **inversion center** *i* seems self-explanatory and is perhaps the easiest element to recognize. But we have to be careful, because the projection through the center to the same distance on the other side must be in three dimensions. Thus, a tetrahedron does *not* have a center of symmetry.

EXAMPLE 17-4: Does methane have a center of symmetry?

Solution: If we look at the usual two-dimensional structural formula of CH_4 we might say yes: The C atom certainly looks like an inversion center for the 4 H atoms. However, we know that in three dimensions the molecule is tetrahedral: If one H is projected through the center, it does not meet another H; so methane does not have a center of symmetry.

The **improper axis of rotation** or **rotation–reflection axis** S_n is probably the most difficult axis to recognize, since it combines rotation and reflection. Such an element may exist when no σ_h plane is present independently, but there is ALWAYS a C axis collinear with an S axis. And, like the C axis, an S_n axis generates additional operations, some of which are identical to existing operations.

EXAMPLE 17-5: Methane has an S_4 axis. Describe the location of this axis.

Solution: Visualize the tetrahedral methane balanced on a desk, so that two H atoms rest on the desk surface and the vertical plane containing these H atoms and the C atom is perpendicular to the surface. (Use a model if you have one!) This vertical plane is a true plane of symmetry and is designated σ_d, because it bisects the angle between the other two H atoms, which are sticking up like two antennae. These H atoms lie in another σ_d plane which is at right angles to the first. The S_4 axis is the line of intersection of these planes; i.e., it is the line that passes through the C atom perpendicular to the desk surface.

To verify this conclusion, rotate the model 90° (360°/n) and consider the reflection through the horizontal plane which contains the C atom. The H atoms that were on top are now in the positions on the desk surface originally occupied by the other H atoms. Note that the S_4 axis is also a C_2 axis (but NOT C_4) and that a tetrahedron has no σ_h plane.

As a general rule, when n is even, an S_n axis requires a C axis with $n/2$ and does not require a σ_h plane. On the other hand, when n is odd, an S_n axis requires a C_n axis and a σ_h plane. Furthermore, when n is even, S_n generates n operations; when n is odd, S_n generates $2n$ operations. (See Problem 17-3.)

EXAMPLE 17-6: Locate all of the symmetry operations of a tetrahedron.

Solution: In Examples 17-4 and 17-5, we saw that there is no center of symmetry and no horizontal plane in a tetrahedron; but we did find an S_4 and a C_2 axis and two σ_d planes for the orientation described in Example 17-5. We take that S_4 axis, which is perpendicular to the surface of the desk, to be the z axis; then we consider the x and y axes. Each of these is also an S_4 axis, with an accompanying C_2 axis and two σ_d planes. Now we have three S_4, three C_2, and six σ_d symmetry elements. Then, because S_n generates additional operations, we also have S_4^1, S_4^2, S_4^3, and S_4^4. And since $S_4^2 = C_2^1$ and $S_4^4 = E$, there are two distinct operations. Thus the three S_4 axes have contributed six new operations.

Now we re-orient the model to put three H atoms on the desk surface and consider the vertical line through the C atom and the H atom above it. This is another symmetry element—a C_3 axis. It generates the operations C_3^1, C_3^2, C_3^3, the last of which is the same as E, so there are two new operations. The other C—H bonds also lie along C_3 axes, so we have a total of four axes and eight distinct operations. Thus the total number of operations is 24, normally written in this manner:

$$E \quad 8C_3 \quad 3C_2 \quad 6S_4 \quad 6\sigma_d$$

Note that C_3^2 is counted with C_3^1, and S_4^3 with S_4^1, because they are in the same class. (See Example 17.2)

B. Matrix representation

A symmetry operation can be replaced mathematically by a **transformation matrix**. Such a matrix, multiplied by the vector of a point, yields the new coordinates—in terms of the original coordinates—after an operation has occurred. A matrix may be represented by its **trace**, which is the sum of the numbers on the main diagonal (upper left corner to lower right).

EXAMPLE 17-7: Consider two points a and b in a horizontal plane, which are equivalent under a C_4^1 rotation. **(a)** What are the coordinates of b in terms of the coordinates of a? **(b)** What matrix will transform the original coordinates into the new coordinates, and what is its trace?

Solution:

(a) The original coordinates are (x_a, y_a) and (x_b, y_b). If a 90° counterclockwise rotation places a on b, then $x_b = -y_a$ and $y_b = x_a$. (Sketch a plot of the situation to help you see this.)

(b) Mathematically, the problem is: What matrix will make the transformation of the vector of point a into the vector of point b?, i.e.,

$$\begin{bmatrix} ? \end{bmatrix} \times \begin{bmatrix} x_a \\ y_a \end{bmatrix} \rightarrow \begin{bmatrix} -y_a \\ x_a \end{bmatrix}$$

$$\begin{bmatrix} ? \end{bmatrix} = \begin{bmatrix} 0 & -1 \\ 1 & 0 \end{bmatrix} \qquad \textit{Check: } [0 \ -1][x_a + y_a] \rightarrow [0 \ -y_a]$$

The matrix $\begin{bmatrix} 0 & -1 \\ 1 & 0 \end{bmatrix}$, which is another way of writing C_4^1, has a trace of $0 + 0 = 0$.

EXAMPLE 17-8: The transformation matrix for E for a point (x_a, y_a) is $\begin{bmatrix} 1 & 0 \\ 0 & 1 \end{bmatrix}$. What is its trace?

Solution: The trace, or sum of the main diagonal, is $1 + 1 = 2$.

17-2. Group Theory and Symmetry

A. Group theory: A mathematical intermission

The symmetry operations of a molecule fit the requirements for a mathematical *group*; thus the power of *group theory* can be brought to bear on problems involving symmetry. We'll define a group here, but we can include only an elementary description of group theory.

note: For an introduction to group theory, see *J. Chem. Ed.* **44**, 128–135 (1967).

- A **group** is a set of members for which a method of combination has been specified and which satisfies the following requirements:

  The combination of any two members is a member.
  The associative law holds.
  There is a unit member E, such that $EX = X = XE$.
  For each member X there must be a member Y which is its *inverse* (reciprocal) $Y = X^{-1}$, such that $XY = E$. (A member may be *its own inverse.*)

- The *order* of the group is the number of members.

EXAMPLE 17-9: Show that the set of integers, which includes all positive and negative integers and zero, meets the requirements of a group. What is the order of the set?

Solution: In examining a set, we must determine (1) the method of combination, (2) the identity member under the method of combination, and (3) the inverse relationship under the method of combination.

(1) For the integers and zero, the method of combination is addition: e.g., $4 + 5 = 9$ where 9 is a member of the set. The associative law holds for addition: e.g., $2 + (3 + 4) = (2 + 3) + 4$.
(2) The identity member, which can be added to a second member to give back that second member, is zero: e.g., $4 + 0 = 4$.
(3) The condition of the inverse relationship, in which a member can be added to a second member to give the identity element zero, is met by the negative of the second member: e.g., $-2 + (+2) = 0$.

The order of the set of integers including zero is infinite.

An assembly of the combinations of all pairs of members of a group is called the **multiplication table**, since the method of combination is usually multiplication. A **representation of a group** is a set of elements that satisfies two requirements: (1) There are elements in the set that can be associated with every member of the group, and (2) the multiplication table of the set is equivalent to that of the group. The elements in the set may be numbers, matrices, or wave functions. There are an infinite number of possible representations of a group, but the more complex representations can be expressed as the sum of a few relatively simple ones called **irreducible representations**. The complex ones are *reducible representations*.

EXAMPLE 17-10: In Example 17-9, the method of combination that makes the set of integers (including zero) a group is addition. Write the multiplication table for this group. (The multiplication table will be infinite for the group of integers, but start it to get the idea of how it goes.)

Solution: Usually, the members are written across the top and down the left side, then the grid is filled in with the combination of each pair (in this case, the sum): $-1 + (-1) = -2$, $-1 + 0 = -1$, etc. Near zero, the table can be written as shown here. Every number in the grid IS a member of the group.

	−1	0	1	2
−1	−2	−1	0	1
0	−1	0	1	2
1	0	1	2	3
2	1	2	3	4

B. Symmetry operations as groups

The complete set of symmetry operations of a molecule satisfies the requirements of a group—more specifically, a **point group**, because one point in the molecule must not move. It's not necessary that an atom be *at* that point, but it *is* necessary that every symmetry element *contain* the point. There are more than 50 point groups, although most are not common.

A system of labeling point groups is based broadly on the axis of rotation, with finer distinctions for planes of symmetry and the others. If there is a principal axis with C_2 axes perpendicular to it, the general type is D; if there are no C_2 axes, the type is C. A subscript number gives the order of the principal axis, e.g., C_3. A subscript letter "h" shows a horizontal plane, e.g., C_{3h}; if there is no horizontal plane but there are vertical planes, a subscript "v" is used, e.g., C_{3v}. The highly symmetrical groups based on a cube have their own symbols, e.g., tetrahedral, T_d, and octahedral, O_h.

EXAMPLE 17-11: What is the point group of (**a**) the $PtCl_4^{2-}$ ion described in Example 17-3? (**b**) of CH_4 in Example 17-6?

Solution: (**a**) The $PtCl_4^{2-}$ ion has a C_4 axis and a σ_h plane; thus it belongs to the C_{4h} point group. (**b**) Methane is tetrahedral, so it belongs to the T_d group.

C. Assigning a molecule to a point group

You must know the point group to which a molecule belongs in order to locate the proper *character table* (described in Section 17-2F), which is needed for applications in spectroscopy, orbital theory, etc. The following procedure for determining the point group will work for most molecules. It's not necessary to locate *all* symmetry operations or even all elements: After determining the point group, you can find a list of all operations at the top of the character table.

Step 1: Is the molecule linear ($C_{\infty v}$ or $D_{\infty h}$), tetrahedral (T_d), octahedral (O_h), or completely asymmetric (C_1)? (Always check against the list in the character table to be certain that *all* operations can be performed on the molecule.) If the molecule is not one of these easy-to-recognize types, go to Step 2.

Step 2: Is there an axis of rotation of type C (not S)?
(*a*) If there is no C axis, the group is C_s (only a plane) or C_i (only a center).
(*b*) If there is a C_n axis, go to Step 3.

Step 3: Are there n C_2 axes perpendicular to the C_n axis?
(*a*) If there are n C_2 axes present, the group is of the general D type.
(*b*) If there are no C_2 axes present, the group is of the general C type (or possibly S_{2n}).
(*c*) If a C axis and an S axis are collinear, use the n of the C axis.

Step 4: Is there a horizontal plane?

(*a*) If a horizontal plane is present, the group is D_{nh} or C_{nh}, and you are finished.

(*b*) If there is no horizontal plane, go to Step 5.

Step 5: Are there *n* vertical planes? [*note*—Only if there is no horizontal plane!]

(*a*) For the *C* type:

(1) If there are *n* vertical planes, the group is C_{nv}.

(2) If there are no vertical planes, the group is C_n.

(*b*) For the *D* type:

(1) If there are *n* vertical planes, the group is D_{nd}.

(2) If there are no vertical planes, the group is D_n.

EXAMPLE 17-12: Assign the water molecule to a point group.

Solution:

Step 1: The water molecule is triatomic and bent; it's not in one of the easy-to-recognize groups.

Step 2: There is a C_2 axis in the plane of the molecule; it passes through the O atom and bisects the angle between the H atoms.

Step 3: There are no C_2 axes perpendicular to the $C_2(z)$, so we've narrowed the choice down to a group of the *C* type.

Step 4: There is no σ_h plane: A σ_h would have to be perpendicular to $C_2(z)$. The group is not C_{2h}.

Step 5: There are 2 σ_v planes: One σ_v is the plane of the molecule; the other σ_v is perpendicular to it. So, the point group is C_{2v}.

D. Multiplication tables

For symmetry groups, the combination process is called multiplication, although it might be considered addition since one operation follows another, adding their individual effects. In the mathematical treatment, however, matrices are multiplied, not added; so it is proper to speak of a *multiplication table* and the *product* of two operations.

EXAMPLE 17-13: Prepare a multiplication table for the four symmetry operations generated by a C_4 axis, similar to the multiplication table in Example 17-10.

Solution: We list the symmetry operations (E, C_4^1, C_2^1, C_4^3) across the top and down the left side, then fill in the grid points with the products of the operations on the column and row which cross at that point. The products will be $C_4^1 \cdot C_4^1 = C_2^1$, $C_4^1 \cdot C_2^1 = C_4^3$, $E \cdot C_2^1 = C_2^1$, etc. Notice that each product is a member of the group, as required.

	E	C_4^1	C_2^1	C_4^3
E	E	C_4^1	C_2^1	C_4^3
C_4^1	C_4^1	C_2^1	C_4^3	E
C_2^1	C_2^1	C_4^3	E	C_4^1
C_4^3	C_4^3	E	C_4^1	C_2^1

EXAMPLE 17-14: The operations in Example 17-13 do constitute a point group. (**a**) What is its symbol? (**b**) What is its order?

Solution:

(**a**) There is a C_4 principal axis but no C_2 axes perpendicular to it, so it is type *C*. There are no planes. The point group is C_4.

(**b**) The order is the number of members—in this case, 4.

E. Representations

Various sets of numbers (e.g., $+1, -1, +1, -1$) have a multiplication table equivalent to that of the C_4 point group shown in Example 17-13 and are therefore representations of the group as defined in Section 17-2A. Moreover, the transformation matrices that express the operations also constitute a representation. Since these matrices can be represented by their traces, the traces constitute a representation of the group—and are more convenient to use than the complete matrix. This set of numbers is called the *character of the representation.*

EXAMPLE 17-15: Referring to Examples 17-7 and 17-8, show how a representation is obtained from transformation matrices.

Solution: In Example 17-7b, we have the transformation matrix of the C_4^1 operation whose trace is 0, denoted $\text{tr}(C_4) = 0$. In Example 17-8, we have $\text{tr}(E) = 2$. We need the matrices for C_2^1 and C_4^3. Following the procedure of Example 17-7, we find

$$C_2^1 = \begin{bmatrix} -1 & 0 \\ 0 & -1 \end{bmatrix} \quad \text{and} \quad C_4^3 = \begin{bmatrix} 0 & 1 \\ -1 & 0 \end{bmatrix}$$

with $\text{tr}(C_2^1) = -2$ and $\text{tr}(C_4^3) = 0$. Now we have one representation of the C_4 group, which is the set of four numbers: 2 0 −2 0. If you look up the C_4 character table, you'll find that 2 0 −2 0 is not one of the representations given. Further explanation, however, is beyond the scope of this book.

F. Character tables

A **character table** is an assemblage of the *irreducible representations* (see Section 17-2A) of the point group. Each representation is assigned a symbol which has a systematic meaning, but which we'll use primarily as a label. These irreducible representations are sometimes called the **symmetry species** of the point group. There must be as many of them as there are *classes* of operations. A **class** is a subgroup of the operations of the group that includes those operations of the same kind that are interchanged by another operation of the group. In Example 17-6 for instance, the eight C_3 operations are one class, the three C_2 operations are another class, etc.

- All members of a class have the same trace.

Character tables usually include additional notations on the right side; these notations tell us which symmetry species is associated with certain molecular properties that depend on the molecular symmetry—in particular, atomic and molecular orbitals and rotational motions.

By way of illustration, we'll look at the short, uncomplicated table of point group C_{2v}, which is shown in Table 17-2:

TABLE 17-2: Character Table for the C_{2v} Point Group

C_{2v}	E	C_2	$\sigma_v(xz)$	$\sigma'_v(yz)$		
A_1	1	1	1	1	z	x^2, y^2, z^2
A_2	1	1	−1	−1	R_z	xy
B_1	1	−1	1	−1	x, R_y	xz
B_2	1	−1	−1	1	y, R_x	yz

note: $\sigma_v(xz)$ and $\sigma'_v(yz)$ belong to different classes and the Cartesian axes of each are given.
The third column contains first-order terms, where x, y, and z represent properties associated with the axes, such as atomic p orbitals or translation in an axial direction (in some tables T_x, T_y, etc., are used); the R's refer to rotational movement about the axis or to the axial component of the angular momentum vector.
The last column contains squares and binary products; for example, xy could represent the d_{xy} atomic orbital.

EXAMPLE 17-16: Describe why y is placed by ("belongs to") the B_2 symmetry species in Table 17-2.

Solution: We have to determine the representation that is developed by applying each operation. First, we imagine a p_y atomic orbital with its positive and negative lobes lying along the y axis; then we consider how it is affected by each symmetry operation:

E No change: This is like multiplying by +1. Thus $\text{tr}(E) = +1$.

C_2^1 A 180° rotation moves the + lobe into the − lobe. Thus $\text{tr}(C_2) = -1$.

$\sigma_v(xz)$ This plane is perpendicular to the y axis, so reflection takes the + lobe into the − lobe. Thus $\mathrm{tr}[\sigma_v(xz)] = -1$.

$\sigma'_v(yz)$ This plane includes the p_y orbital, so reflection takes one half into the other half; there's no sign change. Thus $\mathrm{tr}[\sigma'_v(yz)] = +1$.

Now we have a set of four numbers—the character $1 \ -1 \ -1 \ +1$, which equals the B_2 irreducible representation. Or, we could say that "p_y transforms as B_2."

17-3. Molecular Spectroscopy

A. A bit of background

We know about atomic spectroscopy (introduced in Chapter 14) in conjunction with the development of quantum mechanics and the Bohr theory. Atoms absorb or emit light of specific wavelengths in different regions of the spectrum as electrons move between energy levels. For hydrogen, the Lyman series is in the ultraviolet (uv) region, the Balmer series in the visible region, and others in the infrared (ir) region. The frequency of the radiation is a measure of the energy difference between two energy levels of the atom [Eq. (14-9)]. Spectroscopists now speak of **transitions** between energy levels or between *states of the system.* Limitations on allowed transitions are called **selection rules**. In atoms, a transition is allowed only if the quantum number ℓ of the two states differs by 1, i.e., $\Delta\ell = \pm 1$.

Similar transitions occur between states of molecules. But the experimental evidence is different: For atoms, the spectra are well-spaced, sharp lines, whereas the spectra of molecules are closely spaced lines which blur into a *band* unless an instrument of very high resolution can separate them. The difference in the case of molecules is that energy can be absorbed or emitted in ways other than the movement of electrons. The vibrational and rotational motions of the nuclei are quantized, have energy levels, and undergo transitions between these energy levels. The observed spectrum of a molecule, then, is a combination or superposition of electronic, vibrational, and rotational contributions.

Fortunately, the energy differences between the rotational, vibrational, and electronic energy levels are sufficiently different so that, to a first approximation, each can be treated separately. (Actually, the effect of a lower type of energy is a complication in studying the higher type!) Table 17-3 collects information on the three energy levels which brings out the relationships among them.

B. A bit of theory

As we know (Chapter 16), the Schrödinger equation for a molecule must include nuclear (n) and electronic (e) terms. And the Born–Oppenheimer approximation [Eq. (16-2)] permits separation of the wavefunction into a nuclear and an electronic wavefunction:

$$\psi = \psi_n(R)\psi_e(r, R) \qquad \textbf{[Eq. (16-2)]}$$

Then, by Eq. (16-3), we could write an electronic Hamiltonian operator and an electronic wave equation. But now we need to look at the nuclear part of Eq. (16-2) and write a nuclear wave equation:

$$\mathbf{H}_n\psi_n = E_n\psi_n \tag{17-1}$$

TABLE 17-3: Molecular Energy Levels and Spectral Regions

Energy level	Spectral region	Range of frequency (s^{-1})	Range of energy (kJ mol^{-1})
Rotational	Microwave (and far infrared)	10^9–10^{12}	0.001–0.1
Vibrational (plus rotational)	Infrared	10^{12}–10^{14}	0.1–10
Electronic (plus vibrational and rotational)	Visible and ultraviolet	10^{14}–10^{16}	100–1000

Here, E_n represents **nuclear eigenvalues**, ψ_n represents **nuclear wavefunctions**, and the equation applies to translational, vibrational, and rotational motions of the nuclei. The effects of these motions are not really independent. But, as a first step, we assume that they *are* independent and write

$$\psi_n = \psi_{trans}\psi_{vib}\psi_{rot} \tag{17-2}$$

and

$$E_n = E_{trans} + E_{vib} + E_{rot} \tag{17-3}$$

The translational motion is ignored, because, at ordinary temperatures, the energy levels are so close together that no transitions can be detected. In fact, translational motion of molecules is treated by classical equations, as in the kinetic theory of gases.

We are left with vibrational and rotational motions. Transitions between two vibrational or two rotational states may occur only when there are variations in the dipole moment μ [see Eq. (16-25), $\mu = rq$], which produces an electric effect that may interact with the radiation. Otherwise, the probability of absorption of energy is zero, and transitions are forbidden. For example, rotational transitions do not occur in homonuclear diatomic molecules, such as Cl_2 or H_2, or in linear molecules having a center of symmetry, such as CO_2 or acetylene, because they don't have an electric dipole.

EXAMPLE 17-17: What are the point groups of **(a)** CO_2 and **(b)** HCl? Which molecule will have rotational transitions?

Solution: We see that both molecules are linear, and we know that the principal axis of rotation in a linear molecule is always the internuclear axis, designated C_∞.

(a) In CO_2, there are an infinite number of C_2 axes perpendicular to the principal axis, so the group is the D type. And there is a plane perpendicular to the principal axis, so the group is $D_{\infty h}$. Any linear molecule with a center of symmetry belongs to this point group. And since this linear molecule has a center of symmetry—and hence no dipole—CO_2 will not undergo rotational transitions.

(b) In HCl, there are no C_2 axes and no horizontal plane. But there is an infinite number of vertical planes, so the group is $C_{\infty v}$. Any linear molecule without a center of symmetry belongs to this point group and has a dipole. The dipole moment in a given direction will change during rotation, so HCl will undergo rotational transitions.

In polyatomic molecules, vibrational transitions may occur without the presence of a *permanent* dipole moment: Some of the vibrational modes, when they displace the atoms, create a *temporary* dipole. The magnitude of the dipole moment will continuously change (oscillate) as the atoms move back and forth. Consequently, the linear CO_2 molecule does have a vibrational spectrum (infrared), even though it doesn't have a rotational spectrum (microwave).

17-4. Rotational Spectra

A. The rigid rotor model: A first approximation

A diatomic molecule can be treated, *to a first approximation*, like a rigid rotor. Actually, the centrifugal force will stretch the bond slightly, the extent increasing as the rotational energy increases, but we'll ignore this effect for now. Assuming that the molecule is a rigid rotor, we can apply Eq. (14-41):

$$E_{rot} = E_J = J(J+1)\frac{h^2}{8\pi^2 I} \qquad (J = 0, 1, 2, \ldots)$$

where I is the moment of inertia and J is the rotational quantum number. The second quantum number, M, is given by $M = 0, \pm 1, \pm 2, \ldots, \pm J$ [Eq. (14-42)]. The degeneracy of a rotational level is $2J + 1$, as we saw in Example 14-28.

note: J and M are to the rotational angular momentum of molecules as ℓ and m are to the angular momentum of the hydrogen atom.

1. Rotational constant

If we let $B' = h^2/8\pi^2 I$, we can shorten Eq. (14-41) to

$$E_J = B'J(J+1) \tag{17-4}$$

where $B' = h^2/8\pi^2 I$ is in joules. Then we can calculate the frequency ν of a spectral line from the Bohr expression, Eq. (14-9), rearranged to $\nu = (1/h)|E_f - E_i|$. If we use Eq. (14-41) or Eq. (17-4) for E, we'll have units of s^{-1} for ν. Now, since $\nu = c\tilde{\nu}$, we can convert frequency, Eq. (14-9), to wavenumbers $\tilde{\nu}$ by dividing by c, so that

$$\tilde{\nu} = \frac{\nu}{c} = \frac{1}{hc}\Delta E \tag{17-5}$$

where E is in joules. But it's convenient to incorporate the hc into E—i.e., to express the energy in wavenumbers—since spectroscopists use $\tilde{F}(J)$ for rotational energies in wavenumbers. Then Eq. (17-4) becomes

$$\tilde{F}(J) = \frac{B'}{hc}J(J+1) = BJ(J+1) \tag{17-6}$$

where $$B = \frac{1}{hc}\left(\frac{h^2}{8\pi^2 I}\right) = \frac{h}{8\pi^2 Ic} = (2.7994 \times 10^{-44}\ \text{kg m})(I^{-1}) \tag{17-7}$$

This new B is called the **rotational constant**, which has units of m^{-1} when SI units are used throughout. (In the literature, it is usually given as cm^{-1}.) The value of B is different for every molecule, since the moment of inertia I, which appears in the expression, is different for every molecule. (For example, the hydrohalogens HI to HF have values of B that run from ~ 6 to 21 cm^{-1}.) Moreover, the moment of inertia involves the reduced mass ($I = \mu r_0^2$), so the value of B for a molecule will change when different isotopes are part of its composition.

warning: In practice, the spectral lines measured by infrared instruments are expressed in wavenumbers, $\tilde{\nu}$ (in cm^{-1}), although—to the confusion of beginners and purists—they are often called "frequencies." To add to the confusion, the results of microwave spectroscopy are usually reported as true frequencies, ν (in s^{-1}). If you remember to keep the units of B and E consistent in calculations, you'll avoid this particular confusion.

EXAMPLE 17-18: Calculate the value of the rotational constant B of $H^{79}Br(g)$ whose bond length is 144 pm.

Solution: To obtain B, we need to substitute in Eq. (17-7); but let's find I independently first. To do this, we need the reduced mass μ for $H^{79}Br$:

$$\mu = \frac{m_1 m_2}{m_1 + m_2} = \left(\frac{1(79)}{1+79}\ \text{g mol}^{-1}\right)\left(\frac{1\ \text{mol}}{6.022 \times 10^{23}}\right)\left(\frac{10^{-3}\ \text{kg}}{\text{g}}\right) = 1.640 \times 10^{-27}\ \text{kg}$$

Now $$I = \mu r_0^2 = (1.640 \times 10^{-27}\ \text{kg})(144 \times 10^{-12}\ \text{m})^2 = 3.40 \times 10^{-47}\ \text{kg m}^2$$

Finally, showing all of the constants, we have

$$B = \frac{h}{8\pi^2 Ic} = \frac{6.626 \times 10^{-34}\ \text{J s}}{8\pi^2(3.40 \times 10^{-47}\ \text{kg m}^2)(2.998 \times 10^8\ \text{m s}^{-1})}\left(\frac{\text{kg m}^2\,\text{s}^{-2}}{\text{J}}\right)$$
$$= 8.23 \times 10^2\ \text{m}^{-1} \quad \text{or} \quad 8.23\ \text{cm}^{-1}$$

---

2. Selection rule for a rigid rotor

For rotational transitions in a rigid diatomic molecule, the selection rule is $\Delta J = \pm 1$; any other transitions are forbidden. Since experimental work almost always involves absorption, we usually use only $\Delta J = +1$. This means that $E_f > E_i$.

3. Spacings between energy levels

We can obtain an expression for the energy difference between two levels quickly from Eq. (17-6):

$$\Delta\tilde{F} = \tilde{F}_f - \tilde{F}_i = BJ_f(J_f + 1) - BJ_i(J_i + 1)$$

Taking the usual case of absorption, $J_f = J_i + 1$, we get

$$\Delta\tilde{F} = B(J_i + 1)(J_i + 2) - BJ_i(J_i + 1) = B(J_i^2 + 3J_i + 2 - J_i^2 - J_i) = B(2J_i + 2)$$

So $$\Delta\tilde{F} = 2B(J_i + 1) \quad \textbf{(17-8)}$$

or $$\Delta\tilde{F} = 2BJ_f \quad \textbf{(17-9)}$$

EXAMPLE 17-19: Calculate the energy difference between the $J = 2$ and $J = 3$ levels for $H^{79}Br(g)$, whose rotational constant B was calculated in Example 17-18.

Solution: To obtain $\Delta\tilde{F}$, we could use either Eq. (17-8) or Eq. (17-9). Both give $\Delta\tilde{F} = 2B(3) = 6B$. Substituting $B = 8.23\ \text{cm}^{-1}$, we get $\Delta\tilde{F} = 6(8.23\ \text{cm}^{-1}) = 49.38\ \text{cm}^{-1}$.

EXAMPLE 17-20: Calculate the energies, as a multiple of B, for the ground state and first four rotational excited states of a rigid diatomic molecule. Is there any regularity?

Solution: We merely have to substitute the values of J from 0 to 4 in Eq. (17-6):

State	Energy	Difference
Ground:	$\tilde{F}_0 = B(0)(1) = 0$	
		$2B$
1st:	$\tilde{F}_1 = B(1)(2) = 2B$	
		$4B$
2nd:	$\tilde{F}_2 = B(2)(3) = 6B$	
		$6B$
3rd:	$\tilde{F}_3 = B(3)(4) = 12B$	
		$8B$
4th:	$\tilde{F}_4 = B(4)(5) = 20B$	

We see that the difference interval increases by $2B$ each time.

B. The nonrigid rotor: A complication

The regularity of the increase in separations between consecutive energy levels calculated in Example 17-20 does not appear in the observed spectrum of $H^{79}Br$. In fact, the separation between spectral lines decreases as J increases. This is explained by an elongation of the bond by centrifugal force; in other words, the bond is not really rigid but is slightly elastic. And if r changes, I changes, thus changing B [Eq. (17-7)]. So the rigid rotor model has to be modified to account for nonrigid rotors.

The energy of a nonrigid rotor can be approximated by adding a term to Eq. (17-6):

$$\tilde{F}(J) = BJ(J + 1) - DJ^2(J + 1)^2 \quad \textbf{(17-10)}$$

where D is the **centrifugal distortion constant**, defined, in cm^{-1}, as

$$D = \frac{h^3}{32\pi^4 I^2 r^2 kc} \quad \textbf{(17-11)}$$

The only really new term here is k, the **force constant**, which brings vibrational motion into the picture. The force constant, equivalent to the spring constant in Hooke's law [Eq. (3-10)], is a measure of the *elasticity* of the bond—its tendency to stretch. The smaller the k, the greater is D, and the more the bond will stretch. For a simple harmonic oscillator,

$$k = 4\pi^2 c^2 \mu\tilde{\omega}^2 \quad \textbf{(17-12)}$$

where $\tilde{\omega}$ is the **vibrational wavenumber** (often called, confusingly, *vibrational frequency*.)

EXAMPLE 17-21: Calculate the value of D for the $H^{79}Br$ of Example 17-18 if $k = 412\ \text{kg}\,\text{s}^{-2}$. Compare the value of D with that of B.

Solution: From Example 17-18, we have $I = 3.40 \times 10^{-47}\ \text{kg}\,\text{m}^2$ and $r = 144$ pm. Substituting these and all constants in Eq. (17-11), we get

$$D = \frac{h^3}{32\pi^4 I^2 r^2 kc} = \frac{(6.6262 \times 10^{-34}\ \text{J s})^3}{32\pi^4(3.40 \times 10^{-47}\ \text{kg m}^2)^2(144 \times 10^{-12}\ \text{m})^2(412\ \text{kg s}^{-2})(2.998 \times 10^8\ \text{m s}^{-1})}$$
$$= 3.15 \times 10^{-2}\ \text{m}^{-1} = 3.15 \times 10^{-4}\ \text{cm}^{-1}$$

The value of B from Example 17-18 is $8.23\ \text{cm}^{-1}$, so we see that D is much smaller than B. The effect of nonrigidity will be noticeable only at large J ($J > 10$).

17-5. Vibrational Spectra

A. Harmonic oscillator

A diatomic molecule can be treated, to a first approximation, as a simple harmonic oscillator, whose potential energy is given by $V = \frac{1}{2}kx^2$. We can also write this expression as

$$V = \tfrac{1}{2}k(r_{\text{AB}} - r_e)^2 \qquad \textbf{(17-13)}$$

to emphasize that x represents the displacement of the atoms, or the extension of the bond, away from the equilibrium position.

The proportionality constant k is the force constant, already introduced in Eq. (17-11). We see that it also expresses the proportionality between the vibrational wavenumber $\tilde{\omega}$ and the reduced mass μ for a simple harmonic oscillator if we rewrite Eq. (17-12) as

$$\tilde{\omega} = \frac{1}{2\pi c}\sqrt{\frac{k}{\mu}} \qquad \textbf{(17-14)}$$

or we can express this relationship as a true frequency ω:

$$\omega = \frac{1}{2\pi}\sqrt{\frac{k}{\mu}} \qquad \textbf{(17-15)}$$

EXAMPLE 17-22: If the vibrational wavenumber of $H^{79}Br$ is $2649.7\ \text{cm}^{-1}$, what is its force constant?

Solution: We'll need the reduced mass, which we calculated in Example 17-18 to be 1.640×10^{-27} kg. Since we are given $\tilde{\omega}$ in cm^{-1}, we use Eq. (17-14); solving for k gives back Eq. (17-12):

$$k = (2\pi c\tilde{\omega})^2\mu = [2\pi(2.998 \times 10^8\ \text{m}\,\text{s}^{-1})(2649.7\ \text{cm}^{-1})(100\ \text{m}^{-1}\ \text{cm})]^2(1.640 \times 10^{-27}\ \text{kg})$$
$$= 4.086 \times 10^2\ \text{kg}\,\text{s}^{-2}$$

note: This value doesn't quite agree with the value given in Example 17-21. The discrepancy is probably due to the use of slightly different frequencies.

Now, since the proper SI unit of force is the newton ($1\ \text{N} = 1\ \text{kg}\,\text{m}\,\text{s}^{-2}$),

$$k = (4.086 \times 10^2\ \text{kg}\,\text{s}^{-2})\left(\frac{\text{N}}{\text{kg}\,\text{m}\,\text{s}^{-2}}\right) = 408.6\ \text{N}\,\text{m}^{-1}$$

1. Vibrational energy levels

When the potential energy function V is used in the Schrödinger equation, the eigenvalues are given by Eq. (14-38):

$$E = (v + \tfrac{1}{2})h\nu_0$$

where we use ν_0 for the general ν. The vibrational quantum number v is $0, 1, 2, \ldots$; and when

$v = 0$, the ground-state energy is

$$E_0 = \tfrac{1}{2}h\omega_0$$

where ω_0 is used for the **fundamental vibrational frequency**, rather than ν_0. The fundamental vibrational frequency is the *only* vibrational frequency for a diatomic molecule that truly acts as a simple harmonic oscillator. The vibrational levels, in the simple case, are equally spaced at intervals of $h\omega_0$. At room temperature, this number is large, so most molecules are in the vibrational ground state.

It's customary to convert the energy units from joules to wavenumbers. Thus, dividing by hc to get wavenumbers and using the spectroscopists' symbol $\tilde{G}$ for vibrational energies expressed in wavenumbers, we get

$$\tilde{G} = \left(v + \frac{1}{2}\right)\frac{\omega_0}{c} = \left(v + \frac{1}{2}\right)\tilde{\omega}_0 \qquad \textbf{(17-16)}$$

Now the spacing between levels, $\Delta\tilde{G}$, will equal the wavenumber of the corresponding spectral line.

2. Selection rule for a harmonic oscillator

In order to have a permanent electric dipole moment, a diatomic molecule must be heteronuclear; but in that case, vibrational movement along the internuclear axis will produce an oscillating dipole. The selection rule for transitions between vibrational energy levels in a harmonic oscillator is $\Delta v = \pm 1$. Since nearly all molecules are in the ground state, very few transitions occur, except from $v = 0$ to $v = 1$, and the spectrum consists of only one line, with wavenumber $\tilde{\omega}_0$.

B. Anharmonicity

The correspondence between the parabolic curve of a simple harmonic oscillator and the potential energy curve of a molecule is acceptable only near the bottom. At larger internuclear separations, i.e., greater vibrational energies, the potential energy curve slopes away. The shape of this curve is described rather well by the Morse function. As a result, there is an anharmonic character to the model, evidenced by the existence of weak lines from transitions forbidden by the selection rule. Use of the Morse function in the Schrödinger equation has the effect of adding a term to Eq. (17-16):

$$\tilde{G} = \tilde{\omega}_e(v + \tfrac{1}{2}) - \tilde{\omega}_e x_e(v + \tfrac{1}{2})^2 \qquad \textbf{(17-17)}$$

where x_e is the **anharmonicity constant** having values around $+0.01$. The new wavenumber $\tilde{\omega}_e$ is an *equilibrium vibrational wavenumber* corresponding to vibrations at the equilibrium point, i.e., where v would be $-\tfrac{1}{2}$ in Eq. (17-16). The second term does give smaller spacings between higher energy levels. The constants may be determined empirically from the overtone frequencies, in bands arising from $\Delta v = 2$, 3, etc. (In some tables the product $\tilde{\omega}_e x_e$ is listed rather than the two values.) The selection rule for an anharmonic oscillator is $\Delta v = \pm 1, \pm 2, \pm 3, \ldots$

17-6. Rotational–Vibrational Spectra

A. Superimposed energy levels

Energy that is sufficient to cause a vibrational transition is many times greater than the energy for rotational transitions, so it isn't possible to have pure vibrational spectra. Each vibrational level has a fine structure of many rotational levels superimposed on it, and the energy of the levels is the sum of these separate levels. Staying with the simpler models, we can add Eqs. (17-6) and (17-16) to get a combined expression for vibrational–rotational energies in wavenumbers:

$$\tilde{G}_{v,J} = (v + \tfrac{1}{2})\tilde{\omega}_0 + BJ(J + 1) \qquad \textbf{(17-18)}$$

[For greater accuracy, we should use Eqs. (17-17) and (17-10).]

1. Selection rules for vibration–rotation

Since a change in vibrational level nearly always includes a rotational change, both of the separate selection rules must be followed. Thus, for a harmonic oscillator, only those transitions are allowed for which $\Delta v = \pm 1$ and $\Delta J = \pm 1$. (In a few rare cases, $\Delta J = 0$ does occur, but we won't worry about that here.)

2. Spectral lines

Because ΔJ may be $+1$ or -1, there are two possibilities when we calculate the wavenumber of a spectral line, $\Delta\tilde{G}$. For either $J = +1$ or $J = -1$, the first step is the same—subtraction of the form of Eq. (17-18) for the initial state from the form of Eq. (17-18) for the final state:

$$\Delta\tilde{G} = \tilde{G}_f - \tilde{G}_i = (v_f + \tfrac{1}{2})\tilde{\omega}_0 + BJ_f(J_f + 1) - (v_i + \tfrac{1}{2})\tilde{\omega}_0 - BJ_i(J_i + 1)$$

The $\frac{1}{2}\omega_0$ terms will cancel.

For $\Delta J = +1$, we substitute $J_i + 1$ for J_f [this part is like the derivation of Eq. (17-8)]:

$$\tilde{\omega}_+ = (v_f - v_i)\tilde{\omega}_0 + B[(J_i + 1)(J_i + 2) - J_i(J_i + 1)]$$

Then, since $\Delta v = \pm 1$, $v_f - v_i = 1$, and

$$\tilde{\omega}_+ = \tilde{\omega}_0 + 2B(J_i + 1) \tag{17-19}$$

For $\Delta J = -1$, we substitute $J_i - 1$ for J_f. The first term is the same, and we have

$$\tilde{\omega}_- = \tilde{\omega}_0 - 2BJ_i \tag{17-20}$$

We have assumed that B is the same for both levels, but it will actually be slightly smaller for $v = 1$ than for $v = 0$, because the vibration increases the internuclear distance. Note that, for Eq. (17-20), J_i cannot equal 0. If it did, we could not have $\Delta J = -1$. Also note that $\tilde{\omega}_+$ and $\tilde{\omega}_-$ are the wavenumbers of spectral lines, whereas $\tilde{\omega}_0$ is the wavenumber of the vibration.

EXAMPLE 17-23: Calculate the wavenumbers of the two spectral lines of $H^{79}Br$ produced by transitions from $v = 0$, $J = 3$, to $v = 1$. (See Examples 17-18 and 17-22.)

Solution: To use Eqs. (17-19) and (17-20), we will need B for HBr, which we calculated in Example 17-18 to be 8.23 cm^{-1}. And we know from Example 17-22 that $\tilde{\omega}_0 = 2649.7$ cm^{-1}. The two lines, of course, come from $\Delta J = +1$ and $\Delta J = -1$.

For $\Delta J = +1$, we use Eq. (17-19):

$$\tilde{\omega}_+ = \tilde{\omega}_0 + 2B(J_i + 1) = \tilde{\omega}_0 + 2B(3 + 1) = 2649.7\text{ cm}^{-1} + 8(8.23\text{ cm}^{-1}) = 2715.5\text{ cm}^{-1}$$

For $\Delta J = -1$, we use Eq. (17-20):

$$\tilde{\omega}_- = \tilde{\omega}_0 - 2BJ_i = \tilde{\omega}_0 + 2B(3) = 2649.7\text{ cm}^{-1} - 6(8.23\text{ cm}^{-1}) = 2600.3\text{ cm}^{-1}$$

Notice that the lines are NOT equidistant from the middle.

---

3. *P* and *R* branches

For transitions between two given vibrational levels, the spectral lines form a *band*, spreading out in both directions from $\tilde{\omega}_0$, the **band origin**. The $\Delta J = -1$ set leads to lower energies of transition (spectral line wavenumbers) and gives a group of lines called the ***P* branch**. The $\Delta J = +1$ set leads to higher energies of transition and gives a group of lines called the ***R* branch**.

EXAMPLE 17-24: The first line of a P branch is at $\tilde{\omega}_0 - 2B$, and the first line of an R branch is at $\tilde{\omega}_0 + 2B$. **(a)** What initial and final J values would produce these lines? **(b)** Compare the initial J values, and draw a general conclusion.

Solution:

(a) For the R branch, we look at Eq. (17-19) and see that to get $\tilde{\omega}_0 + 2B$, J_i must be zero. And since $\Delta J = +1$ for R, this transition is $J_i = 0$ to $J_f = 1$. For the P branch, using Eq. (17-20), we get $\tilde{\omega}_0 - 2B$ if $J_i = 1$. And since $\Delta J = -1$ for P, this transition must be $J_i = 1$ to $J_f = 0$.

(b) For R, $J_i = 0$; for P, $J_i = 1$. In general, if any two lines are located at the same multiple of B from the center, the one in the P branch will have originated from the higher J (by 1).

In actual spectra, the spacings between lines in the P branch do not equal those between lines in the R branch. Anharmonicity can cause a difference in the B values for $\Delta J = +1$ and $\Delta J = -1$, so that the interval $2B_P$ is greater than the interval $2B_R$.

B. Polyatomic molecules

For molecules with more than two atoms, the infrared spectra become more complex because there are more vibrational opportunities. A molecule has $3N$ **degrees of freedom** (3 per atom). We subtract 3 for translational motion and either 3 or 2 for rotational motion (3 for nonlinear, 2 for linear), leaving $3N - 6$ (nonlinear) or $3N - 5$ (linear) vibrational degrees of freedom.

Some simplification is achieved by describing all motions in terms of a few simple ones, called the **normal modes of vibration**, which require that all atoms move with the same frequency and in phase. Normal modes of vibration are usually chosen to be oriented along, or perpendicular to, an internuclear axis. (For example, the "symmetric stretch" in linear CO_2 involves simultaneous movement of both O atoms along the internuclear axis—first away from, then toward the C atom.) Usually, normal modes involve stretching or bending movements, but twisting is also possible. Finally, vibrations must cause an oscillating dipole moment if they are to produce a spectral line. (The symmetric stretch mode of CO_2 is "infrared-inactive," because the molecular dipole doesn't change.)

The displacements in the x, y, and z directions of the atoms of a vibrating molecule can be designated symmetric or antisymmetric with respect to each of the molecule's symmetry operations. This process yields a reducible representation of the molecule's point group, which we then reduce to its irreducible components. From these, we subtract the irreducible representations corresponding to translational and rotational displacements, leaving only vibrational representations. Then we can predict which modes will be *infrared-active*. Since the dipole moment has components in the x, y, and z directions, a vibration will cause the dipole to oscillate if its symmetry is the same as one of these components. Thus, if a normal mode belongs to the same representation as the x, y, or z listed on the right side of character tables, that mode will be infrared-active.

EXAMPLE 17-25: The irreducible representations of the three vibrational modes of the water molecule belong to the representations A_1 (two) and B_1. Are any of these infrared-active?

Solution: From the method explained in Section 17-2C and Example 17-12, we know that the point group of H_2O is C_{2v}, whose character table is Table 17-2. To determine if the modes are infrared-active, we look for x, y, and z in the character table. We find z by A_1 and x by B_1, so we predict that all three modes will be infrared-active. These represent one bending motion and two stretches—one symmetric and one antisymmetric. And in fact, three strong bands do appear in the spectrum of water vapor.

17-7. Other Spectroscopies

A. Raman spectra

If visible light is passed through a substance—a liquid or gas—a small amount of the light is scattered by the molecules of the substance. Most of the scattered light has the same frequency as the incident light, but a small portion of it has new frequencies, some larger and some smaller than the incident frequency. The scattering of the incident frequency is called **Rayleigh**

scattering, and the secondary scattering is called **Raman scattering**. In Raman scattering, the gain or loss in frequency is due to a loss or gain of energy in the molecule, corresponding to a rotational or vibrational transition (or combination). The resulting photons are classified as *Stokes' radiation* or *anti-Stokes' radiation*:

- **Stokes' radiation:** Energy gained by a molecule; lower frequency scattered; more intense
- **Anti-Stokes' radiation:** Energy lost by a molecule; higher frequency scattered; less intense

---

EXAMPLE 17-26: Anti-Stokes' lines are weaker than the Stokes' lines. Why?

Solution: Anti-Stokes' radiation requires that a molecule lose energy, and in order to do so, the molecule must be in an excited state before collision with the incident photon. Most molecules, however, are in the ground state, prepared to accept energy and give Stokes' radiation. Since there are fewer molecules in the excited state, undergoing fewer changes to give anti-Stokes' radiation, the line is less intense.

The Raman effect depends on the polarizability of the molecule. In an electric field, a nonpolar molecule may become polar owing to an *induced dipole moment*, whose magnitude is proportional to the field strength E:

$$\mu = \alpha E \tag{17-21}$$

where α is the **polarizability**. The polarizability must change as the molecule vibrates in order for a Raman line to be produced, and this change causes the induced dipole to oscillate at a frequency different from the incident frequency.

The character tables tell us which vibrational modes will be Raman-active. The normal modes of vibration are assigned to irreducible representations as described in Section 17-6B. The components of polarizability involve binary products, so we look in that column of the character table to see if any x^2, xy, $x^2 + y^2$, etc., belong to the same irreducible representation as a mode of vibration. If there is a match, that mode is Raman-active. Sometimes the Raman line is too weak to be observed, as is usually the case for antisymmetric vibrations.

EXAMPLE 17-27: Are the normal modes of water Raman-active?

Solution: The normal modes of water belong to A_1 and B_1. Checking in Table 17-2, we find several binary terms by A_1 and xz by B_1. Therefore all three modes of H_2O are Raman-active as well as ir-active.

Some molecules that are Raman-active are not ir-active. Thus we can use Raman measurements to obtain information such as the internuclear distances in homonuclear diatomic molecules. Another application in structure determination arises from the **rule of mutual exclusion**:

- If a molecule has a center of symmetry, those vibrations that are ir-active will be Raman-inactive and vice-versa.

Thus, if we find both spectra for a vibrational mode, the molecule cannot have a center of symmetry.

EXAMPLE 17-28: What is the arrangement of atoms in the linear molecule N_2O if the same bands are found in the Raman and the ir spectra?

Solution: The molecule cannot have a center of symmetry: Its arrangement must be N—N—O, not N—O—N.

Raman spectra, then, give essentially the same structural information as ir spectra and are useful for molecules that do not give an ir spectrum. And together, the two methods may provide molecular information that neither can separately.

B. Electronic spectra

The third energy level listed in Table 17-3 also gives rise to spectra due to transitions between energy levels. Electronic spectra are the familiar absorption spectra, showing up in the colored solutions of transition-metal ions. The observed spectra are complex because vibrational and rotational transitions are superimposed on electronic transitions. All molecules yield electronic spectra because movement of electrons always changes the dipole moment. Thus, we can obtain molecular structure information from the electronic spectrum of a molecule that does not give rotational or vibrational spectra, such as homonuclear diatomic molecules. Information provided by electronic spectra but not by rotational and vibrational spectra includes the energies of electronic states (especially excited states) and the dissociation energy D_0. [See Eq. (16-1).]

C. Spin resonance spectra

Spinning particles such as electrons and nuclei have a magnetic dipole moment that will interact with an applied magnetic field. The dipole will not be lined up with the direction of the field and will precess around an axis in that direction at a certain frequency. Electromagnetic radiation of the same frequency can interact with the particle, exchanging energy as the magnetic dipole moment interacts with the oscillating magnetic field of the radiation. The two are said to be in resonance. Two important experimental methods: *nuclear magnetic resonance* (nmr) and *electron spin resonance* (esr) are based on this principle.

SUMMARY

In Table S-17, the major equations of this chapter are collected and identified with the model to which they apply. Connections between a few of them are shown. You should determine what conditions are necessary for each equation to be valid or what conditions are imposed to make the connections. Other important things you should know are

1. A symmetry operation is an action performed on an object that leaves the object apparently unchanged, and a symmetry element is a geometric feature about which symmetry operations may occur. (See Table 17-1.)
2. A symmetry operation may be expressed mathematically as a transformation matrix. (See Example 17-7.)
3. The symmetry operations of a molecule constitute a point group and can be treated by group theory. A molecule can be assigned to a point group by following a procedure of elimination; all symmetry operations do not have to be identified.
4. A multiplication table lists all binary combinations of the members of a group. Representations of groups are sets of numbers or matrices which have the same multiplication table as the group.
5. The irreducible representations of a group are known as symmetry species. A character table lists the characters of the irreducible representations, usually with additional information identifying the symmetry properties of rotational and translational motions and of squares and binary products.
6. Spectroscopic lines are produced by energy emitted due to transitions between energy states: electronic, vibrational, and rotational [Eq. (17-3)].
7. Rotational effects are superimposed on vibrational spectra, and both are superimposed on electronic spectra.
8. Vibrational and rotational transitions that are induced by radiation require a fluctuation in the dipole moment.
9. The energy of a rotational transition for a rigid rotor is inversely proportional to the moment of inertia of the molecule and directly proportional to the rotational quantum number [Eq. (17-4)].
10. The selection rule for rotational transitions in a rigid rotor is $\Delta J = \pm 1$.

TABLE S-17: Summary of Important Equations

Born–Oppenheimer Approximation (nuclear part)

$\mathbf{H}_n\psi_n = E_n\psi_n$ (17-1) → $\psi_n = \psi_{trans}\psi_{vib}\psi_{rot}$ (17-2)

→ $E_n = E_{trans} + E_{vib} + E_{rot}$ (17-3)

Rotation

Rigid rotor

$E_J = B'J(J+1)$ (17-4; 14-41) → $\tilde{F}(J) = BJ(J+1)$ (17-6) → $\Delta\tilde{F} = 2B(J_i + 1)$ (17-8)

$B = h/8\pi^2 Ic$ (17-7) → $\Delta\tilde{F} = 2BJ_f$ (17-9)

Nonrigid rotor

$\tilde{F}(J) = BJ(J+1) - DJ^2(J+1)^2$ (17-10) $\qquad D = \dfrac{h^3}{32\pi^4 I^2 r^2 kc}$ (17-11)

$k = 4\pi^2 c^2 \mu\tilde{\omega}^2$ (17-12)

Vibration

Harmonic oscillator

$V = \frac{1}{2}k(r_{AB} - r_e)^2$ (17-13) $\qquad \tilde{\omega} = \dfrac{1}{2\pi c}\sqrt{\dfrac{k}{\mu}}$ (17-14)

$\tilde{G} = \left(v + \dfrac{1}{2}\right)\dfrac{\omega_0}{c} = \left(v + \dfrac{1}{2}\right)\tilde{\omega}_0$ (17-16)

Anharmonic oscillator

$\tilde{G} = \tilde{\omega}_e\left(v + \dfrac{1}{2}\right) - \tilde{\omega}_e x_e\left(v + \dfrac{1}{2}\right)^2$ (17-17)

Vibration–rotation

$\tilde{G}_{v,j} = \left(v + \dfrac{1}{2}\right)\tilde{\omega}_0 + BJ(J+1)$ (17-18, from 17-6 and 17-16)

→ $\tilde{\omega}_+ = \tilde{\omega}_0 + 2B(J_i + 1)$ (17-19)

→ $\tilde{\omega}_- = \tilde{\omega}_0 - 2BJ_i$ (17-20)

Raman

$\mu = \alpha E$ (17-21)

11. If the rotor is not rigid, a term to account for centrifugal distortion is included in the equation. The effect becomes noticeable at large quantum numbers [Eqs. (17-10, 17-11, and 17-12)].
12. As a first approximation, vibrational spectra are explained by treating a diatomic molecule as a simple harmonic oscillator. The energy of a transition depends on the frequency of vibration [Eqs. (17-13) through (17-16)].
13. The selection rule for vibrational transitions in the harmonic oscillator is $\Delta v = \pm 1$. For the anharmonic case, $\Delta v = \pm 1, \pm 2, \ldots$.
14. Since rotational transitions are superimposed on vibrational transitions, a detailed spectrum is a pattern of lines spreading both ways from the center, known as *P* and *R* branches. The spacing between the lines is $2B$, but B is slightly different for each branch.
15. For polyatomic molecules, atomic motions are described using normal modes of vibration.
16. From character tables and the symmetry of the normal modes, we can predict whether a particular normal mode is infrared- and/or Raman-active.
17. In Raman spectroscopy, Stokes' radiation is produced when a molecule gains energy; anti-Stokes' radiation is produced when it loses energy.
18. The Raman effect depends on the polarizability of the molecule [Eq. (17-21)].

SOLVED PROBLEMS

Group Theory and Symmetry

PROBLEM 17-1 (**a**) List the operations generated by a C_6 axis. (**b**) Are any of these distinct from lower-order operations?

Solution:

(**a**) If C_6 is repeated six times, the object will stop at 60°, 120°, 180°, etc. It would be possible to make one move of 120°, one move of 180°, etc., so each of these is an individual operation. The six may be designated

$$C_6^1,\ C_6^2,\ C_6^3,\ C_6^4,\ C_6^5,\ C_6^6$$

This list makes it clear that all six operations arise from a C_6 axis.

(**b**) Some of these are the same as operations that can be described with smaller n: $C_6^2 = C_3^1$, $C_6^3 = C_2^1$, $C_6^4 = C_3^2$, and $C_6^6 = E$. Thus, C_6 produces two distinct operations: C_6^1 and C_6^5. The list of six usually is written as

$$C_6^1,\ C_3^1,\ C_2^1,\ C_3^2,\ C_6^5,\ E$$

PROBLEM 17-2 Locate all of the axes of symmetry in the planar molecule BCl_3.

Solution: Use a model or draw a picture to assist you. The three B—Cl bonds will be arranged at 120° to each other in a plane. The axis perpendicular to this plane is a C_3 axis, since a rotation of 120° will move one Cl atom to the original position of another Cl atom.

Now consider the line joining one of the Cl atoms to the B atom. Do you see that this is a C_2 axis? Rotation of 180° will interchange the other two Cl atoms. What about the other two B—Cl bonds? Each of them also lies on a C_2 axis, so there are three C_2 axes. This result is a necessity of the symmetry: If there is one C_2 axis perpendicular to a C_n axis, there must be n C_2 axes. Even more generally, if there is any feature (axis or atom) that does not lie on a C_n axis, there must be n of those features in the molecule.

Remember that improper axes are possible. In BCl_3, the principal axis also is an S_3 axis. After rotation of 120°, reflection through the plane of the molecule takes the bottom half of each atom into the top half. You might think that specifying this axis is unnecessary, because you already have C_3 and a plane. Mathematically, however, all operations are needed to meet the requirements of group theory.

PROBLEM 17-3 Continue Problem 17-2 by listing all of the operations generated by the symmetry axes of BCl_3.

Solution: You found a C_3, an S_3, and three C_2 axes. The associated operations are

C_3: $C_3^1, C_3^2, E\ (=C_3^3)$

C_2: $C_2^1, E\ (=C_2^2)$; three of them

S_3: $S_3^1, S_3^2, S_3^3, S_3^4, S_3^5, E\ (=S_3^6)$

The six operations for S_3 may surprise you. We are illustrating the difference between even and odd n for improper axes. In Example 17-6 we saw that, for even n, an S_n axis yields n operations. Now we see that an odd n yields $2n$ operations. The BCl_3 molecule may not be the best for checking this assertion, but try the following exercise.

Consider the top half of one Cl atom to be marked. Rotate 120° and reflect ($=S_3^1$). The marked half-atom has moved from above the plane of the molecule to below the plane. The second rotation-plus-reflection move brings it above the plane, the third below. Now you are at 360°, but the configuration is not equivalent to the starting configuration, because the half-atom started above the plane. If you continue to 720° ($=S_3^6$), you will have the half-atom back on top. Thus, $S_3^6 = E$.

What other of the six operations equal a "lesser" operation? From the opposite viewpoint, which are distinct? A list may help:

S_3^1 = 120° and below = distinct S_3^4 = 120° and above = C_3^1

S_3^2 = 240° and above = C_3^2 S_3^5 = 240° and below = distinct

S_3^3 = 360° and below = σ_h S_3^6 = 360° and above = E

Thus the distinct rotation operations for BCl_3 are $C_3^1, C_3^2, C_2^1, C_2^1, C_2^1, S_3^1, S_3^5$, and $E\,(C_3^3, C_2^2, S_3^6)$. Often these are summarized as $2C_3$, $3C_2$, and $2S_3$ operations.

PROBLEM 17-4 (**a**) Continue Problems 17-2 and 17-3 by identifying all planes of symmetry in BCl_3. (**b**) List the complete set of symmetry operations for BCl_3.

Solution:

(**a**) The previous answers have brought out the fact that the plane of the molecule is a plane of symmetry. It's a horizontal plane, σ_h, because it's perpendicular to the principal axis. You've probably seen already that the plane that is perpendicular to σ_h and includes a B—Cl bond is also a plane of symmetry, σ_v. And because there is a C_3 axis, there must be three σ_v planes; and, of course, each B—Cl bond does correspond to one σ_v.

(**b**) Before making the list, for completeness, note that there is no center of symmetry. You can collect the results of this and the previous problems as

$$E,\quad 2C_3,\quad 3C_2,\quad \sigma_h,\quad 2S_3,\quad 3\sigma_v$$

Note that C_3, C_2, and S_3 refer to types of operations here, not axes.

PROBLEM 17-5 An ethane molecule, H_3CCH_3, when viewed along the C—C axis, may be oriented so that the far H atoms appear directly behind the near ones (*eclipsed*), or the far H atoms may appear to be half-way between the near H atoms (*staggered*). Does either of these orientations have an improper axis? If so, describe it. How many distinct operations does it generate?

Solution: Make a ball-and-stick model and turn one of the methyl groups to generate the two forms—or make two models.

In the eclipsed orientation, you can see that the C—C axis is a C_3 rotation axis. There is a σ_h plane, so the combination of C_3 axis and σ_h plane produces an S_3 axis collinear with the C_3 axis. Remember that an odd-order S axis generates $2n$ operations; but in Problem 17-3, you've already found that an S_3 axis yields only two distinct operations: S_3^1 and S_3^5.

In the staggered orientation, there is still a C_3 axis, but there is no σ_h plane. You should see, however, that a 60° rotation followed by reflection does yield an equivalent position of the atoms. So there is an S_6 axis. An even-order S axis generates n operations. List the six operations as in Problem 17-3, and look for those that equal other operations:

$S_6^1 = 60°$ and below = distinct	$S_6^4 = 240°$ and above $= C_3^2$
$S_6^2 = 120°$ and above $= C_3^1$	$S_6^5 = 300°$ and below = distinct
$S_6^3 = 180°$ and below $= i$	$S_6^6 = 360°$ and above $= E$

Thus, there are two distinct operations: S_6^1 and S_6^5. An i has appeared for S_6^3. Do you see from your model that there is in fact a center of symmetry in the staggered form?

PROBLEM 17-6 Construct the transformation matrix for a C_2 operation collinear with the z axis.

Solution: You need some coordinates of points as a starter. You can deduce that a 180° rotation (C_2) can be a symmetry operation only if a point (x_a, y_a, z_a) has an equivalent point at $(-x_a, -y_a, z_a)$. Now, represent each point by a vector and ask what matrix will make the change:

$$\begin{bmatrix} ? \end{bmatrix} \times \begin{bmatrix} x_a \\ y_a \\ z_a \end{bmatrix} \rightarrow \begin{bmatrix} -x_a \\ -y_a \\ z_a \end{bmatrix}$$

To get $-x_a$ from x_a, y_a, z_a takes $-1, 0, 0$. Similarly, $-y_a$ takes $0, -1, 0$; and z_a takes $0, 0, 1$. Making these the rows, you get

$$\begin{bmatrix} -1 & 0 & 0 \\ 0 & -1 & 0 \\ 0 & 0 & 1 \end{bmatrix} = C_2$$

PROBLEM 17-7 In a group, every member must have an inverse. Show that C_2 is its own inverse, using the matrix from Problem 17-6.

Solution: The inverse of X is that member which, multiplied by X, gives the identity E. The matrix for E in two dimensions is $\begin{bmatrix} 1 & 0 \\ 0 & 1 \end{bmatrix}$ (given in Example 17-8). You should see that in three dimensions it is

$$\begin{bmatrix} 1 & 0 & 0 \\ 0 & 1 & 0 \\ 0 & 0 & 1 \end{bmatrix}$$

If C_2 is its own inverse, C_2C_2 should equal E. Multiplying the matrix from Problem 17-6 by itself gives

$$\begin{bmatrix} -1 & 0 & 0 \\ 0 & -1 & 0 \\ 0 & 0 & 1 \end{bmatrix} \times \begin{bmatrix} -1 & 0 & 0 \\ 0 & -1 & 0 \\ 0 & 0 & 1 \end{bmatrix} = \begin{bmatrix} 1 & 0 & 0 \\ 0 & 1 & 0 \\ 0 & 0 & 1 \end{bmatrix}$$

which is E.

PROBLEM 17-8 Transformation matrices are one representation of a group. If the matrices in Problem 17-7 are part of a representation, what would be written down for E and C_2?

Solution: Here you need to see that a matrix may be represented by its trace—the sum of the numbers on the main diagonal. The representation could consist of all of the complete matrices, but much space is saved by writing only the trace: $\mathrm{tr}(E) = 3$ and $\mathrm{tr}(C_2) = -1$.

PROBLEM 17-9 Return to Problem 17-5 and determine the point group of (**a**) the staggered form and (**b**) the eclipsed form of the H_3CCH_3 molecule.

Solution: You've already seen that both orientations of ethane have a C_3 axis. This brings you to Step 3 of the procedure for determining point groups.

(**a**) *Step 3:* For staggered ethane, there are three C_2 axes perpendicular to the C_3 axis. (You may have difficulty locating them: Try looking down the C_3 axis, so you see that a near H and a far H are separated by 60°; a C_2 axis bisects this angle.) Thus the general type is D.

Step 4: There is no σ_h plane.

Step 5: There are three σ_v planes. Thus the point group is D_{3d}.

(**b**) *Step 3:* For eclipsed ethane, again there are three C_2 axes. (This time they lie in the plane of two H atoms and two C atoms.) Again, the general type is D.

Step 4: There is a σ_h plane. Therefore the point group is D_{3h}.

PROBLEM 17-10 What is the point group of BCl_3? (See Problems 17-2, 17-3, and 17-4.)

Solution: Follow the steps of the procedure. Remember that the molecule is planar.

Step 1: BCl_3 does not belong to an easy-to-recognize group.

Step 2: There is a C_3 axis.

Step 3: There are three C_2 axes; therefore the point group is a D type.

Step 4: There is a σ_h plane; therefore the point group is D_{3h}.

note: Eclipsed H_3CCH_3 and BCl_3 belong to the same point group, so they must have similarities in properties that depend on symmetry, such as optical activity and ir-active vibrational modes.

Spectra

PROBLEM 17-11 From an ir spectrum of $^{12}C^{16}O$, the distance between the lines was measured to be 3.84 cm^{-1}. Calculate the bond length of the molecule.

Solution: You must put together several expressions dealing with rotational spectra. First obtain B, then I, then r. The spacing between the lines of a pure rotational spectrum is $2B$, so

$$B = \frac{\text{interval}}{2} = \frac{3.84}{2} = 1.92\ \mathrm{cm^{-1}}$$

Rearranging Eq. (17-7), you get

$$I = \frac{h}{8\pi^2 cB} = \frac{2.7994 \times 10^{-44}\ \mathrm{kg\,m}}{B} = \frac{2.7994 \times 10^{-44}\ \mathrm{kg\,m}}{192\ \mathrm{m^{-1}}} = 1.458 \times 10^{-46}\ \mathrm{kg\,m^2}$$

From the expression for dipole moment, you know that

$$r_0^2 = \frac{I}{\mu}$$

Now you need the reduced mass μ for $^{12}C^{16}O$:

$$\mu = \frac{m_1 m_2}{m_1 + m_2} = \frac{12(16)}{12 + 16}\left(\frac{1}{6.022 \times 10^{23}}\right)(10^{-3}) = 1.139 \times 10^{-26}\ \text{kg}$$

Thus $$r^2 = \frac{1.458 \times 10^{-46}\ \text{kg m}^2}{1.139 \times 10^{-26}\ \text{kg}} \quad \text{and} \quad r = 1.13 \times 10^{-10}\ \text{m} = 113\ \text{pm}$$

PROBLEM 17-12 From the results of Example 17-20, **(a)** calculate the energy difference between the third and fourth rotational excitation levels as a function of B, and **(b)** compare your result with Eq. (17-8).

Solution:

(a) Looking back to Example 17-20, you find that $\tilde{F}_3 = 12B$ and $\tilde{F}_4 = 20B$, so

$$\Delta\tilde{F} = \tilde{F}_4 - \tilde{F}_3 = (20 - 12)B = 8B$$

(b) Equation (17-8) is $\Delta\tilde{F} = 2B(J_i + 1)$. In this case, $J_i = 3$, so $\Delta\tilde{F} = 2B(3 + 1) = 8B$; thus, the results agree.

PROBLEM 17-13 Continue Problem 17-11 by calculating the frequency ν of the radiation absorbed in a rotational transition from the second to the third excited states of $^{12}C^{16}O$.

Solution: Use Eq. (17-9) and $J_f = 3$, so that

$$\Delta\tilde{F} = 2B(J_f) = 2B(3) = 6B$$

Then, using $B = 1.92\ \text{cm}^{-1}$ from Problem 17-11, you get

$$\Delta\tilde{F} = 6(1.92) = 11.52\ \text{cm}^{-1}$$

Remember that the units of energy are cm^{-1} when obtained from the usual rotational constant; consequently, this $\Delta\tilde{F}$ is the wavenumber, $\tilde{\nu}$, and the frequency is

$$\nu = c\tilde{\nu} = (2.998 \times 10^8\ \text{m s}^{-1})(11.52\ \text{cm}^{-1})\left(\frac{100\ \text{m}^{-1}}{\text{cm}^{-1}}\right) = 3.454 \times 10^9\ \text{s}^{-1}$$

PROBLEM 17-14 Continue Problem 7-13 by listing the wavenumbers of the first five lines in the pure rotational spectrum of $^{12}C^{16}O$.

Solution: You *could* use the result of Example 17-20, but pretend you haven't seen it and use Eq. (17-8). The first five lines will be from transitions that start from $J = 0, 1, 2, 3,$ and 4. For these transitions, Eq. (17-8) gives

$$\Delta\tilde{F} = 2B(J_i + 1) = 2B(1) = 2(1.92) = 3.84$$
$$2B(2) = 4(1.92) = 7.68$$
$$\vdots$$

The values will continue to increase by $2B = 3.84$ each time, so you can add indefinitely: 3.84, 7.68, 11.52, 15.36, 19.20, etc., cm^{-1}.

PROBLEM 17-15 If the force constant of $^{14}N^{16}O$ is $1550\ \text{N m}^{-1}$, at what frequency would you expect to find an ir spectrum? Express your answer in wavenumbers.

Solution: You can calculate the frequency using Eq. (17-15), but you need the reduced mass:

$$\mu = \frac{m_1 m_2}{m_1 + m_2} = \frac{14(16)}{14 + 16}\left(\frac{1}{6.022 \times 10^{23}}\right)(10^{-3}) = 1.240 \times 10^{-26}\ \text{kg}$$

Substituting in Eq. (17-15), you get the true frequency:

$$\omega = \frac{1}{2\pi}\left[\left(\frac{1550\ \text{N m}^{-1}}{1.240 \times 10^{-26}\ \text{kg}}\right)\left(\frac{\text{kg m s}^{-2}}{\text{N}}\right)\right]^{1/2} = 5.627 \times 10^{13}\ \text{s}^{-1}$$

To convert to the vibrational wavenumber, divide by c:

$$\tilde{\omega} = \frac{\omega}{c} = \frac{5.627 \times 10^{13}\ \text{s}^{-1}}{2.998 \times 10^{10}\ \text{cm s}^{-1}} = 1.877 \times 10^3\ \text{cm}^{-1}$$

PROBLEM 17-16 Derive expressions for the wavenumbers of the spectral lines that arise from transitions having Δv of (**a**) $+1$ and (**b**) $+2$ for an anharmonic oscillator, starting at $v = 0$.

Solution: You need to calculate $\Delta\tilde{G}$ $(=\tilde{G}_f - \tilde{G}_i)$ using Eq. (17-17).

(**a**) For $\Delta v = +1$, $\Delta\tilde{G} = \tilde{G}_{v=1} - \tilde{G}_{v=0}$; so you have to substitute Eq. (17-17) twice:

$$\Delta\tilde{G} = \tilde{\omega}_e(1 + \tfrac{1}{2}) - \tilde{\omega}_e x_e(1 + \tfrac{1}{2})^2 - [\tilde{\omega}_e(\tfrac{1}{2}) - \tilde{\omega}_e x_e(\tfrac{1}{2})^2]$$
$$= \omega_e[\tfrac{3}{2} - x_e(\tfrac{9}{4}) - \tfrac{1}{2} + x_e(\tfrac{1}{4})] = \tilde{\omega}_e[1 - x_e(\tfrac{8}{4})] = \tilde{\omega}_e(1 - 2x_e)$$

(**b**) For $\Delta v = +2$, $\Delta\tilde{G} = \tilde{G}_{v=2} - \tilde{G}_{v=0}$; so you get

$$\Delta\tilde{G} = \tilde{\omega}_e(2 + \tfrac{1}{2}) - \tilde{\omega}_e x_e(2 + \tfrac{1}{2})^2 - [\tilde{\omega}_e(\tfrac{1}{2}) - \tilde{\omega}_e x_e(\tfrac{1}{2})^2]$$
$$= \tilde{\omega}_e[\tfrac{5}{2} - x_e(\tfrac{25}{4}) - \tfrac{1}{2} + x_e(\tfrac{1}{4})] = \tilde{\omega}_e(2 - 6x_e) = 2\tilde{\omega}_e(1 - 3x_e)$$

PROBLEM 17-17 The ir spectrum of HCl has a strong absorption at 2886 cm^{-1} and a weaker one at 5668 cm^{-1}; these are the *fundamental* and *first overtone*, respectively. Calculate $\tilde{\omega}_e$ and x_e using the equations from Problem 17-16.

Solution: The two wavenumbers given are the $\Delta\tilde{G}$ values of the equations derived in Problem 17-16, so you have two simultaneous equations:

$$2886 = \tilde{\omega}_e(1 - 2x_e) \qquad \text{and} \qquad 5668 = 2\tilde{\omega}_e(1 - 3x_e)$$

Solve one of them for one unknown and substitute the result in the other. The first gives $\tilde{\omega}_e$ easily:

$$\tilde{\omega}_e = \frac{2886}{1 - 2x_e}$$

The second now becomes

$$5668 = 2\left(\frac{2886}{1 - 2x_e}\right)(1 - 3x_e)$$

so that

$$\frac{5668}{2(2886)}(1 - 2x_e) = 1 - 3x_e$$
$$0.9820 - 1.964x_e = 1 - 3x_e$$
$$1.036x_e = 0.0180$$
$$x_e = \frac{0.0180}{1.036} = 0.0174$$

Then

$$\tilde{\omega}_e = \frac{2886}{1 - 2x_e} = \frac{2886}{1 - 2(0.0174)} = \frac{2886}{0.9652} = 2990\ \text{cm}^{-1}$$

PROBLEM 17-18 In deriving Eqs. (17-19) and (17-20), we assumed that B was the same for both vibrational levels. Derive more precise equations using B_1 and B_0, respectively, for the $v = 1$ and $v = 0$ levels.

Solution: You need to find $\Delta\tilde{G}$, using Eq. (17-18), as we did in obtaining Eqs. (17-19) and (17-20). Thus

$$\Delta\tilde{G} = \tilde{G}_1 - \tilde{G}_0 = (1 + \tfrac{1}{2})\tilde{\omega}_0 + B_1 J_f(J_f + 1) - (0 + \tfrac{1}{2})\tilde{\omega}_0 - B_0 J_i(J_i + 1)$$

The $\frac{1}{2}\tilde{\omega}_0$ terms cancel and $1 - 0 = 1$, so you have

$$\Delta\tilde{G} = \tilde{\omega}_0 + B_1 J_f(J_f + 1) - B_0 J_i(J_i + 1)$$

For $\Delta J = +1$, $J_f = J_i + 1$. This is the R branch of the band, so using $\tilde{\omega}_R$ for the lines, you get

$$\tilde{\omega}_R = \tilde{\omega}_0 + B_1(J_i + 1)(J_i + 2) - B_0 J_i(J_i + 1)$$

For $\Delta J = -1$, $J_f = J_i - 1$. This is the P branch, so

$$\tilde{\omega}_P = \tilde{\omega}_0 + B_1(J_i - 1)(J_i) - B_0 J_i(J_i + 1)$$

PROBLEM 17-19 From analysis of the ir spectrum of carbon monoxide ($^{12}C^{16}O$), the center of the P–R band is found to be 2143 cm^{-1}. Also, $B_1 = 1.898$ cm^{-1}, and $B_0 = 1.915$ cm^{-1}. Calculate the wavenumbers of the corresponding P and R lines from $J = 20$ and $J = 19$ (see Examples 17-23 and 17-24), and compare their distances from the center. Refer to Problem 17-18.

Solution: Since B is not constant, you need to substitute in the equations obtained in Problem 17-18 rather than Eqs. (17-19) and (17-20). Thus

$$\tilde{\omega}_R = 2143 + 1.898(20)(21) - 1.915(19)(20) = 2212\ \text{cm}^{-1}$$

and
$$\tilde{\omega}_P = 2143 + 1.898(19)(20) - 1.915(20)(21) = 2060\ \text{cm}^{-1}$$

On the R branch the difference is 69, and on the P branch the difference is 83. (Here, you can see that the spacing between the lines will be slightly larger in the P branch than in the R branch—as it should be in all cases.)

PROBLEM 17-20 **(a)** Continue Problem 17-19 by calculating the internuclear distance r for the molecule in each vibrational level. **(b)** Calculate the percent increase. **(c)** Compare the result to that of Problem 17-11.

Solution:

(a) To calculate r, you have to go back to the definition of B, Eq. (17-7):

$$B = (2.7994 \times 10^{-44}\ \text{kg m})(I)^{-1}$$

Substituting $I = \mu r^2$ and rearranging, you get

$$r^2 = \frac{2.7994 \times 10^{-44}\ \text{kg m}}{\mu B}$$

And in Problem 17-11, you found that $\mu = 1.139 \times 10^{-26}$ kg for $^{12}C^{16}O$, so

$$r^2 = \frac{2.7994 \times 10^{-44}\ \text{kg m}}{1.139 \times 10^{-26}\ \text{kg}}\left(\frac{1}{B}\right) = \frac{2.458 \times 10^{-18}\ \text{m}}{B}$$

For $v = 1$, $$r^2 = \frac{2.458 \times 10^{-18}\ \text{m}}{189.8\ \text{m}^{-1}}$$ $$r = 113.8\ \text{pm}$$

For $v = 0$, $$r^2 = \frac{2.458 \times 10^{-18}\ \text{m}}{191.5\ \text{m}^{-1}}$$ $$r = 113.3\ \text{pm}$$

(b) The present increase is $\frac{0.5}{113.3} \times 100 = 0.44\%$. Obviously, the effect is not very large.

(c) In Problem 17-11, you used $B = 1.92\ \text{cm}^{-1}$ and found $r = 113$ pm. Actually, the $v = 0$ calculation is simply a repeat of Problem 17-11 with a more accurate B value.

Supplementary Exercises

PROBLEM 17-21 Answer the following questions.

(1) How many symmetry operations are generated by a C_5 axis? a σ_v plane? an S_5 axis?
(2) Of the operations generated by an S_4 axis, how many are distinct?
(3) What single operation is equal to performing a reflection in the same plane twice?
(4) Which of these molecules have a center of symmetry: acetylene, ammonia, carbon monoxide, carbon dioxide?
(5) How many planes of symmetry are there in an ammonia molecule?
(6) If a molecule has a C_6 axis, must it also have a center of symmetry?
(7) Give an example of an operation that is its own inverse (other than the identity operation).
(8) List the symmetry operations of a molecule of bromochloroiodomethane.
(9) What geometry does a molecule have if it belongs to the point group $C_{\infty v}$?
(10) What are the SI units of the rotational constant?
(11) If the lines in a far-infrared spectrum are equally spaced at $4.0\ \text{cm}^{-1}$, what is the value of the rotational constant of the molecule?
(12) Which of these molecules would you expect to produce a rotational spectrum: NO, CO_2, HCN, N_2?
(13) For which molecule would the rotational constant be larger: $H^{35}Cl$ or $H^{37}Cl$?

(14) What is the general expression for the interval between the lines of a pure rotational spectrum?
(15) What causes nonrigidity in a rotating diatomic molecule?
(16) In a rotation–vibration spectrum, which branch of a band represents lower energies, *P* or *R*?
(17) Does a molecule gain or lose energy to produce a Stokes' line in a Raman spectrum?
(18) What molecular property must change if a Raman transition is to occur?
(19) Carbon dioxide has ir bands but no corresponding Raman band. What feature must the molecule have?
(20) What molecular information is usually obtained from electronic spectra rather than from vibrational or rotational spectra?

PROBLEM 17-22 List all of the distinct symmetry operations in a regular pentagon.

PROBLEM 17-23 The answer to Problem 17-22 constitutes a point group. What is its order and how many classes are there?

PROBLEM 17-24 What is the inverse of an S_6^1 operation?

PROBLEM 17-25 Write the matrix for the inversion operation in three dimensions.

PROBLEM 17-26 Find the point groups of molecules of **(a)** benzene and **(b)** *m*-dichlorobenzene.

PROBLEM 17-27 Find the point groups of **(a)** ethylene, **(b)** *cis*-dichloroethylene, and **(c)** *trans*-dichloroethylene.

PROBLEM 17-28 If the rotational constant of $H^{127}I$ is 6.51 cm^{-1}, what is the internuclear distance?

PROBLEM 17-29 What is the percent of change in the rotational constant for carbon monoxide if the ^{16}O in $^{12}C^{16}O$ is replaced by ^{18}O? Assume that the internuclear distances are the same. (See Problem 17-11.)

PROBLEM 17-30 If the internuclear distance in $H^{37}Cl$ is 129 pm, what are the wavenumbers of the first three lines in its rotational spectrum?

PROBLEM 17-31 If $H^{127}I$ has a fundamental vibration at 2310 cm^{-1}, what is its force constant?

PROBLEM 17-32 The radical $^{16}OH\cdot$ has a strong ir band at 3568 cm^{-1} and a weaker one at 6967 cm^{-1}. Calculate its equilibrium vibrational wavenumber, its anharmonicity constant, and its force constant.

PROBLEM 17-33 Continuing Problem 17-32, calculate the internuclear distance of the radical, given that the value of *B* was determined to be 18.9 cm^{-1}.

PROBLEM 17-34 Calculate the fundamental vibrational wavenumber of $^{18}OH\cdot$. (See Problem 17-32, and use the fact that isotopic substitution does not affect the force constant.)

Answers to Supplementary Exercises

17-21

(1) 5, 1, 10
(2) 2
(3) *E*
(4) H—C≡C—H, O=C=O
(5) 3
(6) no
(7) $C_2^1, \sigma_h, \sigma_v, \sigma_d, i$
(8) *E* (or C_1)
(9) Linear
(10) m^{-1}
(11) 2.0 cm^{-1}
(12) NO, HCN
(13) $H^{35}Cl$
(14) 2*B*
(15) centrifugal force
(16) *P*
(17) gain
(18) polarizability
(19) center of symmetry
(20) dissociation energy

17-22 $E, \sigma_h, 5\sigma_v, 5C_2^1, S_5^1, S_5^3, S_5^7, S_5^9, C_5^1, C_5^2, C_5^3, C_5^4$ **17-23** 20; 8 **17-24** S_6^5

17-25 $\begin{bmatrix} -1 & 0 & 0 \\ 0 & -1 & 0 \\ 0 & 0 & -1 \end{bmatrix}$ **17-26** **(a)** D_{6h} **(b)** C_{2v} **17-27** **(a)** D_{2h} **(b)** C_{2v} **(c)** C_{2h}

17-28 162 pm **17-29** 4.76% **17-30** 20.8, 41.6, 62.4 cm^{-1} **17-31** 312 N m^{-1}

17-32 3737 cm^{-1}; 0.0226; 775 N m^{-1} **17-33** 97.4 pm **17-34** 3556 cm^{-1}

18 STATISTICAL MECHANICS

THIS CHAPTER IS ABOUT

☑ Statistical Definitions and Probability
☑ Boltzmann Distribution
☑ Partition Functions and Energy
☑ Other State Functions

18-1. Statistical Definitions and Probability

The underlying postulate of statistical mechanics is the assumption that the observed properties of a system (e.g., a gas) are the net result of the properties of the individual molecules constituting that system. We can use thermodynamic methods to deal with the macroscopic properties of a system—temperature, energy, etc.—which can be measured directly in the laboratory. Now we see how to calculate the values of the macroscopic properties of the system from the microscopic properties of molecules—energy states, structure, etc.—which can be measured indirectly or calculated from theory. We use statistics to connect the macroscopic properties to the microscopic properties, because this is the only way to deal with numbers of particles so large as Avogadro's number.

A. Distributions and complexions

A **distribution** (or, sometimes, *configuration*) is one way in which a group of objects may be arranged. A **complexion** is one way of achieving that distribution. The **probability** of finding a particular distribution depends on the number of complexions of that distribution: The more complexions, the more probable the distribution.

EXAMPLE 18-1: Given 3 marked boxes and 4 marbles, describe some ways the marbles could be distributed in the boxes. How many ways could each distribution be achieved?

Solution: If we designate the boxes as 1, 2, and 3, we can see (among others) the following possible distributions and ways of achieving them:

Distribution	Number of ways
(1) All four marbles in box 1	1
(2) Three marbles in box 1, one marble in box 3	4
(3) Two marbles in box 1, one marble in box 2, one marble in box 3	12

Each way of achieving a distribution is a *complexion*. Thus, distribution (3) has 12 complexions.

EXAMPLE 18-2: Assuming that all complexions are equally probable and that distributions (1), (2), and (3) are the only possible distributions, calculate the probability of each distribution given in Example 18-1.

Solution: The probability of each distribution is the ratio of its complexions to all of the possible complexions. The total number of complexions given in Example 18-1 is 17, so

Distribution (1):	$1/17 = 0.059$
Distribution (2):	$4/17 = 0.235$
Distribution (3):	$12/17 = 0.706$
Total:	$17/17 = 1.000$

The number of complexions of a distribution is called the **thermodynamic probability** (or **weight**) W. The thermodynamic probability isn't a true probability, but it is proportional to the true probability.

note: The sum of the true probabilities equals 1, but the sum of the thermodynamic probabilities equals the total number of complexions.

In molecular systems, each distribution is a **state of the system**. The boxes of Example 18-1 are now energy levels and are called electronic or quantum states. (You'll see this word "state" used in different ways in different textbooks.)

B. Distributions over energy levels

The simple system described in Examples 18-1 and 18-2 may be generalized to a large system of N particles for which i energy levels are available. We can describe any distribution by specifying n_0 particles in energy level ε_0, n_1 in ε_1, etc. A particular set of n_i is one distribution (one state of the system). For example, for distribution (2) of Example 18-1, we have $n_0 = 3$, $n_1 = 0$, $n_2 = 1$, where boxes 1, 2, and 3 represent energy levels ε_0, ε_1, and ε_2.

Two important conditions must be met. The total number of particles, N, must be

$$N = \Sigma n_i \qquad \textbf{(18-1)}$$

and the total energy E of the system must be

$$E = \Sigma n_i \varepsilon_i \qquad \textbf{(18-2)}$$

Many possible distributions for a certain N will be eliminated because they do not give the correct E.

We would like to be able to calculate the probability of a particular distribution of molecules over several energy levels (a state of the system) in a system containing a very large number of molecules. This situation is the familiar mathematical problem of finding how many distinguishable ways N objects can be arranged in i boxes, when only a certain number, n_i, may be placed in box i. And the answer to this familiar problem is the thermodynamic probability W:

$$W = \frac{N!}{n_0!n_1!n_2!\cdots n_i!} = \frac{N!}{\Pi n_i!} \qquad \textbf{(18-3)}$$

EXAMPLE 18-3: Calculate the thermodynamic probability for distribution (3) in Example 18-1.

Solution: This distribution would be written as $n_0 = 2$, $n_1 = 1$, $n_2 = 1$, and $N = 4$. Substituting in Eq. (18-3), we have

$$W = \frac{4(3)(2)(1)}{[2(1)](1)(1)} = \frac{24}{2} = 12$$

18-2. Boltzmann Distribution Law

A. Outline of derivation

An important assumption of statistical mechanics is that the most probable distribution of particles in a system is that distribution for which W is a maximum. In Example 18-2, for instance, we easily identified distribution (3) as the most probable, but, with around 10^{23} particles, a different method is needed. The usual procedure is to find those conditions under which ln W is a maximum, because it is simpler than dealing with W. At the maximum, $d(\ln W) = 0$. Equations (18-1) and (18-2) impose constraints on changes in n_i:

$$\Sigma\, dn_i = 0 \qquad \text{and} \qquad \Sigma\, \varepsilon_i\, dn_i = 0$$

Such constraints are handled by Lagrange's method of undetermined multipliers, in which the constraints are multiplied by constants (α and β) and the results are added to the expression for $d(\ln W)$. Now we need Stirling's approximation, which is valid when N is large:

$$\ln N! = N \ln N - N$$

When this is applied to the natural logarithm of Eq. (18-3), we obtain

$$\ln W = N \ln N - \Sigma n_i \ln n_i$$

Eventually, we find that the most probable distribution is that for which the number of particles in each energy level is given by the **Boltzmann distribution law**:

$$n_i = Ae^{-\varepsilon_i/kT} \qquad \textbf{(18-4)}$$

where $\beta = 1/kT$. The Boltzmann distribution law was previously introduced as Eq. (2-46). Remember that k is the *Boltzmann constant*—the gas constant per molecule:

$$k = \frac{R}{N} = 1.3807 \times 10^{-23}\ \text{J K}^{-1}\,\text{molecule}^{-1}$$

$$= 3.3000 \times 10^{-24}\ \text{cal K}^{-1}\,\text{molecule}^{-1} = 0.695\,02\ \text{cm}^{-1}\text{K}^{-1}\,\text{molecule}^{-1}$$

B. Various forms of Boltzmann's distribution law

To evaluate A, we combine Eq. (18-1) and Eq. (18-4):

$$\Sigma n_i = N = A\Sigma e^{-\varepsilon_i/kT}$$

so that

$$A = \frac{N}{\Sigma e^{-\varepsilon_i/kT}}$$

Then the Boltzmann equation (18-4) becomes

BOLTZMANN DISTRIBUTION

$$\frac{n_i}{N} = \frac{e^{-\varepsilon_i/kT}}{\Sigma e^{-\varepsilon_i/kT}} \qquad \textbf{(18-5)}$$

which specifies the fraction of particles at energy level ε_i.

A more general form, with a weighting factor g_i, allows for the possibility that some energy levels may be degenerate:

$$\frac{n_i}{N} = \frac{g_i e^{-\varepsilon_i/kT}}{\Sigma g_i e^{-\varepsilon_i/kT}} \qquad \textbf{(18-6)}$$

Another useful form of the Boltzmann equation, which follows from Eqs. (18-4) or (18-5), is

$$\frac{n_i}{n_0} = e^{-(\varepsilon_i - \varepsilon_0)/kT}$$

which can be simplified by expressing the ε_i relative to the ground state, i.e., by letting $\varepsilon_0 = 0$. Then

$$\frac{n_i}{n_0} = e^{-\varepsilon_i/kT} \qquad \textbf{(18-7)}$$

or, with the weighting factors,

$$\frac{n_i}{n_0} = \frac{g_i}{g_0}\, e^{-\varepsilon_i/kT} \qquad \textbf{(18-8)}$$

EXAMPLE 18-4: Calculate the ratio of the molecules that are in the second vibrational level to those in the first level for HBr(g) at 300 K ($\tilde{\omega}_0 = 2650\ \text{cm}^{-1}$).

Solution: From Eq. (17-16), we know that the spacing between vibrational levels is $\tilde{\omega}_0$, at least at low v. Taking $\varepsilon_0 = 0$, we have $\varepsilon_1 = 2650\ \text{cm}^{-1}$. Vibrational energy levels in HBr are nondegenerate, so we omit weighting factors and use Eq. (18-7). For convenience, let's calculate the exponent first:

$$-\frac{\varepsilon_i}{kT} = -\frac{2650\ \text{cm}^{-1}\ \text{molecule}^{-1}}{(0.6950\ \text{cm}^{-1}\text{K}^{-1}\ \text{molecule}^{-1})(300\ \text{K})} = -12.71$$

Note that the exponent must be dimensionless. Thus,

$$\frac{n_i}{n_0} = e^{-12.71} = 3.0 \times 10^{-6}$$

EXAMPLE 18-5: Use the result of Example 18-4 to calculate the number of molecules in energy level ε_1 (the "first excited state") if there are 10^{23} molecules in ε_0 (the "ground state").

Solution: The n_i/n_0 ratio just calculated can be read as three in one million, so there are three molecules in the excited state for every million molecules in the ground state. Stated this way, it doesn't sound like very much! But if $n_0 = 10^{23}$, then

$$n_1 = (3.0 \times 10^{-6})(10^{23}) = 3.0 \times 10^{17}$$

This number is huge, but let's put it in perspective. When we add n_1 and n_0, we get $0.000\,003 \times 10^{23}$ plus 1×10^{23}, so n_1 is really quite insignificant.

---

note: In Examples 18-4 and 18-5 we have introduced the subject of population of energy levels, which plays a role in determining the intensity of spectral lines. This topic is an application of the Boltzmann distribution appropriate to spectroscopy (see Chapter 17) but we won't develop it further here.

18-3. Partition Functions and Energy

A. Molecular partition function

The denominator of Eq. (18-5) is known as the **molecular partition function** q:

$$q = \Sigma e^{-\varepsilon_i/kT} \tag{18-9}$$

which is a summation over the energy levels (states) of the molecules. The importance of this expression is that we can represent thermodynamic properties using the partition function and the derivatives of its natural logarithm. If the molecular partition function can be determined, then every equilibrium property of a system of independent (distinguishable) particles can be calculated.

In Chapter 16, we saw that the Born–Oppenheimer approximation divides the wavefunction and the energy of a molecule into electronic and nuclear components [Eqs. (16-2) and (16-3)]. And in Chapter 17, we divided the nuclear component into translational, rotational, and vibrational components [Eqs. (17-2) and (17-3)]. If we put the electronic component into Eq. (17-3), using lowercase e for the energy of a molecule, we have

$$e = e_{\text{elec}} + e_{\text{trans}} + e_{\text{vib}} + e_{\text{rot}} \tag{18-10}$$

Similarly, the partition function can be separated into corresponding components,

$$q = q_e q_t q_v q_r \tag{18-11}$$

where each q is given by Eq. (18-9). Then, working with translation only, we obtain the Maxwell distribution law for molecular speeds, also finding that $\beta = 1/kT$, and we obtain the Maxwell–Boltzmann distribution law for molecular energies.

If the partition function is large, many energy levels (states) are available, probably because the separation between levels is small, as in the case of translational motion; but, for electronic energy levels, the separation is often very large, and q_e is close to 1. At $T = 0$, all $q = 1$, and, as temperature increases, q becomes larger, as shown by Eq. (18-9).

B. Canonical partition function

In principle, we could calculate the value of the total energy of a system containing N particles by averaging the fluctuating values of the total energy over time as the energies of individual particles change. But this is not feasible, so we use an approach introduced by Gibbs, based on the *ensemble*. We imagine an **ensemble** to be composed of many identical systems, each of which is a "clone" of the system of concern. If the system is isolated in the thermodynamic sense, neither energy nor matter can enter or leave, so V, E, and N are constant. By definition, then, each member of the ensemble has the same V, E, and N, and this assembly of systems is called a **microcanonical ensemble**.

For a closed system, mass is constant but energy can be exchanged; so N, V, and T are constant only at equilibrium. An ensemble of systems identical to this one is called a **canonical ensemble** and has the important feature that the E values of each member are not quite equal but fluctuate undetectably as the system changes from one complexion to another. At equilibrium, the average value of E is the same for all members and determines T. The canonical ensemble is the appropriate one for the usual thermodynamic expressions, which are restricted to closed systems. For an open system, the ensemble is called a **grand canonical ensemble**.

Instead of measuring one system many times to determine the average value of a property, we can measure the ensemble of identical systems, all at the same time, and average over all members $\mathcal{N}$. Since the number of members is huge, every allowed complexion of the system will occur at least once; thus we assume that

- The average over the ensemble gives the same value as the time average of many measurements of the one system as it fluctuates through all possible complexions.

If the energy of the ensemble is $\mathcal{E}$, the average energy per member is $\bar{E} = \mathcal{E}/\mathcal{N}$. As $\mathcal{N}$ approaches infinity, this average is taken to be the usual internal energy, E, of the system. We also make a second assumption, known as the **principle of equal *a priori* probabilities**:

- Every complexion that meets the conditions is equally probable.

This means that each member of the ensemble contributes equally to the average.

More than one member of the ensemble may have the same energy E; the number having E_i is n_i. A particular set of n_i is one distribution. The most probable distribution is found as described in Section 18-2, but Eq. (18-5) becomes

$$\frac{n_i}{\mathcal{N}} = \frac{e^{-E_i/kT}}{\Sigma e^{-E_i/kT}} \tag{18-12}$$

This expression has been called the **canonical distribution**, analogous to Eq. (18-5), and it tells us the probability p_i that a member of the ensemble has energy E_i. The denominator of Eq. (18-12) is the **canonical partition function** (also called the *system partition function*):

$$Q = \Sigma e^{-E_i/kT} \tag{18-13}$$

which is analogous to Eq. (18-9). Applying Eq. (18-2), we see that the total energy of the ensemble, $\mathcal{E}$, must equal $\Sigma n_i E_i$. The average energy $\bar{E}$ is

$$\bar{E} = \frac{\mathcal{E}}{\mathcal{N}} = \frac{\Sigma n_i E_i}{\mathcal{N}}$$

and substituting Eqs. (18-12) and (18-13) gives

$$\bar{E} = \frac{\Sigma E_i e^{-E_i/kT}}{Q} \tag{18-14}$$

Then, working backward, so to speak, from the result we want, we can evaluate $(\partial Q/\partial T)_V$ from Eq. (18-13) for constant N:

$$\left(\frac{\partial Q}{\partial T}\right)_V = \sum \frac{E_i}{kT^2} e^{-E_i/kT} \tag{18-15}$$

and substituting for the numerator of Eq. (18-14) gives

$$\bar{E} = \frac{kT^2}{Q}\left(\frac{\partial Q}{\partial T}\right)_V = kT^2\left(\frac{\partial \ln Q}{\partial T}\right)_V$$

We take this average to be the internal energy E of each system and write

$$E = kT^2\left(\frac{\partial \ln Q}{\partial T}\right)_V \qquad \textbf{(18-16)}$$

We have seen that the energy of a molecule can be divided into electronic, translational, vibrational, and rotational components [Eqs. (18-10) and (18-11)]. Similarly, for one mole,

$$E = E_{\text{elec}} + E_{\text{trans}} + E_{\text{vib}} + E_{\text{rot}} \qquad \textbf{(18-17)}$$

and Q can be divided into equivalent parts:

$$Q = Q_{\text{elec}}Q_{\text{trans}}Q_{\text{vib}}Q_{\text{rot}} \qquad \textbf{(18-18)}$$

C. Relationship between ensemble and molecular partition functions

One member of the ensemble is a replica of the system, so it will contain N molecules; one molecule will have a range of energy levels available to it. We can rewrite Eq. (18-13) as a summation over all of the energy levels available to the system; then we can factor this expression for Q into N terms each of which is q, i.e., $Q = q^N$. This result applies if the molecules are distinguishable; this situation would occur when, for example, they are held in place in a crystal.

If the molecules are not distinguishable, as in a gas, we have

$$Q = \frac{q^N}{N!} \qquad \textbf{(18-19)}$$

Now we take the logarithm of Eq. (18-19) and apply the Stirling approximation to get

$$\ln Q = \ln q^N - \ln N! = N \ln q - N \ln N + N = N \ln q - N(\ln N - 1) \qquad \textbf{(18-20)}$$

This relationship between Q and q is needed to convert the expressions for the thermodynamic properties derived using Q to expressions involving molecular partition functions.

18-4. Other State Functions

A. Entropy and probability

We can describe entropy as a measure of disorder or randomness—concepts that are related to the number of ways in which a system may be arranged. Further, we can say that a measure of degree of disorder is the probability that a certain arrangement will occur. Thus, choosing the most probable distribution, we can say that the lower its probability, the greater the disorder of the system.

EXAMPLE 18-6: Which distribution in Examples 18-1 and 18-2 represents the greatest disorder? Does this correlate with probability? Is the system more likely to be in an ordered or disordered state?

Solution: Looking back to Example 18-1, we see that in arrangement (1) all the marbles are in one box—quite orderly. In (3), the marbles are spread over three boxes—less orderly. Then, looking at Example 18-2, we find that the probability of distribution (3) is 71% compared to 6% for distribution (1). Thus, the system is much more likely to be in a disordered state than in a perfectly ordered one.

Boltzmann incorporated these ideas into a definition of entropy, saying that entropy S is proportional to the average of the natural logarithm of the probability:

$$S = -k\,\overline{\ln p} \qquad \textbf{(18-21)}$$

The average is given by

$$\overline{\ln p} = \Sigma p_i \ln p_i$$

Then

$$S = -k\Sigma p_i \ln p_i \qquad \textbf{(18-22)}$$

Omitting the proof, we can state that this leads to

$$S = k \ln W \tag{18-23}$$

where k is Boltzmann's constant and W is the number of complexions of the system [(Eq. 18-3)].

EXAMPLE 18-7: How would a system of N particles be arranged in order to have the largest possible entropy?

Solution: From Eq. (18-3) we see that the largest value of W occurs when each n_i equals 1 or 0. Thus the N molecules should be distributed so widely over the energies that each energy level would contain only 1 molecule.

B. Entropy in terms of the partition function

We can bring Q into Eq. (18-22) by substituting an expression for $\ln p_i$ obtained by combining Eqs. (18-12) and (18-13) and taking the logarithm:

$$\ln p_i = \ln e^{-E_i/kT} - \ln Q = -\frac{E_i}{kT} - \ln Q$$

Then
$$S = -k\left(-\sum \frac{p_i E_i}{kT} - \ln Q \, \Sigma p_i\right)$$

It is necessary that $\Sigma p_i = 1$. Then, since $p_i = n_i/\mathcal{N}_i$, we have

$$\Sigma p_i E_i = \sum \frac{n_i E_i}{\mathcal{N}_i} = \bar{E}$$

Again taking $\bar{E}$ to be the usual internal energy E, we have

$$S = \frac{E}{T} + k \ln Q \tag{18-24}$$

Finally, to convert to the molecular partition function, we substitute Eq. (18-20):

$$S = \frac{E}{T} + kN \ln q - kN(\ln N - 1) = \frac{E}{T} + kN(\ln q - \ln N + 1) \tag{18-25}$$

C. Other functions

1. Helmholtz energy

From Eq. (6-9), we know that the Helmholtz energy $A = E - TS$. Using Eq. (18-16) for E and Eq. (18-24) for S, we get

$$A = kT^2\left(\frac{\partial \ln Q}{\partial T}\right)_V - T\left[kT\left(\frac{\partial \ln Q}{\partial T}\right)_V + k \ln Q\right] = -kT \ln Q \tag{18-26}$$

2. Enthalpy and Gibbs energy

For enthalpy and Gibbs energy, we need an expression for the pressure.

EXAMPLE 18-8: Derive an expression for obtaining the pressure from the Helmholtz energy A; then express pressure in terms of the partition function. Write an expression for enthalpy H and Gibbs energy G.

Solution: We must recall the Gibbs equation for dA, Eq. (6-34):

$$dA = -S\,dT - P\,dV$$

And, at constant T,
$$P = -\left(\frac{\partial A}{\partial V}\right)_T$$

So, taking the partial derivative of Eq. (18-26), we find

$$P = kT\left(\frac{\partial \ln Q}{\partial V}\right)_T \quad \textbf{(18-27)}$$

Now we can obtain

$$H = kT\left[T\left(\frac{\partial \ln Q}{\partial T}\right)_V + V\left(\frac{\partial \ln Q}{\partial V}\right)_T\right] \quad \textbf{(18-28)}$$

$$G = -kT\left[\ln Q - V\left(\frac{\partial \ln Q}{\partial V}\right)_T\right] \quad \textbf{(18-29)}$$

---

3. Heat capacity

The heat capacity at constant volume is given by Eq. (3-28):

$$C_V = \left(\frac{\partial E}{\partial T}\right)_V$$

Differentiating Eq. (18-16), we get

$$C_V = kT\left[2\left(\frac{\partial \ln Q}{\partial T}\right)_V + T\left(\frac{\partial^2 \ln Q}{\partial T^2}\right)_V\right] \quad \textbf{(18-30)}$$

D. Calculations

To use any of these expressions in $\ln Q$, we must replace $\ln Q$ by Eq. (18-20) and evaluate q from Eqs. (18-9) and (18-11). This isn't as bad as it looks; usually we want the derivative of $\ln Q$ and, if N is constant, the derivative of Eq. (18-20) is simply

$$\left(\frac{\partial \ln Q}{\partial x}\right)_y = N\left(\frac{\partial \ln q}{\partial x}\right)_y \quad \textbf{(18-31)}$$

EXAMPLE 18-9: Use partition functions to find an expression for E in terms of T for an ideal monatomic gas.

Solution: This is an uncomplicated situation because a single atom does not have vibrational and rotational energies. Near room temperature, we can assume that all atoms are in the ground state and ignore electronic transitions. This leaves only translations, so Eq. (18-17) is reduced to $E = E_t$ and Eq. (18-11) becomes $q = q_t$.

For a particle in a one-dimensional box, $q_t = q_x$, and the energy levels are given by Eq. (14-32):

$$\varepsilon_n = \frac{n^2h^2}{8ma^2}$$

From Eq. (18-9), q_x is

$$q_x = \Sigma e^{-\varepsilon_x/kT}$$

Because the spacings are very small in translational motion, we can replace the summation by integration:

$$q_x = \int_0^\infty e^{-n^2h^2/8ma^2kT}\,dn = \left(\frac{(2\pi mkT)^{1/2}}{h}\right)x$$

Similar expressions are obtained for q_y and q_z. Multiplying them together gives

$$q = q_xq_yq_z = \frac{(2\pi mkT)^{3/2}}{h^3}xyz = (2\pi mkT)^{3/2}\left(\frac{V}{h^3}\right) \quad \textbf{(18-32)}$$

The assumption that the contribution from each direction is the same is a consequence of the principle of **equipartition of energy** (see Section 14-5C).

We will need $\left(\frac{\partial \ln Q}{\partial T}\right)_V$ in order to use Eq. (18-16). From Eq. (18-20), we get an expression like Eq. (18-31):

$$\left(\frac{\partial \ln Q}{\partial T}\right)_V = N\left(\frac{\partial \ln q}{\partial T}\right)_V \tag{18-33}$$

Taking the natural log of Eq. (18-32) and differentiating, we have

$$\left(\frac{\partial \ln q}{\partial T}\right)_V = \left(\frac{\partial}{\partial T}[\ln(CT)^{3/2} + \ln V]\right)_V = \left(\frac{\partial}{\partial T}\left[\frac{3}{2}\ln CT\right]\right)_V + 0$$

where we have let C represent the constants. Then

$$\left(\frac{\partial \ln q}{\partial T}\right)_V = \frac{3}{2}\left[\frac{\partial}{\partial T}(\ln C + \ln T)\right]_V = \frac{3}{2}\left[0 + \frac{\partial \ln T}{\partial T}\right] = \frac{3}{2}\left(\frac{1}{T}\right)$$

Substituting into Eq. (18-33) gives

$$\left(\frac{\partial \ln Q}{\partial T}\right)_V = \frac{3}{2}\left(\frac{N}{T}\right) \tag{18-34}$$

Finally, substituting in Eq. (18-16) gives

$$E = kT^2\left(\frac{3}{2}\right)\left(\frac{N}{T}\right) = \frac{3}{2}NkT \tag{18-35}$$

If $N = L_A$ (Avogadro's constant),

$$E = \frac{3}{2}RT \tag{18-36}$$

which is the familiar expression from kinetic theory, Eq. (2-37).

For molecules, the other energy terms may be significant, and their corresponding q's must be determined by using expressions for vibrational and rotational energy levels (see Chapter 17). Numerical values of energy spacings, force constants, etc., are usually obtained from spectroscopic measurements.

SUMMARY

In Table S-18, the major equations of this chapter are collected, and connections between some of them are shown. You should determine what conditions are necessary for each equation to be valid, what conditions are imposed to make the connections, and/or how the equation is derived. Other important things you should know are

1. Statistical mechanics provides a bridge between the large-scale magnitudes of thermodynamic properties and the small-scale magnitudes of molecular properties.
2. The distribution having the maximum number of complexions is the most probable distribution; it has the greatest thermodynamic probability.
3. One form of the Boltzmann distribution law gives the fraction of the particles in a particular energy level for the most probable distribution and introduces the partition function [Eqs. (18-5) and (18-9)].
4. Another form of the Boltzmann distribution law gives the ratio of the populations of two energy levels for the most probable distribution [Eqs. (18-7) and (18-8)].
5. The molecular partition function permits calculation of all equilibrium properties from molecular properties for a system of distinguishable molecules.
6. The time-dependency of an average value of a property of an actual system is avoided by using the concept of an ensemble of systems.
7. The ensemble distribution [Eq. (18-12)] leads to an ensemble partition function [Eq. (18-13)], which is applicable when the molecules are indistinguishable and from which expressions may be obtained for the thermodynamic properties E and S [Eqs. (18-16) and (18-24)].

TABLE S-18: Summary of Important Equations

Distinguishable molecules

$$N = \sum n_i \quad (18\text{-}1) \qquad E = \sum n_i \varepsilon_i \quad (18\text{-}2) \qquad W = \frac{N!}{n_0! n_1! n_2! \cdots n_i!} \quad (18\text{-}3)$$

$$n_i = A e^{-\varepsilon_i/kT} \quad (18\text{-}4) \longrightarrow \frac{n_i}{N} = \frac{e^{-\varepsilon_i/kT}}{\sum e^{-\varepsilon_i/kT}} \quad (18\text{-}5) \quad \text{or} \quad \frac{n_i}{N} = \frac{g_i e^{-\varepsilon_i/kT}}{\sum g_i e^{-\varepsilon_i/kT}} \quad (18\text{-}6)$$

$$q = \sum e^{-\varepsilon_i/kT} \quad (18\text{-}9)$$

$$\frac{n_i}{n_0} = e^{-\varepsilon_i/kT} \quad (18\text{-}7) \quad \text{or} \quad \frac{n_i}{n_0} = \frac{g_i}{g_0} e^{-\varepsilon_i/kT} \quad (18\text{-}8)$$

Indistinguishable molecules

$$\frac{n_i}{\mathcal{N}} = \frac{e^{-E_i/kT}}{\sum e^{-E_i/kT}} \quad (18\text{-}12) \longrightarrow Q = \sum e^{-E_i/kT} \quad (18\text{-}13) \longrightarrow \left(\frac{\partial Q}{\partial T}\right)_V = \sum \frac{E_i}{kT} e^{-E_i/kT} \quad (18\text{-}15)$$

(18-2)

$$\bar{E} = \frac{\sum E_i e^{-E_i/kT}}{Q} \quad (18\text{-}14) \longrightarrow E = kT^2 \left(\frac{\partial \ln Q}{\partial T}\right)_V \quad (18\text{-}16)$$

State functions

$$S = -k\,\overline{\ln p} \quad (18\text{-}21) \longrightarrow S = -k \sum p_i \ln p_i \quad (18\text{-}22) \longrightarrow S = k \ln W \quad (18\text{-}23)$$

$$S = \frac{E}{T} + k \ln Q \quad (18\text{-}24, \text{ from 18-12 and 18-13})$$

$$Q = \frac{q^N}{N!} \quad (18\text{-}19)$$

$$S = \frac{E}{T} + kN(\ln q - \ln N + 1) \quad (18\text{-}25)$$

$$\ln Q = N \ln q - N(\ln N - 1) \quad (18\text{-}20)$$

$$A = -kT \ln Q \quad (18\text{-}26) \longrightarrow P = kT\left(\frac{\partial \ln Q}{\partial V}\right)T \quad (18\text{-}27)$$

$$H = kT\left[T\left(\frac{\partial \ln Q}{\partial T}\right)_V + V\left(\frac{\partial \ln Q}{\partial V}\right)_T\right] \quad (18\text{-}28) \qquad G = -kT\left[\ln Q - V\left(\frac{\partial \ln Q}{\partial V}\right)_T\right] \quad (18\text{-}29)$$

$$C_V = kT\left[2\left(\frac{\partial \ln Q}{\partial T}\right)_V + T\left(\frac{\partial^2 \ln Q}{\partial T^2}\right)_V\right] \quad (18\text{-}30)$$

Ideal monatomic gas

$$q = \frac{(2\pi mkT)^{3/2} V}{h^3} \quad (18\text{-}32)$$

$$\left(\frac{\partial \ln Q}{\partial T}\right)_V = \frac{3N}{2T} \quad (18\text{-}34) \qquad E = \tfrac{3}{2}NkT \quad (18\text{-}35)$$

$$\left(\frac{\partial \ln Q}{\partial T}\right)_V = N\left(\frac{\partial \ln q}{\partial T}\right)_V \quad (18\text{-}33, \text{ from 18-20})$$

$$E = \tfrac{3}{2}RT \quad (18\text{-}36)$$

8. From the expressions for E and S, expressions may be obtained for the other thermodynamic properties (state functions): A, H, G, and C_V [Eqs. (18-26, 18-28, 18-29, 18-30)].
9. Calculations require evaluation of molecular partition functions using energy level information obtained mainly from spectroscopy [Eqs. (18-10, 18-11, and 18-20)].

SOLVED PROBLEMS

Probability

PROBLEM 18-1 As a quick refresher on probability, find the probabilities of drawing from an ordinary deck of cards (**a**) a particular card, (**b**) a particular number, (**c**) a particular suit.

Solution: The total number of cards, of course, is 52.

(a) When you try to draw a particular card, you know that there are 52 possibilities and only one is the right one, so your chances are (or the probability is) 1/52.
(b) For a particular number, there are 4 possibilities for getting a "correct" card, so the probability is 4/52 or 1/13.
(c) For a particular suit, there are 13 possibilities, so the probability is 13/52 or 1/4.

PROBLEM 18-2 For a system of three marked boxes and two marbles, **(a)** describe the possible distributions and **(b)** find the number of complexions W of each distribution.

Solution: Designate the boxes 1, 2, and 3 and the marbles a and b; then write the possible arrangements of a and b. [*Note:* You might want to make a diagram.]

(a) List the results:

A. a and b in 1	D. a in 1 and b in 2; b in 1 and a in 2
B. a and b in 2	E. a in 1 and b in 3; b in 1 and a in 3
C. a and b in 3	F. a in 2 and b in 3; b in 2 and a in 3

(b) You can see that, in D, E, and F, there are two possibilities; thus, $W = 2$. For A, B, and C, there is only one way to achieve the distribution, so $W = 1$.
(b′) You can also calculate W from Eq. (18-3). For D, E, or F, you would have

$$W = \frac{N!}{n_0!n_1!n_2!} = \frac{2!}{1!1!0!} = 2$$

and for A, B, or C, you would have

$$W = \frac{2!}{2!0!0!} = 1$$

PROBLEM 18-3 Continue Problem 18-2 by calculating the probability of each distribution.

Solution: The probability is the number of complexions divided by the total number of complexions. In Problem 18-2, the total number is $1 + 1 + 1 + 2 + 2 + 2 = 9$; therefore, you can see quickly that the probabilities are 1/9 for A, B, C and 2/9 for D, E, F. To check, find the sum of the probabilities, which should be 1: Sum $= 3(1/9) + 3(2/9) = 9/9 = 1$.

Boltzmann Distribution

PROBLEM 18-4 If the energy of the first excited state of a compound is 10 kcal/mol greater than the ground-state energy, how many molecules will be in the first excited state at 0° C if it is known that there are 10^{20} molecules in the ground state?

Solution: In general, you can calculate the ratio of n_1 to n_0 by using Eq. (18-7): $n_1/n_0 = e^{-\varepsilon_1/kT}$. Then you can multiply the result by n_0 to get n_1. But notice that the energy is given in calories and moles. You can convert the Boltzmann expression to moles simply by replacing k with R. Thus, the exponent of Eq. (18-7) now is

$$-\frac{\varepsilon_1}{RT} = -\frac{10 \times 10^3 \text{ cal mol}^{-1}}{(1.987 \text{ cal K}^{-1}\text{mol}^{-1})(273.15 \text{ K})} = -18.43$$

Thus
$$\frac{n_1}{n_0} = e^{-18.43} = 9.9 \times 10^{-9}$$

And so the actual number in the first excited state is

$$n_1 = (9.9 \times 10^{-9})n_0 = (9.9 \times 10^{-9})(10^{20}) = 9.9 \times 10^{11} \text{ molecules}$$

PROBLEM 18-5 Show how the expression $\ln W = N \ln N - \Sigma n_i \ln n_i$ is obtained from Eq. (18-3) by the use of Stirling's approximation.

Solution: First, you take the natural log of both sides of Eq. (18-3):

$$\ln W = \ln N! - \ln n_0! - \ln n_1! - \cdots = \ln N! - \Sigma \ln n_i! \quad \textbf{(a)}$$

Now you apply Stirling's approximation to both terms on the right-hand side:

$$\ln N! = N \ln N - N \quad \text{and} \quad \Sigma \ln n! = \Sigma n_i \ln n_i - \Sigma n_i$$

And, by Eq. (18-1), $\Sigma n_i = N$

Now substitute all of these in eq. (a) to get

$$\ln W = N \ln N - N - (\Sigma n_i \ln n_i - N) = N \ln N - \Sigma n_i \ln n_i$$

State Functions

PROBLEM 18-6 (**a**) Calculate the entropy of the system for distribution (3) in Example 18-1 using Eq. (18-22), assuming that each marble represents one mole of molecules. (**b**) Calculate the entropy again using Eq. (18-23). See Example 18-3 for W. (**c**) Discuss the discrepancy in the answers.

Solution: Distribution 3 had 2 marbles in one box and 1 each in the other two boxes. The probabilities are 2/4, 1/4, and 1/4.

(**a**) In order to use Eq. (18-22) here, you should change k to R, for moles; then substitute the probability values:

$$S = -k\Sigma p_i \ln p_i = -R(\tfrac{2}{4} \ln \tfrac{2}{4} + \tfrac{1}{4} \ln \tfrac{1}{4} + \tfrac{1}{4} \ln \tfrac{1}{4}) = (-8.314 \text{ J K}^{-1}\text{mol}^{-1})(-0.586)$$
$$= 4.87 \text{ J K}^{-1}\text{mol}^{-1}$$

(**b**) From Example 18-3, you have $W = 12$. Then, from Eq. (18-23),

$$S = R \ln W = R \ln 12 = 20.66 \text{ J K}^{-1}\text{mol}^{-1}$$

(**c**) The discrepancy arises from the fact that the number of particles is small. To derive Eq. (18-22) from (18-23) requires using Stirling's approximation, which assumes that N is very large.

PROBLEM 18-7 Derive an expression for the entropy in terms of temperature for an ideal monatomic gas. Use what you can from Example 18-9.

Solution: Your first inclination would probably be to use Eq. (18-25), but $S = E/T + kN(\ln q - \ln N + 1)$ is awkward to handle. Let's use $S = E/T + k \ln Q$ [Eq. (18-24)] and evaluate $\ln Q$ differently in order to illustrate another form of the Stirling approximation:

$$N! = \frac{N^N}{e^N} \qquad \textbf{(a)}$$

Going back to $Q = q^N/N!$ [Eq. (18-19)], you can substitute eq. (a) for $N!$ Then

$$Q = \frac{q^N e^N}{N^N} = \left(\frac{qe}{N}\right)^N$$
$$\ln Q = N \ln \frac{q}{N} + N \ln e \qquad \textbf{(b)}$$

but $\ln e = 1$, so

$$\ln Q = N \ln \frac{q}{N} + N \qquad \textbf{(c)}$$

Substituting in Eq. (18-24), you get

$$S = \frac{E}{T} + kN \ln \frac{q}{N} + kN \qquad \textbf{(d)}$$

In Example 18-9, you saw that a monatomic gas at room temperature has only translational energy and obtained expressions for q (Eq. 18-32) and for E (Eq. 18-35). Substituting these in eq. (d), you get

$$S = \frac{3}{2}\left(\frac{NkT}{T}\right) + kN \ln\left[(2\pi mkT)^{3/2}\left(\frac{V}{Nh^3}\right)\right] + kN$$

If you assume one mole, $N = L_A$, and you have

$$\tfrac{3}{2}L_A k + kL_A = \tfrac{3}{2}R + R = \tfrac{5}{2}R$$

Then

$$S = \tfrac{5}{2}R + R \ln\left[(2\pi mkT)^{3/2}\left(\frac{V}{L_A h^3}\right)\right] \qquad \textbf{(e)}$$

PROBLEM 18-8 Continue Problem 18-7 by calculating the entropy of one mole of neon at 298.15 K and 1 atm.

Solution: You will use eq. (e) of Problem 18-7. Notice that V must be at 298.15 K. Assuming ideal behavior,

$$V_{298.15} = \frac{(0.08205)(298.15)}{1} = 24.46 \text{ L} = 24.46 \times 10^{-3} \text{ m}^3$$

In eq. (e) the mass m is per molecule and equals M/L_A. It's convenient to evaluate the ln term in eq. (e) before substituting:

$$\ln[\ \ ] = \frac{3}{2}\ln(2\pi mkT) + \ln\frac{V}{L_A h^3}$$

$$= \frac{3}{2}\ln\left[2\pi\left(\frac{20.18 \times 10^{-3}}{6.022 \times 10^{23}}\right)(1.381 \times 10^{-23})(298.15)\right] + \ln\frac{24.46 \times 10^{-3}}{6.022 \times 10^{23}(6.626 \times 10^{-34})^3} = 15.1$$

Now eq. (e) becomes

$$S = 2.5R + 15.1R = 17.6(8.314) = 146 \text{ J K}^{-1}\text{mol}^{-1}$$

PROBLEM 18-9 Derive the ideal gas equation for a monatomic gas.

Solution: You should use $P = kT\left(\frac{\partial \ln Q}{\partial V}\right)_T$ [Eq. (18-27)] for P. You can evaluate the partial derivative by using Eq. (18-31) and the expression for q for a monatomic gas from Example 8-9, Eq. (18-32):

$$q = (2\pi mkT)^{3/2}\left(\frac{V}{h^3}\right)$$

At constant T, q becomes CV and $\ln q = \ln C + \ln V$. Now, from Eq. (18-31),

$$\left(\frac{\partial \ln Q}{\partial V}\right)_T = N\left(\frac{\partial \ln C}{\partial V}\right)_T + N\left(\frac{\partial \ln V}{\partial V}\right)_T = 0 + N\left(\frac{1}{V}\right)$$

Finally, Eq. (18-27) becomes

$$P = kT\left(\frac{N}{V}\right) = \frac{NkT}{V}$$

For 1 mol, $N = L_A$ and $L_A k = R$, so $PV = RT$.

PROBLEM 18-10 Derive an expression for the molecular partition function for rotation for a linear molecule.

Solution: Molecular partition functions are given by Eq. (18-9) with the appropriate expression for ε_i. If you assume the rigid rotor model, you can use Eq. (14-41) [or Eq. (17-4)] for the separation between rotational energy levels in joules, $E(J) = B'J(J + 1)$, where $B' = h^2/8\pi^2 I$. Now you must include the weighting factor, g, because rotational levels have a degeneracy of $2J + 1$. Then Eq. (18-9) becomes

$$q = \Sigma g_i e^{-\varepsilon_i/kT}$$

As in Example 18-9, because the spacings are so small, you can replace the summation with an integration and

$$q_r = \int_0^\infty (2J + 1)e^{-J(J+1)h^2/8\pi^2 IkT}\, dJ$$

This works out nicely because $(2J + 1)\,dJ$ is $d[J(J + 1)]$ and

$$q_r = \frac{8\pi^2 IkT}{h^2}$$

For completeness, it is necessary to recognize a difference between symmetrical and unsymmetrical molecules and divide by a *symmetry factor*, σ:

$$q_r = \frac{8\pi^2 IkT}{\sigma h^2}$$

where σ is 1 for unsymmetrical diatomic molecules (e.g., HCl) and 2 for symmetrical diatomic molecules (e.g., H_2).

Supplementary Exercises

PROBLEM 18-11 The probability of drawing a nine from an ordinary deck of cards is __________. One distribution of three marbles in two marked boxes is two in box 1 and one in box 2: This distribution has __________ complexions. The defining expression for the molecular partition function is __________. The expression that relates the partition function of the system to the molecular partition function is __________. The statistical mechanical definition of entropy is __________.

PROBLEM 18-12 You have 10 marbles and 6 marked boxes. One distribution is 4 marbles in the first box, 2 in the second, and 1 each in the others. How many complexions does this distribution have?

PROBLEM 18-13 The separation between the first two rotational energy levels of HCl is 20 cm^{-1}. Calculate the ratio of the population of the first excited state to the population of the ground state at 25°C.

PROBLEM 18-14 To get some idea of the magnitude of factorials of large numbers, calculate the logarithm to the base 10 of $(10^{23})!$. Remember that this is the power to which 10 must be raised to get the number. [*Hint:* Use Stirling's approximation.]

PROBLEM 18-15 Show that the form of Stirling's approximation $N! = N^N/e^N$ used in Problem 18-7 is equivalent to that used in deriving $\ln N! = N \ln N - N$.

PROBLEM 18-16 Derive Eq. (18-29), $G = -kT\left[\ln Q - V\left(\frac{\partial \ln Q}{\partial V}\right)_T\right]$.

PROBLEM 18-17 Calculate the molar entropy of krypton at 298.15 K and 1 atm.

PROBLEM 18-18 For $H^{79}Br$, B' is found to be 1.63×10^{-24} J. Evaluate the molecular rotational partition function at 298.15 K.

Answers To Supplementary Exercises

18-11 1/13; 3; $q = \Sigma \exp(-\varepsilon_i/kT)$; $Q = q^N/N!$; $S = k \ln W$(or $= -k\Sigma p_i \ln p_i$)
18-12 75 600 **18-13** 0.908 **18-14** 2.26×10^{24} **18-15** *Hint:* $\ln e = 1$
18-16 *Hint:* $G = A + PV$ **18-17** 164 J $K^{-1}mol^{-1}$ **18-18** 2.53×10^3

19 SOLIDS AND SURFACES

THIS CHAPTER IS ABOUT

- ☑ Crystal Shapes and Systems
- ☑ Lattices and Crystal Structures
- ☑ X-Ray Crystallography
- ☑ Surfaces of Solids
- ☑ Surfaces of Liquids

19-1. Crystal Shapes and Systems

A. Similar shapes

In the seventeenth century, certain basic ideas about crystals were developed. The equality of angles between corresponding faces of crystals of the same substance had been recognized, and the suggestion had been made that the visible, outer regularity was due to an inner regular arrangement of some fundamental particles.

- Particles lying in planes account for the flat faces of crystals.

Also, two crystals of the same substance might appear rather different superficially: One may be elongated, another squat; one may have more faces than another. Nevertheless

- Two crystals of the same substance belong to the same crystal system.

B. Crystal systems

Classification of crystals into seven **crystal systems** is achieved by considering the unit lengths along three axes and the angles between the axes. For example, the cubic crystal system has all unit lengths equal and all angles equal to 90°. The triclinic system has all unit lengths and all angles unequal and none of the angles is 90°.

Crystal systems can also be described by their symmetry properties. Only a minimum list—merely the principal axes of symmetry—is sufficient, as Table 19-1 shows.

TABLE 19-1: Minimum Symmetry of Crystal Systems

cubic	four C_3	hexagonal	one C_6
tetragonal	one C_4	orthorhombic	three C_2
trigonal (rhombohedral)	one C_3	monoclinic	one C_2
		triclinic	none

C. Space groups

When the complete symmetry is considered, the point group of the crystal may be determined (see Section 17-2C). Because certain operations that are possible in an unrestricted body such as a molecule of a gas are not possible in an extended body such as a crystal, only 32 point groups are possible for crystals. Each of these **crystallographic point groups** belongs to one of the seven systems listed in Table 19-1.

EXAMPLE 19-1: Give an example of a rotation operation that is impossible in crystals and explain why.

Solution: A five-fold rotation is not possible, because it requires a five-sided figure, which cannot be used to fill all of the space in the bounds of the crystal. Consider, for example, only two dimensions—a planar surface such as a floor—and the shape that pieces of tile must have in order to cover the floor completely. We could use triangles, squares, or hexagons but not pentagons; the pentagon is not a "space-filling" shape.

In an extended object with repeated patterns—most wallpaper, for example—additional symmetry operations are possible: specifically, *glide planes* and *screw axes*. These operations cause a movement, or translation, and are called **space operations**. Space operations expand the 32 point groups to 230 **space groups**.

note: Remember that a point group requires that one point in the body must not move when the symmetry operation occurs. Space operations move *all* points.

19-2. Lattices and Crystal Structures

A. Lattice

The arrangements of groups of atoms within crystals are described by referring to the possible geometric arrangements of single points. An arrangement that gives a repeating pattern of points, such that each point is found in an identical environment, is called a **lattice**. In two dimensions, only five lattices are possible; in three dimensions, there are fourteen lattices possible, and these are known as **Bravais lattices**.

Usually, the lattice is represented by the **unit cell**: the smallest group of points which, when repeated, reproduces the entire lattice. The basic, or primitive, unit cell contains only one lattice point. For example, the cubic point lattice shown in Figure 19-1, may be represented by the cubic unit cell, which is heavily outlined in the figure.

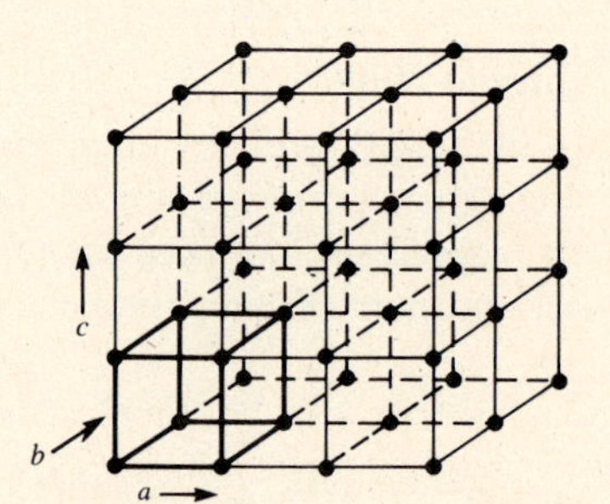

FIGURE 19-1: The cubic point lattice.

EXAMPLE 19-2: Explain how a two-dimensional unit cell in the shape of a rectangle contains only one lattice point.

Solution: In an infinite two-dimensional array of points in columns and rows at 90° to each other, the rectangle formed by joining four points shares each side with a second rectangle adjacent to the first. Each corner is shared by four rectangles; in other words, a given point serves as one corner of four rectangles. We can say that only one-fourth of the point belongs to each rectangle. Thus each of the four corners of a given rectangle contributes one-fourth of a point, and the rectangle contains a total of $4(1/4) = 1$ point.

Sometimes, larger unit cells are used to represent lattices because they show the symmetry of the entire structure better. These larger unit cells are called *non-primitive unit cells* and contain more than one lattice point.

The fourteen Bravais lattices are distributed unequally among the seven crystal systems (see Table 19-1). The triclinic, trigonal, and hexagonal systems have one lattice each, monoclinic and tetragonal two each, cubic three, and orthorhombic four. When we place various unit cells at the lattice points, we obtain the 230 space groups.

There are three cubic lattices, which are known as **simple cubic**, **body-centered cubic** (*bcc*), and **face-centered cubic** (*fcc*). A body-centered cube has, in addition to the eight corner points of the simple cube, another point located at the center of the cube. A face-centered cube has six additional points, one at the center of each face of the cube. The last two cubic lattices occur frequently in crystals of elements, especially the metals, in which the lattice points are occupied by single atoms. The atoms can be treated as simple spheres, and the structures may be considered to be the result of various possibilities for packing identical spherical objects in the smallest volume, i.e., most efficiently.

Finally, we merely will mention the two possible arrangements, which are named for their symmetry: **hexagonal-closest-packed** (*hcp*) and **cubic-closest-packed** (*ccp*). The latter is the same as face-centered cubic.

B. Crystal structures

A lattice provides the geometric framework upon which a crystal structure is built. An atom or group of atoms, called the **basis**, is repeated at every lattice point to produce the **crystal structure**. If the basis is a group of atoms, only one of the group is at the lattice point, the others falling in the spaces between points. For example, if the NO_2^- ion were to form a cubic lattice, an N atom would occupy every point in Figure 19-1. The O atoms would occupy open space, with two near each N. (The crystal structure is often less symmetric than the idealized lattice because of the groups of atoms in the basis.)

EXAMPLE 19-3: Cesium halides appear to fit a body-centered-cubic lattice with one type of ion at the corners and the other in the center. The structure should probably be viewed as two interlocking simple cubic lattices, one of anions and one of cations. Using a simple cubic lattice, describe the structure, taking the CsX unit as the basis.

Solution: Although CsX is ionic, we will imagine the Cs^+ ion and the X^- ion to be joined by a rod, like a dumbbell. We will place one corner of the cube at the origin of a Cartesian coordinate system with the sides along the positive x, y, and z axes. Now we place the Cs^+ ion at the origin and let the connecting rod lie along a diagonal of the cube so that the X^- ion is at the center. If the side of the unit cell equals 1, the coordinates of the center are $(\frac{1}{2}, \frac{1}{2}, \frac{1}{2})$.

Repeating this process by placing a basis unit with Cs^+ at each corner of the cube and the Cs–X axis in the same orientation will place X^- ions at the center of adjoining cubes which share faces, sides, or a corner with the original cube. Continued repetition will produce the entire crystal structure, with Cs^+ ions at the lattice points of a simple cubic lattice. If we look only at the X^- ions, we see that they also fall at the lattice points of a simple cubic lattice.

19-3. X-Ray Crystallography

A. Planes within crystals

The flat faces of crystals suggest a planar aspect of the inner crystal structure. The arrays of lattice points show the source of such planes. If we simplify Figure 19-1 to two dimensions and a simple square array of points, as in Figure 19-2, we can draw a line connecting unit length on the b axis and unit length on the a axis, that is, $a = 0, b = 1$ and $a = 1, b = 0$. Further lines connecting the points (0, 2) and (2, 0) or (0, 3) and (3, 0) produce a set of parallel lines (solid in Figure 19-2). In three dimensions, we would have a set of parallel planes. Another set of lines would be obtained by connecting the points (0, 2) and (1, 0), (0, 3) and $(1\frac{1}{2}, 0)$, (0, 4) and (2, 0), etc. (dashed in Figure 19-2); and another set for (0, 3) to (1, 0), (0, 6) to (2, 0), etc.

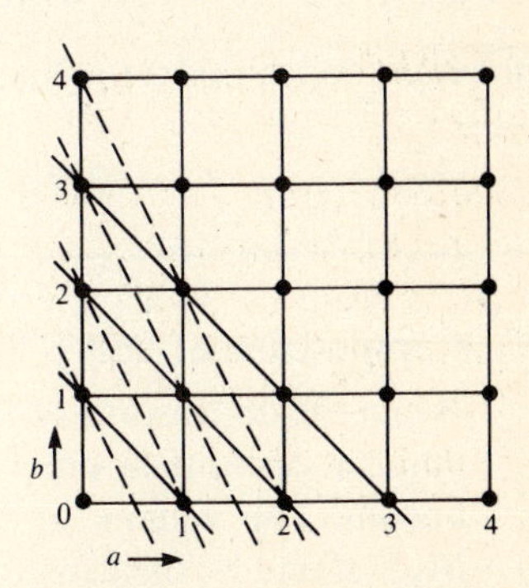

FIGURE 19-2: One *xy* plane of a cubic point lattice.

A set of planes can then be described by the multiples of the unit lengths, such as $(\frac{1}{2}a, b, \infty c)$ for the plane through $(\frac{1}{2}, 0)$ and (0, 1), and parallel to the z axis, but the system known as **Miller indices** uses the reciprocals of the multiples of the unit lengths, *with fractions cleared.*

Thus ($\frac{1}{2}a, b, \infty c$) becomes ($1/\frac{1}{2}, 1/1, 1/\infty$), or (210). (Remember that the larger number means the shorter distance along the axis.)

EXAMPLE 19-4: Give the Miller indices of (**a**) a plane parallel to the ab plane that crosses the c axis at $c = 1$; (**b**) a plane that passes through ($4a, 2b, 1c$).

Solution:

(**a**) Since the plane does not cross a or b, its intercepts on these axes are infinity. The reciprocal of ∞ is 0, so the Miller indices are (001).

(**b**) For ($4a, 2b, 1c$), the reciprocals are ($\frac{1}{4}, \frac{1}{2}, 1$), so the Miller indices are (124).

An entire set of parallel planes can be described by its simplest indices. For (210), parallel planes would be (420), (630), etc., but all could be called the (210) planes. The dashed lines in Figure 19-2 are the intercepts of some of the (210) planes with an ab plane of lattice points.

B. Diffraction of x-rays

Electromagnetic radiation passing through a set of slits will be diffracted if the slit width is about the same as the wavelength of the radiation. The wavelengths of x-rays are about the same as the distance between planes in a crystal, and, in fact, x-rays are diffracted upon passing through a crystal.

Mathematical analysis is simpler if the effect is treated as the result of reflections from parallel planes. Experimentally, if the angle of incidence (the angle between the x-rays and the surface of the crystal) is increased while keeping the wavelength of x-ray constant, a series of strong maxima is observed in the reflected radiation. A maximum can be explained as the result of the constructive interference that occurs when the reflections from two parallel planes are out of phase by an integral number of wavelengths. Each plane in the set of parallel planes will reflect at the same angle, so each reflection from each plane reinforces reflection from all of the others. From such an analysis, Bragg obtained the equation

BRAGG'S LAW
$$n\lambda = 2d \sin \theta \tag{19-1}$$

where n is an integer called the **order of reflection**, λ is the wavelength, d is the separation between adjacent planes, and θ is an angle of incidence for which a maximum of reflection occurs. As θ increases, $n = 1, 2, \ldots$, for the successive maxima.

EXAMPLE 19-5: In NaCl, the first reflection from the (100) planes was at 5.9° when x-rays of wavelength 0.576 Å were used. Calculate the separation between the planes in the (100) set.

Solution: Since this is the first reflection, we assume that $n = 1$; then, from Eq. (19-1),

$$d = \frac{\lambda}{2 \sin \theta} = \frac{0.576 \text{ Å}}{2 \sin 5.9^\circ} = \frac{0.576 \text{ Å}}{2(0.1028)} = 2.80 \text{ Å}$$

C. Composition of unit cells

When the dimensions of the unit cell have been determined from the diffraction pattern, the number of formula units per unit cell can be calculated, *if* the density of the substance is known. The volume of the unit cell may be calculated from the measured dimensions. Multiplying by the density ρ gives the mass of the unit cell:

$$m(\text{cell}) = \rho V \tag{19-2}$$

The mass of one formula unit is the formula weight divided by the Avogadro constant:

$$m(\text{formula unit}) = \frac{\text{FW}}{L_A} \tag{19-3}$$

Dividing Eq. (19-3) by Eq. (19-2) gives the number of formula units per cell:

$$\frac{m(\text{formula unit})}{m(\text{cell})} = \frac{\rho V L_A}{\text{FW}} \tag{19-4}$$

EXAMPLE 19-6: Calculate the number of formula units of NaCl in a unit cell, using data from Example 19-5. The density of NaCl is $2.17 \times 10^3 \text{ kg m}^{-3}$.

Solution: In the complete study of NaCl, it turns out that the unit cell is determined by the (200) planes; thus the length of one side is twice the value obtained in Example 19-5: $a = 2d_{(100)} = 5.60 \text{ Å} = 5.60 \times 10^{-10}$ m. For a cubic structure, $V = a^3$, and Eq. (19-4) becomes

$$\frac{\text{formula units}}{\text{unit cell}} = \frac{(2.17 \times 10^3 \text{ kg m}^{-3})(5.60 \times 10^{-10} \text{ m})^3(6.02 \times 10^{23} \text{ formula units mol}^{-1})}{5.85 \times 10^{-2} \text{ kg mol}^{-1}} = 3.92$$

Since the number must be an integer, we assume some experimental error and round off to 4. In Problem 19-2, you'll find that a face-centered-cubic lattice has 4 points per unit cell, so we conclude that the NaCl structure is *fcc*.

19-4. Surfaces of Solids

A. Adsorption

In the interior of a solid or liquid, a given molecule, atom, or ion is completely surrounded, and the various attractive forces are in balance. At the surface, however, those species on top may not satisfy all of their "attractive ability" with their own kind; frequently, molecules from an adjoining fluid phase will be attracted to these top species and become attached to the surface. The process is called **adsorption**, which is defined very generally as increasing the amount of a substance at a surface. An *adsorbent* attracts an *adsorbate* or *adsorbed phase.*

note: *Ab*sorption differs from *ad*sorption in that the molecules of the *absorbate* penetrate into the interior of the absorbent, mixing more or less completely with the molecules of the solid, rather like water being soaked up in a sponge.

EXAMPLE 19-7: Name some adsorbents and common adsorbates.

Solution: Solids that typically adsorb gases include metals, silica gel, and charcoal. Finely divided Pt and Ni are good catalysts because of their adsorption capability. Gases involved as adsorbates in laboratory studies or industrial processes include H_2, O_2, C_2H_4, C_2H_6, SO_2, and NH_3.

The most significant factor in determining the amount of adsorption is of course surface area. The best adsorbents are porous materials, which have a large exposure of surface. The natures of the adsorbent and adsorbate are also significant. Other factors being equal, the amount of a gas adsorbed depends on temperature and pressure—decreasing at higher temperature and increasing at greater pressure. An interesting correlation, perhaps useful only for qualitative predictions, is: The higher the critical temperature of a gas, the greater the volume which will be adsorbed. Also, a dynamic equilibrium is established between adsorbed gas and the bulk of the gas. This means that, for given conditions, the amount adsorbed is always the same and particles are gained or lost at equal rates.

B. Physical and chemical adsorption

The strength of the binding force between adsorbent and adsorbate species is used to distinguish two types of adsorption. The term **physical adsorption** (also, **physisorption**) refers to forces of the van der Waals type, which are relatively weak. In **chemical adsorption** (also, **chemisorption**), however, it can be considered that a true covalent chemical bond has formed, in which case the binding forces are relatively strong. In the latter case, energy (or enthalpy) changes are much greater than in the former, and only one layer of molecules can be held; the surface becomes saturated. In physical adsorption, it is possible to have two or more layers.

C. Adsorption isotherms

The amount of gas adsorbed usually is expressed per gram of adsorbent, assuming that the surface area is proportional to the mass. Usually, the unit is volume per gram, but mass per gram also is used. Experimentally, the amount adsorbed is measured as a function of the equilibrium pressure at constant temperature. The plot of the data is called an **adsorption isotherm**. Sometimes, the data are stated as the fraction of the surface which is covered, i.e., the actual volume over the volume for a complete monolayer:

$$\theta = \frac{v}{v_m} \tag{19-5}$$

For chemical adsorption, the plot of θ (or v) versus P rises rapidly at first, then levels off, approaching a maximum at $\theta = 1$. For physical adsorption, several types of curves are obtained because additional layers will form. The term "adsorption isotherm" also is applied to the mathematical expressions that have been suggested in attempts to fit the experimental curves. Freundlich proposed a simple empirical equation:

FREUNDLICH ISOTHERM

$$v = kP^a \tag{19-6}$$

where v is the volume adsorbed, P is the equilibrium pressure, and k and a are constants to be evaluated from the data. This isotherm works well at lower pressures but must fail at higher pressures because it does not predict the leveling off.

EXAMPLE 19-8: Show how the Freundlich isotherm can be rearranged and plotted so as to evaluate the constants.

Solution: The exponential term suggests taking the logarithm:

$$\log v = \log k + a \log P \tag{19-7}$$

This is the equation of a straight line with slope a and intercept $\log k$ if $\log v$ is plotted against $\log P$.

Langmuir derived a theoretical expression for the monolayer case:

LANGMUIR ISOTHERM

$$\theta = \frac{bP}{1 + bP} \tag{19-8}$$

Although overly simplified assumptions were made in deriving this equation, Eq. (19-8) does fit a large number of cases of chemical adsorption.

EXAMPLE 19-9 Convert the Langmuir isotherm into an expression for the measured adsorbed volume.

Solution: We can substitute the definition of θ [Eq. (19-5)] into Eq. (19-8) and rearrange:

$$\theta = \frac{v}{v_m} = \frac{bP}{1 + bP}$$

Thus

$$v = \frac{bv_mP}{1 + bP} = \frac{aP}{1 + bP} \tag{19-9}$$

if we let $a = bv_m$.

note: In some treatments, v_m is treated as an empirical constant and a and b are determined from the experimental data.

19-5. Surfaces of Liquids

A. Surface tension

The imbalance of attractive forces on molecules in the surface of a liquid or solid was offered (in Section 19-4A) as an explanation for the phenomenon of adsorption. In the case of liquids, this imbalance also explains the phenomenon of **surface tension**, which is the tendency of a liquid to behave as if there were a thin membrane stretched over it. Since the molecules can move in a liquid, the surface molecules are pulled inward by the attractive forces between them and the molecules in the interior.

EXAMPLE 19-10: Give some illustrations of cases in which a liquid surface behaves like a film or membrane.

Solution: A surface will support objects—a leaf or a needle will float on water; some insects walk on the surface of ponds. A small drop of oil or a soap bubble will assume a spherical shape, attaining the least surface area for a given volume. Liquids rise in capillary tubes.

If a thin soap film is stretched on a suitable wire frame, the stretching force is opposed by the surface tension. Such a force is proportional to the length ℓ of film being stretched:

$$F \propto \ell \qquad \text{and} \qquad F = \gamma \ell \tag{19-10}$$

where γ is the proportionality constant, called the *surface tension*. The units of γ must be force $\cdot\, \ell^{-1}$ or $\mathrm{N\,m^{-1}}$ in SI units. In this particular case,

$$F = \gamma(2\ell) \tag{19-11}$$

where ℓ is the length of the movable side of the frame. The factor 2 arises because there are two surfaces present, one above the wire and one below, so the length of film acted on by F is really 2ℓ.

B. Surface energy

Surface tension may also be expressed in terms of work and energy. In one dimension, work is given by Eq. (3-8). Using dx for the distance moved,

$$dw = F\,dx$$

Substituting Eq. (19-10), we get

$$dw = \gamma \ell\, dx \tag{19-12}$$

The term $\ell\, dx$ is the change in area $\mathcal{A}$, so

$$dw = \gamma\, d\mathcal{A} \tag{19-13}$$

or

$$\gamma = \frac{dw}{d\mathcal{A}} \tag{19-14}$$

Stated in words, surface tension is the work done per unit increase in area, and its SI units are $\mathrm{J\,m^{-2}}$.

Doing this work on the surface increases its Gibbs free energy. For cases in which the surface area is variable, this surface energy must be included in expressions for dG:

$$dG = -S\,dT + V\,dP + \gamma\, d\mathcal{A} \tag{19-15}$$

EXAMPLE 19-11: What is the spontaneous direction of change for surfaces?

Solution: We know that dG must be negative for a spontaneous change. At constant T and P, this requires that $\gamma\, d\mathcal{A}$ be negative, according to Eq. (19-15). In other words, surfaces spontaneously try to decrease their area. This is a thermodynamic statement of what has already been said about surface tension in Example 19-10.

C. Capillarity

One consequence of surface tension is the rise, or fall, of liquids in capillary tubes, a phenomenon that is often used as a method of measuring surface tension. Liquids that "wet glass" rise in a tube until the upward force due to the surface tension is balanced by the downward force due to gravity. The force due to surface tension is given by Eq. (19-10), where, in this case, the length is the inner circumference of the capillary tube, πd or $2\pi r$:

$$F = 2\pi r\gamma \qquad \textbf{(19-16)}$$

When a liquid does not wet the surface completely, the force acts in the direction of the *contact angle* θ, measured from the vertical, and with a vertical component F_v of

$$F_v = F\cos\theta = 2\pi r\gamma\cos\theta \qquad \textbf{(19-17)}$$

The force due to gravity F_g is the acceleration of gravity times the mass of liquid in the tube (= density times volume):

$$F_g = g\rho(\pi r^2 h) \qquad \textbf{(19-18)}$$

where h is measured from the level of liquid outside the tube. At equilibrium, $F_v = F_g$, or

$$2\pi r\gamma\cos\theta = g\rho\pi r^2 h$$

Solving for γ gives

$$\gamma = \frac{g\rho r h}{2\cos\theta} \qquad \textbf{(19-19)}$$

Usually liquids *do* wet glass almost completely, and θ can be taken to be 0°, or $\cos\theta = 1$.

EXAMPLE 19-12: How high will water at 0°C rise in a capillary tube 2 mm in diameter? At 0°, $\gamma = 7.56 \times 10^{-2}\ \mathrm{N\,m^{-1}}$ and the density is almost $1.00\ \mathrm{g\,cm^{-3}}$. Assume that $\theta = 0°$.

Solution: This problem is an exercise in applying Eq. (19-19). Rearranging to get h, we have

$$h = \frac{2\gamma\cos\theta}{g\rho r}$$

Before substituting, let's put the terms in the proper units:

$$r = d/2 = 1\ \mathrm{mm} = 10^{-3}\ \mathrm{m}$$

$$\rho = (1.00\ \mathrm{g\,cm^{-3}})\left(\frac{1\ \mathrm{kg}}{10^3\ \mathrm{g}}\right)\left(\frac{10^6\ \mathrm{cm^3}}{\mathrm{m^3}}\right) = 10^3\ \mathrm{kg\,m^{-3}}$$

$$\gamma = (7.56 \times 10^{-2}\ \mathrm{N\,m^{-1}})\left(\frac{\mathrm{kg\,m\,s^{-2}}}{\mathrm{N}}\right) = 7.56 \times 10^{-2}\ \mathrm{kg\,s^{-2}}$$

Now

$$h = \frac{2(7.56 \times 10^{-2}\ \mathrm{kg\,s^{-2}})(1)}{(9.81\ \mathrm{m\,s^{-2}})(10^3\ \mathrm{kg\,m^{-3}})(10^{-3}\ \mathrm{m})} = 1.54 \times 10^{-2}\ \mathrm{m}$$

D. Effects of temperature and composition

The surface tension of a liquid decreases with increase in temperature, finally becoming zero at the critical temperature. At the critical point, there is no surface, no distinction between liquid and vapor.

The surface tension of solutions depends on the nature of the solute. For solutes that increase γ, the effect is small, so increasing the concentration has little additional effect. Examples, in the solvent water, are salts and some nonelectrolytes. For solutes that decrease γ, the effect may increase drastically with increased concentration. Examples for water are most organic compounds, especially soaps and detergents.

SUMMARY

In Table S-19, the major equations of this chapter are collected, and connections between them are shown. You should determine what conditions are necessary for each equation to be valid or what conditions are imposed to make the connections. Other important things you should know are

1. Natural crystals exhibit regularities in positions of the faces and in angles between faces.
2. Crystals are classified into seven systems, based on unit lengths along three axes and on angles between the axes. Each crystal system has a unique minimum symmetry.
3. Crystallographic point groups are limited to 32 because certain operations are excluded by the requirement that all space must be filled.
4. Two symmetry operations that translate all points are possible in infinite structures: screw axes and glide planes. These permit 230 space groups.
5. Fourteen Bravais lattices constitute all of the ways in which points can be arranged in a regular manner and fill all space.
6. Lattices are represented by unit cells; repetition of the unit cell produces the complete lattice. A primitive unit cell contains only one lattice point.
7. Spheres may be packed efficiently in two ways: hexagonal-closest-packing and cubic-closest-packing.
8. A crystal structure may be viewed as a lattice with an atom or group of atoms (known as the basis) at each lattice point.
9. Planes of points in a lattice or of atoms in a crystal structure are identified by Miller indices.
10. X-rays are diffracted upon passing through a crystal, and the results are used to determine crystal structure.
11. Adsorption is the phenomenon in which one substance becomes more concentrated at the surface of another. Physical (or van der Waals) adsorption involves weak attractive forces; chemical adsorption involves the strong forces of chemical bonding.
12. The amount of gas adsorbed by a solid depends on (**a**) the nature of the adsorbent, (**b**) the nature of the adsorbate, (**c**) the surface area of the adsorbent, (**d**) the temperature, and (**e**) the pressure of the gas.
13. In adsorption, a true dynamic equilibrium is reached in which the rate of adsorption equals the rate of desorption.
14. An adsorption isotherm is the relationship between the amount of gas adsorbed and the equilibrium pressure at constant temperature.
15. The Langmuir isotherm holds well for most cases of chemical adsorption.
16. The surface tension of a liquid can be defined as the force per unit length which resists an

TABLE S-19: Summary of Important Equations

Crystals

$n\lambda = 2d \sin\theta$ (19-1)

$\dfrac{m(\text{formula units})}{m(\text{unit cell})} = \dfrac{\rho V N_A}{\text{FW}}$ (19-4)

Surfaces

$v = kP^a$ (19-6)

$\theta = \dfrac{bP}{1 + bP}$ (19-8) $\qquad v = \dfrac{aP}{1 + bP}$ (19-9)

$F = \gamma \ell$ (19-10) $\rightarrow$ $dG = -S\,dT + V\,dP + \gamma\, d\mathscr{A}$ (19-15)

Capillarity

$F = \gamma(2\pi r)$ (19-16) $\rightarrow$ $\gamma = \dfrac{g\rho r h}{2\cos\theta}$ (19-19) increase in the area of a surface. Its effect is that the liquid acts as if it were covered by a thin membrane.

17. The surface energy is the surface tension expressed as Gibbs energy per unit area. Numerically, they are equal: $N\,m^{-1} = J\,m^{-2}$.
18. Liquids rise or fall in capillary tubes, depending on whether they wet or do not wet the surface, because of surface tension.

SOLVED PROBLEMS

Lattices and Crystal Structures

PROBLEM 19-1 (**a**) Describe a three-dimensional unit cell, analogous to the illustration in Example 19-2. (**b**) How many lattice points does it contain? Explain your response.

Solution:

(**a**) In Example 19-2, we used a rectangle to describe a two-dimensional unit cell. In three dimensions, you would have a block or, in the special case of a square rectangle, a cube. Either has 8 corners.

(**b**) A given point would be the corner for 8 blocks, thus contributing $\frac{1}{8}$ of a point to a particular block. Then, 8 corners, contributing $\frac{1}{8}$ each, are equivalent to 1 point per block.

PROBLEM 19-2 How many lattice points are contained by the unit cell of a face-centered-cubic lattice? Explain.

Solution: Each corner of the cube contributes $\frac{1}{8}$ of a point, as in the Problem 19-1, for a total of 1 point. A point in the center of a face is shared by 2 cubes, thus contributing $\frac{1}{2}$ to each cube. There are 6 faces; at $\frac{1}{2}$ each, they contribute 3 points. The total for the *fcc* unit cell is 4 points.

X-Ray Crystallography

PROBLEM 19-3 If the separation between two (100) planes is unity, what is the separation between two (110) planes of the same crystal?

Solution: Visualize the two (100) planes as opposing vertical sides of a square box: One (110) plane (also perpendicular to the base) will include the diagonal through the base, connecting two opposite corners; parallel planes in the (110) set will pass through the other two corners of the base. The distance between the (110) planes, $d_{(110)}$, is that from the corner of a square to its center. This distance is one side of the right triangle formed by taking one side of the square (= 1) as the hypotenuse and the center as the opposite corner. Thus, by the Pythagorean theorem,

$$1^2 = d^2 + d^2 = 2d^2 \qquad d^2 = \tfrac{1}{2} \qquad d_{(110)} = 1/\sqrt{2}$$

PROBLEM 19-4 If you are using x-rays of wavelength $\lambda = 1.54 \times 10^{-10}$ m and the distance between planes is $d = 1.54 \times 10^{-10}$ m, at what angle would you find the first reflection?

Solution: Use Bragg's law, Eq. (19-1). Rearranging it, you get

$$\sin\theta = \frac{n\lambda}{2d}$$

Since $\lambda = d$ in this case and $n = 1$, $\sin\theta = 0.5$ and $\theta = 30°$.

PROBLEM 19-5 Cesium halides are known to have simple cubic crystal structures. If the density of CsCl is 4.0 g cm^{-3}, what is the length of one side of the unit cell?

Solution: You can find the length of one side of a cube, a, by taking the cube root of the volume, which you can calculate by using Eq. (19-4), rearranged to

$$V = \frac{N(\mathrm{FW})}{\rho L_A} = a^3$$

The formula weight of CsCl is 168.4, and N is 1 (see Example 19-3), so

$$a^3 = \frac{(1 \text{ molecule})(168.4 \text{ g mol}^{-1})}{(4.0 \text{ g cm}^{-3})(6.02 \times 10^{23} \text{ molecule mol}^{-1})} = 69.9 \times 10^{-24} \text{ cm}^3$$

$$a = 4.1 \times 10^{-8} \text{ cm}$$

Surfaces of Solids

PROBLEM 19-6 Describe, from the molecular viewpoint, the equilibrium situation when a gas is adsorbed on a solid.

Solution: This is another example of a dynamic equilibrium, similar to evaporation and condensation. In the case of adsorption, before equilibrium is reached, gas molecules are adsorbed on the surface, rapidly at first but more slowly as the surface area becomes occupied. Some of the adsorbed molecules will acquire sufficient kinetic energy to escape or *desorb*. The rate will be small at first but will increase as more molecules are adsorbed. Eventually, the two rates will become equal, and a state of dynamic equilibrium will have been reached.

PROBLEM 19-7 Rearrange the Langmuir isotherm into a form suitable for plotting and for evaluating the constants.

Solution: For either of the forms of the isotherm, $\theta = bP/(1 + bP)$ [Eq. (19-8)] or $v = aP/(1 + bP)$ [Eq. (19-9)], the procedure is to take the reciprocal. This gets rid of the unwieldy denominator. For Eq. (19-8),

$$\frac{1}{\theta} = \frac{1 + bP}{bP} = \frac{1}{bP} + 1 \qquad \textbf{(a)}$$

This is the equation of a straight line; a plot of $1/\theta$ versus $1/P$ will have a slope of $1/b$ and an intercept of 1.
For Eq. (19-9), you get

$$\frac{1}{v} = \frac{1 + bP}{aP} = \frac{1}{aP} + \frac{b}{a} \qquad \textbf{(b)}$$

Now the slope is $1/a$ and the intercept is b/a when $1/v$ is plotted against $1/P$.
A variation often used in either case is to multiply through by P. For the second case, eq. (b),

$$\frac{P}{v} = \frac{1}{a} + \frac{b}{a}P \qquad \textbf{(c)}$$

Now a plot of P/v versus P has a slope of b/a and an intercept of $1/a$.

PROBLEM 19-8 In a study of the adsorption of oxygen on silica, two of the data points were $v = 8.47 \text{ cm}^3 \text{ g}^{-1}$ at 0.0160 atm and $v = 25.0 \text{ cm}^3 \text{ g}^{-1}$ at 0.0560 atm. Assuming that these points fall on the line drawn through all of the data points, evaluate the constants a and b of the Langmuir adsorption isotherm, Eq. (19-9). [*Hint:* See eq. (b) of Problem 19-7.]

Solution: You need to rewrite Eq. (19-9) in a form that gives a straight line when plotted. Either eq. (b) or eq. (c) of Problem 19-7 could be used, but let's choose (b). To calculate the slope ($= 1/a$), you need

$$m = \frac{y_2 - y_1}{x_2 - x_1}$$

where $y = 1/v$ and $x = 1/P$. Calculating the reciprocals,

v	$1/v$	P	$1/P$
8.47	0.118	0.0160	62.5
25.0	0.0400	0.0560	17.9

Now $$\frac{1}{a} = m = \frac{0.118 - 0.0400}{62.5 - 17.9} = \frac{0.078 \text{ g cm}^{-3}}{44.6 \text{ atm}^{-1}}$$

And $$a = \frac{1}{m} = \frac{44.6}{0.078} = 5.7 \times 10^2 \text{ cm}^3 \text{ g}^{-1} \text{atm}^{-1}$$

To calculate b, you need the intercept ($I = b/a$). This is the value of y when $x = 0$. Using the equation for m again,

$$m = \frac{0.078}{44.6} = 0.001\,75 = \frac{0.118 - I}{62.5 - 0}$$

$$0.118 - I = 0.109$$

$$I = 0.009 \text{ g cm}^{-3}$$

$$b = Ia = (0.009 \text{ g cm}^{-3})(572 \text{ cm}^3 \text{ g}^{-1}\text{atm}^{-1}) = 5.1 \text{ atm}^{-1}$$

PROBLEM 19-9 Continue Problem 19-8 by calculating the fraction of the surface covered at $P = 0.0160$ atm.

Solution: The fraction covered is given by Eq. (19-8), and from Problem 19-8, $b = 5.1 \text{ atm}^{-1}$, so

$$\theta = \frac{bP}{1 + bP} = \frac{5.1(0.0160)}{1 + 5.1(0.0160)} = \frac{0.082}{1.082} = 0.076$$

PROBLEM 19-10 **(a)** If a certain surface is 25% covered at a pressure of 2.0 atm, at what pressure would it just become completely covered? **(b)** What does the result imply?

Solution:

(a) The relationship between fraction of surface covered and pressure is the Langmuir isotherm, Eq. (19-8). You could set up expressions for the two conditions and eliminate b, but it is more instructive—and perhaps simpler—to proceed in two steps. First, calculate b after rearranging Eq. (19-8):

$$bP = \theta + \theta bP \qquad \textbf{(a)}$$

$$bP - \theta bP = b(P - \theta P) = \theta$$

$$b = \frac{\theta}{P(1 - \theta)} \qquad \textbf{(b)}$$

Substituting $\theta = 0.25$ and $P = 2.0$ atm,

$$b = \frac{0.25}{(2.0 \text{ atm})(0.75)} = 0.17 \text{ atm}^{-1}$$

Then, for the case of complete coverage, $\theta = 1$. Now solve eq. (b) for P:

$$P = \frac{\theta}{b(1 - \theta)} \qquad \textbf{(c)}$$

Substituting $\theta = 1$ gives zero in the denominator, and $P = \infty$.

(b) This result suggests that the surface can never be completely covered. This may be correct—or perhaps the Langmuir isotherm is not valid when coverage nears saturation.

PROBLEM 19-11 Continue Problem 19-10 by calculating the pressure that would give 90% coverage.

Solution: You can use eq. (c) obtained in Problem 19-10. Now, $\theta = 0.90$, and

$$P = \frac{\theta}{b(1 - \theta)} = \frac{0.90}{(0.17 \text{ atm}^{-1})(0.10)} = 53 \text{ atm}$$

Surfaces of Liquids

PROBLEM 19-12 Would $\gamma = g\rho rh/(2 \cos \theta)$ [Eq. (19-19)] be altered if the liquid were mercury, which does not wet glass? Explain.

Solution: The equation would not be altered; its derivation does not depend on the nature of the liquid or whether the liquid is attracted to or repelled by the glass. The level of mercury inside the capillary would be lower than the outside, so h would be negative.

PROBLEM 19-13 In a certain capillary tube at 0°C, the mercury level was 12.8 mm below the surface of the liquid outside the tube. What is the radius of the tube? Known values for mercury at 273 K are $\gamma = 0.476 \text{ N m}^{-1}$, $\rho = 13.6 \text{ g cm}^{-3}$, and $\theta = 140°$.

Solution: You need to solve Eq. (19-19) for r and make substitutions in the proper units. [See Example 19-12.]

$$r = \frac{2\gamma \cos \theta}{g\rho h} \qquad \rho = 13.6 \times 10^3 \text{ kg m}^{-3} \qquad h = -12.8 \times 10^{-3} \text{ m}$$

Then
$$r = \frac{2(0.476 \text{ N m}^{-1})(\cos 140°)}{(9.81 \text{ m s}^{-2})(13.6 \times 10^3 \text{ kg m}^{-3})(-12.8 \times 10^{-3} \text{ m})} = 4.27 \times 10^{-4} \text{ m} = 0.427 \text{ mm}$$

PROBLEM 19-14 In an experiment to determine how well benzene wets glass, benzene was found to rise to 1.71 mm in a certain capillary tube at 20°C. In the same tube, water rose 4.90 mm at 20°C. Calculate the contact angle for benzene. Known values at 20°C are $\rho_{\text{benzene}} = 0.879 \text{ g cm}^{-3}$, $\rho_{\text{water}} = 0.998 \text{ g cm}^{-3}$, $\gamma_{\text{benzene}} = 2.89 \times 10^{-2} \text{ N m}^{-1}$, $\gamma_{\text{water}} = 7.28 \times 10^{-2} \text{ N m}^{-1}$.

Solution: This situation is slightly complicated because the radius of the capillary is not given but must be found from the calibration with water. You could derive one large expression, but it is simpler to find r first, at least as a set-up calculation. First, rearrange Eq. (19-19) to

$$r = \frac{2\gamma \cos \theta}{g\rho h}$$

Then from the data for water,
$$r = \frac{2(7.28 \times 10^{-2})(\cos 0)}{9.81(0.998 \times 10^3)(4.90 \times 10^{-3})}$$

Don't calculate those terms that will cancel when you substitute in Eq. (19-19) for benzene, i.e., 10^{-2}, g, and 2.

Then
$$r = \left(\frac{2 \times 10^{-2}}{g}\right)(1.49)$$

Now rearrange Eq. (19-19) to get $\cos \theta$ and put r on the end:

$$\cos \theta = \left(\frac{g\rho h}{2\gamma}\right) r$$

Thus, for benzene

$$\cos \theta = \frac{g(0.879 \times 10^3)(1.71 \times 10^{-3})}{2(2.89 \times 10^{-2})}\left[\frac{(2 \times 10^{-2})(1.49)}{g}\right] = \frac{0.879(1.71)(1.49)}{2.89} = 0.775$$

So
$$\theta = 39.2°$$

Supplementary Exercises

PROBLEM 19-15 The pentagon is a "non-space-filling" figure (see Example 19-1). Two other common regular geometric shapes (two-dimensional) that are not space-filling are ________ and ________. The non-primitive unit cell of a body-centered cubic lattice contains ________ lattice points. Which Bravais lattice is *ccp*? ________ The space operations are ________ and ________. In which type of adsorption would greater energy be required to remove the adsorbed substance? ________ Which type of adsorption stops when a single layer of adsorbed species has been completed? ________ Adsorption (increases, decreases) as pressure increases. A good way to de-gas a solid at constant pressure is to (heat, cool) it. The SI units of surface energy are ________. The surface tension of a liquid (increases, decreases) if the temperature increases. A common type of substance that decreases the surface tension (at room temperature) of water is ________. If a certain liquid rises to 8 mm in a capillary of 0.6 mm diameter, it will rise to ________ in a capillary of 0.3 mm diameter.

PROBLEM 19-16 If the distance between two (100) planes in a cubic lattice is assigned the value 1, what is the distance between two (111) planes?

PROBLEM 19-17 With x-rays having a wavelength of 0.587 Å, a reflection was obtained from a crystal at 10.5°. If this is a first-order reflection, what is the distance between planes in the set causing the reflection?

PROBLEM 19-18 A NaCl crystal is set up to measure x-ray wavelengths by using the planes that are separated by 2.80×10^{-10} m. What is the wavelength of an x-ray that gives a second-order reflection at 14.0°?

PROBLEM 19-19 If the unit cell of KCl is *fcc* and the density is 1.99 g cm^{-3}, what is the length of one side of the unit cell?

PROBLEM 19-20 What would be the adsorbed volume per gram at infinite pressure according to (**a**) the Freundlich isotherm? (**b**) the Langmuir isotherm?

PROBLEM 19-21 When the pressure is 0.50 atm, a particular surface is 60% covered. What pressure is required to get 80% coverage?

PROBLEM 19-22 In a study of the adsorption of a gas on a certain substance, the following data were obtained:

Pressure (atm)	1.30	1.81	2.92	4.88	6.52
Volume (cm^3 g^{-1})	5.57	7.01	8.82	11.10	13.09

Determine a and b of Eq. (19-9), one form of the Langmuir isotherm, from an appropriate plot.

PROBLEM 19-23 In a laboratory study of an unknown liquid, a student determined its density to be 0.79 g cm^{-3} at 20°C. When a capillary tube of diameter 0.80 mm was placed vertically in the liquid, the liquid rose to 14.4 mm. Calculate the surface tension, assuming that the contact angle is zero.

PROBLEM 19-24 What will the depression of mercury be in a 0.50-mm-diameter capillary at 20°? At this temperature, $\rho = 13.5$ g cm^{-3} and $\gamma = 47.2 \times 10^{-2}$ N m^{-1}. The contact angle for mercury is 140°.

Answers to Supplementary Exercises

19-15 octagon (any polygon above the hexagon), circle; two; *fcc*; screw axis, glide plane; chemical; chemical; increase; heat; J m^{-2}; decreases; soap; 16 mm
19-16 $1/\sqrt{3}$ **19-17** 1.61 Å **19-18** 6.77×10^{-11} m **19-19** 6.29 Å **19-20** (**a**) ∞ (**b**) v_m
19-21 1.3 atm **19-22** $a = 6.0$ cm^3 g^{-1}atm^{-1} $b = 0.33$ atm^{-1} **19-23** 2.2×10^{-2} N m^{-1}
19-24 22 mm

FINAL EXAM (Chapters 10–19)

Time: 100 minutes

Total Points: 200

Some physical constants:

Boltzmann	$k = 1.38 \times 10^{-23}\ \mathrm{J\,K^{-1}}$	electron charge	$e = 1.60 \times 10^{-19}\ \mathrm{C}$
Planck	$h = 6.62 \times 10^{-34}\ \mathrm{J\,s}$	velocity of light	$c = 3.00 \times 10^{8}\ \mathrm{m\,s^{-1}}$
Rydberg	$R_H = 1.10 \times 10^{7}\ \mathrm{m^{-1}}$		

A. Short Answers (2 points per answer = 20)

1. Name the scientist who proposed that all electromagnetic radiant energy is quantized.

____

2. Write the de Broglie expression for the wavelength of a particle.

____

3. What is the range of the angular coordinate θ? ____________________

4. In the self-consistent field method, what is the name for the difference between the calculated value and the true value? ____________________

5. What is the name of the procedure for obtaining molecular orbitals from atomic orbitals?

____

6. What hybridization is needed to explain the structure of PF_5?

____

7. What do we call the number of members in a group? ____________________

8. Name the term that attempts to account for nonrigidity in a rotor.

____

9. In rotational–vibrational spectra, what sign of ΔJ leads to the R branch?

____

10. What factor is most important in determining the amount of adsorption?

____

B. Long Answers (Total points = 128)

1. **(a)** Sketch a typical phase diagram of a system of two solids A and B that has a congruent melting point at $X_B = 0.75$.
(b) What is the formula of the compound formed?
(c) Put on your diagram an x to show a point where the system will consist of solid A and a melt. (5)

2. The Debye–Hückel theory proposes that an atmosphere of ions of the opposite sign surrounds each ion in solutions of electrolytes. At a given temperature, what properties of solvent and solute are significant in determining the thickness of the ionic atmosphere? (6)

3. Describe briefly the method that must be used to measure accurately the emf of a cell. (5)

4. Define the term "mechanism of a reaction" as used in chemical kinetics. (2)

5. Write the rate equation for a first-order reaction, then derive from it an expression which, when plotted, gives a straight line. (5)

6. **(a)** Briefly describe the situation known as the ultraviolet catastrophe.
(b) Why was it important in the development of the theory of matter? (6)

7. **(a)** What condition must be satisfied for an equation to be an eigenequation?
(b) Give an example of an eigenequation and identify the eigenfunction and the eigenvalue. (6)

8. One of the eigenfunctions for a free particle moving in one dimension is $\psi_1 = Ae^{ibx}$.
(a) Why is this eigenfunction not suitable for a particle contained in a one-dimensional box?
(b) Describe how to obtain a suitable eigenfunction for the particle in a one-dimensional box.
(c) The eigenfunction obtained may be expressed as a sine function. What is the significant feature of the term for the angle in this expression? (10)

9. Describe briefly in words the major steps in solving the Schrödinger equation to obtain the wave function. (6)

10. What is the "penetration effect"? Include a sketch of the plot that explains this effect. (6)

11. **(a)** Write the general expression for the expectation value of the ground-state energy and then describe the variation method for obtaining approximate wave functions.
(b) What is the major disadvantage of this method? (6)

12. Write the complete electron configuration, showing the distribution of electrons in each orbital in unfilled levels, of **(a)** an atom of ^{23}V and **(b)** the ion $^{28}Ni^{2+}$. (4)

13. **(a)** Write the expression that relates the two dissociation energies.
(b) Give the full name of each term. (6)

14. State the Born–Oppenheimer approximation. (3)

15. What is meant by an "ionic term" in the wave function for a molecular orbital for the hydrogen molecule? (4)

16. Describe an antibonding molecular orbital from the point of view of **(a)** probability density and **(b)** electron charge density. (5)

17. **(a)** Write the complete molecular electron configuration for CN.
(b) What is the bond order of this species?
(c) Is this species paramagnetic? Justify your answer. (9)

18. Predict the shape of the species SCl_6, describing your reasoning by use of VSEPR. (5)

19. **(a)** List the symmetry operations generated by a S_3 axis. Use the usual symbol and give the angle in degrees.
(b) Which of the operations are distinct?
(c) Is one of the operations equivalent to the σ_v operation? (9)

20. Determine the point group and describe the location of the principal axis of **(a)** dichloromethane (Cl_2H_2C) and **(b)** naphthalene ($C_{10}H_8$, planar) (6)

21. **(a)** Write the mathematical expression that defines the canonical partition function without the weighting factor.
(b) Derive an expression for the energy of an ensemble in terms of the natural logarithm of the canonical partition function. (6)

22. How many lattice points are in a unit cell (nonprimitive) of a body-centered cubic lattice? Describe briefly how you deduce the number. (4)

23. Using thermodynamic reasoning, determine which process should occur naturally: an increase in surface area or a decrease in surface area. (4)

C. Problems (Total points = 52). To save time, do not do the arithmetic; just *set up* the final calculation. Solve all equations for the unknown, and substitute numerical values for all symbols. Put the proper units on the final answer.

1. The conductivity of a saturated solution of silver chloride was measured to be 1.849 ×

$10^{-4}\ \Omega^{-1}\ m^{-1}$ at 298 K. The water used had a conductivity of $5.8 \times 10^{-6}\ \Omega^{-1}\ m^{-1}$. Calculate the solubility product constant of AgCl. For AgCl, the molar conductivity at infinite dilution is $0.01382\ \Omega^{-1}\ m^2\ mol^{-1}$. (5)

2. If the ions are at unit activity, it is not possible to reduce Sn^{2+} with Pb at 298 K, i.e., $E°$ is negative for the reaction

$$Pb + Sn^{2+} \rightarrow Pb^{2+} + Sn \qquad E° = -0.010\ V$$

Calculate what the activity of Sn^{2+} would have to be to make the $E = +0.010$ V if the Pb^{2+} were kept at unit activity. (5)

3. For a second-order reaction of the type 2A → products, the rate constant has been found to be $0.36\ M^{-1}\ min^{-1}$. How long will it take for an initial concentration of 0.50 M to become 0.25 M? (4)

4. If a reaction has an activation energy of $35\ kJ\ mol^{-1}$ and the rate constant at 20°C is $1.8\ M\ min^{-1}$, calculate the rate constant at 35°C, all other factors being equal. (4)

5. Calculate the energy of the photon of light that gives the third line in the Balmer series of the hydrogen spectrum. (6)

6. If the percent ionic character of HI(g) is 5.0% and the measured dipole moment is 1.27×10^{-30} C m, what is the internuclear distance? (6)

7. In a pure rotational spectrum of $H^{19}F$, the average difference between the measured wave numbers of the lines was 42 cm^{-1}. Calculate the internuclear distance of the molecule. (8)

8. Calculate the force constant of $H^{35}Cl$ if its "vibrational frequency" has been found to be 2991 cm^{-1}. (6)

9. For a certain gas at 300 K, the energy of the first excited state is $5.0\ kcal\ mol^{-1}$ greater than the energy of the ground state. If there are 10^{25} molecules in the ground state, calculate how many molecules would be expected in the first excited state. Assume that the energy levels are not degenerate. (4)

10. In an x-ray crystallographic study, the wavelength of the x-rays was 1.31×10^{-10} m. If the first reflection for a certain orientation of the crystal is at an angle of 23.7°, what is the separation between the atomic planes? (4)

Solutions to Final Exam

A. Short Answers

1. Einstein [Section 14-1G]
2. $\lambda = h/mv$ [Section 14-3A]
3. 0 to π [Section 15-1A]
4. correlation energy [Section 15-5A]
5. linear combination of atomic orbitals (LCAO) [Section 16-2B]
6. sp^3d [Section 16-4D]
7. order [Section 17-2A]
8. centrifugal distortion constant [Section 17-4B]
9. positive [Section 17-6A]
10. surface area [Section 19-4A]

B. Long Answers

1. **(a)** and **(c)** See Figure E-3. **(b)** AB_3 [Section 10-5A]

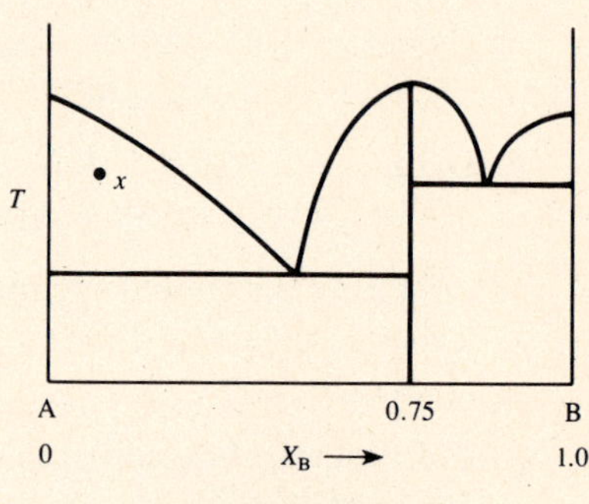

FIGURE E-3

2. The dielectric constant of the solvent, the concentration of the solute, and the charges on the ions. The last two could be grouped together as the ionic strength [Section 11-6A].
3. One must use a potentiometer that opposes the cell's emf by a known emf so that, at the balance point, there is no deflection of the galvanometer needle. The known emf is supplied by a standard cell such as one of the calomel electrodes [Section 12-2B].
4. The mechanism of a reaction is a series of steps (each an elementary reaction) that add up to equal the overall reaction [Section 13-1C].
5. The rate equation must be integrated, then rearranged. The first-order rate equation is

$$\text{rate} = k_1 c = -\frac{dc}{dt}$$

Rearranging gives
$$\frac{dc}{c} = -k_1\, dt = d \ln c$$

And integrating without limits gives

$$\ln c = -k_1 t + I$$

If ln c is plotted against t, a straight line results [Section 13-4A].

6. **(a)** The laws of classical physics predicted that the amount of energy of a given frequency radiated by a black body should increase as the frequency increases. Aside from the fact that the energy would be impossibly large at high frequencies, measurements showed that the energy dropped back towards zero at high frequencies after passing through a maximum. Something very serious was wrong with the classical theory.
(b) This problem led Planck to his quantum hypothesis, which underlies the present theory of matter [Section 14-1F].
7. **(a)** The operation of the operator on a function gives back the function but multiplied by a constant.
(b)
$$\frac{d}{dx} e^{2x} = 2e^{2x}$$
The eigenfunction is e^{2x}, and the eigenvalue is 2 [Section 14-4B].
8. **(a)** At the edges of the box, ψ_1 must = 0. For this to occur, A must = 0, but then $\psi = 0$ for every x.
(b) You need a second eigenfunction such as $\psi_2 = Be^{-ibx}$. Then ψ_1 and ψ_2 can be combined by a linear combination to give two new eigenfunctions, both of which are satisfactory. Each eigenfunction is multiplied by a weighting factor:

$$\psi_3 = c_1\psi_1 + c_2\psi_2 \qquad \psi_4 = c_1\psi_1 - c_2\psi_2$$

(c) The term for the angle contains $n\pi$, where n may have only integral values, because $\sin\theta = 0$ only for integral multiples of π. Thus the concept of an integral quantum number arises naturally from the mathematics [Section 14-5B].

9. The equation is transformed to spherical coordinates: r, θ, ϕ. The resulting equation can be expressed as the product of three parts, one for each of the coordinates. Each of these parts can be solved separately. The total wave function is obtained by multiplying together the three solutions [Section 15-1A].

10. The penetration effect is the appreciable probability that the electron in an orbital is closer to the nucleus than the value of n for that orbital would suggest. Your sketch should be the plot of the radial probability for the s orbital for an n greater than 1.The plot should have as many maxima as n, growing larger away from the origin. If $n = 2$, your sketch should look like the dashed curve in Figure 15–4. The small "inner" maxima represent the probability that the electron will be near the corresponding distance from the nucleus [Section 15-1H].

11. **(a)** The expectation value is given by Eq. (14-31). For energy,

$$E_g = \int \psi^* \mathbf{H} \psi \, d\tau \qquad \text{[Eq. (15-33)]}$$

In the variation method, a reasonable trial wave function is used to calculate a value of E which always will be greater than or equal to E_g. Then values in the expression are adjusted to give a different trial function, and another E is calculated. The process is continued, with the values being adjusted in successive approximations to bring E as close to E_g as possible.

(b) The disadvantage is that you must know the value of E_g [Sections 14-4D and 15-3A].

12. **(a)** $1s^2\ 2s^2\ 2p^6\ 3s^2\ 3p^6\ 4s\ \underline{\uparrow\downarrow}\ 3d\ \underline{\uparrow}\ \underline{\uparrow}\ \underline{\uparrow}\ \underline{\ \ }\ \underline{\ \ }$

(b) $1s^2\ 2s^2\ 2p^6\ 3s^2\ 3p^6\ 3d\ \underline{\uparrow\downarrow}\ \underline{\uparrow\downarrow}\ \underline{\uparrow\downarrow}\ \underline{\uparrow}\ \underline{\uparrow}$ [Section 15-5B]

13. **(a)**

$$D_{eq} = D_0 + E_0 \qquad \text{or} \qquad D_{eq} = D_0 + h\nu_0/2 \qquad \text{[Eq. (16-1)]}$$

(b) D_{eq} is the equilibrium dissociation energy, D_0 is the spectroscopic dissociation energy, E_0 is the energy of the ground state of a harmonic oscillator, h is Planck's constant, and ν_0 is the frequency of vibration of the atoms in the ground state [Section 16-1A].

14. The Born–Oppenheimer approximation is the assumption that the nuclei of the atoms in a molecule are fixed in position relative to the motion of the electrons. In other words, the motion of the nuclei is ignored [Section 16-1B].

15. Ionic terms arise in combinations of atomic orbitals in which both electrons are on the same nucleus, e.g., $1s_A(1)\ 1s_A(2)$. Here both electrons would be on atom A [Section 16-3A].

16. **(a)** An antibonding orbital has a node in the plot of ψ^2, i.e., there is a region between the two nuclei where the probability of finding the electrons is zero or very small.

(b) We say that there is a minimum of electron charge density between the nuclei, permitting the repulsion between the nuclei to prevent bond formation [Section 16-3B].

17. **(a)** Write the symbols for the MOs, then fill in electrons. For CN, there are 13 electrons.

$$(1\sigma_g)^2(1\sigma_u^*)^2(2\sigma_g)^2(2\sigma_u^*)^2(1\pi_u)^4(3\sigma_g)^1 \qquad \text{[Section 16-3H]}$$

(b) Bond order $= \frac{1}{2}$(electrons in bonding orbitals − electrons in antibonding orbitals) $= \frac{1}{2}(7 - 4) = 1.5$ [Section 16-3F].

(c) Yes, because there is one unpaired electron in one of the two π_u orbitals [Section 16-3G].

18. Octahedral. The S atom has 6 outer-shell electrons, and there are 6 Cl atoms, each of which must contribute one electron, making a total of 6 electron pairs. The repulsion between the 6 pairs forces them to the 6 equivalent positions at the corners of an octahedron with the S atom at its center [Section 16-4D].

19. **(a)** There are six operations: $S_3^1(120°)$, $S_3^2(240°)$, $S_3^3(360°)$, $S_3^4(120°)$, $S_3^5(240°)$, $S_3^6(360°)$.

(b) $S_3^2 = C_3^2$, $S_3^3 = \sigma_h$, $S_3^4 = C_3^1$, and $S_3^6 = E$, so the only distinct operations are S_3^1 and S_3^5.

(c) No [Section 17-1A and Problem 17-3].

20. **(a)** Dichloromethane has a tetrahedral shape. Visualize it with the two Cl atoms on the desk and the Cl—C—Cl plane perpendicular to the desk. The principal axis passes through C, perpendicular to the desk, bisecting the Cl—C—Cl angle. Of course, it also bisects the H—C—H angle. This axis is C_2. There are no C_2 axes perpendicular to it nor is there a horizontal plane of symmetry. Therefore the point group is C_{2v}.

(b) There is a C_2 axis perpendicular to the plane of the molecule, which normally is taken to be the principal axis. There are two C_2 axes perpendicular to it, and the plane of the molecule is a horizontal plane of symmetry. Therefore the point group is D_{2h}. Any of the three C_2 axes could be chosen as the principal axis, with the same result [Section 17-2C].

21. **(a)**

$$Q = \Sigma e^{-E_i/kT} \qquad \text{[Eq. (18-13)]}$$

(b) The probability that the ith member of the ensemble has energy E_i is

$$p_i = \frac{e^{-E_i/kT}}{Q}$$

The average value of the energy of the ensemble is

$$\bar{E} = \Sigma p_i E_i = \Sigma E_i \frac{e^{-E_i/kT}}{Q}$$

Since

$$\left(\frac{\partial Q}{\partial T}\right)_V = \Sigma \frac{E_i}{kT} e^{-E_i/kT},$$

$$\bar{E} = \frac{kT^2}{Q}\left(\frac{\partial Q}{\partial T}\right)_V = kT^2\left(\frac{\partial \ln Q}{\partial T}\right)_V \qquad \text{[Section 18-3B]}$$

22. Each of the eight corners of the cube contributes $\frac{1}{8}$ of a point, for a net contribution of one point. The point at the center belongs entirely to the cell; thus the total is two [Section 19-2A].

23. From Eq. (19-13), the work done to change the area of a surface is $\gamma\, d\mathscr{A}$. Work done on a surface increases its Gibbs free energy or $dG = \gamma\, d\mathscr{A}$. For a spontaneous process, dG must be negative, so $d\mathscr{A}$ must be negative. Thus the surface area will decrease if given the opportunity [Section 19-5B].

C. Problems

1. Determine the concentration of the solution, then $K_{sp} = [Ag^+][Cl^-] = [Ag^+]^2$. Find the concentration by using Eq. (11-18), $C = \kappa/\Lambda_0$, where

$$\kappa\,(\text{salt}) = \kappa\,(\text{soln}) - \kappa\,(\text{water}) \qquad \text{[Eq. (11-22)]}$$
$$= (1.849 \times 10^{-4} - 0.058 \times 10^{-4})\,\Omega^{-1}\,\text{m} = 1.791 \times 10^{-4}\,\Omega^{-1}\,\text{m}^{-1}$$

$$C = \frac{\kappa}{\Lambda_0} = \frac{1.791 \times 10^{-4}\,\Omega^{-1}\,\text{m}^{-1}}{0.01382\,\Omega^{-1}\,\text{m}^2\,\text{mol}^{-1}}\left(\frac{\text{m}^3}{10^3\,\text{dm}^3}\right)$$

$$K_{sp} = C^2 = \left(\frac{1.791}{138.2} \times 10^{-3}\,\text{mol}\,\text{m}^{-3}\right)^2$$

2. Use the Nernst equation, Eq. (12-12), at 298 K.

$$E = E^\circ - \frac{0.05916}{n}\log\frac{a(Pb^{2+})}{a(Sn^{2+})}$$

Substituting, and letting x represent the activity of Sn^{2+},

$$+0.010 = -0.010 - \frac{0.05916}{2}\log\frac{1}{x} \qquad 0.020 = +\frac{0.05916}{2}\log x$$

$$x = \text{antilog}\left[\frac{2(0.020)}{0.05916}\right]$$

3. Since the concentration is reduced in half, the time sought is the half-life. And since the reaction is of Type I [Section 13-4B], you can use Eq. (13-26),

$$t^{1/2} = \frac{1}{k_2 c_0} = \frac{1}{0.36(0.50)}\,\text{min}$$

If you missed the one-half relationship, you could have used Eq. (13-17),

$$\frac{1}{c} - \frac{1}{c_0} = k_2 t$$

$$t = \frac{1}{k_2}\left(\frac{1}{c} - \frac{1}{c_0}\right) = \frac{1}{0.36}\left(\frac{1}{0.25} - \frac{1}{0.50}\right)\text{min}$$

4. Use the Arrhenius equation in its integrated form, Eq. (13-31):

$$\ln\frac{k_2}{k_1} = \frac{E_a}{R}\left(\frac{T_2 - T_1}{T_1 T_2}\right)$$

$$\ln k_2 = \ln k_1 + \frac{E_a}{R}\left(\frac{T_2 - T_1}{T_1 T_2}\right)$$

$$k_2 = \text{antiln}\left[\ln 1.8 + \frac{35 \times 10^3}{8.314}\left(\frac{308 - 293}{(308)(293)}\right)\right] M\,\text{min}^{-1}$$

5. A photon of light has energy $h\nu$ or $\varepsilon = h\nu = hc/\lambda$. To calculate the wavelength, use the Ritz–Rydberg equation, Eq. (14-1). The Balmer series is generated when $n = 2$, so the third line arises from the transition $n_2 = 5 \rightarrow n_1 = 2$.

$$\frac{1}{\lambda} = R_H\left(\frac{1}{n_1^2} - \frac{1}{n_2^2}\right) = \frac{\varepsilon}{hc}$$

$$\varepsilon = hcR_H\left(\frac{1}{n_1^2} - \frac{1}{n_2^2}\right) = (6.62 \times 10^{-34})(3.00 \times 10^8)(1.10 \times 10^7)\left(\frac{1}{4} - \frac{1}{25}\right)\,\text{J}$$

6. Percent ionic character is given by Eq. (16-26): $I = 100\mu_{exp}/\mu_{ion}$, where μ_{ion} is the dipole moment for the completely ionized molecule, given by Eq. (16-25), $\mu = rq$. Rearranging, $r = \mu_{ion}/q$. Rearranging Eq. (16-26), $\mu_{ion} = 100\mu_{exp}/I$. Then

$$r = \frac{100\mu_{exp}}{Iq} = \frac{100(1.27 \times 10^{-30})}{5(1.60 \times 10^{-19})}\ \text{m}$$

7. The interval is twice the rotational constant B, so $B = 42/2 = 21\ \text{cm}^{-1}$. You need to calculate I from B, then r from I, using Eq. (17-7):

$$B = \frac{h}{8\pi^2 Ic} \qquad \text{where} \qquad I = \mu r_0^2$$

$$r_0^2 = \frac{I}{\mu} = \frac{1}{\mu}\frac{h}{8\pi^2 cB}$$

$$\mu = \frac{m_1 m_2}{m_1 + m_2} = \frac{(1)(19)}{1 + 19}\ \text{g mol}^{-1}\left(\frac{1\ \text{mol}}{6.02 \times 10^{23}}\right)\left(\frac{10^{-3}\ \text{kg}}{\text{g}}\right)$$

$$r = \left(\frac{20(6.02 \times 10^{26})(6.63 \times 10^{-34}\ \text{J s})}{(19\ \text{kg})(8\pi^2)(3.00 \times 10^8\ \text{m s}^{-1})(21 \times 10^2\ \text{m}^{-1})}\right)^{1/2}\ \text{m}$$

8. Use Eq. (17-14) or the more convenient form of Eq. (17-12), $k = 4\pi^2 c^2 \mu \tilde{\omega}^2$. Remember that $\tilde{\omega}$ often is called a frequency even though it is a wave number and must be expressed in reciprocal meters. You need μ, the reduced mass:

$$\mu = \frac{m_1 m_2}{m_1 + m_2} = \frac{(1)(35)}{1 + 35}\ \text{g mol}^{-1}\left(\frac{1}{6.02 \times 10^{23}}\ \text{mol}\right)(10^{-3}\ \text{kg g}^{-1})$$

Then
$$k = 4\pi^2(3.0 \times 10^8)^2\left(\frac{35 \times 10^{-3}}{36(6.02 \times 10^{23})}\right)(2991 \times 10^2)^2\ \text{kg s}^{-2}$$

9. Use the Boltzmann distribution in the form of Eq. (18-7). For moles, use R instead of k:

$$\frac{n_i}{n_0} = e^{-\varepsilon_i/RT}$$

The exponent is
$$-\frac{5.0 \times 10^3\ \text{cal mol}^{-1}}{1.99\ \text{cal mol}^{-1}\text{K}^{-1}(300\ \text{K})}$$

$$n_i = 10^{25} e^{-5/(1.99)(300)}$$

10. Solve the Bragg equation, Eq. (19-1), for the separation d between the atomic planes.

$$n\lambda = 2d \sin\theta \qquad d = \frac{n\lambda}{2 \sin\theta}$$

At the first reflection, $n = 1$, so

$$d = \frac{1(1.31 \times 10^{-10})}{2 \sin(23.7^\circ)}$$

APPENDIX

TABLE A-1: Standard Molar Properties at 298.15 K

Substance	ΔH_f° kJ mol^{-1}	S° J K^{-1}mol^{-1}	ΔG_f° kJ mol^{-1}
Inorganic			
$H_2O(g)$	−241.82	188.83	−228.57
$H_2O(l)$	−285.83	69.91	−237.13
$H_2(g)$	0	130.68	0
$O(g)$	249.17	161.06	231.73
$O_2(g)$	0	205.14	0
$O_3(g)$	142.7	238.93	163.2
$Br_2(g)$	30.91	245.46	3.11
$Br_2(l)$	0	152.23	0
$HBr(g)$	−36.40	198.70	−53.45
$Cl_2(g)$	0	223.07	0
$HCl(g)$	−92.31	186.91	−95.30
$I_2(g)$	62.44	260.69	19.33
$I_2(s)$	0	116.14	0
$HI(g)$	26.48	206.59	1.70
$CaO(s)$	−635.09	39.75	−604.03
$Ca(OH)_2(s)$	−986.1	83.4	−898.6
$CaCO_3$(calcite)	−1206.92	92.9	−1128.79
$NO(g)$	90.25	210.76	86.57
$NO_2(g)$	33.18	240.06	51.31
$N_2O_4(g)$	9.16	304.29	97.89
$NH_3(g)$	−46.11	192.45	−16.45
$Na_2O(s)$	−414.2	75.1	−375.5
$NaOH(s)$	−425.61	64.46	−379.49
$Na_2SO_4(s)$	−1387.1	149.6	−1270.2
$Na_2SO_4 \cdot 10H_2O$	−4327.3	592	−3647.4
$PCl_3(g)$	−287.0	311.78	−267.8
$PCl_5(g)$	−374.9	364.58	−305.0
$SO_2(g)$	−296.83	248.22	−300.19
$SO_3(g)$	−395.72	256.76	−371.06
$H_2S(g)$	−20.63	205.79	−33.56
$H_2SO_4(l)$	−813.99	156.90	−690.00
$SnO(s)$	−285.8	56.5	−256.9
$SnO_2(s)$	−580.7	52.3	−519.6
Carbon and Compounds			
C (graphite)	0	5.74	0
C (diamond)	1.90	2.38	2.90
$CO(g)$	−110.53	197.67	−137.17
$CO_2(g)$	−393.51	213.74	−394.36
$CH_4(g)$	−74.81	186.26	−50.72
$C_2H_4(g)$	52.26	219.56	68.15
$CH_3CO_2H(l)$	−484.5	159.8	−389.9
$CH_3CO_2H(aq)$	−485.76	178.7	−396.46
$C_2H_6(g)$	−84.68	229.60	−32.82
$C_2H_5OH(g)$	−235.10	282.70	−168.49
$C_2H_5OH(l)$	−277.69	160.70	−174.78
$C_3H_8(g)$	−103.89	270.02	−23.38
$C_4H_{10}(g)$	−126.15	310.23	−17.03
$C_6H_6(g)$	82.93	269.31	129.72
$C_6H_{12}(g)$	−123.14	298.35	31.91

TABLE A-1: (continued)

Substance	ΔH_f° kJ mol^{-1}	S° J K^{-1}mol^{-1}	ΔG_f° kJ mol^{-1}
Ions (in solution at infinite dilution)			
H^+	0	0	0
Ca^{2+}	−542.83	−53.1	−553.58
Na^+	−240.12	59.0	−261.91
$CH_3CO_2^-$	−486.01	86.6	−369.31
Cl^-	−167.16	56.5	−131.23
OH^-	−229.99	−10.75	−157.24
SO_4^{2-}	−909.27	2.01	−744.53

TABLE A-2: Coefficients for the Heat-Capacity Equation

$C_{P,m} = a + bT + cT^2$ in J K^{-1}mol^{-1}, at 1 atm

Substance	a	$b \times 10^3$	$c \times 10^7$
Gases			
Br_2	35.241	4.07	−14.86
Cl_2	31.696	10.14	−40.37
H_2	29.066	−0.83	20.12
O_2	25.723	12.98	−38.61
HBr	27.521	4.00	6.61
HCl	28.166	1.81	15.47
H_2O	30.359	9.61	11.84
H_2S	28.719	16.12	32.84
SO_2	32.217	22.2	−34.7
CO	26.861	6.97	−8.19
CO_2	25.999	43.50	−148.31
CH_4	14.146	75.50	−179.90
C_2H_4	11.839	119.66	−365.07
C_2H_6	9.184	160.17	−460.27
Solid			
C (graphite)	−5.296	58.60	−432.24

INDEX

Page numbers in italics refer to Solved Problems.

Resistance measurements, 252
Resonance integral, 400
Reversible engines, 97
Reversible process, 53
  entropy change, 100, *113*
  isothermal expansion, *72*
Reversible reaction, 319
Reversible work, calculations, 54